PHASE DIAGRAMS OF
BINARY TITANIUM ALLOYS

Joanne L. Murray, Ph.D.

Editor

Metals Park, Ohio 44073

The National Standard Reference Data System (NSRDS) was established in 1963 for the purpose of promoting the critical evaluation and dissemination of numerical data of the physical, chemical, and materials sciences. The NSRDS is coordinated by the Office of Standard Reference Data of the National Bureau of Standards, but involves the efforts of many groups in universities, government laboratories, and private industry. The primary aim of the NSRDS is to provide compilations of critically evaluated physical and chemical property data needed by the scientific and engineering community. Activities carried out under the NSRDS emphasize the evaluation of published data by scientists who are expert in the research areas involved.

The output of the International Data Program for Alloy Phase Diagrams contained in this monograph has been reviewed by the Office of Standard Reference Data and accepted as a product of the National Standard Reference Data System.

David R. Lide, Jr., Director
Standard Reference Data
National Bureau of Standards

Library of Congress Catalog Card Number 87-70152
ISBN: 0-87170-248-7
SAN: 204-7586

Managing Editor: Hugh Baker
Production Supervisor: Linda Kacprzak

PRINTED IN THE UNITED STATES OF AMERICA

FOREWORD

The Monograph Series on Alloy Phase Diagrams represents a compendium of completed detailed evaluations of binary and higher-order systems carried out in separate categories of the International Programme for Alloy Phase Diagram Data. When the U.S. Phase Diagram Evaluation Program was initiated by the joint action of ASM INTERNATIONAL and the National Bureau of Standards several years ago, the first major step was the recognition that, with the mushrooming amount of phase diagram information being published in the literature, the very large task of data assessment and phase diagram evaluations had to be suitably subdivided to become manageable. Accordingly, all binary systems were divided into categories, each representing a major metal or a group of related metals. In each category (of which now more than thirty are active in several different countries) the available published data on binary systems is being compiled and evaluated under the guidance and responsibility of an appointed Category Editor, or Editors, each for a given finite set of binary systems. The output from this very large effort is channeled to the worldwide scientific and technological community in a variety of ways. Individual evaluations are being published in the *Bulletin of Alloy Phase Diagrams* following a peer review process. Information on phase diagram and crystal structure data is being utilized for the development of a computerized data base. Abbreviated extracts from individual evaluations were published in 1986 in a two-volume reference set entitled "Binary Alloy Phase Diagrams." The detailed record of all individual evaluations completed for a given metal, or related groups of metals, will be published as a monograph in the Monograph Series on Alloy Phase Diagrams.

Evaluations included in a given monograph may come from a number of different binary categories, depending on which category editor initiated the study of a given system. In some cases, the evaluations may represent joint efforts by two categories and by several authors. Naturally, the style and approach by the different authors may differ, and the same applies regarding the degree and form with which the various parts in each evaluation have been treated.

This large effort would not have been possible without the commitment, devotion and very conscientious work by a large number of people. Special thanks are due to all authors, to the professional staffs at ASM and NBS, to the members of the Binary Phase Diagram Advisory Committee and to members of various sponsoring organizations and their committees who have all contributed to this joint effort. Their work, their professional backgrounds and experience and their enthusiastic commitment to the Phase Diagram Program have made this monograph possible. May it serve all those who have need for it now and in the future and who have encouraged us all to persist.

T. B. Massalski
Editor-in-Chief
Binary Alloy Phase Diagram Program

ACKNOWLEDGMENTS

The following evaluations of titanium alloy phase diagrams were performed under the auspices of the ASM/NBS Data Program for Alloy Phase Diagrams. Work was jointly funded by the Office of Naval Research and the National Bureau of Standards through the Office of Standard Reference Data.

The Editor owes a particular debt of gratitude to Robert Mehrabian. Without his encouragement and support, neither the ambition nor the resources would have existed to complete this project.

Joyce F. Harris prepared the bibliographies and typescripts, with considerable initiative and independence.

David F. Redmiles and James S. Sims wrote the software for the computer graphics and the database. Their responsiveness to the special needs of the evaluator have made the software a powerful tool for examining and evaluating constitution data.

Roseann Hayes maintains the computerized graphics database and was largely responsible for the orderly state of the vast number of data files that went into the preparation of the computer-drawn figures.

Software for thermodynamic optimizations was graciously provided by H.L. Lukas and E.-Th. Henig of the Max-Planck Institute, Stuttgart, Federal Republic of Germany. Valuable suggestions on the thermodynamic modeling of several systems were made by N. Saunders, P. Miodownik and I. Ansara. The systematic analysis of the thermodynamics of binary titanium systems by L. Kaufman has been a valuable groundwork for the present re-evaluations. John Cahn has been a patient teacher of all aspects of phase equilibria. A.J. McAlister and R.D. Shull provided unpublished experimental data to resolve some of the more inexplicable discrepancies in the literature.

This project has required the cooperation of many people at NBS and ASM, and I would like to thank the following for ideas, encouragement and help: H. Baker, L.H. Bennett, K.J. Bhansali, T. Bise, W. Boettinger, B. Burton, L. Charlton, R.V. Drew, F.C. Johnson, L. Kacprzak, D.J. Kahan, T.B. Massalski, D. Orser, J. Perepezko, C. Qualey, M.R. Read, H. Rubin, W.W. Scott, Jr., C.E. Sirofchuck, and L.J. Swartzendruber.

I would also like to thank authors and co-authors, C. Bale, A. Pelton, K. Spear, H. Wriedt, and G. Weatherly for valuable contributions to this volume.

J.L. Murray
August 1987

INTRODUCTION

Scope

The ASM/NBS Alloy Phase Diagram Data Program has set for itself the ambitious goal of not only updating Hansen's standard work on the constitution of binary alloys, but of widening its scope to include more crystal structure data, metastable phase equilibria, and modern thermodynamic treatments of stable and metastable equilibria. Many of the following evaluations were published previously in the *Bulletin of Alloy Phase Diagrams* during the early stages of the data program. The early efforts reflected the slow, experimental development of scope, methods, conventions, style, and format of the data program output. All the evaluations have undergone substantial revision and updating, to provide a uniform treatment of data and format of presentation.

As far as possible within the limitations of the available experimental data, each evaluation attempts to be a complete account of the phase equilibria of the system: stable, metastable, and constrained. Evaluation of the experimental literature is the primary goal, but whenever possible, a thermodynamic analysis of the system is given. Each evaluation is divided into four parts: equilibrium diagram, metastable phases, crystal structures, and thermodynamics.

In the section on the equilibrium diagram, a summary is given of the experimental results and their probable accuracy, experimental techniques, and sample purity. All experimental data are presented in figures or according to guidelines proposed by Hume-Rothery et al.[1] Discussions are given of the selection of the most reliable data, the basis for the selection, and the construction of the assessed diagram.

Topics covered under metastable equilibria are metastable extensions of equilibrium boundaries, massive and martensitic transformations, amorphous and other nonequilibrium phases formed during rapid solidification and coherent equilibria.

Crystal structures and lattice parameter data are compiled, but a thorough critical evaluation of these data has not been attempted.

The coverage of thermodynamic data is limited to heats of formation of intermetallic phases, high-temperature heats of mixing, and partial Gibbs energies. Previous calculations of the diagram are described and compared to the results of the present evaluations and optimizations.

Literature Coverage and Reference Style

Essentially all experimental phase equilibrium investigations are described, including those results that reflect the effects of contamination and results which were misinterpreted, so that experimental pitfalls may be avoided in future work. The complete literature on each system has therefore been covered, starting with references cited by Hansen.[2] Because high-purity Ti has been available only since about 1950, some very early or cursory work has been omitted when clearly superseded by modern work. Similarly, some published results had to be abandoned if they were no longer obtainable.

Reference citations are composed of five-character codes: the year of publication and the first three letters of the first author's surname. Exceptions include frequently cited compilations, which are listed under "General References." Although the referencing style may appear odd and uncomfortable at first, it facilitates continuous update of evaluations. The reference lists also contain annotations, which include an asterisk (*) preceding the reference code of a key paper, a number symbol (#) indicating the presence of a phase diagram in the paper, and section headings and document classifications.

Style of Presentation

The assessed line diagram is shown as Fig. 1 and is provided in atomic percent, with a secondary weight percent scale. Only the single-phase fields are labeled, space permitting. Invariant temperatures and special compositions are often labeled for ease of using the diagram, but the primary source of these numerical data is the table entitled "Special Points of the Diagram." Experimental data are shown in atomic percent on a set of succeeding detailed figures plotted on scales suitable to the accuracy of the original data. No satisfactory graphical method has been found to compare metallographic data of various investigations; these data are shown as estimated points on the phase boundaries, and it is recommended that the original literature be consulted if greater precision is required.

Calculation of the Phase Diagrams

Where both phase diagram data and thermodynamic data exist, nonlinear least-squares optimization of Gibbs energies with respect to both types of data permit one to verify the mutual consistency of the various data, to resolve minor discrepancies among investigators, and in some instances to extrapolate phase boundaries beyond their equilibrium limits. Therefore, phase diagram calculations have been attempted for almost all the evaluated systems.

The systems of Ti with the early transition metals, rare earths, and actinides exhibit positive, but not very large, excess Gibbs energies and exhibit nearly subregular solution behavior. For these systems, gaps in the experimental data can be filled by the calculations in at least a qualitative way. For example, liquid phase miscibility gaps have been predicted for several Ti-rare earth systems.

The systems of Ti with other elements (except Ag) show either almost complete immiscibility or strongly negative excess Gibbs energies. For systems that are relatively well characterized, not only can the phase diagram be calculated within experimental uncertainty, but also the calculation can be correlated with nonequilibrium properties such as glass-forming ability. Examples are the Ti-Cu and Ti-Ni systems. For most Ti-based systems, however, the absence or inaccuracy of experimental data limits one to unrealistically simple models for the Gibbs energies, and phase diagram calculations are of limited utility. Many calculations for these systems have been omitted altogether. For calculations that are included, an attempt has been made to define a specific goal or inquiry of the calculation and to define a range of validity of the Gibbs energy functions.

Optimization calculations were performed using software provided by H. Lukas and E.-Th. Henig. The optimization techniques and equations of error have been thoroughly documented by those authors.[3] Definitions of the parameters of the Gibbs energy expressions are provided in the text of each evaluation.

The so-called "lattice stability parameters" or Gibbs energy of the pure elements in both stable (measured) and unstable forms are taken chiefly from the work of Kaufman.[4] These parameters are currently undergoing re-examination and revision by several groups of investigators. The effect of revision of these quantities must be a first order of business for future systematic calculations of Ti systems. However, for the present, it has been judged necessary to use a self-consistent and generally accepted set of parameters.

Accuracy of the Diagrams

Because of the susceptibility of Ti to contamination and its relatively high melting temperature, true binary Ti phase diagrams are notoriously difficult to characterize. The systematic examination of many Ti systems often permits one, by comparison with other systems, to conclude that sample purity or the method of examination was not adequate for determination of the binary phase equilibria. It is hoped that these evaluations will stimulate renewed interest in experimental phase diagram work for high-purity Ti alloys.

References

1. W. Hume-Rothery, J.W. Christian, and W.B. Pearson, *Metallurgical Equilibrium Diagrams*, Chapman and Hall, Ltd., London, England (1952).
2. M. Hansen and K. Anderko, *Constitution of Binary Alloys*, McGraw-Hill, New York, or General Electric Co., Business Growth Services, Schenectady, NY (1958).
3. H.L. Lukas, E.T. Henig, and B. Zimmerman, "Optimization of Phase Diagrams by a Least Squares Method Using Simultaneously Different Types of Data," *Calphad*, *1*(3), 225–236 (1977).
4. L. Kaufman and H. Bernstein, *Computer Calculations of Phase Diagrams*, Academic Press, New York (1970).

GENERAL REFERENCES

[**Hansen**]: M. Hansen and K. Anderko, *Constitution of Binary Alloys*, McGraw-Hill, New York, or General Electric Co., Business Growth Services, Schenectady, NY (1958).

[**Elliott**]: R.P. Elliott, *Constitution of Binary Alloys*, First Supplement, McGraw-Hill, New York, or General Electric Co., Business Growth Services, Schenectady, NY (1969).

[**Shunk**]: F.A. Shunk, *Constitution of Binary Alloys*, Second Supplement, McGraw-Hill, New York, or General Electric Co., Business Growth Services, Schenectady, NY (1969).

[**Pearson2**]: W.B. Pearson, *Handbook of Lattice Spacings and Structures of Metals and Alloys*, Vol. 1, Pergamon Press, New York (1967).

[**Hultgren, E**]: R. Hultgren, P.D. Desai, D.T. Hawkins, M. Gleiser, K.K. Kelley, and D.D. Wagman, *Selected Values of the Thermodynamic Properties of the Elements*, American Society for Metals, Metals Park, OH (1973).

[**Hultgren, B**]: R. Hultgren, P.D. Desai, D.T. Hawkins, M. Gleiser, and K.K. Kelley, *Selected Values of the Thermodynamic Properties of Binary Alloys*, American Society for Metals, Metals Park, OH (1973).

[**Ti**]: "Melting Points of the Elements," *Bull. Alloy Phase Diagrams*, 2(1), 145–146 (1981).

ABBREVIATIONS

Common abbreviations in phase diagram work may be used without definition. These include abbreviations for all elements, crystal structure terms, and units. The abbreviations used in these volumes include:

atmosphere	atm
atomic percent	at.%
austenitic finish	A_f
austenitic start	A_s
body-centered cubic	bcc
body-centered tetragonal	bct
boiling point	B.P.
Celsius	°C
close-packed hexagonal	cph
cubic centimeters	cm^3
Curie temperature	T_C
degree (angular)	°
differential scanning calorimetry	DSC
differential thermal analysis	DTA
double close-packed hexagonal	dcph
electromotive force	emf
enthalpy	H
entropy	S
face-centered cubic	fcc
face-centered tetragonal	fct
Fahrenheit	°F
gas	G or g
Gibbs energy	G
gram	g
Guinier-Preston	GP
heat capacity	C_p
hexagonal	hex
high temperature	HT
hour	h
joule	J
Kelvin	K
liquid	L
logarithm (base 10)	log
(base e)	ln

low temperature	LT
martensitic finish	M_f
martensitic start	M_s
maximum	max
megapascal	MPa
melting point	M.P.
millimicron (nanometer)	nm
minimum	min
minute (time)	min
(angular)	'
mole	mol
nanometer	nm
Néel temperature	T_N
parts per billion	ppb
parts per million	ppm
percent	%
pressure	P
radio frequency	RF
rare earth	RE
room temperature	RT
second (time)	s
(angular)	"
small angle X-ray scattering	SAXS
solid	S or s
sublimation point	S.P.
temperature	T
transmission electron microscopy	TEM
transformation temperature for partitionless solidification	T_0
triple point	T.P.
unknown	*
vapor	v
versus	vs
weight percent	wt.%
X-ray diffraction	XRD

TABLE OF CONTENTS

Ti	1	N-Ti	176
Ag-Ti	6	Na-Ti	187
Al-Ti	12	Nb-Ti	188
As-Ti	25	Nd-Ti	195
Au-Ti	27	Ni-Ti	197
B-Ti	33	O-Ti	211
Ba-Ti	38	Os-Ti	229
Be-Ti	40	P-Ti	234
Bi-Ti	44	Pb-Ti	237
C-Ti	47	Pd-Ti	239
Ca-Ti	52	Pt-Ti	247
Cd-Ti	54	Pu-Ti	253
Ce-Ti	56	Rb-Ti	258
Co-Ti	59	RE-Ti	259
Cr-Ti	68	Re-Ti	260
Cs-Ti	79	Rh-Ti	263
Cu-Ti	80	Ru-Ti	270
Er-Ti	96	S-Ti	275
Eu-Ti	98	Sb-Ti	282
Fe-Ti	99	Sc-Ti	284
Ga-Ti	112	Se-Ti	287
Gd-Ti	115	Si-Ti	289
Ge-Ti	118	Sn-Ti	294
H-Ti	123	Sr-Ti	300
Hf-Ti	136	Ta-Ti	302
Hg-Ti	140	Tc-Ti	307
In-Ti	143	Te-Ti	309
Ir-Ti	145	Th-Ti	311
K-Ti	150	Ti-U	313
La-Ti	151	Ti-V	319
Li-Ti	153	Ti-W	328
Mg-Ti	156	Ti-Y	333
Mn-Ti	159	Ti-Zn	336
Mo-Ti	169	Ti-Zr	340

Ti (Titanium)

47.88

By J.L. Murray and H.A. Wriedt

Melting Point

A summary of melting point determinations since 1950 is given in Table 1. Experiments of [54Ori], [67Rud], [74Ber], and [77Cez] were designed to prevent contamination of samples by refractory containers. [53Sch], [54Ori], [67Rud], [74Ber], and [77Cez] claimed to have achieved blackbody conditions. [54Ori] and [74Ber] did not document the purity of their samples; other work used iodide titanium. [67Rud] verified that their samples were not contaminated during the melting experiments by checking for agreement of the βTi ⇄ αTi transformation temperatures before and after the melting point determination.

The high melting point values determined by [51Han], [52Ade], and [53May] are discredited by the later work. Of the other determinations, that of [67Rud] particularly recommends itself for the attention to accuracy of the temperature measurement and for the demonstration that contamination had been avoided. The values reported by [53Sch], [54Ori], [56Dea], [59Bic], [74Ber], and [77Cez] agree with that of [67Rud] within the reported uncertainties.

The previous assessment of the melting point [Melt] of 1670 °C (IPTS-68) is therefore accepted, with an uncertainty of about ±5 °C. In other recent assessments, [84Des] adopted 1772 ± 5 °C from [77Cez], and [79JAN] adopted 1666 ± 10 °C from [74Ber].

Allotropic Transformation βTi ⇄ αTi

At atmospheric pressure, Ti has the cph $A3$ structure at low temperature and transforms at 882 ± 2 °C to the high-temperature bcc $A2$ structure.

The transformation temperature value reported by [78Cez] is in serious conflict with the others (Table 2). The discrepancy is attributable to the use of pulse heating (2500 to 2700 °C/s); superheating of as much as 90 °C was reported for experiments where very high heating rates were used [74Mar].

The assessed transformation temperature of 882 ± 2 °C is based on phase diagram data for binary Ti systems as well as on direct measurements. When alloys of high purity are used, (αTi)/(βTi) phase boundary data are consistently found to extrapolate to 882 °C at pure Ti. The effect of O or N contamination is to raise the temperature of the transformation. Values as high as 900 °C have been reported in studies of binary systems, and the existence of an appreciable two phase αTi + βTi region in apparently pure Ti was due to contamination by O and/or N.

In other recent assessments, [84Des] adopted 883 °C (basis unknown) and [79JAN] adopted 893 °C from [78Cez]. The difference between the present value and that adopted by [84Des] is slight. The value adopted by [79JAN] is considered incorrect.

Thermodynamic Properties

The recent assessments of [84Des] and [79JAN] included tabulations of heat capacities, enthalpies, entropies, and Gibbs energies of the equilibrium Ti phases as a function of temperature. For the enthalpies of transformation and melting, the following values were chosen:

$\Delta_{trs}H = 4\,171 \pm 126$ J/mol [79JAN]

$\Delta_{trs}H = 3\,825 \pm 400$ J/mol [84Des]

$\Delta_{fus}H = 13\,818 \pm 500$ J/mol [84Des]

$\Delta_{fus}H = 14\,146 \pm 126$ J/mol [79JAN]

For binary phase diagram calculations, the "lattice stability parameters" of the pure elements are required. These are the enthalpies and entropies of transformation between various structures, including hypothetical forms such as fcc Ti. The approximation of zero heat capacity differences among phases is most often used, partly because of the often negligible effect of C_p differences on phase diagrams and partly because estimation of C_p differences among strictly unstable phases is usually impractical. The lattice stability parameters estimated by [70Kau], which are still in general use, are adopted:

$\Delta G(\text{bcc} \rightarrow \text{L}) = 16\,234 - 8.368\,T$

$\Delta G(\text{cph} \rightarrow \text{L}) = 20\,585 - 12.134\,T$

$\Delta G(\text{fcc} \rightarrow \text{L}) = 17\,238 - 12.134\,T$

Table 1   Experimental Determinations of the Melting Point of Pure βTi

Reference	Reported melting point, °C	Melting point (IPTS-68), °C
[51Han]	1720 ± 10	1723
[52Ade]	1700 ± 15	1703
[53May]	1680 ± 10	1683
[53Sch]	1660 ± 10	1663
[54Ori]	1672 ± 4	1675
[56Dea]	1660 ± 10	1663
[59Bic]	1667 ± 8	1670
[67Rud]	1668 + 4	1671
[74Ber]	1666 ± 4	1667
[77Cez]	1672 ± 5	1672
Assessed		1670 ± 5

Table 2   Experimental Determinations of the βTi ⇄ αTi Transformation Temperature

Reference	Transformation temperature (IPTS-68), °C	Experimental technique
[51Mcq]	883	Hydrogen pressure
[51Duw]	882 ± 4	Cooling curve
[52Kot]	881	Drop calorimetry
[57Sco]	883 ± 2	Adiabatic calorimetry
[67Rud]	882	DTA
[74Cor] .... ~ 882		Electrical resistivity
[76Etc]	882 ± 2	Dilatometry
[78Cez]	893 ± 6	Pulse heating and resistivity
[84Mca]	883 ± 2	DTA (cooling, heating)
Assessed	882 ± 2 °C	

1

Fig. 1---Effect of Temperature and Hydrostatic Pressure on the Equilibrium Crystal Structure of Pure Ti

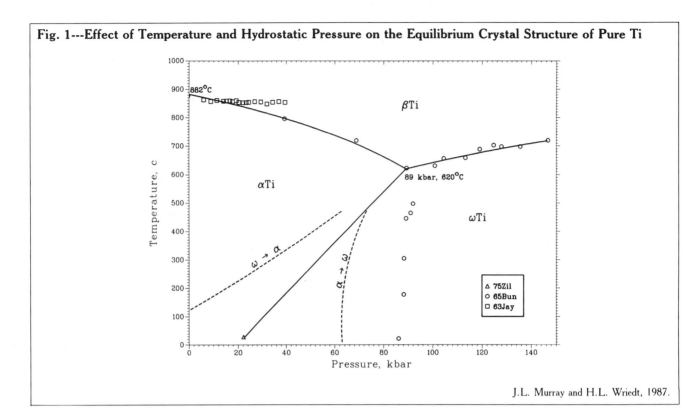

J.L. Murray and H.L. Wriedt, 1987.

Table 3 Room-Temperature Lattice Parameters of αTi

Reference	Lattice parameters, nm		Temperature, °C
	a	c	
[49Cla]	0.29504	0.46833	25 ± 2
[49Gre]	0.29450	0.46845	25
[50Fin]	0.29504	0.46834	RT
[55Sza]	0.29506	0.46788	25
[59Spr]	0.29506	0.46797	22
[62Woo]	0.29511	0.46843	25 ± 2
[68Paw]	0.29508	0.46855	28
[68Sch]	0.29503	0.46810	20
[77Dec]	0.29512	0.46826	21 ± 1

Table 4 Coefficients of Linear Thermal Expansion for αTi

Reference	Coefficient		Temperature range, °C
	$\frac{10^6}{a}\frac{da}{dT}$, °C^{-1}	$\frac{10^6}{c}\frac{dc}{dT}$, °C^{-1}	
[59Spr]	9.55 ± 0.5	10.65 ± 0.7	25 to 700
[53Mch] ...	11.0	8.8	25 to 225
[68Paw] ...	9.5	5.6	28 to 155
[53Ber]	11.03	13.37	25 to 700
[68Sch]	~6	~8	−200
	~10	~12.7	400
	~16	~18.7	882

Estimates of the Gibbs energies of Ti in other structural forms, such as the double cph structures encountered in the rare earth systems, are listed in the appropriate binary system evaluations.

Pressure-Temperature Diagram

The effect of pressure on the temperature of the αTi/βTi equilibrium was determined for purified Ti by [63Jay] and [65Bun]. The data of [63Jay] extrapolate to 860 °C at zero pressure and are therefore inconsistent with the accepted transition point of 882 °C. These low values of transition temperatures are attributed, tentatively, to alloying of the Ti with Mo and Ta containers.

High hydrostatic pressure causes αTi to transform to the hexagonal ω structure, which persists metastably after reduction of the pressure to 0.1 MPa [63Jam]. A first-order transition, presumably αTi → ωTi, was found in purified Ti at 8.5 ± 0.5 GPa at room temperature [63Jay]. From his own experimental data, [65Bun] constructed a pressure-temperature equilibrium diagram

showing the boundaries between regions of stability for the αTi, βTi, and ωTi phases. The triple point was shown at about 9.4 GPa at 625 °C; coexistence of ωTi with αTi at 0 °C was shown at 8.5 GPa, in agreement with the [63Jay] result. [73Zil] found that, at room temperature, the equilibrium pressure for coexistence of αTi and ωTi is approximately 2.0 GPa; the discrepancy from earlier work was explained by the sensitivity of the equilibrium to departures from purely hydrostatic stressing.

The only available data on the pressure-temperature relationship for the equilibrium coexistence of βTi and ωTi phases are those of [65Bun], which show the transition temperature increasing with increasing pressure above the triple point, initially with a slope of about 22 °C/GPa.

The assessed temperature-pressure diagram (Fig. 1) is based on the work of [65Bun] for the αTi/βTi and βTi/ωTi equilibria and on [73Zil] for the αTi/ωTi equilibrium. Another transformation of αTi, caused by high pressure applied dynamically at room temperature, was reported by [70Ger]. Shock waves producing 35 GPa pres-

Fig. 2---Effect of Temperature on the Lattice Parameters of Ti

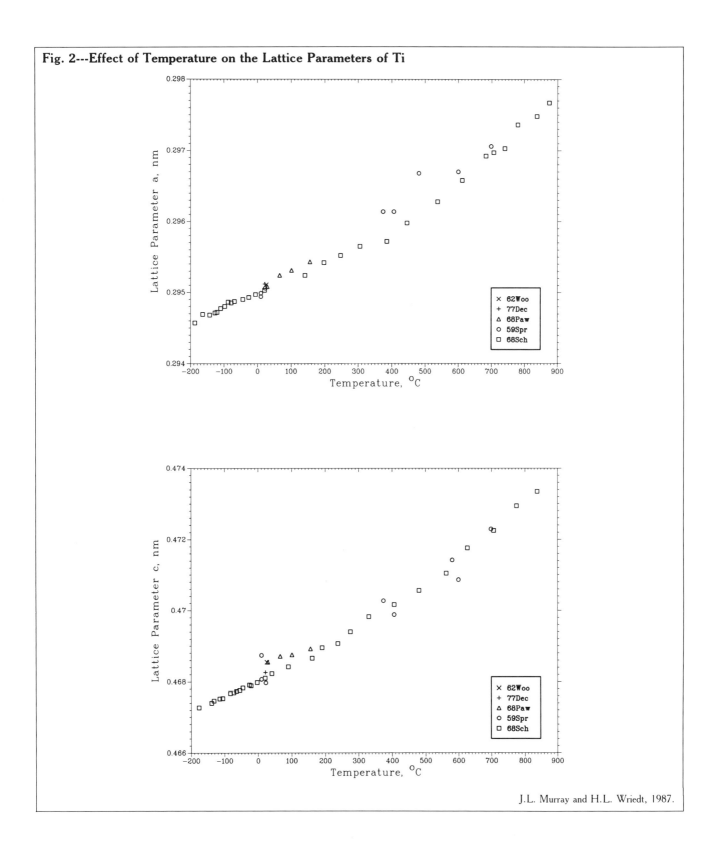

J.L. Murray and H.L. Wriedt, 1987.

sure effected a transformation to a "new," possibly bcc phase. Although not indicated by [70Ger], it seems possible that the "new" phase is βTi, both from the reported magnitude of the lattice parameter and consistency with the trend of the metastable temperature-pressure relationship for the αTi/βTi equilibrium.

Lattice Parameters

Room-temperature lattice parameter data for αTi are listed in Table 3. [62Woo] used samples of very high purity compared to that in any previous measurement; the [62Woo] values were adopted by [Pearson2]. More re-

3

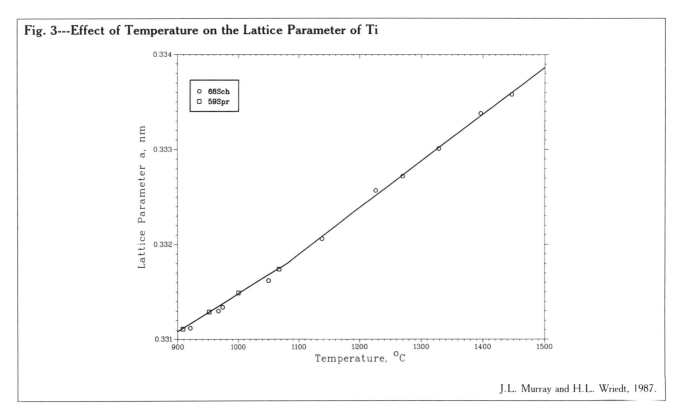

Fig. 3---Effect of Temperature on the Lattice Parameter of Ti

J.L. Murray and H.L. Wriedt, 1987.

Table 5 βTi Lattice Parameters

Reference	Lattice parameter, nm	Temperature, °C
[50Epp]	0.33065	900
[59Spr]	0.33111	908
	0.33129	951
	0.33149	1000
	0.33174	1067
[68Sch]	0.33112	920
	0.33130	966
	0.33134	973
	0.33162	1050
	0.33206	1138
	0.33257	1226
	0.33272	1270
	0.33301	1329
	0.33338	1397
	0.33358	1447

cently, [68Sch] and [77Dec] measured lattice parameters of exceptionally high-purity Ti. The measurements by [77Dec] were made as a function of O content on samples with 90 to 4000 at. ppm O, and results were extrapolated to zero O content. These results are accepted. Measurements of the coefficient of linear thermal expansion are summarized in Table 4; and lattice parameters are shown as a function of temperature in Fig. 2. The work of [68Sch] is preferred; the thermal expansion coefficient is strongly temperature dependent.

Considerably less work has been done on the lattice parameter of βTi, because, for an accurate measurement, samples must be maintained above 882 °C without contamination. The measurements of [50Epp], [59Spr], and [68Sch] are listed in Table 5 and plotted in Fig. 3. [59Spr]

noted some contamination of their specimens. Based on [59Spr] and [68Sch], the linear thermal expansion coefficient is about 10.8×10^{-6}/K at 900 to 1100 °C and 14.7×10^{-6}/K at 100 to 1500 °C. These expressions lead to a lattice parameter of 0.32763 nm at room temperature. This value is comparable to values obtained by extrapolation of lattice parameters of quenched binary alloys (Ti-Cr and Ti-V) to 100% Ti. The lattice parameter at 900 °C is 0.33106 nm, which differs slightly from the value of 0.33065 nm [Pearson2] from [50Epp].

Lattice parameters of quenched ωTi are $a = 0.2813$ nm and $c = 0.4625$ nm [63Jam].

Cited References

49Cla: H.T. Clark, "The Lattice Parameters of High Purity αTi; and the Effects of Oxygen on Them," *Metall. Trans., 185,* 588-589 (1949). (Crys Structure; Experimental)

49Gre: E.S. Greiner and W.E. Ellis, "Thermal and Electrical Properties of Ductile Titanium," *Trans. AIME, 180, 657-665 (1949). (Crys Structure; Experimental)*

50Epp: D.S. Eppelsheimer and R.R. Penman, "Accurate Determination of the Lattice of βTi at 900 °C," *Nature, 166,* 960 (1950). (Crys Structure; Experimental)

50Fin: W.L. Finlay and J.A. Snyder, "Effects of Three Interstitial Solutes (Nitrogen, Oxygen, and Carbon) on the Mechanical Properties of High-Purity, αTi," *Trans. AIME, 188,* 277-286, 1368-1369 (1950). (Experimental)

51Duw: P. Duwez, "Effect of Rate of Cooling on the Alpha-Beta Transformation in Titanium and Titanium-Molybdenum Alloys," *J. Met., 3,* 765-771 (1951). (Allotropic Transformation; Experimental)

51Han: M.A. Hansen, E.L. Kaman, H.D. Kessler, and D.J. McPherson, "Systems Titanium-Niobium," *J. Met., 3,* 881-888 (1951). (Meta Phases; Experimental)

51Mcq: A.D. McQuillan, "Some Observations on the α ⇄ β Transformation in Titanium," *J. Inst. Met., 78,* 249-257 (1951). (Allotropic Transformation; Experimental)

52Ade: H.K. Adenstedt, J.R. Pequingnot, and J.M. Raymer, "The Titanium-Vanadium System," *Trans. ASM, 44,* 980-1003 (1952). (Meta Phases; Experimental)

52Kot: C.W. Kothen, "The High Temperature Heat Contents of Molybdenum and Titanium and the Low Temperature Heat Capacities of Titanium," Diss., Ohio State Univ., Publ. No. 52-23697, 1-89 (1952). (Allotropic Transformation; Experimental)

53Ber: L.P. Berry and G.V. Raynor, "A Note on the Lattice Spacings of Titanium at Elevated Temperatures," *Research, 6,* 21S-23S (1953). (Crys Structure; Experimental)

53May: D.J. Maykuth, H.R. Ogden, and R.I. Jaffee, "Titanium-Tungsten and Titanium-Tantalum Systems," *Trans. AIME, 197,* 231-237 (1953). (Meta Phases; Experimental)

53Mch: C.J. McHargue and P. Hammond, "Deformation Mechanisms in Titanium at Elevated Temperatures," *Acta Metall., 1,* 700-705 (1953). Crys Structure; Experimental)

53Sch: T.H. Schofield and A.E. Bacon, "The Melting Point of Titanium," *J. Inst. Met., 82,* 167-169 (1953-1954). (Meta Phases; Experimental)

54Ori: R.A. Oriani and T.S. Jones, "An Apparatus for the Determination of the Solidus Temperatures of High-Melting Alloys," *Rev. Sci. Instr., 25*(3), 248-250 (1954). (Meta Phases; Experimental)

55Sza: I. Szanto, "On the Determination of High-Purity αTi Lattice Parameters," *Acta Tech. Acad. Sci. Hung., 13,* 363-372 (1955). (Crys Structure; Experimental)

56Dea: D.K. Deardorff and E.T. Hayes, "Melting Point Determination of Hafnium, Zirconium, and Titanium," *Trans. AIME, 206,* 509-511 (1956). (Meta Phases; Experimental)

57Sco: J.L. Scott, "Calorimetric Investigation of Zirconium, Titanium, and Zirconium Alloys from 60 to 860 °C," U.S. Atom. Energy Comm., ORNL-2328 (1957). (Allotropic Transformation; Experimental)

59Bic: R.L. Bickerdike and G. Hughes, "An Examination of Part of the Titanium-Carbon System," *J. Less-Common Met., 1,* 42-49 (1959). (Meta Phases; Experimental)

59Spr: J. Spreadborough and J.W. Christian, "The Measurement of the Lattice Expansions and Debye Temperatures of Titanium and Silver by X-Ray Methods," *Proc. Phys. Soc., (London), 74,* 609-615 (1959). (Crys Structure; Experimental)

62Woo: R.M. Wood, "The Lattice Constants of High Purity αTi," *Proc. Phys. Soc., 80,* 783-786 (1962). (Crys Structure; Experimental)

63Jam: J.C. Jamieson, "Crystal Structures of Titanium, Zirconium, and Hafnium at High Pressures," *Science, 140,* 72-73 (1963). (Pressure; Experimental)

63Jay: A. Jayaraman, W. Klement, and G.C. Kennedy, "Solid-Solid Transitions in Titanium and Zirconium at High Pressures," *Phys. Rev., 131*(2), 644-649 (1963). (Pressure; Experimental)

65Bun: F.P. Bundy, "Formation of New Materials and Structures by High Pressure Treatment," ASTM Special Technical Publication No. 374, *Irreversible Effects of High Pressure and Temperature on Materials,* (ASTM Matls. Sci. Ser. 7), 52-64 (1965). (Pressure; Experimental)

67Rud: E. Rudy and J. Progulski, "A Pirani Furnace for the Precision Determination of the Melting Temperatures of Refractory Metallic Substances," *Plansee Pulver., 15,* 13-45 (1967). (Meta Phases; Experimental)

68Paw: R.R. Pawar and V.T. Deshpande, "The Anisotropy of the Thermal Expansion of αTi," *Acta Crystallogr., A24,* 316-317 (1968). (Crys Structure; Experimental)

68Sch: N. Schmitz-Pranghe and P. Dunner, "Crystal Structure and Thermal Expansion of Scandium, Titanium, Vanadium and Manganese," *Z. Metallkd., 59,* 377-382 (1968) in German. (Crys Structure; Experimental)

70Ger: V.N. German, A.A. Bakanova, L.A. Tarasova, and Yu.N. Sumulov, "Phase Transformation of Titanium and Zirconium in Shock Waves," *Fiz. Tverd. Tela, 12*(2), 637-639 (1970) in Russian; TR: *Sov. Phys. Solid State, 12*(2), 490-491 (Pressure; Experimental)

70Kau: L. Kaufman and H. Bernstein, *Computer Calculation of Phase Diagrams,* Academic Press, New York (1970). (Thermo; Theory)

73Zil: V.A. Zil'Bershteyn, G.I. Nosova, and E.I. Estrin, "α ⇄ ω Transformation in Titanium and Zirconium," *Fiz. Metal. Metalloved., 35*(3), 584-589 (1973) in Russian; TR: *Phys. Met. Metallogr., 35*(3), 128-133 (1973). (Pressure; Experimental)

74Ber: B.Ya. Berezin, S.A. Kats, M.M. Kenisarin, and V.Ya. Chekhovskoi, "Heat and Melting Temperature of Titanium," *Tepf. Vy. Temp., 12*(3), 524-529 (1974) in Russian; TR: *High Temp., 12*(3), 450-455 (1974). (Meta Phases; Experimental)

74Cor: M. Cormier and F. Claisse, β ⇄ α Phase Transformation in Ti and Ti-O Alloys," *J. Less-Common Met., 34*(2), 181-189 (1974). (Allotropic Transformation; Experimental)

74Mar: M.M. Martynyuk and V.I. Tsapkov, "The Resistivity, Enthalpy and Phase Transitions of Titanium, Zirconium and Hafnium During Pulse Heating," *Izv. Akad. Nauk SSSR, Met.,* (2), 181-188 (1974) in Russian. (Allotropic Transformation; Experimental)

76Etc: E. Etchessahar and J. Debuigne, "Study of the Allotropic Transformations of High-Purity Ti and of Zr by Dilatometry in an Ultra-High Vacuum," *C.R. Acad. Sci. Paris (Ser. C), 283,* 63-66 (1976) in French. (Allotropic Transformation; Experimental)

77Cez: A. Cezairliyan and A.P. Miller, "Melting Point, Normal Spectral Emittance (at the Melting Point), and Electrical Resistivity (above 1900 K) of Titanium by a Pulse Heating Method," *J. Res. Nat. Bur. Standards, 82*(2), 119-122 (1977). (Meta Phases; Experimental)

77Dec: M. Dechamps, A. Quivy, G. Bauer, and F. Lehr, "Influence of the Distribution of the Interstitial Oxygen Atoms on the Lattice Parameters in Dilute H.C.P. Titanium-Oxygen Solid Solutions (90-4000 ppm at)," *Scr. Metall., 11,* 941-945 (1977). (Crys Structure; Experimental)

78Cez: A. Cezairliyan and A.P. Miller, "Thermodynamic Study of the α ⇄ β Phase Transformation in Titanium by a Pulse Heating Method," *J. Res. Nat. Bur. Stand., 83*(2), 127-132 (1978). (Allotropic Transformation; Experimental)

84Mca: A.J. McAlister, private communication (1984). (Allotropic Transformation; Experimental)

84Des: P.D. Desai, "Thermodynamic Properties of Titanium," CINDAS Report 77 (1984). (Thermo; Compilation)

85Cha: M.W. Chase, Jr., C.A. Davis, J.R. Downey, Jr., D.J. Frurip, R.A. McDonald, and A.N. Syverud, "JANAF Thermochemical Tables," 3rd ed., *J. Phys. Chem. Ref. Data, 14,* Suppl. No. 1 (1985). (Thermo, Compilation)

The Ag-Ti (Silver-Titanium) System

107.8682   47.88

By J.L. Murray and K.J. Bhansali

Equilibrium Diagram

The first experimental studies of the Ti-Ag system led to the supposition that the two components are completely immiscible in the liquid state [52Rau] and that no intermediate phases are formed [43Wal]. Later studies revealed a wide two-phase [(βTi) + L] region, which makes it difficult to prepare homogeneous alloys. The existence of two compounds and the outlines of the diagram are now well established (see Table 1).

The major studies of the system were conducted by [53Ade], [53Wor], [60Mcq], [69Ere], and [78Pli]. [59Mor] also made a comprehensive study of the Ti-Ag system; however, the alloys used for this study were contaminated by interstitials, probably oxygen. The evidence for this is the wide melting range of nominally pure Ti and the brittleness of the alloys. Therefore, the work of [59Mor] has not been considered in the present assessment.

The assessed diagram (Fig. 1) is drawn from the present thermodynamic calculations; the calculations are

Table 1   Special Points of the Assessed Ti-Ag Phase Diagram

Reaction	Compositions of the respective phases, at.% Ag			Temperature, °C	Reaction type
Assessed					
(βTi) + L ⇌ TiAg	15.5	~ 94	~ 48	1020 ± 5	Peritectic
(βTi) + TiAg ⇌ Ti₂Ag	~ 12	~ 48	33.3	940 ± 5	Peritectoid
(βTi) ⇌ (αTi) + Ti₂Ag	7.6	~ 4.7	33.3	855 ± 5	Eutectoid
L ⇌ TiAg + (Ag)	~ 95	~ 50	~ 95	959 ± 1	Eutectic
(βTi) ⇌ (αTi)		0		882	Allotropic transformation
L ⇌ (βTi)		0		1670	Melting point
L ⇌ (Ag)		100		961.93	Melting point
Calculated					
(βTi) + L ⇌ TiAg	15.5	92.4	50	1019	Peritectic
(βTi) + TiAg ⇌ Ti₂Ag	11.8	50	33.3	941	Peritectoid
(βTi) ⇌ (αTi) + Ti₂Ag	7.5	4.6	33.3	856	Eutectoid
L ⇌ TiAg + (Ag)	94.4	50	94.6	960	Eutectic

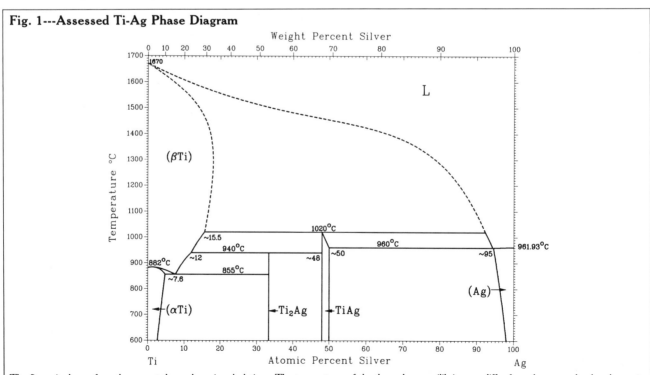

Fig. 1---Assessed Ti-Ag Phase Diagram

The figure is drawn from the present thermodynamic calculations. The temperatures of the three-phase equilibria may differ from the assessed values by up to 1 °C, which is within the accuracy of the data (see Table 1).

J.L. Murray and K.J. Bhansali, 1987.

Fig. 2---Ti-Rich Portion of the Diagram vs Experimental Data

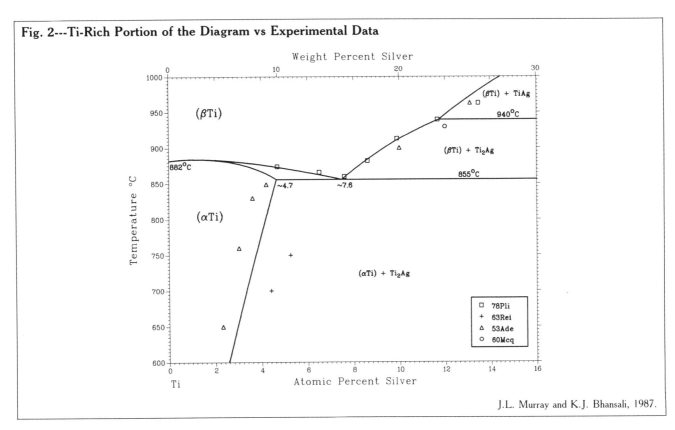

J.L. Murray and K.J. Bhansali, 1987.

Table 2 Peritectic Reaction (βTi) + L ⇄ TiAg

Reference	(βTi) composition, at.% Ag	Liquid composition, at.% Ag	Temperature, °C	Method
[53Dec]	...	...	1000	Melting of TiAg
[53Ade]	16.3 ± 0.7	...	1038 ± 3	Metallography
	12 ± 0.5	...	1427	(incipient melting)
	6.7	...	~ 1608	
[60Mcq]	~ 15	~ 90	1017 to 1030	Metallography
	~ 15	...	1400	
	> 15	...	1300	
[69Ere]	...	~ 94	1020 ± 5	Thermal analysis

consistent with the assessment of the experimental data, within the experimental accuracy. The Ti-rich region of the diagram is shown in Fig. 2.

The equilibrium solid phases of the Ti-Ag system are:

- bcc (βTi) solid solution, with a homogeneity range of 0 to 15.5 at.% Ag
- cph (αTi) solid solution, with a homogeneity range of 0 to about 4.7 at.% Ag
- fcc (Ag) solid solution, shown in Fig. 1 with a range of 95 to 100 at.% Ag
- Essentially stoichiometric Ti_2Ag, with the $MoSi_2$ structure
- TiAg with the γTiCu structure and a composition range of approximately 48 to 50 at.% Ag

(βTi) Solidus and Liquidus.

Addition of Ag to (βTi) lowers the melting point; a peritectic reaction L + (βTi) ⇄ TiAg occurs at 1020 ± 5 °C [53Ade, 60Mcq, 69Ere]. The compositions of (βTi) and liquid in equilibrium at 1020 °C are 15.5 ± 0.5 and 94 ±

2 at.% Ag. The liquidus has the shape associated with a metastable liquid miscibility gap; it is steep at Ti- and Ag-rich compositions, but is nearly flat at about 1400 °C.

A difficulty in experimental work on Ti-Ag is the loss of Ag by volatilization during melting. This has prevented direct quantitative determination of the liquidus. Flattening of the liquidus was observed by [53Ade] as a change in volume fraction of liquid as a function of the temperature of the heat treatment. The only quantitative data on the liquid/(βTi) phase relationships are incipient melting and metallographic determinations of the (βTi) solidus [53Ade] and the composition of the liquid in the peritectic equilibrium (Table 2).

The present estimate for the liquidus is based on thermodynamic optimization of the Gibbs energies with respect to experimental data on the solidus and peritectic reaction. The liquidus and solidus are drawn in Fig. 1 with dashed curves to indicate their uncertainty.

Solidus data of [53Ade] and [60Mcq] are mutually consistent; they reported the maximum solubility of Ag in (βTi) at the peritectic temperature as 16.3 and about 15

Table 3 (αTi) and (βTi) Boundaries of the Ti-Ag System

Reference	Compositions, at.% Ag		Temperature, °C	Experimental method
	(αTi) solvus	β transus		
[78Pli]	< 4.7	7.6	860 ± 5(a)	Microprobe, metallography
	...	6.5	866	
	...	4.7	874	
[69Ere]	...	...	851 ± 5(a)	Thermal analysis
[53Ade]	4.2 ± 0.5	8.3 ± 0.2	849 ± 4(a)	Metallography
	3.6	...	830	
	3.0	...	760	
	~ 2.3	...	650	
[53Wor]	7	11.5	855 ± 5(a)	Metallography, lattice parameters
[63Rei]	5.25	...	750	Microhardness
	4.4	...	700	
	2.55	...	600	

(a) Eutectoid temperature.

at.%, respectively. Additional solidus data are summarized in Table 2. Both experimental studies suggested a bending of the solidus in the vicinity of 1400 °C; however, [60Mcq] pointed out that the high melting temperatures of 12 at.% Ag alloys may be a spurious effect resulting from the volatilization of Ag. Although present thermodynamic calculations reproduce the retrograde solidus and the available liquidus data, in the absence of thermochemical data, the calculated liquidus and solidus are not unique, and they must be understood to be uncertain.

The liquidus composition at the peritectic temperature was determined by [69Ere], who estimated the liquid composition to be about 94 at.% Ag, where peritectic arrests were observed by thermal analysis. Because of the small volume fraction of (βTi), this may underestimate the Ti content of the liquid.

Eutectic Reaction and Solubility of Ti in (Ag).

The type of the reaction involving (Ag), L, and TiAg is difficult to determine experimentally. [53Ade], [53Dec], and [69Ere] reported that the reaction is probably of the eutectic type and occurs close to the melting point of pure Ag. [60Mcq], however, reported that the reaction is of the peritectic type and occurs at 1017 °C. The eutectic reaction L ⇌ (Ag) + TiAg at 959 °C [69Ere] is accepted here.

[60Mcq] heat treated and quenched alloys from successively higher temperatures, seeking metallographic evidence that melting had begun. They noted that during heat treatment alloys were distorted and in some instances had lost more than half their weight. The present evaluators conclude that any liquid formed during heat treatment was lost, until the peritectic temperature (1020 °C) was reached. At this temperature, the volume fraction of liquid increased and the liquid could be observed metallographically.

[69Ere], on the other hand, used differential thermal analysis with pure Ag as the reference alloy. They determined that a Ag-Ti alloy melted approximately 1 °C below pure Ag and that therefore the reaction is of the eutectic type. The method used by [69Ere] does not suffer the same difficulties as the metallographic technique.

Based on thermal analysis of a series of alloys, [69Ere] estimated the eutectic composition to be about 95 at.% Ag. Because the eutectic temperature is so close to the melting point of pure Ag, van't Hoff's law requires that the two-phase region be very narrow and the solubility of Ti in (Ag) must be about 5 at.%. This conflicts with an observation by [53Ade] of two phases in an as-cast 98.9

at.% Ag alloy. Because [69Ere] used purer starting materials, their results are preferred. Further work is needed to determine the (Ag) solvus.

(αTi)/(βTi) Boundaries and the Eutectoid Reaction.

Ag is a slightly β-stabilizing addition to Ti; the eutectoid reaction (βTi) → (αTi) + Ti$_2$Ag occurs at 855 ± 5 °C and 7.6 ± 0.2 at.% Ag [53Ade, 78Pli]. Experimental data on the (αTi)/(βTi) and the (αTi)/(αTi) + Ti$_2$Ag boundaries are summarized in Table 3. For the eutectoid temperature, reports range between 849 ± 5 °C and 860 ± 5 °C and are consistent with 855 ± 5 °C.

For the assessed (βTi) transus and eutectoid reaction, the work of [78Pli] is chosen, because of the high purity of the materials used to prepare alloys. Conflicting results all tend to place the phase boundaries at higher Ag content. This can be understood as the effect of oxygen contamination, oxygen being a stabilizer of the cph structure. The work of [53Ade] is in best agreement with that of [78Pli]; therefore, the present evaluators conclude that oxygen contamination was minimized in this work and accept it for the (αTi) solvus.

The phase boundaries (Fig. 1) are drawn from the present thermodynamic calculations, which have some interesting features. The experimental data on the (αTi)/(βTi) phase field deviate significantly from van't Hoff's law, which must be strictly observed in the limit of infinite dilution. Regular solution models do not reproduce these phase boundaries accurately, rather they approximate the van't Hoff relation down to the eutectoid temperature. Subregular solution models that do reproduce the data predict a congruent (αTi)/(βTi) transformation very near the transformation of pure Ti. A congruent transformation is not ruled out by the data of [53Ade], or by any other observations.

Compounds.

In the literature, there is some confusing nomenclature on the compound phases, which stems from early misidentifications of the structure (see below). For example, [60Mcq] reported a phase at 32 at.% Ag that they designated Ti$_3$Ag. This nomenclature persists in the recent literature. In fact, the correct stoichiometries are Ti$_2$Ag and TiAg, and no observations have been found in the literature of other compounds.

[69Ere] reported that TiAg has a homogeneity range of 2 at.%, but no further details were given. [60Mcq] reported that 46.8 at.% Ag alloys were single-phase TiAg. [59Wit] identified the composition of TiAg formed from

Table 4 Peritectoid Reaction and β/(β + Ti₂Ag) Boundary

Reference	(βTi) composition, at.% Ag	Temperature, °C	Method
[78Pli]	13.5	963	Metallography, microprobe
	11.7	940	
	9.9	913	
	8.6	882	
[53Ade]	~ 10	900	Metallography
	13.1	963 ± 10	
[60Mcq]	12	930(a)	Metallography
[69Ere]	...	945 ± 5(a)	Thermal analysis, heating

(a) Peritectoid temperature.

Table 5 Crystal Structures of the Ti-Ag System

Phase	Homogeneity range, at.% Ag	Pearson symbol	Space group	Strukturbericht designation	Prototype	Reference
(αTi)	0 to ~4.7	hP2	$P6_3/mmc$	A3	Mg	[Pearson2]
(βTi)	0 to 15.5	cI2	Im3m	A2	W	[Pearson2]
Ti₂Ag	33.3	tI6	I4/mmm	$C11_b$	MoSi₂	[64Sch, 65Sch]
TiAg	48 to 50	tP4	P4/nmm	B11	γCuTi	[64Sch, 65Sch]
(Ag)	95 to ~100	cF4	Fm3m	A1	Cu	[Pearson2]

Table 6 Ti-Ag Lattice Parameters

Phase	Composition, at.% Ag	Lattice parameters, nm		Reference
		a	c	
(αTi)	0	0.2951	0.4685	[53Wor]
	2.75	0.2945	0.4692	
	6.6	0.29405	0.47007	
Ti₂Ag	33.3	0.295	1.185	[64Sch, 65Sch]
	33.3	0.295	1.182	[66Ere]
	33.3	0.296(a)	1.185(a)	[53Wor]
TiAg	50	0.290	0.814	[64Sch, 65Sch]
	50	0.290	0.812	[66Ere]
	50	0.290(a)	0.815(a)	[53Thy]

(a) Converted to the accepted unit cell (see text).

molten Ag at 1000 °C (in a nonequilibrium experiment) as 49.88 at.% Ag by microprobe analysis. Based on these findings, the maximum composition range of TiAg is drawn as 48 to 50 at.% Ag, and an uncertainty of ± 1 at.% Ag is estimated for each of these compositions.

[60Mcq] and [69Ere] reported that the composition range of Ti₂Ag is narrow. [53Wor] placed 28 and 30 at.% Ag alloys in the single-phase Ti₂Ag field, but these data were rejected in the present assessment because the alloys were made with commercial titanium and the effect of interstitials is evident throughout the study.

The peritectoid reaction (βTi) + TiAg ⇄ Ti₂Ag occurs at 945 ± 5 °C according to [69Ere] (thermal analysis, heating), or at 930 °C [60Mcq] (metallography). Data on the peritectoid temperature and β/(β ± Ti₂Ag) boundary are summarized in Table 4. For the present assessment, 940 ± 5 °C is used.

Metastable Phases

The bcc (βTi) phase can transform to cph (αTi) partitionlessly during quenching from the (βTi) field. The mechanism of transformation is massive or martensitic,

depending on cooling rate and composition. [77Pli, 78Pli] demonstrated the existence of the massive transformation by microstructural and microprobe analysis. Higher cooling rates or higher Ag content favored the martensitic over the massive mechanism. Nearly 100% of a 4.7 at.% Ag alloy transformed to massive (αTi) upon an iced brine quench, whereas 10 to 50% of a 13.5 at.% Ag alloy transformed massively, the remainder transforming martensitically. [60Sat] determined the start temperature of the martensitic transformation (M_s) to be about 705 and 710 °C for 5.3 and 8.3 at.% Ag alloys, respectively.

[78Pli] also made thermal analysis measurements as a function of cooling rate. They extrapolated thermal arrest temperatures to zero cooling rate using a model for nucleation and growth kinetics to estimate T_0 temperatures for the (αTi)/(βTi) transition. The uncertainty in the T_0 thus derived is too large to permit a quantitative comparison with the phase diagram calculations. The temperature found by this method in one instance lies above the metastable (βTi) transus and near it for another alloy.

Phase Diagrams of Binary Titanium Alloys

Crystal Structures and Lattice Parameters

The crystal structures of the equilibrium phases are summarized in Table 5 and the lattice parameters are listed in Table 6. The compounds, Ti_2Ag and TiAg, originally were assigned ordered fct structures, $L1_2$ and $L1_0$, respectively [53Wor, 53Thy]. Later investigation revealed that the structures are bct [65Sch, 69Ere], a finding which parallels work on the structures of the Ti-Cu compounds. For Ti_2Ag, the unit cells are related by:

$$a_{bct}^2 = 2\, a_{fct}^2$$

$$c_{bct} = 3\, c_{fct}$$

For TiAg, the unit cells are related by:

$$a_{bct}^2 = 2\, a_{fct}^2$$

$$c_{bct} = 2\, c_{fct}$$

In Table 6, the lattice parameters are, where necessary, corrected to give the correct unit cell. Lattice parameter data are not available for (βTi), because the bcc structure cannot be retained during quenching. Lattice parameters of (αTi) are given in Table 6.

Thermodynamics

From the phase diagram, estimates can be made of Gibbs energies of the equilibrium phases. On the one hand, only rough estimates can be expected from an analysis based on phase diagram data alone, especially if much uncertainty is attached to it. On the other hand, no experimental thermodynamic data have yet been obtained, nor have theoretical calculations of Gibbs energies been performed. Therefore, the present thermodynamic analysis represents the first estimate of the thermodynamic properties of this system.

In this calculation, the Gibbs energies of the solution phases are represented by a subregular solution polynomial expansion:

$$G(i) = (1 - x)\, G^0(Ti,i) + x\, G^0(Ag,i) + RT\,[x \ln x + (1 - x)$$
$$\ln (1 - x)] + x(1 - x)\,[B(i) + C(i)\,(1 - 2x)]$$

where i designates the phase; x is the atomic fraction of Ag; $G^0(i)$ designates the Gibbs energies of the pure metals; and $B(i)$ and $C(i)$ are the interaction parameters. Lattice stability parameters are from [70Kau], except for that of cph Ag, which was estimated. Ti_2Ag and TiAg are represented as line compounds.

For the liquid/(βTi) boundaries, the input data are as follows. The maximum solubility of Ag in (βTi) is 15.5 to 16.3 at.% at the peritectic temperature 1020 °C, and the composition of the liquid in equilibrium is approximately 94 at.% Ag. The experimental data suggest that the solidus is retrograde. The liquidus descends steeply on the Ti- and Ag-rich sides of the diagram, but tends to flatten between 40 and 70 at.% Ag. The wide two-phase region and inflection in the liquidus, together with the probably retrograde solidus, strongly suggest that despite the existence of compounds, metastable miscibility gaps exist in both the liquid and (βTi) solutions. This was verified by preliminary calculations. An upper boundary on the interaction Gibbs energies is imposed by the requirement that there be no monotectic equilibrium separation of two liquid phases. This means that regular solution interaction parameters must be less than about 27 000 J/mol for each phase. Least-squares optimizations of liquid and bcc phase parameters led to the values given in Table 7. From

Table 7 Thermodynamic Properties of the Ti-Ag System

Gibbs energies of the pure components

$G^0(Ti,L)$	$= 0$
$G^0(Ag,L)$	$= 0$
$G^0(Ti,bcc)$	$= -16\,234 + 8.368\,T$
$G^0(Ag,bcc)$	$= -8\,180 + 7.782\,T$
$G^0(Ti,cph)$	$= -20\,585 + 12.134\,T$
$G^0(Ag,cph)$	$= -10\,000 + 9.665\,T$
$G^0(Ti,fcc)$	$= -17\,238 + 12.134\,T$
$G^0(Ag,fcc)$	$= -11\,945 + 9.665\,T$

Solution phases

$B(L)$	$= 24\,873$
$C(L)$	$= 327$
$B(bcc)$	$= 24\,432$
$B(cph)$	$= 47\,950$
$C(cph)$	$= -25\,000$

Note: Values are given in J/mol and J/mol · K.

the available data, it is not possible to estimate excess entropies, and they are assumed to be zero.

The excess Gibbs energy of the fcc phase was determined from the constraint of a eutectic reaction of TiAg, L, and (Ag) at approximately 959 °C. Except for the eutectic reaction temperature and approximate composition, nothing is known about the phase boundaries of the fcc (Ag) phase. Therefore, the fcc Gibbs energy given in Table 7 should only be used in the dilute solution limit.

Next, the Gibbs energy of the cph phase (αTi) is considered. The melting entropy of cph Ag was roughly estimated as equal to that of fcc Ag. The melting enthalpy was set to $-10\,000$ J/mol; this approximately equals other Ag melting enthalpies and satisfies the constraint that the melting temperature of cph Ag must lie below the melting point of the stable fcc solid. Using regular solution models for the bcc and cph phases, it was found that the compositions of coexisting phases at the eutectoid temperature could not be reproduced by thermodynamic calculations, the width of the (βTi) + (αTi) field being about 2 at.% too small. A subregular solution term is needed to represent that Gibbs energies; however, it was also found that a congruent (βTi) $\rightarrow$ (αTi) transformation appears for large subregular contributions. The congruent transformation in the present calculations occurs very near the $\beta \rightarrow \alpha$ transformation of Ti and would not have been observed in any of the experiments performed to date. The Gibbs energies of Ti_2Ag and TiAg were optimized with respect to the phase boundary data and the temperatures of the three-phase equilibria. These data are reproduced within experimental error.

In summary, the Gibbs energies of Table 7 reproduce what is known about this diagram quite accurately. They are, however, first approximations and in need of further refinement.

Cited References

43Wal: H.J. Wallbaum, "Transition Metal Alloys," *Naturwissenschaften, 31*, 91-92 (1943). (Equi Diagram; Experimental)

52Rau: E. Raub, P. Walter, and M. Engel, "Alloys of Titanium with Copper, Silver and Gold," *Z. Metallkd., 43*, 112-118 (1952). (Equi Diagram; Experimental)

*53Ade: H.K. Adenstedt and W. Freeman, "The Tentative Titanium-Silver Binary System," WADC Tech. Rept., Part 1, 53-109 (1953). (Equi Diagram; Experimental)

The Ag-Ti System

53Dec: N.A. Dececco and J.M. Parks, "The Brazing of Titanium," *Welding J.*, 1071-1073 (1953). (Equi Diagram; Experimental)

53Thy: R.J. Van Thyne, W. Rostoker, and H.D. Kessler, "Observations on the Phase TiAg," *J. Met.*, 670-671 (1953). (Crys Structure; Experimental)

53Wor: H.W. Worner, "The Structure of Titanium-Silver Alloys in the Range 0-30 at.% Silver," *J. Inst. Met.*, 82, 222-226 (1953). (Equi Diagram; Experimental)

59Mor: T. Moringa, I. Miura, and T. Takuai, "On the Phase Diagram of the Titanium-Silver System," *J. Jpn. Inst. Met.*, 23(2), 117 (1959). (Equi Diagram; Experimental)

59Wit: D.B. Wittry, "Metallurgical Applications of Electron Probe Microanalysis," Advances in X-Ray Analysis, Proc. 8th Annual Conf. on Applied X-ray Analysis, W.M. Mueller Ed., 197-207 (1959). (Equi Diagram; Experimental)

*60Mcq: M.K. McQuillan, "A Study of the Titanium-Silver System," *J. Inst. Met.*, 88, 235-239 (1960). (Equi Diagram; Experimental)

60Sat: T. Sato, S. Hukai, and Y. Huang, "The M_s Points of Binary Titanium Alloys," *J. Aust. Inst. Met.*, 5(2), 149-153 (1960). (Meta Phases; Experimental)

63Rei: R. Reinbach and D. Fischmann, "Diffusion in the Titanium-Silver System," *Z. Metallkd.*, 54(5), 314-316 (1963). (Equi Diagram; Experimental)

65Sch: K. Schubert, "On the Constitution of the Titanium-Copper and Titanium-Silver Systems," *Z. Metallkd.*, 56(3), 197-199 (1965). (Crys Structure; Experimental)

*69Ere: V.N. Eremenko, Y.I. Buyanov, and N.M. Panchenko, "Constitution Diagram of the System Titanium-Silver," *Porosh. Metall.*, 7(79), 55-59 (1969) in Russian; TR: *Sov. Powder Met. Metal. Ceram.*, 7(79), 562-566 (1969). (Equi Diagram; Experimental)

70Kau: L. Kaufman and H. Bernstein, *Computer Calculation of Phase Diagrams*, Academic Press, New York (1970). (Thermo; Theory)

77Pli: M.R. Plichta, J.C. Williams, and H.I. Aaronson, "On the Existence of the β → α-m Transformation in the Alloy Systems Ti-Ag, Ti-Au, and Ti-Si," *Metall. Trans. A*, 8, 1885-1892 (1977). (Meta Phases; Experimental)

*78Pli: M.R. Plichta and H.I. Aaronson, "The β → α-m Transformation in Three Ti-X Systems," *Acta Metall.*, 26, 1293-1305 (1978). (Equi Diagram, Meta Phases; Experimental)

The Al-Ti (Aluminum-Titanium) System

26.98154   47.88

By J.L. Murray

Equilibrium Diagram

The assessed Ti-Al phase diagram is shown in Fig. 1 and the special points of the diagram are summarized in Table 1. This assessment differs greatly from previous assessments [Hansen, Elliott, Shunk]. The outlines of the diagram were determined by [52Bum] in the only comprehensive study of the system. The $(\alpha Ti)/Ti_3Al$ boundaries are based on a critical reassessment of literature data through 1983 and work performed in this laboratory [84Shu]. The seminal works in the resolution of the many controversies surrounding Ti_3Al are [67Bla], [70Bla], and [84Shu]. The present diagram includes three additional phases, $TiAl_2$, δ, and $\alpha TiAl_3$, discovered by [64Sch], and [65Ram]. Recent work on long-period superlattice structures [82Mii, 84Loi1] showed that Al-rich alloys exhibit

Table 1   Special Points of the Assessed Ti-Al Phase Diagram

Reaction	Compositions of the respective phases, at.% Al			Temperature, °C	Reaction type
$L \rightleftarrows (\beta Ti)$		~ 11		~ 1710	Congruent
$L + (\beta Ti) \rightleftarrows TiAl$	53	47.5	51	~ 1480	Peritectic
$(\beta Ti) + TiAl \rightleftarrows (\alpha Ti)$	43	49	45	~ 1285	Peritectoid
$(\alpha Ti) \rightleftarrows Ti_3Al$		~ 32		~ 1180	Congruent
$(\alpha Ti) \rightleftarrows Ti_3Al + TiAl$	40	39	48	~ 1125	Eutectoid
$L + TiAl \rightleftarrows \delta$	73.5	69.5	71.5	~ 1380	Peritectic
$L + \delta \rightleftarrows TiAl_3$	80	72.5	75	~ 1350	Peritectic
$TiAl + \delta \rightleftarrows TiAl_2$	65	70	67	~ 1240	Peritectoid
$\delta \rightleftarrows TiAl_2 + TiAl_3$	71.5	68	75	~ 1150	Eutectoid
$L + TiAl_3 \rightleftarrows (Al)$	99.9	75	99.3	665	Peritectic
$TiAl_3 \rightleftarrows \alpha TiAl_3$	...	75	...	~ 600	Unknown
$L \rightleftarrows (\beta Ti)$		0		1670	Melting point
$(\beta Ti) \rightleftarrows (\alpha Ti)$		0		882	Allotropic transformation
$L \rightleftarrows (Al)$		100		660.452	Melting point

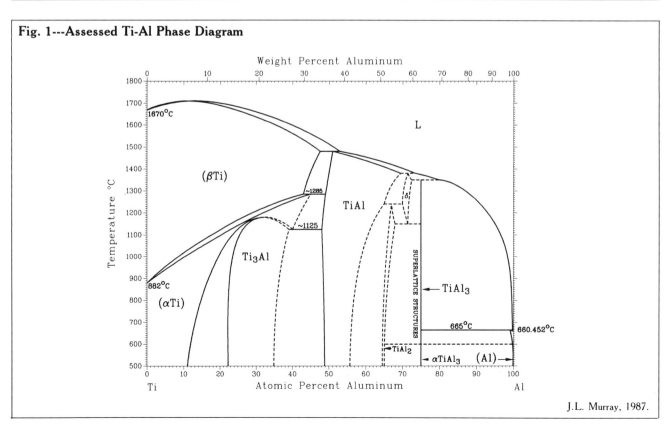

Fig. 1---Assessed Ti-Al Phase Diagram

J.L. Murray, 1987.

Fig. 2---Experimental Data on the Ti-Al Liquidus

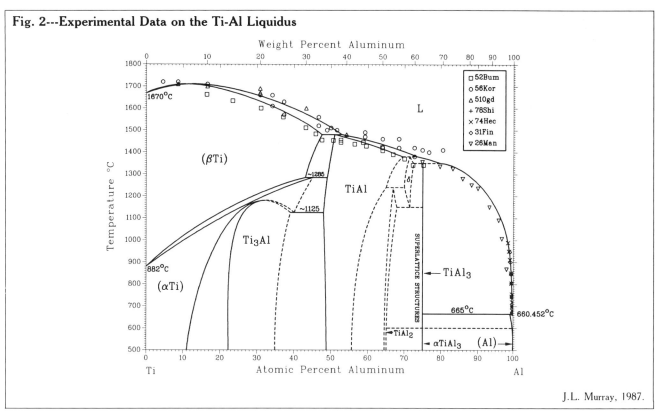

J.L. Murray, 1987.

considerably greater complexity than had previously been thought. The details of this part of the phase diagram, however, are not yet known.

The established equilibrium solid phases of the Ti-Al system are:

- The bcc (βTi) and cph (αTi) solid solutions. The addition of Al stabilizes (αTi) relative to (βTi). The maximum solubilities of Al in (βTi) and (αTi) are about 48 and 45 at.%, respectively.
- Ti_3Al, an ordered hexagonal structure based on (αTi). Ti_3Al is also designated α_2 in the literature. The (αTi)/Ti_3Al phase boundaries were at one time hotly debated, but the essential features of the ordering transition are now established. It is probable that the congruent transformation to the ordered structure is reached below the (αTi)/((αTi) + (βTi)) boundary.
- TiAl, an ordered $L1_0$ fcc phase of homogeneity range approximately 48 to 68 at.% Al. TiAl forms from the melt by a peritectic reaction with (βTi).
- $TiAl_2$ and δ phases with ordered fcc structures. The existence of these phases is established, but the phase boundaries have not been determined. Moreover, two related structural variants of $TiAl_2$ have been reported, but details of the transition are not yet known. The δ and $TiAl_2$ phase fields are therefore shown in Fig. 1 with dashed curves.
- $TiAl_3$, a stoichiometric phase with the $D0_{22}$ ordered fcc structure, and a low-temperature form, $\alpha TiAl_3$.
- The fcc (Al) solid solution, in which the maximum Ti solubility is about 0.7 at.%.

Ti-Rich Liquidus (0 to 80 at.% Al).

At least three peritectic reactions occur in this system: the two reactions L + (βTi) ⇄ TiAl [52Bum, 56Kor]

and L + $TiAl_3$ ⇄ (Al) [31Fin, 72Max, 74Cis, 74Ker, 78Shi] are well established. Near 75 at.% Al, peritectic microstructures are also seen--either the single reaction L + TiAl ⇄ $TiAl_3$ or the two reactions L + TiAl ⇄ δ and L + $TiAl_2$ ⇄ $TiAl_3$ (as in Fig. 1) are plausible interpretations of the data.

Liquidus data are plotted in Fig. 2. On the Ti-rich side, no investigation [51Ogd, 52Bum, 56Kor] is consistent with the melting point of pure Ti, 1670 °C [Melt]. The data of [52Bum] and [56Kor] are preferred to the those of [51Ogd], but discrepancies of 50 to 60 °C remain between the values of [52Bum] and [56Kor]. [52Bum] reported an uncertainty of ± 15 °C in thermal analysis and incipient melting data, but this is probably an underestimate, because it does not include the effect of contamination on the melting temperatures.

In the assessed diagram (Fig. 1), (βTi) is shown as having a congruent melting point at 1710 °C; this is suggested not only by the melting data, but also by rough thermodynamic calculations of the diagram. The L + (βTi) ⇄ TiAl and L + TiAl ⇄ $TiAl_3$ peritectic temperatures are shown as 1490 ± 30 °C and 1370 ± 30 °C, respectively. Temperatures between those of [56Kor] and [52Bum] were chosen to maximize consistency with the liquidus data for $TiAl_3$. Data reported for the peritectic temperatures are given in Table 2.

Al-Rich Alloys.

The Al-rich part of the diagram is shown in greater detail in Fig. 3. The peritectic reaction L + $TiAl_3$ ⇄ (Al) was established by [31Fin]; the peritectic temperature is 665 ± 0.5 °C [31Fin, 72Max, 74Cis, 74Ker, 78Shi]. The $TiAl_3$ liquidus was studied over its whole range of 80 to 100 at.% Al only by [23Erk] and [26Man]. For Al-rich alloys (less than 1 at.% Ti), [31Fin] showed that undercooling makes thermal analysis an unsuitable technique to determine the liquidus; chemical analysis of the separated liquid is a more accurate technique, despite problems in achieving equilibrium compositions [70Dav]. The assessed diagram is based on analyses of the composition of the liquid in equilibrium with TiAl$_3$ [31Fin, 74Hec, 78Shi], and in particular on the recent work of [78Shi], who used diffusion couples. Liquidus data from these studies are given in Table 3. It is expected that undercooling decreases with increasing Ti content, because of the greater driving force for precipitation of TiAl$_3$. The liquidus is drawn above the thermal analysis data, but approaches it at the upper peritectic temperature of 1350 °C. The peritectic nature of the reaction and its approximate temperature were established by [52Bum] by microstructural analysis. (Some recent data [84Abd] on

the TiAl$_3$ liquidus curve have been included in Fig. 3. The technique of electromagnetic phase separation was used, and the results agree well with previous work on which the assessed boundary is based.)

Table 2   Literature Reports of Peritectic Temperatures

Reference	Reaction	Temperature, °C	Technique/ comments
[510gd]	L + β ⇄ α	~1630	Incipient melting
	L + γ ⇄ TiAl$_3$	~1350	
[52Bum] ...	L + β ⇄ γ	1460	Thermal analysis, cooling
	L + γ ⇄ Tial$_3$	1340	Incipient melting
[56Kor] ...	β + L ⇄ γ	1510	Thermal analysis, cooling
	L + γ ⇄ TiAl$_3$	1400	
[62Pot] ...	γ + L ⇄ TiAl$_2$	~1400	Interpretation of [52Bum, 56Kor]
	TiAl$_2$ + L ⇄ TiAl$_3$	1340	based on additional X-ray data

Table 3   Liquidus Data for Al-Rich Alloys

Reference	Composition, at.% Al	Temperature, °C
[78Shi]	99.55	850
	99.69	800
	99.81	750
	99.88	700
	99.91	670
[74Hec]	99.893	680
	99.876	700
	99.853	720
	99.802	745
	99.786	755
	99.672	805
	99.545	850
	99.235	912
	99.000	950
	98.696	990
[31Fin]	99.21	948
	99.4	901
	99.6	846
	99.7	806
	99.82	742
	99.88	700
	99.91	675
[26Man]	98.3	868
	97.0	1007
	96.1	1090

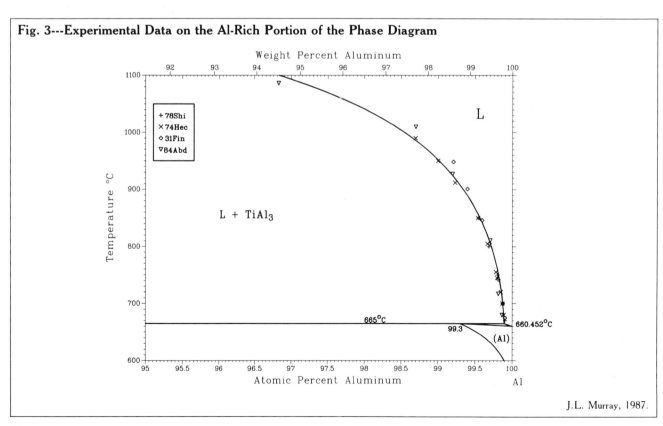

Fig. 3---Experimental Data on the Al-Rich Portion of the Phase Diagram

J.L. Murray, 1987.

Based on the disappearance of the peritectic reaction and the appearance of a single-phase microstructure at 80 at.% Al, [23Erk] erroneously identified the first equilibrium Al-rich compound as a congruently melting phase, TiAl$_4$. Metastable ordered fcc phases appear during solidification [73Oha, 74Cis, 74Ker], which have the approximate composition 80 at.% Al [74Cis, 74Ker], but the equilibrium phase is TiAl$_3$ [26Man, 31Fin, 74Cis, 74Ker]. By microstructural analysis, [52Bum] established that TiAl$_3$ forms by a peritectic reaction near 1350 °C. [52Bum] stated that Ti$_3$Al has a homogeneity range, also based on microstructure, but this homogeneity range has never been determined. The homogeneity ranges referred to in [52Bum] are reinterpreted as those of δ and TiAl$_2$, and TiAl$_3$ is therefore drawn as a line compound in Fig. 1.

Reported values of the maximum solubility of Ti in (Al) differ greatly. [46Buc] found the low value of 0.1 at.% Ti using parametric microhardness data. [74Cis] and [74Ker] reported 0.7 at.% Ti. Unpublished research ([48Fin], 0.7 at.% Ti; [72Max], 0.65 at.% Ti) supports this value. Thermodynamic calculations show that only the higher estimates give a partition coefficient in reasonable agreement with the known heat of fusion of pure Al. The maximum solubility is therefore placed at 0.7 at.% Ti.

(αTi)/(βTi) Boundaries.

Addition of Al to Ti stabilizes (αTi) relative to (βTi). Data on the (αTi)/(βTi) boundaries are shown in Fig. 4. The assessed phase boundaries are based on [83Mca], [61Enc], [62Cla], and [70Bla]. The two-phase field is drawn inside the scatter of the select data, because the effect of contamination is to broaden the two-phase field [54Thy, 56Sch].

Most investigators agree concerning the average position of the (αTi)/(βTi) boundaries; exceptions are [60Sat1], [61Enc], [73Loo], and [54Mcq]. In the first two references, the discrepancies are probably due to the effect of Mo and gaseous impurities, respectively. In the experiments of [73Loo], equilibrium was not reached in the diffusion couples studied; the effect of the composition of the couple was shown by [80Ouc].

Results of the hydrogen pressure experiments of [54Mcq] suggest that the (αTi)/(βTi) boundaries show a minimum at about 1 at.% Al. Because of the possible effect of hydrogen on the phase equilibria and the discrepancy between this method and more conventional techniques in the range 2 to 7 at.% Al, the minimum is not accepted.

Finally, the reason for the discrepancy between the thermal analysis work of [65Kor] and other work is unknown. Beyond 20 at.% Al, these (αTi)/(βTi) data conflict with the selected data on the (αTi)/Ti$_3$Al boundary and with most other work on the (αTi)/(βTi) equilibrium.

The high-temperature range 1200 to 1450 °C was examined only by [52Bum] and [61Enc]. [52Bum] annealed samples for only to 10 min in this range to avoid sample contamination; [61Enc] used longer annealing times, but undoubtedly introduced significant interstitial impurities. Gaseous impurities tend to stabilize (αTi) relative to (βTi), and the [61Enc] observations can be rationalized in this manner.

The temperature of the peritectoid equilibrium among (βTi), (αTi), and TiAl has not been determined experimentally, except that [79Col] suggested that the apparent failure to achieve (αTi)/TiAl equilibrium may indicate that the three phases are in equilibrium at 1315 °C. The assessed peritectoid temperature (1285 °C) is also based on the requirement of consistency with the data on the (αTi)/(βTi) boundaries.

Fig. 4---Select Experimental Data on the (αTi)/(βTi) Boundaries

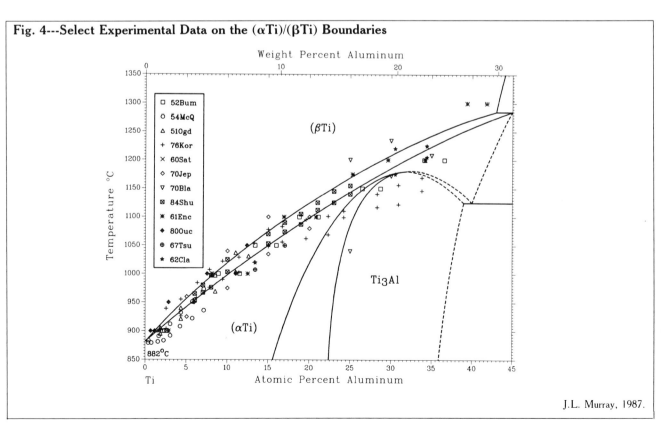

J.L. Murray, 1987.

(αTi)/Ti₃Al Boundaries.

The (αTi)/Ti₃Al phase boundaries are based on data selected from several sources [61Enc, 62Cla, 65Kor, 66Tsu, 67Bla, 70Lut, 71Sam, 73Nam, 76Kor1, 76Kor2, 76Zel, 77Sas, 77Bag, 79Erm, 79Muk, 80Sha, 83Nam, 84Shu] (Fig. 5). The essential features of the phase boundaries were clarified by the TEM work of [67Bla] and [70Bla], which resolved several hotly debated controversies [66Cro, 68Mar]. The assessed boundaries are now established within about 1 at.% in the range 0 to 25 at.% Al. Below about 850 °C, the assessed (αTi)/((αTi) + Ti₃Al) boundary represents the coherent, rather than the stable incoherent, equilibrium because the most reliable experiments use techniques that determine the coherent equilibrium. [67Bla] estimated that the coherency stresses are small and that the incoherent phase boundary does not differ greatly from the coherent boundary.

A peritectoid reaction (αTi) + (βTi) ⇌ Ti₃Al near 1080 °C was proposed by several investigators [56Sag, 61Enc, 60Sat1, 62Cla, 65Kor, 66Tsu, 67Tsu, 81Swa], and it was thought that this reaction ended the extent of the disordered (αTi) phase. It is now established, however, that the (αTi) + Ti₃Al field narrows with increasing temperature, passes through a maximum, and ends in a three-phase equilibrium with TiAl [70Bla, 84Shu]. The (αTi)/Ti₃Al boundaries pass very close to the (αTi)/(βTi) boundaries, and it is not certain whether the (αTi)/Ti₃Al and (αTi)/Ti₃Al and (αTi)/(βTi) phase fields intersect.

The data of [67Bla] and [70Bla] were consistent with much of the previous work [62Cla, resistivity; 61Enc, metallography] and were later confirmed by TEM [70Lut, 73Nam]. The (αTi)/Ti₃Al boundary was also confirmed by DSC and TEM measurements in this laboratory [84Shu]. The continuation of the order/disorder transition into the range 30 to 40 at.% Al was confirmed by [77Bag], [79Muk], and [84Shu] using TEM and DSC. Because the ordering transition cannot be suppressed during quenching, the microscopic evidence is based on the nature of antiphase boundaries.

This part of the phase diagram was at one time very controversial; phases both to the Ti-rich [61Gru, 62Fed, 65Gla] and the Al-rich [56Sag, 60Sat1, 61Yao, 62Cla, 65Gla] side of Ti₃Al have been reported. The spurious appearance of a second phase was due to (1) etching effects due to hydrides [61Enc], (2) anomalous effects in physical properties not due to phase changes 56Sag, 61Gru, 61Yao, 65Kor, 76Kor1], (3) contamination [60Sat1], or (4) failure to reach equilibrium [62Cla, lattice parameters]. Most of these results have since been rejected by their authors [61Enc, 56Sag, 65Kor]. More recently, [83Loi2] reported a phase of ideal stoichiometry Ti₂Al. Based on the volume fractions of the phases, this phase was a nonequilibrium structure. The only established stable equilibrium phases in the range 0 to 40 at.% Al are the disordered cph and the D0₁₉ structures.

There are also striking discrepancies in the quantitative placement of the (αTi)/Ti₃Al boundaries. All the experimental work on these boundaries are not plotted in Fig. 5, because they are extremely scattered. Microprobe analysis of annealed diffusion couples, for example, located the boundaries as much as 5 at.% to the Ti-side of the assessed boundaries [80Ouc, 73Loo, 72Kor]. The lattice parametric technique is not reliable for this coherent order/disorder transformation [62Cla, 76Kor1]. Magnetic susceptibility measurements [81Swa] appear to be a reliable technique, but linear tie lines must be rigorously established, and this may be difficult where the two-phase field is very narrow. Optical metallography as a probe for this phase transition contains fatal pitfalls, primarily because of hydride formation [61Enc, 66Cro].

Fig. 5---Select Experimental Data on the (αTi)/Ti₃Al Boundaries

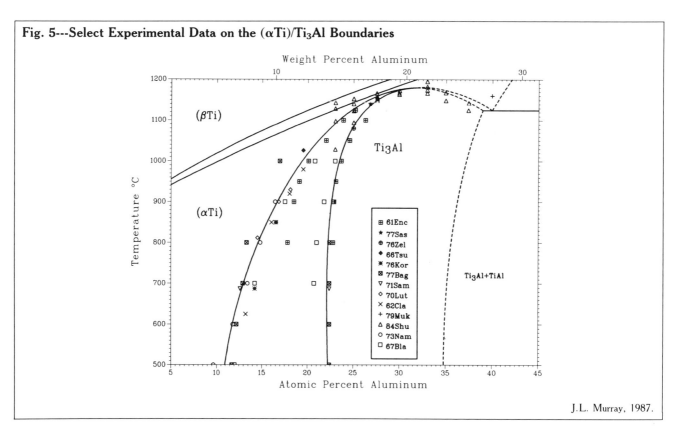

J.L. Murray, 1987.

The eutectoid reaction $(\alpha Ti) \rightleftarrows Ti_3Al + TiAl$ is tentatively drawn at 1125 °C, based on extrapolation of ordering temperatures to the $Ti_3Al/TiAl$ boundary and some preliminary work at this laboratory [84Shu].

Further work remains to be done on this portion of the diagram. Both the difference between the coherent and incoherent boundaries and the effect of impurities, other than oxygen, remain to be determined. Oxygen stabilizes Ti_3Al relative to (αTi) [76Lim], but oxygen contamination does not explain, for example, the slight discrepancy between [67Bla] (high oxygen) and other TEM work (low oxygen).

TiAl (Ordered fcc).

Data on the TiAl/(TiAl + Ti_3Al) boundary were reported by [76Col] and [79Col] through magnetic susceptibility, metallography, and hardness measurements, [51Ogd], [52Bum], and [56Kor] via metallography of heat treated and quenched alloys, [71Sam] by emf measurements and [73Loo] and [80Ouc] by microprobe analysis of annealed diffusion couples. The work of [52Gru] is not considered, because of the high impurity content of their alloys. Select data are shown in Fig. 6; they are scattered in a band about 2 at.% wide. The work of [79Col] is preferred, because a careful study was made of the approach to equilibrium and several reliable experimental techniques were used.

On the Al-rich side, the extent of the TiAl field is difficult to determine because of residual coring in the alloys [52Bum]. Data over a wide temperature range were provided by [52Bum] and [73Loo], and the homogeneity range at 900 °C was reported by [56Kor]. [52Bum] and [73Loo] found that the homogeneity range extends to nearly 70 at.% Al at high temperatures. The difficulty of equilibrating samples at high temperatures, without se-

vere contamination, restricts both the temperature range and accuracy of the data. The uncertainty in the Al-rich phase boundary is at least ± 2 at.%.

[52Duw] stated that X-ray diffraction showed that 62 at.% Al alloys were single phase after being annealed at 750 °C and quenched. Using electron diffraction, [82Mii] observed no $TiAl_2$ at 63 at.% Al, in agreement with [52Duw], but [82Mii] did observe superlattice reflections that intensified upon annealing at 700 °C. Because of the experimental difficulties in preparing equilibrium alloys, these structural data were not used to determine the phase boundary. The boundary above 1200 °C is based on the data of [52Bum]; the smooth decrease in the homogeneity range with decreasing temperature, as shown in Fig. 1, is based on the diffusion couple data of [73Loo].

Stoichiometric TiAl remains ordered up to 1350 °C, and the absence of antiphase domains in cast alloys supports the supposition that the ordered phase is formed from the melt [54Ell, 75Tey].

A tetragonal phase, Ti_3Al_5, with a superlattice structure based on $L1_0$, was observed by [82Mii]. Based on the amounts of phases observed [82Mii] as a function of composition, Ti_3Al_5 was verified by [83Loi1, 83Loi2] and [84Loi1, 84Loi2], who presented further evidence that it is not an equilibrium structure. Even at the stoichiometric composition and after long annealing times, the Ti_3Al_5 was always found in a matrix of (equilibrium) TiAl and $TiAl_2$. The nomenclature Ti_7Al_{11} pertains to the structure of the interface between Ti_3Al_5 and the matrix [84Loi2]; Ti_7Al_{11} is certainly not an equilibrium bulk phase.

$TiAl_2$ and δ.

The diagram of [52Bum] showed no Al-rich phases other than TiAl and $TiAl_3$. Recent work clearly indicates

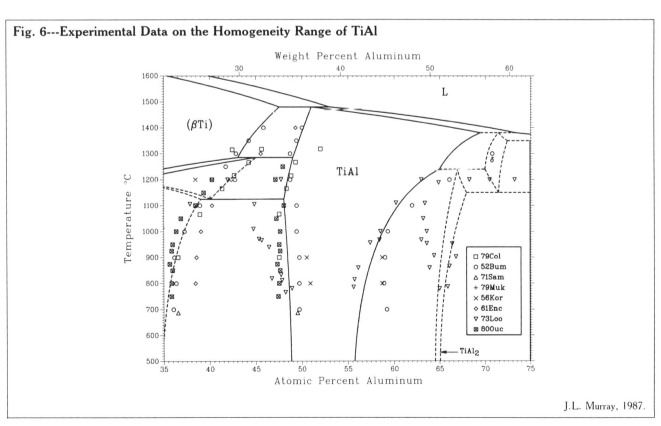

Fig. 6---Experimental Data on the Homogeneity Range of TiAl

J.L. Murray, 1987.

that several additional phases must be taken into account [65Ram, 73Loo, 80Mii, 81Mii, 82Mii, 83Loi2, 84Loi1]. In particular, the electron microscopy and electron diffraction of [82Mii], [83Loi1], [83Loi2], [84Loi1], and [84Loi2] demonstrated that long-period superlattice structures and other complex structural arrangements are formed in the range 50 to 75 at.% Al. There are, however, reasons to avoid making full use of these results to construct the equilibrium diagram: First, the experimental work is not yet complete. Second, this work has so far been primarily concerned with identifying the structures present and not with demonstrating that these are bulk phases or that equilibrium was achieved.

By X-ray diffraction, [65Ram] found a phase $TiAl_2$ isomorphous with Ga_2Hf and estimated that it is stable to about 1250 °C. The existence of $TiAl_2$ was verified by [73Loo], [80Mii], [83Loi2], and [84Loi1]. According to [80Mii], there are two forms of $TiAl_2$: a Ga_2Hf-type structure and also a Ga_2Zr-type structure ($\alpha TiAl_2$). [80Mii] did not explore the composition or temperature ranges of stability of these forms. According to [83Loi2], the Ga_2Hf-type phase is the high-Al form. [84Loi] made the opposite identification. In the assessed diagram, a single phase "$TiAl_2$" is shown as stable up to 1240 °C, with a homogeneity range roughly based on the microprobe work of [73Loo]. Details of the possible phase transformation between the two forms of $TiAl_2$ are unresolved.

The phase designated δ in this assessment was called Ti_5Al_{11} by [65Ram] and Ti_2Al_5 by [73Loo]. [65Ram] found δ only in alloys annealed above 950 °C and attributed to it a faulted DO_{23} structure. [73Loo] found a phase with a homogeneity range of 70.5 to 73.5 at.% Al in diffusion couples annealed at 1200 °C, but not in couples annealed at 1100 °C. The present author therefore includes a high-temperature phase δ in the assessed diagram; δ but not $TiAl_2$ was found by [65Ram] in as-cast alloys, and therefore a peritectic reaction involving δ is shown, with peritectoid decomposition of $TiAl_2$. The peritectoid temperature is unknown.

In alloys annealed between 1100 and 1200 °C, [82Mii] observed one-dimensional antiphase structures (APS) and argued that the structure observed by [65Ram] should be reinterpreted in terms of an APS. According to [82Mii], the superlattice structures can be viewed as intimate mixtures of the DO_{22} and DO_{23} structures, with the more DO_{23}-like structures stable at high temperature. These findings were verified in detail by [84Loi1]. [84Loi1] drew a "phase diagram" of the APS structures in which high-temperature structures were shown to decompose by a eutectoid reaction at about 1000 °C, and the low-temperature structures were shown as stable up to a peritectoid reaction at about 900 °C. In the assessed diagram, the approximate composition and temperature range in which these structures are found is designated as "Long Period Superlattice Structures."

Finally, [65Ram] observed a phase that existed only below 780 °C, which they placed to the Ti-side of $TiAl_3$. [73Loo] also found an allotropic change in $TiAl_3$ at some temperature below 638 °C, but they detected no composition difference between the two forms. The assessed diagram, with a dashed line indicating a reaction at 600 °C, tentatively follows the work of [73Loo]; however, further experimental work is needed.

Metastable Phases

Metastable equilibria and nonequilibrium transitions have been investigated in two areas: (1) the martensitic

(βTi) → (αTi) transformation and (2) solidification of Al-rich alloys.

Start temperatures for the martensitic transformation on cooling, M_s were measured by [70Jep] and [60Sat2] as functions of composition and cooling rate. [60Sat2] found that M_s temperatures lie in the range 800 to 840 °C and that, for rates between 600 and 2000 °C/s, M_s was temperature independent. [70Jep] used cooling rates of 10^{-2} to 10^5 °C/s and found that M_s decreased with increasing cooling rate. [70Jep] listed M_s and the reversion temperature on heating, β_s, for cooling rates of 1000 to 2000 °C/s. These are more than 100 °C higher than results obtained by [60Sat2]. Data are summarized in Table 4.

For higher Al content, ordered Ti_3Al rather than disordered (αTi) may be formed during quenching from the (βTi) range. [76Kor2] reported that homogeneous α alloys are formed only up to 20 at.% Al, beyond which a two-phase microstructure appeared.

Rapid solidification produces (Al) solid solutions containing up to 0.2 at.% Ti [34Boh, 52Fal]. Interest in solidification of dilute Al alloys is due to the effectiveness of Ti as a grain refiner. [74Cis] and [74Ker] examined the metastable extensions of the (Al) liquidus and solidus, by measuring nucleation temperatures as a function of cooling rate and by microprobe analysis of the (Al) dendrites. Their estimates of the extended boundary are shown as Fig. 7. [74Cis] and [74Ker] also made microprobe analyses of compound phases nucleated by metastable peritectic reactions and concluded that, in addition to equilibrium $TiAl_3$, two metastable compounds appear. Based on the similarities between microstructures and orientational relationships that are observed in the Ti-Al system and those that occur in the Al-Zr system, [73Oha] suggested that the metastable phase has the $L1_2$ structure.

Crystal Structures and Lattice Parameters

The crystal structures of phases found in Ti-Al alloys are summarized in Table 5. Lattice parameters of the compound phases are listed in Table 6. Lattice parameters of the hexagonal phases (αTi) and Ti_3Al are shown graphically in Fig. 8.

The long-period superlattice structures found near $TiAl_3$ are described by one-dimensional arrays of antiphase boundaries along the c axis of a hypothetical $L1_2$ structure [82Mii, 84Loi]. The D_{22} structure has a period M of one cell, the DO_{23} structure has M equals to 2. More complex arrangements described by $1 < M < 2$ correspond to periodic arrays of the DO_{22}-type and DO_{23}-type configurations. They can be described by integer ratios $M = p/q$, where p is the periodicity of the structure and q is the

Table 4 Data on the Martensite (βTi) → (αTi) Transformation

Reference	Composition, at.% Al	Start temperature (M_s), °C	Reversion temperature (A_s), °C
[60Sat2]	3.5	835	...
	5.2	800	...
	6.9	800	...
	11.8	840	...
[70Jep]	5	918	940
	10	960	1005
	15	1015	1045
	20	1060	1110

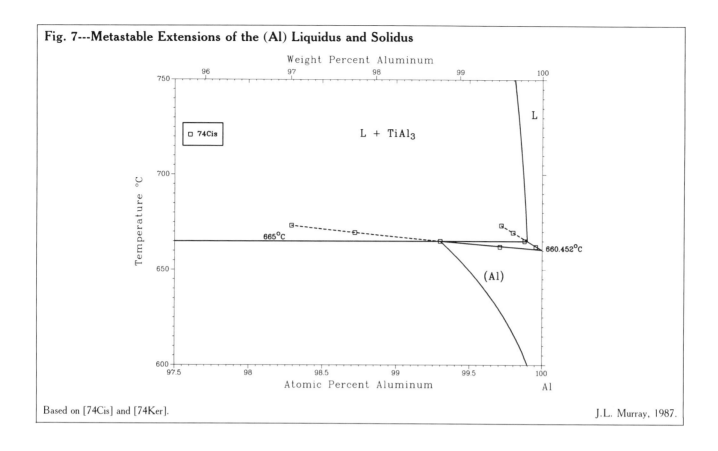

Fig. 7---Metastable Extensions of the (Al) Liquidus and Solidus

Weight Percent Aluminum

L + TiAl₃

□ 74Cis

665°C

660.452°C

(Al)

Temperature °C

Atomic Percent Aluminum

Al

Based on [74Cis] and [74Ker].

J.L. Murray, 1987.

Table 5 Crystal Structures of the Ti-Al System

Phase	Homogeneity range, at.% Al	Pearson symbol	Space group	Struktur-bericht designation	Prototype	Reference
(αTi)	0 to 45	$hP2$	$P6_3/mmc$	$A3$	Mg	[Pearson2]
(βTi)	0 to 47.5	$cI2$	$Im3m$	$A2$	W	[Pearson2]
Ti₃Al	22 to 39	$hP8$	$P6_3/mmc$	$D0_{19}$	Ni₃Sn	[55Cla, 61Gol, 61Pie, 73Geh]
TiAl	48 to 69.5	$tP4$	$P4/mmm$	$L1_0$	AuCu	[52Duw]
Ti₃Al₅	58 to 63(a)	$tP32$	$I4/mbm$	...	...	[82Mii] 84Loi2]
TiAl₂	65 to 68	$tI24$	$I4_1/amd$	...	Ga₂Hf	[62Pot, 62Sch, 65Ram]
αTiAl₂	(b)	$oC12$	$Cmmm$	...	Ga₂Zr	[80Mii, 83Loi2, 84Loi1]
δ	70 to 72.5	(c)	...	...	...	[82Mii, 84Loi]
TiAl₃	75	$tI8$	$I4/mmm$	$D0_{22}$	Al₃Ti	[31Fin, 39Bra]
αTiAl₃	75	(b,d)	...	...	...	[65Ram, 64Sch, 73Loo1]
(Al)	99.3 to 100	$cF4$	$Fm3m$	$A1$	Cu	[Pearson2]

(a) Not an equilibrium phase. (b) Not shown on the assessed diagram. (c) Long-period superlattice. See text for a description. (d) Tetragonal. A superstructure of the $D0_{22}$ lattice.

Table 6 Lattice Parameters of the Ti-Al System

Phase	Composition, at.% Al	Lattice parameters, nm		Reference
		a	*c*	
$\alpha TiAl_3$	75	0.3875	3.384	[73Loo]
	69 to 72	0.389	3.392	[80Mii]
	72(a)	0.384	3.346	[65Ram]
$TiAl_3$	75	0.3843	0.8591	[31Fin]
	75	0.3844	0.8596	[39Bra]
	75	0.3851	0.8608	[64Dag]
	75	0.384	0.859	[65Ram]
	75	0.3849	0.8610	[73Loo]
	75	0.384	0.856	[80Mii]
	98(a)	0.3853	0.8618	[82Tsu]
TiAl	(b)	0.40155	0.40625	[510gd]
	(c)	0.3976	0.4049	
	(d)	0.3999	0.4080	[52Gru]
	46	0.4011	0.4069	[52Duw]
	50	0.4004	0.4071	
	55	0.3997	0.4075	
	60	0.3990	0.4074	
	62	0.3988	0.4081	
	50	0.3998	0.4076	[64Dag]
	52.13	0.3992	0.4073	[52Bum]
	55.23	0.3980	0.4075	
	60.22	0.3975	0.4093	
	68.47	0.3957	0.4097	
$TiAl_2$	67	0.3976	2.436	[62Sch, 62Pot]
	67 to 72	0.398	2.436	[80Mii]
	69	0.3917	1.6524	[65Ram]
	69	0.392	1.648	[80Mii]
(Al)	100	0.40497	...	[52Fal]
	99.915	0.40487	...	
	99.91	0.40487	...	
	99.8	0.40477	...	

(a) Two-phase alloys. (b) At Ti_3Al boundary. (c) At $TiAl_2$ boundary. (d) All compositions examined.

number of antiphase boundaries in a period. According to [82Mii], the high-temperature phase described by [65Ram] has $M = 8/6$, and in general at high-temperatures, values of $1.6 < M < 1.91$ are found. According to [84Loi1], the values $M = 4/3$, $3/2$, and $11/7$ are found at low temperatures and more complex configurations are found at high temperature. [84Loi1] also observed composition modulations that accounted for the nonstoichiometry of the observed structures.

Thermodynamics

Thermodynamic measurements on the Ti-Al system include heats of mixing in the liquid phase [74Esi]; Al and Ti activities in the bcc phase [71Hoc, 73Geg]; heats of formation of (αTi), TiAl, and $TiAl_3$ at 300 to 350 °C [55Kub, 60Kub]; and heat capacity of $TiAl_3$ [74Stu]. Emf measurements were also made to determine Al partial Gibbs energies at 980 K [71Sam], but because only integrated quantities were reported, the emf data were used only as phase boundary data in the present assessment. Heats of mixing and heats of formation are given in Table 7. The compilation of thermodynamic data for Ti-Al made by [83Lia] includes tables of other thermodynamic properties, smoothed using regular and subregular solution models.

Calculations of the Phase Diagram.

The Ti-Al system presents several difficulties in determining Gibbs energy functions: (1) Heats and entropies

of transformation are needed for pure Al in the metastable bcc and cph forms to separate the excess contribution to the bcc and cph Gibbs energies. (2) Much of the phase diagram data, particularly for the liquidus and solidus and Al-rich compounds, are not sufficiently accurate to decide whether or not a given Gibbs energy function agrees with the phase diagram. (3) The high-temperature thermochemical data are of unknown accuracy. (4) To make quantitative comparison between phase boundary data and calculations, the broad homogeneity ranges of the compounds Ti_3Al and TiAl must be modeled. Based on recent experimental work, the region between TiAl and $TiAl_3$ will require substantial further revision, and there is no phase diagram data presently available on which one can base a reliable phase calculation.

[73Kau] and [78Kau] calculated the equilibria involving the solution phases and Ti_3Al, TiAl, and $TiAl_3$; the intermetallic phases were modeled as line compounds, and equal regular solution excess entropies were attributed to all the solution phases. Within the limitation of the line compound model for TiAl and Ti_3Al, the phase diagram calculated by [78Kau] agrees with the experimental data. The earlier calculation of [73Kau] is very similar.

The purposes of the present calculations are to refine the Gibbs energies of the solution phases using least-squares optimization with respect to thermochemical and assessed phase diagram data, and in particular, to examine excess entropy contributions and make an estimate of their accuracy.

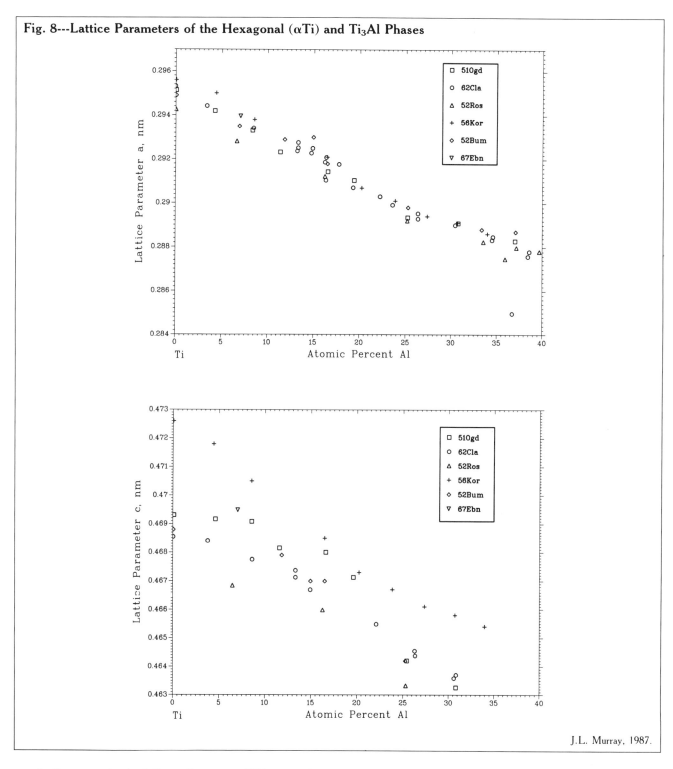

Fig. 8---Lattice Parameters of the Hexagonal (αTi) and Ti$_3$Al Phases

J.L. Murray, 1987.

In the present calculations, the excess Gibbs energies of the solution phases are represented as:

$$G^{ex}(i) = x(1 - x) [B(i) + C(i) (1 - 2x)]$$

where i designates the phase; x is the atom fraction of Al; and $B(i)$ and $C(i)$ are interaction parameters. Gibbs energies for the pure elements were taken from [70Kau] and were not varied in the optimizations. TiAl$_3$ was modeled as a line compound. The presence of Ti$_3$Al was ignored.

Optimizations of Gibbs energy parameters were made with respect to assessed phase boundary data, heats of mixing [74Esi], and partial Gibbs energies [71Hoc, 73Geg]. Optimizations of different subsets of thermodynamic parameters were compared; the regular solution model with an excess entropy was selected. Thermodynamic parameters resulting from the present calculations are listed in Table 8. These parameters may be used to calculate the (βTi) and TiAl$_3$ liquidus curves and the (βTi)/(αTi) boundaries.

Table 7 Experimental Heats of Mixing and of Formation

Reference	Heat of mixing ($\Delta_{mix}H$), J/mol	Heat of formation(a) ($\Delta_f H$), J/mol	Composition, at.% Al
[74Esi]	− 8 870	...	10
	− 15 480	...	20
	− 20 120	...	30
	− 25 770	...	40
	− 29 660	...	50
	− 29 830	...	60
	− 26 990	...	70
	− 20 590	...	80
	− 11 270	...	90
[55Kub]	...	− 9 460	8.55
	...	− 18 400	16.45
	...	− 23 900	23.85
	...	− 29 000	30.75
	...	− 30 900	35.90
	...	− 32 900	37.2
	...	− 39 400	48.9
	...	− 40 200	54.2
	...	− 40 250	58.05
	...	− 41 800	59.25
	...	− 37 400	75
	...	− 36 300	75
[60Kub]	...	− 25 300	25
	...	− 30 000	35
	...	− 34 900	45
	...	− 36 500	50
	...	− 36 200	50
	...	− 37 400	55
	...	− 38 900	60

(a) From fcc Al and (αTi).

The order/disorder transformation (αTi) → Ti₃Al was treated by [81Kik] using the cluster variation method. Cluster variation calculations provide a "generic" phase diagram, in which dimensionless energy parameters must be fixed. If this is done by fixing the temperature and composition of the congruent point, good agreement with experiment can be achieved for this system, particularly in the vicinity of the congruent point. It is therefore suggested that a quantitative analysis of the (αTi) → Ti₃Al transition could profitably be further pursued by the cluster variation method.

Cited References

23Erk: E. van Erkelenz, "Aluminum-Titanium Alloys and the Effect of Titanium on Aluminum," *Metall. Erz,* 20(11), 206-210 (1923) in German. (Equi Diagram; Experimental)

26Man: W. Manchot and A. Leber, "On the Alloying of Titanium with Aluminum," *Z. Anorg. Chem, 150,* 26-34 (1926) in German. (Equi Diagram; Experimental)

*31Fin: W.L. Fink, K.R. Van Horn, and P.M. Budge, "Constitution of High-Purity Aluminum-Titanium Alloys," *Trans. AIME, 93,* 421-439 (1931). (Equi Diagram; Experimental)

34Boh: H. Bohner, "Undercooling of the High Melting Compounds in Aluminum Alloys," *Z. Metallkd., 26*(12), 268-271 (1934) in German. (Meta Phases; Experimental)

39Bra: G. Brauer, "The Crystal Structures of TiAl₃, NbAl₃, TaAl₃ and ZrAl₃," *Z. Anorg. Chem., 242*(1), 1-22 (1946) in German. (Equi Diagram; Experimental)

46Buc: H. Buckle, "Solubility Determinations Using Microhardness Measurements," *Z. Metallkd., 37,* 43-47 (1946) in German. (Equi Diagram; Experimental)

48Fin: W.L. Fink and L.A. Willey, *Metals Handbook,* American Society for Metals, Metals Park, OH, 1167 (1948). (Equi Diagram; Experimental)

Table 8 Gibbs Energies for the Ti-Al System

Thermodynamic properties of the pure elements and TiAl₃

$G^0(Ti,L) = 0$

$G^0(Al,L) = 0$

$G^0(Ti,cph) = - 20\,585 + 12.134\,T$

$G^0(Al,cph) = - 5\,230 + 9.707\,T$

$G^0(Ti,bcc) = - 16\,234 + 8.368\,T$

$G^0(Al,bcc) = - 628 + 6.694\,T$

$G^0(Ti,fcc) = - 17\,238 + 12.134\,T$

$G^0(Al,fcc) = - 10\,740 + 11.506\,T$

$G^0(TiAl_3) = - 82\,500 + 30.209\,T$

Interaction parameters for the solution phases

$B(L) = - 99\,987$		$B(bcc) = - 115\,653$
$C(L) = - 66\,140$		$C(bcc) = - 63\,137$
$B(cph) = - 118\,239$		$B(fcc) = - 121\,496$
$C(cph) = - 61\,415$		$C(fcc) = - 56\,715$

Note: Values are in J/mol and J/mol · K.

51Ogd: H.R. Ogden, D.J. Maykuth, W.L. Finlay, and R.I. Jaffee, "Constitution of Titanium-Aluminum Alloys," *Trans. AIME, 191,* 1150-1155 (1951). (Equi Diagram; Experimental)

*52Bum: E.S. Bumps, H.D. Kessler, and M. Hansen, "Titanium-Aluminum System," *Trans. AIME, 194,* 609-614 (1952). (Equi Diagram; Experimental)

52Duw: P. Duwez and J.L. Taylor, "Crystal Structure of TiAl," *Trans. AIME, 194,* 70-71 (1952). (Crys Structure; Experimental)

52Fal: G. Falkenhagen and W. Hofmann, "The Effect of Very High Cooling Rates on the Solidification and Structure of Binary Alloys," *Z. Metallkd., 43*(3), 69-81 (1952) in German. (Meta Phases; Experimental)

52Gru: W. Gruhl, "Investigation of Ti-Rich Aluminum Titanium Alloys," *Metall, 6*(5/6), 134-135 (1952) in German. (Equi Diagram; Experimental)

52Ros: W. Rostoker, "Observations on the Lattice Parameters of the Alpha Solid Solution in the Titanium-Aluminum System," *Trans. AIME, 194",* 212-213 (1952). (Crys Structure; Experimental)

54Ell: R.P. Elliott and W. Rostoker, "The Influence of Aluminum on the Occupation of Lattice Sites in the TiAl Phase," *Acta Metall., 2,* 884-885 (1954). (Crys Structure; Experimental)

54Mcq: A.D. McQuillan, "A Study of the Behaviour of Titanium-Rich Alloys in the Titanium-Tin and Titanium-Aluminium Systems," *J. Inst. Met., 834f1,* 181-184 (1954-55). (Equi Diagram; Experimental)

54Thy: R.J. Van Thyne and H.D. Kessler, "Influence of Oxygen, Nitrogen, and Carbon on the Phase Relationships of the Ti-Al System," *Trans. AIME, 200,* 193-199 (1954). (Equi Diagram; Experimental)

55Cla: D. Clark and J.C. Terry, "Superlattice Formation in αTi-Aluminium Solid Solutions," *Met. Mater., 3,* 116 (1956). (Equi Diagram; Experimental)

55Kub: O. Kubaschewski and W.A. Dench, "The Heats of Formation in the Systems Titanium-Aluminium and Titanium-Iron," *Acta Metall., 3,* 339-346 (1955). (Thermo; Experimental)

56Kor: I.I. Kornilov, E.N. Pylaeva, and M.A. Volkova, "Phase Diagram of the Titanium-Aluminum Binary System," *Izv. Akad. Nauk SSSR, Otd. Khim. Nauk, 7,* 771-780 (1956) in Russian. (Equi Diagram; Experimental)

56Sag: K. Sagel, E. Schulz, and U. Zwicker, "Investigation of the Titanium-Aluminum System," *Z. Metallkd., 47*(8), 529-534 (1956) in German. (Equi Diagram; Experimental)

56Sch: T.H. Schofield and A.E. Bacon, "The Constitution of the Titanium-Rich Alloys of Titanium, Aluminum, and Oxygen," *J. Inst. Met., 85,* 193-196 (1956-1957). (Equi Diagram; Experimental)

60Kub: O. Kubaschewski and G. Heymer, "Heats of Formation of Transition-Metal Aluminides," *Trans. Faraday Soc., 56,* 473-478 (1960). (Thermo; Experimetnal)

60Sat1: T. Sato and Y.C. Huang, "The Equilibrium Diagram of the Ti-Al System," *Trans. Jpn. Inst. Met., 1,* 22-27 (1960). (Equi Diagram; Experimental)

60Sat2: T. Sato, S. Hukai, and Y.C. Huang, "The M_s Points of Binary Titanium Alloys," *J. Aust. Inst. Met., 5*(2), 149-153 (1960). (Meta Phases; Experimental)

61Enc: E. Ence and H. Margolin, "Phase Relations in the Titanium-Aluminum System," *Trans. AIME, 221,* 151-157 (1961). (Equi Diagram; Experimental)

61Gol: A.J. Goldak and J.G. Parr, "The Structure of Ti_3Al," *Trans. AIME, 221,* 639-640 (1961). (Crys Structure; Experimental)

61Gru: N.V. Grum-Grzhimailo, I.I. Kornilov, E.N. Pylaeva, and M.A. Volkova, "Metallic Compounds in the Region of α-Solid Solutions of the System Titanium-Aluminum," *Dokl. Akad. Nauk SSSR, 137*(3), 599-602 in Russian. (Equi Diagram; Experimental)

61Pie: P. Pietrokowsky, "Occurrence of an Ordered Phase in an As-Cast Titanium-Aluminium Alloy," *Nature, 190,* 77-78 (1961). (Crys Structure; Experimental)

61Yao: Y.L. Yao, "Magnetic Susceptibilities of Titanium-Rich Titanium-Aluminum Alloys," *Trans. ASM, 54,* 241-246 (1961). (Equi Diagram; Experimental)

62Cla: D. Clark, K.S. Jepson, and G.I. Lewis, "A Study of the Titanium-Aluminium System up to 40 at.% Aluminium," *J. Inst. Met., 91,* 197-203 (1962-1963). (Equi Diagram; Experimental)

62Fed: S.G. Fedotov, T.T. Nartova, and E.P. Sinodora, "Elastic Properties of Titanium Aluminum Alloys," *Dokl. Akad. Nauk SSSR, 146*(6), 1377-1379 (1972) in Russian. (Equi Diagram; Experimental)

62Pot: M. Potzschke and K. Schubert, "The Structures of Some T4-B3 Homologous and Quasi-Homologous Systems," *Z. Metallkd., 53*(8), 548-561 (1962) in German. (Crys Structure; Experimental)

62Sch: K. Schubert, H.G. Meissner, M. Potzschke, W. Rossteutscher, and E. Stolz, "Structure Data on Metallic Phases," *Naturwissenschaften, 49*(3), 57 (1962) in German. (Crys Structure; Experimental)

64Sch: K. Schubert, H.G. Meissner, A. Raman, and W. Rossteutscher, "Structure Data on Metallic Phases," *Naturwissenschaften, 51*(12), 287 (1964) in German. (Crys Structure; Experimental)

65Gla: V.V. Glazova, "Reaction of Titanium with Aluminum," *Dokl. Akad. Nauk SSSR, 160,* 109-111 (1965) in Russian; TR: *Dokl. Chem., 160,* 11-13 (1965). (Equi Diagram; Experimental)

65Kor: I.I. Kornilov, E.N. Pylaeva, M.A. Volkova, P.I. Kripyakevich, and V.Y. Markiv, "Phase Structure of Alloys in the Binary Ti-Al System Containing 0-30% Al," *Dokl. Akad. Nauk SSSR, 161,* 843-846 (1965); TR: *Dokl. Chem., 161,* 332-335 (1965). (Equi Diagram; Experimental)

65Ram: A. Raman and K. Schubert, "The Constitution of Some Alloy Series Related to $TiAl_3$. II. Investigations in Some T(4)-Al-Si and T(4).(6)-In System," *Z. Metallkd., 56,* 44-52 (1965) in German. (Crys Structure; Experimental)

66Cro: F.A. Crossley, "Ti-Rich End of the Ti-Al Equilibrium Diagram," *Trans. AIME, 236,* 1174-1185 (1966). (Equi Diagram; Experimental)

66Tsu: T. Tsujimoto and M. Adachi, "Reinvestigation of the Ti-Rich Region of the Ti-Al Equilibrium Diagram," *J. Inst. Met., 94,* 358-363 (1966). (Equi Diagram; Experimental)

67Bla: M.J. Blackburn, "The Ordering Transformation in Ti-Al Alloys Containing up to 25 at.% Al," *Trans. AIME, 239,* 1200-1208 (1967). (Equi Diagram; Review)

67Ebn: A.E. Ebneter, thesis, Air Force Institute of Technology, Wright-Patterson AFB (1967). (Crys Structure; Experimental)

67Tsu: T. Tsujimoto and M. Adachi, "Study of the Ti-Rich Region of the Ti-Al," *Trans. Nat. Res. Inst. Met. (Tokyo), 9*(2), 68-81 (1967). (Equi Diagram; Experimantal)

68Mar: H. Margolin, "The Ordering Transformation in Ti-Al Alloys Containing up to 25 at.% Al," *Trans. AIME, 242, 742-743 (1968). (Equi Diagram; Experimental)*

70Bla: M.J. Blackburn, "Some Aspects of Phase Transformations in Titanium Alloys," *Sci. Technol. Appl. Titanium,* Proc. Int. Conf., R.I. Jaffee, Ed., 633-643 (1970). (Equi Diagram; Experimental)

70Dav: I.G. Davies, J.M. Dennis, and A. Hellawell, "The Nucleation of Aluminum Grains in Alloys of Aluminum with Titanium and Boron," *Metall. Trans., 1,* 275-279 (1970). (Equi Diagram; Experimental)

70Jep: K.S. Jepson, A.R.G. Brown, and J.A. Gray, "The Effect of Cooling Rate on the β Transformation in Titanium-Niobium and Titanium-Aluminium Alloys," *Sci. Technol. Appl. Titanium,* Proc. Int. Conf., R.I. Jaffee, Ed., 677-690 (1970). (Meta Phases; Experimental)

70Kau: L. Kaufman and H. Bernstein, *Computer Calculations of Phase Diagrams,* Academic Press, New York (1970). (Thermo; Theory)

70Lut: G. Lutjering and S. Weissmann, "Mechanical Properties of Age-Hardened Titanium-Aluminum Alloys," *Acta Metall., 18,* 785-795 (1970). (Equi Diagram; Experimental)

71Hoc: M. Hoch and R.J.J. Usell, "Thermodynamics of Titanium Alloys. Pt. 2., Titanium and Aluminum Activities in the BCC β Phase of the Ti-Al System," *Metall. Trans., 2*(2), 2627-2632 (1971). (Thermo; Experimental)

71Sam: V.V. Samokhval, P.A. Poleshchuk, and A.A. Vecher, "Thermodynamic Properties of Aluminium-Titanium and Aluminium-Vanadium Alloys," *Zh. Fiz. Khim., 45*(8), 2071-2078 (1971) in Russian; TR: *Russ. J. Phys. Chem., 45*(8), 1174-1176 (1971). (Thermo; Experimental)

72Kor: I.I. Kornilov, F.N. Tavadze, G.N. Ronami, K.M. Konstantinov, T.A. Peradze, and Yu.A. Maksimov, "The Influence of Stabilizing α- and β-Elements on the Position of Phase Boundaries in the Titanium-Aluminum System," *Soobshch. Akad. Nauk Gruz. SSR, 66,* 133-136 (1972) in Russian. (Equi Diagram; Experimental)

72Max: I. Maxwell and A. Hellawell, "Constitution of the System Al-Ti-B with Reference to Al-Base Alloys," *Metall. Trans., 3*(6), 1487-1493 (1972). (Equi Diagram; Experimental)

73Geg: H. Gegel and M. Hoch, "Thermodynamics of α-Stabilized Ti-X-Y Systems," *Sci. Technol. Appl. Titanium,* Proc. Int. Conf., R.I. Jaffee, Ed., 2, 923-933 (1973). (Thermo; Experimental)

73Geh: P.C. Gehlen, "The Crystallographic Structure of Ti_3Al," *Sci. Technol. Appl. Titanium, Proc. Int. Conf., R.I. Jaffee, Ed.,* 2, 923-933 (1973). (Crys Structure; Experimental)

73Kau: L. Kaufman and H. Nesor, "Phase Stability and Equilibria as Affects by the Physical Properties and Electronic Structure of Titanium Alloys," *Sci. Technol. Appl. Titanium,* Proc. Int. Conf., R.I. Jaffee, E.D., 2, 773-800 (1973). (Thermo; Theory)

73Loo: F.J.J. van Loo and G.D. Rieck, "Diffusion in the Titanium-Aluminum System-II. Interdiffusion in the Composition Range Between 25 and 100 at.% Ti," *Acta Metall., 21,* 73-84 (1973). (Equi Diagram; Experimental)

73Nam: T.K.G. Namboodhiri, C.J. McMahon, and H. Herman, "Decomposition of the α-Phase in Titanium-Rich Ti-Al Alloys," *Metall. Trans., 4,* 1323-1331 (1973). (Equi Diagram; Experimental)

73Oha: T. Ohashi and R. Ichikawa, "Grain Refinement in Aluminum-Zirconium and Aluminum-Titanium Alloys by Metastable Phases," *Z. Metallkd., 64*(7), 517-521 (1973) in German. (Meta Phases; Experimental)

74Cis: J. Cisse, H.W. Kerr, and G.F. Bolling, "The Nucleation and Solidification of Al-Ti Alloys," *Metall. Trans., 5,* 633-641 (1974). (Equi Diagram, Meta Phases; Experimenal)

74Esi: Yu.O. Esin, N.P. Bobrov, M.S. Petrushevskiy, and P.V. Gel'd, "Enthalpy of Formation of Liquid Aluminum Alloys with Titanium and Zirconium," *Izv. Akad. Nauk SSR, Met.,* (5), 104-109 (1974) in Russian; TR: *Russ. Metall.,* (5), 86-89 (1974). (Equi Diagram; Experimental)

74Hec: M. Heckler, "Solubility of Ti Liquid in Al," *Aluminium (Dusseldorf), 50*(6), 405-407 (1974) in German. (Equi Diagram; Experimental)

Phase Diagrams of Binary Titanium Alloys

74Ker: H.W. Kerr, J. Cisse, and G.F. Bolling, "Equilibrium and Non-Equilibrium Peritectic Transformation," *Acta Metall.*, 22(6), 677-686 (1974). (Equi Diagram; Experimental)

74Stu: J.M. Stuve and M.J. Ferrante, "Low-Temperature Heat Capacity and High-Temperature Enthalpy of TiAl$_3$," Rept. Invest. No. 7834, U.S. Dept. of the Interior, Bureau of Mines, Washington, DC, 9 p (1974). (Thermo; Experimental)

75Esi: Yu.O. Esin, N.P. Borov, M.S. Petrushevskii, and P.V. Geld, "Concentration Dependence of the Enthalpies of Formation of Melts of Ti with Al," *Tepl. Yyok. Temp.*, 13(1), 84-88 (1975) in Russian. (Thermo; Experimental)

75Tey: Ye.I. Teytel and E.S. Yakovleva, "Investigation of the Fine Structure of the Cast Alloy Ti-Al," *Fiz. Met. Metalloved.*, 40(1), 129-134 (1975) in Russian; TR: *Phys. Met. Metallogr.*, 40(1), 109-114 (1975). (Equi Diagram; Experimental)

76Col: E.W. Collings, "Magnetic Investigations of Electronic Bonding and α_2 Through γ Phase Equilibria in the Titanium Aluminum System," *Titanium and Titanium Alloys, Scientific and Technological Aspects*, Moscow, USSR, 2, 1391-1402 (1976). (Equi Diagram; Experimental)

76Kor1: I.I. Kornilov, T.T. Nartova, and S.P. Chernyshova, "The Ti-Rich Range of the Ti-Al Phase Diagram," *Izv. Akad. Nauk SSSR, Met.*, (6), 192-198 (1976) in Russian; TR: *Russ. Metall.*, (6), 156-161 (1976). (Equi Diagram; Experimental)

76Kor2: I.I. Kornilov, T.T. Nartova, S.P. Chernyshova, and V.M. Naidan, "Structure and Properties of TiAl Alloys Quenched from the β Region," *Izv. Akad. Nauk SSSR, Met.*, (5), 189-192 (1976) in Russian; TR: *Russ. Metall.*, (5), 163-166 (1976). (Meta Phases; Experimental)

76Lim: J.Y. Lim, C.J. McMahon, D.P. Pope, and J.C. Williams, "The Effect of Oxygen on the Structure and Mechanical Behavior of Aged Ti−8 wt.% Al," *Metall. Trans. A*, 7, 139-144 (1976). (Equi Diagram; Experimental)

76Zel: I.A. Zelenkov and E.N. Osokin, "Effects of Phase Transitions Upon Some Physical Properties of the Compound Ti$_3$Al and Its Alloys," *Porosh. Metall.*, 2(158), 44-48 (1976) in Russian; TR: *Sov. Powder Metall. Met. Ceram.*, 15(2), 1543-1552 (1977). (Equi Diagram; Experimental)

77Bag: R.G. Baggerly, "X-Ray Analysis of Ti$_3$Al Precipitation in Ti-Al Alloys," *Advances in X-Ray Analysis*, Plenum Press, New York and London, Vol. 18, 1543-1552 (1977). (Equi Diagram; Experimental)

77Sas: S.M.L. Sastry and H.A. Lipsitt, "Ordering Transformations and Mechanical Properties of Ti$_3$Al and Ti$_3$Al-Nb Alloys," *Metall. Trans. A.*, 8, 1543-1552 (1977). (Equi Diagram; Experimental)

78Kau: L. Kaufman and H. Nesor, "Coupled Phase Diagrams and Thermochemical Data for Transition Metal Binary Systms - V," *Calphad*, 2(4), 325-348 (1978). (Thermo; Theory)

78Shi: K. Shibata, T. Sato, and G. Ohira, "The Solute Distributions in Dilute Al-Ti Alloys During Undirectional Solidification," *J. Cryst. Growth*, 44, 435-445 (1978). (Equi Diagram; Experimental)

79Col: E.W. Collings, "Magnetic Studies of Phase Equilibria in Ti-Al (30-57 at.%) Alloys," *Metall. Trans. A*, 10(4), 463-474 (1979). (Equi Diagram; Experimental)

79Erm: M.I. Ermolova, E.I. Guskova, and O.P. Solomina, *Mitom*, (3), 55 (1979) in Russian. (Equi Diagram; Experimental)

79Muk: P. Mukhopadhyay, "Dissolution of TiAl Plates in a Ti-40% Al Alloy," *Metallography*, 12, 119-123 (1979). (Equi Diagram; Experimental)

80Mii: R. Miida, M. Kasahara, and D. Watanabe, "Long-Period Antiphase Domain Structures of Al-Ti Alloys near Composition Al$_3$Ti," *Jpn. J. Appl. Phys.*, 19(11), L707-L710 (1980). (Equi Diagram; Experimental)

80Ouc: K. Ouchi, Y. Iijima, and K. Hirano, "Interdiffusion in the Ti-Al System," *Titanium '80, Science and Technology*, Vol. 1, Proc. Fourth Int. Conf. on Titanium, 19-22 May 1980, Kyoto, Japan, H. Kimura and O. Izumi, Ed., TMS-AIME, Warrendale, PA, 559-568 (1980). (Equi Diagram; Experimental)

80Sha: R.E. Shalin and Yu.K. Kovneristy, "Phase Stability and Phase Equilibrium in Titanium Alloys," *Titanium '80, Science and Technology* H. Kimura and O. Izumi, Ed., TMS-AIME, Warrendale, PA, Vol. 1, Proc. Fourth Int. Conf. on Titanium, 19-22 May 1980, Kyoto, Japan, 277-293 (1980). (Equi Diagram; Experimental)

81Kik: R. Kikuchi, "Equivalence of H.C.P. an F.C.C. Phase Diagrams and Phase Diagram of Hexagonal Binary Alloy with Anisotropic Interactions," HRL Rep. NB8ONAAE0188, Hughes Research Laboratories (1981). (Thermo; Theory)

81Mii: H. Miida, S. Hashimoto, and D. Watanabe, "Long-Period Structures of the IIIb-IVa Alloys Near Composition A$_3$B," *Sci. Rept. Res. Inst. Tohoku Univ.*, A 29, Suppl. (1), 19-24 (1981). (Equi Diagram; Experimental)

81Swa: L.J. Swartendruber, L.H. Bennett, L.K. Ives, and R.D. Shull, "The Ti-Al Phase Diagram: the α_2 Phase Boundary," *Mater. Sci. Eng.*, 51, P1-P9 (1981). (Equi Diagram; Experimental)

82Mii: R. Miida, R. Hashimoto, and D. Watanabe, "New Type of A$_5$B$_3$ Structure in Al-Ti and Ga-Ti Systems; Al$_{15}$Ti$_3$ and Ga$_5$Ti$_3$," *Jpn. J. Appl. Phys.*, 21(1), L59-L61 (1982). (Equi Diagram, Crys Structure; Experimental)

82Tsu: S. Tsunekawa and M.E. Fine, "Lattice Parameters of Al$_{13}$(Zr$_x$i$_{1-x}$) vs x in Al−2 at.% (Ti + Zr) Alloys," *Scr. Metall.*, 16, 391-392 (1982). (Crys Structure; Experimental)

83Lia: W.W. Liang, "A Thermodynamic Assessment of the Aluminum-Titanium System," *Calphad*, 7(1), 13-20 (1983). (Thermo; Theory)

83Loi1: A. Loiseau and A. Lasalmonie, "New Ordered Structures in the Non-Stoichiometric Compound TiAl," *Acta Crystallogr. B*, 39, 580-587 (1983) in French. (Equi Diagram; Experimental)

83Loi2: A. Loiseau, A. Lasalmonie, G. Van Tendeloo, and J. Van Landuyt, "Ordered Structures in the Composition Range from AB to AB$_3$," Proc. Conf. on Phase Transformations, Maleme (1983). (Equi Diagram; Experimental)

83Nam: T.K.G. Namboodhiri, "On the Ti-Al Phase Diagram," *Mater. Sci. Eng.*, 57, 21-22 (1983). (Equi Diagram; Experimental)

84Abd: A. Abdel-Hamid, C.H. Allibert, and F. Durand, "Equilibrium Between TiAl$_3$ and Molten Al: Results from the Technique of Electromagnetic Phase Separation," *Z. Metallkd.*, 76, 455-458 (1984).

84Loi1: A. Loiseau, A. Lasalmonie, G. Van Tendeloo, J. Van Landuyt, and S. Amelincks, "New Investigation of the Ti-Al Phase Diagram by High Resolution Electron Microscopy," Fifth Int. Conf. on Titanium, Lutjering, U. Zwicker and W. Bunk, Ed., Munich, Sep 10-14 (1984). (Equi Diagram; Experimental)

84Loi2: A. Loiseau and A. Lasalmonie, "High Resolution Electron Microscopy Study of Al$_{5-x}$Ti$_{3+x}$ Alloys," *Acta Crystallogr., B*, to be published (Equi Diagram; Experimental)

84Shu: R.D. Shull, A.J. McAlister, and R. Reno, "Phase Equilibria in the Titanium-Aluminum System," Proc. Fifth Int. Conf. on Titanium, G. Lutjering, U. Zwicker, and W. Bunk, Ed., Munich, Sep 10-14 (1984). (Equi Diagram; Experimental)

The As-Ti (Arsenic-Titanium) System

74.9216 47.88

By J.L. Murray

Equilibrium Diagram

Essentially nothing is known about the Ti-As phase diagram. The only phase equilibrium investigation for this system was done by [59Hay] for Ti-rich alloys. Heat treatments were made in air, and therefore, the diagram is certainly not a correct binary diagram. A schematic of the [59Hay] results (Fig. 1) is included in this assessment only as a guide to future experiment. The existence of a eutectic reaction involving (Ti) can probably be extrapolated to the binary system. The compound Ti_4As was assigned its stoichiometry on the basis of its volume fraction in two-phase alloys. Its structure was not determined, but it was neither TiAs nor DO_{19}.

Crystal Structures and Lattice Parameters

Crystal structures and lattice parameters of other intermetallic phases of the Ti-As system are listed in Table 1. These phases are based on stoichiometries TiAs and $TiAs_2$.

[64Wen] observed that single crystals of $TiAs_2$ formed only when excess As was present and above 820 °C, and they concluded that liquid As participated in the growth of single crystals. They also described unpublished vapor pressure studies by [54Trz], which implied that $TiAs_2$ has a wide homogeneity range. However, the homogeneity range was not reflected in any variation of lattice parameters with composition. From X-ray and density measure-

ments and from "visual inspection of the samples," [72Kje] concluded that $TiAs_2$ is strictly stoichiometric; [64Hul] could not find any evidence for a homogeneity range. It is concluded that $TiAs_2$ probably has a narrow homogeneity range.

$\alpha TiAs$ has a superstructure of NiAs (TiAs). [55Bac] believed TiAs to be the high-temperature (and possibly Ti-rich) form. The heat of formation of TiAs from αTi and solid As is -74.9 J/mol, determined from the difference between the heat of combustion of the compound and that of the pure components [59Sch].

Cited References

54Luk: K. Lukaszewicz and W. Trzebiatowski, "Crystal Structure of Titanium Arsenide, TiAs," *Bull. Acad. Polon. Sci., Classe III, 2*(6), 277-279 (1954). (Crys Structure; Experimental)

54Trz: W. Trzebiatowski and K. Lukaszewicz, "The Structure of Titanium Arsenide," *Roczniki Chem., 28* 150-151 (1954) in German. (Crys Structure; Experimental)

55Bac: K. Bachmayer, H. Nowotny, and A. Kohl, "The Structure of TiAs," *Monatsh. Chem., 86,* 39-43 (1955) in German. (Crys Structure; Experimental)

***59Hay:** R. Haynes, "Commercially Pure Titanium-Arsenic Alloys," *J. Inst. Met., 88,* 277-279 (1959-1960). (Equi Diagram; Experimental)

59Sch: S.A. Schukarev, V.P. Morozova, and Li Miao-hsiu, "Enthalpy of Formation of Compounds of Titanium with Elements of the Main Subgroup of Group V," *Zh. Obshch. Khim., 29,* 2465-2467 (1959) in Russian; TR: *J. Gen. Chem. USSR, 29,* 2427-2429 (1959). (Crys Structure; Experimental)

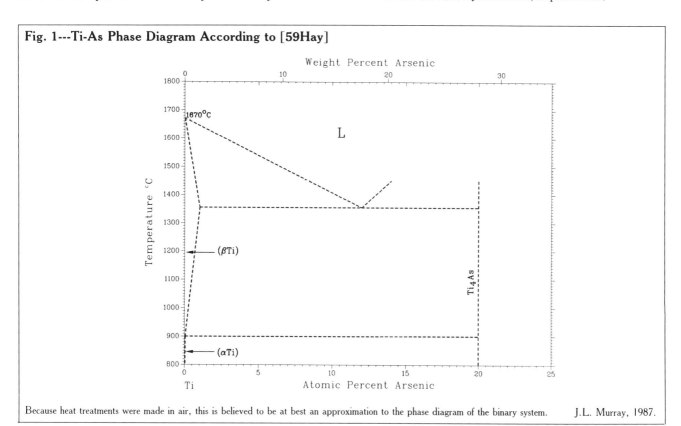

Fig. 1---Ti-As Phase Diagram According to [59Hay]

Because heat treatments were made in air, this is believed to be at best an approximation to the phase diagram of the binary system. J.L. Murray, 1987.

Table 1 Crystal Structures and Lattice Parameters of the Equilibrium Phases of the Ti-As System

Phase	Homogenity range, at.% As	Pearson symbol	Space group	Struktur-bericht designation	Prototype	Lattice parameters, nm	Reference
(βTi)	0 to ?	cI2	Im3m	A2	W	...	
(αTi)	0 to ?	hP2	P6₃/mmm	A3	Mg	...	
Ti₄As	~ 19.7	(a)	...	...	...	...	[59Hay]
βTiAs	50	hP4	P6₃/mmc	B8₁	AsNi	$a = 0.364$ $c = 0.615$	[55Bac]
αTiAs	50	hP8	P6₃/mmc	B_i	AsTi	$a = 0.365$ $c = 1.230$	[54Luk]
						$a = 0.3642$ $c = 1.2064$	[54Trz]
						$a = 0.3647$ $c = 1.2305$	[55Bac]
TiAs₂	66.7	oP24	Pnnm	...	TiAs₂	$a = 1.327$ $b = 0.896$ $c = 0.350$	[64Wen]
						$a = 1.3220$ $b = 0.8915$ $c = 0.3478$	[64Hul]
						$a = 1.32303$ $b = 0.89148$ $c = 0.34793$	[72Kje]
(As)	100	hR2	R3̄m	A7	As	...	

(a) Unknown.

64Hul: F. Hullinger, "TiAs₂-Type Phases," *Nature, 204,* 991 (1964). (Crys Structure; Experimental)

64Wen: S. Wenglowskii, G.B. Bokii, and E.A. Pobedimskaya, "The Crystal Structure of Titanium Diarsenide, TiAs₂, *Zh. Strukt. Khim.,* 5(1), 64-69 (1964) in Russian; TR: *J. Struct.* *Chem. USSR,* 5(1), 55-59 (1964). (Crys Structure; Experimental)

72Kje: A. Kjekshus, "On the Crystal Structures of ZrSb₂ and αHfSb₂," *Acta Chem. Scand., 26*(4), 1633-1639 (1972). (Crys Structure; Experimental)

The Au-Ti (Gold-Titanium) System

196.9665   47.88

By J.L. Murray

Equilibrium Diagram

The main outlines of the Ti-Au system are the result of two studies [56Pie, 62Pie]. Additional data from [54Mcq], [62Sto], [63Hah], [70Don], and [78Pli] refine, but essentially corroborate, the findings of [56Pie, 62Pie]. Thermodynamic calculations of the diagram agree very well with the experimental data, within uncertainties expected for a Ti-based system for which there is no experimental thermodynamic data, i.e., within ± 35 °C for the three-phase equilibria. In the assessed phase diagram (Fig. 1), the (Au) boundaries were drawn from the thermodynamic calculation, because they agreed with the assessment within 2 °C; the (βTi) liquidus and solidus were based on calculations as well as experiment because the experimental observations do not conform with thermodynamic constraints. The remainder of the diagram was drawn from the experimental data.

The equilibrium solid phases of the Ti-Au system are:

- The bcc (βTi) solid solution, with a maximum Au solubility of about 15 at.%.
- The cph (αTi) solid solution, with a maximum Au solubility of about 1.7 at.%.
- The fcc (Au) solid solution, with a maximum Ti solubility of 12 at.%.
- The essentially stoichiometric compound Ti_3Au, with the $A15$ structure and a congruent melting point.

- The TiAu compounds, with a maximum homogeneity range of 38 to 52 at.% Au. The allotropic forms with the CsCl, AuCd, and γTiCu structures are designated γTiAu, βTiAu, and αTiAu, respectively.
- The essentially stoichiometric compound $TiAu_2$, with the $MoSi_2$ structure and a congruent melting point.
- The compound $TiAu_4$, with a homogeneity range of 79 to 82 at.% Au. $TiAu_4$ has the Ni_4Mo structure and melts by a peritectic reaction with (Au).

Liquidus and Solidus.

The three-phase equilibria and congruent melting points listed in Table 1 were determined by [56Pie, 62Pie]. The temperature uncertainties are based on the temperature intervals used by [56Pie, 62Pie] for heat treatments. [56Pie, 62Pie] and [52Rau] made the only investigations of the liquidus for this system. [56Pie, 62Pie] used metallographic and incipient melting techniques. In addition to the invariant temperatures and congruent melting temperatures, [56Pie] obtained some information on the (βTi) and Ti_3Au solidus curves. [52Rau] used thermal analysis to investigate the Au-rich part of the liquidus. The data of [52Rau] are not considered to be uniformly reliable because of the possibility of contamination by alumina crucibles. This effect may account for their identification of the Au-rich compound as $TiAu_6$ and for unexplained thermal arrests in the composition range 50

Fig. 1---Assessed Ti-Au Phase Diagram

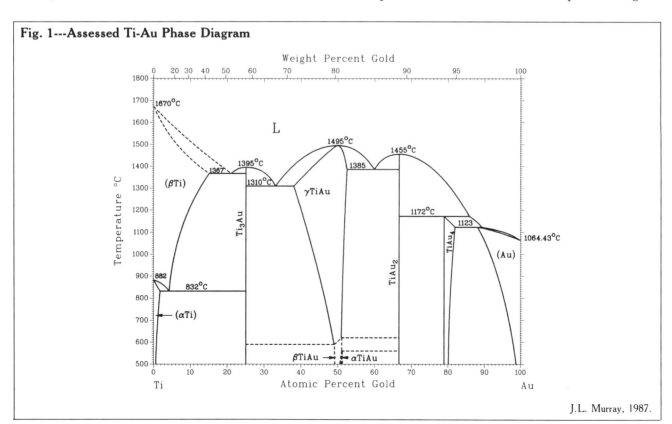

J.L. Murray, 1987.

Fig. 2---Assessed Ti-Au Phase Diagram vs Select Experimental Data

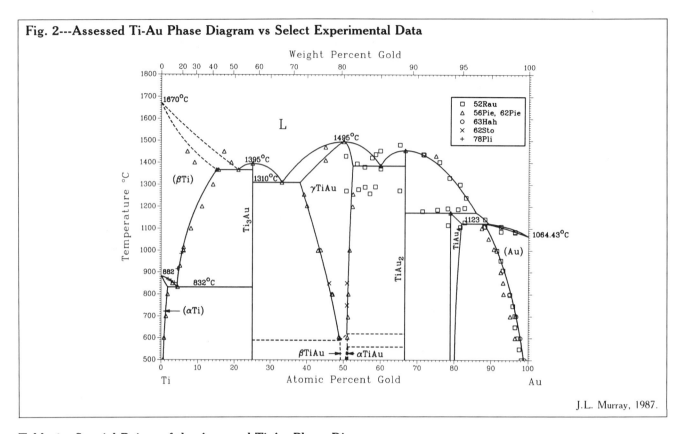

J.L. Murray, 1987.

Table 1   Special Points of the Assessed Ti-Au Phase Diagram

Reaction	Compositions of the respective phases, at.% Au			Temperature, °C	Reaction type
L ⇌ (βTi) + Ti₃Au	21 ± 1	15 ± 1	25	1367 ± 5	Eutectic
(βTi) ⇌ (αTi) + Ti₃Au	4.2	1.7 ± 0.3	25	832 ± 5	Eutectoid
L ⇌ Ti₃Au + γTiAu	33 ± 1	25	38	1310 ± 10	Eutectic
L ⇌ γTiAu + TiAu₂	60 ± 1	52	66.7	1385 ± 5	Eutectic
TiAu₄ + TiAu₂ ⇌ L	80	66.7	86	1172 ± 5	Peritectic
TiAu₄ + L ⇌ (Au)	82	89	88	1123 ± 3	Peritectic
L ⇌ Ti₃Au		25		1395 ± 10	Congruent
L ⇌ γTiAu		50		1495 ± 25	Congruent
L ⇌ TiAu₂		66.7		1455 ± 5	Congruent
L ⇌ (βTi)		0		1670	Melting point
(βTi) ⇌ (αTi)		0		882	Allotropic transformation
L ⇌ (Au)		100		1064.43	Melting point

to 70 at.% Au. The Au-rich TiAu₂ liquidus and the peritectic temperatures, however, agree quite well with the [56Pie] and [62Pie] data (Table 1).

The incipient melting technique is not without special difficulties in this system; [56Pie] noted that microscopic examination was "not always rewarding" and was strongly influenced by cooling rate. In particular, the (βTi) solidus as drawn by [56Pie] approaches the eutectic temperature with nearly zero slope, that is to say, it approaches a metastable congruent point. This would imply that the metastable extension of the (βTi) boundaries also contains a congruent maximum. Construction of a thermodynamically valid metastable extension based on there data would require an implausible Gibbs energy function, which is not justified in view of the uncertainties in the experimental data. Likewise, extrapolating the

liquidus and solidus to the pure Ti melting point leads to difficulties in satisfying the van't Hoff relation. The assessed liquidus and solidus are therefore based on rough thermodynamic calculations, as well as on the experimental observations.

(αTi), (Ti), and Ti₃Au.

The eutectoid reaction (βTi) ⇌ (αTi) + Ti₃Au occurs at 832 ± 2 °C, with (αTi) and (βTi) compositions of 1.7 ± 0.1 and 4.2 ± 0.2 at.% Au, respectively. This region of the diagram was examined by [54Mcq], [56Pie], and [78Pli] by metallographic methods, supplemented with hydrogen pressure experiments [54Mcq] or microprobe analysis of equilibrated alloys [78Pli]. The work of [78Pli] was least likely to be influenced by interstitial impurities. Data on

The Au-Ti System

Table 2 (αTi) Solvus

Reference	(αTi) composition, at.% Au	Temperature, °C	Experimental method
[56Pie] ...	1.75 ± 0.2	800	Metallography
	1.2 ± 0.4	700	
	0.7 ± 0.15	600	
[54Mcq] ...	2 to 3	810 to 830	Metallography
	< 2	840	
[68Pli]	< 1.7	832	Metallography, microprobe

Table 3 Martensitic Transformation of γTiAu [70Don]

Composition, at.% Au	Transformation temperature, °C	
	Cooling, M_s	Heating, A_s
45	575	535
47.4	605	560
50	620	615
52.6	630	630

Note: Compositions at which γTiAu is supersaturated are included.

the (βTi) boundaries are shown in Fig. 2; data on the (αTi) solvus are also listed in Table 2.

The (βTi) transus is difficult to determine metallographically because of the massive transformation; (βTi) that has transformed to (αTi) during quenching differs very little in appearance from equilibrium (αTi). The agreement among various authors on the eutectoid point is nevertheless very good, and reported values for the (βTi) transus agree within ± 1 at.%.

[56Pie] determined three points of the (αTi) solvus, which extrapolate to a maximum solubility of about 1.7 ± 0.2 at.% Au at the eutectoid temperature. This agrees well with the work of [78Pli], but not with [54Mcq], who found 2 at.% Au alloys annealed between 810 and 830 °C to be single-phase (αTi). Because it was supplemented by microprobe analyses, the work of [78Pli] is preferred. [54Mcq], [56Pie], and [78Pli] did not make structural determinations to verify that the compound in equilibrium with (Ti) had the A15 structure.

Ti₃Au was observed by electron diffraction in thin-film diffusion couples [72Tis] and by X-ray diffraction [70Don] as the first precipitate from supersaturated Ti-rich TiAu. After prolonged heating, the equilibrium A15 structure transforms to another, interstitially stabilized structure [70Don]. Determinations of Ti₃Au boundaries by [54Mcq], [56Pie], and [78Pli] did not use structural determinations to verify that the compound in equilibrium with (βTi) or (αTi) had the A15 structure. The work of [78Pli] was least likely to be influenced by interstitial impurities.

TiAu Allotropes.

The congruent melting temperature of γTiAu is 1490 ± 15 °C [56Pie]. The maximum extent of the single-phase γTiAu field is 38 to 52 at.% Au, [56Pie, 62Sto] at 1310 °C. There are at least two, and perhaps three, equilibrium allotropic forms of TiAu; the forms are designated γTiAu, βTiAu, and αTiAu, proceeding from high- to low-temperature and high- to low-Ti content (see Fig. 1). There are some discrepancies concerning the structures of the allotropic forms. However, investigators agree that the room-temperature phase on the Ti-rich side of the phase field has the B19 AuCd structure.

[56Pie] found metallographic evidence of a martensitic transformation in a 54 at.% Au alloy; however, crystal structures were not determined. [62Sto], in an X-ray study of alloys annealed between 500 and 550 °C, observed the B19 structure for Ti-rich compositions, but the γCuTi structure for Au-rich compositions. On the other hand, [62Sto] however, observed the martensitic structure for all near-equiatomic alloys quenched from 800 °C. These observations led them to suggest that the high-temperature phase γTiAu has the B19 structure and is stable to

low temperature only at Ti-rich compositions, but that the γCuTi structure is stable below about 600 °C on the Au-side. They also suggested the possibility that the high-temperature phase has yet a third structure, based on the observation of martensitic microstructures.

[70Don] splat quenched a series of alloys between 40 and 52.6 at.% Au to prepare metastable single-phase TiAu alloys over the widest possible composition range and examine the martensitic transformation. Thermal analysis and high-temperature X-ray diffraction were used. They found that the low-temperature phase had the B19 structure for all compositions studied and that it had transformed martensitically from the high-temperature CsCl structure. The martensite transformation temperatures for compositions in the single-phase field are above 600 °C. M_s and A_s temperatures are listed in Table 3.

[70Don] did not observe αTiAu (γTiCu structure) even in single-phase 52.6 at.% Au alloys. It is possible that it may be a ternary phase; [62Sto] observed that oxide phases form during the relatively long anneals required to produce the γTiCu structure. αTiAu is tentatively included as an equilibrium binary phase whose composition and temperature range of stability are known only approximately.

Au-Rich Alloys: TiAu₂, TiAu₄, and (Au) Solvus.

The compound TiAu₂ was observed in all investigations of this composition range [62Pie, 62Sto, 52Rau], as well as in thin-film diffusion couples [72Tis], and as the first precipitate from TiAu supersaturated with Au [70Don]. TiAu₂ melts congruently at 1445 °C.

TiAu₄, however, has been variously identified as TiAu₃ [43Wal] and as TiAu₆ [52Rau]. The latter identification was based on interpretation of a microstructure as single phase at that composition; the former identification is clearly the result of large interstitial impurity content. The observation of a two-phase TiAu₂ + TiAu₄ assemblage at 75 at.% Au [62Pie, 69Sin] appears to preclude the occurrence of TiAu₃ as an equilibrium binary phase. The composition range of TiAu₄ was deduced by [62Pie] from metallographic observations and lattice parameter measurements.

Determinations of the (Au) solvus were made by [52Rau], [62Pie], and [63Hah]. [52Rau] and [62Pie] used metallographic methods. [63Hah] used electrical conductivity measurements between 500 and 800 °C. The lattice parameter data [63Hah] imply slightly lower (by 0.5 to 1 at.%) solubilities at low temperatures, and these are preferred. At higher temperatures, the data of [52Rau] and [62Pie] appear to be mutually consistent.

Liquidus 50 to 100 at.% Au.

Thermal analysis (cooling) data were reported for Au-rich alloys by [52Rau], and incipient melting data were reported by [62Pie]. For alloys containing more than 67 at.% Au, reported liquidus and invariant temperatures

29

Table 4 Crystal Structures of the Ti-Au System

Phase	Homogeneity range, at.% Au	Pearson symbol	Space group	Struktur-bericht designation	Proto-type	Reference
(αTi)	0 to 1.7	$hP2$	$P6_3/mmc$	$A3$	Mg	[Pearson2]
(βTi)	0 to 15	$cI2$	$Im3m$	$A2$	W	[Pearson2]
Ti$_3$Au	25	$cP8$	$Pm3n$	$A15$	Cr$_3$Si	[Pearson2]
γTiAu	38 to 52	$cP2$	$Pm3m$	$B2$	CsCl	[68Reu]
βTiAu	49 to 50	$oP4$	$Pmma$	$B19$	AuCd	[70Don]
αTiAu	50	$tP4$	$P4/nmm$	$B11$	CuTi	[62Sto]
TiAu$_2$	66.7	$tI6$	$I4/mmm$	$C11_b$	MoSi$_2$	[62Sto]
TiAu$_4$	79 to 82	$tI10$	$I4/m$	$D1_a$	Ni$_4$Mo	[62Pie]
(Au)	88 to 100	$cF4$	$Fm3m$	$A1$	Cu	[62Sto]

are in agreement. Between 50 and 67 at.%, [52Rau] found two arrest temperatures apparently corresponding to invariant reactions; they attributed arrests at 1284 °C to the eutectic reaction L $\rightleftarrows$ γTiAu + TiAu$_2$. Arrests at higher temperatures correspond to the eutectic reaction observed by [62Pie] at 1385 °C. Other invariant reactions listed in Table 1 and shown in Fig. 1 are also based on the work of [62Pie], primarily because it was supported by metallography.

Thermal analysis (cooling) data for the (Au) liquidus and solidus were reported by [52Rau]. The liquidus and solidus were constructed for Fig. 1 from thermodynamic calculations that reproduced the liquidus and peritectic temperature. The calculated two-phase region is somewhat narrower than that based on the experimental data, as would be expected when solidus data are derived from cooling curves.

Metastable Phases

The (βTi) structure cannot be retained metastably during quenching for alloys containing 0 to 6 at.% Au [54Mcq]. For cooling rates less than approximately 10^3 °C/s, the mechanism of the transformation was shown to be of the massive type, by microstructural and microprobe analysis [78Pli]. At faster cooling rates, the transformation is martensitic [78Pli]. The absence of a martensitic microstructure for normal quenching rates creates difficulties in determining the (βTi) transus [54Mcq, 56Pie].

[78Pli] reported arrest temperatures as a function of cooling rate for three alloys of compositions 1.8, 2.6, and 3.5 at.% Au. For the 1.8 and 3.5 at.% Au alloys, a martensite transformation "plateau" occurred, and the M$_s$ temperatures were 770 and 745 °C, respectively. Using a kinetic model of the massive transformation, they also extrapolated to zero cooling rate to define a "T_0 temperature" for the massive transformation. These temperatures, based on the assessed phase diagram, do not bear any apparent relation to the temperature of equal Gibbs energies for the (αTi) and (βTi) phases.

Crystal Structures and Lattice Parameters

Crystal structures are given in Table 4 and lattice parameter data are given in Table 5. Lattice parameters of (αTi) and (βTi) have not been measured as a function of composition. The structures reported for αTiAu, βTiAu, and γTiAu were discussed above; reported structures of the other compounds will now be reviewed.

Ti$_3$Au.

The original identification of the structure of Ti$_3$Au as AuCu$_3$-type [39Lav] was due to the effect of impurities.

[52Duw] determined by X-ray diffraction that Ti$_3$Au has the $A15$ structure, and this structure was verified by [62Sto]. [68Reu] measured order parameters and fractional occupancies of each atomic site. The stabilization of the AuCu$_3$ structure by interstitial impurities was verified by [70Don], who observed a transformation from $A15$ to AuCu$_3$ upon prolonged heating.

TiAu$_2$.

Structure determinations of TiAu$_2$ were made by [62Pie] and [62Sto] and verified (in thin film diffusion couples) by [72Tis]. The observation by [52Rau] of a hexagonal TiAu$_2$ structure is attributed to impurities.

TiAu$_4$.

[43Wal] attributed to TiAu$_4$ an ordered distorted cph "Cu$_3$Ti-type" structure. This is incorrect for both the Au-Ti and the Cu-Ti compounds: the stoichiometries of the phases are Cu$_4$Ti and Au$_4$Ti. There is no equilibrium compound of stoichiometry Au$_3$Ti [62Pie, 69Sin]. Similarly, the observation by [52Rau] of an Au$_6$Ti compound is attributed to the effect of impurities. TiAu$_4$ has the NiMo$_4$ structure.

Thermodynamic Modeling

No experimental thermodynamic data are available on the Ti-Au system. We nevertheless find it possible to reproduce many features of the experimental diagram by calculations, and it is thus possible to make some tentative estimates of relative Gibbs energies of the phases.

In this calculation, the Gibbs energies of the solution phases are represented as:

$$G(i) = (1 - x) G^0(\text{Ti},i) + G^0(\text{Au},i) x + RT [x \ln x + (1 - x) \ln (1 - x)] + B(i) x(1 - x) + C(i) x(1 - x)(1 - x)$$

where i designates the phase; x the atom fraction of Au; G^0, the Gibbs energies of the pure components; and $B(i)$ and $C(i)$ are temperature-dependent coefficients defining the excess quantities.

TiAu is modeled by a Wagner-Schottky Gibbs energy:

$$G = G^0(\text{TiAu}) (1 + v) + RT[x_{\text{Ti}} \ln x_{\text{Ti}} + x_{\text{Au}} \ln x_{\text{Au}} + v \ln v + s \ln s - (1 + v) \ln (1 + v)] + C(i) s + D(i) v$$

Wagner-Schottky compounds are conceptually resolved into appropriate Ti and Au sublattices, where v is the concentration of vacancies on the Ti sublattice; s is the concentration of substitutional Ti atoms on the Au sublattice; X_{Ti} and X_{Au} are the concentrations of Ti and Au on their respective sublattices; and x is the overall composi-

Table 5 Lattice Parameters of the Ti-Au System

Phase	Composition, at.% Au	Lattice parameters, nm			Reference
		a	b	c	
(βTi)	0	0.33065	...	...	[Pearson]
	0	0.33149	...	...	
(αTi)	0	0.29511	...	0.46843	[Pearson]
Ti₃Au	25	0.50947	...	...	[68Reu]
	25	0.5096	...	...	[52Duw]
	25	0.5089	...	...	[62Sto]
γTiAu	50	0.3254	...	...	[70Don]
βTiAu	48(a)	0.463	0.2944	0.4880	[62Sto, 62Sch]
	50	0.460	0.293	0.485	[70Don]
αTiAu	50(a)	0.333	...	0.6030	[62Sto, 62Sch]
TiAu₂	67	0.343	...	0.8538	[62Sto]
	67	0.3419	...	0.8514	[62Pie]
TiAu₄	80	0.6485	...	0.4002	[62Sto]
	80	0.6553	...	0.4043	[58Sch]
	78(a)	0.6460(8)	...	0.3970(6)	[62Pie](b)
	80	0.6459(8)	...	0.3976(3)	
	82	0.6458(3)	...	0.3986(3)	
(Au)	100	0.4078(2)	...	...	[62Pie]
	97.8	0.4076(7)	...	...	
	96.2	0.4974(8)	...	...	
	93.9	0.4072(0)	...	...	
	91.9	0.4070(6)	...	...	
	90.3	0.4068(2)	...	...	
	88.0	0.4065(3)	...	...	

(a) Two-phase alloys. (b) Averaged from several annealing temperatures.

Table 6 Estimated Gibbs Energies of Ti-Au

Gibbs energies of the pure components

G^0 (Ti,L) $= 0$
G^0 (Au,L) $= 0$

G^0 (Ti,bcc) $= -16\,234 + 8.363\ T$
G^0 (Au,bcc) $= -8\,786 + 7.510\ T$
G^0 (Ti,cph) $= -20\,585 + 12.134\ T$
G^0 (Au,cph) $= -9\,372 + 7.510\ T$
G^0 (Ti,fcc) $= -17\,238 + 12.134\ T$
G^0 (Au,fcc) $= -12\,564 + 9.393\ T$

Gibbs energies of compounds (referenced to L)

G (Ti₃Au) $= -32\,692 + 7.566\ T$
G^0 (TiAu) $= -33\,424 + 5.037\ T$
C (TiAu) $= 43\,978$
D (TiAu) $= 72\,318$
G (TiAu₂) $= -25\,330 + 2.072\ T$
G (TiAu₄) $= -31\,963 + 11.953\ T$

Excess Gibbs energies of solution phases

B (L) $= -63\,607 + 2.114\ T$ B(bcc) $= -47\,413 + 10.531\ T$
C (L) $= -10\,497$ C(bcc) $= -14\,224$
B (cph) $= -21\,813$ B(fcc) $= -57\,791$
C (cph) $= -15\,000$ C(fcc) $= -1\,000$

Note: Values are given in J/mol and J/mol · K.

tion. All concentrations are referred to as the total number of atoms. The concentrations of vacancies and substitutions are determined by minimizing the Wagner-Schottky Gibbs energy with respect to the variables s and v. The three allotropes (γTiAu, βTiAu, and αTiAu) are not distinguished in these calculations. Ti₃Au, TiAu₂, and TiAu₄ are represented as line compounds.

The excess Gibbs energy of the liquid was initially estimated to be approximately equal to that of the Ti-Cu liquid, based on the phase diagram. The choice given in Table 6 is arbitrary in the sense that a range of at least 20 kJ/mol in B(L) can give equally accurate phase diagrams. We estimate that the values given in Table 6 probably err in the direction of being too negative. Least-squares optimizations of the Gibbs energies of the various phases were performed, always fixing the excess Gibbs energy of the liquid.

Two strategies were compared for optimization calculations. In the first, optimizations were performed in the following sequence. Gibbs energies of the four compounds were simultaneously optimized with respect to the congruent melting points and three-phase equilibria. Then, the excess Gibbs energies of the solid solution phases were optimized with respect to the three-phase equilibria and solvus curves. The entire phase diagram except for the (βTi) solidus and eutectoid composition could be reproduced essentially exactly, but these calculations led to very large excess entropies for the solid solution phases and very low melting entropies for the compounds. This strategy was therefore rejected.

In the second strategy, the optimizations involving the compounds and the solid solution phases were performed in the reverse order. The entropies were reduced to physically reasonable values, and the errors in the phase diagram increased to magnitudes expected of a rough calculation. The most important discrepancy is still in the (αTi)/(βTi) boundaries and eutectoid composition. The results of these optimizations are reported in Table 6, and the calculated phase diagram is compared to the input data in Fig. 3.

Fig. 3---Calculated Ti-Au Phase Diagram vs Select Experimental Data

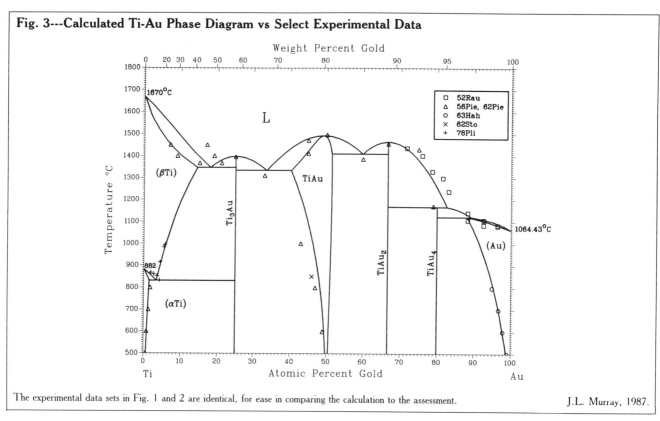

The experimental data sets in Fig. 1 and 2 are identical, for ease in comparing the calculation to the assessment.

J.L. Murray, 1987.

Cited References

39Lav: F. Laves and H.J. Wallbaum, "Crystal Chemistry of Titanium Alloys," *Naturwissenschaften, 27,* 674-675 (1939) in German. (Equi Diagram; Experimental)

43Wal: H.J. Wallbaum, "Transition Metal Alloys," *Naturwissenschaften, 31,* 91-92 (1943) in German. (Equi Diagram; Experimental)

52Duw: P. Duwez and C.B. Jordon, "The Crystal Structure of Ti_3Au and Ti_3Pt," *Acta Crystallogr., 5,* 213-214 (1952). (Crys Structure; Experimental)

52Rau: E. Raub, P. Walter, and M. Engel, "Alloys of Titanium with Copper, Silver and Gold," *Z.Metallkd., 43,* 112-118 (1952) in German. (Equi Diagram; Experimental)

54Mcq: M.K. McQuillan, "Note on the Constitution of the Titanium-Gold System in the Region 0-6 at.% Gold," *J. Inst. Met., 82,* 511-512 (1954). (Equi Diagram; Experimental)

*56Pie: P. Pietrokowsky, E.P. Frink, and P. Duwez, "Investigation of the Partial Constitution Diagram Ti-TiAu$_2$," *Trans. AIME, 206,* 930-935 (1956). (Equi Diagram; Experimental)

58Sch: K. Schubert, H. Breimer, R. Gohle, H.L. Lukas, H.G. Meissner, and E. Stolz, "Some Structure Results for Intermetallic Phases III," *Naturwissenschaften, 45,* 360-361 (1958) in German. (Crys Structure; Experimental)

62Pie: P. Pietrokowsky, "The Partial Constitutional Diagram TiAu$_2$-Au: Lattice Parameters of the α Solid Solutions and the Intermediate Phase TiAu$_4$," *J. Inst. Met., 90,* 434-438 (1962). (Equi Diagram; Experimental)

62Sch: K. Schubert, H.G. Meissner, M. Potzschke, W. Rossteutscher, and E. Stolz, "Structure Data for Intermetallic Phases (7)," *Naturwissenschaften, 49,* 57 (1962) in German. (Crys Structure; Experimental)

62Sto: E. Stolz and K. Schubert, "Structural Studies of Several T^4-B Homologous and Quasihomologous Systems," *Z. Metallkd., 43,* 433-444 (1962) in German. (Equi Diagram; Experimental)

63Hah: H.D. Hahlbohm, "Electrical, Magnetic, and Galvanomagnetic Measurements on Copper-Titanium and Gold-Titanium Solid Solutions," *Z. Metallkd., 54,* 515-518 (1963) in German. (Equi Diagram; Experimental)

68Reu: E.C. Van Reuth, "Atomic Ordering in Binary A15-Type Phases," *Acta Crystallogr. B, 24,* 186-196 (1968). (Crys Structure; Experimental)

69Sin: A.K.Sinha, "Close-Packed Ordered AB_3 Structures in Binary Transition Metal Alloys," *Trans. AIME, 245,* 237-240 (1969). (Equi Diagram; Experimental)

70Don: H.C. Donkersloot and J.H.N. Van Vucht, "Martensitic Transformations in Gold-Titanium Alloys Near the Equiatomic Composition," *J. Less-Common Met., 20,* 83-91 (1970). (Equi Diagram; Experimental)

72Tis: T.C. Tisone and J. Drobek, "Diffusion in Thin Film Ti-Au, Ti-Pd, and Ti-Pt Couples," *J. Vacuum Sci. Technol., 9(1),* 271-275 (1972). (Equi Diagram; Experimental)

78Pli: M.R. Plichta, H.I. Aaronson, and J.H. Perepezko, "The Thermodynamics and Kinetics of the β to α$_m$ Transformation in Three Ti-X Systems," *Acta Metall., 26,* 1293-1305 (1978). (Meta Phases; Experimental)

The B-Ti (Boron-Titanium) System

10.81   47.88

By J.L. Murray, P.K. Liao and K.E. Spear

Equilibrium Diagram

The assessed Ti-B phase diagram is shown in Fig. 1, and its special points are summarized in Table 1. The liquidus was calculated from Gibbs energy functions optimized with respect to thermochemical and phase diagram data, as described below.

The equilibrium phases of the system are:

- Terminal solid solutions—high-temperature bcc (βTi), low-temperature cph (αTi), and rhombohedral (βB).

- Intermetallic compounds, TiB and TiB$_2$, whose structures and melting mechanisms are well established.
- Ti$_3$B$_4$, which forms from the melt only in a narrow temperature range. Ti$_3$B$_4$ was reported by [64Fen] and [66Fen], but not by [66Rud]. Its existence was confirmed by a recent study [86Spe].

There is disagreement about the eutectic composition. [64Fen] and [66Rud] gave <1 and 7 ± 1 at.% B, respectively. An unambiguous eutectic microstructure is difficult to obtain because TiB rapidly segregates to the grain

Table 1 Special Points of the Assessed Ti-B Phase Diagram

Reaction	Compositions of the respective phases, at.% B			Temperature, °C	Reaction type
L $\rightleftarrows$ (βTi) + TiB	7 ± 1	<1	~ 50	1540 ± 10	Eutectic
L + TiB$_2$ $\rightleftarrows$ Ti$_3$B$_4$	42 ± 3	~ 65.5	58.1	2200 ± 25	Peritectic
L + Ti$_3$B$_4$ $\rightleftarrows$ TiB	~ 39	58.1	50	2180	Peritectic
L $\rightleftarrows$ TiB$_2$	...	66.7	...	3225 ± 25	Congruent
L $\rightleftarrows$ (βB) + TiB$_2$	~ 98	~ 100	~ 66.7	2080 ± 20	Eutectic
(βTi) + TiB $\rightleftarrows$ (αTi)	~ 0.1	49	~ 0.2	884 ± 2	Peritectoid
L $\rightleftarrows$ (βTi)		0		1670	Melting point
(βTi) $\rightleftarrows$ (αTi)		0		882	Allotropic transformation
L $\rightleftarrows$ (βB)		100		2092	Melting point

Fig. 1---Assessed Ti-B Phase Diagram

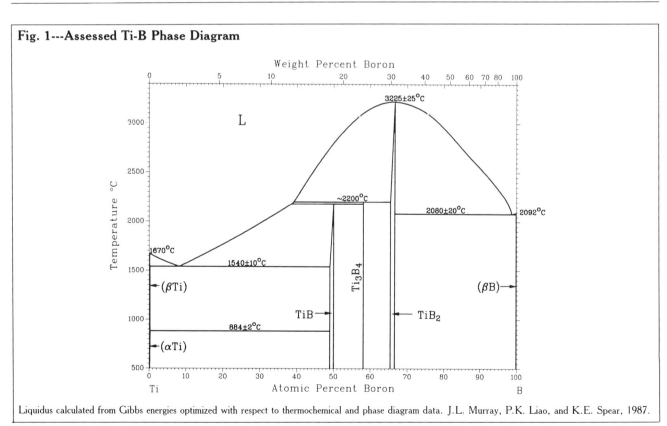

Liquidus calculated from Gibbs energies optimized with respect to thermochemical and phase diagram data. J.L. Murray, P.K. Liao, and K.E. Spear, 1987.

uators therefore use the composition given by [66Rud], who performed studies of the microstructure as a function of quenching rate. It is also noted that the eutectic composition given by [64Fen] would require the liquidus to descend more than 200 °C in less than a 1 at.% interval, and this is thermodynamically implausible.

The solubility of B in (βTi) and (αTi) was investigated by [49Ehr] and [51Bre], who utilized X-ray diffraction, and by [51Ogd], [54Pal], and [66Rud], who utilized metallography. The assessed diagram is based on the microscopic work only. The solubility of B in (αTi) near the pure metal transformation temperature of 882 °C was reported as less than 1.7 at.% B [51Ogd] and as less than 0.2 at.% B [54Pal]. The solubility of B in (βTi) is less than 1 at.% at the eutectic temperature 1540 °C [66Rud].

The temperature of the three-phase reaction of (βTi), (αTi), and TiB is 882 °C, which, within experimental uncertainty, is identical to that of the pure metal transformation [51Ogd, 54Pal, 66Rud]. Microscopic work, however, suggests that the reaction is of the peritectic type [54Pal, 66Rud].

There are major discrepancies in the literature concerning the structures and stability ranges of the equilibrium phases, in part because many of the reported binary phases are derived from studies of ternary systems [51Gre, 52Gla]. Major studies of the binary system are those of [54Pal], [64Fen], and [66Rud]. The work of [54Pal] was not used to construct the assessed diagram in the high-temperature range because of the evident effect of contamination on their specimens. Both [64Fen] and [66Rud] exercised considerable caution to avoid contamination; nevertheless, there are substantial discrepancies between these two studies, primarily concerning the existence of Ti_3B_4 and the temperatures of the invariant reactions. For the invariant temperatures, the data of [66Rud] are preferred because their technique for determining incipient melting temperatures has been successfully tested for a number of high-temperature systems and because results were verified by differential thermal analysis for a large number of samples.

Ti-Rich Alloys

Boron is only slightly soluble (<1 at.%) in (βTi) and (αTi). Solidification occurs by the eutectic reaction, L $\rightleftarrows$ (βTi) + TiB, with eutectic temperature 1540 $\pm$ 10 °C [64Fen, 66Rud] and composition 7 at.% B [66Rud]. The eutectic temperature was reported as 1540 $\pm$ 10 °C [66Rud], 1530 $\pm$ 10 °C [64Fen], ~1670 °C [54Pal]. The latter value can be disregarded because it was clearly affected by contamination of the Ti, as evidenced by the very high melting point attributed to pure Ti. The work of [66Rud] is preferred, and 1530 °C agrees with the assessment within experimental uncertainty.

In the assessed diagram, the peritectoid reaction is placed at 884 $\pm$ 2 °C, with (αTi) and (βTi) compositions 0.15 and 0.1 at.%, respectively. These values are chosen only to indicate the reaction type and do not represent assessed solubility data.

Effect of Contamination on Ti-Rich Alloys.

Conflicting reports of the solubility of B in (αTi) and (βTi) are attributed to [49Ehr] and [51Bre]. Based on X-ray diffraction patterns, [49Ehr] concluded that cph (αTi) was found between 0 and 28.6 at.% B, and superlattice lines indicated that an ordered hexagonal phase was formed by a higher order transformation. [51Bre] confirmed these observations, but interpreted the "superlatt-

ice" lines as possibly due to nitrogen or oxygen contamination. [51Ogd] established that there is no extended solubility, but confirmed the superlattice lines and tentatively assigned them to an ordered hexagonal compound, Ti_2B. [52Gla], [52Pos], [53Sch], and [54Pal] found a tetragonal phase and assigned it stoichiometry Ti_2B, based partly on the previous speculation of [51Ogd]. Based on metallographic work, [54Pal] proposed that Ti_2B is a high-temperature phase. [64Fen] and [66Rud] established that the lowest boride is TiB and that neither Ti_2B nor any other sub-boride exists in the binary system.

Monoboride TiB.

The existence of the monoboride was established by structural studies [54Dec, 60Wit, 64Fen, 66Rud] and by melting point and metallographic studies [64Fen, 66Rud]. It was also identified in B fiber–Ti composite materials [76The].

TiB forms from the melt by a peritectic reaction. According to [66Rud], the reaction is L + TiB_2 $\rightleftarrows$ TiB and occurs at 2190 $\pm$ 20 °C. According to [64Fen], the reaction is L + Ti_3B_4 $\rightleftarrows$ TiB and occurs at 2000 °C. In the assessed diagram, two reactions, L + TiB_2 $\rightleftarrows$ Ti_3B_4 and L + Ti_3B_4 $\rightarrow$ TiB, are shown at 2200 and 2180 °C, respectively.

TiB has a narrow homogeneity range of about 49 to 50 at.% B, as determined by metallography and the small variation of lattice parameters with composition [54Dec, 64Fen, 66Rud].

Ti_3B_4.

[64Fen] and [66Fen] observed an additional phase, Ti_3B_4. The crystal structure was reported as isomorphous with that of Ta_3B_4. The composition of this phase is 58.1 at.% B, as determined by chemical extraction, and its homogeneity range is small. A second phase, thought to be possibly a high-temperature form of Ti_3B_4, was found to have a structure completely different from any other phase in the system. During heating, Ti_3B_4 is transformed at about 2010 °C, but the transformation is not reversible. High- and low-temperature forms of Ti_3B_4 are not distinguished in the assessed diagram, because the irreversibility of the transformation strongly suggests that it is a contamination effect.

[64Fen] indicated that Ti_3B_4 forms from the melt by a peritectic reaction at 2020 °C, which is 20 °C above the peritectic reaction involving TiB. In the assessed diagram, the peritectic reactions are separated by 20 °C, but are positioned at 2180 and 2200 °C, in accordance with [66Rud].

Using electron microscopy, [81Ner] observed Ti_3B_4 in the course of an investigation of the interaction of B-Ti layers in the presence of high-temperature gradients. Very recently, [86Spe] confirmed the existence of Ti_3B_4 by arc-melting and annealing studies. The absence of Ti_3B_4 in the diagram proposed by [66Rud] may be explained by the narrow composition and temperature range at which Ti_3B_4 forms from the melt. [66Rud] also noted that TiB had different etching properties depending on whether it was formed by the peritectic reaction with TiB_2, or by primary solidification. This may also be explainable in terms of an additional equilibrium phase. However, further verification for the stability range of this phase is needed.

Diboride TiB_2.

Diboride TiB_2 melts congruently at 3225 $\pm$ 25 °C [66Rud]. Reported melting points range between 2790 and 3225 °C.

The B-Ti System

Reference	Congruent melting temperature, °C
[57Lsk]	2850 ± 50
[52Kie]	2900 ± 80
[53Sch, 54Pos]	2920 ± 30
[52Gla]	2790
[64Fen]	> 2880
[66Rud]	3225 ± 25

[64Fen] noted that large volumes of vapor prevented the accurate determination of the TiB₂ melting point. The major difficulty is reaction with the crucible material, which results in the detection of a ternary eutectic temperature lower than the melting point of the binary compound. The highest reported congruent temperature was therefore used to construct the assessed diagram.

The homogeneity range of TiB₂ was determined metallographically as 65.5 to 67 at.% B [64Fen]. [66Rud] found that samples containing 65.2 and 66.3 at.% B contained excess TiB and (βB), respectively, at the grain boundaries and estimated the range as less than 2 at.%--about 66 at.% B over the entire solidus range. The small variation of lattice parameters also confirms the narrow homogeneity range (see Table 2). In the assessed diagram, the range is taken as 65.5 to 66.7 at.% B.

B-Rich Alloys

Two additional phases which are not included in the assessed diagram were reported: Ti₂B₅ [52Gla, 52Pos, 53Sch] and TiB_x (also referred to as TiB₁₀ or TiB₁₂) [51Gre]. Ti₂B₅ was reported to be stable only in the range 1700 to 2100 °C [53Sch] and to have the hexagonal W₂B₅ structure. The reflectivity of the phase [51Gre] and the carbon content in chemical analysis [52Gla, 52Pos] indicate that these phases are not stabilized by carbon. Although the two phases were prepared in the same temperature range, [51Gre] did not observe Ti₂B₅, and [52Gla], [52Pos], and [53Sch] did not observe TiB_x.

Evidence against the existence of Ti₂B₅ or TiB_x was given by [60Sey], [64Fen], and [66Rud]. In these studies, only TiB₂ and (βB) were found in proportions consistent with the phase rule. [64Fen] heated alloys to 2050 and 2150 °C and heat treated specimens at 1820 and 1920 °C, but found no evidence for Ti₂B₅. The eutectic reaction L ⇌ (βB) + TiB₂ occurs at 2080 ± 20 °C, near pure B (>98 at.% B) [66Rud]. [64Fen] reported incipient melting at 2040 °C. The results of [66Rud] are preferred. Eutectic microstruc-

tures were not observed by [64Fen] or [66Rud], and significant quantities of TiB₂ are found in 95 at.% B alloys. It is therefore assumed that the eutectic composition is close to pure B and that the solubility of Ti in (βB) is very low.

Metastable Phases

[81Tav] reported that by rapid (10⁶ K/s) quenching of Ti-rich eutectic alloys from the liquid state, extension of the solubility of B in Ti can be obtained. An amorphous phase was also detected. Phases were identified by X-ray diffraction, but other details of the experiments were not given.

Crystal Structure and Lattice Parameters

Crystal structure and lattice parameter data are summarized in Table 2.

The AlB₂ structure of the diboride TiB₂ was verified by numerous studies [49Ehr, 49Nor, 54Pal, 64Fen, 66Rud]. Ti₃B₄ structure data is taken from [64Fen], [66Fen], and [86Spe].

The structure of the monoboride has been the subject of controversy. Binary TiB has the orthorhombic FeB structure [54Dec, 60Wit, 64Fen, 66Rud]. Other researchers proposed the cubic NaCl [50And] and the zincblende structures [49Ehr]. The sample preparation techniques used by [54Dec], [60Wit], [64Fen], and [66Rud] were most likely to produce the binary phase. It is now established that the NaCl-type phase is a ternary boronitride [81Leb].

Thermodynamics

Experimental Data.

Experimental thermodynamic data for the binary titanium boride phases are available only for TiB₂. The low-temperature heat capacity and 298 K entropy value were reported by [63Wes]. Although this entropy value has not been confirmed, the reported value of 28.49 J/mol·K is quite consistent with values for other transition metal diborides [69Spe]. A reasonable estimate for the TiB(s) entropy value at 298 K is 29.6 J/mol·K. It is obtained by assuming a reaction entropy of zero for the formation of TiB from TiB₂ and αTi metal.

High-temperature heat content values for the diboride have been measured by [57Wal], [59Kre], [62Mez], [63Mcd], [63Nee], and [64Kir]. Assessments of these data by [71Stu] and [Hultgren, B] are in good agreement. The heat capacity values of [71Stu] can be represented as:

Table 2 Crystal Structures and Lattice Parameters of the Ti-B System

Phase	Homogeneity range, at.% B	Pearson symbol	Space group	Strukturbericht designation	Prototype	Lattice parameters, nm			Reference
						a	b	c	
(αTi)	0 to < 0.2	hP2	P6₃/mmc	A3	Mg	...	...	...	[Pearson2]
(βTi)	0 to < 0.2	cI2	Im3m	A2	W	...	...	...	[Pearson2]
TiB	49 to 50	oP8	Pnma	B27	FeB	0.612	0.306	0.456	[54Dec]
						0.6105	0.3048	0.4542	[60Wit]
TiB₂ ..	65.5 to 66.7	hP3	P6/mmm	C32	AlB₂	0.3028 to 0.3040	...	0.3228 to 0.3234	[65Geb]
						0.3028	...	0.3230	[51Bre]
						0.3030	...	0.3228	[65Kau]
Ti₃B₄ ...	56.1	oI14	Immm	D7_b	Ta₃B₄	0.3259	1.373	0.3042	[66Fen]
						0.3260	1.372	0.3041	[86Spe]
(βB)	~ 100	hR108	R3m	...	βB	1.09251	...	2.3814	[77Cal]

Table 3 Ti-B Thermodynamic Properties

Gibbs energies of the pure elements and compounds

$G^0(\text{Ti,L})$ = 0
$G^0(\text{B,L})$ = 0
$G^0(\text{Ti,bcc})$ = $-16\,234\,+\,8.368\,T$
$G^0(\text{Ti,cph})$ = $-20\,585\,+\,12.134\,T$
$G^0(\text{B},\beta)$ = $-50\,208\,+\,21.23\,T$
$G^0(\text{TiB})$ = $-142\,942\,+\,28.03\,T$
$G^0(\text{Ti}_3\text{B}_4)$ = $-149\,674\,+\,27.833\,T$
$G^0(\text{TiB}_2)$ = $-145\,637\,+\,22.434\,T$

Excess Gibbs energy of the liquid

$B_0 = -249\,809\,+\,19.183\,T$
$B_1 = 44\,973$
$B_2 = 45\,409$
$B_3 = -15\,284$

$A_0 = -272\,514\,+\,19.183\,T$
$A_1 = -67\,899$
$A_2 = -68\,114$
$A_3 = -38\,210$

Note: Values are given in J/mol and J/mol · K; the entity for mol is an atom.

$$C_p = 556.4 + 25.9 \times 10^{-3}T - 17.5 \times 10^5/T^2$$
$$- 2.5 \times 10^{-6}\,T^2 \text{ J/mol} \cdot \text{K}$$

where mol is mole of TiB_2.

Various values for the enthalpy of formation of TiB_2—ranging from -134 to -324 kJ/mol—have been reported. Critical assessments by [71Stu] and [Hultgren, B] both yielded 298 K values of about -280 kJ/mol. A more recent paper by [80Yur] discusses these values as well as the original data [55Bre, 58Epe, 61Low, 61Wil, 62Sch, 63Fes, 64Kib, 66Hub, 75Akh]. [80Yur] also presented experimental measurements of equilibrium reactions of TiB_2 with nitrogen gas to produce Ti and B nitrides. These equilibrium measurements yield a 298 K enthalpy of formation for TiB_2 of -304.2 ± 3.8 kJ/mol [66Hub]. The reason for the disagreement between the results from the equilibrium and calorimetric measurements is not known, but it may be related to formation of ternary boronitride phases or compositional uncertainties in the samples studied calorimetrically. In the oxygen bomb calorimetry study by [66Hub], combustion of only 99% of the amount of sample believed to have been burned reduced the calculated enthalpy of formation from -323.8 kJ/mol to -305.0 kJ/mol.

Phase Diagram Calculations.

A check for consistency between the thermodynamic data discussed above and the phase diagram data of [66Rud] was made by least-squares optimization. This process also provides a means for calculating unmeasured phase boundaries and thermodynamic functions. The liquidus curves, excess Gibbs energy for the liquid phase, and the Gibbs energies of formation of the TiB(s) and Ti_3B_4(s) phases were calculated.

The phase diagram in Fig. 1 and the thermodynamic data given in Table 3 provide the resulting optimized set of thermochemical data for the binary Ti-B system. The reference states of the elements were chosen as the pure liquids and the excess Gibbs energy of the liquid can be represented by either a Redlich-Kister polynomial equation:

$$\Delta G = x(1 - x)\,[A_0 + A_1(1 - 2x) + A_2(1 - 2x)^2$$
$$+ A_3(1 - 2x)^3]$$

or a Legendre polynomial expansion:

$$\Delta G = x(1 - x)\,[B_0\,P_0(1 - 2x) + B_1\,P_1(1 - 2x)$$
$$+ B_2\,P_2(1 - 2x) + B_3\,P_3(1 - 2x)]$$

where x is the mole fraction of B, $P_i\,(1 - 2x)$ are the Legendre polynomials, and A_i and B_i are the temperature-dependent polynomial coefficients (as listed in Table 3). At $x = 0.5$, the excess enthalpy and entropy of the liquid are:

$$\Delta H = -68\,128 \text{ J/mol}$$

and

$$\Delta S^{\text{ex}} = -4.80 \text{ J/mol·K}$$

The inadequacy of the subregular model for describing the liquid phase Gibbs energy is evident from the calculation reported by [84Kau]. In that calculation, the invariant temperatures differ from observation by over 100 °C.

The present calculations simplified the Gibbs energy expressions for the solid phases by assuming that they have fixed stoichiometric compositions. The Gibbs energy of formation expressions contained a constant enthalpy and entropy term:

$$\Delta_f G^0 = \Delta_f H^0 - \Delta_f S^0\,T$$

If the data in Table 3 are used to calculate the Gibbs energy for the chemical reaction:

$$\beta\text{Ti(s)} + 2\,\beta\text{B(s)} = \text{TiB}_2\text{(s)}$$

the following equation is obtained:

$$\Delta_f G^0\,(\text{TiB}_2) = -320\,261 + 16.47\,T \text{ J/mol}$$

or

$$\Delta_f G^0\,(\text{TiB}_2) = -106\,754 + 5.49\,T \text{ J/mol}$$

Strictly speaking, because the thermodynamic values obtained when optimizing the thermochemical properties of the system correspond to temperatures at which the liquid phase is stable, the enthalpy and entropy values from the above equations ($-320\,261$ J/mol and -16.47 J/mol·K, respectively) refer to temperatures in the range of 2500 K.

The optimized high-temperature thermodynamic functions for the formation of TiB_2 can be compared with the 298 K data by using the thermodynamic data for βTi and βB from [79Cha] and the above C_p equation for TiB_2(s). The calculated entropy change for the formation of the diboride from βTi and βB at 2500 K is -23.51 J/mol·K, as compared with the optimized value, -16.47 J/mol·K. The respective 298 K values reported for the enthalpy of formation of TiB_2 by [66Hub], [75Akh], and [80Yur] become $-314\,000$, $-328\,260$, and $-333\,770$ J/mol at 2500 K, as compared with the optimized value of $-320\,261$ J/mol.

The optimization process provides values for the Gibbs energies of formation of the TiB(s) and Ti_3B_4(s) phases that are consistent with the phase diagram given in Fig. 1 and the above Gibbs energy equation for TiB_2(s). For TiB(s), the data in Table 3 is:

$$\beta\text{Ti(s)} + \beta\text{B(s)} = \text{TiB(s)}$$

$$\Delta_f G^0(\text{TiB}) = -219\,442 + 26.46\,T \text{ J/mol}$$

or

$$\Delta_f G^0(\text{TiB}) = -109\,721 + 13.23\,T \text{ J/mol}$$

For Ti$_3$B$_4$(s), these data yield:

$$3 \ \beta\text{Ti(s)} + 4 \ \beta\text{B(s)} = \text{Ti}_3\text{B}_4\text{(s)}$$

$$\Delta_f G^0(\text{Ti}_3\text{B}_4) = -798 \ 184 + 84.81 \ T \ \text{J/mol}$$

or

$$\Delta_f G^0(\text{Ti}_3\text{B}_4) = -114 \ 026 + 12.12 \ T \ \text{J/mol}$$

Cited References

49Ehr: P. Ehrlich, "Binary Systems of Titanium with Nitrogen, Carbon, Boron, and Beryllium," *Z. Anorg. Chem., 259*, 1-41 (1949) in German. (Equi Diagram; Experimental)

49Nor: J.T. Norton, H. Blumenthal, and S.J. Sindeband, "Structure of Diborides of Titanium, Zirconium, Columbium, Tantalum and Vanadium," *Metall Trans., 185*, 749-751 (1949). (Crys Structure; Experimental)

50And: L.H. Andersson and R. Kiessling, "Investigations on the Binary Systems of Boron with Chromium, Columbium, Nickel, and Thorium, Including a Discussion of the Phase TiB in the Titanium-Boron System," *Acta Chem. Scand., 24*, 160-164 (1950). (Crys Structure; Experimental)

51Bre: L. Brewer, D.L. Sawyer, D.H. Templeton, and C.H. Dauben, "A Study of the Refractory Borides," *J. Am. Ceram. Soc., 34*, 173-179 (1951). (Equi Diagram, Crys Structure; Experimental)

51Gre: H.M. Greenhouse, O.E. Accountius, and H.H. Sisler, "High-Temperature Reactions in the System Titanium Carbide-Boron Carbide," *J. Am. Ceram. Soc., 73*, 5086-5087 (1951). (Equi Diagram; Experimental)

51Ogd: H.R. Ogden and R.I. Jaffee, "Titanium-Boron Alloys," *Trans. AIME, 191*, 335-336 (1951). (Equi Diagram; Experimental)

52Gla: F.W. Glaser, "Contribution to the Metal-Carbon-Boron Systems," *Trans. AIME, 194*, 391-396 (1952). (Equi Diagram; Experimental)

52Kie: R. Kieffer, F. Benesovsky, and E.R. Honak, "A New Method of Preparing Borides of the Transition Metals, Particularly of Titanium and Zirconium Borides," *Z. Anorg. Chem., 268*, 191-200 (1952) in German. (Equi Diagram; Experimental)

52Pos: B. Post and F.W. Glaser, "Borides of Some Transition Metals," *J. Chem. Phys., 20*, 1050-1051 (1952). (Crys Structure; Experimental)

53Sch: P. Schwarzkopf and F.W. Glaser, "Structure and Chemical Properties of Borides of Transition Metals of the Fourth, Fifth and Sixth Group," *Z. Metallkd., 44*, 353-358 (1953) in German. (Crys Structure; Experimental)

54Dec: B.F. Decker and J.S. Kasper, "The Crystal Structure TiB," *Acta Crystallogr., 7*, 77-80 (1954). (Crys Structure; Experimental)

54Pal: A.E. Palty, H. Margolin, and J.P. Nielsen, "Titanium-Nitrogen and Titanium-Boron Systems," *Trans. ASM, 46*, 312-328 (1954). (Equi Diagram; Experimental)

54Pos: B. Post, F.W. Glaser, and D. Moskowitz, "Transition Metal Diborides," *Acta Metall., 2*, 20-25 (1954). (Equi Diagram; Experimental)

55Bre: L. Brewer and H. Haraldsen, "The Thermodynamic Stability of Refractory Metal Borides," *J. Electrochem. Soc., 102*, 399-406 (1955). (Thermo; Experimental)

57Lsk: I.I. Lskoldsky and L.R. Bogorodskaya, "The Possibility of Production of Metalloceramic (Sintered) Hard Alloys Based on Chromium, Titanium, and Tungsten Borides," *Zh. Prikl. Khim., 30*, 177-185 (1957) in Russian; TR: *J. Appl. Chem. USSR, 30*, 181-188 (1957). (Equi Diagram; Experimental)

57Wal: B. Walker, C. Ewing, and R. Miller, "Heat Capacity of Titanium Diboride from 30 to 700 °C," *J. Phys. Chem., 61*, 1682-1683 (1957). (Thermo; Experimental)

58Epe: V.A. Epelbaum and M.I. Starostina, "Thermochemical Investigations of Boron and Certain Borides," *Bor. Trudy Konf. Khim., Bora Ego Soedinenii*, 97 (1958) in Russian. (Thermo; Experimental)

59Kre: A.N. Krestovnikov and M.S. Vendrikh, "Thermodynamics of Titanium Diborides," *Izv. V.U.Z. Tsvetn. Metall., 2*, 54-57 (1959) in Russian. (Thermo; Experimental)

60Sey: A.U. Seybolt, "An Exploration of High Boron Alloys," *Trans. ASM, 52*, 971-989 (1960). (Equi Diagram; Experimental)

60Wit: A. Wittmann, H. Nowotny, and H. Boller, "Investigation of the Ternary System Titanium-Molybdenum-Boron," *Monatsh. Chem., 91*, 608-615 (1960) in German. (Crys Structure; Experimental)

61Low: C.E. Lowell and W.S. Williams, "High Temperature Calorimeter for the Determination of Heats of Formation of Refractory Compounds," *Rev. Sci. Instrum., 32*, 1120-1123 (1961). (Thermo; Experimental)

61Wil: W. Williams, "The Heat of Formation of Titanium Diboride: Experimental and Analytical Resolution of Literature Conflict," *J. Phys. Chem., 65*, 2213-2216 (1961). (Thermo; Experimental)

62Mez: R. Mezaki, E. Tilleux, D.W. Barnes, and J.L. Margrave, "Thermodynamics of Nuclear Materials (1962)," IAEA, Vienna, 775-788 (1962). (Thermo; Experimental)

62Sch: P. Schissel and O. Trulson, "Mass Spectrometric Study of the Vaporization of the Titanium-Boron System," *J. Phys. Chem., 66*, 1492-1496 (1962). (Thermo; Experimental)

63Fes: V.V. Fesenko and A.S. Bolgar, "Evaporation Rate and Vapor Pressure of Carbides, Silicides, Nitrides, and Borides," *Sov. Powder Metall. Met. Ceram., 1*, 11 (1963) in Russian. (Thermo; Experimental)

63Mcd: R.A. McDonald, F.D. Oetting, and H. Prophet, Rep. N64-18824, Dow Chemical Co., Midland, MI (1963). (Thermo; Experimental)

63Nee: D.S. Neel, C.D. Pears, and S. Oglesby, Tech. Doc. Rep. No. ASD-TDR-62-765, Southern Res. Inst., Birmingham, AL (1963). (Thermo; Experimental)

*63Wes: E.F. Westrum, "Heat Capacity," Chapt. XV, Tech. Doc. Rep. No. RTD-TDR-63-4096, Part I, L. Kaufman and E.Y. Clougherty, Ed., Man Labs, Inc., Cambridge, MA, 239-261 (1963). (Thermo; Experimental)

64Fen: R.G. Fenish, "Phase Relationships in the Titanium-Boron System," NRM-138, 1-37 (1964). (Equi Diagram; Experimental)

64Kib: G.M. Kibler, T.F. Lyon, M.I. Linevsky, and V.J. Desantis, Tech. Rep. No. WADD-TR-60-646, Part III, Vol. 2, General Electric Co., Evansdale, OH (1964). (Thermo; Experimental)

64Kir: V.A. Kirilin, A.E. Sheindlin, V.Ya. Chekhovskoi, and V.I. Tyukaev, "Enthalpy and Heat Capacity of Titanium Diboride at 273.15-2600 °K," *Teplofiz. Vys. Temp., 2*, 710-715 (1964) in Russian. (Thermo; Experimental)

65Geb: J.J. Gebhardt and R.F. Cree, "Vapor-Deposited Borides of Group IVA Metal," *J. Am. Ceram. Soc., 48*, 262-267 (1965). (Crys Structure; Experimental)

65Kau: L. Kaufman and E.V. Clougherty, "Investigation of Boride Compounds for High Temperature Applications," *Metals for the Space Age*, Plansee Proceedings 1964, Plansee Metallwerk, Reutte, Austria, F. Benesovsky, Ed., 722-758 (1965). (Thermo; Experimental)

66Fen: R.G. Fenish, "A New Intermediate Compound in the Titanium-Boron System, Ti$_3$B$_4$," *Trans. AIME, 236*, 804 (1966). (Crys Structure; Experimental)

*66Hub: E.J. Huber, "The Heat of Formation of Titanium Diboride," *J. Chem. Eng. Data, 11*(3), 430-431 (1966). (Thermo; Experimental)

66Rud: E. Rudy and St. Windisch, "Ternary Phase Equilibria in Transition Metal-Boron-Carbon-Silicon Systems Part I. Related Binary System Volume VII. Ti-B System," Technical Rep. No. AFML-TR-65-2, Part I, Vol. VII (1966). (Equi Diagram; Experimental)

69Spe: K.E. Spear, H. Schafer, and P.W. Gilles, "Thermodynamics of Vanadium Borides," in *High Temperature Technology*, Butterworths, London, 201-212 (1969). (Thermo; Experimental)

71Stu: D.R. Stull and H. Prophet, *JANAF Thermochemical Tables*, 2nd ed., Nat. Stand. Ref. Data Ser., Nat. Bur. Stand. (U.S.), 37 (1971). (Thermo; Compilation)

*75Akh: V.V. Akhachinskij and N.A. Chirin, "Enthalpy of Formation of Titanium Diboride," Thermodynamics of Nuclear

Materials 1974, Vol. II, IAEC, Vienna, 467-476 (1975). (Thermo; Experimental)

76The: J. Thebault, R. Pailler, G. Bontemps-Moley, M. Bourdeau, and R. Naslain, "Chemical Compatibility in Boron Fiber-Titanium Composite Materials," *J. Less-Common Met., 47,* 221-233 (1976). (Equi Diagram; Experimental)

77Cal: B. Callmer, "An Accurate Refinement of the β-Rhombohedral Boron Structure," *Acta Crystallogr. B, 33,* 1951-1954 (1977). (Crys Structure; Experimental)

79Cha: M.W. Chase, *JANAF Thermochemical Data* (Boron, March 31, 1979; Titanium, June 30, 1979), Dow Chemical Co., Midland, MI (1979). (Thermo; Experimental)

80Yur: T.J. Yurick and K.E. Spear, "Thermodynamics of TiB₂ from Ti-B-N Studies," *Thermodynamics of Nuclear Materials 1979,* Vol. I, IAEA-SM-236/53, 73-90 (1980). (Thermo; Experimental)

81Leb: A. Lebugle, R. Nyholm, and N. Martensson, "Electron Spectroscopy Studies of Titanium Boronitrides," *J. Less-*

Common Met., 82(1/2), 269-275 (1981). (Equi Diagram, Crys Structure; Experimental)

81Ner: V.A. Neronov, M.A. Korchagin, V.V. Aleksandrov, and S.N. Gusenko, "Investigation of the Interaction Between Boron and Titanium," *J. Less-Common Met., 82,* 125-129 (1981). (Equi Diagram; Experimental)

81Tav: G.F. Tavadze, O.Sh. Okrostsvaridze, F.N. Tavadze, G.V. Tasagareishvili, and G.A. Mazmishvili, "Crystallization of Eutectic Systems Ti-B, V-B, Zr-B, and Hf-B at Ultrahigh Rates of Cooling," *J. Less-Common Met., 82,* 368 (1981). (Meta Phases; Experimental)

84Kau: L. Kaufman, B. Uhrenius, D. Birnie, and K. Taylor, "Coupled Pair Potential, Thermochemical and Phase Diagram Data for Transition Metal Binary Systems-VII," *Calphad, 8*(1), 25-66 (1984). (Equi Diagram; Theory)

*86Spe: K.E. Spear, P. McDowell, and F. McMahon, "Experimental Evidence for the Existence of the Ti₃B₄ Phase," *J. Am. Ceram. Soc., 69*(1), C4-C5 (1986). (Equi Diagram, Crys Structure; Experimental)

The Ba-Ti (Barium-Titanium) System

137.33   47.88

By J.L. Murray

[78Ali] measured solubilities of Ti in liquid Ba at several temperatures and found them to be very low. From the limited liquid solubility, it is deduced that mutual solid solubilities are extremely low and that no intermetallic compounds occur.

[78Ali] held Ti isothermally in liquid Ba for 1.5 to 3 h under pressure in helium. The solubility of Ti in liquid Ba was determined by chemical analysis of several samples of

the liquid layer. Microstructures were also reported to have been examined. Alloys were prepared from "degassed Ti used as a getter, and barium . . . of analytical purity."

[78Ali] reported solubilities of Ti in liquid Ba and Sr at 800, 900, and 1000 °C in the form of equations and figures. However, values of the two do not correspond. Assuming the figures to be correctly labeled, this evalua-

Fig. 1---Assessed Ti-Ba Phase Diagram

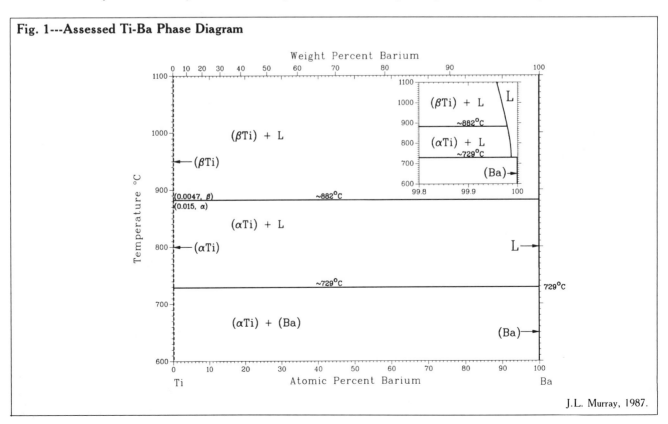

J.L. Murray, 1987.

The Ba-Ti System

Fig. 2---Solubility of Ti in Liquid Ba

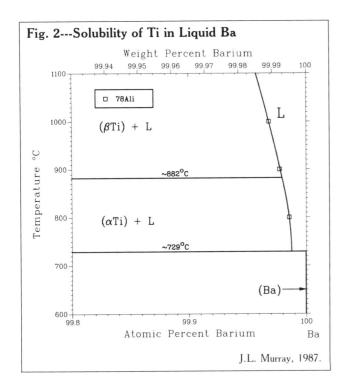

J.L. Murray, 1987.

tor concluded that the equations for Ba and Sr had been reversed and that in the equation temperature was given in °C and composition x in atomic fraction of Ba. Then, the equation for the solubility is:

$$\log x = -2.1151 - 1388/T$$

Using 7120 J/mol for the heat of fusion of Ba [84Cha] and the van't Hoff relation, it is estimated that the temperature of the eutectic reaction $L \rightleftarrows (\alpha Ti) + (Ba)$ lies about 0.1 °C below the melting point of Ba, 729 °C [Melt].

[78Ali] found the solubility of Ba in (αTi) and (βTi) to be 0.015 and 0.0047 at.% Ba respectively, using a method that was not described. These values are shown in the assessed phase diagram, Fig. 1, but it is not recommended that a great deal of reliance be placed on solid solubility values obtained by an unknown method from unconfirmed liquid solubility data. Experimental data are shown in Fig. 2.

Special points of the phase diagram are listed in Table 1, and crystal structure data for the pure components are listed in Table 2.

Cited References

78Ali: F.N. Alidzhanov, A.V. Vakhobov, and T.D. Dushanbe, "Ba-Ti and Sr-Ti Phase Diagrams," *Izv. Akad. Nauk SSSR, Met.,* (2), 223-224 (1978); TR: *Russ. Metall.,* (2), 177-178 (1978). (Equi Diagram; Experimental)

84Cha: M.W. Chase, "Heats of Transformation of the Elements," *Bull. Alloy Phase Diagrams,* 4(1), 123-124 (1984). (Thermo; Compilation)

Table 1 Special Points of the Assessed Ti-Ba Phase Diagram

Reaction	Compositions of the respective phases, at.% Ba			Temperature, °C	Reaction type
$(\beta Ti) + (LBa) \rightleftarrows (\alpha Ti)$	(0.0047)	99.98	(0.015)	~ 882	Peritectic
$(LBa) \rightleftarrows (\alpha Ti) + (Ba)$	99.988	(<0.015)	~100	~ 729	Eutectic
$(\beta Ti) \rightleftarrows (\alpha Ti)$		0		882	Allotropic transformation
$L \rightleftarrows (Ba)$		100		729	Melting point

Table 2 Crystal Structures of the Ti-Ba System

Phase	Homogeneity range, at.% Ba	Pearson symbol	Space group	Strukturbericht designation	Prototype	Reference
(βTi)	0 to ~ 0.0047	cI2	Im3m	A2	W	[Pearson2]
(αTi)	0 to 0.015	hP2	P6$_3$/mmc	A3	Mg	[Pearson2]
(Ba)	~ 100	cI2	Im3m	A2	W	[Pearson2]

The Be-Ti (Beryllium-Titanium) System

9.01218   47.88

By J.L. Murray

Equilibrium Diagram

There is considerable interest in glass formation in Ti-rich Ti-Be alloys [78Tan] and in Be-rich alloys for high-temperature applications [65Bea]. The only previous assessment of the diagram is that of [Elliott], which is a rough estimate based on limited experimental work. The present assessment (Fig. 1), although incorporating a somewhat larger set of experimental data and thermodynamic calculations, is not a substantial improvement, particularly on the Be-rich side. The assessed phase diagram is in every detail based on uncertain hypotheses and on either conflicting or crude experimental data. Further experimental phase diagram work is clearly needed.

The equilibrium solid phases of the system are:

- The high-temperature bcc (βTi,βBe) and low-temperature cph (αTi,αBe) solid solutions based on the pure elements. Because mutual solid solubilities are limited, these phases will be designated (βTi), (βBe), (αTi), and (αBe).
- $TiBe_2$, a $C15$ Laves phase.
- $TiBe_3$.
- Two allotropic forms; αTi_2Be_{17} and βTi_2Be_{17}. αTi_2Be_{17} is found at the Be-rich side of stoichiometry, βTi_2Be_{17} at the Ti-rich side. Because there is no quantitative data on the homogeneity ranges and no data available on the temperature ranges of

stability, the two allotropic forms are not distinguished in the phase diagram (Fig. 1) or in the reaction table (Table 1). They are however distinguished in the discussion of crystal structure.
- $TiBe_{12}$.

The major experimental investigations of this system are [60Bed], [62Obi], and [66Hun]. The work of [62Obi] covered the range 0 to 60 at.% Be at 775 to 1000 °C by microscopy, X-ray diffraction, and measurement of electrical and magnetic properties. The original paper is not available, and the data must be assessed on the basis of purity of materials and the phase diagram itself; 99.6% Ti and 99.5% Be were used. TiBe was observed as the first Ti-rich intermetallic phase, rather than $TiBe_2$. Other equiatomic TiBe phases are known to be metastable or ternary phases [78Tan, 79Gie]. Based on observations of the approach to the equilibrium (αTi) + $TiBe_2$ assemblage in rapidly solidified alloys [79Tan], the first stable Ti-rich compound is $TiBe_2$, not TiBe. [62Obi] showed two invariant reactions, $L \rightleftarrows (\beta Ti)$ + TiBe and $(\beta Ti) \rightleftarrows (\alpha Ti)$ + TiBe, which are tentatively identified with the corresponding reactions involving $TiBe_2$. [62Obi] placed the eutectic composition at 28.6 at.% Be and the eutectoid composition at about 3 at.% Be, based on microscopic work. The maximum solubility of Be in (βTi) was given as 9.8 at.%. Invariant temperatures, also based on microscopy, are listed in Table 2.

Fig. 1 ---Assessed Ti-Be Phase Diagram

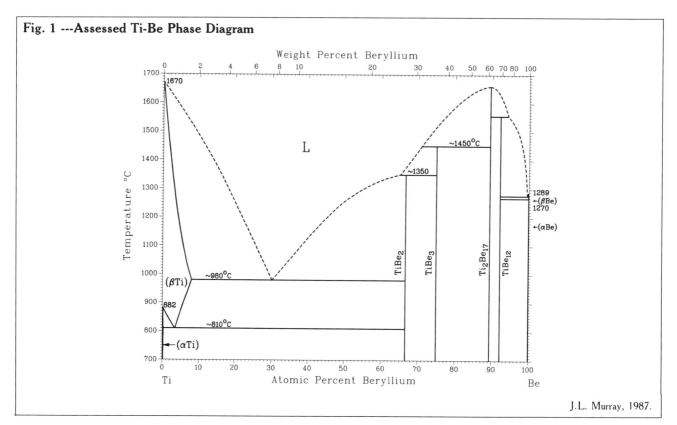

J.L. Murray, 1987.

Table 1 Special Points of the Assessed Ti-Be System

Reaction	Compositions of the respective phases, at.% Be			Temperature, °C	Reaction type
L $\rightleftarrows$ (βTi) + TiBe$_2$	~ 30	~ 7.5	66.7	~ 980	Eutectic
(βTi) $\rightleftarrows$ (αTi) + TiBe$_2$	~ 3	~ 0	66.7	810	Eutectoid
L + TiBe$_3$ $\rightleftarrows$ TiBe$_2$	~ 65	75	66.7	~ 1350	Peritectic
L + Ti$_2$Be$_{17}$ $\rightleftarrows$ TiBe$_3$	~ 71	89.5	75	~ 1450	Peritectic
L + Ti$_2$Be$_{17}$ $\rightleftarrows$ TiBe$_{12}$	~ 94	89.5	92.3	~ 1570	Peritectic
L $\rightleftarrows$ Ti$_2$Be$_{17}$		89.5		~ 1660	Congruent
L $\rightleftarrows$ (βBe) + TiBe$_{12}$	~ 100	~ 100	92.3	~ 1285	Eutectic
(βBe) $\rightleftarrows$ (αBe) + TiBe$_{12}$	~ 100	~ 100	92.3	~ 1254	Eutectoid
L $\rightleftarrows$ (βTi)		0		1670	Melting point
(βTi) $\rightleftarrows$ (αTi)		0		882	Allotropic transformation
L $\rightleftarrows$ (βBe)		100		1289	Melting point
(βBe) $\rightleftarrows$ (αBe)		100		1256	Allotropic transformation

Table 2 Experimental Determinations of Invariant Temperatures

Reaction	Temperature, °C	Comment	Reference
L $\rightleftarrows$ (βTi) + TiBe$_2$	980 ± 5	Microscopy	[62Obi]
	1030 ± 50	Thermal analysis	[60Bed](a)
	1031 ± 6	Microscopy	[66Hun]
(βTi) $\rightleftarrows$ (αTi) + TiBe$_2$	815 ± 3	Microscopy	[62Obi]
	810 ± 5	Microscopy	[66Hun]
L + TiBe$_3$ $\rightleftarrows$ TiBe$_2$	1350 ± 50	Thermal analysis	[60Bed]
L → Ti$_2$Be$_{17}$ $\rightleftarrows$ TiBe$_3$	1450 ± 50	Thermal analysis	[60Bed]
L + Ti$_2$Be$_{17}$ $\rightleftarrows$ TiBe$_{12}$	1550 ± 60	Thermal analysis	[60Bed]
	1593	"Melting of TiBe$_{12}$"	[65Bea]
L $\rightleftarrows$ Ti$_2$Be$_{17}$	1670 ± 50		[60Bed]
	1632	"Melting of Ti$_2$Be$_{17}$"	[65Bea]

(a) Data of [60Bed] are from [Elliott].

[66Hun] investigated the range 0 to 22 at.% Be at 800 to 1150 °C. The master Ti-Be alloy contained 1.05 wt.% Fe. The eutectic composition was bracketed between 21 and 28.9 at.% Be, the eutectoid composition was placed at 2.6 at.% Be, and the maximum solubility of Be in (βTi) was placed at 5 at.%. Based on lattice parameter measurements, [62Amo] reported very low solubility of Be in (αTi).

[60Bed] examined several alloys ranging between 60 and 100 at.% Be and reported invariant temperatures involving the liquid, with uncertainties of 50 to 60 °C. It is not known whether metallographic work was done to determine the types of the invariant reactions [Elliott].

On the Be-rich side of the diagram, some additional data are available [50Kau, 63Hin, 65Bea]. [50Kau] determined metallographically that there is a eutectic reaction L $\rightleftarrows$ TiBe$_{12}$ + (βBe) and that the solubility of Ti in (βBe) is low. By X-ray diffraction, [63Hin] verified that the phase identified by [52Rae] as TiBe$_{12}$ is present in an as-solidified 99.8 at.% Be alloy.

[65Bea] listed melting points of Ti$_2$Be$_{17}$ and TiBe$_{12}$; that of Ti$_2$Be$_{17}$ is shown as a congruent melting point and that of TiBe$_{12}$ is identified with the peritectic reaction L + Ti$_2$Be$_{17}$ $\rightleftarrows$ TiBe$_{12}$.

In this assessment, the topology of the diagram is the same as that shown by [Elliott], except that the melting point of TiBe$_3$ is interpreted as a peritectic reaction. This interpretation is based on the results of previous and present thermodynamic calculations of the diagram, which give the topology shown in Fig. 1 for any reasonable choice of enthalpies and entropies of fusion for the compounds. For the Ti-rich eutectic temperature, the lower value [62Obi] is preferred, based on sample purity. The solubility of Be in (βTi) is placed midway between the determinations of [62Obi] (10 at.%, 2 wt.%) and [66Hun] (5 at.%, 1 wt.%). Other invariant temperatures are taken from [60Bed], or lie between determinations of [60Bed] and [65Bea].

Metastable Phases

[78Tan], [79Gie], and [79Tan] produced amorphous structures in the range 37 to 41 at.% Be by rapid solidification (10^7 to 10^8 °C/s). The solubility of Be in (βTi) was extended to about 15 at.% by this technique. Other phases observed to nucleate from the liquid were TiBe$_2$ and a metastable TiBe phase with the CsCl structure. It was tentatively proposed that, with increasing Be content, TiBe replaces (βTi) as the primary nucleant from the liquid. The glass-forming range lies, as theoretically expected, between the intermetallic compound and the eutectic point.

During quenching, (βTi) transforms martensitically to the cph structure (α″Ti) in the range 0 to 15 at.% Be

[79Tan]. The start temperature of the martensite transformation has not been measured.

Crystal Structure and Lattice Parameters

The crystal structures and lattice parameters of equilibrium and metastable phases are listed in Table 3.

The only equilibrium phase for which conflicting structure reports occur is $TiBe_{12}$. [52Rae] reported that $TiBe_{12}$ has a structure of symmetry $P6/mmm$ with 624 atoms in the unit cell. [63Hin] verified that this structure is present in as-solidified Be-rich alloys. [61Zal] and [64Gil], however, found the tetragonal structure listed in Table 3. [64Gil] suggested that the complex hexagonal structure may correspond to Ti_2Be_{17}, but this conflicts with the finding of [63Hin] and the discrepancy remains unresolved.

[62Obi] found a phase TiBe of hexagonal symmetry, and this does not correspond with the structure of any stable or metastable phase of similar composition.

[78Tan], [79Gie], and [79Tan] found two metastable phases of approximate stoichiometry TiBe--a metastable binary CsCl phase and a ternary Fe_3W_3C, $E9_3$-type phase stabilized by oxygen or nitrogen.

Thermodynamics

There are no experimental thermochemical data on this system. [73Kau] and [79Kau] estimated Gibbs energies based on the phase diagram. [79Kau] used a phase diagram based partially on [Elliott] and partially on [66Hun]; the Ti-rich eutectic composition was taken to be about 38 at.% Be. The major difference between the calculations of [73Kau] and [79Kau] is that [73Kau] used the regular solution approximation for the liquid and bcc phases and [79Kau] added substantial subregular contributions. It is thought that the subregular contributions were added to obtain better agreement with the (βTi) eutectic composition 37.5 at.% Be, as evaluated by [Elliott]. According to the [73Kau] calculation, $TiBe_2$ and $TiBe_3$ melt by peritectic reactions; according to the [79Kau] calculations, they melt congruently. The present calculations were undertaken because it is thought that

the (βTi) eutectic composition can be placed nearer to 28 at.% Be and smaller subregular contributions to the excess Gibbs energies are required.

The Gibbs energies of the solution phases are represented as:

$$G(i) = G^0(Ti,i)(1-x) + G^0(Be,i)x + RT[x \ln x + (1-x) \ln(1-x)] + x(1-x)[B(i) + C(i)(1-2x)]$$

where i designates the phase; x is the atomic fraction of Be; G^0 is the Gibbs energy of the pure metals; and $B(i)$ and $C(i)$ are expansion coefficients of the excess Gibbs energies. Lattice stabilities are from [79Kau]. $TiBe_2$, $TiBe_3$, Ti_2Be_{17}, and $TiBe_{12}$ are assumed to be line compounds.

Results of the present calculation are listed in Table 4; they are close to those of [73Kau], which are also listed. The topology of the calculated diagram is shown in Fig. 1. The calculated peritectic temperatures are 1356, 1461, and 1500 °C, compared to 1350, 1450, and 1550 °C, as shown in Fig. 1. The calculated eutectic temperature however is 921 °C; it is not adjusted to agree with the assessed diagram because the adjustment would require that $TiBe_3$ decompose relative to $TiBe_2 + Ti_2Be_{17}$.

In view of the absence of thermochemical data or accurate phase diagram data, more elaborate calculations are believed to be unjustified. Either the Gibbs energies of the present calculations or those of [73Kau] can be used for rough estimates of thermochemical quantities.

Acknowledgment.

The author gratefully acknowledges contributions to the thermodynamic analysis of this system made by Peter Miodownik and Nigel Saunders.

Cited References

50Kau: A.R. Kaufman, P. Gordon, and D.W. Lillie, "The Metallurgy of Beryllium," *Trans. ASM, 42*, 785-844 (1950). (Equi Diagram; Experimental)
52Rae: R.F. Raeuchle and R.E. Rundle, "The Structure of $TiBe_{12}$," *Acta Crystallogr., 5*, 85-93 (1952). (Crys Structure; Experimental)

Table 3 Crystal Structures and Lattice Parameters of the Ti-Be System

Phase	Homogeneity range, at.% Be	Pearson symbol	Space group	Strukturbericht designation	Prototype	Lattice parameters, nm		Reference
						a	c	
(βTi,βBe)	0 to ~7.5	cI2	Im3m	A2	W	...	...	[Pearson2]
(αTi,αBe)	~0	hP2	$P6_3/mmc$	A3	Mg	...	...	[Pearson2]
						0.6451	...	[59Pai]
$TiBe_2$	66.7	cF24	Fd3m	C15	$MgCu_2$	0.64532	...	[79Gio]
						0.644	...	[79Tan]
						0.6450	...	[75Stu]
$TiBe_3$	75	hR12	$R\bar{3}m$	...	$NbBe_3$	0.449	2.132	[61Zal]
αTi_2Be_{17}	89.5	hR19	$R\bar{3}m$	...	Nb_2Be_{17}	0.7392(a)	1.079	[61Zal]
						0.740	1.084	[61Gla]
						0.734	1.074	[60Pai]
βTi_2Be_{17}	89.5	hP38	$P6_3/mmc$	...	Th_2Ni_{17}	0.736	0.730	[61Zal]
						0.735	0.726	[64Gil]
$TiBe_{12}$	92.3	tI26	I4/mmm	$D20_b$?	$Mn_{12}Th$	0.736	0.4195	[64Gil]
						0.735	0.419	[61Zal]
TiBe	~50	cP2	Pm3m	B2	CsCl	0.2940	...	[78Tan]

(a) For lattice parameters of αTi_2Be_{17}, hexagonal axes are used.

The Be-Ti System

Table 4 Thermodynamic Properties of the Ti-Be System

Gibbs energies of the pure elements [79Kau]

$G^0(\text{Ti,L}) = 0$

$G^0(\text{Be,L}) = 0$

$G^0(\text{Ti,bcc}) = -16\,234 + 8.368\,T$

$G^0(\text{Be,bcc}) = -10\,418 + 6.694\,T$

$G^0(\text{Ti,cph}) = -20\,585 + 12.134\,T$

$G^0(\text{Be,cph}) = -15\,020 + 9.706\,T$

Excess Gibbs energies and intermetallic compounds

Present work		[73Kau]	
$B(\text{L})$	$= -54\,000$	$B(\text{L})$	$= -37\,656$
$C(\text{L})$	$= 12\,000$		
$B(\text{bcc})$	$= -21\,996$	$B(\text{bcc})$	$= 0$
$C(\text{bcc})$	$= 12\,000$		
$B(\text{cph})$	$= 20\,000$	$B(\text{cph})$	$= 41\,840$
$G(\text{TiBe}_2)$	$= -44\,000 + 13.563\,T$	$G(\text{TiBe}_2)$	$= -38\,851 + 11.715\,T$
$G(\text{TiBe}_3)$	$= -44\,300 - 14.33\,T$	$G(\text{TiBe}_3)$	$= -41\,572 + 14.330\,T$
$G(\text{Ti}_2\text{Be}_{17})$	$= -44\,711 - 17.51\,T$	$G(\text{Ti}_2\text{Be}_{17})$	$= -42\,279 + 17.510\,T$
$G(\text{TiBe}_{12})$	$= -41\,237 - 17.429\,T$	$G(\text{TiBe}_{12})$	$= -39\,356 + 17.429\,T$

Note: Values are given in J/mol and J/mol · K.

59Pai: R.M. Paine, A.J. Stonehouse, and W.W. Beaver, "An Investigation of Intermetallic Compounds for Very High Temperature Applications," WADC Tech. Rept., Part I (1959). (Crys Structure; Experimental)

60Bed: R.G. Bedford, U.S. Atomic Energy Comm., UCRL-5991-T, 8 p (1960). (Equi Diagram; Experimental; #)

60Pai: R.M. Paine and J.A. Carrabine, "Some New Intermetallic Compounds of Beryllium," *Acta Crystallogr.*, 13, 680-681 (1960). (Crys Structure; Experimental)

61Gla: E.I. Gladyshevskii, P.I. Kripyakevich, M.Yu. Teslyuk, O.S. Zarechnyuk, and Yu.B. Kuz'ma, "Crystal Structures of Some Intermetallic Compounds," *Kristallografiya*, 6(2), 267-268 (1961) in Russian; TR: *Sov. Phys. Crystallogr.*, 6, 207-208 (1961). (Crys Structure; Experimental)

61Zal: A. Zalkin, D. Sands, R.G. Bedford, and O.H. Krikorian, "The Beryllides of Ti, V, Cr, Zr, Nb, Mo, Hf, Ta," *Acta Crystallogr.*, 14, 63-64 (1961). (Crys Structure; Experimental)

62Amo: V.M. Amonenko, V.Ye. Ivano, G.F. Tikhinskiy, and V.A. Finkel, "X-Ray Diffraction Study of the Solubility of Impurities in Beryllium," *Fiz. Met. Metalloved.*, 14(6), 852-856 (1962) in Russian; TR: *Phys. Met. Metallog.*, 14, 47-51 (1962). (Equi Diagram; Experimental)

62Obi: I. Obinata, K. Kurikar, and M. Simura, "A Study of Ti-Be Alloys," *Titanium*, 10(7), 160-166 (1962) in Japanese. (Equi Diagram; Experimental)

63Hin: E.D. Hindle and G.F. Slattery, "A Metallographic Survey of Some Dilute Beryllium Alloys," *Inst. Met. Mon. Rept. Ser.*, (28), 651-664 (1963). (Equi Diagram; Experimental)

64Gil: E. Gillam and H.P. Rooksby, "Structural Relationships in Beryllium-Titanium Alloys," *Acta Crystallogr.*, 17, 762-763 (1964). (Crys Structure; Experimental)

65Bea: W.W. Beaver, A.J. Stonehouse, and R.M. Paine, "Development of Intermetallic Compounds for Aerospace Applications," *Plansee Proceedings (Metals for the Space Age), Metallwerk Plansee AG, Reutte/Tirol*, 682-700 (1963). (Equi Diagram; Experimental)

66Hun: D.B. Hunter, "The Titanium-Beryllium Phase Diagram up to 10 wt.% Be," *Trans. AIME*, 236, 900-902 (1966). (Equi Diagram; Experimental)

73Kau: L. Kaufman and H. Nesor, "Phase Stability and Equilibria as Affected by the Physical Properties and Electronic Structure of Titanium Alloys," Intl. Conf. on Titanium, 2nd *Titanium Science and Technology*, R.I. Jaffee and H.M. Burte, Ed., Inst. of Techn., Cambridge, MA, 773-800 (1973). (Equi Diagram; Experimental)

75Stu: M. Stumke and G. Petzow, "Crystal Structure and Lattice Constants of Transition Metal-Diberyllides and Diborides in Ternary Solid Solutions," *Z. Metallkd.*, 66(5), 292-297 (1975) in German. (Crys Structure; Experimental)

78Tan: L.E. Tanner and B.C. Giessen, "Structure and Formation of the Metastable Phase m-TiBe," *Metall. Trans. A*, 9, 67-68 (1978). (Meta Phases; Experimental)

79Gie: B.C. Giessen, J.C. Barrick, and L.E. Tanner, "Formation and Structure of a New Eta Phase Ti$_3$Be$_3$(O,N)x," *Mater. Sci. Engr.*, 38, 211-216 (1979). (Meta Phases; Experimental)

79Gio: A.L. Giorgi, B.T. Matthias, G.R. Stewart, F. Acker, and J.L. Smith, "Itinerant Ferromagnetism in the C_{15} Laves Phase-TiBe$_{2-x}$Cu$_x$," *Solid State Comm.*, 32, 455-458 (1979). (Crys Structure; Experimental)

79Kau: L. Kaufman and L.E. Tanner, "Coupled Phase Diagrams and Thermochemical Descriptions of the Titanium-Beryllium, Zirconium-Beryllium and Hafnium-Beryllium Systems," *Calphad*, 3(2), 91-107 (1979). (Thermo; Experimental; #)

79Tan: L.E. Tanner and R. Ray, "Metallic Glass Formation and Properties in Zr and Ti Alloyed with Be--I. The Binary Zr-Be and Ti-Be Systems," *Acta Metall.*, 27, 1727-1747 (1979). (Meta Phases; Experimental)

The Bi-Ti (Bismuth-Titanium) System

208.9804   47.88

By J.L. Murray

Equilibrium Diagram

The Ti-Bi phase diagram has been examined experimentally only for Ti-rich and Bi-rich alloys and is known only approximately. The assessed phase diagram is shown in Fig. 1. Special points of the diagram are summarized in Table 1, and crystal structures of the equilibrium phases are described in Table 2. According to [58Aue], the absorption of oxygen by pyrophoric Ti-Bi alloys is so rapid that contamination is unusually difficult to avoid. Moreover, they noted that free Bi was almost always found in their samples, indicating that it is very difficult to achieve equilibrium in this system.

The equilibrium solid phases of the Ti-Bi system are:

- The bcc (βTi), cph (αTi), and (Bi) solid solutions. The solubility of Ti in (Bi) is less than 5×10^{-3} at.%, assuming that the solubility in solid (Bi) is less than in the Bi-rich liquid. The maximum solubilities of Bi in (βTi) and (αTi) are 10 and 0.5 at.%, respectively [60Obi].
- The most Ti-rich intermetallic phase was designated Ti$_4$Bi by [51Now] and [58Aue], based on a structural similarity to Ti$_4$Pb. According to the metallographic and X-ray work of [60Obi], how-

Table 1 Special Points of the Assessed Ti-Bi Phase Diagram

Reaction	Compositions of the respective phases, at.% Bi			Temperature, °C	Reaction type
L + (βTi) $\rightleftharpoons$ Ti$_3$Bi	~ 26	10	~ 25	1340	Peritectic
(βTi) $\rightleftharpoons$ (αTi) + Ti$_3$Bi	8.6	~ 0.5	~ 25	725	Eutectoid
L + Ti$_3$Bi + Ti$_2$Bi	?	~ 25	~ 33	?	Unknown
L $\rightleftharpoons$ Ti$_2$Bi + (Bi)	~ 100	~ 33	100	271	Eutectic
L $\rightleftharpoons$ (βTi)		0		1670	Melting point
(βTi) $\rightleftharpoons$ (αTi)		0		882	Allotropic transformation
L $\rightleftharpoons$ (Bi)		100		271.442	Melting point

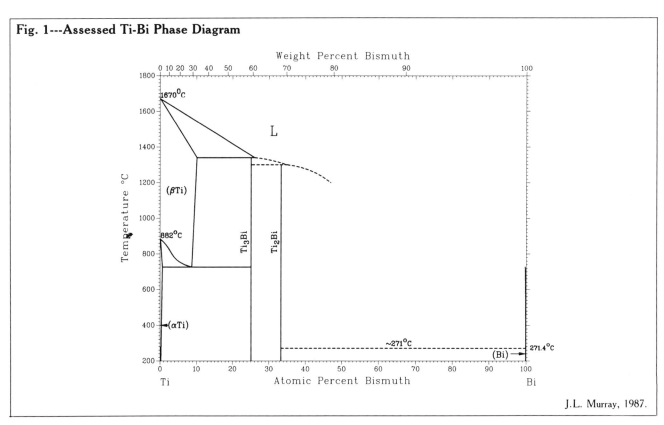

Fig. 1---Assessed Ti-Bi Phase Diagram

J.L. Murray, 1987.

44

The Bi-Ti System

Fig. 2---Experimental Data for Ti-Rich Alloys

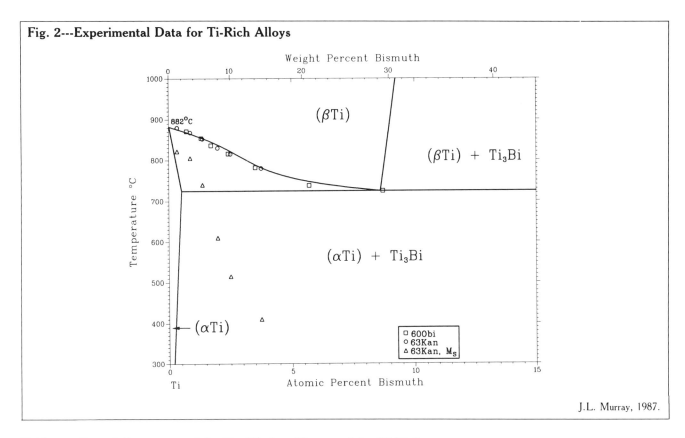

J.L. Murray, 1987.

Table 2 Crystal Structures of the Equilibrium Phases of the Ti-Bi System

Phase	Homogeneity range, at.% Bi	Pearson symbol	Space group	Strukturbericht designation	Prototype	Lattice parameters, nm a	c	Reference
(αTi)	0 to 0.5	$hP2$	$P6_3/mmc$	$A3$	Mg	...	...	[Pearson2]
(βTi)	0 to 10	$cI2$	$Im3m$	$A2$	W	...	...	[Pearson2]
Ti_3Bi	20	(a)	(b)	...	...	0.6020	0.8204	[60Obi]
Ti_2Bi	33 to 37	$tP12$	$P4_2/mmc$	...	Ti_2Bi	0.4048 to 0.4078	1.453 to 1.457	[58Aue]
(Bi)	100	$hR2$	$R\bar{3}m$	$A7$	Bi	...	...	[Pearson2]

(a) Tetragonal. (b) Identified by [51Now] as similar to $D0_{19}$, but with lower than hexagonal symmetry.

ever, the phase has the stoichiometry Ti_3Bi. In this evaluation the Ti_3Bi designation is accepted.

• The most Bi-rich intermetallic phase is Ti_2Bi [58Aue]. Ti_2Bi was also found in co-sputtered thin films [70Sho] and another phase "γ" (in addition to Ti_3Bi) was found by [60Obi] in a 30 at.% Bi alloy. This phase is designated with Ti_2Bi in this evaluation. Ti_2Bi has the composition range 33 to 37 at.% Bi, according to X-ray diffraction and density measurements [58Aue].

Ti_3Bi and Ti_2Bi were the only intermetallic phases found in a survey of the whole system [58Aue].

Ti-Rich Alloys.

For Ti-rich alloys, the only phase diagram study was made by [60Obi], using microscopic examination, X-ray diffraction, and differential thermal analysis. Two reactions were found:

$L + (\beta Ti) \rightleftarrows Ti_3Bi$ at 1340 ± 20 °C $(\beta Ti) \rightleftarrows (\alpha Ti) + Ti_3Bi$ at 725 ± 20 °C

The (βTi) solvus was determined up to 1000 °C and extrapolated to the peritectic temperature to obtain a maximum solubility of 10 at.% Bi in (βTi). The solubility of Bi in (αTi) is about 0.5 at.%; the shape of the phase boundary was not determined. [63Kan] made a continuous cooling study of the $(\beta Ti) \rightarrow (\alpha Ti)$ transformation. They verified the diagram of [60Obi] up to 3.74 at.% Bi and determined the start temperature, M_s, of the martensite transformation. M_s is illustrated in Fig. 2 with the equilibrium data. [78Fra] studied the kinetics of the eutectoid decomposition, and this work does not conflict with the eutectoid temperature and composition as determined by [60Obi].

Bi-Rich Alloys.

The solubility of Ti in liquid Bi was determined by [65Wee] by chemical analysis of filtered samples of the equilibrated melt. The liquidus was described by the expression:

$$\log_{10} (\text{at. ppm Ti}) = 3.4075 - 2230/T$$

45

Fig. 3---Detail of the Bi-Rich Portion of the Phase Diagram

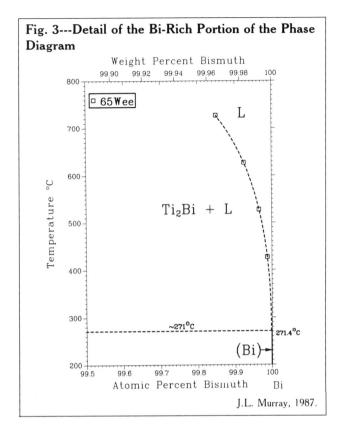

J.L. Murray, 1987.

liquidus of [65Wee], and the emf data of [78Alg], the present evaluator attempted to construct a subregular solution excess Gibbs energy for the liquid and the Gibbs energy of Ti_2Bi relative to the liquid phase. The aim was to interpolate the liquidus between 35 and 99 at.% Bi. The calculation predicted a very asymmetric excess Gibbs energy and a miscibility gap in the liquid phase. Some slight evidence to suggest that there may be a miscibility gap in the liquid was mentioned by [58Aue]; free Bi was always found in melted samples. However, the present evaluator does not believe that there is sufficient evidence at this time to justify a thermodynamic prediction of a miscibility gap in Ti-Bi, and therefore thermodynamic functions are not tabulated.

Cited References

*51Now: H. Nowotny and J. Pesl, "Investigation of the Titanium-Lead System," *Monatsh. Chem., 82*, 344-347 (1951) in German. (Equi Diagram; Experimental)

*58Aue: H. Auer-Welsbach, H. Nowotny, and A. Kohl, "Friction-Pyrophoric Titanium Alloys; Ti_2Bi, A New Structure Type," *Monatsh. Chem., 89*, 154-159 (1958) in German. (Equi Diagram; Experimental)

*60Obi: I. Obinata, Y. Takeuchi, and S. Saikawa, "The System Titanium-Bismuth," *Trans. ASM, 52*, 1059-1070 (1960). (Equi Diagram; Experimental)

63Kan: H. Kaneko and Y.C. Huang, "Continuous Cooling Transformation Characteristics of Titanium Alloys of Eutectoidal Type (Part l)," *J. Jpn. Inst. Met., 27*, 393-397 (1963) in Japanese. (Equi Diagram; Experimental)

*65Wee: J.R. Weeks, "Liquidus Curves of Nineteen Dilute Binary Alloys of Bismuth," *Trans. ASM, 58*, 303-322 (1965). (Equi Diagram; Experimental)

70Sho: R.T. Shoemaker, C.E. Anderson, and G.L. Liedl, "Sputtering of Bismuth-Titanium Two-Phase Cathodes," *J. Electrochem. Soc., 117*(11), 1438-1439 (1970). (Equi Diagram; Experimental)

78Alg: M.M. Alger, "The Thermodynamics of Highly Solvated Liquid Metal Solutions," *Diss. Abstr. Int., 42*(11), 251 (1982). (Equi Diagram; Experimental)

78Fra: G.W. Franti, J.C. Williams, and H.I. Aaronson, "A Survey of Eutectoid Decomposition in Ten Ti-X Systems," *Metall. Trans. A, 9*, 1641-1649 (1978). (Equi Diagram; Experimental)

where T is in K in the temperature range 300 to 700 °C. Several liquidus points calculated from this expression are plotted in Fig. 3.

Titanium partial Gibbs energies were measured by the emf method [78Alg] for alloys containing up to 11 at.% Ti at 856 and 873 °C. Based on the data of [65Wee], all these alloys were two-phase. Using a roughly estimated melting point for Ti_2Bi based on [60Obi], the Bi-rich

The C-Ti (Carbon-Titanium) System

12.011   47.88

By J.L. Murray

Equilibrium Diagram

The condensed Ti-C system has been characterized over the composition range 0 to about 70 at.% C; in addition to the bcc (βTi) and cph (αTi) terminal solid solutions, there is a stable equilibrium carbide, TiC, with a homogeneity range of 32 to 48.8 at.% C. The carbide has another form, designated Ti_2C, in which vacancies are ordered on the carbon sublattice [67Gor]. This ordered phase is not the same "Ti_2C" phase as that mentioned by [Hansen] as having been proposed but not verified; the latter "Ti_2C" was proposed by analogy to V_2C, which has a hexagonal crystal structure. The most important phase

diagram investigations are those of [53Cad], [55Ogd], [56Wag], [59Bic], and [65Rud]. At the higher temperatures, the assessed diagram (shown in Fig. 1) is based primarily on [65Rud]; at lower temperatures, it is based primarily on [53Cad] and [56Wag]. Some additional data are contained in reports not generally available; the review of the system by [67Sto] is recommended as a summary of these data. Special points of the assessed phase diagram are presented in Table 1.

Liquidus.

The Ti-rich liquid solidifies by the eutectic reaction L ⇌ (βTi) + Ti_2C at 1648 ± 5 °C, based on reported eutectic

Table 1   Special Points of the Assessed Ti-C Phase Diagram

Reaction	Compositions of the respective phases, at.% C			Temperature, °C	Reaction type
L ⇌ (βTi) + Ti_2C	1.8	0.6	32	1648 ± 5	Eutectic
L ⇌ TiC		44		3067 ± 15	Congruent
TiC ⇌ Ti_2C		33.3		~ 1900	Congruent
L ⇌ TiC + C	63	48.8	~ 100	2776 ± 6	Eutectic
(βTi) + TiC ⇌ (αTi)	0.6	38	1.6	920 ± 3	Peritectoid
L ⇌ (βTi)		0		1670	Melting point
(βTi) ⇌ (αTi)		0		882	Allotropic transformation

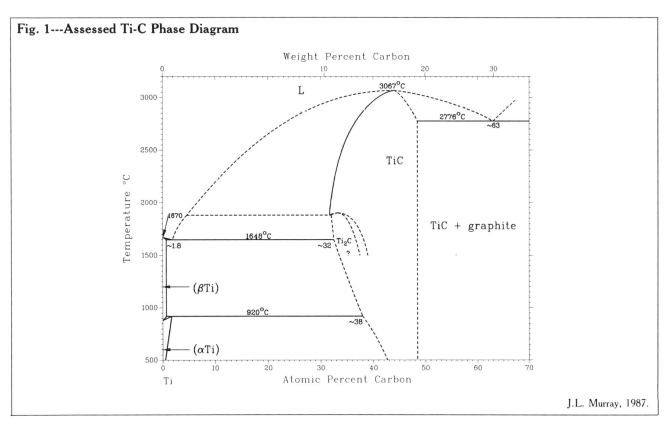

Fig. 1---Assessed Ti-C Phase Diagram

J.L. Murray, 1987.

Fig. 2---Experimental Data on the Ti-C Phase Diagram

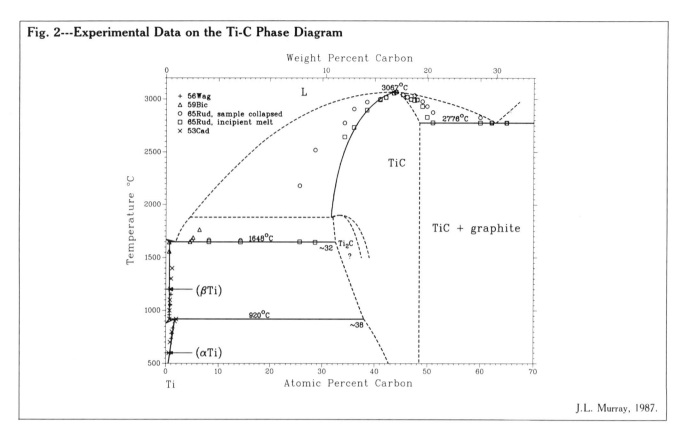

J.L. Murray, 1987.

Table 2   Reported Congruent Melting Temperatures of TiC

Reference	Melting point, °C
[25Fri]	3160 ± 1000
[31Agt]	3140
[53Sch]	3250
[55Gea]	3030
[60Eng]	2940
[65Rud]	3067

temperatures of 1650 ± 5 °C [65Rud] and 1645 °C [59Bic]. [65fRud] used the Pirani technique to measure melting temperatures, and these data are judged to be quite accurate based on the results of a series of measurements made for Ti-, Zr-, and Hf-based alloys. [59Bic] used a quenching and metallographic technique that makes use of the reaction of Ti with graphite crucibles. Both [59Bic] and [65Rud] measured the melting point of pure Ti as 1667 ± 8 °C, in close agreement with the accepted value of 1670 °C. Melting point and liquidus data of [59Bic] and [65Rud] are shown in Fig. 2. [53Cad] previously reported a peritectic reaction L + TiC ⇌ (βTi) at 1750 ± 20 °C based on quenching and metallographic studies; they observed the melting point of pure Ti to be 1725 °C. This indicates that the high or melting temperatures should be attributed to oxygen or nitrogen contamination of the specimens during heat treatment and that the peritectic reaction does not occur in the binary system. Similarly, peritectic reactions reported by [61Kur] at 1700 °C and by [56Nis] at 1750 °C were not used to construct the assessed diagram.

[59Bic] estimated liquidus compositions by chemical analysis of the liquid held at temperature until the melt composition was changing only very slowly. The temperature range 1650 to 1763 °C was examined. These data imply a eutectic composition of about 4.5 at.% C. [65Rud] roughly estimated the liquidus to be somewhat above the temperature at which the samples collapsed. They placed the eutectic composition at slightly less than 2 at.% C, based on metallographic examination of as-cast alloys and noted that the low carbon content of the eutectic and structural changes that occurred during quenching made the metallographic work difficult.

The assessed liquidus was constructed with the aid of thermochemical calculations by [84Uhr]. These calculations are thought to be quite accurate regarding the limiting slope of the (βTi) liquidus. Linear extrapolation of the calculated liquidus to the assessed eutectic temperature predicts a eutectic composition of 1.8 at.% C, in agreement with the metallographic results of [65Rud]. The TiC liquidus must lie above the temperature at which the sample collapsed, but there is no quantitative basis for the exact value selected by [65Rud]. Therefore, the thermochemical calculation of [84Uhr] was used to estimate the liquidus between the eutectic point and the congruent melting point of TiC.

Reported congruent melting points of TiC range between 1940 and 3250 °C (see Table 2). Based on considerations discussed above, the value 3067 ± 15 °C [65Rud] is chosen for the assessed diagram. Similarly, on the C-rich side of stoichiometry, the eutectic reaction L ⇌ TiC + C (graphite) is placed at 2776 °C with the eutectic composition about 63 at.% C [65Rud]. [56Nis] was qualitatively in agreement with [65Rud], but placed the eutectic point at 58 at.% C and 3050 °C. [63Lee] also estimated the eutectic composition as 68 at.%, based on microscopy and mechanical-property determination of cast samples. [61Por] showed the eutectic point at 85 at.% C and 3088 ± 50 °C,

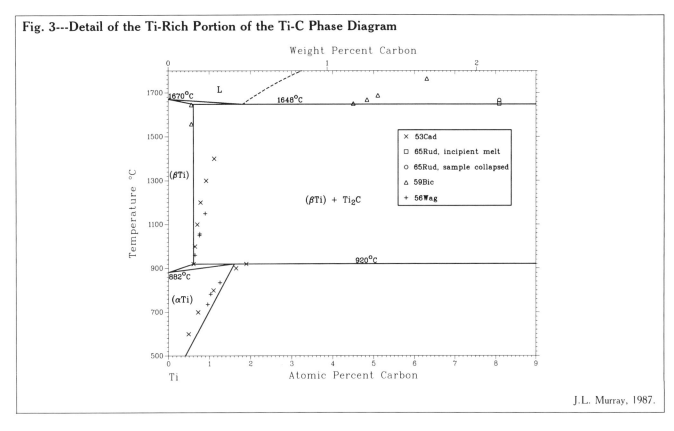

Fig. 3---Detail of the Ti-Rich Portion of the Ti-C Phase Diagram

J.L. Murray, 1987.

Table 3 Crystal Structures of the Ti-C System

Phase	Homogeneity range, at.% C	Pearson symbol	Space group	Struktur-bericht designation	Proto-type	Reference
(βTi)	0 to 0.6	cI2	Im3m	A2	W	[Pearson2]
(αTi)	0 to 1.6	hP2	P6₃/mmc	A3	Mg	[Pearson2]
Ti₂C .... ~	32 to 36	cF48	Fd3m	...	...	[67Gor]
TiC .... ~	32 to 48.8	cF8	Fm3m	B1	NaCl	[Pearson2]

but gave no documentation of experiments. The value 63 at.% C is preferred; eutectic microstructures observed in cast samples at 58 at.% C lie to the side of the faceted phase (TiC), as expected for nonequilibrium solidification.

The Ti-rich limit of the TiC phase field was determined metallographically by [53Cad] (38 at.% C at 920 °C). These values are accepted in the present evaluation. The C-rich limit of the TiC phase field was estimated as 48.5 at.% C at 2776 °C by [65Rud], also based on the melting point results and metallography. Lattice parameter data are available for alloys of greater carbon content, however, and the TiC phase boundaries are therefore considered uncertain on the C-rich side.

(αTi)/(βTi) Equilibria.

There is general agreement that (βTi) transforms by the peritectoid reaction (βTi) + TiC ⇄ (αTi) [53Cad, 55Ogd, 56Wag, 61Kur]. Data for the reported peritectoid temperatures and solid solubilities are shown in Fig. 3. Agreement among investigations is good. The peritectoid temperature is 920 °C, based on [53Cad], who bracketed it ±3 °C by metallographic examination of annealed alloys.

Ordering in the Carbide.

No particular distinction has yet been made between equilibria involving TiC and those involving the ordered

Ti₂C phase, because most investigators did not distinguish the two structures and very little information is available on the stability range of the ordered form. [62Bit] found discontinuities in the magnetic susceptibility of TiC as a function of composition. [67Gor] performed neutron diffraction studies of alloys annealed at temperatures between 1100 and 2000 °C. Superlattice lines were found in the concentration range 32 to 36 at.% C. At 33 at.% C, Ti₂C was estimated to be stable up to about 1900 °C. On this basis, an ordering reaction has been loosely sketched in Fig. 1.

Crystal Structure and Lattice Parameters

Table 3 summarizes the crystal structure data for the equilibrium phases. Room temperature lattice parameters of TiC are plotted as a function of composition in Fig. 4 and those of (αTi) are listed in Table 4.

The NaCl (B1) structure of TiC was first determined by [24Ark] and has been verified by all subsequent investigations. The existence of a Ti₂C carbide (mentioned by [Hansen]) was hypothesized by [33Jac], based on analogy with the V-C and Sc-C systems, but in fact no such compound exists in Ti-C. The structure of the phase designated Ti₂C in Fig. 1 is derived from that of TiC by ordering of carbon atoms and vacancies in the octahedral

Fig. 4---Lattice Parameters of TiC

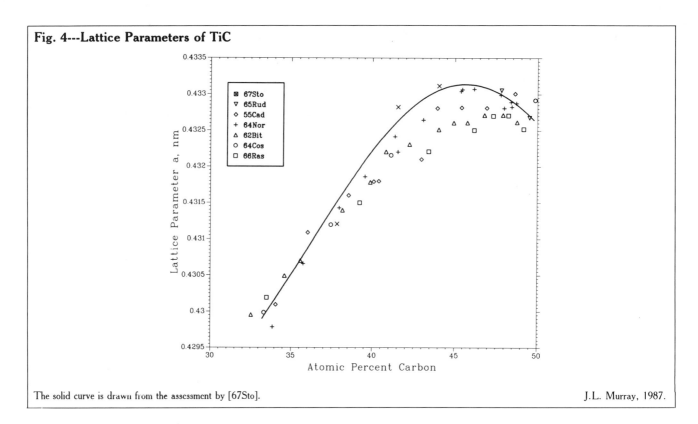

Legend:
- ⊠ 67Sto
- ▽ 65Rud
- ◇ 55Cad
- + 64Nor
- △ 62Bit
- ○ 64Cos
- □ 66Ras

Lattice Parameter a, nm (y-axis)
Atomic Percent Carbon (x-axis)

The solid curve is drawn from the assessment by [67Sto].

J.L. Murray, 1987.

Table 4   Room-Temperature Lattice Parameters of (αTi) [53Cad]

Composition, at.% C	Lattice parameters, nm	
	a	c
0	0.29511	0.46831
0.4	0.29517	0.46872
0.8	0.29525	0.46828
1.2	0.29539	0.46885
1.6	0.29552	0.46823

sites of the Ti lattice [67Gor]. The lattice parameter of the Ti$_2$C cell is twice that of the TiC cell, $a = 0.86$ nm at 33 at.% C [67Gor].

Extensive data are available on the lattice parameter of TiC, particularly for the stoichiometric composition (see Hansen). The more recent work, summarized in Fig. 4, is considered to render obsolete the earlier measurements cited by [Hansen]. The solid curve (Fig. 3) is based on the assessment of [67Sto].

Linear thermal expansion coefficients of TiC were reported by [58Ell], [63Kri], and [64Hou]:

Reference	Linear expansion coefficient, °C^{-1}	Temperature range, °C
[58Ell]	7.42×10^{-6}	24 to 538
[63Kri]	3.4×10^{-6}	−190 to 26
[64Hou]	7.5×10^{-6} to 9.9×10^{-6}	25 to 2080

Thermodynamics

A thermodynamic assessment of the Ti-C system was presented by [84Uhr]. The liquid, bcc, and cph phases

were described as ordinary substitutional solutions of Gibbs energies:

$$G(i) = (1 - x) G^0(\text{Ti},i) + x G^0(\text{C},i) + RT[x \ln x + (1 - x) \ln (1 - x)] + x(1 - x) [A(i) + B(i) (1 - 2x)]$$

where i indicates the phase; x is the atomic fraction of C; G is the Gibbs energy of the pure components; and A and B are temperature-dependent parameters of the subregular expansion of the excess Gibbs energy.

The carbide TiC is described by a sublattice model in which the partial Gibbs energies of TiC are described as:

$$G_{\text{Ti}} = G^0_{\text{TiV}} + RT \ln (1 - y) + y^2 L_{\text{CV}}$$

$$G_{\text{C}} = G^0_{\text{TiC}} - G^0_{\text{TiV}} + RT \ln y(1 - y) + (1 - 2y) L_{\text{CV}}$$

$$y = x/(1 - x) = \text{site fraction of C}$$

$$G^0_{\text{TiV}} = G^0(\text{Ti,fcc})$$

$$G^0_{\text{TiC}} = G^0(\text{TiC, stoichiometric})$$

L_{CV} parametrizes the interaction between carbon and vacancy (V) on their sublattice when the Ti sublattice is completely full.

[84Uhr] evaluated the Gibbs energy parameters of TiC using heats of formation tabulated in [Hultgren] and data [67Sto] on the vapor pressure of Ti over TiC as a function of composition. The parameters derived by [84Uhr] were:

$$L_{\text{CV}} = -173\,640 \text{ J/mol}$$

$$^0G_{\text{TiV}} = {^0G}^{\text{gas}}_{\text{TiV}} - 151\,600 \text{ J/mol}$$

$$^0G_{\text{TiC}} = G^{\text{cph}}_{\text{Ti}} + G^{\text{graphite}}_{\text{C}} = 184\,560 + 11.058\,T$$

Thermodynamic data obtained by equilibration of CH_4/H_2 mixtures over TiC [67Gri, 68Ale] and emf data [71Mal, 73Koy] were also examined by [84Uhr]. The data of [67Sto], however, were preferred, primarily because they were more consistent with accepted heat of formation results.

No other thermodynamic data are available for this system, and the Gibbs energies of the L, cph, and bcc solutions were determined from the phase diagram.

The calculated diagram is characterized by the reactions:

L $\rightleftharpoons$ (βTi) + TiC at 1654 °C and 1.3, 0.7, and 32 at.% C

L $\rightleftharpoons$ TiC at 3067 °C and 44 at.% C

L $\rightleftharpoons$ TiC + graphite at 2776 °C and 65, 48, and 100 at.% C

where the compositions are those of the respective phases in the reactions. Agreement with the experimental diagram is good, and as mentioned above, the assessed TiC liquidus is drawn from the calculations of [84Uhr].

Cited References

24Ark: A.E. van Arkel, *Physica, 4,* 286-301 (1924). (Crys Structure, Experimental)

31Agt: C. Agte and K. Moers, "Methods for Preparing High-Purity Refractory Carbides, Nitrides and Borides and Description of Some of Their Properties," *Z. Anorg. Chem., 198,* 233-275 (1931) in German. (Equi Diagram; Experimental)

33Jac: B. Jacobson and A. Westgren, "Nickel Carbide and Its Relation to the Other Carbides of the Series of Elements Scandium-Nickel," *Z. Phys. Chem.,* B20, 361-367 (1933) in German. (Equi Diagram; Experimental)

53Cad: I. Cadoff and J.P. Nielsen, "Titanium-Carbon Phase Diagram," *Trans. AIME, 197,* 248-252 (1953). (Equi Diagram; Experimental)

55Ogd: H.R. Ogden, R.I. Jaffee, and F.C. Holden, "Structure and Properties of Ti-C Alloys," *Trans. AIME, 203,* 73-80 (1955). (Equi Diagram; Experimental)

56Nis: N. Nishimura and H. Kimura, "On the Equilibrium Diagram Titanium-Oxygen-Carbon System," *Nippon Kinzoku Gakkai-Shi, 20,* 528-531 (1956) in Japanese. (Equi Diagram; Experimental)

56Wag: F.C. Wagner, E.J. Bucur, and M.A. Steinberg, "The Rate of Diffusion of Carbon in α and β Titanium," *Trans. ASM, 48,* 742-761 (1956). (Equi Diagram; Experimental)

58Ell: R.O. Elliott and C.P. Kempter, "Thermal Expansion of Some Transition Metal Carbides," *J. Phys. Chem., 62,* 630-631 (1958). (Crys Structure; Experimental)

59Bic: R.L. Bickerdike and G. Hughes, "An Examination of Part of the Titanium-Carbon System," *J. Less-Common Met., 1,* 42-49 (1959). (Equi Diagram; Experimental)

60Eng: J.L. Engelke, F.A. Halden, and E.P. Farley, "Synthesis of New High Temperature Materials," Tech. Rep. WADC-TR-59-654 (PB 161720) (1960). (Equi Diagram; Experimental)

61Kur: N.N. Kurnakov and M.Ya. Troneva, "Equilibrium Diagram of the Carbon-Titanium Binary System up to 50% Carbon," *Zh. Neorg. Khim., 6,* 1347-1350 (1961) in Russian; TR: *Russ. J. Inorg. Chem., 6*(6), 690-694 (1961). (Crys Structure; Experimental)

61Por: K.I. Portnoi, Yu.V. Levinskii, and V.I. Fadeeva, "Reaction with Carbon of Some Refractory Carbides and Their Solid Solutions," *Izv. Akad. Nauk SSSR, Otd. Tekhn. Nauk, Met. Top.,* (2), 147-149 (1961) in Russian. (Thermo; Experimental)

62Bit: H. Bittner and H. Goretzki, "Magnetic Investigations of the Carbides TiC, ZrC, HfC, VC, NbC and TaC," *Monatsh Chem., 93,* 1000-1004 (1962) in German. (Equi Diagram; Experimental)

63Kri: N.H. Krikorian, T.C. Wallace, and J.L. Anderson, "Low-Temperature Thermal Expansion of the Group 4a Carbides," *J. Electrochem. Soc., 110,* 587-588 (1963). (Crys Structure; Experimental)

63Lee: D.H. Leeds, E.G. Kendall, and J.F. Ward, "Preparation and Properties of Arc-Cast TiC-C Alloys," Tech. Rep. SSD-TDR-63-216 (AD 422165) (1963). (Equi Diagram; Experimental)

64Cos: P. Costa and R.R. Conte, "Properties of the Carbides of the Transition Metals," Int. Symp. on Compounds of Interest in Nuclear Reactor Technology, J.T. Waber, P. Chiotti, and W.N. Miner, Ed., The Metallurgy Society of the AIME, Nuclear Metallurgy, 10 (1964). (Crys Structure; Experimental)

64Hou: C.R. Houska, "Thermal Expansion and Atomic Vibration Amplitudes for TIC, TiN, ZrC, ZrN, and Pure Tungsten," *J. Phys. Chem. Solids, 25,* 359-366 (1964). (Crys Structure; Experimental)

64Nor: J.T. Norton and R.K. Lewis, "Properties of Non-Stoichiometric Metallic Carbides," final report, Advanced Metals Research Corp. for Hdgtrs., NASA under Contract No. NASw-663 (1964).

65Rud: E. Rudy, D.P. Harmon, and C.E. Brukl, "Ternary Phase Equilibria in Transition Metal-Boron-Carbon-Silicon Systems," AFML-TR-65-2, Part I, Volume II (1965). (Equi Diagram; Experimental)

66Ras: H. Rassaerts, F. Benesovsky, and H. Nowotny, "Investigation of the Systems Titanium- and Hafnium-Chromium-Carbon," *Plansee Pulvermetall., 14,* 23-28 (1966) in German. (Crys Structure; Experimental)

67Gor: H. Goretzki, "Neutron Diffraction Studies on Titanium-Carbon and Zirconium-Carbon Alloys," *Phys. Status Solidi, 20,* K141-K143 (1967). (Equi Diagram; Experimental)

67Sto: E.K. Storms, "The Refractory Carbides," Refractory Materials, A Series of Monographs, J.L. Margrave, Ed., Academic Press, New York (1967). (Equi Diagram, Crys Structure, Thermo; Assessment)

68Ale: V.I. Alekseev, A.S. Panov, Ye.V. Fiveiskii, and L.A. Shvartsman, "Thermodynamic Properties of Nonstoichiometric Vanadium and Titanium Carbide," Thermodynamics of Nuclear Materials, 1967, Int. At. Energy Agency, Vienna, 435-447 (1968). (Thermo; Experimental)

71Mal: V.I. Malkin and V.V. Pokidyshev, "Emf Study of the Thermodynamic Characteristics of TiC Within the Range of Homogeneity," *Zh. Fiz. Khim., 45*(8), 2044-2046 (1971) in Russian; TR: *Russ. J. Phys. Chem., 45*(8),1159-1161 (1971). (Thermo; Experimental)

73Koy: K. Koyama and Y. Hashimoto, *Nippon Kinzoku Gakkai-Shi, 37,* 406 (1973). (Thermo)

84Uhr: B. Uhrenius, "Calculation of the Ti-C, W-C and Ti-W-C," *Calphad, 8*(2), 101-119 (1984). (Equi Diagram; Experimental)

The Ca-Ti (Calcium-Titanium) System

40.08   47.88

By J.L. Murray

The investigation by [60Obi] is the only reported experimental study of the Ti-Ca system. [60Obi] prepared solutions of Ti in liquid Ca and of Ca in bcc (βTi) and cph (αTi), by heating either Ca or Ca + Ti powders in sealed Ti containers.

Equilibration temperatures of 845 to 1300 °C were used. The mutual solubilities were measured by chemical analysis of the annealed Ca or of the Ti powders from which excess Ca had been removed. Metallographic and X-ray examination of the Ti container and the Ti powders revealed no evidence of intermetallic compounds.

A schematic phase diagram is shown in Fig. 1, and details of the data for the Ca-rich liquidus and the (βTi) and (αTi) solvus curves are shown in Fig. 2 and 3, respectively. The discontinuity in the observed solubility of Ca in (βTi) at 1150 °C suggests that equilibrium was probably not reached below 1050 °C. If the solubilities of Ca in (βTi) are approximately correct, then a peritectic reaction Ca(L) + (βTi) ⇄ (αTi) occurs near 882 °C, the temperature of the allotropic transformation of pure Ti. The solubility data, however, are not precise enough to

permit an estimate of the peritectic temperature. Special points on the phase diagram are given in Table 1, and crystal structure data are given in Table 2.

No determination was made of the solubility of Ti in fcc (βCa), nor were thermal analysis measurements made. Therefore, it is not known whether addition of Ti to Ca raises or lowers the Ca melting point.

Considering the very low mutual solubilities of Ti and Ca, it is likely that Ti and Ca do not mix even in the liquid state at high temperatures, i.e., above the melting point of pure Ti, 1670 °C. The available phase diagram data, however, are not complete or accurate enough to permit a thermodynamic analysis of the system, and such an analysis would be required to provide even a speculative completion of the phase diagram.

Cited Reference

60Obi: I. Obinata, Y. Takeuchi, and S. Saikawa, "The System Titanium-Cadmium," *Trans. ASM, 52,* 1072-1083 (1960). (Experimental)

Fig. 1---Assessed Ti-Ca Phase Diagram

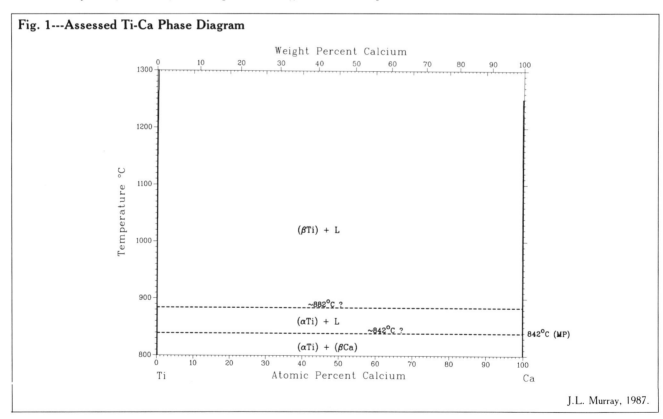

J.L. Murray, 1987.

The Ca-Ti System

Fig. 2---Detail of the Ca-Rich Portion of the Diagram

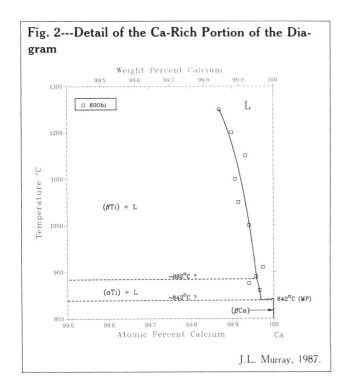

J.L. Murray, 1987.

Fig. 3---Detail of the Ti-Rich Portion of the Diagram

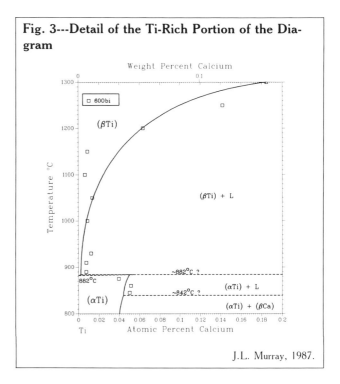

J.L. Murray, 1987.

Table 1   Special Points of the Assessed Ti-Ca Phase Diagram

Reaction	Compositions of the respective phases, at.% Ca			Temperature, °C	Reaction type
L + (βTi) ⇌ (αTi)	99.96	0.002	0.05	~ 882	Peritectic
L ⇌ (αTi) + (βCa)	99.97	0.004	~ 100	~ 840	Eutectic (?)
L ⇌ (βTi)		0		1670	Melting point
(βTi) ⇌ (αTi)		0		882	Allotropic transformation
L ⇌ (βCa)		100		840	Melting point

Table 2   Crystal Structures of the Ti-Ca System

Phase	Homogeneity range, at.% Ca	Pearson symbol	Space group	Struktur-bericht designation	Prototype	Reference
(αTi)	0 to ~0.05	hP2	$P6_3/mmc$	A3	Mg	[Pearson2]
(βTi)	?	cI2	Im3m	A2	W	[Pearson2]
(βCa)	~ 100	cF4	Fm3m	A1	Cu	[Pearson2]

The Cd-Ti (Cadmium-Titanium) System

114.41   47.88

By J.L. Murray

Equilibrium Diagram

The boiling point of Cd is 767 °C, and the melting point of Ti is 1670 °C. It is therefore extremely difficult to prepare or heat treat Ti-Cd alloys of fixed composition. The phase diagram is nevertheless of interest because of the embrittling effect of liquid Cd on Ti. The present knowledge of the diagram (Fig. 1) is also summarized in Table 1.

The equilibrium solid phases of Ti-Cd are:

- The cph (αTi) solid solution, which dissolves about 6.5 at.% Cd; nothing is known about the homogeneity range of the cph (Cd) solid solution.
- The bcc (βTi) solid solution, which dissolves at least 22 at.% Cd.
- Ti_2Cd, stable up to about 850 ± 50 °C.
- TiCd, stable up to 620 °C.

[62Cha] determined the solubility of Ti in liquid Cd and showed that there are two intermetallic compounds. The saturated liquid and the solid in equilibrium with it were separated and chemically analyzed. The composition

Fig. 1---Assessed Ti-Cd Phase Diagram

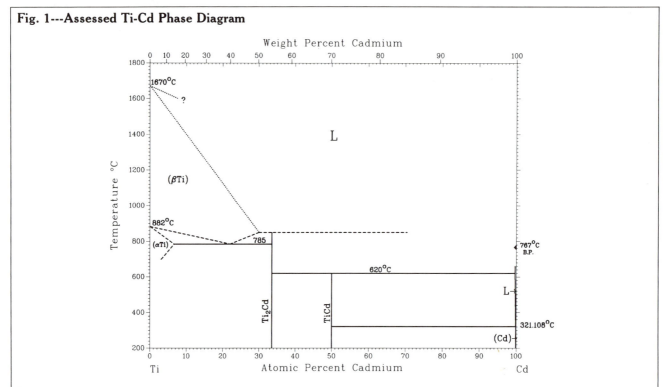

Phase boundaries drawn with dashed curves are uncertain, but have been investigated experimentally. Dotted curves are drawn for the sake of completeness only.

J.L. Murray, 1987.

Table 1   Special Points of the Assessed Ti-Cd Phase Diagram

Reaction	Compositions of the respective phases, at.% Cd			Temperature, °C	Reaction type
L + (βTi) $\rightleftarrows$ Ti_2Cd	~ 99.7	~ 30	33.3	850 ± 50	Peritectic
L + Ti_2Cd $\rightleftarrows$ TiCd	99.88	33.3	50	620	Peritectic
L $\rightleftarrows$ TiCd + (Cd)	~ 100	50	~ 100	321	Eutectic
(βTi) $\rightleftarrows$ (αTi) + Ti_2Cd	22	~ 6.5	33.3	785 ± 5	Eutectoid
L $\rightleftarrows$ (βTi)		0		1670	Melting point
(βTi) $\rightleftarrows$ (αTi)		0		882	Allotropic transformation
L $\rightleftarrows$ (Cd)		100		321.108	Melting point
L(Cd) $\rightleftarrows$ V(Cd)		100		677	Boiling point

Table 2   Cd-Rich Liquidus [62Cha]

| Temperature, °C | Composition | |
	at. ppm Ti	wt. ppm Ti
651.5	1501	640
628.9	1290	550
603.9	1196	510
573.4	915	390
547.8	919	392
523.4	699	298
502.5	680	290
474.4	554	236
449.6	467	199
447.4	479	204
424.4	380	162
399.6	350	149
394.2	335	143
378.6	312	133
349.6	282	120
339.2	279	119

of the saturated liquid was obtained by dissolving the entire sample, analyzing the entire sample eliminated any effects of segregation on cooling.

Liquid solubility data are listed in Table 2. [62Cha] provided the following expression to represent the Cd-rich liquidus between 340 and 650 °C:

$$\log N_{Ti} = 1.690 - 5.960 \times 10^3\, T^{-1} + 1.683 \times 10^6\, T^{-2}$$

where N_{Ti} is the atomic percent of Ti, and T is temperature in K. The data are plotted in Fig. 2.

Thermal analyses and chemical and X-ray analyses of crystals isolated from furnace-cooled melts showed that the phases in equilibrium with the liquid are Ti_2Cd above 620 °C and TiCd below 620 °C.

[72Rob] investigated the Ti-rich portion of the diagram using metallographic and X-ray techniques and chemical analysis. The results of that study are summarized in Table 1. Heat treatments of 1 to 48 h were performed. As expected, significant amounts of Cd were lost, and the surface layer of the alloys was entirely (αTi).

Table 3   Crystal Structures and Lattice Parameters of the Ti-Cd System

| Phase | Composition range, at.% Cd | Pearson symbol | Space group | Struktur-bericht designation | Proto-type | Lattice parameters, nm | | Reference |
						a	c	
(αTi)	0 to ~6.5	hP2	P6₃/mmc	A3	Mg	0.29511	0.46843	[Pearson2]
(βTi)	0 to ~30	cI2	Im3m	A2	W	0.33065	...	[Pearson2]
Ti_2Cd	33.3	tI6	P4/mmm	C11_b	MoSi₂	0.2865	1.342	[62Sch]
TiCd	50	tP4	P4/nmm	B11	γTiCu	0.2904	0.8954	[62Sch]
(Cd)	~100	hP2	P6₃/mmc	A3	Mg	0.29788	0.56167	[Pearson2]

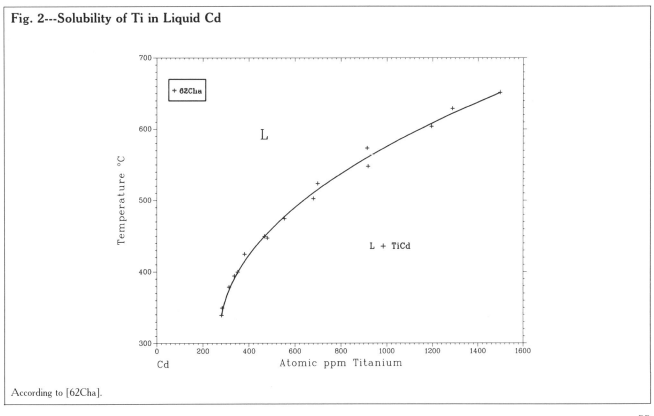

Fig. 2---Solubility of Ti in Liquid Cd

According to [62Cha].

The loss of Cd makes the attainment of equilibrium at a fixed composition extremely uncertain. For this reason, it is considered that these experiments, although carefully conducted, are most accurate for determining the temperatures of three-phase reactions. They are approximate for estimating the compositions of the three-phase reactants and contain no information on the shapes of the phase boundaries. The eutectoid temperature appears to be located to within ±10 °C. The peritectic temperature for the reaction L + (βTi) ⇌ Ti₂Cd is located between 800 and 900 °C. During the 1-h anneal at 900 °C, the sample lost considerable Cd and became porous, but the microstructure was identified as (βTi) plus liquid.

Crystal Structures and Lattice Parameters

The structures and lattice parameters for the Ti-Cd system are listed in Table 3. The compounds Ti₂Cd and TiCd have the same structures as the corresponding compounds in the Ti-Ag system. Details of the structures were given by [62Sch]. Lattice parameters listed for the

solution phases are the lattice parameters of the appropriate pure components [Pearson2].

Thermodynamics

Thermodynamic data are not available for the Ti-Cd system, and the prognosis is pessimistic, not only for experimental thermodynamic data, but also for theoretical estimates of Gibbs energies derived from the phase diagram.

Cited References

62Cha: M.G. Chasanov, P.D. Hunt, I., Johnson, and H.M. Feder, "Solubility of 3-*d* Transition Metals in Liquid Cadmium," *Trans. AIME,* 224, 935-939 (1962). (Equi Diagram; Experimental)

62Sch: R.V. Schablaske, B.S. Tani, and M.G. Chasanov, "The Crystal Structures of TiCd and Ti₂Cd," *Trans. AIME,* 224, 867-868 (1962). (Crys Structures; Experimental)

72Rob: W.M. Robertson "The Titanium-Rich End of the Ti-Cd Phase Diagram," *Metall. Trans.,* 3, 1443-1445 (1972). (Equi Diagram; Experimental)

The Ce-Ti (Cerium-Titanium) System

140.12   47.88

By J.L. Murray

This assessment is based on the same experimental work as [74Gsc], but adds a thermodynamic calculation of the system. Special points of the diagram are summarized in Table 1, and crystal structure data are given in Table 2.

Experimental work was done on the Ti-rich part of the system by [57Sav], [57Tay], and [62Sav], and unpub-

lished research was cited by [74Gsc]. All investigators agreed that no intermetallic phases are formed and that there is a miscibility gap in the liquid phase.

The equilibrium solid phases are the solid solutions based on the pure elements: (1) The bcc (βTi) and cph (αTi), the high- and low-temperature forms, respectively; and (2) bcc (δCe) and fcc (γCe), high- and low-temperature

Table 1 Special Points of the Assessed Ti-Ce Phase Diagram

Reaction	Compositions of the respective phases, at.% Ce			Temperature, °C	Reaction type
L ⇌ (LTi) + (LCe)		50		~ 1650	Critical point
(LTi) ⇌ (βTi) + (LCe)	~ 23	~ 1.7	~ 77	1450 ± 25	Monotectic
(LCe) + (βTi) ⇌ (αTi)	98	0.35	1.2	910 ± 10	Peritectic
(LCe) ⇌ (αTi) + (δCe)	~ 98.8	0.8	99.8	790 ± 20	Eutectic
(δCe) ⇌ (γCe) + (αTi)	99.9	~ 100	0.6	710 ± 20	Eutectoid
L ⇌ (βTi)		0		1670	Melting point
(βTi) ⇌ (αTi)		0		882	Allotropic transformation
L ⇌ (δCe)		100		798	Melting point
(δCe) ⇌ (γCe)		100		726	Allotropic transformation
(γCe) ⇌ (βCe)		100		61	Allotropic transformation

The Ce-Ti System

Table 2 Crystal Structures and Lattice Parameters of the Ti-Ce System

Phase	Composition range, at.% Ce	Pearson symbol	Space group	Struktur- bericht designation	Proto- type	Lattice parameters, nm _a_	_c_	Reference
(αTi)	0 to 1.2	hP2	P6₃/mmc	A3	Mg	0.29511	0.46843	[Pearson2]
(βTi)	0 to 1.7	cI2	Im3m	A2	W	0.33065(a)	...	[Pearson2]
(βCe)	99.8 to 100	cI2	Im3m	A2	W	0.412(b)	...	[86Gsc]
(γCe)	100	cF4	Fm3m	A1	Cu	0.51610	...	[86Gsc]
(βCe)	100	hP4	P6₃/mmc	A3′	La	0.36810	1.1857	[86Gsc]

Note: Lattice parameters are for the pure elements. (**a**) At 900 °C. (**b**) At 757 °C.

forms of Ce in the temperature range under consideration (600 to 1800 °C).

[57Tay] examined alloys containing up to 7.2 at.% Ce. The solubility of Ce in (αTi) and (βTi) was less than 0.4 at.% in the range 750 to 980 °C, based on microscopy and X-ray diffraction. The (nonequilibrium) (βTi) → (αTi) transformation temperature increased with increasing Ce content. Samples of unspecified compositions exhibited incipient melting when heated to 1343 °C, but not at 1316 °C, and the phase diagram therefore showed solid (Ce) in equilibrium with (βTi) up to 1330 °C, i.e., 590 °C above the melting point of pure δCe.

The diagram of [57Sav], determined by microscopy, X-ray diffraction, dilatometry, and thermal analysis, is qualitatively the same diagram, accounting for the phases (βTi), (αTi), Ce, and the liquid. The maximum solubility of Ce in (αTi) was given as 1.7 at.% and the lowest melting temperature as 1450 °C.

Both [57Sav] and [57Tay] stated that at about 910 °C there is a peritectoid reaction (βTi) + (Ce) ⇌ (αTi) and

that melting occurs only above at least 1330 °C. Because the melting point of pure Ce is 798 °C, postulation that incipient melting occurs only at about 1400 °C requires (1) that the melting point of Ce is raised by about 600 °C by small additions of Ti and (2) that there is significant solubility of Ti in (Ce). In other words, it is not possible to extrapolate these phase relationships to Ce-rich compositions in any thermodynamically plausible way. It may be that [57Tay] and [57Sav] observed the melting of a cerium oxide phase at 1350 to 1450 °C. Therefore, neither [57Tay] nor [57Sav] are used as the quantitative basis for the assessed binary phase diagram.

The work of [62Sav] corrects the difficulties discussed above and illustrates the reactions given in Table 1. In the [62Sav] diagram, the melting point of Ce is lowered by small additions of Ti, and three-phase reactions at temperatures above 800 °C involve the Ce-rich liquid rather than a solid phase. Invariant reactions at 1450 °C and 910 °C were determined by thermal analysis and microscopic studies and those at 790 °C and 710 °C apparently

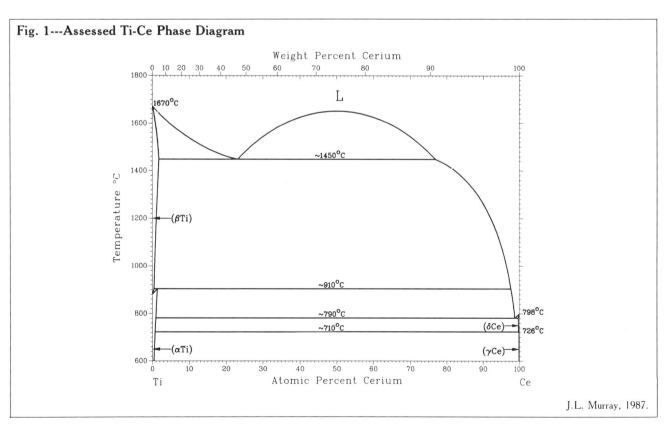

Fig. 1---Assessed Ti-Ce Phase Diagram

J.L. Murray, 1987.

57

Phase Diagrams of Binary Titanium Alloys

Table 3 Gibbs Energies of the Ti-Ce System

Properties of the pure elements

$G^0(\text{Ti,L}) = 0$
$G^0(\text{Ce,L}) = 0$

$G^0(\text{Ti,bcc}) = -16\,234 + 8.368\,T$
$G^0(\text{Ce,bcc}) = -5\,460 + 5.0982\,T$

$G^0(\text{Ti,hex}) = -20\,585 + 12.134\,T$
$G^0(\text{Ce,hex}) = -8\,510 + 8.509\,T$

$G^0(\text{Ti,fcc}) = -17\,238 + 12.134\,T$
$G^0(\text{Ce,fcc}) = -8\,452 + 8.092\,T$

Interaction parameters of the solution phases

$B(\text{L}) = 32\,000$
$B(\text{bcc}) = 55\,087$
$B(\text{hex}) = 42\,322$
$B(\text{fcc}) = 90\,404$

Note: Values are given in J/mol and J/mol · K.

---

by microscopic methods only. The work of [62Sav] was used to construct the assessed diagram (Fig. 1). Uncertainties given in Table 1 are roughly estimated, based on the temperature of the reaction and the amount of data provided by [62Sav].

The assessed diagram is the result of a thermodynamic calculation. Input experimental data were the temperatures of the three-phase reactions and the compositions of (αTi) and (βTi) at the three-phase equilibria [62Sav]. Because of the limited input data, the calculation was limited to a regular solution approximation.

Gibbs energies of the solution phases are represented as:

$$G(i) = (1 - x)\,G^0(\text{Ti},i) + x\,G^0(\text{Ce},i) + RT\,[x \ln x + (1 - x) \ln (1 - x)] + x\,(1 - x)\,[B(i)\,(1 - 2x)]$$

where i designates the phase; x is the atom fraction of Ce; $G^0(i)$, the Gibbs energies of the pure components; and $B(i)$, the regular solution interaction parameter. The pure component parameters of Ce are based on [Hultgren, E].

The method for optimizing regular solution parameters is as follows. Fixing the Gibbs energy of the liquid phase, the parameters of the bcc, fcc, and hexagonal solutions were varied to optimize agreement with the experimental invariant temperatures. The homogeneity ranges of the solid phases depend on the fixed Gibbs energy of the liquid phase. This parameter was determined by trial and error, so that both the solubility limits and the invariant temperatures would be reproduced. The results of these calculations are given in Table 3. Calculated and experimental invariant temperatures agree well within estimated uncertainties, with agreement better for the high temperatures.

The present calculation roughly predicts the solubility of Ti in solid (Ce), but this prediction assumes the regular solution approximation. Based on analogy with the Ti-Sc and Ti-Y systems, the deviation from the regular solution approximation is not expected to be large enough to make a qualitative difference in the phase diagram.

Cited References

57Sav: E.M. Savitskii and G.S. Burkhanov, "Phase Diagrams of Titanium-Lanthanum and Titanium-Cerium Alloys," *Zh. Neorg. Khim.*, 2(11), 2609-2616 (1957) in Russian; TR: *J. Inorg. Chem.*, 2, 199-219 (1957). (Equi Diagram; Experimental)

57Tay: J.L. Taylor, "Preliminary Investigation of the Ti-Ce System," *Trans. AIME*, 209, 94-96 (1957). (Equi Diagram; Experimental)

62Sav: E.M. Savitskii and G.S. Burkhanov, "Phase Diagrams of Titanium Alloys with the Rare Earth Metals," *J. Less-Common Met.*, 4, 301-314 (1962) in German. (Equi Diagram; Experimental)

74Gsc: K.A. Gschneidner, Jr. and M.E. Verkade, "Selected Cerium Phase Diagrams," IS-RIC-7, Rare-Earth Information Center, Iowa State University, Ames, IA (1974). (Equi Diagram; Experimental)

81Gsc: K.A. Gschneidner, Jr. and F.W. Calderwood, "Critical Evaluation of Binary Rare Earth Phase Diagrams," IS-RIC-PR-1, Rare-Earth Information Center, Iowa State University, Ames, IA (1981). (Crys Structure; Review)

The Co-Ti (Cobalt-Titanium) System

58.9332   47.88

By J.L. Murray

Equilibrium Diagram

Certain features of the phase equilibria in the Ti-Co system have recently been the subject of experimental work. The phase diagram contains two "deep eutectics," in which amorphous alloys can be formed; the higher-order ferromagnetic transition has a pronounced effect on the solvus, and it is possible that the ordering transformation in supersaturated (βCo) proceeds by a spinodal mechanism. However, many important features of the diagram have received only cursory examination, or are the subject of controversy. Most notably, liquidus temperatures have not been measured in the range 0 to 20 at.% Co, and there are serious discrepancies in the range 20 to 80 at.% Co. There are conflicting reports about which of the Laves phases TiCo$_2$ are stable phases, and it has been suggested

Table 1 Special Points of the Assessed Ti-Co Phase Diagram

Reaction	Compositions of the respective phases, at.% Co			Temperature, °C	Reaction type
L ⇌ (βTi) + Ti$_2$Co	23.2	14.5	32.9	1020	Eutectic
(βTi) ⇌ (αTi) + Ti$_2$Co	7.0	0.86	33.2	685	Eutectoid
L + TiCo ⇌ Ti$_2$Co	27.1	49.0	33.1	1058	Peritectic
L ⇌ TiCo		50		1325	Congruent
TiCo + L ⇌ TiCo$_2$(c)	55.2	67.2	66.5	1235	Peritectic
TiCo$_2$(c) + L ⇌ TiCo$_2$(h)	67.0	71.0	68.75	1210	Peritectic
L ⇌ TiCo$_2$(h) + TiCo$_3$	75.8	72.0	77.2	1170	Eutectic
L + (αCo) ⇌ TiCo$_3$	79.3	85.9	80.7	1210	Peritectic
L ⇌ (βTi)		0		1670	Melting point
(βTi) ⇌ (αTi)		0		882	Allotropic transformation
L ⇌ (αCo)		100		1495	Melting point
(αCo) ⇌ (εCo)		100		421	Allotropic transformation

Fig. 1---Assessed Ti-Co Phase Diagram

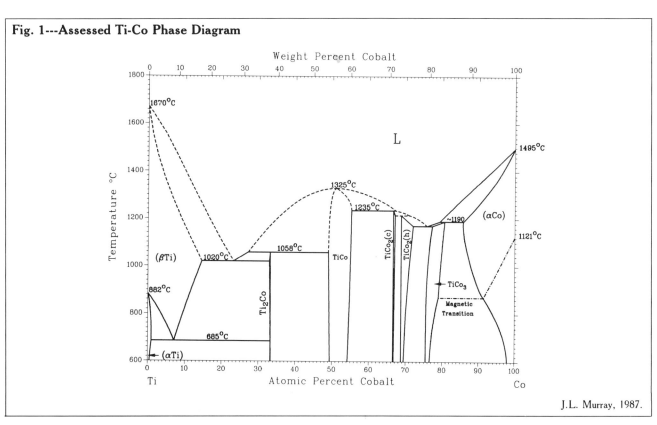

J.L. Murray, 1987.

(based on observation of polytypism) that this part of the diagram may be considerably more complex than previously assumed.

The assessed diagram is shown in Fig. 1, and special points of the diagram are listed in Table 1.

The equilibrium solid phases of the system are:

- The cph solid solutions, (αTi) and (εCo). (αTi) is stable below 882 °C, and (αCo) is stable below approximately 422 °C. The temperature range of Fig. 1 has not been extended to include the (εCo)/(αCo) transition, because the equilibrium phase relations among TiCo₃, (αCo), and (εCo) have not been determined experimentally.
- The bcc solid solution, (βTi), stable in pure Ti above 882 °C. The maximum solubility of Co in (βTi) is 14.5 at.% at 1020 °C.
- The fcc solid solution, (αCo), stable in pure Co above 422 °C. The maximum solubility of Ti in (αCo) is 14.1 at.% at 1190 °C.
- Ti₂Co, an ordered fcc structure containing 96 atoms per unit cell. The homogeneity range of Ti₂Co is no more than about 0.3 at.% about stoichiometry.
- TiCo, with the CsCl structure. TiCo melts congruently at 1325 °C. Its homogeneity range is 49 ± 1 to 55 ± 0.5 at.% Co at 1200 °C.
- Cubic (C15) and hexagonal (C36) Laves phases of approximate stoichiometry TiCo₂, here distinguished as TiCo₂(c) and TiCo₂(h), respectively. TiCo₂(h) is slightly richer in Co; the homogeneity ranges of TiCo₂(c) and TiCo₂(h) are approximately 66.5 to 67.0 at.% Co and 68.75 to 72 at.% Co, respectively.

- TiCo₃, with the ordered fcc AuCu₃ structure. The maximum homogeneity range of TiCo₃ is 75.5 to 80.7 at.% Co.

Ti-Rich Liquidus and Solidus.

[37Kro] reported that the addition of 4.1 at.% Co lowers the melting point of Ti to approximately 1500 °C. By metallographic examination of as-cast samples, [55Orr] estimated that the solidus and liquidus curves meet the eutectic isotherm at 1020 ± 5 °C, with compositions of approximately 14.5 and 23.2 at.% Co, respectively. No other experimental work on the (βTi) solidus and liquidus has been reported. Because of the incompleteness of the experimental data on which the calculations are based, and the large temperature range over which the interpolation is made, the estimated liquidus and solidus should be viewed with considerable skepticism.

(αTi) and (βTi) Phase Equilibria.

The solubility of Co in (αTi) was estimated to be less than 0.8 at.% Co by [55Orr]. (βTi) decomposes by the eutectoid reaction (βTi) ⇌ (αTi) + Ti₂Co at 685 °C [55Orr, 63Kan]. Experimental data on the phase equilibria involving (βTi) are shown in Fig. 2.

The data of [55Orr] for the (βTi) transus are based on the original metallographic findings rather than their diagram, because their two-phase field includes several alloys that they identified as single phase. The diffusion data of [76Str] also lie to the Co-side of the other determinations. The same slight discrepancy between findings in diffusion couples and bulk alloys was noted in the Ti-Ni system. The results of diffusion experiments were therefore considered to be qualitative verification of the metallographic results, but were not used to draw the

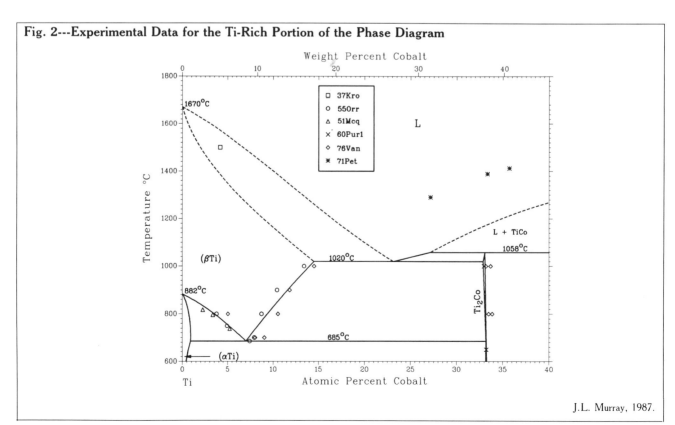

Fig. 2---Experimental Data for the Ti-Rich Portion of the Phase Diagram

J.L. Murray, 1987.

evaluated phase diagram. The data of [51Mcq] were obtained by the hydrogen pressure technique.

Ti₂Co.

[41Wal] identified Ti₂Co as a complex fcc structure with 96 atoms per unit cell. This structure is a ternary phase in several related Ti systems (for example, Ti-Fe-O) [52Ros]. He claimed that Ti₂Co is not an equilibrium phase in Ti-Co; however, at 33 at.% Co, [52Ros] appears to have identified the Laves phase, indicating that there may have been an error in the sample composition. The same author [55Ros] later confirmed that Ti₂Co is a stable equilibrium binary phase. Ti₂Co was also observed by [71Pet] and [76Str] to have the complex fcc structure.

Ti₂Co is formed by the peritectic reaction TiCo + L ⇄ Ti₂Co [55Orr, 71Pet]. [55Orr] determined the peritectic temperature to be 1055 ± 5 °C by examination of as-cast structures and metallographic determination of incipient melting. [71Pet] observed the peritectic solidification by differential thermal analysis at 1060 ± 10 °C. By high temperature X-ray diffraction, [68Ian] observed Ti₂Co in equilibrium with TiCo only to 1050 °C. In Fig. 1, the peritectic temperature is shown as 1058 ± 5 °C.

[60Pur1] made a metallographic and X-ray study of the homogeneity range of Ti₂Co and found that Ti₂Co is not exactly stoichiometric, but contains excess Ti at higher temperatures. From the metallographic data, it was found that the center of the single-phase field lies at 33.00 at.% Co at 1000 °C and between 33.15 and 33.33 at.% Co at 650 °C and that the single-phase field is no more than 0.3 at.% wide. Based on lattice parameter data at different temperatures, [60Pur1] estimated the homogeneity range of Ti₂Co to be less than 0.03 at.%.

Diffusion data [76Str] agreed qualitatively with the metallography of [60Pur1], in that Ti₂Co shifts slightly to Ti-rich concentrations with increasing temperature. The single-phase field was reported to extend from 33.2 to 33.7 at.% Co at 1000 °C and from 33.5 to 33.9 at.% Co at 800 °C. Because the precision of the metallographic study was greater than that obtainable by electron probe microanalysis, the metallographic results of [60Pur1] are preferred as the basis of the assessed diagram. At 1020 °C, the range is shown as 32.9 to 33.1 at.% Co in Fig. 1 and 2.

TiCo.

[39Lav] first identified the CsCl structure of TiCo. [50Duw] found no evidence of ordering even after annealing at 650 and 800 °C, but this work was inconclusive, possibly because of the similarity of the atomic X-ray scattering factors of Ti and Co. [57Phi] and [60Pie] established by X-ray studies that TiCo is indeed ordered. [67Dor] and [75Hut] used neutron scattering to verify the existence of ordering; [75Hut] observed no incipient disordering even up to the melting temperature. Similarly, [73Ara] found no evidence for any transformation up to 1127 °C in electrical resistivity and magnetic susceptibility data.

Select experimental data on the homogeneity range of TiCo are compared in Fig. 3. There is general agreement that the maximum homogeneity range is 49 ± 1 to 55 ± 0.5 at.% Co over the range 900 to 1200 °C. Determinations of the homogeneity range were based on metallographic studies [69Aok, 71Pet, 72Suz] and on high-temperature X-ray diffraction [68Ian], as well as on lattice parameter measurements [60Stu] and microprobe analysis of annealed diffusion couples [76Str]. The metallographic data of [71Pet] are omitted from Fig. 3, because the failure to

Fig. 3---Experimental Data for the TiCo Phase Equilibria

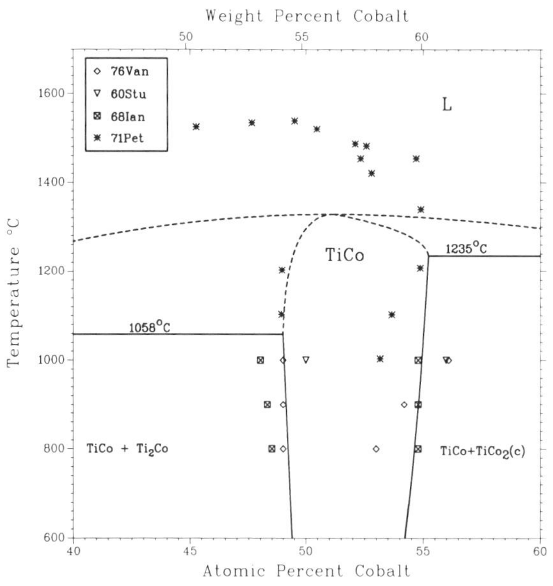

J.L. Murray, 1987.

observe any melting until 1495 °C on the Ti-side of stoichiometry points to the influence of contamination.

There are large inconsistencies in observed melting temperatures of TiCo. [69But] and [71Pet] reported melting points of 1500 and 1520 °C, respectively, for nearly stoichiometric alloys. [68Ian] and [75Hut] reported melting points of 1325 ± 50 °C and approximately 1350 °C, respectively. [69But] reported that above 1200 °C the microstructure "definitely degenerates in most of the compositions studied." On the Ti-side of stoichiometry, [71Pet] observed thermal arrests at approximately 1450 °C, which they attributed to a reaction involving a high-temperature compound X. High-temperature X-ray diffraction [68Ian] showed that Ti₂Co disappeared above 1050 °C, consistent with the peritectic melting reaction.

Laves Phases TiCo₂(c) and TiCo₂(h).

The Laves phases, a cubic variant TiCo₂(c) and a hexagonal variant TiCo₂(h), were recognized by [41Wal]. Both the cubic and hexagonal forms were also observed by [59Dwi], [68Nak], [70Nak], [71Pet], and [76Str]; all investigators who reported both forms found the hexagonal phase to be the stable form at higher Co concentration. Homogeneity ranges were determined by microprobe analyses of annealed diffusion couples [76Str], composition dependence of lattice parameters and magnetic properties [68Nak, 70Nak], and metallography of heat treated and quenched samples [71Pet]. The results, shown in Fig. 4, are in reasonable agreement. [71Pet] determined by differential thermal analysis that TiCo₂(c) and TiCo₂(h) are formed from the melt by peritectic reactions at 1235 ± 10 and 1215 °C, respectively.

However, there are also conflicting reports that suggest that only the hexagonal variant [50Duw] or only the cubic variant [83Uhr] is stable. [50Duw], by X-ray diffraction, found only TiCo₂(h) in a series of alloys containing 63 to 68 at.% Co. [83Uhr] found only TiCo₂(c) formed from the liquid in a 71 at.% Co sample. [59Fou] found only TiCo₂(h)

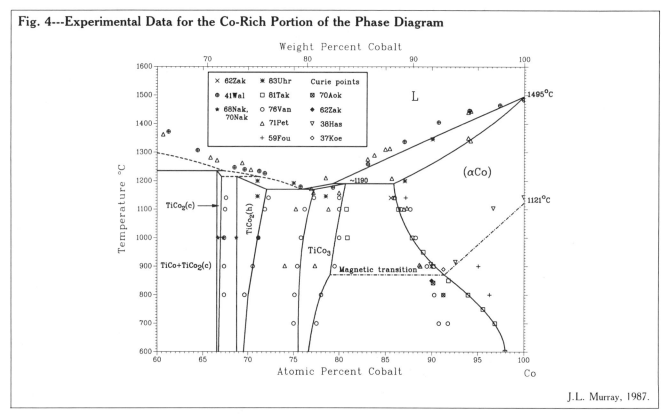

Fig. 4---Experimental Data for the Co-Rich Portion of the Phase Diagram

J.L. Murray, 1987.

in an as-cast 66.1 at.% Co sample. Finally, [72All], by electron microscopy, found that this system exhibits polytypism; the stable hexagonal form can be described as a 4H stacking, and in addition 6H and 12-layer stackings were found. These contradictory observations are at present unresolvable. In the assessed diagram, both forms are tentatively shown as equilibrium phases.

TiCo₃.

[59Fou] first identified $TiCo_3$ as an equilibrium phase of the system; [59Bib] and [61Fou] found it to be ordered with the fcc $CuAu_3$ structure. $TiCo_3$ has approximately a 5 at.% homogeneity range, to the Co-side of stoichiometry. [59Fou] found that the single-phase field narrows and shifts to higher Co content as the temperature increases and that $TiCo_3$ forms by a peritectic reaction between the liquid and (αCo). In the diagram of [62Zak], however, the $TiCo_3$ single-phase field is vertical and ends in a peritectoid reaction with (αCo) and $TiCo_2$(h). [62Zak] gave no details of their experimental procedures. Because the topology of the diagram as given by [59Fou] was verified by [71Pet] and [83Uhr], it is assumed that the phase boundaries shown by [62Zak] were extrapolated from other composition or temperature regions. In this evaluation, the peritectic temperature is taken to be 1210 °C [71Pet], and the peritectic composition of $TiCo_3$ is taken to be ~81 at.% Co.

Solubility of Ti in (αCo).

The diagram of Fig. 1 is based primarily on the data of [81Tak], who recognized the effect of the magnetic transition on the solvus. Drawing an equilibrium solvus from experimental phase stability data is complicated by the intervention of the higher-order ferromagnetic phase transition. The addition of Ti lowers the ferromagnetic transition in (αCo) from 1121 °C at pure Co to 870 °C at the end of the single-phase region (9.5 at.% Ti). As a

result, the solvus curve is "kinked" at the intersection with the line of higher-order transitions. Models based on a mean-field approximation succeed in modeling the overall change in solubility [81Tak]. Theories of critical phenomena that go beyond mean-field theory predict different details for the behavior of the (αCo)/(αCo)+$TiCo_3$) boundary at the ferromagnetic transition; the boundary becomes horizontal below the transition and has a discontinuity in slope.

Phase diagrams based on metallographic data should be re-interpreted in view of this effect, because most investigators have drawn smooth curves between single- and two-phase regions such as would ordinarily occur in the absence of the magnetic transition. For example, the metallographic data of [59Fou] are consistent with the data of [81Tak], though the phase boundaries they drew are not.

Above the ferromagnetic transition, the data of [76Str], [71Pet] and [62Zak] agree. Below the Curie temperature, [62Zak] and [76Str] reported a much higher solubility than did [59Fou] or [81Tak]. [76Str] had difficulty producing large enough precipitates for a composition determination of the (αCo) phase, and discrepancies may be due to the larger uncertainies of the diffusion experiments in this region. [62Zak] did not describe their experimental procedure or tabulate the experimental data.

The (εCo)/(αCo) Transformation.

The allotropic transition in pure Co has associated with it a very low enthalpy change and therefore takes place sluggishly. The mechanism of transformation in pure Co is martensitic [79Bet]. The transformation temperatures on cooling and heating (M_s and A_s, respectively) depend strongly on heating and cooling rates, purity of the material, and other factors such as grain size and plastic

The Co-Ti System

Table 2 M_s and A_s Temperatures for the $(\alpha Co)/(\epsilon Co)$ Transformation

Reference	Composition, at.% Co	Heating (A_s), °C	Cooling (M_s), °C	$T_0 = (M_s + A_s)/2$, °C
[37Koe]	100	425	380	402
	96.3	410	300	355
	93.9	370	260	315
[38Has]	100	...	...	446
	97.9	...	...	437
	95.4	...	...	433
	92.6	...	...	427
[72Jea]	100	...	...	422
	99	...	...	395

deformation. Thus, the equilibrium transition temperature even in the pure metal is not certain, and addition of Ti gives rise to different results depending on the experimental circumstances [59Bib, 79Bet]. The temperature range of Fig. 1 has not been extended to include the $(\epsilon Co)/(\alpha Co)$ transition, because the equilibrium phase relations among TiCo$_3$, (αCo), and (ϵCo) have not been determined experimentally.

The equilibrium transition temperature in pure Co is approximately 422 °C, the average of the martensitic transformation temperatures on heating and cooling—(M_s + A_s)/2 [67Ada]—as quoted in [79Bet], [72Jea], and other references cited in [79Bet]. [37Koe], [62Zak], [72Jea], and [77Nik] reported that both M_s and A_s decrease with increasing Ti content. [38Has] reported that the addition of Ti increases A_s and decreases M_s. Numerical data are given in Table 2. [71Sha] reported that their experiments were not consistent with the diagram of [Hansen], but further details are not available. Because the equilibrium transition of pure Co is in the best agreement with current evaluations of pure Co data [79Bet], the results of [72Jea] are preferred.

Co-Rich Liquidus and Solidus.

The melting temperature of pure Co is 1495 °C [Melt]. The peritectic reaction L + (αCo) $\rightleftarrows$ TiCo$_3$ occurs at about 1200 °C [59Fou], at 1210 °C [71Pet], or at about 1190 °C [83Uhr]. The results of [83Uhr] are preferred. Thermal analysis data for the (αCo) liquidus and solidus were reported by [41Wal], [71Pet], and [83Uhr]. [59Fou] reported the solidus composition at the peritectic temperature as 87.2 at.% Co and estimated the liquidus composition as 80 at.% Co. Experimental uncertainties of approximately ±30 °C should be attached to the high-temperature liquidus and solidus data of [71Pet] and additional uncertainties of 15 °C because of inaccuracy of reading the data from the published graph.

[41Wal] showed the liquidus and solidus as extending down to a eutectic equilibrium between (αCo) and Ti-Co$_2$(h) at 1135 °C, in disagreement with later work, which takes into account TiCo$_3$. The high-temperature data agree reasonably with the later work; however, and it is speculated that the measured liquidus and solidus of [41Wal] represent metastable extensions of the equilibrium phase boundaries. Similarly, [83Uhr] found a metastable reaction at 1145 °C; the as-cast microstructure consisted of coarse precipitates of TiCo$_2$ in a matrix of TiCo$_3$. [83Uhr] speculated that because of slow nucleation of TiCo$_3$ the liquid had solidified without partitioning and that the arrest temperature and the composition of TiCo$_3$ corresponded to a point on the metastable extension of the TiCo$_2$ liquidus.

Table 3 Martensite Transformation Temperatures

Reference	Composition, at.% Co	M_s temperature, °C
[60Sat]	2.6	620
	3.5	520
	4.2	430
[63Kan]	0.81	765
	1.63	670
	2.45	585
	3.27	472
	4.10	415

Metastable Phases

In Ti-rich alloys, the cph $(\alpha' Ti)$ phase can form martensitically from (βTi) during quenching. At sufficiently high Co content, metastable (βTi) can be retained after quenching. The ω phase also appears as an intermediate metastable phase in the decomposition of (βTi) into equilibrium (αTi) + (βTi). In Co-rich alloys, the (αCo) to (ϵCo) transformation takes place martensitically, with large hysteresis and formation of metastable structures. Other metastable transformations in Ti-Co include formation of amorphous alloys during rapid solidification and the formation of modulated microstructures during decomposition of supersaturated (αCo) solutions.

Metastable Phases in Ti-Rich Alloys.

The minimum Co content for which (βTi) can be retained metastably during quenching was reported by [55Orr], [58Rau], [58Swa], and [59Yak]; observed values range from 4.5 [59Yak] to 10 at.% Co [58Swa]. The composition reported depends on the investigators' judgment of whether (βTi) was fully retained, as well as on the quenching rate.

At lower Co content, cph $(\alpha' Ti)$ forms martensitically from (βTi) during quenching. Martensitic transformation temperatures on cooling (M_s) are summarized in Table 3.

In alloys in which the martensitic transition does not occur, the ω phase is formed either during quenching from the (βTi) region, or during aging at temperatures between 300 and 400 °C. Higher annealing temperatures cause the direct precipitation of equilibrium (αTi). [60Pur2] found metallographic evidence of ω coexisting with stress-induced martensite in tempered alloys of 4.5 to 10 at.% Co. [58Rau] produced significant amounts of ω in alloys of 8.3 and 10 at.% Co by annealing at 400 °C and found traces of ω in other alloys. ω phase precipitation involves Co enrichment of the (βTi) phase, with ω phase composition approaching a "critical" or "saturation" composition. [79Shc] associated anomalies in the concentration depen-

63

dence of the superconducting critical temperature and magnetic and electrical properties with a critical concentration of approximately 5 at.% Co. [63Bor] found traces of ω in quenched 3.3 and 4.9 at.% Co alloys. ω phase precipitates are cubic in shape [73Ika], indicating substantial mismatch with the matrix.

(εCo)/(αCo) Martensitic Transformation.

[77Nik] found that, although cph martensite is formed in the composition range 100 to 98 at.% Co, a faulted structure appeared between 98 and 96 at.% Co and that from 96 to 93 at.% Co the martensite has a 126-layer rhombohedral structure. The rhombohedral structure was found to be unstable relative to the cph martensite, to which it transformed under plastic deformation. The M_s for the transformation fell to room temperature at 96 at.% Co, beyond which the multilayer martensite was formed.

Amorphous Alloys

[80Ino] produced a mixed structure of crystalline and amorphous phases near the two eutectic compositions at 23 and 88 at.% Co by rapid solidification. In the composition range 87 to 89 at.% Co, only the amorphous phase was present without any trace of the crystalline phase. Formation of the amorphous phase was attributed to the combination of the deep eutectic and the presence of the compound $TiCo_3$ in the equilibrium assemblage. The crystallization temperature of the 88 at.% Co sample was 504 °C, as determined by differential thermal analysis. The first phase to precipitate from the amorphous phase was not equilibrium $TiCo_3$, as might be expected, but rather $TiCo_2(c)$. Later $TiCo_3$ began to form. The composition of neither precipitate corresponded to a phase boundary in the equilibrium diagram.

Decomposition of Supersaturated (αCo).

In the course of decomposition of supersaturated (αCo) to equilibrium (αCo) + $TiCo_3$, satellites have been found around the fundamental diffraction peaks and modulated microstructures have been observed [59Bib, 61Fou, 66Ber, 70Tka, 72Zak]. This suggests two possible mechanisms for the decomposition. (1) The system first undergoes spinodal decomposition into solute-lean and solute-enriched regions, which then order. The modulated microstructure and satellite peaks would in this case be characteristic of spinodal decomposition and appear in the very early stages of decomposition. (2) The ordered

phase may nucleate in the ordinary way, with precipitates randomly distributed through the matrix. If this is the decomposition mechanism, the modulations are caused by the alignment of the randomly distributed particles along directions determined by the elastic properties of the matrix, and they appear during coarsening. Because the two mechanisms for decomposition give the same microstructure in the later stages of aging, one must examine very rapidly quenched samples to distinguish them. In the Ti-Ni and Ti-Cu systems, spinodal decomposition has been shown to occur.

[72Zak] examined an 88 at.% Co sample quenched from 1150 °C. Satellite peaks were observed, and precipitates were found, some randomly distributed and some lined up in <100> directions. [70Tka] examined a 91 at.% Co sample quenched from 1200 °C. Anomalous point contrast and satellite peaks were observed. These findings are consistent with, but do not prove, the spinodal character of the decomposition process.

Crystal Structures and Lattice Parameters

The crystal structures of the equilibrium and metastable phases are summarized in Table 4. Lattice parameters of the cubic phases are listed in Table 5; lattice parameters of the hexagonal phases are listed in Table 6. The effect of addition of Co to (αTi) on the lattice parameter has not been reported, neither has the effect of addition of Ti to (εCo). A typographical error in [71Pet]—the lattice parameter of $TiCo_2(h)$—has been corrected in Table 5. The rhombohedral martensite was reported to have a lattice parameter of 8.653 nm with α = 140″ [77Nik].

Thermodynamics

[80Esi] measured enthalpies of mixing of liquid alloys containing 45 to 100 at.% Co at 2000 K by high-temperature calorimetry. The experimental data are fit within reported experimental uncertainties by the polynomial expansion:

$$\Delta_{mix}H (x, T = 2000 \text{ K}) = x(1 - x) [-145\,840 + -19\,933 \\ (1 - 2x)] \text{ J/mol}$$

where x is the atomic fraction of Co. No experimental data have been reported on partial Gibbs energies or heats of formation of the compounds.

Table 4 Crystal Structures of the Ti-Co System

Phase	Homogeneity range, at.% Co			Pearson symbol	Space group	Struktur- bericht designation	Prototype	Reference
(αTi)	0	to	0.8	hP2	$P6_3/mmc$	A3	Mg	[Pearson2]
(βTi)	0	to	14.5	cI2	$Im3m$	A2	W	[Pearson2]
Ti_2Co	32.9	to	33.3	cF96	$Fd3m$	$E9_3$	Fe_3W_3C	[41Wal]
TiCo	49	to	55	cP2	$Pm3m$	B2	CsCl	[75Hut]
$TiCo_2(c)$	66.5	to	67	cF24	$Fd3m$	C15	$MgCu_2$	[41Wal]
$TiCo_2(h)$	68.75	to	72	hP24	$P6_3/mmc$	C36	$MgNi_2$	[41Wal]
$TiCo_3$	75.5	to	80.7	cP4	$Pm3m$	$L1_2$	$CuAu_3$	[59Fou]
(αCo)	85.6	to	100	cF4	$Fm3m$	A1	Cu	[Pearson2]
(εCo)	~ 99	to	100	hP2	$P6_3/mmc$	A3	Mg	[Pearson2]
ω	(a)			hP3	$P6/mmm$	...	ωMnTi	(b)
(α″Co)	(a)			(c)	...	...	...	[77Nik]

(a) Metastable. (b) The "ideal" ω structure is hexagonal, but a distorted trigonal form has also been observed in some Ti systems. The structure of ω in Ti-Co has not been definitively established. (c) Rhombohedral.

[70Kau], [75Kau], and [78Kau] presented Gibbs energies from which the phase diagram was calculated. [70Kau] used the regular solution approximation for the solution phases; [75Kau] and [78Kau] introduced excess entropy contributions. Compounds were represented as stoichiometric. [83Uhr] presented a calculation of the Co-rich part of the diagram based on the regular solution approximation for the liquid and fcc Gibbs energies and on a sublattice representation for TiCo$_3$ and TiCo$_2$. Gibbs energies used by [83Uhr] and [78Kau] differ substantially (see below and Table 7).

[79Lio] applied a theory of radiation induced order-disorder transitions to the TiCo$_3$/(αCo) transition. Based on the phase diagram calculation of [78Kau], they calculated modified phase diagrams showing shifts in the solvus under various radiation dosages. TiCo$_3$ was predicted not to completely disorder, and the solvus was predicted to shift to higher Ti content.

Although optimization techniques are available by which Gibbs energies can be constructed to essentially reproduce the assessed phase diagram, it appears that further calculation is not justified at this time. This judgment is based on the lack of heat of formation and partial Gibbs energy data and on qualitative uncertainties and discrepancies in the phase diagram. For example, the liquidus is determined, even approximately, only for the most Co-rich alloys. Also, the question of the relative stabilities of TiCo$_2$(c) and TiCo$_2$(h) remains to be resolved. The Gibbs energies calculated by [78Kau] are consistent with the enthalpy of mixing data of [80Esi]; however, the calculated melting point of TiCo was greater than 1500 °C, which is believed to be too high. The Gibbs energies calculated by [83Uhr] imply phase boundaries in quantitative agreement with data for Co-rich alloys; but phase equilibria over the whole composition range were not calculated and the regular solution Gibbs energy for the liquid is not consistent with the measurements of [80Esi].

Parameters of the Gibbs energy expansions presented by [78Kau] and [83Uhr] are listed in Table 7. For the solution phases, the expansions are defined by:

$$G(i) = (1 - x) G^0(\text{Ti},i) + xG^0(\text{Co},i) + RT [x \ln x + (1 - x) \ln (1 - x)] + x(1 - x) [B(i) + C(i) (1 - 2x)]$$

where x is the atomic fraction of Co; G^0 is the Gibbs energies of the pure elements; and B and C are temperature-dependent coefficients of the expansion in Legendre polynomials.

For the sublattice representation of the ordered phases TiCo$_3$ and TiCo$_2$ [83Uhr], the Gibbs energy is defined by:

$$G_m^\gamma = \varphi [y_j G_j^{0,\gamma} + y_j y_k L_{jk}^\alpha + RT/4 (y_j \ln y_j)]$$

where the y_j are the site fractions of component i in the sublattices of the ordered phase Co$_a$(Co,Ti)$_b$.

Magnetic Properties

(αCo).

Pure (αCo) is ferromagnetic with a Curie temperature (T_C) of 1121 °C; T_C is lowered by the addition of Ti. Reported values of T_C are given in Table 8. These values are mutually consistent and lie on a straight line. The alloy examined by [70Aok] had been annealed at 1000 °C. The 10 at.% Ti alloy examined by [62Zak], although represented in his phase diagram as lying on the (αCo) solvus at 850 °C, should be within the two-phase (αCo) + TiCo$_3$ region at that temperature. In this evaluation, the intersection of the line of magnetic transitions and the (αCo) solvus is chosen as 870 °C and 9.5 at.% Ti (see Fig. 1 and 3). The line of magnetic transitions is drawn as a straight line.

Table 5 Lattice Parameters of Cubic Phases

Phase	Reference	Composition, at.% Co	Lattice parameter, nm
(βTi)	[Pearson]	0.0	3.3132
Ti$_2$Co	[50Duw]	33.3	1.1283
	[60Pur1]	33.1	1.129
	[61Mat]	33.3	1.130
	[71Pet]	33.3	1.130
TiCo	[57Phi]	50	0.2991 ± 0.0001
	[69Aok]	50	0.2995
		55.5	0.2970
	[50Duw]	50	0.2988
	[71Pet]	54.5	0.2965 ± 0.0001
		49.3	0.3002 ± 0.0001
		53	0.2965 ± 0.0001
		49.3	0.2997 ± 0.0001
	[69But]	50	0.29960 ± 0.00003
		52	0.29877 ± 0.00003
TiCo$_2$(c) ..	[39Wal]	66.7	0.6691 ± 0.0006
	[59Dwi]	66.7	0.6706
	[71Pet]	66.7	0.6716 ± 0.0001
TiCo$_3$	[59Fou]	82.4	0.3603
		79.0	0.3609
		75.0	0.3613
	[70Aok]	79.5	0.3602
		75	0.3614
	[71Pet]	79	0.3603 ± 0.0001
		77.5	0.3606 ± 0.0001
		75.5	0.3621 ± 0.0001
		74.5	0.3618 ± 0.0001
(αCo)	[Pearson2]	100	0.35441
	[59Fou]	100	0.3554
		93.1	0.3573
		89.2	0.3582

Table 6 Lattice Parameters of Hexagonal Phases

Phase	Reference	Composition, at.% Co	Lattice parameters, nm	
			a	c
(αTi)	[Pearson2]	0	0.29564	0.46928
TiCo$_2$(h)	[39Wal]	67.7	0.4715 ± 0.0007	1.537 ± 0.002
	[50Duw]	66.7	0.472	1.5392
	[59Fou]	66.7	0.472	1.540
	[71Pet]	69	4.732	1.5427
(αCo)	[Pearson2]	100	0.25054 to 0.25073	0.40892 to 0.40698

Table 7 Thermodynamic Parameters of the Ti-Co System

Lattice stability parameters of Ti and Co [77Kau, 78Kau]

G^0(Ti,cph) = − 20 585 + 12.134 T
G(Co,cph) = − 17 591 + 10.334 T

G^0(Ti,L) = 0
G^0(Co,L) = 0

G^0(Ti,bcc) = − 16 234 + 8.468 T
G^0(Co,bcc) = − 11 339 + 6.568 T

G^0(Ti,fcc) = − 17 238 + 12.134 T
G^0(Co,fcc) = − 17 155 + 9.706 T

Compounds [78Kau]

G(Ti$_2$Co) = −39 662 + 8.36 T (referenced to fcc)
G(TiCo) = −92 154 + 6.224 T (referenced to bcc)
G(TiCo$_2$(h)) = −36 410 + 7.152 T (referenced to cph)
G(TiCo$_3$) = −31 223 + 6.504 T (referenced to cph)

Ti-Co interaction parameters [78Kau]

B(L) = −154 808 + 41.84 T B(cph) = −111 503 + 48.12 T
C(L) = 12 552 − 12.55 T C(cph) = 47 488 − 18.82 T

B(bcc) = −126 357 + 41.84 T B(fcc) = −133 260 + 41.84 T
C(bcc) = 16 736 − 12.55 T C(fcc) = 21 966 − 12.55 T

Gibbs energies according to [83Uhr]

G^0(Ti,fcc) = −17 238 + 12.134 T B(L) = 60 800
G^0(Co,fcc) = −17 155 + 9.706 T B(fcc) = 45 300

G^0(TiCo$_2$) = −6435 − 8.03 T + 0.29 G(Ti,fcc) + 0.71 G(Co,fcc)
G^0(TiCo$_3$) = −15 900 + 0.25 G(Ti,fcc) + 0.75 G(co,fcc)

L(TiCo$_3$) = 9450 − 4.2 T

Note: Values are given in J/mol and J/mol·K.

Intermetallic Compounds

[69Aok] and [70Aok] studied the magnetic properties of alloys of compositions between 79.5 and 75 at.% Co annealed at 1000 °C. They found stoichiometric TiCo$_3$ to be paramagnetic and found that nonstoichiometric TiCo$_3$ becomes ferromagnetic, with T_C increasing with increasing Co content and equal to 38 K at 78.6 at.% Co.

[60Nev] found no measurable magnetization in TiCo at 7 K. Stoichiometric TiCo was found by [69But] to follow a Curie-Weiss law, with a large negative Weiss constant (−499 K) from 50 to 400 K. The samples used by [69But], however, showed evidence of contamination; the melting points were very high (approximately 1500 °C) and none of the alloys were single phase. [73Ara] extended the temperature range studied and found significant deviations from Curie-Weiss behavior. [69Aok] found the susceptibility of TiCo to be represented by a modified Curie-Weiss expression:

$$X = X_0 - \alpha T + C/(T - \theta)$$

The Curie constant C varies strongly with composition, being approximately zero at the stoichiometric composition. The strong composition dependence of the magnetic properties means that conflicting results are inevitable, unless the magnetic properties are studied as a function of composition.

[70Nak] found TiCo$_2$(c) to be ferromagnetic with a Curie temperature of 44 K at 71 at.% Co and 18 K at 68.7 at.% Co. [70Nak] identified TiCo$_2$(h) as antiferromagnetic (with a Nèel temperature of 43 K); but later work at the same laboratory [77Ike] on purer specimens showed that TiCo$_2$(h) is Pauli paramagnetic.

Cited References

37Koe: W. Koester and E. Wagner, "The Influence of Aluminum, Titanium, Vanadium, Copper, Zinc, Tin and Antimony on the Polymorphic Transformation of Cobalt," *Z. Metallkd., 29,* 230-232 (1937) in German. (Equi Diagram; Experimental)

37Kro: W. Kroll, "Titanium Alloys," *Z. Metallkd., 29,* 189-192 (1937) in German. (Equi Diagram; Experimental)

38Has: U. Haschimoto, "The Effect of Various Elements on the α/β Allotropic Transformation Point of Cobalt," *Nippon Kinzoku Gakkai-Shi, 2,* 67-77 (1938) in Japanese. (Equi Diagram; Experimental)

39Lav: F. Laves and H.J. Wallbaum, "Crystal Chemistry of Titanium Alloys," *Naturwissenschaften, 27,* 674 (1939) in German. (Crys Structure; Experimental)

39Wal: H.J. Wallbaum and H. Witte, "The Crystal Structure of TiCo$_2$," *Z. Metallkd., 31,* 185-187 (1939) in German. (Equi Diagram, Crys Structure; Experimental)

41Wal: H.J. Wallbaum, "The Systems of the Iron-Group Metals with Titanium, Zirconium, Niobium, and Tantalum," *Arch. Eisenhuttenwes., 10,* 521-526 (1941) in German. (Equi Diagram; Experimental)

50Duw: P. Duwez and J.L. Taylor, "The Structure of Intermediate Phases in Alloys of Titanium with Iron, Cobalt, and Nickel," *Trans. AIME, 188,* 1173-1176 (1950). (Crys Structure; Experimental)

51Mcq: A.D. McQuillan, "The Effect of the Elements of the First Long Period on the α/β Transformation in Titanium," *J. Inst. Met., 80,* 363-368 (1951-1952). (Experimental)

52Ros: W. Rostoker, "Observations on the Occurrence of Ti2X Phases," *Trans. AIME, 194,* 209-210 (1952). (Equi Diagram; Experimental)

55Orr: F.L. Orrell and M.G. Fontana, "The Titanium-Cobalt System," *Trans. ASM, 47,* 554-564 (1955). (Equi Diagram; Experimental)

57Phi: T.V. Philip and B.A. Beck, "CsCl-Type Ordered Structures in Binary Alloys of Transition Elements," *Trans. AIME, 209,* 1269-1271 (1957). (Crys Structure; Experimental)

58Rau: E. Raub and H. Beeskow, "Tempering and Phase Changes Connected with β/α-Titanium Transformation in Titanium-Cobalt Alloys," *Z. Metallkd., 49*(4), 185-190 (1958). (Meta Phases; Experimental)

58Swa: P.R. Swann and J.G. Parr, "Phase Transformations in Titanium-Rich Alloys of Titanium and Cobalt," *Trans. AIME, 212,* 276-279 (1958). (Meta Phases; Experimental)

59Bib: H. Bibring and J. Manenc, "Structure of Phases in the System Co-Ti," *C.R. Hebd. Seances Acad. Sci., 249,* 1508-1510 (1959) in French. (Equi Diagram; Experimental)

59Dwi: A.E. Dwight, "CsCl-Type Equiatomic Phases in Binary Alloys of Transition Elements," *Trans. AIME, 215,* 283-286 (1959). (Crys Structure; Experimental)

59Fou: R.W. Fountain and W.D. Forgeng, "Phase Relations and Precipitation in Cobalt-Titanium Alloys," *Trans. AIME, 215,* 998-1008 (1959). (Equi Diagram; Experimental)

59Yak: F.W. Yakymyshyn, G.R. Prudy, R. Taggart, and J.G. Parr, "The Relationship Between the Constitution and Mechanical Properties of Titanium-Rich Alloys of Titanium and Cobalt," *Trans. ASM, 53,* 283-294 (1959). (Meta Phases; Experimental)

60Nev: M.V. Nevitt, "Magnetization of the Compound TiFe," *J. Appl. Phys., 31*(1), 155-157 (1960). (Equi Diagram; Experimental)

Table 8 Curie Temperatures of (αCo)

Reference	Composition, at.% Co	Temperature, °C
[37Koe]	91.3	890
[38Has]	100	1142
	96.7	1103
	92.6	916
	89.9	908
[62Zak]	90	850
[70Aok]	90.17	842

The Co-Ti System

60Pie: P. Pietrokowsky and F.G. Youngkin, "Ordering in the Intermediate Phases TiFe, TiCo, and TiNi," *J. Appl. Phys.*, *31*(10), 1763-1766 (1960). (Meta Phases; Experimental)

60Pur1: G.R. Purdy and R. Taggart, "A Substructure in ω-Hardened Alloys of Cobalt in Titanium," *Trans. AIME, 218*, 186-187 (1960). (Meta Phases; Experimental)

60Pur2: G.R. Purdy and J.G. Parr, "The Composition Range of Ti₂Co," *Trans. AIME, 218*, 225-227 (1960). (Equi Diagram; Experimental)

60Sat: T. Sato, S. Hukai and Y.C. Huang, "The Ms Points of Binary Titanium Alloys," *J. Aust. Inst. Met.*, *5*(2), 149 (1960). (Meta Phases; Experimental)

60Stu: H.P. Stuewe and Y. Shimomura, "Lattice Constants of the Cubic Phases FeTi, CoTi, NiTi," *Z. Metallkd.*, *3*, 180-181 (1960) in German. (Meta Phases; Experimental)

61Fou: R.W. Fountain, G.M. Faulring, and W.D. Forgeng, "Structural Relationships Between Precipitate and Matrix in Cobalt-Rich Cobalt-Titanium Alloys," Trans. AIME, 221, 747-751 (1961). (Meta Phases; Experimental)

61Mat: B.T. Matthias, V.B. Compton, and E. Corenzwit, "Some New Superconducting Compounds," *J. Phys. Chem. Solids*, *19*(1-2), 130-133 (1961). (Crys Structure; Experimental)

62Zak: E.K. Zakharov and B.G. Livshitz, "Phase Diagram of the System Cobalt-Chromium-Titanium," *Izv. Akad. Nauk SSSR, OTN*, (5), 143-150 (1962) in Russian. (Equi Diagram; Experimental)

63Bor: B.A. Borok, E.K. Novikova, L.S. Golubeva, N.A. Ruch-'eva, E.K. Novikova, and R.P. Shchegoleva, "Dilatometric Study of Binary Alloys of Titanium," *Metalloved. Term. Obrab. Met.*, (2), 32-36 (1963) in Russian; TR: *Met. Sci. Heat Treat.*, (2), 94-98 (1963). (Equi Diagram; Experimental)

63Kan: H. Kaneko and Y.C. Huang, "Continuous Cooling Transformation Characteristics of Titanium Alloys of Eutectoidal Type (i)," *J. Jpn. Inst. Met.*, *27*, 393-397 (1963) in Japanese. (Equi Diagram, Meta Phases; Experimental)

66Ber: A.L. Berezina and K.V. Chuistov, "Nature of the Structural Changes at Early Stages of Disintegration of the Supersaturated Solid Solution of Titanium in Cobalt," *Fiz. Met. Metalloved.*, *22*(3), 404-409 (1966) in Russian; TR: *Phys. Met. Metallogr.*, *22*(3), 84-89 (1966). (Meta Phases; Experimental)

67Ada: R. Adams and C. Altstetter, "Enthalpy Changes in the Allotropic Phase Changes in Cobalt," paper presented to the Metallurgical Society of AIME, Oct (1967). (Equi Diagram; Experimental)

67Dor: A.V.Doroshenko, S.A. Nemnonov, and S.K. Sidorov, "Neutron Diffraction Analysis of the Structure of the Alloys TiFe and TiCo," *Fiz. Met. Metalloved.*, *23*(3), 562-563 (1967) in Russian; TR: *Phys. Met. Metallogr.*, *23*(3), 168-169 (1967). (Crys Structure; Experimental)

68Ian: A. Iannucci, A.A. Johnson, E.J. Hughes, and P.W. Barton, "Study of the Solubility Limits of the Compound TiCo Using High-Temperature X-Ray Diffractometry," *J. Appl. Phys.*, *39*(5), 2222-2224 (1968). (Equi Diagram; Experimental)

68Nak: T. Nakamichi, "Ferro- and Antiferromagnetism of the Laves Phase Compound in Fe-Ti Alloy System," *J. Phys. Soc. Jpn.*, *25*, 1189 (1968). (Equi Diagram; Experimental)

69Aok: Y. Aoki, T. Nakamichi, and M. Yamamoto, "Magnetic Properties of Cobalt-Titanium Alloys with the CsCl-Type Structure," *J. Phys. Soc. Jpn.*, *27*(6), 1455-1458 (1969). (Equi Diagram; Experimental)

69But: S.R. Butler, J.E. Hanlon, and R.J. Wasilewski, "Electric and Magnetic Properties of TiCo and Related Compositions," *J. Phys. Chem. Solids*, *30*, 281-286 (1969). (Equi Diagram; Experimental)

70Aok: Y. Aoki, "Magnetic Properties of the Intermetallic Compound with the Cu₃Au-Type Structure in Cobalt-Titanium Alloy System," *J. Phys. Soc. Jpn.*, *28*(6), 1451-1456 (1970). (Equi Diagram; Experimental)

70Kau: L. Kaufman and H. Bernstein, *Computer Calculation of Phase Diagrams*, Academic Press, New York (1970). (Thermo; Theory)

70Nak: T. Nakamich, Y. Aoki, and M. Yamamoto, "Ferromagnetic Properties of the Intermetallic Compound with the Hexagonal Laves-Phase Structure in Cobalt-Titanium System," *J.*

Phys. Soc. Jpn., *28*(3), 590-595 (1970). (Equi Diagram; Experimental)

70Tka: O.Y. Tkachenko and K.V. Chuistov, "A Study of the Morphology of Homogeneous Precipitation in the Alloy Co–9 at.% Ti," *Fiz. Met. Metalloved.*, *29*(4), 834-840 (1970) in Russian; TR: *Phys. Met. Metallogr.*, *29*(4), 159-165 (1970). (Meta Phases; Experimental)

71Pet: V.V. Pet'kov and M.V. Kireev, "Intermediate Phases in the Titanium-Cobalt System," *Metallofizika*, (33), 107-115 (1971) in Russian. (Equi Diagram; Experimental)

71Sha: I.M. Sharshakov, D.E. Soldatenko, L.V. Nikiforova, and V.N. Belko, "Phase Transformations in Cobalt-Molybdenum and Cobalt-Titanium Alloys," *Trudy Asp. Fiz.-Tekn. Fak. Voron. Polit. Inst.*, (2), 48-52 (1971) in Russian. (Meta Phases; Experimental)

72All: C.W. Allen, P. Delavignette, and Amelinckx, "Electron Microscopic Studies of the Laves Phases TiCr₂ and TiCo₂," *Phys. Status Solidi A*, *9*, 237-246 (1972). (Metas Phases; Experimental)

72Jea: R. Jeanjean, J. Dubois, Y. Fetiveau, and R. Riviere, "Enthalpy Measurements on the Allotropic Transformation in Cobalt Alloys," *Mem. Sci. Rev. Metall.*, *69*(2), 165-169 (1972) in French. (Thermo; Experimental)

72Suz: T. Suzuki and K. Masumoto, "Composition Dependence of Density in NiTi and CoTi," *Metall. Trans.*, *3*, 2009-2010 (1972). (Crys Structure; Experimental)

72Zak: M.I. Zakharova and N.A. Vasil'yeva, "A Study of the Decomposition of the Solid Solution in Cobalt-Titanium, Iron-Cobalt-Titanium-Aluminium and Iron-Nickel-Titanium-Aluminium Alloys," *Fiz. Met. Metalloved.*, *33*(5), 1027-1033 (1972) in Russian. (Meta Phases; Experimental)

73Ara: S. Arajs, A.A. Stelmach, and M.C. Martin, "Magnetic Susceptibility and Electrical Resistivity of TiCo at Elevated Temperatures," *J. Less-Common Met.*, *32*, 178-180 (1973). (Equi Diagram; Experimental)

73Ika: H. Ikawa, S. Shin, M. Miyagi, and M. Morikawa, "Some Fundamental Studies on the Phase Transformation from β Phase to α Phase in Titanium Alloys," Sci. Technol. Appl. Titanium, Proc. Int. Conf., R. I. Jaffee, Ed., 1545 (1973). (Meta Phases; Experimental)

75Hut: H. Huthmann and G. Inden, "High-Temperature Neutron Diffraction on FeTi and CoTi," *Phys. Status Solidi A*, *28*, K129 (1975). (Equi Diagram, Thermo; Experimental)

75Kau: L. Kaufman and H. Nesor, "Calculation of Superalloy Phase Diagrams Part III," *Metall. Trans. A*, *6*, 2115-2122 (1975). (Thermo; Theory)

76Str: P.J. Van der Straten, G.F. Bastin, F.J. Van Loo, and G.D. Rieck, "Phase Equilibria and Interdiffusion in the Cobalt-Titanium System," *Z. Metallkd.*, *67*(3), 152-157 (1976). (Equi Diagram; Experimental)

77Ike: K. Ikeda, "Electrical Resistivity of Laves Phase Compounds Containing Transition Elements. II. Co₂A (A = Ti, Y, Zr, and Nb)," *J. Phys. Soc. Jpn.*, *42*(5), 1541-1546 (1977). (Equi Diagram; Experimental)

77Nik: B.I. Nikolin, "Formation of Multilayer (α') and Imperfect Martensitic Phases in Cobalt-Titanium Alloys," *Dokl. Akad. Nauk SSSR*, *233*, 587-590 (1977) in Russian; TR: *Sov. Phys. Dokl.*, *22*(4), 226-228 (1977). (Meta Phases; Experimental)

78Kau: L. Kaufman, "Coupled Phase Diagrams and Thermochemical Data for Transition Metal Binary Systems - III," *Calphad*, *2*(2), 117 (1978). (Equi Diagram, Thermo; Theory)

79Bet: W. Betteridge, "The Properties of Metallic Cobalt," *Prog. Mater. Sci.*, *24*(2), 51-142 (1979). (Equi Diagram; Experimental)

79Lio: K.Y. Liou and P. Wilkes, "The Radiation Disorder Model of Phase Stability," *J. Nucl. Mater.*, *87*, 317-330 (1979). (Equi Diagram, Thermo; Theory)

79Shc: A.S. Shcherbakov, A.F. Prekul, N.V. Volkenshtein, and A.L. Nikolaev, "Electronic Structure and Anomalies of the Electrical and Magnetic Properties of Alloys of Ti with Co and Ni," *Fiz. Tverd. Tela*, *21*, 676-681 (1979); TR: *Sov. Phys. Solid State*, *21*(3), 398-401 (1979). (Equi Diagram; Experimental)

80Esi: Yu.O. Esin, M.G. Valishev, P.B. Gel'd, and M.S. Petrush-evskii, "Calculation of Heat of Formation of Binary Alloys of

Iron, Cobalt, and Nickel, with Titanium," *Zh. Fiz. Khim., 54*(9), 2267-2270 (1980). (Equi Diagram; Experimental)

80Ino: A. Inoue, K. Kobayashi, C. Suryanarayana, and T. Masumoto, "An Amorphous Phase in Co-Rich Co-Ti Alloys," *Scr. Metall., 14*, 119-123 (1980). (Meta Phases; Experimental)

81Tak: T. Takayama, M.Y. Way, and T. Nishizawa, "Effect of Magnetic Transition on the Solubility of Alloying Elements in BCC Iron and FCC Cobalt," *Trans. Jpn. Inst. Met., 22*(5), 315-325 (1981). (Experimental)

83Uhr: B. Uhrenius and K. Forsen, "On the Co-Ti System," *Z. Metallkd., 74*(9), 610-615 (1983). (Thermo; Experimental)

The Cr-Ti (Chromium-Titanium) System

51.996   47.88

By J.L. Murray

Equilibrium Diagram

The equilibrium solid phases of the Ti-Cr system are:

- The cph (αTi) solid solution, in which Cr has a small solubility.
- The bcc (βTi,Cr) solid solution. In a narrow temperature range below the congruent melting point, Ti and Cr are completely miscible in the bcc phase. For the sake of brevity, the bcc solid solution will be designated β, unless a particular composition range is indicated. In this case, the bcc phase will be called (βTi), (Cr), or (βTi,Cr), as required.

- αTiCr$_2$, βTiCr$_2$, and γTiCr$_2$ are Laves phases with the *C*15, *C*14, and *C*36 structures, respectively. αTiCr$_2$ has the lowest Cr content, and βTiCr$_2$ and γTiCr$_2$ are low- and high-temperature forms of the hexagonal Laves phase.

The present reassessment of the diagram incorporates recent work on the Laves phases and a new thermodynamic analysis of the system. The assessed diagram is shown in Fig. 1, and special points of the system are summarized in Table 1.

Fig. 1---Assessed Ti-Cr Phase Diagram

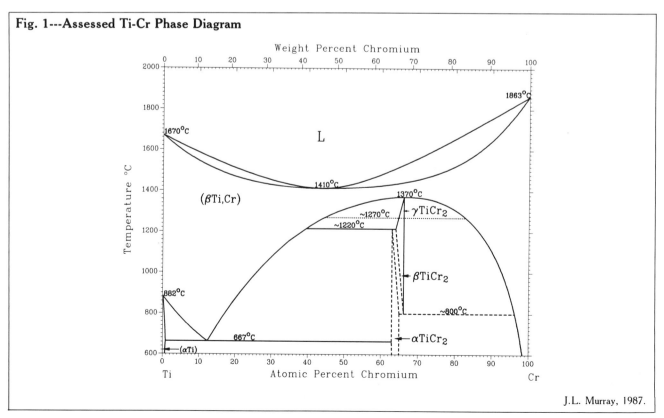

J.L. Murray, 1987.

The Cr-Ti System

Fig. 2---Detail of the Liquidus and Solidus Showing Experimental Data

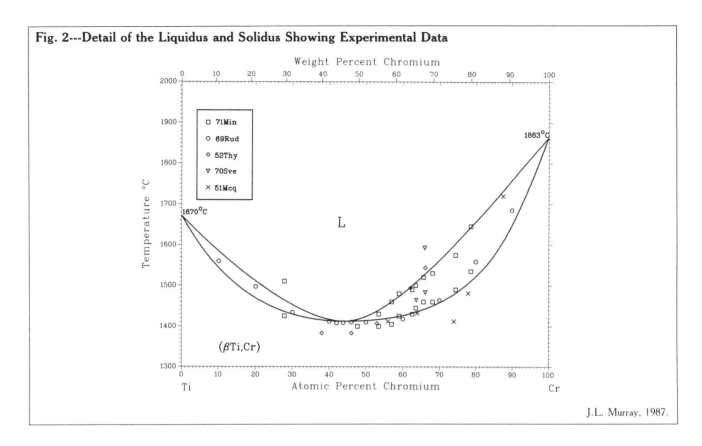

J.L. Murray, 1987.

Table 1 Special Points of the Ti-Cr Phase Diagram

Reaction	Compositions of the respective phases, at.% Cr			Temperature, °C	Reaction type
L ⇌ (βTi,Cr)		44		1410 ± 5	Congruent
(βTi) ⇌ γTiCr$_2$		~ 66		1370 ± 10	Congruent
(βTi) ⇌ (αTi) + αTiCr$_2$	12.5 ± 0.5	0.6	~ 63	667 ± 10	Eutectoid
(βTi) + βTiCr$_2$ ⇌ αTiCr$_2$	39	~ 63	~ 65	~ 1220	Peritectoid
γTiCr$_2$ ⇌ βTiCr$_2$		~ 65 to 66		~ 1270	Unknown
βTiCr$_2$ ⇌ (Cr) + αTiCr$_2$	~ 65	96	~ 66	~ 800	Eutectoid
L ⇌ (βTi)		0		1670	Melting point
(βTi) ⇌ (αTi)		0		882	Allotropic transformation
L ⇌ (Cr)		100		1863 ± 20	Melting point

and [69Rud] determined solidus temperatures by the incipient melting technique, and [62Sve] and [71Min] determined solidus temperatures by thermal analysis. The experimental data are compared to the assessed diagram in Fig. 2, and the reported congruent points are summarized in Table 2. The reported congruent temperatures range between 1380 °C [52Thy] and 1412 ± 5 °C [69Rud]. The assessed value is 1410 ± 5 °C.

Notable discrepancies in the data are high liquidus temperatures for some Ti-rich alloys (e.g., [62Sve]) and low melting points for Cr-rich alloys [51Mcq]. The latter observations can be attributed to inhomogeneity of the alloys. The former discrepancy is attributable to contamination and experimental uncertainty in the high-temperature measurement. The assessed diagram is based on least-squares optimization of thermodynamic functions with respect to the experimental data. Because of the good agreement with chosen experimental data over most of

the composition range, the calculated diagram is preferred where there are slight discrepancies.

(βTi)/(αTi) Equilibrium and the Eutectoid Reaction.

Experimental determinations of the β transus and temperature of the eutectoid reaction β ⇌ (αTi) + αTiCr$_2$ are summarized in Table 3. Experimental data on the β transus are shown in Fig. 3. In the present assessment, the eutectoid point is drawn at 12.5 ± 0.5 at.% Cr and 66

Liquidus and Solidus.

[52Thy], [62Sve], and [71Min] determined the liquidus using high-temperature thermal analysis; [51Mcq] ± 10 °C, based primarily on [61Erm] and also on [52Cuf] and [52Thy].

Contamination of alloys by gaseous impurities stabilizes (αTi) with respect to β and raises the temperature of the β transus. Data from several sources [52Duw, 58Bag1, 59Gol, 62Mik1] illustrate this effect, and they were there-

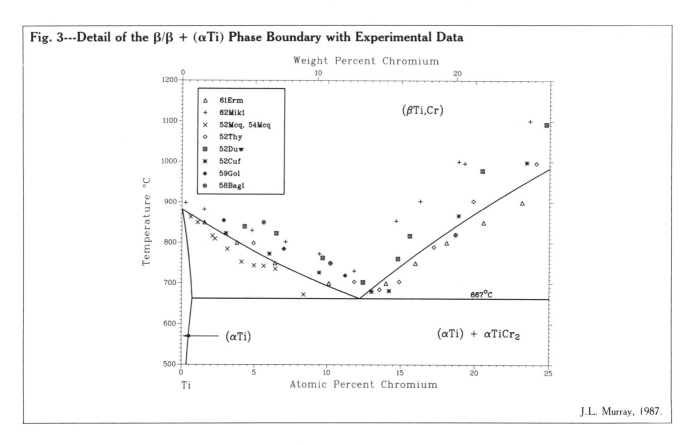

Fig. 3---Detail of the β/β + (αTi) Phase Boundary with Experimental Data

J.L. Murray, 1987.

Table 2 Experimental Studies of the Liquidus and Solidus

	Congruent melting		
Reference	Composition, at.% Cr	Temperature, °C	Method
[51Mcq]	48.7 to 53.5	1400 ± 10	Incipient melting (solidus)
[52Thy]	43	1380	Thermal analysis (liquidus)
[57Kor, 62Mik1]		~ 1400	Thermal analysis
[62Sve]	53	~ 1400	Thermal analysis (liquidus/solidus)
[69Rud]	44	1412 ± 5	Incipient melting (solidus)
[71Min]	48	1410	Thermal analysis (liquidus/solidus)

Table 3 Experimental Determinations of the (βTi) Eutectoid Temperature

Reference	Temperature, °C	Technique/comment
[52Thy]	685 ± 15	Metallography
[52Duw]	660 ± 10	Metallography
[52Cuf]	670 ± 5	Dilatometry
[54Fro]	675 to 700	Appearance of β in (αTi) + αTiCr$_2$ alloys on heating
[54Mcq]	~ 20 to 550	Hydrogen pressure, metallography (rapid quenching)
[57Kor]	728 ± 10	Thermal analysis
[58Bag1]	~ 670	Metallography
[59Gol]	670 ± 5	Metallography
[61Erm]	667 ± 10	Appearance of β in (αTi) + αTiCr$_2$ alloys on heating, metallography (rapid quenching)
[62Mik1]	690 to 700	Resistivity

fore excluded from the calculation of the assessed boundary.

[54Mcq] has argued that the eutectoid reaction occurs below 550 °C. She claimed that apparent two-phase microstructures were due to precipitation during quenching, and therefore, an extremely rapid quench is required. She also found that rapid quenching experiments were corroborated by hydrogen pressure experiments. These conclusions are rejected in the present evaluation for the following reasons. First, the author of [54Mcq] appears to have failed to work the alloys to hasten equilibration. Second, the hydrogen pressure technique rather consistently underestimates the temperature of the β transus in several Ti alloy systems. Finally, there is no evidence of anomalies in thermodynamic properties of the bcc phase, which would be required by the diagram of [54Mcq].

The determination of the eutectoid temperature by [61Erm] was made by testing for the disappearance of αTiCr$_2$ in equilibrated (αTi) + αTiCr$_2$ alloys, thus avoiding the difficulties noted by [54Mcq]. For determination of the β transus, [61Erm] also used a rapid quenching technique, but found only a minor discrepancy with data obtained by conventional metallographic techniques.

Experimental data on the solubility of Cr in (αTi) are summarized in Table 4. Although the details of the phase boundaries have not been studied, there is agreement that the maximum solubility of Cr in (αTi) is about 0.6 at.%.

The β/(αTi) boundaries of the assessed diagram are drawn from thermodynamic calculations in which Gibbs energies were optimized with respect to select thermodynamic and phase diagram data. The assessed composition and temperature of the eutectoid point are 12.5 ± 0.5 at.% Cr and 667 ± 10 °C. The calculated values shown in Fig. 1, 3, and 4 are 12.2 at.% Cr and 663 °C.

Laves Phases.

Concerning the number and homogeneity ranges of equilibrium intermetallic phases, there is considerable discrepancy. Three Laves phases α-, β-, and γTiCr$_2$ are found in this system. The existence of the cubic C15 Laves phase (αTiCr$_2$) was established by [52Cuf] and [52Duw]. The existence of the higher temperature, higher Cr content βTiCr$_2$ phase was established by [53Lev] and verified [60Gro1, 60Gro2, 63Far, 70Sve, 71Min]. γTiCr$_2$ (with the hexagonal MgNi$_2$ structure) may not be an equilibrium

Table 4 Experimental Data on Solubility of Cr in (αTi)

Reference	Temperature, °C	Composition, at.% Cr
[52Cuf]	~ 670	0.42 to 0.56
[63Luz]	650	0.4 to 0.6
	500	0.3 to 0.5
[62Mik2]	450	0.2 to 0.5
	660	~ 0.6

phase. It was reported by [62Sve] and [70Sve] to be the equilibrium form above about 1270 °C. [71Min], however, attributed thermal arrests found at this temperature to the formation of a ternary Ti-Cr-O phase diagram rather than a third equilibrium Laves phase; however, they did not identify the structure of the high-temperature phase with any known ternary phase. A transition between βTiCr$_2$ and γTiCr$_2$ is indicated on the assessed diagram by a dotted line.

The determination of the homogeneity ranges and invariant reactions of the Laves phases is hindered by sluggishness of the reactions and metastable persistence of phases [60Gro1, 63Far]. Moreover, these structures can be described as different stacking sequences of the same basic cell; small amounts of one phase may be interpreted as stacking fault variants of another and a number of defect structures or polytypes have been observed [79All, 81Lia]. Experimental data on homogeneity ranges and invariant reactions are summarized in Tables 5 and 6, respectively. In drawing the assessed diagram, we have placed greatest reliance on the metallographic work of [63Far].

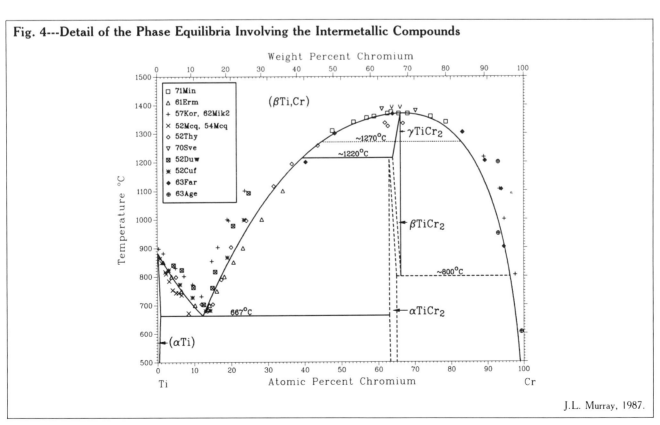

Fig. 4---Detail of the Phase Equilibria Involving the Intermetallic Compounds

J.L. Murray, 1987.

Phase Diagrams of Binary Titanium Alloys

Data on the extent of the (βTi,Cr) field are in good agreement, as shown in Fig. 4. Exceptions are [52Duw] and [62Mik1, 62Mik2], who observed a less extensive single-phase β region. This result is probably caused by contamination of alloys by gaseous impurities and can also be seen in their data for the β transus. The (βTi,Cr) field of the assessed diagram was determined by optimization of Gibbs energies with respect to the β/TiCr₂ phase equilibrium data. The calculated congruent transformation β ⇌ γTiCr₂ occurs at 1371 °C, compared to the assessed value of 1370 ± 10 °C.

Magnetic Transition

Pure Cr is antiferromagnetic, with a Nèel temperature (T_N) about 312 K. [73Ara] measured the effect on T_N of alloying with a series of elements and found that the addition of 0.7 at.% Ti to Cr lowers T_N to 190 K.

Metastable Phases

Decomposition of metastable (βTi) can involve one or more of several processes, depending on alloy composition and heat treatment. In low-Cr alloys, supersaturated (αTi) can form martensitically during quenching (α'); in higher Cr alloys, coherent ω phase forms either during quenching or during low-temperature heat treatment. The metastable β miscibility gap can be observed if α' and ω can be prevented from forming. Other metastable structures besides α', ω, and β have been identified [58Spa2, 60Gro2, 69Eri1]. [60Gro2] and [69Eri1] tentatively identified the phases as transition structures related to the equilibrium Laves phases.

(βTi,Cr) Miscibility Gap.

[71Nar] and [73Cha] reported that, if a metastable bcc alloy (14 at.% Cr) is heat treated in the range 300 to

Table 5 Homogeneity Ranges of the Laves Phases

Reference	Phase	Maximum composition range, at.% Cr	Technique/comment
[51Mcq]	αTiCr₂	~ 60	Metallography, X-ray diffraction, no variation of lattice parameter with composition was observed
[52Cuf]	αTiCr₂	58 to 63.1	X-ray diffraction, metallography, microhardness (using 99.7 wt.% Ti)
[52Duw]	αTiCr₂	~ 60	X-ray (alloys were not in equilibrium)
[57Kor]	αTiCr₂	57 to 60	
[63Far]	αTiCr₂	~ 63 to 66	X-ray diffraction, metallography
	βTiCr₂	~ 63 to 66	
[61Gol]	αTiCr₂	66.7	Chemical analysis of separated compound
[70Sve, 62Sve]	αTiCr₂	60 ± 1	X-ray diffraction, metallography, and
	βTiCr₂	64 ± 1	high-temperature physical and
	γTiCr₂	64 ± 1	mechanical properties
[71Min]	αTiCr₂	63 to 65	Metallography, X-ray diffraction
	βTiCr₂	63 to 65	

Table 6 Three-Phase Equilibria Involving Laves Phases

Reaction	Reference	Temperature, °C	Note
β ⇌ γTiCr₂	[71Min]	1360 ± 15	Thermal analysis
	[63Kor]	1340	Thermal analysis, heating
	[51Mcq]	1360	Metallography
	[52Cuf]	1350	X-ray diffraction, metallography, dilatometry
	[52Thy]	1350	Metallography
	[62Sve, 70Sve, 73Sve]	1360 ± 15	High-temperature physical properties, thermal analysis
γTiCr₂ ⇌ βTiCr₂	[71Min]	1270	Interpreted as ternary phase
	[70Sve, 62Sve]	1235 ± 5	Thermal analysis, cooling;
		1250 ± 5	thermal analysis, heating
(βTi) + βTiCr₂ ⇌ αTiCr₂	[70Sve, 73Sve]	1195 ± 10	Thermal analysis, heating; thermal analysis, cooling
	[71Min]	1225 ± 10	Thermal analysis, heating
	[63Far]	1150 ± 5	Metallography, X-ray
	[60Gro]	1160 ± 30	X-ray diffraction
	[63Kor]	1220	Thermal analysis, heating
βTiCr₂ ⇌ αTiCr₂ + (Cr)	[63Far]	850 ± 50	Metallography, X-ray
	[71Min]	~ 800	Metallography, X-ray

Fig. 5---Start Temperatures for the Martensite and Reverse Martensite Transitions

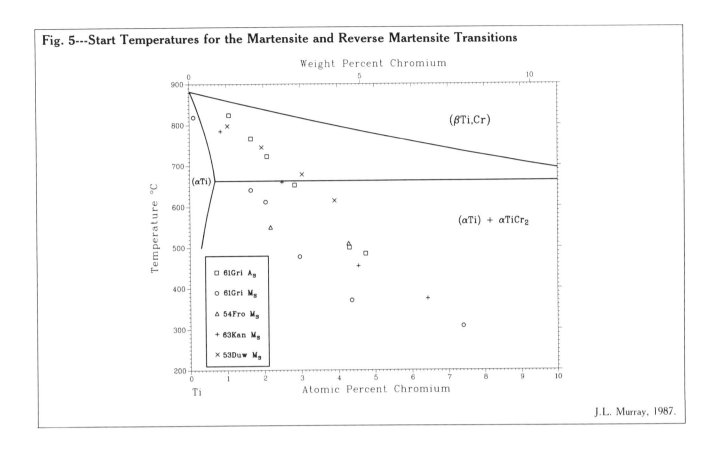

J.L. Murray, 1987.

450 °C and precipitation of ω or stable (αTi) is avoided, then two β phases are formed. The β particles are associated with (αTi) precipitates, but they do not act as preferred nucleation sites for ω. [71Nar] and [73Cha] also stated that the β precipitate contains less Cr than the matrix, based on the association of β precipitates and equilibrium (αTi) plates. A metastable β miscibility gap is consistent with the present knowledge of the thermodynamics of the system, but only if the β precipitate is solute rich. In the thermodynamic calculations to be described below, a metastable monotectoid reaction (βTi) ⇄ (αTi) + (Cr) occurs at 574 °C, and the relative stability of (αTi) in metastable equilibrium with solute-enriched β provides a more conventional explanation for the observed precipitation of (αTi) in association with β precipitates. The effect of coherency energy on the miscibility gap remains to be examined.

Martensitic Transformation.

The cph phase (α'Ti) can form martensitically during the quench from the β field. The β phase cannot be fully retained during quenching unless the Cr content exceeds about 6 at.% [52Duw, 52Thy, 66Kol, 69Eri1]. Start temperatures for the martensitic β → α' transformation (M_s) and for the reverse transformation (A_s) were measured by [53Duw, 54Fro, 61Gri, 63Kan (see also 60Sat) or [70Hua]; they are plotted in Fig. 5.

ω Phase.

The ω phase, a ubiquitous metastable phase in Ti-transition metal systems, was discovered in the Ti-Cr system by [54Fro]. [57Aus] and [58Spa1] identified the structure of ωTiCr as cubic γ brass-type and orthorhombic, respectively. However, the structure is now definitely established in many Ti and Zr alloys as based on a hexagonal distortion of the bcc structure. In Ti-Cr, this structure has undergone further distortion, and ω phase has trigonal symmetry $P\bar{3}m1$ [55Bag, 58Bag2, 58Sil, 62Bag].

The ω phase occurs as a coherent precipitate in the bcc matrix; coherent ω particles take the form of cubes [73Cha, 73Ika] because of the large misfit between the lattice parameters of Ti and Cr. These cubes are 4 to 5 nm in size [73Cha].

In Ti-Cr, ω appears in metastable β alloys either during cooling from the β field, or during low-temperature aging. As-quenched ω forms in alloys with compositions between approximately 3 and 9 at.% Cr [61Age, 63Bor, 66Kol, 69Eri2, 69Luh].

Temperatures of ω formation on cooling and reversion on heating were measured for alloys containing 5.1 to 9.3 at.% Cr [61Gri, 62Bag, 72Ika, 73Ika, 73Luz]. Results are somewhat disparate, partly because of the range of cooling rates and alloy compositions used. For a 5.1 at.% Cr alloy and high cooling and heating rates, [62Bag] found that β → ω at 355 °C and ω → β at 455 °C. [72Ika] studied the transformation β → ω as a function of cooling rate and found that the start temperature can be depressed to at least 200 °C, but at lower cooling rates a plateau is reached in the β → ω temperature at about 390 to 410 °C. This suggests that both supercooling and superheating occur. During tempering, ω + β is formed below about 450 °C, and ω redissolves in the range 400 to 450 °C [58Spa2, 62Yam, 69Hic, 71Miy].

[69Hic] determined the compositions of ω and β after aging for various times and at temperatures between 300 and 400 °C. The composition of ω approaches 6.5 at.% Cr, and the composition of β approaches approximately 20 at.% Cr. The tie-line compositions and cooling data do not

Fig. 6---Lattice Parameters of Single-Phase bcc Alloys

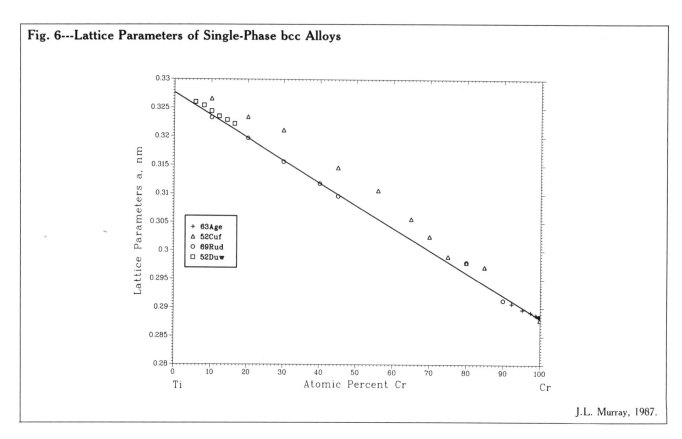

J.L. Murray, 1987.

Table 7   Crystal Structures and Lattice Parameters of the Ti-Cr System

Phase	Homogeneity range, at.%	Pearson symbol	Space group	Struktur-bericht designation	Proto-type	Lattice parameters, nm		Reference
						a	c	
(αTi)	0 to 0.2	$hP2$	$P6_3/mmc$	$A3$	Mg	...	...	[Pearson2]
(βTi,Cr)	0 to 9	$cI2$	$Im3m$	A_2	W	...	...	[Pearson2]
αTiCr$_2$	63 to 65	$cF24$	$Fd3m$	$C15$	MgCu$_2$	0.6957	...	[52Duw]
						0.691	...	[52Cuf]
						0.69442(a)	...	[62Sve]
						0.6918(b)	...	[62Sve]
βTiCr$_2$	64 to 66	$hP12$	$P6_3/mmc$	$C14$	MgZn$_2$	0.4931	0.7961	[53Lev]
γTiCr$_2$	64 to 66	$hP24$	$P6_3/mmc$	$C36$	MgNi$_2$	...	...	[70Sve]
ω	(c)	$hP3$	$P\bar{3}m1$	...	ωCrTi	0.4616	0.2827	[58Bag2]
						0.4609	0.2826	[55Bag]

(a) On the Ti-side. (b) On the Cr-side. (c) Metastable.

form a consistent picture of the metastable β-ω phase boundaries, presumably because of the effect of elastic energies on the thermodynamics of the transition.

Crystal Structures and Lattice Parameters

The crystal structures of equilibrium and metastable phases and lattice parameters of the compounds and ω phases are summarized in Table 7. Lattice parameters of quenched β alloys are shown in Fig. 6, and lattice parameters of supersaturated (martensitic) (αTi) are shown in Fig. 7.

Thermodynamics

[67Poo] obtained activity data for Cr and Ti in the β phase at 1250, 1290, 1360, and 1380 °C by the Knudsen

cell technique. [72Rol] used a Knudsen cell and mass spectrometer to measure activities in bcc Ti-Cr-V ternary alloys and fitted the excess Gibbs energies to a temperature-dependent regular solution model, finding the excess Gibbs energy in the β phase to be $1172(1 - T/1840) x(1 - x)$ kJ/mol, where x is the atomic fraction of Cr. This does not conflict with the work of [67Poo].

[62Ger] measured the latent heat of the cph $\rightarrow$ bcc transformation in a 6 at.% Cr alloy, finding 1882 ± 210 J/mol. [70Rub] reported activity measurements for several alloys in the temperature range 750 to 850 °C, but according to the authors, the values are to be taken as approximate only.

Calculated Phase Boundaries.

Previous calculations of the phase diagram [75Mol, 78Kau] used regular and subregular solution models,

Fig. 7---Lattice Parameters of Single-Phase cph Alloys

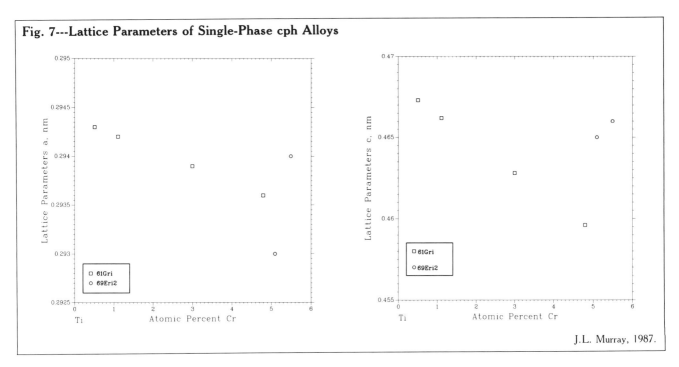

J.L. Murray, 1987.

Fig. 8---Experimental Data on Partial Cr Gibbs Energies of the bcc Phase vs Calculated Values (Solid Curves)

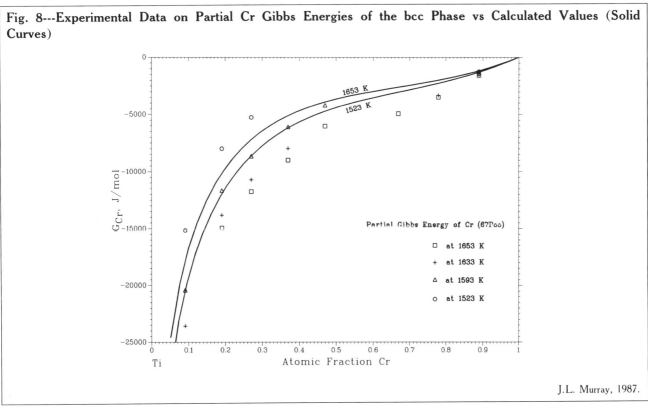

J.L. Murray, 1987.

respectively, to describe the solution phases. The [78Kau] calculation used incorrect solidus data, and as a result, the critical temperature of the metastable bcc miscibility gap is considerably higher than in the calculation of [75Mol], or in the present calculations. The present calculations employ least-squares optimization of Gibbs energies with respect to thermodynamic data and assessed phase diagram data.

The Gibbs energies of the solution phases are represented as:

$$G(i) = G^0(\text{Ti},i)\,(1 - x) + G^0(\text{Cr},i)\,x + RT[x \ln x + (1 - x) \ln(1 - x)]\,x(1 - x)\,[B(i) + C(i)(1 - 2x)]$$

where i designates the phase; x is the atomic fraction of Cr; G^0 the Gibbs energies of the pure components; and B(i)

Table 8 Ti-Cr Thermodynamic Properties

Properties of the pure components

$G^0(\text{Ti,L})$ = 0
$G^0(\text{Cr,L})$ = 0

$G^0(\text{Ti,bcc})$ = $-16\,234 + 8.368\,T$
$G^0(\text{Cr,bcc})$ = $-18\,200 + 8.368\,T$

$G^0(\text{Ti,cph})$ = $-20\,585 + 12.134\,T$
$G^0(\text{Cr,cph})$ = $-26\,568 + 12.134\,T$

Solution phases and compounds

	Present evaluation	[78Kau]	[75Mol]
$B(\text{L})$	5 598	6 776	4 184
$C(\text{L})$	$-2\,528$	6 776	...
$B(\text{bcc})$	17 346	18 828	16 318
$C(\text{bcc})$	$-3\,195$	$-6\,000$	...
$B(\text{cph})$	40 000	45 690	30 543
$C(\text{cph})$	...	$-12\,553$	...
$G(\text{TiCr}_2)$	$-19\,959 + 7.1414\,T$n(a)	$5\,480 - 2.879\,T$(b)	...
$G(\text{"Ti}_2\text{Cr}_3\text{"})$	$-17\,808 + 7.7873\,T$(a)	...	...

(a) Referred to liquid. (b) Referred to bcc, see [78Kau].

and $C(i)$ are parameters in the expansion of the excess Gibbs energy. The pure element Gibbs energies are from [70Kau], but were adjusted to reproduce the melting point of Cr. Because of the uncertainties in the experimentally determined homogeneity ranges, a line compound representation for the Gibbs energies was found most appropriate. The Laves phases were modeled as two line compounds with initial guesses of stoichiometries TiCr$_2$ (βTiCr$_2$ and γTiCr$_2$) and Ti$_2$Cr$_3$ (αTiCr$_2$). Because the composition of αTiCr$_2$ appears to be closer to 63 at.% Cr, calculations were also performed using that stoichiometry. Very little difference was found in either the Gibbs energy of the compound or the phase boundaries of the bcc phase in equilibrium with it, and these calculations were not pursued further.

This calculation makes use of partial Gibbs energy data for the bcc phase and assessed phase diagram data already discussed. In preliminary calculations, Gibbs energies of all the phases were simultaneously varied. Gibbs energies of liquid and solid phases were then separately fine-tuned, keeping the Gibbs energy of the bcc phase fixed. Calculated values are compared to experimental data on partial Gibbs energies of the bcc phase in Fig. 8. Excess entropies could not be determined by this technique using the experimental data available, and they were therefore set to zero.

The calculated reaction temperatures, compared to the assessed values are:

Reaction	Reaction temperatures, °C	
	Calculated	Assessed
L $\rightleftarrows$ β	1413	1410 ± 5
β $\rightleftarrows$ γTiCr$_2$	1371	1370 ± 10
(βTi) $\rightleftarrows$ (αTi) + αTiCr$_2$	663	667 ± 10
(βTi) + γTiCr$_2$ $\rightleftarrows$ αTiCr$_2$	1215	~1220

Numerical values of the thermodynamic parameters are summarized in Table 8.

Cited References

51Mcq: M.K. McQuillan, "A Provisional Constitutional Diagram of the Chromium-Titanium System," *J. Inst. Met., 80*, 379-390 (1951). (Equi Diagram; Experimental)

52Cuf: F.B. Cuff, N.J. Grant, and C.F. Floe, "Titanium-Chromium Phase Diagram," *Trans. AIME, 194*, 848-853 (1952). (Equi Diagram; Experimental)

52Duw: P. Duwez and J.L. Taylor, "A Partial Titanium-Chromium Phase Diagram and the Crystal Structure of TiCr$_2$," *Trans. ASM, 44*, 495-513 (1952). (Equi Diagram, Crys Structure; Experimental)

52Mcq: A.D. McQuillan, "The Effect of the Elements of the First Long Period on the α-β Transformation in Titanium," *J. Inst. Met., 80*, 363-368 (1951-1952). (Equi Diagram; Experimental)

52Thy: R.J. van Thyne, H.D. Kessler, and M. Hansen, "The Systems Titanium-Chromium and Titanium-Iron," *Trans. ASM, 44*, 974-989 (1952). (Equi Diagram; Experimental)

53Duw: P. Duwez, "The Martensite Transformation Temperature in Titanium Binary Alloys," *Trans. ASM, 45*, 934-940 (1953). (Meta Phases; Experimental)

53Lev: B.W. Levinger, "High Temperature Modification of TiCr$_2$," *Trans. AIME, 197*, 196 (1953). (Crys Structure; Experimental)

54Fro: P.D. Frost, W.M. Parris, L.L. Hirsch, J.R. Doig, and C.M. Schwartz, "Isothermal Transformation of Titanium-Chromium Alloys," *Trans. ASM, 46*, 231-256 (1954). (Meta Phases; Experimental)

54Mcq: M.K. McQuillan, "A Redetermination and Interpretation of the Titanium-Rich Region of the Titanium-Chromium System," *J. Inst. Met., 82*, 433-439 (1953-54). (Equi Diagram; Experimental)

55Bag: Yu.A. Bagariatskii, G.I. Nosova, and T.V. Tagunova, "Crystallographic Structure and Properties of ω Phase in Titanium Chromium Alloys," *Dokl. Akad. Nauk SSSR, 105*(6), 1225-1228 (1955) in Russian. (Meta Phases, Crys Structure; Experimental)

57Aus: A.E. Austin and J.R. Doig, "Structure of the Transition Phase Omega in Ti-Cr Alloys," *Trans. AIME, 209*, 27-30 (1957). (Meta Phases, Crys Structure; Experimental)

57Kor: I.I. Kornilov, V.S. Mikheyev, and T.S. Chernova, "Constitution Diagram of Ti-Cr," *Tr. Inst. Metall. Akad. Nauk SSSR, 2*, 126-134 (1957) in Russian. (Equi Diagram; Experimental)

The Cr-Ti System

58Bag1: Yu.A. Bagariatskii, G.I. Nosova, and T.V. Tagunova, "Study of the Phase Diagrams of the Alloys Titanium-Chromium, Titanium-Tungsten, and Titanium-Chromium-Tungsten, Prepared by the Method of Powder Metallurgy," *Zh. Neorg. Khim., 3*(3), 777-785 (1958) in Russian; TR: *Russ. J. Inorg. Chem., 3*(3), 330-341 (1958). (Equi Diagram; Experimental)

58Bag2: Yu.A. Bagariatskii and G.I. Nosova, "Exact Determination of the Atomic Coordinates in the Metastable ω-Phase of Ti-Cr Alloys," *Kristallografiya, 3*(1), 17-28 (1958) in Russian; TR: *Sov. Phys. Crystallogr., 3*, 15-26 (1958). (Meta Phases, Crys Structure; Experimental)

58Sil: J.M. Silcock, "An X-Ray Examination of the ω Phase in TiV, TiMo and TiCr Alloys," *Acta Metall., 6*, 481-493 (1958). (Meta Phases, Crys Structure; Experimental)

58Spa1: S.A. Spachner, "Comparison of the Structure of the ω Transition Phase in Three Titanium Alloys," *Trans. AIME, 212*, 57-59 (1958). (Meta Phases, Crys Structure; Experimental)

58Spa2: S.A. Spachner and W. Rostoker, "Transformation Kinetics of Two Titanium Alloys in the Transition Phase Region," *Trans. AIME, 212*, 765-769 (1958). (Meta Phases, Crys Structure; Experimental)

59Gol: A.W. Goldenstein, A.G. Metcalfe, and W. Rostoker, "The Effect of Stress on the Eutectoid Decomposition of Titanium-Chromium Alloys," *Trans. ASM, 51*, 1036-1053 (1959). (Equi Diagram; Experimental)

60Gro1: K.A. Gross and I.R. Lamborn, "Allotropic Modifications of TiCr₂," *J. Inst. Met., 88*, 416 (1959-1960). (Equi Diagram; Experimental)

60Gro2: K.A. Gross and I.R. Lamborn, "Some Observations on Decomposition of Metastable β-Phase in Titanium-Chromium Alloys," *J. Less-Common Met., 2*, 36-41 (1960). (Meta Phases; Experimental)

60Sat: T. Sato, S. Hukai, and Y.C. Huang, "The M_s Points of Binary Titanium Alloys," *J. Aust. Inst. Met., 5*(2), 149-152 (1960). (Meta Phases; Experimental)

61Age: N.V. Ageev, O.G. Karpinskii, and L.A. Petrova, "Stability of the β-Solid Solution of Titanium-Chromium Alloys," *Zh. Neorg. Khim., 6*, 1976 (1961) in Russian; TR: *Russ. J. Inorg. Chem., 6*, 1011-1012 (1961). (Meta Phases; Experimental)

61Erm: F. Ermanis, P.A. Farrar, and H. Margolin, "A Reinvestigation of the Systems Ti-Cr and Ti-V," *Trans. AIME, 221*, 904-908 (1961). (Equi Diagram; Experimental)

61Gri: V.N. Gridnev, V.I. Trefilov, D.V. Lotsko, and N.F. Chernenko, "Mechanisms of Phase Transformations in Ti-Cr Alloys," *Akad. Nauk URSR Kiev Instytut Metallof., Sbornik Nauchn. Rabot, 12*, 37-44 (1961) in Russian. (Meta Phases; Experimental)

62Bag: Yu.A. Bagariatskii and G.I. Nosova, "The β-ω Transformation in Titanium Alloys During Quenching: A Singular Kind of Martensitic Transformation," *Fiz. Met. Metalloved., 13*, 415-425 (1962) in Russian; TR: *Phys. Met. Metallogr., 13*(3), 92-101 (1962). (Meta Phases; Review)

62Ger: S.D. Gertsriken and B.P. Slyusar, "Determination of the Heat of Melting of Titanium, Zirconium and the Alloy Ti6.5% Cr," *Ukr. Fiz. Zh., 7*(4), 339-442 (1962) in Russian. (Thermo; Experimental)

62Mik1: V.S. Mikheev and V.S. Alekasashin, "Electrical Volume Resistivity of Alloys of the Titanium-Chromium System up to Temperatures of 1100 °C," *Fiz. Met. Metalloved., 14*, 231-237 (1962) in Russian; TR: *Phys. Met. Metallogr., 14*(2), 62-67 (1962). (Equi Diagram; Experimental)

62Mik2: V.S. Mikheev and T.S. Chernova, "Solubility of Chromium in αTi and Mechanical Properties of the Binary System Titanium-Chromium," *Titan Splav. AN SSSR Inst. Met., 7*, 68-73 (1962) in Russian. (Equi Diagram; Experimental)

62Sve: V.N. Svechnikov, Yu.A. Kocherzhinskii, and V.I. Latysheva, "Constitution Diagram of Chromium-Titanium," *Problems in the Physics of Metals and Metallurgy,* (16), 132-135 (1962) in Russian. (Equi Diagram; Experimental)

62Yam: T. Yamane and J. Ueda, "Tempering Behavior of a Ti-Cr Alloy Quenched from the β Region," *Trans. Jpn. Inst. Met., 6*, 151-154 (1962). (Meta Phases; Experimental)

63Bor: B.A. Borok, E.K. Novikova, L.S. Golubeva, R.P. Shchegoleva, and N.A. Ruch'eva, "Dilatometric Investigation of Binary Alloys of Titanium," *Metalloved. Term. Obrab. Met.,* (2), 32-36 (1963) in Russian; TR: *Met. Sci. Heat Treat.,* (2), 94-98 (1963). (Meta Phases; Experimental)

*63Far: P.A. Farrar and H. Margolin, "A Reinvestigation of the Chromium-Rich Region of the Titanium-Chromium System," *Trans. AIME, 227*, 1342-1345 (1963). (Equi Diagram; Experimental)

63Kan: H. Kaneko and Y.C. Huang, "Continuous Cooling Transformation Characteristics of Titanium Alloys of Eutectoidal Type (Part 1)," *J. Jpn. Inst. Met., 27*, 393-397 (1963) in Japanese. (Meta Phases; Experimental)

63Kor: I.I. Kornilov, K.I. Shakhova, P.B. Budberg, and N.A. Nedumova, "Phase Diagram of the System TiCr₂-NbCr₂," *Dokl. Akad. Nauk SSSR, 149*(6), 1340-1342 (1963) in Russian; TR: *Dokl. Chem., Proc. Acad. Sci. USSR, 149*, 362-364 (1963). (Equi Diagram; Experimental)

63Luz: L.P. Luzhnikov, V.M. Novikova and A.P. Mareev, "Solubility of β-Stabilizers in αTi," *Metalloved. Term. Obrab. Met.,* (2), 13-16 (1963) in Russian; TR: *Met. Sci. Heat Treat.,* (2), 78-81 (1963). (Equi Diagram; Experimental)

66Kol: B.A. Kolachev and V.S. Lyastoskaya, "Metastable Phase Diagram of the Titanium-Chromium System," *Izv. VUZ Metall.,* (2), 123-128 (1966) in Russian. (Meta Phases; Experimental)

67Poo: M.J. Pool, R. Speiser, and G.R. St. Pierre, "Activities of Chromium and Titanium in Binary Chromium-Titanium Alloys," *Trans. AIME, 239*, 1180-1186 (1967). (Thermo; Experimental)

69Eri1: R.H. Ericksen, R. Taggart, and D.H. Polonis, "The Martensite Transformation in Ti-Cr Binary Alloys," *Acta Metall., 17*, 553-564 (1969). (Meta Phases; Experimental)

69Eri2: R.H. Ericksen, R. Taggart, and D.H. Polonis, "The Characteristics of Spontaneous Martensite in Thin Foils of Ti-Cr Alloys," *Trans. AIME, 245*, 359-363 (1969). (Meta Phases; Experimental)

69Hic: B.S. Hickman, "ω Phase Precipitation in Alloys of Titanium with Transition Metals," *Trans. AIME, 245*, 1329-1336 (1969). (Meta Phases; Experimental)

69Luh: T.S. Luhman, R. Taggart, and D.H. Polonis, "Correlation of Superconducting Properties with the β to ω Phase Transformation in Ti-Cr Alloys," *Scr. Metall., 3*, 777-782 (1969). (Meta Phases; Experimental)

69Rud: E. Rudy, "Compilation of Phase Diagram Data," Tech. Rep. AFML-TR-65-2, Part V, Wright Patterson Air Force Base (1969). (Equi Diagram; Experimental)

70Hua: Y.C. Huang, S. Suzuki, H. Kaneko, and T. Sato, "Thermodynamics of M_s Points in Titanium Alloys," Sci., Technol. Appl. Titanium, Proc. Int. Conf., R.I. Jaffee, Ed., 691-693 (1970). (Meta Phases; Experimental)

70Kau: L. Kaufman and H. Bernstein, *Computer Calculation of Phase Diagrams,* Academic Press, New York (1970). (Thermo; Theory)

70Rub: A.N. Rubtsov, Yu.G. Olesov, V.I. Cherkashin, and A.B. Suchkov, "Activity of Aluminum, Vanadium and Chromium in Binary Alloys with Titanium," *Izv. Akad. Nauk SSSR, Met.,* (6), 84 (1970) in Russian; TR: *Russ. Metall.,* (6), 56-58 (1970). (Thermo; Experimental)

70Sve: V.N. Svechnikov, M.Yu. Teslyuk, Yu.A. Kocherzhinsky, V.V. Petkov, and E.V. Dabizha, "Three Modifications of TiCr₂," *Dopov. Akad. Nauk Ukr. RSR, 32A*(9), 837-841 (1970) in Russian. (Equi Diagram; Experimental)

*71Min: S.A. Minayeva, P.B. Budberg, and A.L. Gavze, "Phase Structure of Ti-Cr Alloys," *Izv. Akad. Nauk SSSR, Met.,* (4), 205-209 (1971); TR: *Russ. Metall.,* (4), 144-147 (1971). (Equi Diagram; Experimental)

71Miy: M. Miyagi and S. Shin, "Isothermal Transformation Characteristic of Metastable β-type Titanium Alloys," *J. Jpn. Inst. Met., 35*(7), 716-722 (1971) in Japanese. (Meta Phases; Experimental)

71Nar: G.H. Narayanan, T.S. Luhman, T.F. Archbold, R. Taggart, and D.H. Polonis, "A Phase Separation Reaction in a

Binary Titanium-Chromium Alloy," *Metallography, 4*, 343-358 (1971). (Meta Phases; Experimental)

72Ika: H. Ikawa, S. Shin, and M. Morikawa, "Continuous Cooling Transformation in Binary, α and β and Metastable βTi Alloys," *Yosetsu Gakkai-Shi, 41*(4), 394-402 (1972) in Japanese. (Meta Phases; Experimental)

72Rol: E.J. Rolinski, M. Hoch, and C.J. Oblinger, "Determination of Thermodynamic Interaction Parameters in Solid V-Ti-Cr Alloys Using the Mass Spectrometer," *Metall. Trans., 3*, 1413-1418 (1972). (Thermo; Experimental)

73Ara: S. Arajs, K.V. Rao, H.U. Astrom, and T.F. DeYoung, "Determination of Nèel Temperatures of Binary Chromium Alloys from Electrical Resistivity Data," *Phys. Scr., 8*(3), 109-112 (1973). (Equi Diagram; Experimental)

73Cha: V. Chandrasekaran, R. Taggart, and D.H. Polonis, "Decomposition Processes Prior to Detection of the ω Phase in Aged Ti-Cr Alloys," *Metallography, 6*(4), 313-322 (1973). (Meta Phases; Experimental)

73Ika: H. Ikawa, S. Shin, M. Miyagi, and M. Morikawa, "Some Fundamental Studies on the Phase Transformation from β Phase to α Phase in Titanium Alloys," Sci. Technol. Appl. Titanium, Proc. Int. Conf., R.I. Jaffee, Ed., 1545-1556 (1973). (Meta Phases; Experimental)

73Luz: L.P. Luzhnikov and V.M. Novikowa, "Reversion at the Ageing of Titanium Alloys," Sci. Technol. Appl. Titanium, Proc. Int. Conf., R.I. Jaffee, Ed., 1535-1543 (1973). (Meta Phases; Experimental)

73Sve: V.N. Svechnikov, Yu.A. Kocherzhinskii, V.V. Petkov, A.V. Polenur, and L.S. Guzei, "Phase Transformations in Compounds of Cr with Ti," *Akad. Nauk Ukr. SSR, Metallofiz., 46*, 72-75 (1973) in Russian. (Equi Diagram; Experimental)

75Mol: V.V. Molokanov, P.B. Budberg, and S.P. Alisova, "Phase Diagram of the Titanium-Chromium System," *Dokl. Akad. Nauk SSSR, 233*(5), 1184-1186 (1975); TR: *Dokl. Phys. Chem., 223*, 847-849 (1975). (Thermo; Theory)

78Kau: L. Kaufman and H. Nesor, "Coupled Phase Diagrams and Thermochemical Data for Transition Metal Binary Systems-I," *Calphad, 2*(1), 55-80 (1978). (Thermo; Experimental)

79All: C.W. Allen and K.C. Liao, "Shear Transformations in the Laves Phase TiCr$_{-2}$," Proc. ICOMAT, 1979, Martensitic Transformations, Cambridge, MA, 124-129 (1979).

81Lia: K.C. Liao and C.W. Allen, "Shear Transformation in the Ti-Cr Laves Phases," *Solid to Solid Phase Transformations*, Met. Soc. AIME, Warrendale, PA, 1493-1497 (1981).

The Cs-Ti (Cesium-Titanium) System

132.9054 47.88

By C.W. Bale

Equilibrium Diagram

The assessed Cs-Ti phase diagram is shown in Fig. 1. As part of a project to study the corrosive attack of refractory alloys and superalloys by liquid Cs, [64Tep] measured the solubility of Mo–1/2 wt.% Ti alloys in liquid Cs held in a Nb–1 wt.% Zr container at 1371 °C. The results were <0.0014 at.% Ti and 0.0014 at.% Mo after 5 min, 0.028 at.% Ti and 0.0035 at.% Mo after 14 h, and 0.042 at.% Ti and 0.0014 at.% Mo after 110 h. [64Tep] concluded that the Ti had not equilibrated and that diffusion from the Mo–0.5 wt.% Ti was probably the rate-controlling step. [66Mck] reported that in earlier experiments [63Tep] obtained a solubility of from <0.0017 to 0.025 at.% Ti at the same temperature for Mo–0.5 wt.% Ti alloys held in a Mo–0.5 wt.% Ti container. Any oxygen impurity that leads to the formation of highly stable cesium titanates [82Koh] and the synergistic solubility

effects of the solutes would probably have a marked effect on the results.

In a subsequent experiment, [64Tep] successfully re-fluxed liquid Cs held in Mo–0.5 wt.% Ti capsules—one capsule at 1371 °C for 255 h and two capsules at 1149 °C for 292 and 1000 h, respectively. The capsules were examined, and some dissolution was detected. Oxygen was also added and it affected the solubility.

If the measurements of [63Tep] and [64Tep] represent true equilibrium conditions, the actual solubility of pure Ti in Cs is much higher, because the results correspond to Ti in a dilute solid solution of Mo.

There are no other published data for the Cs-Ti system. From the results of [63Tep] and [64Tep] and by analogy with other alkali metal–Group IVA systems, it is probable that the Cs-Ti system is almost completely immiscible in both the solid and liquid states.

If the solubilities are very limited, it follows from

Fig. 1---Assessed Cs-Ti Phase Diagram

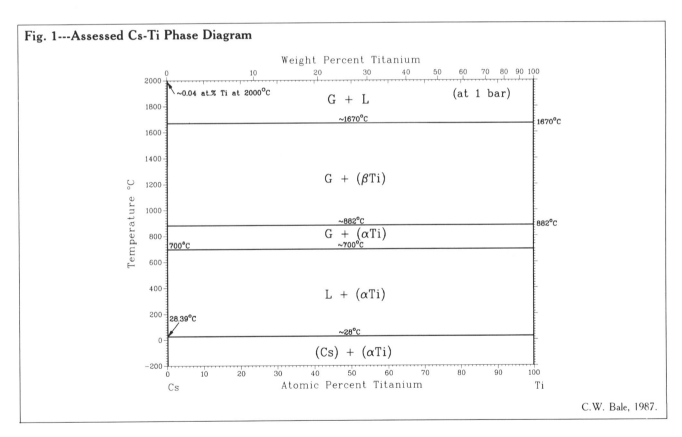

C.W. Bale, 1987.

Table 1 Crystal Structures and Lattice Parameters of the Cs-Ti System

Phase	Composition, at.% Ti	Pearson symbol	Space group	Strukturbericht designation	Prototype	Lattice parameters, nm a	c	Comment	Reference
(Cs)	0	cI2	Im3m	A2	W	0.6141	...	At RT	[King1]
(αTi)	100	hP2	P6₃/mmc	A3	Mg	0.29503	0.55263	At RT	[King2]
(βTi)	100	cI2	Im3m	A2	W	0.33065	...	Above 900 °C	[King2]

thermodynamic considerations that the univariant temperatures in the phase diagram are virtually identical with the transition temperatures of the pure components.

From vapor-pressure calculations, it is estimated that the composition of the gas phase at 2000 °C and 1 atm is approximately 0.04 at.% Ti, in equilibrium with almost pure liquid Ti.

Crystal structure data for the elements are summarized in Table 1.

Cited References

63Tep: F. Tepper and J. Greer,"Factors Affecting the Compatibility of Liquid Cesium with Containment Metals," MSA Research Corp. Tech. Rep. ASD-TDR-63-824, Part I (1963). (Equi Diagram; Experimental)

64Tep: F. Tepper and J. Greer,"Factors Affecting the Compatibility of Liquid Cesium with Containment Metals," MSA Research Corp. Tech. Rep. AFML-TR-64-327, Contract No. AF 33(657)-9168 (1964). (Equi Diagram; Experimental)

66Mck: R.L. McKisson, R.L. Eichelberger, R.C. Dahleen, J.M. Carborough, and G.R. Argue, "Solubility Studies of Ultra Pure Transition Elements in Ultra Pure Alkali Metals," N. American Aviation Inc., Tech. Rep., NASA Cr-610, for Lewis Research Center (1966). (Equi Diagram; Compilation, Experimental)

82Koh: R. Kohli,"A Thermodynamic Assessment of the Behavior of Cesium and Rubidium in Reactor Fuel Elements," Material Behavior and Physical Chemistry in Liquid Metal Systems, Karlsruhe, FR Germany, 24-26 Mar 1981; Plenum Press, New York 345-350 (1982). (Equi Diagram; Theory)

The Cu-Ti (Copper-Titanium) System

63.546   47.88

By J.L. Murray

Equilibrium Diagram

The Ti-Cu phase diagram shows a depression of the melting points of both elements, so that the liquidus has a minimum near the center of the diagram. In the composition range 50 to 67 at.% Cu, a series of closely spaced, structurally related compounds appear. There are also the Ti-rich and Cu-rich compounds Ti_2Cu and $TiCu_4$. Experimental phase diagram data are uneven in accuracy, in the sense that the structures and compositions of the compounds have been carefully determined and verified, but many of the phase boundaries involving liquid and solid solution phases were examined in only one study [52Jou]. Metastable phase equilibria have been thoroughly investigated; studies cover the decomposition of supersaturated

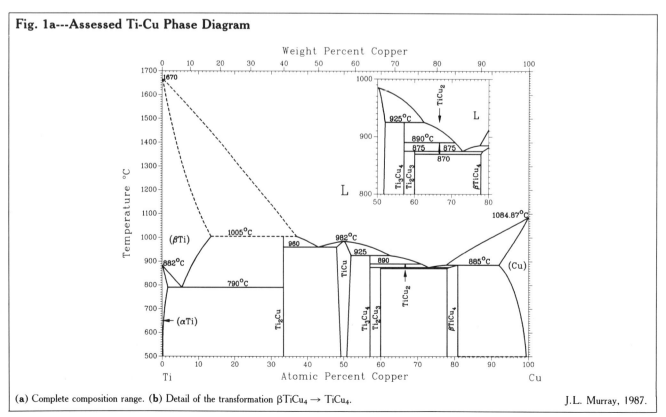

Fig. 1a---Assessed Ti-Cu Phase Diagram

(a) Complete composition range. (b) Detail of the transformation $\beta TiCu_4 \rightarrow TiCu_4$.

J.L. Murray, 1987.

The Cu-Ti System

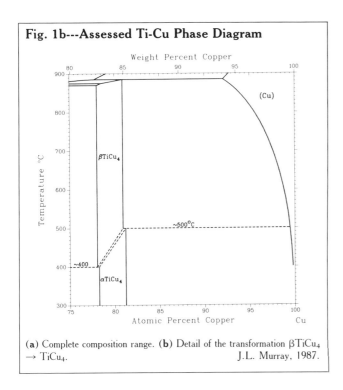

Fig. 1b---Assessed Ti-Cu Phase Diagram

(a) Complete composition range. (b) Detail of the transformation βTiCu₄ → TiCu₄.

J.L. Murray, 1987.

- The essentially stoichiometric compounds Ti_3Cu_4, Ti_2Cu_3, and Ti_2Cu, with related crystal structures.
- High- and low-temperature allotropes $\beta TiCu_4$ and $\alpha TiCu_4$, each with the approximate homogeneity range 78 to 80.9 at.% Cu. Metastable ordered structures can form in this composition range before the appearance of equilibrium $\beta TiCu_4$.

The assessed phase diagram is shown in Fig. 1, and numerical values of the special points of the diagram are summarized in Table 1.

Ti_2Cu and (βTi)/L Boundaries.

There has been disagreement about whether the most Ti-rich compound has the stoichiometry Ti_3Cu or Ti_2Cu. The existence of Ti_2Cu with the bct $MoSi_2$ structure and the absence of Ti_3Cu are now established. [51Kar] found a nearly pure compound at 25 and 27.5 at.% Cu, but the alloys had been severely contaminated. The stoichiometry Ti_3Cu was also assumed by [70Lut] and [52Rau], although without any direct evidence. [67Gar] claimed to have established the existence of both Ti_3Cu and Ti_2Cu by means of X-ray diffraction and optical microscopy, but they did not give enough details of the experiments to assess the accuracy of their work.

Evidence for the existence of Ti_2Cu was found by [52Jou], [61Enc], [62Mue], [63Mue], [70Ble], and [71Wil]. [52Jou], [61Enc], [62Mue], and [63Mue] were able to prepare essentially single-phase Ti_2Cu alloys. [61Enc], [62Mue], and [63Mue] related the structure of Ti_2Cu to that proposed by [51Kar] for "Ti_3Cu", which strongly suggests that the same phase was observed in all of these studies. The work of [61Enc], [62Mue], and [63Mue] established the existence of Ti_2Cu.

The work of [71Wil] and [70Ble] on age-hardening in supersaturated (αTi) showed that Ti_2Cu is the phase in equilibrium with (αTi) below the eutectoid temperature of 790 °C. [71Wil] used select area electron diffraction to identify the structure of the precipitates from supersaturated (αTi); [70Ble] used X-ray and microprobe analysis to measure the compositions of precipitates. The identification of Ti_2Cu [70Ble, 71Wil] as the phase in equilibrium with (αTi) implies that there is no additional equilibrium phase between them.

The homogeneity range of Ti_2Cu was examined by [63Mue] and [66Ere]; [66Ere] considered the range to be

cph and fcc solid solutions and the formation of noncrystalline alloys in the composition range 30 to 75 at.% Cu.

The equilibrium solid phases of the Ti-Cu system are:

- The solid solutions based on the pure components—cph (αTi), the stable form of Ti below 882 °C; bcc (βTi), the stable form of Ti between 882 °C and the melt; and fcc (Cu). The maximum solubilities of Cu in (αTi) and (βTi) are 1.6 and 13.5 at.% at 790 and 1005 °C, respectively. The maximum solubility of Ti in (Cu) is 8 at.% at 885 °C.
- The essentially stoichiometric compound Ti_2Cu with the $MoSi_2$ structure.
- TiCu ($B11$ structure), which has a homogeneity range of 48 to 52 at.% Cu and melts congruently at 985 °C.

Table 1 Special Points of the Ti-Cu Phase Diagram

Reaction	Compositions of the respective phases, at.% Cu			Temperature, °C	Reaction type
(βTi) + L $\rightleftarrows$ Ti_2Cu	13.5	36.5	33.3	1005 ± 10	Peritectic
(βTi) $\rightleftarrows$ (αTi) + Ti_2Cu	5.4	1.6	33.3	790 ± 10	Eutectoid
L $\rightleftarrows$ Ti_2Cu + TiCu	43	33.3	48	960 ± 5	Eutectic
L $\rightleftarrows$ TiCu		50		985 ± 10	Congruent
TiCu + L $\rightleftarrows$ Ti_3Cu_4	52	62.5	57.1	925 ± 10	Peritectic
Ti_3Cu_4 + L $\rightleftarrows$ $TiCu_2$	57.1	71	66.7	890 ± 10	Peritectic
Ti_3Cu_4 + $TiCu_2$ $\rightleftarrows$ Ti_2Cu_3	57.1	66.7	60	875 ± 10	Peritectoid
L $\rightleftarrows$ $TiCu_2$ + $\beta TiCu_4$	73	66.7	78	875 ± 10	Eutectic
$TiCu_2$ $\rightleftarrows$ Ti_2Cu_3 + $\beta TiCu_4$	66.7	60	78	870 ± 10	Eutectoid
L + (Cu) $\rightleftarrows$ $\beta TiCu_4$	77	92	80.9	855 ± 10	Peritectic
$\beta TiCu_4$ $\rightleftarrows$ Ti_2Cu_3 + $\alpha TiCu_4$	~ 78	60	~ 78	~ 400	Eutectoid
$\beta TiCu_4$ $\rightleftarrows$ $\alpha TiCu_4$ + (Cu)	~ 80.9	~ 80.9	99.5	~ 500	Peritectoid
(βTi) $\rightleftarrows$ (αTi)		0		882	Allotropic transformation
L $\rightleftarrows$ (βTi)		0		~ 1670	Melting point
L $\rightleftarrows$ (Cu)		100		1084.87	Melting point

Fig. 2---Detail of Experimental Liquidus Data for Ti-Rich Alloys

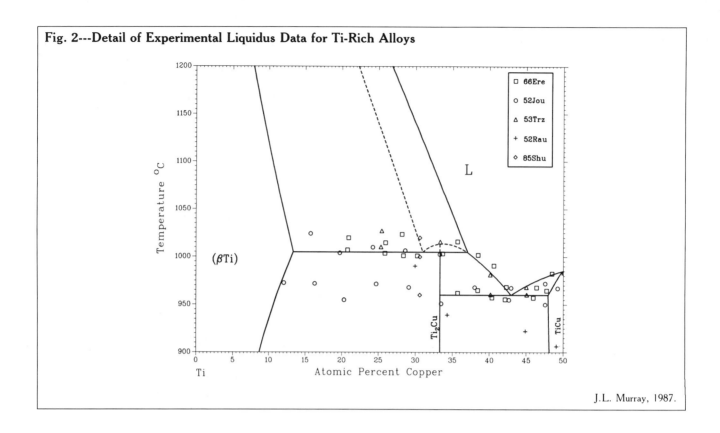

J.L. Murray, 1987.

30 to 33 at.% Cu at 800 °C based on the microstructural work. [63Mue], however, used closely spaced alloys for X-ray structure determination and found a 33.3 at.% Cu alloy to be most nearly single phase, with greater amounts of second phase in 32.8 and 33.8 at.% Cu alloys. The X-ray data are preferred, and the compound is shown as stoichiometric in the assessed diagram.

The solid lines in Fig. 2 represent the present best judgment of the correct melting reaction, and the dashed lines represent an alternate version that has been proposed several times [53Trz, 66Ere]. Also shown in Fig. 2 are all of the available thermal analysis data. Unfortunately, there is very little documentation on details of the experiments; in some references, cooling and heating curves were not differentiated.

Additional arrests above the one at about 1005 °C were seen by most investigators, and the appropriate interpretation of these observations is unclear. The liquidus apparently could not be detected in any of the alloys. There is, therefore, only microstructural evidence to distinguish between alternate versions of the diagram. According to [52Jou], all alloys in the composition range 15 to 33 at.% Cu were only partially melted slightly above 1005 °C, which supports the peritectic construction in Fig. 1. According to [66Ere], the as-cast microstructure of a 30 at.% alloy supports a eutectic construction. The microstructure of an as-cast 31 at.% alloy (produced in this laboratory [85Shu]) supports the peritectic construction that is tentatively accepted in this assessment.

Extrapolation of the liquidus from higher temperatures would allow one to distinguish between the two reaction types; there are, however, no experimental liquidus data between 1670 and 1005 °C. Any assessment of the liquidus must therefore be a hypothetical construction, except at the melting point of Ti and the melting

reaction at 1005 °C. Whether the reaction is of the eutectic or peritectic type changes the position of the liquidus by at least 5 at.% at 1005 °C.

Few data [52Jou] are available on the (βTi) solidus. Based on solidus and (βTi)/[(βTi) + Ti$_2$Cu] data from [52Jou], the maximum solubility of Cu in (βTi) is 13.5 ± 1.5 at.% Cu. The composition of the liquid at the peritectic isotherm is taken to be approximately 35.5 ± 2 at.% Cu. Thermodynamic calculations [83Mur] were used to interpolate the (βTi) liquidus and solidus between the melting point of Ti and the peritectic temperature (see "Thermodynamics").

(αTi)/(βTi) Boundaries.

The eutectoid reaction (βTi) $\rightleftarrows$ (αTi) + Ti$_2$Cu occurs at 790 ± 10 °C; the compositions of (αTi) and (βTi) at 790 °C are 1.6 ± 0.1 and 5.4 ± 0.3 at.% Cu, respectively. Experimental data are given in Table 2 and plotted in Fig. 3.

At any given composition, [51Mcq] and [52Jou] disagreed by approximately 10 °C in the (βTi) transus, with the hydrogen pressure data of [51Mcq] lying at lower temperatures. The thermoelectric power data of [79Vig] and [80Pel] are in better accord with [52Jou] than with [51Mcq]. This discrepancy between the hydrogen pressure data and classical metallographic determinations exists in all Ti systems for which such data exist, and therefore, the data of [52Jou] and [79Vig] are preferred.

The boundaries of the (αTi) region are based on data from [52Jou], [70Ble], [71Wil], and [79Vig]. The metallographic data [62Bor, 63Luz] do not disagree significantly, but are less precise than the lattice parameter and thermoelectric power data. The most restricted solubilities of Cu in (αTi) [79Vig] are also preferred, because of the possibility that small precipitates might escape detection by optical microscopy.

Measured eutectoid temperatures vary between 776 to 780 °C [51Mcq, 79Vig] and 798 to 800 °C [52Jou, 66Ere, thermal analysis]. [70Ble] reported lattice parameters of equilibrium (αTi) as a function of temperature; the minimum lattice parameter is expected to occur at the eutectoid temperature. In this evaluation, the eutectoid temperature of 790 °C was chosen to bring about the best compromise among the (αTi) solubility data, thermal analyses, and lattice parameter data.

Equiatomic TiCu.

[51Kar] and [52Rau] found two compounds near the equiatomic composition—a tetragonal $B11$ phase at compositions containing excess Ti and the tetragonal CuAuI structure at compositions containing excess Cu. The existence of two compounds was not verified by subsequent investigations [65Sch, 66Ere], and therefore a single compound is shown in Fig. 1. If TiCu had two allotropes, a diffusionless transformation would probably occur at some composition and should be revealed in the microstructure of an equiatomic alloy; however, this has not been ob-

Table 2   Solubility of Cu in (αTi)

Reference	Composition, at.% Cu	Temperature, °C	Method/note
[51Mcq]	0.5	830	Hydrogen pressure
	1.0	790	
[79Vig]	1.4	780 ± 5	Thermoelectric power
	1.29	775	
	1.12	765	
	0.78	735	
	0.47	690	
	0.17	600	
	0.11	550	
	0.05	500	
	0.03	475	
[63Luz]	0.4 ± 0.1	550	Metallography, hardness, electrical resistivity
	0.45 ± 0.1	700	
[62Bor]	0.8 to 1.1	400 to 750	Metallography
[52Jou]	1.4	798	Metallography, thermal analysis

Fig. 3---Detail of Experimental Data on the (βTi)/(αTi) Phase Equilibria

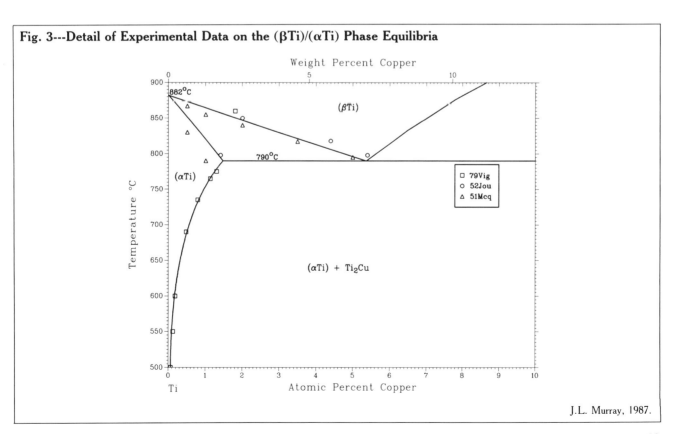

J.L. Murray, 1987.

Table 3 Reaction Temperatures Between 60 and 76 at.% Cu

Reaction	Reference	Temperature, °C	Reaction type
$TiCu + L \rightleftarrows Ti_3Cu_4$	[66Ere]	918	Peritectic
	[52Jou]	916	Peritectic
	[53Trz]	935	Peritectic
$Ti_3Cu_4 + L \rightleftarrows TiCu_2$	[66Ere]	878	Peritectic
	[52Jou]	880	Undetermined
	[53Trz]	892	Peritectic
	[66Zwi]	895	Eutectic
$L \rightleftarrows TiCu_2 + \beta TiCu_4$	[66Ere]	~ 860	Eutectic
	[52Jou]	~ 873	Eutectic
	[53Trz]	880	Eutectic
	[66Zwi]	898	Eutectic
	[52Rau]	875	Eutectic
$Ti_3Cu_4 + TiCu_2 \rightleftarrows Ti_2Cu_3$	[66Ere]	865	Peritectoid
	[53Trz]	885	Peritectoid
	[52Jou]	~ 875	(a)
	[65Sch]	> 870	Peritectoid
$TiCu_2 \rightleftarrows \beta TiCu_4 + Ti_2Cu_3$	[66Ere]	850	Eutectoid
	[53Trz]	872	Eutectoid
	[66Zwi]	865	Eutectoid
	[65Sch]	> 860	Eutectoid
	[69Sin]	> 850	Eutectoid
$L + (Cu) \rightleftarrows \beta TiCu_4$	[66Ere]	870	Peritectic
	[66Zwi]	~ 925	Peritectic
	[53Trz]	895	Eutectic
	[52Jou]	885	Peritectic
	[52Rau]	890	Peritectic
	[58Vig]	870	Eutectic

(**a**) Uninterpreted thermal arrests.

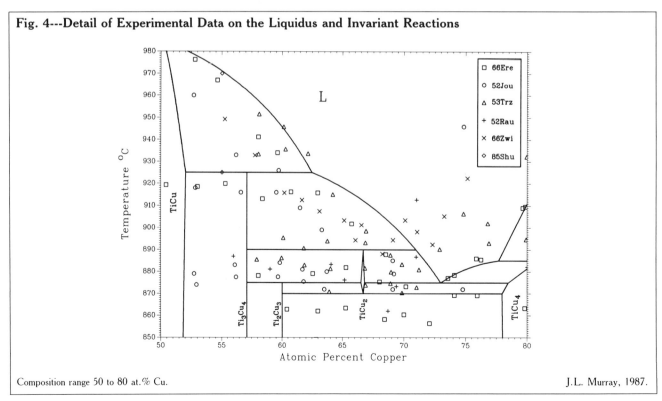

Fig. 4---Detail of Experimental Data on the Liquidus and Invariant Reactions

Composition range 50 to 80 at.% Cu.

J.L. Murray, 1987.

served. It is possible that the second compound observed by [51Kar] was Ti₃Cu₄.

TiCu melts congruently at 985 ± 10 °C [52Jou, 66Ere, 53Trz] (Table 3). Liquidus data are mutually consistent within the experimental uncertainty.

The TiCu homogeneity range was examined by [51Kar] and [66Ere]. On the Ti-rich side, the two reports are consistent at 48 at.% [66Ere] and 47 at.% [51Kar]. On the Cu-rich side, the data of [51Kar] must be discarded because reaction with the crucible caused contamination. Therefore, the data of [66Ere] are used, and the maximum TiCu range is set equal to 48 to 52 at.% Cu.

Composition Range 55 to 75 at.% Cu.

Between the congruent melting point of TiCu and the eutectic reaction at 73 at.% Cu, there occurs a cascade of peritectic reactions involving the compounds Ti₃Cu₄, Ti₂Cu₃, TiCu₂, and TiCu₄. These peritectic reactions (and additional solid-state reactions) obscure the existence of several intermetallic compounds and complicate the interpretation of microstructures. A number of quite different phase diagram constructions based primarily on thermal analysis have been presented (e.g., [53Trz, 66Ere, 66Zwi]). Thermal analysis data are compared with the assessed diagram in Fig. 4.

[65Sch] and [66Ere] made mutually consistent structural identifications of the phases Ti₃Cu₄, Ti₂Cu₃, and TiCu₂. The phase initially identified as TiCu₃ [39Lav] was shown to exist over a range of compositions near TiCu₄ [79Eco, 63Pie, 66Ere]. [62Zwi] postulated a congruently melting compound, Ti₃Cu₇, based on optical observations of melting points and metallographic data. This compound was not subsequently verified, and postulated reactions involving Ti₃Cu₇ are not listed in Table 3.

By etching studies and microprobe analysis of as-cast samples, [63Pie] estimated the compositions of the liquids in the series of three-phase equilibria. Peritectic reactions occurred at 62.5, 71, and 77 at.% Cu, and a eutectic reaction occurred at 73 at.% Cu. Between 62.5 and 73 at.% Cu, therefore, two compounds form from the melt. TiCu₄ was shown as melting congruently by [58Vig] and [53Trz], based on thermal analysis data; however, microprobe and metallographic work [63Pie, 52Jou, 66Ere] showed more definitively that TiCu₄ is formed by a peritectic reaction.

[53Trz] and [66Ere] agreed that a solid-state transformation occurs about 20 °C below the eutectic temperature and that another solid-state transformation occurs at about the same temperature as the eutectic reaction, between about 60 and 67 at.% Cu. Finally, [65Sch] found that TiCu₄ is in equilibrium with Ti₂Cu₃ at 850 °C.

Based on the reaction types, liquidus compositions, and compositions of the intermetallic compounds, the phase diagram must have the topology given by [65Sch] and [66Ere]. Other constructions (e.g., [53Trz]) require the omission of a compound. The reaction temperatures reported by [53Trz] are consistently higher than those reported by [66Ere], but DSC results of [85Shu] favor the higher reaction temperatures. Errors of ± 10 °C are attached to the temperatures given in Table 1, but the differences between temperatures of the various reactions are better known.

TiCu₄.

At the stoichiometry TiCu₄, two phases exist in this system. βTiCu₄ is an established stable equilibrium phase with the ZrAu₄ structure. TiCu₄ has the MoNi₄ structure and was once thought to be a metastable coherent phase, which after long aging, even at low temperature, is succeeded by βTiCu₄ (see "Metastable Phases"). Recent TEM work, however, has established that βTiCu₄ transforms during cooling to αTiCu₄ and therefore that αTiCu₄

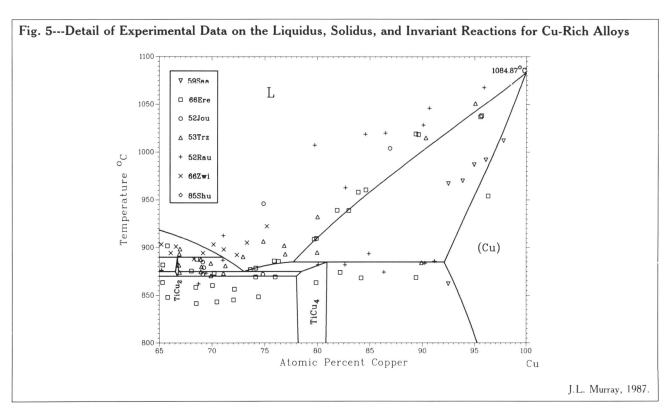

Fig. 5---Detail of Experimental Data on the Liquidus, Solidus, and Invariant Reactions for Cu-Rich Alloys

J.L. Murray, 1987.

Fig. 6---Experimental Data on the Stable Equilibrium (Cu) Solvus, Coherent Solvus, and Spinodal

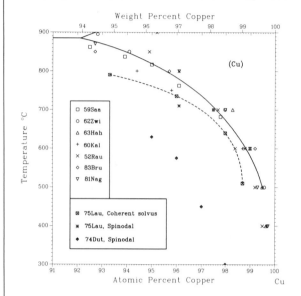

The peritectoid reaction $\alpha TiCu_4$ + (Cu) $\rightleftarrows$ $\beta TiCu_4$ is omitted for clarity.

J.L Murray, 1987.

is the stable phase at low temperature [83Bru]. $\alpha TiCu_4$ is stable up to about 500 °C on the Cu-rich side and up to about 400 °C on the Ti-rich side.

The stoichiometry $TiCu_3$ was first attributed to the phase designated $\beta TiCu_4$ in the present evaluation [39Lav, 51Kar, 52Jou, 53Trz]. [51Kar] attributed to this phase a homogeneity range of 75 to 79 at.% Cu. Later X-ray work showed that 75 at.% Cu alloys contained two phases—Ti_2Cu_3 and $TiCu_4$ [66Vuc, 69Sin]. [66Ere] reported that the composition range of $TiCu_4$ is 78 to 80 at.% Cu, in good agreement with a microprobe determination of the range as 78.5 to 80 at.% Cu [83Bru]. [62Zwi] reported that the composition of $TiCu_4$ in equilibrium with Ti_2Cu_3 at all temperatures is 77.8 at.% Cu (Ti_2Cu_7); [66Vuc] stated that this composition was "presumably" 78.9 at.% Cu (Ti_4Cu_{15}) at an unknown temperature. [79Eco] determined the composition of $\beta TiCu_4$ in equilibrium with (Cu) at 760 °C as 80.9 at.% Cu. The homogeneity range of $\beta TiCu_4$ is therefore estimated as 78 to 80.9 at.% Cu. The homogeneity range of $\alpha TiCu_4$ has not been determined.

[66Zwi] found that contamination by oxygen tends to stabilize $TiCu_4$ at compositions that are enriched in Ti. Because the alloys used by [51Kar] underwent significant reaction with silica tubes, this may explain their attribution of the stoichiometry $TiCu_3$ to this phase.

(Cu) Liquidus, Solidus and Solvus.

Pure Cu melts at 1084.84 °C [Melt]. The assessed liquidus and solidus are approximately straight lines between the melting point of Cu and the end compositions of the peritectic reaction at 890 ± 10 °C, at 77 and 92 at.% Cu. The peritectic composition, 77 at.% Cu, is based on the metallographic and microprobe examination of as-cast alloys by [63Pie]. [63Pie] determined the composition at which the primary phase changes from (Cu) to $TiCu_4$.

Thermal analysis data on the (Cu) liquidus were given by [52Jou], [52Rau], [53Trz], and [66Ere] (Fig. 5). As

shown, there are discrepancies of up to 70 °C among the liquidus data. The highest observed liquidus temperatures [52Rau] are inconsistent with the more accurately known peritectic point and are therefore rejected.

The maximum solubility of Ti in (Cu) is 8 ± 1 at.%, and the solubility decreases to less than 0.1 at.% at 400 °C. Data on the (Cu) solvus are shown in Fig. 6 and listed in Table 4 [52Rau, 53Trz, 58Vig, 59Saa, 60Kal, 62Zwi, 63Hah, 81Nag, 83Bru]. At temperatures below 500 °C, lower solubilities [62Zwi, 63Hah] are preferred.

Solidus temperatures [52Rau, 59Saa, 66Ere] exhibit discrepancies of up to 100 °C. The assessed solidus is therefore based primarily on the peritectic reaction temperature and the (Cu) solvus. The interpolation between the Cu melting point and the peritectic temperature is made by thermochemical calculations ([83Mur], see "Thermodynamics").

Metastable Phases

Cu-Rich Alloys.

Cu-rich alloys have long been known to age-harden [31Kro, 31Sch, 59Saa]. According to [75Lau], the accepted description of the kinetics of decomposition of supersaturated (Cu) solid solutions is as follows. During quenching, a solid solution containing more than about 4 at.% Ti begins to undergo spinodal decomposition into Ti-enriched and Ti-depleted disordered phases [73Cor, 75Lau]. In the early stages of aging, sidebands (or satellite reflections) in electron or X-ray diffraction data indicate the presence of a periodic lattice strain due to oriented coherent particles, i.e., composition waves in the <100> direction [62Heu, 66Sat, 67Sai, 76Dat, 81Ciz]. After a critical composition of the Ti-rich clusters is reached, ordering begins. [75Lau] tentatively ascribed the $L1_0$ structure to an early transitory precipitate; $\alpha TiCu_4$ forms as a persistent metastable precipitate [71Hak, 73Kni, 74Lau1, 74Lau2, 75Lau]. The metastable ordered precipitates are coherent with the fcc matrix and form *in situ* from the disordered precipitate. Finally, incoherent $\beta TiCu_4$ is formed as the equilibrium phase above 500 °C [72Mic, 73Cor, 74Lau1]. Ti-rich clusters are also observed in alloys containing less than 4 at.% Ti, but ordering does not occur in the early stages of decomposition.

Experimental data on the coherent solvus and spinodal are summarized in Table 5 and are plotted in Fig. 6. There are no reported disagreements with the coherent solvus data of [71Hak], [74Lau1], and [75Lau]. There are, however, discrepancies in the data on the spinodal.

[75Lau] placed the spinodal about 35 °C below the coherent solvus. They bracketed the spinodal temperature by comparing microstructures of alloys aged at a temperature that was reached by up-quenching from room temperature to microstructures of alloys aged at the same temperature, but approached from a previous aging temperature above the coherent solvus. [75Lau] used the coarseness of the modulated microstructure as an indication of whether the alloy had passed through the spinodal before the final quench. [74Dut] and [75Vai], however, used only aging treatments at single temperatures. [74Dut], [77Dut], and [78Dut] distinguished between a plate-like and a modulated microstructure to roughly estimate the spinodal curve. [75Vai] considered the absence of precipitate-free zones near grain boundaries a test of whether spinodal decomposition had occurred. The work of [75Lau] is preferred as the best test of whether an alloy has passed through the spinodal. Using small-angle X-ray scattering, [73Tsu1] and [73Tsu2] placed the spi-

Table 4 Experimental Data on the (Cu) Solvus

Reference	Temperature, °C	Solubility, at.% Ti	Method/note
[59Saa]	675 to 700	2.2	Metallography
	750 to 775	3.9	
	805 to 830	5.0	
	825 to 850	6.1	
	850 to 875	7.5	
[53Trz]	300	1.4	Lattice parameters
	600	2.5	
	870	6.5	
[60Kal]	800	5.6	Metallography
	750	4.2	
	600	1.3	
	500	~ 0.4	
[52Rau]	400	0.5	X-ray (a)
	500	0.8	
	600	1.6 (1.2)	
	700	2.3 (2.5)	
	800	3.9 (3.9)	
	850	5.1 (4.8)	
[63Hah]	700	1.7	Electrical resistivity
	600	1.0	
	500	0.5	
	400	0.35	
[62Zwi]	895	7.2	Electrical conductivity, metallography
	850	5.9	
	700	2.0	
	500	0.4	
[81Nag]	870	7.3	See [83Bru]
	800	3.9	
	600	1.0	
	500	0.7	
	400	0.27	
[83Bru]	850	7.3	Microprobe analysis, equilibrated alloys
	800	4.3	
	700	2.5	
	600	0.8 ± 0.1	
	400	0.3 ± 0.1	

(a) Values in parentheses were obtained by microscopy.

nodal at considerably lower temperatures than did [75Lau]. The discrepancy between the results of [73Tsu1, 73Tsu2], and [75Lau] may be caused either by double Bragg scattering, or by failure of the linear spinodal theory to correctly model the experiment done by [73Tsu1] and [73Tsu2].

Ti-Rich Alloys.

Ti-rich (βTi) transforms martensitically to the cph structure during quenching. The addition of Cu to (βTi) does not cause the bcc structure to be retained after quenching at any composition [73Zan, 70Wil1, 70Wil2]. This feature is unusual in alloys of Ti with β-stabilizing elements; it occurs because of diffusional decomposition processes that are rapid in bcc (βTi). The martensite has a massive microstructure for compositions between 0 and 3.8 at.% Cu and an acicular morphology at greater Cu content [70Wil1, 70Wil2]. [60Sat] measured the M_s (start temperatures) of the martensitic (βTi) $\rightarrow$ (αTi) transformation as:

Composition, at.% Cu	M_s temperature, °C
3.6	700
5.6	600
7.4	500
9.0	400

Age hardening occurs in supersaturated (αTi) produced either by quenching from the single phase (αTi) region or by the martensitic transformation. Precipitation of Ti_2Cu takes place by two processes: (1) heterogeneous nucleation and growth of Ti_2Cu at dislocations and (2) uniform nucleation of thin plates or discs of a coherent Cu-rich intermediate precipitate [71Wil]. The fully coherent precipitate has the cph structure. As the particles grow without thickening, coherency is partially lost first on the edge of the disc, then on the flat faces. At this stage, the precipitate can be identified as Ti_2Cu by select area

Phase Diagrams of Binary Titanium Alloys

Table 5 Coherent Solvus and Spinodal (Cu-Rich Alloys)

Reference	Coherent solvus		Spinodal		Method/note
	Composition, at.% Ti	Temperature, °C	Composition, at.% Ti	Temperature, °C	
[74Lau, 75Lau]	1.3	510	3.9	710	Reversion (TEM)
	2.0	635	...	...	
	2.9	675	...	...	
	4.1	735	...	...	
	5.0	746	...	...	
	6.8	790	...	...	
[75Vai]	5.2	720	...	...	TEM
[76Nis]	...	...	4.2	> 500	...
[76Dat]	...	...	5.2	400 to 500	...
[73Tsu1, 73Tsui2]	...	...	5.2	350	SAXS
[74Dut, 77Dut]	...	...	2	~ 300	TEM
	...	...	3	~ 450	
	...	...	4	~ 575	
	...	...	5	~ 630	

Table 6 Crystallization Temperatures of Amorphous Alloys

Reference	Composition, at.% Cu	Crystallization temperature, °C
[81Sak]	31	357
	35	366
	39	384
	43	407
	50	407
	58	428
	66	424
	70	419
	72	404
	75	354
[82Woy]	35	380
	43	395
	50	430
	60	420
	66.7	360
[79Sak]	66.7	403(a)

(a) Glass transition temperature.

diffraction. The ordering reaction occurs *in situ* rather than by redissolution of the intermediate phase [70Lue, 71Wil].

Rapid Solidification.

[68Ray] reported the formation of a noncrystalline phase in splat-quenched alloys in the composition range 65 to 70 at.% Cu. [73Pol] and [75Pol] showed that by splat quenching glassy alloy films could also be prepared in the composition ranges 36 to 44 and 59 to 67 at.% Cu. It was later observed that glasses could be formed over the range 30 to 75 at.% Cu using the chill-block melt-spinning technique [81Sak, 82Woy]. Crystallization temperatures are listed in Table 6. The local structure of the glass was studied primarily by neutron diffraction and also by X-ray diffraction and EXAF techniques [79Sak, 80Rao, 80Sak, 81Fuk, 81Sak]. An unusual "pre-peak" in the structure factor was observed by all investigators and attributed to chemical short-range order of the glass. [81Sak] also observed similar, but less intense, ordering in the liquid phase. On the basis of correlation distances, [80Rao] compared the short-range ordered structure to that of $TiCu_2$.

Table 7 Crystal Structures of the Ti-Cu System

Phase	Homogeneity range, at.% Cu		Pearson symbol	Space group	Strukturbericht designation	Prototype	Reference
(αTi)	0	to 1.6	$hP2$	$P6_3/mmc$	A3	Mg	[Pearson2]
(βTi)	0	to 13.5	$cI2$	$Im3m$	A2	W	[Pearson2]
Ti_2Cu	33.3		$tI6$	$I4/mmm$	$C11_b$	$MoSi_2$	[63Mue]
TiCu	48	to 52	$tP4$	$P4/nmm$	B11	γCuTi	[51Kar]
Ti_3Cu_4	57.1		$tI14$	$I4/mmm$	...	Ti_3Cu_4	[65Sch]
Ti_2Cu_3	60.0		$tP10$	$P4/nmm$	...	Ti_2Cu_3	[65Sch]
$TiCu_2$	66.7		$oC12$	$Amm2$	...	VAu_2	[65Sch]
$TiCu_4$	78	to 80.9	$oP20$	$Pnma$	...	Au_4Zr	[68Pfe]
$αTiCu_4$	~ 78	to ~ 80.9	$tI10$	$I4/m$	$D1_a$	$MoNi_4$	[75Lau]
$TiCu_3$	(a)		$oP8$	$Pmnm$	$D0_a$	$βCu_3Ti$	[71Gie]
β"	(a)		$tP2$	$P4/mmm$	$L1_0$	AuCuI	[75Lau]
(Cu)	100	to 92	$cF4$	$Fm3m$	A1	Cu	[Pearson2]

(a) Metastable.

Metastable crystalline phases are also observed in rapidly solidified alloys. [71Gie] found a metastable phase $TiCu_3(m)$, which is distinct from the equilibrium compounds. The solid solubility of Ti in (Cu) can be extended to 20.2 at.% Ti according to [73Pol], or to 17 ± 2 at.% Ti according to [71Gie]. [73Pol] found that the bcc solid solution is the first crystalline phase to form during heating of amorphous alloys.

Crystal Structures and Lattice Parameters

The crystal structures of equilibrium and metastable phases are summarized in Table 7; more detailed structural information and lattice parameters for the compounds are given in Tables 8 and 9, respectively. The crystal structures fall into the following classes: the orthorhombic structures $TiCu_2$ and $TiCu_4$ approximating

Table 8 Detailed Structures of Ti-Cu Compounds

Phase	Atom positions	Coordinates	z parameter	Reference
Ti_3Cu_4	2 Ti (a)	(0 0 0)		[65Sch, 66Ere]
	4 Ti (e)	(0 0 z) (0 0 −z)	0.295	
	4 Cu (e)	(0 0 z) (0 0 −z)	0.135	
	4 Cu (e)	(0 0 z) (0 0 −z)	0.430	
Ti_2Cu_3	4 Ti (c)	(0.25 0.25 z) (0.75 0.75 −z)	0.145	[65Sch]
		(0.25 0.25 z) (0.75 0.75 −z)	0.565	
	6 Cu (c)	(0.25 0.25 z) (0.75 0.75 −z)	0.955	
		(0.25 0.25 z) (0.75 0.75 −z)	0.755	
		(0.25 0.25 z) (0.75 0.75 −z)	0.335	
TiCu	2 Cu (e)	(0 0.25 z) (0.5 0 −z)	0.1	[51Kar, 66Ere]
	2 Ti (e)	(0 0.5 z) (0.5 0 −z)	0.65	
Ti_2Cu	2 Cu (a)	(0 0 0) (0.5 0.5 0.5)		[62Mue, 63Mue]
	4 Ti (e)	(0 0 z) (0.5 0.5 0.5 + z)	0.339	
		(0 0 −z) (0.5 0.5 0.5 − z)		
$TiCu_3(m)$	2 Ti (a)	± (0.25 0.5 + z 0.25)	0.19 ± 0.02	[71Gie]
	2 Cu (b)	± (0.75 0.5 − z 0.25)		
	4 Cu (f)	± (xz 0.25) ± x + 0.5 −z 0.25)	x = ∼ 0	

Table 9 Lattice Parameters of Ti-Cu Compounds

Phase	Reference	Lattice parameters, nm		
		a	b	c
Ti_2Cu	[66Ere]	0.2943	...	1.0784
	[62Mue]	0.2944	...	1.0786
	[70Ble]	0.295	...	1.078
	[61Enc]	0.2944(a)	...	1.083
	[63Mue]	0.29438 ± 0.00000	...	1.07861 ± 0.00001
TiCu	[66Ere]	0.3107	...	0.5919
	[51Kar]	0.3108	...	0.5887
Ti_3Cu_4	[66Ere]	0.3126	...	1.9964
	[65Sch]	0.312	...	1.994
Ti_2Cu_3	[65Sch]	0.313	...	1.395
	[66Ere]	0.3140	...	1.3962
	[68Pfe]	0.3137	...	1.4024
$TiCu_2$	[65Sch]	0.438	0.797	0.449
	[66Ere]	0.4363	0.7997	0.4478
	[68Pfe]	0.437	0.795	0.448
$TiCu_4$	[66Ere]	0.4522	0.4344	1.2897
	[79Eco]	0.4530 ± 1	0.4342 ± 1	1.2930 ± 3
	[66Vuc]	0.4531	0.4343	1.2929
	[51Kar](b)	0.4503	0.4313	1.2860
	[62Heu](b)	0.450	0.436	1.28
	[69Sin]	0.4526	0.4345	1.2924
	[68Pfe]	0.4530	0.4342	1.2930
	[76Vai](b)	0.4503	0.4313	1.2860
$TiCu_3(m)$	[71Gie]	0.5450	0.4426	0.4307
$\alpha TiCu_4$	[74Lau2]	0.584	...	0.362

(a) Converted from fct to bct cell. (b) Converted from [51Kar] cell.

Table 10 Lattice Parameters of (Cu)

Reference	Composition, at.% Ti	Lattice parameter, nm
[53Trz]	0.48	0.3616
	1.0	0.3617
	3.15	0.3625
	4.0	0.3627
	5.78	0.3640
[51Kar]	0	0.3609
	5.0	0.3616
	6.5	0.3620
[70Kru]	0	0.36150
	1.19	0.36179
	2.46	0.36255
	3.51	0.36306
	4.65	0.36352
	6.25	0.36389
	7.77	0.36421
[62Heu]	4.3	0.3636
	4.5	0.3634
	5.0	0.3633
[52Rau]	0.97	0.36158
	1.35	0.36160
	2.05	0.36163
	2.85	0.36169
	3.25	0.36171
[60Nes]	5.9	0.3620
[56Sim]	0.6	0.3618
	0.9	0.3620
	2.5	0.3624
	3.8	0.3629
	4.0	0.3630
	5.1	0.3635
[73Pol]	20.2	0.3703

12-fold coordination and the tetragonal structures Ti_2Cu, $TiCu$, Ti_3Cu_4, and Ti_2Cu_3 approximating 8-fold coordination.

Ti_2Cu.

Three problems that originally confused the identification of the structure of Ti_2Cu were: (1) identification of the stoichiometry as Ti_3Cu [52Jou, 67Gar]; (2) the appearance of a ternary oxide phase; and (3) the difficulty of recognizing the equilibrium structure using X-ray diffraction. [39Lav] reported a phase Ti_2Cu with the ordered fcc Fe_3W_3C structure. [51Kar] reported an fct phase in the binary alloy, and the Fe_3W_3C structure as the ternary phase Ti_3Cu_3O with a large solubility of Ti. [52Ros], however, concluded that the Fe_3W_3C structure belongs to a true binary phase. [61Enc] and [70Lue] verified the fct structure; the X-ray data of [52Jou] are consistent with the same fct lattice [61Enc]. [62Mue] and [63Mue] discovered additional low-angle forward reflecting lines in the X-ray diffraction pattern that could not be attributed to the fct structure. Because the X-ray scattering amplitudes of Ti and Cu are not very far apart, the structure of Ti_2Cu is difficult to determine by X-ray diffraction. Neutron scattering was used to unambiguously identify the structure as $MoSi_2$. [71Wil] verified the $MoSi_2$ structure. The bct and fct cell parameters are related as:

$$c_{bct} = 3 \ c_{fct} \quad a^2 = 2 \ a_{fct}^2$$

TiCu.

[51Kar] found two phases of near equiatomic composition: the Ti-rich structure was $B11$, and the Cu-rich structure was $L1_0$. The $L1_0$ structure was not verified by any subsequent investigation; the $B11$ structure is generally accepted as the equilibrium structure [66Ere].

Ti_3Cu_4, Ti_2Cu_3, and $TiCu_2$.

The structures of Ti_3Cu_4 and Ti_2Cu_3 have been fully determined. A disagreement between [65Sch] and [66Ere] on Ti_2Cu_3 stemmed partly from a misinterpretation of the abbreviated description of the structure given by [65Sch]. [65Sch] detected diffraction lines that would not be allowed if the structure chosen by [66Ere] were correct. Therefore, Table 7 lists the structure reported by [65Sch]. [65Sch], [66Ere], and [68Pfe] all attributed to Ti_2Cu the orthorhombic VAu_2 structure; no complete determination of atomic positions has yet been made.

$\beta TiCu_4$ and $TiCu_3(m)$.

[51Kar] thought that the equilibrium compound richest in Cu had the stoichiometry $TiCu_3$ and composition range 75 to 79 at.% Cu. [71Gie] discovered that a metastable phase $TiCu_3(m)$ formed during rapid solidification. The symmetry is the same as that found by [51Kar], but the structure has been separately tabulated, because the unit cell found by [71Gie] differs significantly in shape from the [51Kar] cell.

The structure determinations of $\beta TiCu_4$ [68Pfe, 69Sin] are given in Tables 8 and 9. The unit cell upon which the structure is indexed differs from the one that was originally proposed by [51Kar], which is still used to describe the lattice [72Mic, 76Vai, 77Dut]. The unit cells are related by:

$$c_{[51Kar]} = 0.4 \ c_{[68Pfe]}$$

Using several electron microscopy techniques, [79Eco] verified that the structure should be indexed on the larger unit cell. Modifications to the positions of atoms within the unit cell are required to fully explain the [79Eco] TEM data. [80Eco] gave additional information on intrinsic disorder and complex faults within the $\beta TiCu_4$ structure, verifying the original observation by [51Kar] of different degrees of order depending on the heat treatment.

$\alpha TiCu_4$.

[73Kni] attributed an $L1_2$ structure to the intermediate ordered precipitate from the supersaturated (Cu) solution. [71Hak], [74Lau2], and [75Lau] demonstrated that the correct structure is $D1_a$, and that the observations of [73Kni] are also consistent with this structure. [60Nes] and [62Heu] reported an fct intermediate phase with quite different lattice parameters.

Terminal Solid Solution, (Cu).

Lattice parameters of the (Cu) solid solution are listed in Table 10. Lattice parameter measurements for (αTi) were made by [70Ble] as a function of solution temperature; (βTi) lattice parameters are not available because (βTi) cannot be retained metastably to room temperature.

Thermodynamics

Experimental Work.

[81Yok] and [82Kle] made calorimetric investigations of the enthalpy of mixing of liquid alloys at 1373 K in the

Fig. 7---Heats of Mixing for the Liquid Phase at 1100 °C

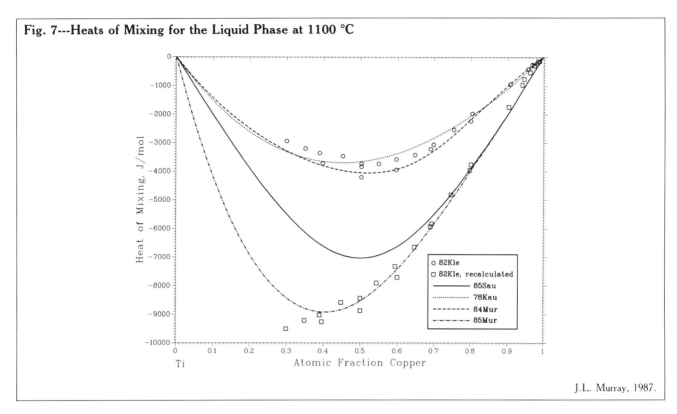

J.L. Murray, 1987.

composition range 30 to 100 at.% Cu. The two studies differ in two respects. The enthalpies of mixing of the Ag-Cu system were used for calibrations by [81Yok]; [82Kle] remeasured the Ag-Cu system and corrected the Ti-Cu results of [81Yok]. After recalculation of the heats measured by [81Yok], there were still unexplained discrepancies between [81Yok] and [82Kle] at compositions away from 50 at.% Cu. [82Som] measured Cu activity in liquid alloys at 1496 K using the Knudsen effusion technique and made calorimetric measurements of heats of mixing. The latter results show greater scatter than those of [82Kle] because of reaction of oxygen with Ti and incomplete mixing of Ti with the liquid alloy.

[80Gac] estimated the excess Gibbs energy of the fcc solution at 98 at.% Cu (1140 K) using the solid cell emf method. The formation of oxides made the G^{ex}(fcc) value very uncertain; comparison of the measured G^{ex}(fcc) with that of the liquid and the phase diagram indicate that this value is too low.

[71Hak] used the Knudsen cell technique to measure Cu activities in bcc (βTi) and interpreted the results in terms of a regular solution model for the excess Gibbs energy, with $G^{ex} = x(1 - x)$ 8033 J/mol of atoms, where x is the atomic fraction of Cu. This excess Gibbs energy is somewhat higher than those derived from the properties of the liquid and the phase diagram.

[79Ari] measured the pressure of hydrogen in equilibrium with Ti-Cu compounds and their hydrides to determine activities in the hydrogen-free alloys. Standard enthalpies and entropies of $TiCu_4$, $TiCu$, and Ti_2Cu were reported. Zero entropy of formation was reported for $TiCu_4$ and $TiCu$.

Thermodynamic Calculations.

Thermodynamic calculations of the Ti-Cu system were made by [70Kau], [78Kau], [83Mur], [85Sau], and [85Mur]. [70Kau] used the regular solution approxima-

tion for all of the solution phases;[78Kau] used additional asymmetry terms in the interaction Gibbs energies. The calculations of [83Mur] were made primarily to predict the (βTi) liquidus, which has not been determined experimentally. The calculated phase boundaries incorporated in the assessed diagram by [83Mur] have been retained in the present version. As a further application of thermodynamic calculations, Gibbs energy functions provide the starting point for modeling the metastable phase equilibria. For theoretical work on the formation of amorphous alloys, thermodynamic calculations provide extrapolations of T_0 curves of equal Gibbs energies to low temperatures.

[85Mur] provided a discussion of the differences among several of the previous calculations of the diagram in terms of the underlying assumptions about the thermodynamics of metastable phases and the treatment of experimental data. The most important difference is in the treatment of the heat of mixing data of [82Kle], who calculated excess enthalpies of mixing of the liquid using an experimentally derived estimate for the heat of fusion of bcc Ti at 1373 K. For the purpose of thermodynamic calculations, however, it is desirable to calculate the excess enthalpy using the constant heat of fusion at the melting point [85Sau]. In Fig. 7, the excess enthalpies, as calculated by [82Kle] and recalculated by [85Sau], are compared with the results of the calculations by [78Kau], [83Mur], [85Sau], and [85Mur]. Thermodynamic functions are compared in Table 11. All calculations reproduce experimental phase diagram data approximately within the experimental uncertainty. The Gibbs energies by [85Sau] are tentatively recommended as incorporating the qualitative characteristics of the thermodynamics of this system and make a reasonable and simple fit to both thermodynamic and phase diagram data.

In Table 11, Gibbs energies of the solution phases are represented as:

Table 11   Thermodynamic Models of Ti-Cu

Properties of the pure components

$G^0(\text{Ti,L})$ = 0	$G^0(\text{Cu,L})$ = 0	
$G^0(\text{Ti,cph})$ = $-20\,585 + 12.134\,T$	$G^0(\text{Cu,cph})$ = $-12\,426 + 10.878\,T$	
$G^0(\text{Ti,fcc})$ = $-17\,238 + 12.134\,T$	$G^0(\text{Cu,fcc})$ = $-13\,054 + 9.623\,T$	
$G^0(\text{Ti,bcc})$ = $-16\,234 + 8.368\,T$	$G^0(\text{Cu,bcc})$ = $-9\,498 + 8.786\,T$ (a)	
	$G^0(\text{Cu,bcc})$ = $-9\,037 + 8.368\,T$ (b)	

Properties of the solution phases

Reference	Gibbs energy parameter (c)
[78Kau]	$A(\text{L})$ = $-14\,644$, $B(\text{L})$ = -2929
	$A(\text{bcc})$ = $5\,858$
	$A(\text{fcc})$ = $4\,602$
	$A(\text{cph})$ = $18\,410$
[83Mur]	$A(\text{L})$ = $-12\,000$, $B(\text{L})$ = -3539, $C(\text{L})$ = 8278 (50 to 100 at.% Cu)
	$A(\text{L})$ = $-16\,009$, $B(\text{L})$ = 566 (0 to 50 at.% Cu)
	$A(\text{bcc})$ = $4\,300$
	$A(\text{fcc})$ = $4\,600$
	$A(\text{cph})$ = $16\,500$
[85Sau]	$A(\text{L})$ = $-24\,667 + 7.593\,T$, $C(\text{L})$ = $-6667 - 2.093\,T$
	$A(\text{bcc})$ = $5\,000$
	$A(\text{fcc})$ = $4\,100$
	$A(\text{cph})$ = $17\,500$
[85Mur]	$A(\text{L})$ = $-34\,056 + 15.295\,T$, $B(\text{L})$ = $-15\,359 + 9.133\,T$
	$C(\text{L})$ = $-14 + 1.094\,T$
	$A(\text{bcc})$ = $3\,917$
	$A(\text{fcc})$ = $8\,097$
	$A(\text{cph})$ = $18\,248$

Properties of the compounds(d)

Reference	
[83Mur]	$G(\text{Ti}_2\text{Cu})$ = $-24\,702 + 11.388\,T$
	$G(\text{TiCu})$ = $-20\,368 + 7.297\,T$
	$G(\text{TiCu}_4)$ = $-23\,428 + 14.130\,T$
	$G(\text{Ti}_3\text{Cu}_4)$ = $-26\,741 + 13.176\,T$
[85Sau](e)	$G(\text{Ti}_2\text{Cu})$ = $-25\,725 + 12.072\,T$
	$G(\text{TiCu})$ = $-26\,220 + 11.921\,T$
	$G(\text{TiCu}_4)$ = $-19\,650 + 10.768\,T$
	$G(\text{Ti}_3\text{Cu}_4)$ = $-40\,278 + 24.237\,T$

Note: Values are given in J/mol and J/mol · K; mol is mol of atoms. (**a**) [70Kau, 84Mur]. (**b**) [85Sau, 85Mur]. (**c**) See Eq 1. (**d**) Only results of [83Mur] and [85Sau] are listed, because they are of most interest for reproducing calculations of the phase. (**e**) [85Sau] listed only heats of formation. The tabulated Gibbs energies are deduced from the heats of formation and the phase diagram.

$$G(i) = G^0(\text{Ti},i)\,(1 - x) + G^0(\text{Cu},i)\,x + RT[x \ln x + (1 - x) \ln (1 - x)] + x(1 - x)\,[B(i) + C(i)\,(1 - 2x) + D(i)\,(x^2 - 6x + 1)]$$

where i designates the phase; x is the atomic fraction of Cu, G^0 the Gibbs energies of the pure metals; and $B(i)$, $C(i)$, and $D(i)$ are coefficients in the polynomial expansion of the excess Gibbs energy.

Two compounds (Ti_2Cu_3 and TiCu_2) were not modeled, because there is neither any thermodynamic data nor sufficient phase diagram data to determine both enthalpies and entropies. Ti_2Cu, TiCu, Ti_3Cu_4, and TiCu_4 were modeled as stoichiometric compounds.

Metastable Equilibria.

The decomposition of the supersaturated cph and fcc solid solutions into two disordered phases before the precipitation of the metastable or equilibrium compound phases strongly suggests that both of these phases exhibit metastable miscibility gaps. In particular, evidence has been found for spinodal decomposition of the fcc solid solution, indicating that an alloy of about 5 at.% Ti is within the metastable spinodal even at about 973 K. Based on present understanding of the thermodynamics of this system, it appears improbable that spinodal decomposition is the effect of large positive excess Gibbs energies such as those which give rise to approximately symmetrical equilibrium miscibility gaps in Ag-Cu, Cd-Zn, or Al-Zn, for example. Rather, spinodal decomposition appears to be the result of subtle variations of the curvature of the Gibbs energy as a function of composition, and the contribution of short-range ordering may have to be included in a thermodynamic analysis which accounts for spinodal decomposition.

T_0 curves, the locus of points where the Gibbs energies of two phases are equal, are of interest for guiding interpretations of the results of rapid solidification experiments, because when the temperature is depressed below T_0, it becomes thermodynamically favorable for a metastable liquid to crystallize without compositional segregation. The Gibbs energies of all three phases—liquid, bcc, and fcc—are extrapolated beyond the temperature range in which they can be tested against either thermochemical or phase diagram data. The Gibbs energies of the solid phases are also extrapolated far beyond the composition ranges over which they have been determined. Therefore, the T_0 curves depend strongly on implicit assumptions

The Cu-Ti System

Fig. 8---T_0 Curves from Recent Calculations of the Ti-Cu Phase Diagram

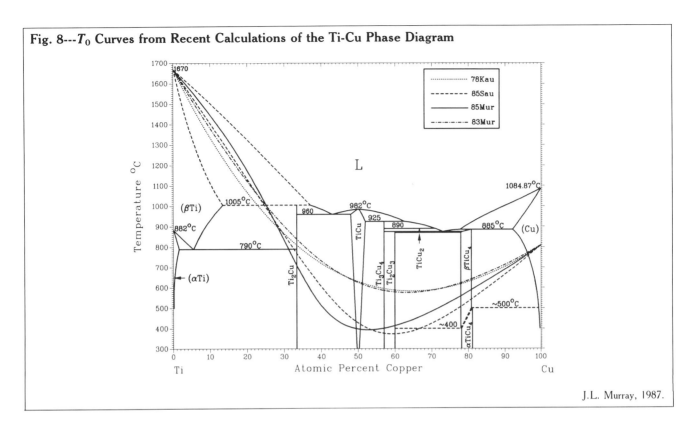

J.L. Murray, 1987.

underlying their calculation. The liquid/bcc T_0 curves from recent calculations are compared in Fig. 8.

Cited References

31Kro: W. Kroll, "Tempered Titanium-Copper Alloys," *Z. Metallkd., 23*, 33-39 (1931) in German. (Equi Diagram; Experimental)

31Sch: E.E. Schumacher and W.C. Ellis, "Age Hardening of Titanium Alloys," *Metals and Alloys, 2*, 111 (1931). (Meta Phases; Experimental)

39Lav: F. Laves and H.J. Wallbaum, "On the Crystal Chemistry of Titanium Alloys," *Naturwissenschaften, 27*, 674-675 (1939) in German. (Crys Structure; Experimental)

51Kar: N. Karlsson, "An X-Ray Study of the Phases in the Copper-Titanium System," *J. Inst. Met., 79*, 391-405 (1951). (Crys Structure; Experimental)

51Mcq: A.D. McQuillan, "The Application of Hydrogen Equilibrium-Pressure Measurements to the Investigation of Titanium Alloy Systems," *J. Inst. Met., 79*, 73-88 (1951). (Equi Diagram; Experimental)

*52Jou: A. Joukainen, N.J. Grant, and C.F. Floe, "Titanium-Copper Binary Phase Diagram," *Trans. AIME, 194*, 766-770 (1952). (Equi Diagram; Experimental)

52Rau: E. Raub, P. Walter, and M. Engel, "Alloys of Titanium with Copper, Silver and Gold" *Z. Metallkd., 43*, 112-118 (1952) in German. (Equi Diagram; Experimental)

52Ros: W. Rostoker, "Observations of the Occurrence of Ti₂X Phases," *Trans. AIME, 194*, 209-210 (1952). (Equi Diagram; Experimental)

*53Trz: W. Trzebiatowski, J. Berak, and T. Romotowski, "The Copper-Titanium System," *Rocznik. Chem., 27*, 426-437 (1953) in Polish. (Equi Diagram; Experimental)

56Sim: W. Simon, Dissertation Work, Berg Akad. Clausthal, (1956) in German. (Crys Structure; Experimental)

58Vig: V.N. Vigdorovitch, A.N. Krestovnikov, and M.V. Malitsev, "Phase Diagram of the Copper-Titanium System," *Izv. Akad. Nauk SSR Otd. Tech. Nauk, (2)*, 145-148 (1958) in Russian. (Equi Diagram; Experimental)

59Saa: M.J. Saarivirta and H.S. Cannon, "Copper-Titanium Alloys Have High Strength," *Met. Prog., 72(2)*, 81-84 (1959). (Equi Diagram; Experimental)

60Kal: K.P. Kalinin and M.Z. Spiridonova, "Investigation of the Properties of Copper-Titanium Alloys," *Tr. Gos. Nauchn-Issled. Proektn. Inst. Obrab. Isvetn. Met., 18*, 46-57 (1960) in Russian. (Equi Diagram; Experimental)

60Nes: E.G. Nesterenko and K.V. Chuistov, "Structural Transformations in the Aging of Copper-Titanium Alloy," *Fiz. Met. Metalloved., 9(3)*, 415-421 (1960) in Russian; TR: *Phys. Met. Metallogr., 9(3)*, 81-86 (1960). (Meta Phases; Experimental)

60Sat: T. Sato, S. Hukai, and Y.C. Huang, "The Mₛ Points of Binary Titanium Alloys," *J. Aust. Inst. Met., 5(2)*, 149-153 (1960). (Meta Phases; Experimental)

61Enc: E. Ence and H. Margolin, "A Study of the Ti-Cu-Zr System and the Structure of Ti₂Cu," *Trans. AIME, 221*, 320-322 (1961). (Crys Structure; Experimental)

62Bor: N.G. Boriskina and K.P. Myasnikova, "Solubility of Iron, Manganese and Copper in α Titanium," *Titan Ego Splavy, Akad. Nauk SSR Inst. Met., 7*, 61-67 (1962) in Russian. (Equi Diagram; Experimental)

*62Heu: U. Heubner and G. Wasserman, "Investigation of Precipitation and Tempering Behavior of Supersaturated Cu-Ti Solid Solutions," *Z. Metallkd., 53*, 153-154 (1962) in German. (Meta Phases; Experimental)

62Mue: M.H. Mueller, M.V. Nevitt, and H.W. Knott, "A Study of the Ti-Cu-Zr System and the Structure of Ti₂Cu," *Trans. AIME, 224*, 611-612 (1962). (Crys Structure; Experimental)

*62Zwi: U. Zwicker, "Tempering and Mechanical Properties of Copper-Titanium Alloys," *Z. Metallkd., 53*, 709-714 (1962) in German. (Equi Diagram; Experimental)

63Luz: I.P. Luzhikov, V.M. Novikova, and A.P. Mareev, "Solubility of β-Stabilizers in α-Titanium," *Metalloved. Term. Obrab. Metallov., (2)*, 13-16 (1963) in Russian; TR: *Met. Sci. Heat Treat., (2)*, 78-81 (1963). (Equi Diagram; Experimental)

63Hah: H.D. Hahlbohm, "Electrical, Magnetic and Galvanomagnetic Measurements of Copper-Titanium and Gold-Titanium," *Z. Metallkd., 5*, 515-518 (1963) in German. (Equi Diagram; Experimental)

63Mue: M.H. Mueller and H.W. Knott, "The Crystal Structures of Ti$_2$Cu, Ti$_2$Ni, Ti$_4$Ni$_2$O, and Ti$_4$Cu$_2$O," *Trans. AIME, 227,* 674-678 (1963). (Crys Structure; Experimental)

*63Pie: P. Pietrokowsky and J.R. Maticich, "The Use of the Electron-Microprobe Analyzer in the Study of Binary Metal Alloy Systems," X-Ray and X-Ray Microanalysis, 3rd Int. Symp., 591-602 (1963). (Equi Diagram; Experimental)

*65Sch: K. Schubert, "On the Constitution of the Titanium-Copper and Titanium-Silver Systems," *Z. Metallkd., 56*(3), 197-198 (1965) in German. (Equi Diagram; Experimental)

*66Ere: V.N. Eremenko, Y.I. Buyanov, and S.B. Prima, "Phase Diagram of the System Titanium-Copper," *Porosh. Met., Akad. Nauk Ukr. SSR, 6*(6), 77-87 (1966) in Russian; TR: *Sov. Powder Met.,* (6), 494-502 (1966). (Equi Diagram; Experimental)

66Sat: K. Sato, "Direct Observations on Precipitation in Copper-Titanium Alloys," *Trans. Jpn. Inst. Met., 7,* 267-272 (1966). (Meta Phases; Experimental)

66Vuc: J.H.N. van Vucht, "Influence of Radius Ratio on the Structure of Intermetallic Compounds of the AB_3 Type," *J. Less-Common Met., 11,* 308-322 (1966). (Crys Structure; Experimental)

66Zwi: U. Zwicker, E. Kalsch, T. Nishimura, D. Ott, and H. Seilstorfer, "Effect of Impurities on the Phase Equilibria of Cu-Rich Copper-Titanium Alloys," *Metall, 20*(12), 1252-1255 (1966) in German. (Equi Diagram; Experimental)

67Gar: Yu.V. Gardina, L.T. Gordeeva, L.G. Timonina, and N.R. Zifferman, "Metallic Compounds in the Ti-Cu System," *Metalloved. Term. Obrab. Metall.,* (2), 10-11 (1967); TR: *Met. Sci. Heat Treat.,* (2), 85-86 (1967). (Equi Diagram; Experimental)

67Sai: K. Saito, K. Iida, and R. Watanabe, "The Ageing Behaviour Cu–4 wt.% Ti Alloy," *Trans. Nat. Res. Inst. Met., 9*(5), 21-27 (1967). (Meta Phases; Experimental)

68Pfe: H.U. Pfeifer, S. Bhan, and K. Schubert, "On the Structures of Ti-Ni-Cu and Quasihomologous Alloys," *J. Less-Common Met., 14,* 291-302 (1968) in German. (Crys Structure; Experimental)

68Ray: R. Ray, B.C. Geissen, and N.J. Grant, "New Non-Crystalline Phases in Splat Cooled Transition Metal Alloys," *Scr. Metall., 2,* 357-359 (1968). (Meta Phases; Experimental)

69Sin: A.K. Sinha, "Close-Packed Ordered AB_3 Structures in Binary Transition Metal Alloys," *Trans. AIME, 245,* 237-240 (1969). (Crys Structures; Experimental)

70Ble: P.A. Blenkinsop and R.E. Goosey, "A Study of the Age Hardening Reaction in Titanium–2 1/2 Copper," Sci. Technol. Appl. Titanium, Proc. Int. Conf., R.I. Jaffee and N.E. Promisel, Ed., 783-793 (1970). (Equi Diagram; Experimental)

70Kau: L. Kaufman and H. Bernstein, *Computer Calculations of Phase Diagrams,* Academic Press, New York (1970). (Thermo; Theory)

70Kru: W.E. Krull and R.W. Newman, "The Lattice Parameter of the αCopper-Titanium Solution," *J. Appl. Crystallogr., 3,* 519-521 (1970). (Crys Structure; Experimental)

70Lut: G. Lutjering and S. Weissmann, "Mechanical Properties and Structure of Age-Hardened Ti-Cu Alloys," *Metall. Trans., 1,* 1641-1649 (1970). (Equi Diagram; Experimental)

70Wil1: J.C. Williams, D.H. Polonis, and R. Taggart, "An Electron Microscopy Study of Phase Transformations in Titanium-Copper Alloys," Sci. Technol. Appl. Titanium, Proc. Int. Conf., R.I. Jaffee and N.E. Promisel, Ed., 733-743 (1970). (Meta Phases; Experimental)

70Wil2: J.C. Williams, R. Taggart, and D.H. Polonis, "The Morphology and Substructure of Ti-Cu Martensite," *Metall. Trans., 1,* 2265-2270 (1970). (Meta Phases; Experimental)

71Gie: B.C. Giessen and D. Szymanski, "A Meta Phase Ti-Cu$_3$(m)," *J. Appl. Crystallogr., 4,* 257-259 (1971). (Meta Phases; Experimental)

71Hak: T. Hakkarainen, "Formation of Coherent Cu$_4$Ti Precipitates in Cu-Rich Cu-Ti Alloys," thesis, Helsinki University of Technology (1971). (Meta Phases; Experimental)

71Wil: J.C. Williams, R. Taggart, and D.H. Polonis, "An Electron Microscopy Study of Modes of Intermetallic Precipitation in Ti-Cu Alloys," *Metall. Trans., 2,* 1139-1148 (1971). (Meta Phases; Experimental)

72Mic: H.T. Michels, I.B. Cadoff, and E. Levine, "Precipitation-Hardening in Cu–3.6 Wt% Ti," *Metall. Trans., 3,* 667-674 (1972). (Meta Phases; Experimental)

73Cor: J.A. Cornie, A. Datta, and W.A. Soffa, "An Electron Microscopy Study of Precipitation in Cu-Ti Sideband Alloys," *Metall. Trans., 4,* 727-733 (1973). (Meta Phases; Experimental)

73Kni: R. Knights and P. Wilkes, "The Precipitation of Titanium in Copper and Copper-Nickel Base Alloys," *Acta Metall., 21,* 1503-1514 (1973). (Meta Phases; Experimental)

73Pol: A.F. Polesya and L.S. Slipchenko, "Formation of Amorphous Phases and Meta Solid Solutions in Binary Ti and Zr Alloys with Fe, Ni, and Cu," *Akad. Nauk SSSR Metally,* (6), 173-178 (1973); TR: *Russ. Metall.,* (6), 103-107 (1973). (Meta Phases; Experimental)

73Tsu1: T. Tsujimoto, K. Saito, and K. Hashimoto, "X-Ray Small-Angle Scattering Study on the Early Stages of the Decomposition in a Cu–4 wt.% Ti Alloy," *J. Jpn. Inst. Met., 37*(1), 67-72 (1973) in Japanese. (Meta Phases; Experimental)

73Tsu2: T. Tsujimoto, K. Hashimoto, and K. Saito, "X-Ray Small Angle Scattering Study on the Later States of the Decomposition in a Cu–4 wt.% Ti Alloy," *J. Jpn. Inst. Met., 37*(1), 61-67 (1973) in Japanese. (Meta Phases; Experimental)

73Zan: A. Zangvil, S. Yamamoto, and Y. Murakami, "Electron Microscopic Determination of Orientation Relationship and Habit Plane for Ti-Cu Martensite," *Metall. Trans., 4,* 467-475 (1973). (Meta Phases; Experimental)

74Dut: J. Dutkiewicz, "The Influence of Temperature on the Aging of Copper-Titanium Alloys," *Bull. Acad. Polon. Sci., 22*(4), 323-328 (1974). (Meta Phases; Experimental)

74Lau1: D.E. Laughlin and J.W. Cahn, "Ordering in Copper-Titanium Alloys," *Metall. Trans., 5,* 972-974 (1974). (Meta Phases; Experimental)

74Lau2: D.E. Laughlin and J.W. Cahn, "The Crystal Structure of the Metastable Precipitate Copper-Based Copper-Titanium Alloys," *Scr. Metall., 8,* 75-78 (1974). (Meta Phases; Experimental)

75Lau: D.E. Laughlin and J.W. Cahn, "Spinodal Decomposition in Age Hardening Copper-Titanium Alloys," *Acta Metall., 23,* 329-339 (1975). (Meta Phases; Experimental)

75Pol: A.F. Polesya and L.S. Slipchenko, "Metastable Phase with bcc Lattice in the Copper-Titanium System," *Izv. V.U.Z. Tsvetn. Metall.,* (6), 144-147 (1975) in Russian. (Meta Phases; Experimental)

75Vai: T.K. Vaidyanathan and K. Mukherjee, "Continuous Precipitation in Cu-Rich Cu-Ti Binary and Cu-Ti-Al Ternary Alloys," *J. Mater. Sci., 10,* 1697-1710 (1975). (Meta Phases; Experimental)

76Dat: A. Datta and W.A. Soffa, "The Structure and Properties of Age Hardened Cu-Ti Alloys," *Acta Metall., 24,* 987-1001 (1976). (Meta Phases; Experimental)

76Vai: T.K. Vaidyanathan and K. Mukherjee, "Precipitation in Cu-Ti and Cu-Ti-Alloys, Discontinuous and Localized Precipitation," *Mater. Sci. Eng., 24,* 143-152 (1976). (Meta Phases; Experimental)

77Dut: J. Dutkiewicz, "Mechanism of Spinodal Decomposition. Discontinuous Precipitation and Ordering in Aged Alloys with F.C.C. Lattice," *Sci. Bull. Stan. Stas. Univ. Min. Met.,* 675, Bull. 80, 56-99 (1977) in Polish. (Meta Phases; Experimental)

78Dut: J. Dutkiewicz, "Spinodal Decomposition, Ordering, and Discontinuous Precipitation in Deformed and Aged Copper-Titanium Alloys," *Met. Technol., 5*(10), 333-340 (1978). (Meta Phases; Experimental)

78Kau: L. Kaufman, "Coupled Phase Diagrams and Thermochemical Data for Transition Metal Binary Systems-III," *Calphad, 2*(5), 117-146 (1978). (Thermo; Experimental)

*79Ari: M. Arita, R. Kinaka, and M. Someno, "Application of the Metal-Hydrogen Equilibration for Determining Thermodynamic Properties in the Ti-Cu System," *Metall. Trans. A, 10,* 529-534 (1979). (Thermo; Experimental)

79Eco: R.C. Ecob, J.V. Bee, and B. Ralph, "The Structure of the β-Phase in Dilute Copper-Titanium Alloys," *Phys. Status Solidi (a), 52,* 201-210 (1979). (Crys Structure; Experimental)

79Sak: M. Sakata, N. Cowlam, and H.A. Davies, "Neutron Diffraction Measurement of the Structure Factor of a CuTi

Metallic Glass," *J. Phys. F. Met. Phys.*, 9(12), L235-L240 (1979). (Meta Phases; Experimental)

*79Vig: G. Vigier, J.M. Pelletier, and J. Merlin, "Determination of Copper Solubility in Titanium and Study of Ti-Cu Solid Solution Stability by Thermoelectric Power Measurements," *J. Less-Common Met., 64*, 175-183 (1979). (Equi Diagram; Experimental)

80Eco: R.C. Ecob, J.V. Bee, and B. Ralph, "Some Ordering Phenomena in the βCu$_4$Ti Phase," *J. Microscopy, 19*, Pt. 1, 153-161 (1980). (Crys Structure; Experimental)

80Gac: J.C. Gachon, J.P. Hilger, M. Notin, and J. Hertz, "Enthalpies and Free-Enthalpies of Formation for Dilute Binary Solid Solutions Cu-M (M in the First Long Period)," *J. Less-Common Met., 72*, 167-192 (1980). (Thermo; Experimental)

80Pel: J.M. Pelletier, G. Vigier, R. Borrelly, and J. Merlin, "Isothermal Decomposition of Solid Solutions in a Ti-2.5 wt.% Cu Alloy," *Titanium '80, Science and Technology* Proc. 4th Int. Conf. on Titanium, H. Kimura and O. Izumi, Ed., Met. Soc. AIME, Warrendale, PA, 1408-1417 (1980). (Meta Phases; Experimental)

80Rao: D. Raoux, J.F. Sadoc, P. Lagarde, A. Sadoc, and A. Fontaine, "Local Structure in a Cu$_2$Ti Amorphous Alloy by EXAFS and X-Ray Scattering," *J. Phys. Colloq.*, C8, *41*(8), C-207-C-210 (1980). (Meta Phases; Experimental)

80Sak: M. Sakata, N. Cowlam, and H.A. Davies, "Measurements of Compositional Order in Binary Metallic Glasses," *J. Phys. Colloq.*, C8, C-190-C-193 (1980). (Meta Phases; Experimental)

81Ciz: P. Cizinsky and J. Pesicka, "The Spinodal Decomposition of Cu-Ti Alloys," *Czech J. Phys. B, 31*, 752-755 (1981). (Meta Phases; Experimental)

81Fuk: T. Fukunaga, K. Kai, N. Masaaki, N. Watanabe, and K. Suzuki, "High Resolution Short-Range Structure of Ni-Ti and Cu-Ti Alloy Glasses by Pulsed Neutron Total Scattering," Proc. 4th Int. Conf. on Rapidly Quenched Metals (Sendai), 347-350 (1981). (Meta Phases; Experimental)

81Nag: K. Nagata and S. Nishikawa, "Aging and Reversion Phenomena of Cu-Ti Alloy," *Rep. Inst. Ind. Sci. Univ. Tokyo, 29*(4), 99-138 (1981).

81Sak: M. Sakata, N. Cowlam, and H.A. Davies, "Chemical Short-Range Order in Liquid and Amorphous Cu$_{66}$Ti$_{34}$ Alloys," *J. Phys. F. Met. Phys., 11*, L157-L162 (1981). (Meta Phases; Experimental)

81Yok: H. Yokokawa and O.J. Kleppa, "Thermochemistry of Liquid Alloys of Transition Metals. II.—Copper and Titanium at 1372 K," *J. Chem. Thermodynam., 13*(8), 703-715 (1981). (Thermo; Experimental)

82Kle: O.J. Kleppa and S. Watanabe, "Thermochemistry of Alloys of Transition Metals: Part III. Copper-Silver, -Titanium, -Zirconium, and -Hafnium at 1373 K," *Metall. Trans. B, 13*, 391-401 (1982). (Thermo; Experimental)

82Som: F. Sommer, K.H. Klappert, I. Arpshofen, and B. Predel, "Thermodynamic Investigations of Liquid Copper-Titanium Alloys," *Z. Metallkd., 73*(9), 581-584 (1982). (Thermo; Experimental)

82Woy: C. Woychik and T.B. Massalski, private communication (1982). (Meta Phases; Experimental)

83Bru: J.Y. Brun, J. Sylvaine, T. Hamar, and H.A. Colette, "Cu-Ti and Cu-Ti-Al Solid State Phase Equilibria in the Cu-Rich Region," *Z. Metallkd., 74*(8), 525-529 (1983). (Equi Diagram; Experimental)

83Mur: J. Murray, "The Cu-Ti (Copper-Titanium) System," *Bull. Alloy Phase Diagrams, 4*(1), 81-95 (1983). (Equi Diagram; Experimental)

85Mur: J. Murray, "Assessment and Calculation of the Ti-Cu Phase Diagram," *Noble Metal Alloys*, T.B. Massalski, W.B. Pearson, L.H. Bennett, and Y.A. Chang, Ed., Proc. Symp. on the Noble Metals, The Metallurgical Society, Warrendale, PA (1985). (Thermo)

85Sau: N. Saunders, "Phase Diagram Calculations for Eight Glass Forming Alloys," *Calphad*, to be published (1985). (Thermo; Theory)

85Shu: R.D. Shull, A.J. McAlister, and M. Kaufman, private communication (1985). (Equi Diagram; Experimental)

The Er-Ti (Erbium-Titanium) System

167.26   47.88

By J.L. Murray

Equilibrium Diagram

There is general agreement that the Ti-rare earth phase diagrams are characterized by the absence of intermetallic compounds, limited mutual solubilities in the solid state, and, often, miscibility gaps in the liquid phase. The Ti-Er system conforms to this pattern, although it appears that the liquid miscibility gap is a purely metastable feature of this system. The assessed phase diagram for the Ti-Er system is shown in Fig. 1.

[60Lov] measured liquidus and solidus temperatures by differential thermal analysis and optical pyrometry for alloys containing 61 to 99 at.% Er. Experimental data are shown in Fig. 2. Probably because of impurities, the melting point of Er was observed to be 1560 °C, compared to the accepted value of 1529 °C [86Gsc]. The eutectic temperature was reported to be 1320 ± 20 °C, based primarily on optical pyrometric data. The eutectic composition was estimated as 74 at.% Er, based on metallographic results. [60Bec] reported some metallographic evidence of liquid immiscibility, but the calculations presented below suggest that, although the miscibility gap is very close to the liquidus, it is metastable in this system.

Lattice parameters of (Er) and (αTi) showed no detectable variation with composition, indicating that mutual solubilities are low. The solubility of Ti in (Er) was estimated as 3.4 at.%, based "on all the available data" [60Lov].

The invariant reaction associated with the βTi → αTi transformation was found to occur at about 890 °C, based on differential thermal analysis [60Lov]. The transformation in pure Ti occurs at 882 °C, indicating that the reaction is peritectoid in character.

The assessed diagram (Fig. 1 and Table 1) is drawn from thermodynamic calculations performed as part of the present evaluation. The data of [60Lov] are fitted to well within the experimental uncertainty, and the excess Gibbs energies are in line with those found for other Ti-rare earth systems.

Metastable Phases

[76Wan] used TEM to examine splat-cooled Ti-Er alloys for evidence of metastable extension of the solid solubility of Ti in (Er). The observed maximum solubility was about 5 at.%, representing a possible slight (~1 at.%) extension from the equilibrium value.

Crystal Structure

Crystal structure data for Ti-Er are listed in Table 2. Lattice parameter data for (αTi) and (Er) were listed by [60Lov]; there is no detectable alloying effect.

Thermodynamics

There are no experimental thermodynamic data for the Ti-Er system. The calculations presented below are

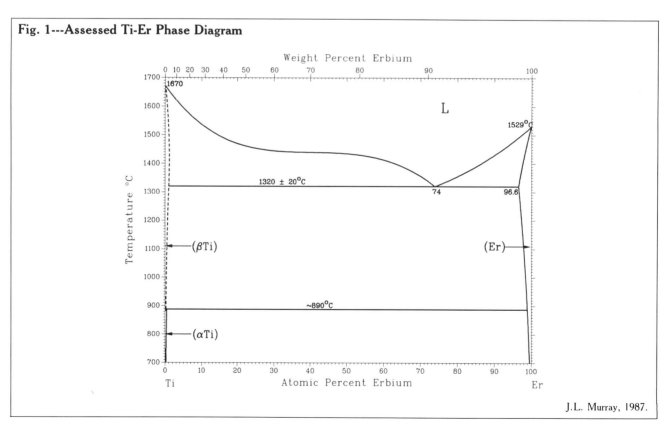

Fig. 1---Assessed Ti-Er Phase Diagram

J.L. Murray, 1987.

The Er-Ti System

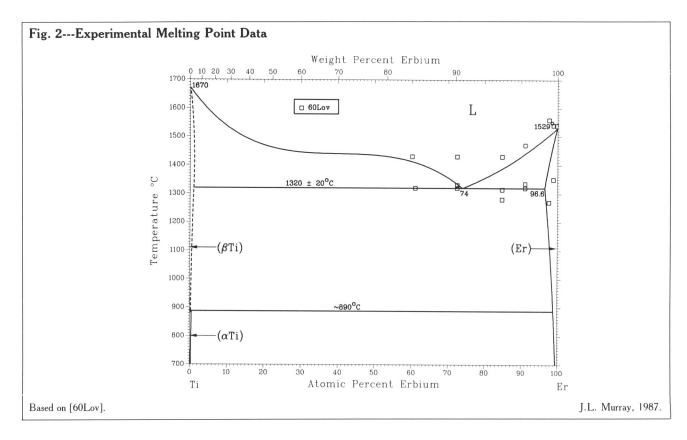

Fig. 2---Experimental Melting Point Data

Based on [60Lov].

J.L. Murray, 1987.

Table 1   Special Points of the Assessed Ti-Er Phase Diagram

Reaction	Compositions of the respective phases, at.% Er			Temperature, °C	Reaction type
L ⇌ (βTi) + (Er) .... ~74		~0.9	~96.6	1320	Eutectic
(βTi) + (Er) ⇌ (αTi) .... ~0.1		~99	~0.3	~890	Peritectoid
L ⇌ (βTi)		0		1670	Melting point
(βTi) ⇌ (αTi)		0		992	Allotropic transformation
L ⇌ (Er)		100		1529	Melting point

Table 2   Crystal Structures of the Ti-Er System

Phase	Homogeneity range, at.% Er		Pearson symbol	Space group	Struktur-bericht designation	Prototype	Reference
(αTi)	0 to	~ 0.3	hP2	P6₃/mmc	A3	Mg	[Pearson2]
(βTi)	0 to	~ 0.9	cI2	Im3m	A2	W	[Pearson2]
(Er)	~ 99 to	100	hP2	P6₃/mmc	A3	Mg	[86Gsc]

based on the phase diagram only. It is thought that the excess properties derived in this way are fairly realistic, first because the miscibility gaps occurring in all phases can be directly related to excess Gibbs energies and second because the derived Gibbs energies are consistent with the series of Ti-rare earth systems.

Parameters of the Gibbs energy expansions are listed in Table 3. For the solution phase i, the expansion is defined by:

$$G(i) = (1 - x)\, G^0(Ti,i) + x\, G^0(Er,i) + RT\,[x \ln x + (1 - x)\ln(1 - x)] + x(1 - x)\,[B(i) + C(i)\,(1 - 2x)]$$

where x is the atomic fraction of Er; G^0 is the Gibbs energy of the pure elements; and B and C are temperature-independent coefficients of the expansion in Legendre polynomials. The Gibbs energies of Ti are taken from [70Kau]; those of Er are based on the emthalpy of fusion from [Hultgren, E] and the melting temperature [86Gsc]. The Gibbs energy of hypothetical bcc Er is roughly estimated, based on the systematics of the rare earths.

Least-squares optimizations were based on invariant temperature data and on estimated solid solubilities. The calculated diagram is shown in Fig. 1 and 2. These Gibbs energies probably give a reasonably accurate qualitative

Table 3 Ti-Er Thermodynamic Parameters

Properties of the pure components

G^0(Ti,L)   = 0
G^0(Er,L)   = 0

G^0(Ti,bcc) = − 16 234 + 8.368 T
G^0(Er,bcc) = − 12 000 + 7.647 T

G^0(Ti,cph) = − 20 585 + 12.134 T
G^0(Er,cph) = − 19 903 + 11.045 T

Excess functions of solution phases

B(L)   = 26 914
C(L)   = 4 149

B(bcc) = 61 000

B(cph) = 50 299
C(cph) = 4 724

Note: Values are given in J/mol and J/mol · K.

picture of the thermodynamics of the system. It should be understood that a careful experimental redetermination of the phase diagram will result in revision of excess Gibbs engergies, because they are based on the phase diagram.

Cited References

60Bec: R. Beck, U.S. Atom. Energy Comm. LAR-10, 93 p (1960). (Equi Diagram; Experimental)

60Lov: B. Love, "The Metallurgy of Yttrium and the Rare-Earth Metals, Part I, Phase Relationships," WADD Tech. Report 60-74, Part I (1960).

70Kau: L. Kaufman and H. Bernstein, *Computer Calculation of Phase Diagrams*, Academic Press, New York (1970). (Thermo; Theory)

76Wan: R. Wang, "Solubility and Stability of Liquid-Quenched Metastable cph Solid Solutions," *Mater. Sci. Eng.*, 7(3), 135-140 (1976). (Meta Phases; Experimental)

86Gsc: K.A. Gschneidner, Jr. and F.W. Calderwood, in *Handbook of the Physics and Chemistry of Rare Earths*, Vol. 8, K.A. Gschneidner, Jr. and L. Eyring, Ed., North-Holland Physics Publishing, Amsterdam (1986). (Crys Structure; Review)

The Eu-Ti (Europium-Titanium) System

151.96   47.88

By J.L. Murray

There is general agreement that Ti-(trivalent)rare earth phase diagrams are characterized by the absence of intermetallic compounds, limited mutual solubilities in the solid state, and, often, miscibility gaps in the liquid phase. Based on very limited experimental work, the Ti-Eu system appears to conform to this pattern [60Kat]. Eu, however, is a divalent metal; metallurgically and chemically it behaves like the alkaline-earth metals (Ca, Sr, and Ba), which also have the above characteristics, except the liquid solubilities are much smaller in the alkaline earths than in the rare-earth metals.

According to [60Kat] (see also [Elliott]), Ti-Eu alloys are difficult to prepare because of the high vapor pressure of Eu. They examined an aggregate liquid phase sintered at 1150 °C by chemical, X-ray, spectrographic, and metallographic analysis and concluded that the solubility of Ti in the liquid at 1150 °C was about 8.9 to 11.7 at.%. They also found that there is slight solubility of Eu in (Ti). Crystal structure data are listed in Table 1.

Cited References

60Kat: H. Kato and C.E. Armantrout, U.S. Atom. Energy Comm. USBM-U-745 (QPR 7), 41-43 (1960).

86Gsc: K.A. Gschneidner, Jr. and F.W. Calderwood, in *Handbook of the Physics and Chemistry of Rare Earths*, Vol. 8, K.A. Gschneidner, Jr. and L. Eyring, Ed., North-Holland Physics Publishing, Amsterdam (1986). (Crys Structure; Review)

Table 1 Crystal Structures of the Ti-Eu System

Phase	Composition, at.% Eu	Pearson symbol	Space group	Strukturbericht designation	Prototype	Reference
(βTi)	0	cI2	Im3m	A2	W	[Pearson2]
(αTi)	0	hP2	P6$_3$/mmc	A3	Mg	[Pearson2]
(Eu)	100	cI2	Im3m	A2	W	[86Gsc]

The Fe-Ti (Iron-Titanium) System

55.847   47.88

By J.L. Murray

Equilibrium Diagram

The equilibrium solid phases of the Ti-Fe system are:

- The cph (αTi) solid solution, based on the equilibrium phase of pure Ti below 882 °C. The (αTi) solvus is retrograde in form, and the maximum solubility of Fe in (αTi) is less than 0.05 at.%.
- The bcc (βTi) and (αFe) solid solutions. The bcc phase is the stable solid phase of pure Ti above 882 °C and the stable solid phase of pure Fe below 912 °C and above 1394 °C.
- The bcc (βTi) and (αFe) solid solutions. The bcc phase is the stable solid phase of pure Ti above 882 °C and the stable solid phase of pure Fe below 912 °C and above 1394 °C.
- The fcc (γFe) solid solution, based on the stable form of pure Fe between 912 and 1394 °C. The (αFe)/(γFe) phase boundaries take the form of a γ loop. The maximum solubility of Ti in (γFe) is 0.8 at.% at approximately 1150 °C.
- The compound TiFe, with the CsCl structure. TiFe forms from the melt by the peritectic reaction L + TiFe$_2$ ⇌ TiFe at 1317 °C. Its homogeneity range is 47.5 to 50.3 at.% Fe at the peritectic temperature.
- The compound based on TiFe$_2$ stoichiometry, the MgZn$_2$-type Laves phase. TiFe$_2$ melts congruently

at 1427 °C. The homogeneity range of TiFe$_2$ is 64.5 to 72.4 at.% Fe.

[14Lam] and [36Jel] reported a phase TiFe$_3$, with the $D0_{22}$ (TiAl$_3$-type) structure [36Jel]. [71Shi], however, investigated alloys near the TiFe$_3$ composition by diffusion experiments and X-ray examination of suitable alloys, but did not find TiFe$_3$. The results of [14Lam] and [36Jel] are attributed to contamination of the alloys, particularly by Al and Si.

[39Lav], [50Duw], and [56Enc] reported a phase Ti$_2$Fe with a structure of the Ti$_2$Ni type, and there has been some controversy about whether it is a true binary phase or is stabilized by impurities [55Thy]. [54Gru] found an fcc (Ti,Al)$_2$Fe phase and suggested that Ti$_2$Fe occurs in the binary system by a peritectic reaction that is not observed because of kinetic factors, but which is facilitated by the presence of impurities. [56Enc] found incomplete evidence suggesting that TiFe$_2$ is a true binary phase formed by a peritectoid reaction at about 1000 °C. However, this structure belongs to a ternary oxide phase ubiquitous in Ti-transition metal alloys [52Ros], and subsequent work strongly suggests that this phase does not appear in the binary system [55Thy, 56Kor, 65Bor, 80Dew].

The assessed Ti-Fe phase diagram (Fig. 1) is topologically the same as those of [Shunk] and [Hultgren, B]. The major changes made in this evaluation are to the homo-

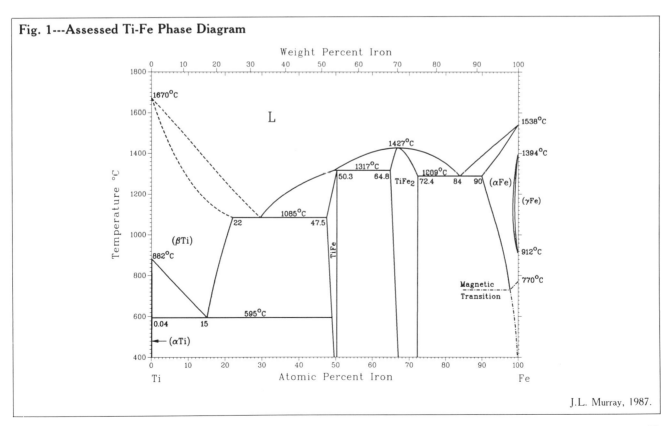

Fig. 1---Assessed Ti-Fe Phase Diagram

J.L. Murray, 1987.

Phase Diagrams of Binary Titanium Alloys

Table 1   Special Points of the Ti-Fe Phase Diagram

Reaction	Compositions of the respective phases, at.% Fe			Temperature, °C	Reaction type
L ⇌ (βTi) + TiFe	29.5	21	47.5	1085	Eutectic
(βTi) ⇌ (αTi) + TiFe	15	0.004	49	595	Eutectoid
L + TiFe$_2$ ⇌ TiFe	49.5	64.8	50.3	1317	Peritectic
L ⇌ TiFe$_2$ + (αFe)	84	72.4	90	1289	Eutectic
L ⇌ TiFe$_2$		66.7		1427	Congruent
(βTi) ⇌ (αTi)		0		882	Allotropic transformation
L ⇌ (βTi)		0		1670	Melting point
L ⇌ (αFe)		100		1536	Melting point
(αFe) ⇌ (γFe)		100		911	Allotropic transformation
(γFe) ⇌ (αFe)		100		1392	Allotropic transformation

---

Fig. 2---Experimental Data on the Ti-Fe Phase Diagram

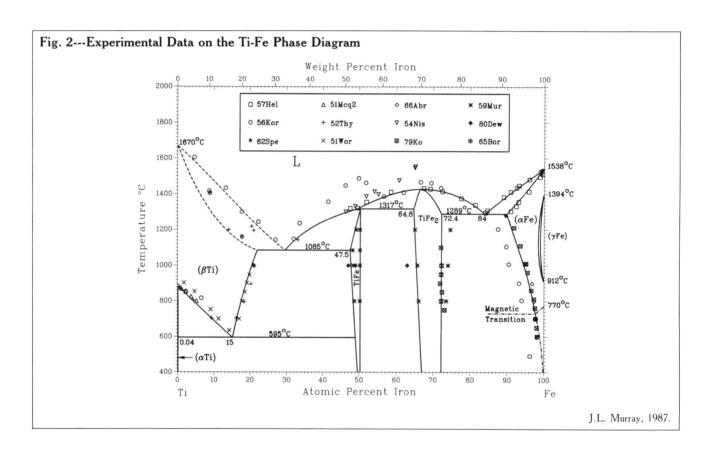

J.L. Murray, 1987.

---

geneity ranges of (γFe), TiFe$_2$, and (αTi). Special points of the phase diagram are summarized in Table 1. The assessed diagram is compared to the experimental data in Fig. 2; details of the (αTi) solvus and the γ loop are shown in Fig. 3 and 4, respectively.

Ti-Rich Liquidus, Solidus, and (βTi)/(βTi) + TiFe Boundary.

The temperature of the eutectic reaction L ⇌ (βTi) + TiFe is 1085 ± 5 °C, and the solidus and liquidus compositions are 21 and 29.5 ± 0.5 at.% Fe, respectively [52Thy, 56Kor, 65Bor]. Experimental liquidus and solidus data are compared with the assessed diagram in Fig. 2. Reported values of the eutectic temperature are:

Reference	Eutectic temperature, °C	Experimental technique
[51Wor]	~ 1060	Optical observation
[52Thy]	1080	Thermal analysis (cooling), metallography
[54Nis]	1080	Metallography
[56Kor]	1100	Thermal analysis
[65Bor]	1090	Thermal analysis (heating)

The Ti-rich liquidus and solidus are the least well-known phase boundaries of the system. Contamination, either in the starting materials [51Wor] or during melting

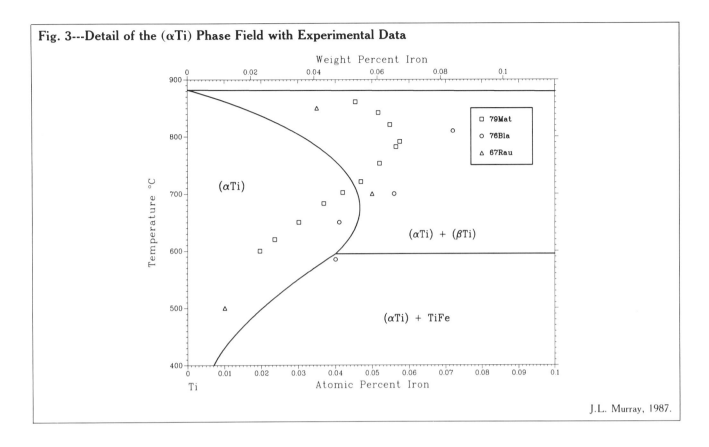

Fig. 3---Detail of the (αTi) Phase Field with Experimental Data

J.L. Murray, 1987.

[52Thy, 56Kor], is a source of inaccuracy in all the liquidus and solidus data. [56Kor] found the melting point of pure Ti to be 1725 °C, compared to the evaluated melting point of 1670 °C, probably because the alloys were contaminated during melting.

[52Thy] estimated the eutectic composition as the composition at which only one arrest could be detected by thermal analysis. The assessed eutectic composition (29.5 at.% Fe) is based on [52Thy] and is also consistent with the thermal analysis data of [56Kor]. The maximum solubility of Fe in (βTi) is 22 at.% at 1085 °C, based primarily on extrapolation of the (βTi)/(βTi) + TiFe boundary to 1085 °C and is also consistent with solidus data.

The assessed (βTi)/(βTi) + TiFe boundary is based on metallographic data [51Wor, 52Thy, 56Kor] and on a microprobe analysis [80Dew] of an alloy equilibrated at 1000 °C. Based on the tie-line data, the results of [80Dew] are roughly estimated to be accurate to 0.5 at.%.

(βTi) Transus.

(βTi) decomposes by the eutectoid reaction (βTi) ⇄ (αTi) + TiFe. Experimental determinations of the (βTi) transus were made by [51Wor], [51Mcq1], [51Mcq2], [52Thy], and [56Kor]. [51Wor] and [52Thy] used quenching and metallographic techniques, whereas [51Mcq1] and [51Mcq2] used the hydrogen pressure method and metallography, and [56Kor] used thermal analysis. Contamination of alloys or insufficiently rapid quenching can lead to erroneously high temperatures for the (βTi) transus; however, in the series of Ti-based systems, the hydrogen pressure technique tended to underestimate the temperature of the transus.

Moreover, the eutectoid reaction is very sluggish [67Rau]. The eutectoid temperature was reported as

−600 °C [51Wor], 585 °C [52Thy], 615 °C [56Kor], 625 ± 10 °C [54Pol], and 595 °C [67Rau]. The [67Rau] value is considered most reliable because thermal analysis, X-ray diffraction, and metallography were used in conjunction. The assessed value (595 °C) is based on the work of [67Rau].

Solubility of Fe in (αTi).

A detail of the (αTi) solvus is given in Fig. 3. [62Bor] and [63Luz] found that solubilities were between 0.2 and 0.5 at.% Fe. They used short anneals and reported that equilibrium was not achieved in every alloy. [67Rau], however, found that the maximum solubility was only 0.05 at.% Fe at approximately 700 °C; the low solubility was confirmed [74Stu, 76Bla, 79Mat, 79Nas, 80Mat].

[67Rau] used very long heat treatments and cold working to achieve equilibrium. The assessed solvus is based primarily on the data of [67Rau] because of the care taken by these investigators to homogenize and equilibrate alloys.

[79Mat] and [80Mat] used very pure starting materials and took special care not to contaminate the alloys; they annealed their samples 5 h and reported that longer anneals gave similar results. [79Mat, 80Mat] found a retrograde solubility, with maximum solubility of 0.057 at.% Fe at 790 °C.

[76Bla] and [79Nas] used quantitative analysis of Mössbauer spectra to determine the (αTi) solvus. Several alloys within the two-phase (βTi) + (αTi) region were annealed and the composition of (αTi) was determined. This method would overestimate the solubility if equilibrium were not achieved, and therefore, for each annealing temperature, the lowest reported solubility is shown in Fig. 3.

Table 2 Isothermal Melting and Three-Phase Equilibria of TiFe$_2$

Reference	Invariant temperatures, °C			Method
	Congruent melt	Peritectic reaction	Eutectic reaction	
[54Nis]	1530	1317	1340	Optical observation of melting
[57Hel]	1427	1316	1289	Thermal analysis (heating, cooling)
[56Kor]	1480	...	1298	Thermal analysis (cooling)
[65Bor]	...	1315	1325	Thermal analysis (heating)

The low solubilities were also verified by [74Stu] and [77Stu]. By means of Mössbauer spectroscopy, [74Stu] determined that a metastable 0.19 at.% Fe single-phase cph alloy formed (θ phase) during aging at 320 °C. θ is a metastable phase from which TiFe ultimately precipitates, and therefore, the 0.19 at.% Fe alloy lies within the two-phase region.

The assessed (αTi) solvus shown in Fig. 3 is based on a least-squares optimization of Gibbs energies using select experimental data [67Rau, 76Bla, 79Mat, 79Nas, 80Mat]. The calculated solvus agrees better with the data of [67Rau] than with those of [76Bla, 79Nas] or [79Mat, 80Mat].

Equiatomic Compound TiFe.

TiFe has the ordered CsCl structure [67Dor, 75Hut], with maximum homogeneity range of 47.5 to 50.3 ± 0.5 at.% Fe at the eutectic temperature (1085 °C).

[59Mur] determined the homogeneity range by a combination of lattice parametric and metallographic methods for several annealing temperatures. The variation of lattice parameters with composition is consistent with the microstructural data.

[80Dew] annealed 35 and 57 at.% Fe alloys at 1000 °C (within the two-phase regions on both sides of stoichiometry) and used microprobe analysis to determine the compositions of equilibrium TiFe. The Ti-rich phase boundary deviates from other studies in the direction that would be expected if the alloy had not fully equilibrated during annealing. The value of 40 at.% Fe for the composition of the Fe-rich boundary cannot be explained in the same way. However, the ternary isothermal sections constructed by [80Dew] suggest that uncertainties in the compositions are about 1 at.%. Crystal structure studies of TiFe have established that the stoichiometric alloy is single phase.

Early studies of the phase diagram [41Wal, 56Kor] indicated that TiFe melts congruently at a temperature above 1500 °C. Based on the thermal analysis study of [57Hel], however, TiFe forms from the liquid at 1317 °C by the peritectic reaction L + TiFe$_2$ ⇌ TiFe. The peritectic character of this reaction was also verified microstructurally for single-crystal TiFe specimens grown by the Bridgeman technique [80Lie].

TiFe$_2$.

TiFe$_2$ is a Laves phase of the hexagonal MgZn$_2$ type; it was discovered in the Ti-Fe system by [38Wit] and [41Wal]. Alloys of composition near TiFe$_2$ are very reactive. [57Hel] reported that alloys of less than 70 at.% Fe were contaminated by alumina crucibles and that thoria or thoria-lined crucibles had to be used. [67Bru] reported that 1 at.% Al and 1 vol.% TiO were present in a 72 at.% Fe sample because of the same problem. The high values of the congruent melting temperatures observed by [38Wit], [41Wal], and [54Nis] were thus probably due

entirely to contamination of the alloys. The assessed eutectic, peritectic, and congruent melting temperatures (1289, 1317, and 1427 °C, respectively) are based on thermal analysis of [57Hel] (heating and cooling). Other reported invariant temperatures are listed in Table 2.

On the Ti-rich side, metallographic and lattice parametric results agree that the TiFe$_2$ composition lies between approximately 64.6 and 65.1 at.% Fe near the peritectic temperature [59Mur]. [68Nak] used magnetic (Nèel temperature) data to determine the composition range, but various investigators disagreed about the composition dependence of T_N (see Magnetic Properties); therefore, T_N should not be considered an accurate indicator of composition. By microprobe determinations of the composition of TiFe$_2$ in a 57 at.% Fe alloy annealed for 500 h at 1000 °C, [80Dew] obtained 63 at.% Fe.

The composition of TiFe$_2$ in equilibrium with (αFe) is approximately 72.4 at.% Fe between 750 and 1100 °C, based on microprobe analysis of equilibrated alloys and lattic parameter measurements [79Ko, 81Tak]. [59Mur] proposed a somewhat larger homogeneity range for TiFe$_2$, based on lattice parameter measurements (see Fig. 2).

(αFe) Liquidus, Solidus, and Solvus.

The melting point of pure Fe is 1538 °C [82Swa]. The assessed liquidus and solidus are based on the thermal analyses of [57Hel]. Contamination of the alloys was a serious problem in early studies [14Lam, 38Tof, 38Wit], and they are not considered in the present assessment.

The (αFe) solvus was determined by [66Abr] using lattice parameter data and by [79Ko] and [81Tak] using both lattice parameters and microprobe analyses of equilibrated samples. Analysis of the volume fractions of coexisting phases [62Spe] corroborates the work of [66Abr] and [79Ko]. Based on metallographic work, [56Kor] and [65Bor] reported higher solubilities. The lower solubilities are preferred, because equilibrium was more likely to have been achieved.

The shape of the (αFe) solvus is expected to be influenced by the second-order ferromagnetic phase transition. Theoretical models for this effect were discussed by [79Ko] and [81Tak]. In the Ti-Fe system, this effect is small because the Curie temperature of (αFe) is changed very little by the addition of Ti (see Magnetic Properties). Thus, the (αFe) solvus is shown as a single, continuous curve through the magnetic transition.

Solubility of Ti in (γFe).

Pure Fe undergoes a transition from the low-temperature bcc structure to the fcc structure at 912 °C and back to the high-temperature bcc form at 1394 ± 2 °C [82Swa]. A detail of the γ loop is shown in Fig. 4, with select experimental data. The maximum solubility of Ti in (γFe) is 0.8 at.% at approximately 1100 to 1150 °C, and the two-phase (γFe) + (αFe) region is approximately 0.6 at.% wide at the same temperature [57Hel, 59Mol, 63Wad,

Fig. 4---Detail of the Ti-Fe γ Loop with Experimental Data

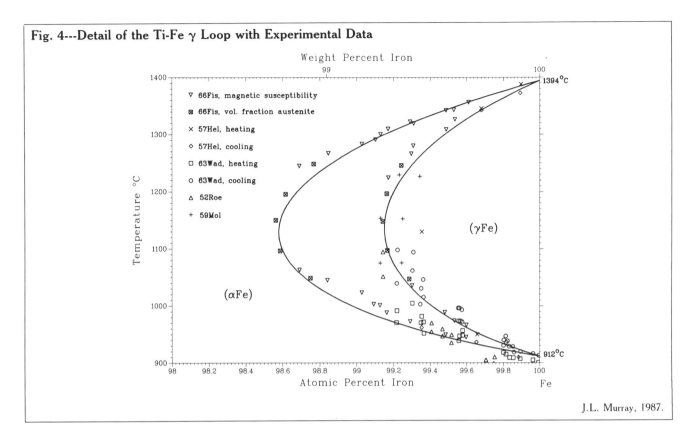

J.L. Murray, 1987.

66Fis]. [Shunk] reported a much lower solubility of 0.2 at.% Ti; [Shunk] seems, however, to have misinterpreted comments by [63Wad] on unpublished work. The early work [30Mic, 35Sve, 38Tof, 48Pet] was not used in the present assessment.

[52Roe] reported that the maximum extent of the outer loop is approximately 0.8 at.% Ti, from dilatometric measurements on alloys containing up to 1.1 at.% Ti. [52Roe] also reported a minimum in the (γFe) + (αFe) boundaries near 912 °C; this has not been verified, although the phase boundaries were found by later investigations to be nearly horizontal near 912 °C.

The assessed γ loop is based primarily on magnetic susceptibility data [66Fis] and is also consistent with thermal analysis data [57Hel] and dilatometry [63Wad].

Metastable Phases

Martensitic Transformations of (βTi).

In Ti-rich alloys, cph (αTi) can form martensitically from (βTi) during quenching. In some Ti-based alloy systems (e.g., Ti-Mo), depending on the alloy composition, either a cph (α'Ti) or an orthorhombic (α″Ti) structure may form martensitically during quenching from (βTi). In Ti-Fe alloys, the orthorhombic martensite has not been observed; however, an fcc martensite has been reported in this system.

Quenching of (βTi) produces martensite in alloys containing less than approximately 3 to 4 at.% Fe [52Thy, 55Pol, 63Bor, 66Nis, 76Stu]. With greater Fe content, (βTi) can be retained after quenching, and ω phase is also observed.

Figure 5 shows experimental data on the start temperatures of the martensitic and reverse martensitic transformations (M_s and β_s, respectively) [50Duw], 60Gri,

63Kan, 80Yam]. Additional qualitative results were given by [54Pol], [55Pol], and [60Sat]. [54Pol] claimed that M_s increases with increasing Fe content; this is not generally the case for the Ti–transition metal alloys. The M_s data of [60Gri], [63Kan], and [80Yam] are mutually consistent; those of [50Duw] lie somewhat higher in temperature. The dashed lines shown in Fig. 5 are based primarily on [80Yam].

Using TEM, [66Nis] and [67Nak] found three phases in a quenched 2.6 at.% Fe alloy—retained (βTi), (α'Ti) martensite, and an fcc martensite. The fcc martensites found by TEM are often thin foil artifacts and have an orthorhombic structure in the bulk [74Spu]. [69Gus] found an fcc martensite in addition to the metastable (α'Ti) and ω phases using X-ray diffraction; the type of phase present depended on the cooling rate. The fcc martensite was only found in surface layers of the samples, which suggests that the phase was stabilized by interstitial impurities.

[74Stu], [76Stu], and [77Ron] found a metastable phase, θ, which formed from the α' martensite during aging above 280 °C. The θ phase is indistinguishable from the (α″Ti) martensite by metallographic or X-ray diffraction methods and was detected only by Mössbauer spectroscopy.

Metastable ω Phase.

The ω phase is an intermediate phase in the decomposition of metastable (βTi) into equilibrium (αTi) + (βTi). Its structure in Ti-Fe alloys was originally reported to be complex bcc [56Yos], but is now generally acknowledged to be hexagonal (P6/mmm) [58Sil, 69Hic, 69Osh]. ω precipitates are cubic in shape [69Hic, 73Ika]; the cubic morphology is associated with a large misfit between the bcc and ω lattices.

Fig. 5---Start Temperatures (M_s) and Reversion Temperatures (β_s) of the Martensitic (βTi) → ($\alpha' Ti$) Transformation

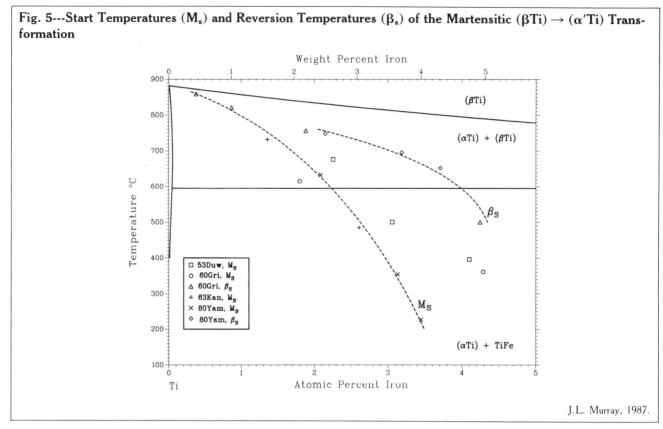

J.L. Murray, 1987.

The ω phase forms from metastable (βTi) phase either at fixed composition during quenching (athermal ω), or during aging at temperatures between approximately 300 to 450 °C [63Bor, 69Hic, 76Stu]. During aging, the ω phase composition approaches 4 to 5 at.% Fe, and the (βTi) composition approaches 10 to 14 at.% Fe, depending upon the aging temperature [69Hic, 78Stu]. If alloys are aged above about 450 °C to 500 °C, equilibrium (αTi) is formed directly from metastable (βTi) [69Hic, 71Miy].

The ω phase can also be formed in supersaturated (αTi) alloys (0.19 to 0.22 at.% Fe) as a result of plastic deformation or coherency stresses between (αTi) and θ phases, as detected by Mössbauer spectroscopy [77Ron, 77Stu].

As-quenched (athermal) ω phase is found only in a limited composition range near the saturation composition of aged ω, 4 at.% Fe [71Odi, 75Fed, 76Stu]. [74Gus] and [78Gus] found (β + ω) in the range 4 to 8 at.% Fe; according to [59Age], the range in which (β + ω) is formed depended on the temperature from which the alloy was quenched. As-quenched ω precipitates in a brine-quenched 4.3 at.% Fe alloy were too small to be resolved by TEM [75Fed, 60Osh]. [72Kha] found athermal ω in a 16.7 at.% Fe alloy. The phase was designated as ω' because the structure contained ⟨111⟩ displacements from the "perfect" ω structure. "ω-like" diffuse streaking in electron diffraction patterns occurs in an alloy containing 0.5 at.% Fe [71Pat]. The diffuse streaking indicates short-range ω-like distortions of the crystal structure.

Low-Temperature Transformation of TiFe₂.

[71Ike], [72Ike1], [72Ike2], and [72Ike3] postulated the existence of a martensitic transformation in $TiFe_2$ below room temperature based on the occurrence of strong thermo-hysteresis in the electrical resistivity. The struc-ture of the martenite was not determined, but [72Ike1], [72Ike2], and [72Ike3] suggested that it may be another Laves phase variant of the MgZn₂-type. The start temperature of the transformation was −8 °C at stoichiometry and decreased away from stoichiometry.

Rapid Solidification.

By splat-cooling of the liquid at rates estimated as 10^6 to 10^8 °C/s, [72Ray] produced extended (βTi) solutions containing up to 35 at.% Fe and TiFe in the range 35 to 50 at.% Fe. The lattice parameters of the ordered and disordered phases followed a single linear trend with composition, from which [72Ray] concluded that the metastable ordering transition may be second order in character. Using cooling rates of 10^7 to 10^8 °C/s, [73Pol] produced amorphous alloys in the range 28 to 30 at.% Fe (near the Ti-rich eutectic point). [80Ino] did not produce amorphous alloys by melt spinning in the range 77 to 87 at.% Fe (near the Fe-rich eutectic point).

Crystal Structures and Lattice Parameters

Crystal structure data for the Ti-Fe phases are summarized in Table 3. TiFe was first reported to have the CsCl structure by [39Lav]. Because the X-ray scattering factors of Ti and Fe are nearly equal, it is difficult to unambiguously determine whether the structure is ordered by X-ray diffraction [50Duw, 51Fre, 56Kor, 57Phi, 59 Dwi]. [60Pie] and [72Ray] did find X-ray evidence of ordering; [72ray] measured order parameters as a function of composition for a wide range of metastable splat-cooled alloys. Neutron scattering [67Dor, 74Ike1, 74Ike2, 74Ike3, 75Hut] definitively established that TiFe has the CsCl structure up to a melting point. Mössbauer spectroscopy can also be used to measure ordering in TiFe [67Wer,

Table 3 Crystal Structures of the Ti-Fe System

Phase	Homogeneity range, at.% Fe	Pearson symbol	Space group	Struktur-bericht designation	Prototype	Reference
(αTi)	0 to 0.04	$hP2$	$P6_3/mmc$	$A3$	Mg	[Pearson2]
(βTi)	0 to 22	$cI2$	$Im3m$	$A2$	W	[Pearson2]
TiFe	48 to 50.2	$cP2$	$Pm3m$	$B2$	CsC1	[67Dor, 75Hut]
TiFe$_2$	64.5 to 72.4	$hP12$	$P6_3/mmc$	$C14$	MgZn2	[50Duw]
(αFe)	90 to 100	$cI2$	$Im3m$	$A2$	W	[Pearson2]
(γFe)	99.4 to 100	$cF4$	$Fm3m$	$A1$	Cu	[Pearson2]
ω	(a)	$hP3$	$P6/mmm$	...	ωMnTi	[69Osh]

(a) Metastable.

Table 4 Room-Temperature Lattice Parameters of TiFe$_2$

Reference	Composition, at.% Fe	Lattice parameters, nm	
		a	c
[50Duw]	66.7	0.4769	0.7745
[59Mur]	65.2	0.4804	0.7849
	66.3	0.4804	0.7843
	66.7	0.4797	0.7836
	68.7	0.4785	0.7816
	68.9	0.4783	0.7814
	69.9	0.4783	0.7815
	70.9	0.4783	0.7811
	72.8	0.4777	0.7807
[68Bru]	64.6	0.4777	0.7807
	65.2	0.4781	0.7795
	69.5	0.4786	0.7810
	71.4	0.4781	0.7795

Table 5 Lattice Parameters of FeTi

Reference	Composition, at.% Fe	Lattice parameter, nm
[50Duw]	50	0.2969
[59Dwi]	50	0.2976
[59Mur]	48	0.2988
	50.3	0.2976
[60Stu]	50	0.2976
	47	0.2986
[72Ray]	37.5	0.3039
	40	0.3023
	43	0.3011
	45	0.2995
	47.5	0.2992
	50	0.2976
[74Ike3]	50.5	0.29735
	47.5	0.29780
[80Dew]	47	0.2986
	49	0.2980
[80Lie]	~ 50	0.2978 ± 0.0002

68Swa, 76Stu]. Lattice parameter data for TiFe are listed in Table 4.

Room-temperature lattice parameters of TiFe$_2$ are listed in Table 5. The lattice parameters of TiFe$_2$ exhibit a change in slope at the stoichiometric composition; the slope change is correlated with a change in the magnetic properties at the same composition [68Bru].

The lattice parameters of (αTi) are not changed by the addition of the small amount of Fe that can be dissolved in it [52Thy, 67Rau, 79Mat]. The lattice parameters of (βTi) are shown in Fig. 6. The lattice parameter data of [72Ray] were obtained from metastable splat-cooled bcc alloys. These lattice parameters showed no discontinuity between the ordered TiFe and disordered (βTi) phases. [72Ray] concluded that the ordering transition may be second order in character.

The effect of Ti on the lattice parameters of (αFe) was studied by [55Sut], [63Arr], [66Abr], and [72Zwe]. [63Arr] and [72Zwe] showed that the marked discrepancy between data for filings and bulk specimens was probably caused by loss of Ti (through oxidation) during cold working. The data of [63Arr], [66Abr], and [72Zwe], were obtained from bulk specimens. [66Abr] summarized lattice parameter data (for bulk specimens heat treated at 700 °C and furnace cooled) by:

$$a(nm) = 0.28662 + 31.30 \, (10^{-5}) \, x$$

where x is the Fe content in at.%. Lattice parameters reported by [63Arr] were derived from density measurements. [63Arr] and [72Zwe] reported only the difference between the lattice parameters of the alloys and that of pure αFe; the lattice parameters shown in Fig. 7 were

calculated using 0.28662 nm [66Abr] as the lattice parameter of pure αFe.

Thermodynamics

Experimental Data.

[55Kub] measured the heat of formation of TiFe from the two bcc metals by reaction calorimetry and found $-20\,300 \pm 630$ J/mol of atoms. They could not measure the heat of formation of TiFe$_2$, because the reaction was incomplete.

[74Wag] and [75Fur] measured activities of Ti and Fe in the liquid phase at 1545 and 1600 °C, respectively, using Knudsen cell mass spectrometry. [74Wag] reported that alloys wetted the thoria crucible walls and tended to creep through the effusion orifice during the experiment. [75Fur] avoided this problem by the addition of small amounts of Th. [74Wag] analyzed their activity data in terms of the regular solution model and found an interaction parameter ($B(L)$, see below) of $-40\,600$ J/mol of atoms. [75Fur] reported that at infinite dilution the heats of solution of Fe and Ti are $-65\,689$ and $-67\,362$ J/mol of atoms, respectively, and that a maximum in the heat of mixing of $-15\,481$ J/mol of atoms occurs at 35 at.% Fe. The work of [75Fur] is preferred, and their tabulated Ti activity values were used in the thermodynamic calculations described below.

Phase Diagrams of Binary Titanium Alloys

Fig. 6---Room Temperature Lattice Parameters of bcc (βTi)

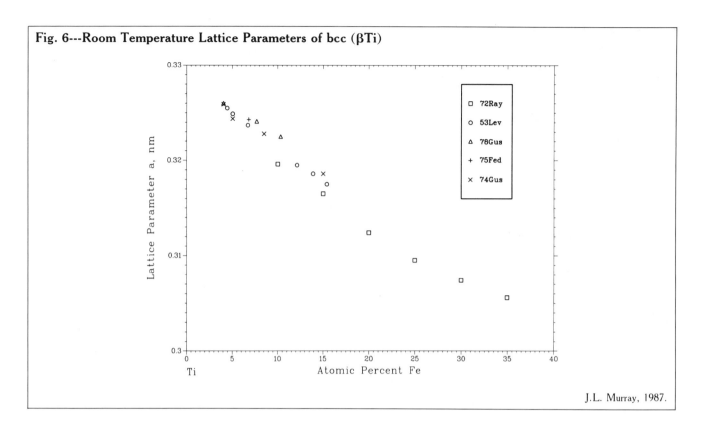

J.L. Murray, 1987.

Ti activities in dilute Fe alloys have also been derived from studies of the Ti-Fe-O system, notably by [70Fru] using the emf method. Interpretation of the data requires assumptions about the identity of the equilibrium oxide phase, and therefore, only data pertaining to the binary system are used in this evaluation.

Previous Theoretical Work.

[79Ko] gave a theoretical discussion of the effect of the magnetic transition on the (αFe) solvus for several Fe and Co alloy systems.

[61Wad] modeled the thermodynamics of the γ loop; based on the now obsolete data of [52Roe], he calculated very large Gibbs energy differences between the (γFe) and (αFe) solutions. [66Fis] calculated the difference between the enthalpy of solution of Ti in (αFe) and (γFe) as -6163 J/mol (of atoms), in good agreement with this assessment which is based on essentially the same data.

[75Ind] modeled the transformation from the disordered bcc to the ordered CsCl structure, based on the phase diagram and the experimental heat of formation of TiFe. [75Ind] assumed that the heat of formation referred to cph Ti and ferromagnetic Fe, whereas [78Kau] and the present author assumed that the reference states are bcc Ti and Fe, because the reaction was reported to occur above 1000 °C [55Kub]. The calculations of [75Ind] therefore appear to overestimate the stability of the ordered phase relative to the disordered phase.

[78Kau] modeled the thermodynamics of the entire Ti-Fe system. The compounds were approximated as stoichiometric phases. The enthalpy of TiFe was taken from [55Kub]. The disordered phases were modeled by the subregular solution approximation, with the liquid phase Gibbs energy based on the experimental data of [74Wag]. The calculation of [78Kau] is in qualitative agreement with the assessed diagram, but quantitative errors in the

Ti-rich eutectic and eutectoid temperatures and the congruent melting point of TiFe$_2$ are about 50 °C, outside the experimental uncertainty.

Calculation of the Phase Diagram.

The present calculations were undertaken to update the thermodynamic functions of [78Kau], by considering recent thermodynamic data [75Fur] and this assessment of the diagram. The coefficients of excess Gibbs energies of the solution phases and the Gibbs energies of the compounds (Table 6) were determined by least-squares optimizations with respect to thermochemical data [55Kub, 75Fur] and assessed phase diagram data.

The Gibbs energies of the solution phases are represented as:

$$G(i) = G^0(\text{Ti},i)(a - x) + G^0(\text{Fe}, i) x + RT[x \ln x + (1 - x) \ln (1 - x)] + x(1 - x)(B(i) + C(i)(1 - 2 x))$$

where i designates the phase; x is the atomic fraction of Fe; G^0 is the Gibbs energies of the pure elements; and $B(i)$ and $C(i)$ are coefficients in the polynomial expansion of the excess Gibbs energy. TiFe and TiFe$_2$ were approximated as line compounds, because of their limited homogeneity ranges and the lack of detailed experimental data with which to compare a defect model of the Gibbs energy.

The pure element Gibbs energies and initial parametrization of the excess Gibbs energies were taken from [70Kau], [77Kau], and [78Kau]. They modeled the Gibbs energy of pure Fe in the temperature ranges 300 to 1100 K and 1100 to 1800 K. In each temperature regime, the Gibbs energies were given as cubic polynomials in temperature. The temperature coefficients of the (γFe) phase were adjusted slightly to match the transformation temperatures of pure Fe [82Swa]. Thermodynamic parameters used in this evaluation are compared in Table 6 together with the parameters of [78Kau].

Table 6   Thermodynamic Properties of Ti-Fe Alloys

Properties of the pure elements

$G^0(\text{Ti,L})$ = $16\,234 - 8.368\ T$
$G^0(\text{Fe,L})$ = $13\,807 - 7.624\ T$

$G^0(\text{Ti,bcc})$ = 0
$G^0(\text{Fe,bcc})$ = 0

$G^0(\text{Ti,cph})$ = $-4\,351 + 3.766\ T$
$G^0(\text{Fe,cph})$ = $4\,280 + 1.224\ T + 0.7472\ 10^{-2}\ T^2 + 0.5124\ 10^{-5}\ T^3$ (300 to 1100 K)

$G^0(\text{Ti,fcc})$ = $1\,004 + 3.766\ T$
$G^0(\text{Fe,fcc})$ = $5\,251 - 9.441\ T + 0.5295\ 10^{-2}\ T^2 + 0.9222\ 10^{-5}\ T^3$ (1100 to 1800 K)

Compounds

	This evaluation	[78Kau]
TiFe	$-19\,000\ -\ 0.6405\ T$	$-20\,292 + 0.188\ T$(a)
TiFe$_2$	$-45\,417 + 14.82\ \ T$	$-6\,874 + 3.53\ \ T$(b)

Excess Gibbs energies of the solution phases

	Present evaluation	[78Kau]
B(L)	$-80\,411 + 17.515\ T$	$-54\,392$
C(L)	$16\,120$	$8\,368$
B(bcc)	$-56\,016 + 14.465\ T$	$-34\,518$
C(bcc)	$20\,089$	$11\,506$
B(cph)	$-5\,532 + 21.558\ T$	0
B(fcc)	$-23\,571 + 10.152\ T$	$-21\,966$
C(fcc)	$44\,392$	$11\,506$

Note: Values are given in J/mol; T in K. (**a**) bcc reference state. (**b**) cph reference state.

Fig. 7---Room Temperature Lattice Parameters of bcc (αFe)

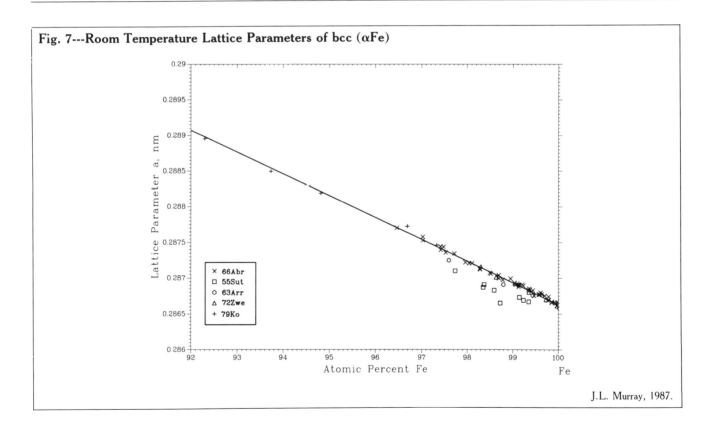

J.L. Murray, 1987.

The liquidus, the invariant reactions involving the liquid, and the (αFe) solvus are reproduced within the experimental uncertainty by the calculations. However, the (βTi)/(αTi) and (βTi)/TiFe phase boundaries (in par-ticular the eutectoid temperature) and the heat of forma-tion of TiFe are reproduced only qualitatively.

The present Gibbs energies are consistent with the experimental results of [55Kub] and [75Fur]. The calcu-

Fig. 8---Curie Temperatures and Nèel Temperatures for Ferromagnetic and Antiferromagnetic TiFe$_2$

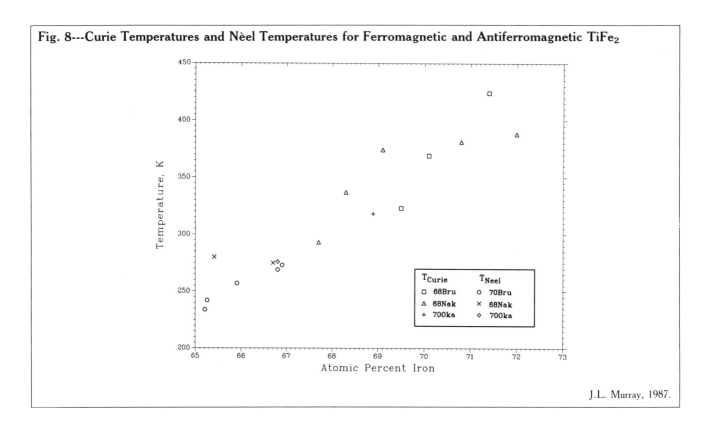

J.L. Murray, 1987.

lated excess entropy of the liquid is also consistent with the excess entropy estimated by Gibbs-Duhem integration of the activity data [75Fur].

Magnetic Transitions

The Curie temperature of pure αFe is 770 $\pm$ 2 °C [82Swa]. The effect of Ti on the ferromagnetic phase transition in (αFe) was studied by [09Por], [59Arr], and [73Pal]. [09Por] found that the Curie temperature (T_C) was lowered to 690 °C by the presence of 3 at.% Ti; the alloy also contained 0.1 wt.% C. [68Nak] reported that the critical temperature of (αFe) is 730 °C in the two-phase (αFe) + TiFe$_2$ region. [59Arr] found that T_C decreases with increasing Ti content, with a slope of 3.7 K per at.% Ti. Beyond 2 at.% Ti, he reported an anomalous drop in T_C. Sample preparation and heat treatment were not described, but it is probable that the explanation for the anomaly is that the alloys beyond 2 at.% Ti were two-phase. [73Pal], who made metastable single-phase alloys with up to 5 at.% Ti, verified the results of [59Arr] out to 2 at.% Ti, but found T_C to be continuous with no anomaly. This result is indicated in Fig. 1.

At stoichiometry and in alloys containing excess Fe, TiFe$_2$ is ferromagnetic, but at Ti-rich compositions, it is antiferromagnetic [68Bru, 68Nak, 70Bru, 70Wer1, 70Wer2]. T_C increases with Fe content [68Nak, 68Bru]. Magnetic transition temperatures are shown in Fig. 7.

Stoichiometric TiFe is superparamagnetic; away from stoichiometry, TiFe follows a Curie-Weiss Law with a paramagnetic T_C of -10 K. Away from stoichiometry, TiFe displays a resistance minimum, with a Kondo temperature of 47 K [73Ike]. Anomalies in the electrical resistivity and magnetic properties of TiFe as functions of composition were studied by [71Ike], [72Ike1], [72Ike2], [72Ike3], and [73Ike].

Cited References

09Por: M.A. Portevin, "Contribution to the Study of Special Ternary Steels," *Rev. Met.*, 6, 1354-1362 (1909) in French. (Equi Diagram; Experimental)

14Lam: J. Lamort, "On Titanium Alloys," *Ferrum*, 11(8), 225-234 (1914) in German. (Equi Diagram; Experimental)

30Mic: A. Michel and P. Benazet, "Influence of Titanium on the Transformation Points of Steel," *Rev. Met.*, 27, 326-333 (1930) in French. (Equi Diagram; Experimental)

35Sve: V.N. Svechnikov and V.N. Gridnev, "Effect of Titanium on Polymorphic Transformations in Iron," *Domez*, (2), 41-43 (1935) in Russian; TR: *Chem. Abst.*, 29, 6547 (1935). (Equi Diagram; Experimental)

36Jel: W. Jellinghaus, "The Crystal Structure of Fe$_3$Ti," *Z. Anorg. Chem.*, 227, 62-64 (1936) in German. (Crys Structure; Experimental)

38Tof: V.W. Tofaute and A. Buttinghaus, "The Fe-Rich Corner of the Fe-Ti-C System," *Arch. Eisenhuetten.*, 12, 33-37 (1938) in German. (Equi Diagram; Experimental)

38Wit: H. Witte and H.J. Wallbaum, "Thermal and X-Ray Study of the Iron-Titanium System," *Z. Metallkd.*, 30(3), 100-102 (1938) in German. (Equi Diagram; Experimental)

39Lav: F. Laves and H.J. Wallbaum, "Crystal Chemistry of Titanium Alloys," *Naturwissenschaften*, 27, 674-675 (1939) in German. (Crys Structure; Experimental)

41Wal: H.J. Wallbaum, "The Systems of Titanium, Zirconium, Niobium, and Tantalum with Iron," *Arch. Eisenhuetten.*, 14(10), 521-526 (1941) in German. (Equi Diagram; Experimental)

48Pet: W. Peter and W.A. Fischer, "Relation Between the Phase Diagrams and Mechanical Properties, Especially Durability, of Fe-Nb and Fe-Ti Alloys," *Arch. Eisenhuetten.*, 19, 161-168 (1948) in German. (Equi Diagram; Experimental)

50Duw: P. Duwez and J.L. Taylor, "The Structure of Intermediate Phases in Alloys of Titanium with Iron, Cobalt, and Nickel," *Trans. AIME*, 188, 1173-1176 (1950). (Crys Structure; Experimental)

51Fre: W.I. Fretague, C.S. Barker, and E.A. Peretti, Air Force Technical Report 6597, Parts I and II (1951-1952). (Experimental)

51Mcq1: A.D. McQuillan, "The Effect of the Elements of the First Long Period on the α → β Transformation in Ti," *J. Inst. Met.,* 80, 363 (1951). (Equi Diagram; Experimental)

51Mcq2: A.D. McQuillan, "The Application of Hydrogen Equilibrium-Pressure Measurements to the Investigation of Titanium Alloy Systems," *J. Inst. Met.,* 79, 73-88 (1951). (Equi Diagram; Experimental)

51Wor: H.W. Worner, "The Constitution of Titanium-Rich Alloys of Iron and Titanium," *J. Inst. Met.,* 79, 173-188 (1951). (Equi Diagram; Experimental)

52Roe: W.P. Roe and W.P. Fisherl, "γ Loop Studies in the Fe-Ti, Fe-Cr, and Fe-Ti-Cr Systems," *Trans. ASM,* 44, 1030-1041 (1952). (Crys Structure; Experimental)

52Thy: R.J. van Thyne, H.D. Kessler, and M. Hansen, "The Systems Titanium-Chromium and Titanium-Iron," *Trans. ASM,* 44, 974-989 (1952). (Equi Diagram; Experimental)

53Lev: B.W. Levinger, "Lattice Parameter of βTi at Room Temperature," *Trans. AIME,* 197, 195 (1953). (Crys Structure; Experimental)

54Gru: W. Gruhl and D. Ammann, "On the Occurrence of Ti$_2$Fe," *Arch. Eisenhuetten.,* 25(11/12), 599-560 (1954) in German. (Equi Diagram; Experimental)

54Nis: H. Nishimura and K. Kamei, "Investigation of the System Ti-Fe-Al Alloys. Part I. Studies on Ti-Fe System," *Bull. Eng. Res. Inst., Kyoto Univ.,* 6, 38-42 (1954) in Japanese. (Equi Diagram; Experimental)

54Pol: D.H. Polonis and J.G. Parr, "Phase Transformations in Titanium-Rich Alloys of Iron and Titanium," *Trans. AIME,* 200, 1148-1154 (1954). (Equi Diagram; Experimental)

55Kub: O. Kubaschewski and W.A. Dench, "The Heats of Formation in the Systems Titanium-Aluminum and Titanium," *Acta Metall.,* 3, 339-346 (1955). (Equi Diagram; Experimental)

55Pol: D.H. Polonis and J.G. Parr, "Martensite Formation in Powders and Lump Specimens of Ti-Fe Alloys," *Trans. AIME,* 203, 64 (1955). (Meta Phases; Experimental)

55Sut: A.L. Sutton and W. Hume-Rothery, "The Lattice Spacings of Solid Solutions of Titanium, Vanadium, Chromium, Manganese, Cobalt and Nickel in αFe," *Philos. Mag.,* 46, 1295-1309 (1955). (Crys Structure; Experimental)

55Thy: R.J. van Thyne and L.D. Jaffe, "Discussion of Phase Transformations in Titanium-Rich Alloys of Iron and Titanium," *Trans. AIME,* 203, 718 (1955). (Equi Diagram; Experimental)

56Enc: E. Ence and H. Margolin, "Re-Examination of Ti-Fe and Ti-Fe-O Phase Relations," *Trans. AIME,* 206, 572-577 (1956). (Equi Diagram; Experimental)

56Kor: I.I. Kornilov and N.G. Boriskina, "Phase Diagram of the System Ti-Fe," *Dokl. Akad. Nauk SSSR,* 108(6), 1083 (1956) in Russian. (Equi Diagram; Experimental)

56Yos: H. Yoshida, "On the Crystal Structure of the Intermediate Phase ω in Titanium-Iron," *J. Jpn. Inst. Met.,* 20, 292-294 (1956). (Meta Phases, Crys Structure; Experimental)

57Hel: A. Hellawell and W. Hume-Rothery, "The Constitution of Alloys of Iron and Manganese with Transition Elements of the First Long Period," *Philos. Trans. Roy. Soc., London,* 249, 417-459 (1957). (Equi Diagram; Experimental)

57Phi: T.V. Philip and P.A. Beck, "CsCl-Type Ordered Structures in Binary Alloys of Transition Elements," *Trans. AIME,* 209, 1269-1271 (1957). (Crys Structure; Experimental)

58Sil: J.M. Silcock, "An X-Ray Examination of the Omega Phase in TiV, TiMo and TiCr Alloys," *Acta Metall.,* 6, 481 (1958). (Crys Structure; Experimental)

59Age: N.V. Ageev and L.A. Petrova, "The Stability of β-Phases in Alloys of Titanium with Iron and Nickel," *Zh. Neorg. Khim.,* 4(5), 1092-1099 (1954); TR: *Russ. J. Inorg. Chem.,* 4(5), 496-499 (1959). (Equi Diagram; Experimental)

59Arr: A. Arrott and J.E. Noakes, "Saturation Magnetization and Curie Points in Dilute Alloys of Iron," *J. Appl. Phys., Suppl.,* 30(4), 97S-98S (1959). (Magnetism; Experimental)

59Dwi: A.E. Dwight, "CsCl-Type Equiatomic Phases in Binary Alloys of Transition Elements," *Trans. AIME,* 215, 283-286 (1959). (Crys Structure; Experimental)

59Mol: S.H. Moll and R.E. Ogilvie, "Solubility and Diffusion of Titanium in Iron," *Trans. AIME,* 215, 613-618 (1959). (Equi Diagram; Experimental)

59Mur: Y. Murakami, H. Kimura, and Y. Nishimura, "An Investigation on the Titanium-Iron-Carbon System," *Trans. Nat. Res. Inst. Met. (Tokyo),* 1(1), 7-21 (1959). (Equi Diagram; Experimental)

60Gri: Y.N. Gridnev, Yu.N. Petrov, V.A. Rafalovskiy, and V.I. Trefilov, "Investigation of ω Phase Formation in Titanium Alloys," *Vopr. Fiz. Met. Metalloved., An Ukr SSR Sb. Nauchn. Rabot,* (11), 82-86 (1960) in Russian. (Meta Phases; Experimental)

60Pie: P. Pietrokowsky and F.G. Youngkin, "Ordering in the Intermediate Phases TiFe, TiCo, and TiNi," *J. Appl. Phys.,* 31(10), 1763-1766 (1960). (Crys Structure; Experimental)

60Sat: T. Sato, S. Hukai, and Y.C. Huang, "The M$_s$ points of Binary Titanium Alloys," *J. Aust. Inst. Met.,* 5(2), 149-153 (1960). (Meta Phases; Experimental)

60Stu: H.P. Stuewe and Y. Shimomura, "Lattice Constants of Cubic Phases FeTi, CoTi, NiTi," *Z. Metallkd.,* 3, 180-181 (1960) in German. (Crys Structure; Experimental)

61Wad: T. Wada, "Thermodynamic Studies on the α-γ Transformation of Iron Alloys," *Sci. Rep. RITU, A13,* 215-224 (1961). (Thermo; Theory)

62Bor: N.G. Boriskina and K.P. Myasnikova, "Solubility of Iron, Manganese and Copper in αTi, Titanium and its Alloys," *Akad. Nauk SSSR 7,* 61-67 (1962) in Russian. (Equi Diagram; Experimental)

62Spe: G.R. Speich, "Precipitation of Laves Phases from Iron-Niobium (Columbium) and Iron-Titanium Solid Solutions," *Trans. AIME,* 224, 850-858 (1962). (Equi Diagram; Experimental)

63Arr: A. Arrott and J.E. Noakes, "Thermal, Electrical, and Magnetic Properties of Iron and Its Dilute Alloys," *Iron and Its Dilute Solid Solutions,* C.W. Spencer and F.E. Werner, Ed., Interscience, New York, 81-97 (1963). (Crys Structure; Experimental)

63Bor: B.A. Borok, E.K. Novikova, L.S. Golubeva, N.A. Rucheva, and R.P. Shchegoleva, "Dilatometric Study of Binary Alloys of Titanium," *Metalloved. Term. Obrab. Met.,* (2), 32-36 (1963) in Russian; TR: *Met. Sci. Heat Treat.,* (2), 94-98 (1963). (Meta Phases; Experimental)

63Kan: H. Kaneko and Y.C. Huang, "Continuous Cooling Transformation Characteristics of Titanium Alloys of Eutectoidal Type (I)," *J. Jpn. Inst. Met.,* 27, 393-397 (1963) in Japanese. (Equi Diagram, Meta Phases; Experimental)

63Luz: L.P. Luzhinkov, V.M. Novikova and A.P. Mareev, "Solubility of β-Stabilizers in αTi," *Metalloved. Term. Obrab. Met.,* (2), 13-16 (1963) in Russian; TR: *Met. Sci. Heat Treat.,* (2), 78-81 (1963). (Equi Diagram; Experimental)

63Wad: T. Wada, "Austenite Loop in Iron-Titanium System," *J. Jpn. Inst. Met.,* 27(3), 119 (1963) in Japanese; TR: *Trans. Nat. Res. Inst. Met. (Tokyo),* 6(2), 43-46 (1963). (Equi Diagram; Experimental)

65Bor: N.G. Boriskina and I. I. Kornilov, "Investigation of the Alloy Systems Ti-Fe and Ti-Cr-Fe," *Sb. Nouyve Titanov. Splavov, Moscow, Izd. Nauk, 6,* 61-75 (1965) in Russian. (Equi Diagram; Experimental)

66Abr: E.P. Abrahamson and S.L. Lopata, "The Lattice Parameters and Solubility Limits of Alpha Iron as Affected by Some Binary Transition-Element Additions," *Trans. AIME,* 236, 76-87 (1966). (Crys Structure, Equi Diagram; Experimental)

66Fis: W.A. Fischer, K. Lorenz, H. Fabritius, A. Hoffman, and G. Kalwa, "Investigation of Phase Transformations in Iron Alloys Using a Magnetic Balance," *Arch. Eisenhuetten.,* 37, 79-86 (1966) in German. (Equi Diagram; Experimental)

66Nis: Z. Nishiyama, M. Oka, and H. Nakagawa, "Transmission Electron Microscope Study of the Martensites in a Titanium−3 wt.% Iron Alloy," *J. Jpn. Inst. Met.,* 30, 16-21 (1966) in Japanese. (Meta Phases; Experimental)

67Bru: W. Bruckner, K. Kleinstuck, and G.E.R. Schulze, "Atomic Arrangement in the Homogeneity Range of the Laves Phases $ZrFe_2$ and $TiFe_2$," *Phys. Status Solidi, 23,* 475-480 (1967). (Equi Diagram; Experimental)

67Dor: A.V. Doroshenko, S.A. Nemnonov, and S.K. Sidorov, "Neutron Diffraction Analysis of the Structure of the Alloys TiFe and TiCo," *Fiz. Met. Metalloved., 23*(3), 562-563 (1967) in Russian; TR: *Phys. Met. Metallogr., 23*(3), 168-169 (1967). (Crys Structure; Experimental)

67Nak: H. Nakagawa, S. Sato, and Z. Nishiyama, "Transmission Electron Microscope Study of the Martensites in a Titanium–3 wt.% Iron Alloy, Supplement," *Trans. Jpn. Inst. Met., 31,* 525-527 (1967) in Japanese. (Meta Phases; Experimental)

67Rau: E. Raub, Ch.J. Raub and E. Roschel, "The αTi-Fe Solid Solution and Its Superconducting Properties," *J. Less-Common Met., 12,* 36-40 (1967). (Equi Diagram; Experimental)

67Wer: G.K. Wertheim and J.H. Wernick, "Mössbauer Effect Study of B.C.C. Structure Alloys, FeAl and FeTi," *Acta Metall., 15,* 297-302 (1967). (Crys Structure; Experimental)

68Bru: W. Bruckner, R. Perthel, K. Kleinstuck, and G.E.R. Schultze, "Magnetic Properties of $ZrFe_2$ and $TiFe_2$ Within Their Homogeneity Range," *Phys. Status Solidi, 29,* 211-216 (1968). (Equi Diagram; Experimental)

68Nak: T. Nakamichi, "Ferro- and Antiferromagnetism of the Laves Phase Compound in Fe-Ti Alloy System," *J. Phys. Soc. Jpn., 25,* 1189 (1968). (Equi Diagram; Experimental)

68Swa: L.J. Swartzendruber and L.H. Bennett, "Line Profiles in the Nuclear Magnetic Resonance and Mössbauer Effect of $(TiFe)_{1-x}$ $(Co)_x$ Alloys," *J. Appl. Phys., 39*(5), 2215-2220 (1968). (Crys Structure; Experimental)

69Gus: L.N. Guseva, L.A. Petrova, and I.A. Ogloblina, "The Martensitic Phase with Face-Centered Cubic Lattice in Titanium Alloy Containing 5.9 % Alloy," *Dokl. Akad. Nauk SSSR, 185*(4), 799-801 (1969) in Russian; TR: *Sov. Phys. Dokl., 14*(4), 367-370 (1969). (Experimental)

69Hic: B.S. Hickman, "ω Phase Precipitation in Alloys of Titanium with Transition Metals," *Trans. AIME, 245,* 1329-1336 (1969). (Meta Phases; Experimental)

69Osh: E. Oshio, Y. Yoshiga, and M. Adachi, "Transmission Electron Microscope Observations of ω Phase in Titanium–5 wt.% Iron Alloy," *J. Jpn. Inst. Met., 33,* 437-442 (1969) in Japanese. (Meta Phases; Experimental)

70Bru: W. Bruckner, K. Kleinstuck, and G.E.R. Schulze, "Mössbauer Study of the Laves Phase (Ti) $1 - x$ (Fe)$2 + x$," *Phys. Status Solidi (a), 1,* K1-K4 (1970). (Equi Diagram; Experimental)

70Fru: R.J. Fruehan, "Activities in Liquid Fe-Al-O and Fe-Ti-O Alloys," *Metall. Trans., 1,* 3403-3408 (1970). (Thermo; Experimental)

70Kau: L. Kaufman and H. Bernstein, *Computer Calculation of Phase Diagrams,* Academic Press, New York (1970). (Thermo; Theory)

70Oka: M. Okazaki, "Thermomagnetic Study of Several Fe-Ti Alloys of Compositions near Fe_2Ti," *C.R. Acad. Sci., Paris, 270B,* 254-256 (1970) in French. (Equi Diagram; Experimental)

70Wer1: G.K. Wertheim, J.H. Wernick, and R.C. Sherwood, "Model for the Composition-Dependent Ferromagnetic to Antiferromagnetic Transition in Fe_2Ti," *J. Appl. Phys., 41*(3), 1325-1326 (1970). (Experimental)

70Wer2: G.K. Wertheim, D.N.E. Buchanan, and J.H. Wernick, "Magnetic Properties of Inequivalent Iron Atoms in Fe_2Ti," *Solid State Comm., 8,* 2173-2176 (1970). (Magnetism; Experimental)

71Ike: K. Ikeda, T. Nakamichi, and M. Yamamoto, "Thermohysteresis Phenomena of the Electrical Resistivity in the Laves Phase Compounds in Fe-Ti Systems," *J. Phys. Soc. Jpan., 30,* 1504-1505 (1971). (Equi Diagram; Experimental)

71Miy: M. Miyagi and S. Shin, "Isothermal Transformation Characteristics of Metastable β-type," *J. Jpn. Inst. Met., 35,* 716-722 (1971) in Japanese. (Meta Phases; Experimental)

71Odi: L.P. Odinokova and B.A. Brusilovskiy, "Decomposition of the β Phase in Titanium-Iron Alloys During Continuous Cooling," *Fiz. Met. Metallovd.,31*(4), 713-718 (1971) in Russian; TR:

Phys. Met. Metallogr., 31(4), 41-45 (1971). (Meta Phases; Experimental)

71Pat: N.E. Paton, D. de Fontaine, and J.C. Williams, "Director Observation of the Diffusionless $\beta \rightarrow \beta + \omega$ Transformation in Titanium Alloys," Proc. Electron Microscopy Society of American, 29th Annual EMSA Meeting, 122-123 (1971). (Meta Phases; Experimental)

71Shi: A. Ya Shinyayev, "Diffusion of Iron in Fe-Ti Alloys," *Izv. Akad. Nauk SSSR, Met.,* (4) 263-267 (1971) in Russian; TR: *Russ. Metall.,* (4), 185-188 (1971). (Equi Diagram; Experimental)

72Ike1: K. Ikeda, T. Nakamichi, and M. Yamamoto, "Thermo-Hysteresis Phenomenon of the Electrical Resistivity of Fe_2Ti Suggesting Its Martensitic Transformation ," *Phys. Stat. Sol. (a), 12,* 595-603 (1972). (Meta Phases; Experimental)

72Ike2: K. Ikeda, T. Nakamichi, and M. Yamamoto, "Resistance Minimum in Iron-Titanium Compounds $(Fe)_{1-x}$ $(Ti)_{1+x}$ with the CsCl-Type Structure," *J. Phys. Soc. Jpn., 32,* 280 (1972). (Equi Diagram; Experimental)

72Ike3: K. Ikeda, T. Nakamichi, K. Noto, Y. Muto, and M. Yamamoto, "Influence of Non-Stoichiometry on the Resistance Minimum and Superparamagnetism in the CsCl-type Compounds $(Fe)_{1-x}(Ti)_{1+x}$," *Phys. Status Solidi (b), 51,* K39-K42 (1972). (Equi Diagram; Experimental)

72Kha: N.A. Khatanova, A.G. Timushev, and M.I, Zakharova, "Decomposition of the Solid Solution in the Alloy Ti–10 wt.% Fe," *Fiz. Met. Metalloved.,34*(4), 892-894 (1972) in Russian; TR: *Phys. Met. Metalogr.,34*(4), 218-220 (1972). (Meta Phases, Crys Structure; Experimental)

72Ray: R. Ray, B.C. Giessen, and N.J. Grant, "The Constitution of Metastable Titanium-Rich Ti-Fe Alloys: An Order-Disorder Transition," *Metall. Trans., 3,* 627-629 (1972). (Meta Phases, Crys Structure)

73Ika: K. Ikeda, "Anomalous Thermoelectric Power in the Cs-Cl Type Compounds $(Fe)_{1-x}(Ti)_{1+x}$," *J. Phys. Soc. Jpn. 34,* 272 (1973). (Equi Diagram; Experimental)

73Pal: R.J. Palma and K. Schroeder, "High Temperature Specific Heats of Iron-Rich Titanium Alloys Between 600 and 1150 K," *J. Less-Common Met., 31,* 249-253 (1973). (Equi Diagram; Experimental)

73Pol: A.F. Polesya and L.S. Slipchenko, "Formation of Amorphous Phases and Metastable Solid Solutions in Binary Ti and Zr Alloys with Fe, Ni, and Cu," *Izv. Akad. Nauk SSSR, Met.,* (6), 173-178 (1973) in Russian; TR: *Russ. Metall.,* (6), 103-1097 (1973). (Meta Phases; Experimental)

74Gus: L.N. Guseva and L.K. Dolinskaya, "Metastable Phases in Titanium Alloys with Group VIII Elements Quenched from the β-Region," *Izv. Akad. Nauk SSSR, Met.,* (6), 195-202 (1974) in Russian; TR: *Russ. Metall.,* (6), 155- 159 (1974). (Meta Phases; Experimental)

74Ike1: K. Ikeda, T. Nakamichi, and M. Yamamoto, "Origin of Superparamagnetism in the CsCl-Type Compounds, $(Fe)_{1-x}(Ti)_{1+x}$, near the Stoichiometric Composition," *J. Phys. Soc. Jpn., 37,* 652-659 (1974) in Japanese. (Equi Diagram; Experimental)

74Ike2: K. Ikeda, "Kondo Effect in the Transport Properties of the CsCl-Type Compounds $Fe_{1-x}Ti_{1+x}$ I. Their Anomalous Behaviors in the Titanium-Rich Compounds," *Phys. Status Solidi (b),* 655-663 (1974). (Equi Diagram; Experimental)

74Ike3: K. Ikeda, "Kondon Effect in the Transport Properties of the Cs-Cl-Type Compounds $(Fe)_{1-x}(Ti)_{1+x}$, II. Magnetic Scattering Center due to Atomic Disordering," *Phys. Stat. Sol. (b), 63,* 361-270 (1974). (Equi Diagram; Experimental)

74Spu: R.A. Spurling, C.G. Rhodes and J.C. Williams, "The Microstructure of Ti Alloys as Influenced by Thin-Foil Artifacts," *Metall. Trans., 5,* 2597-2600 (1974). (Meta Phases; Experimental)

74Stu: M.M. Stupel, M. Ron, and B.Z. Weiss, "A Metastable Phase in α-Ti-Fe Revealed by Mössbauer Analysis," *J. Phys. (Paris) Colloq., 36*(6), C6-483-C6-485 (1974). (Meta Phases; Experimental)

74Wag: S. Wagner and G.R. St. Pierre, "Thermodynamics of the Liquid Binary Iron-Titanium by Mass Spectrometry," *Metall. Trans., 5,* 887-889 (1974). (Thermo; Experimental)

The Fe-Ti System

75Fed: S.G. Fedotov, N.F. Kvasova, and E.P. Sinodova, "Phase Transformations in Unstable Alloys of Titanium and Iron," *Izv. Akad. Nauk SSSR, Met.*, (3), 193 (1975) in Russian; TR: *Russ. Metall.*, (3), 162-165 (1975). (Meta Phases; Experimental)

75Fur: T. Furukawa and E. Kato, "Thermodynamics of Binary Liquid Iron-Titanium Alloys by Mass Spectrometry," *Tetsu-to-Hagane, 16*, 382-387 (1975). (Thermo; Experimental)

75Hut: H. Huthmann and G. Inden, "High-Temperature Neutron Diffraction on FeTi and CoTi," *Phys. Status Solidi (a), 28,* K129-K130 (1975). (Crys Structure; Experimental)

75Ind: G. Indlen, "Determination of Chemical and Magnetic Interchange Energies in BCC Alloys, II. Applications to Non-Magnetic Alloys," *Z. Metallkd., 66*(11), 648-653 (1975). (Thermo; Theory)

76Bla: A. Blaesius and U. Gonser, "Precision Phase Analysis," *J. Phys. Colloq., 37*(12), C6-397-C6-399 (1976). (Equi Diagram; Experimental)

76Stu: M.M. Stupel, M. Ron, and B.Z. Weiss, "Phase Identification in Titanium-Rich Ti-Fe Systems by Mossbauer Spectroscopy," *J. Appl. Phys., 47*(1), 6-12 (1976). (Meta Phases; Experimental)

77Kas: N.G. Kasumzade, A.Yu. Nadzhafov, and A.A. Zhukhovitskii, "Certain Aspects of the Stability of Intermetallic Compounds with the Laves-Phase Structure," *Izv. VUZ Chernaya Metall.*, (7), 133-135 (1977) in Russian. (Equi Diagram; Experimental)

77Kau: L. Kaufman, "Proceedings of the 4th Calphad Meeting–Workshop on Computer Based Coupling of Thermochemical and Phase Diagram Data," *Calphad, 1*(1), 7-89 (1977). (Thermo; Experimental)

77Ron: M. Ron, M.M. Stupel, and B.Z. Weiss, "The α to ω Transformation in a Plastically Deformed αTi(Fe) Alloy," *Acta Metall., 25*, 1355-1362 (1977). (Meta Phases; Experimental)

77Stu: M.M. Stupel, B.Z. Weiss, and M. Ron, "Formation of an ω-Phase from αTi(Fe) During Aging," *Acta Metall., 25*, 667-671 (1977). (Meta Phases; Experimental)

78Gus: L.N. Guseva and L.K. Dolinskaya, "Formation Conditions of Athermal ω Phase in Alloys of Titanium with Transition Elements," *Krist. Strukt. Svoistva. Met. Splavov*, 59-63 (1978) in Russian. (Meta Phases, Crys Structure; Experimental)

78Kau: L. Kaufman, "Coupled Phase Diagrams and Thermochemical Data for Transition Metal Binary Systems—III," *Calphad, 2*(2), 117-146 (1978). (Thermo; Theory)

78Stu: M.M. Stupel, M. Ron, and B.Z. Weiss, "Formation of the ω-Phase During the Aging of β, Ti(7.1 wt.% Fe) — A Mössbauer Study," *Metall. Trans., A, 9*, 249-252 (1978). (Meta Phases; Experimental)

79Ko: M. Ko and T. Nishizawa, "Effect of Magnetic Transition on the Solubility of Alloys Elements in α Fe," *J. Jpn. Inst. Met., 43*(2), 118-126 (1979) in Japanese. (Equi Diagram; Experimental)

79Mat: J. Matyka, F. Faudot, and J. Bigot, "Study of Iron Solubility in αTi," *Scr. Metall., 13*, 645-648 (1979). (Equi Diagram; Experimental)

79Nas: S. Nasu, U. Gonser, A. Blasium, and F.E. Fujita, "Phase Analysis in Metals and Alloys by Mössbauer Spectroscopy," *J. Phys. Colloq., C2*(3), C2-619-C2-620 (1979). (Equi Diagram; Experimental)

80Dew: D. Dew-Hughes, "The Addition of Mn and Al to the Hydriding Compound FeTi: Range of Homogeneity and Lattice Parameters," *Metall. Trans. A, 11*, 1219-1225 (1980). (Equi Diagram; Experimental)

80Ino: A. Inoue, K. Kobayashi, C. Suryanarayana, and T. Masumoto, "An Amorphous Phase in Co-Rich Co-Ti Alloys," *Scr. Metall., 14*, 119-123 (1980). (Meta Phases; Experimental)

80Lie: J. Liebertz, S. Stahr, and S. Haussuhl, "Growth and Properties of Single Crystals of FeTi," *Kristall. Tech., 15*(11), 1257-1260 (1980). (Crys Structure; Experimental)

80Mat: J. Matyka, F. Faudot, and J. Bigot, "Iron Retrograde Solubility in α Ti and Precipitation Phenomenon," *Titanium '80, Sci. and Tech., Proc. 4th Int. Conf. Ti, Kyoto, Japan, May 198-22*, H. Kimura and O. Izumu, Ed., 2941-2947 (1980). (Equi Diagram; Experimental)

80Yam: T. Yamane and M. Ito, "M_s and β(s) Temperature of Ti-Fe Alloys," *Titanium 80, Sci. and Tech., Proc. 4th Int. Conf. Ti, Kyoto, Japan, May 19-22*, H. Kimura and O. Izumu, Ed., 1513-1520 (1980). (Meta Phases; Experimental)

81Tak: T. Takayama, M.Y. Way, and T. Nishizawa, "Effect of Magnetic Transition on the Solubility of Alloying Elements in BCC Iron and FCC Cobalt," *Trans. Jpn. Inst. Met., 22*(5), 315-325 (1981). (Equi Diagram; Experimental)

82Swa: L.J. Swartzendruber, "The Fe (Iron) System," *Bull. Alloy Phase Diagrams, 3*(2), 161-165 (1982). (Equi Diagram; Experimental)

The Ga-Ti (Gallium-Titanium) System

69.72   47.88

By J.L. Murray

Equilibrium Diagram

[Shunk] constructed a Ti-Ga phase diagram based on thermal analysis, X-ray diffraction, and examination of cast alloys by [62Pot].* This is the only complete Ti-Ga phase diagram in the literature. Nevertheless, it must be rejected on the following grounds:

- The published microstructures are inconsistent with the phase diagram as drawn.**
- An oxide phase (or phases) appears in the composition range 35 to 67 at.% Ga, and assumptions must be made about the ternary Ti-Ga-O system in order to interpret the results for the binary system.

Table 1   Special Points of the Assessed Ti-Ga Phase Diagram

Reaction	Compositions of the respective phases, at.% Ga			Temperature, °C	Reaction type
L $\rightleftarrows$ (βTi) + Ti$_2$Ga	31	24	33.3	...	Eutectic
L $\rightleftarrows$ Ti$_2$Ga		33.3		...	Congruent
(βTi) + Ti$_2$Ga $\rightleftarrows$ Ti$_3$Ga	16	33.3	25	1030 ± 20	Peritectoid
(βTi) + Ti$_3$Ga $\rightleftarrows$ (αTi)	13.5	23	14	940 ± 10	Peritectoid
L $\rightleftarrows$ (βTi)		0		1670	Melting point
(βTi) $\rightleftarrows$ (αTi)		0		882	Allotropic transformation
L $\rightleftarrows$ (Ga)		100		29.7741	Melting point

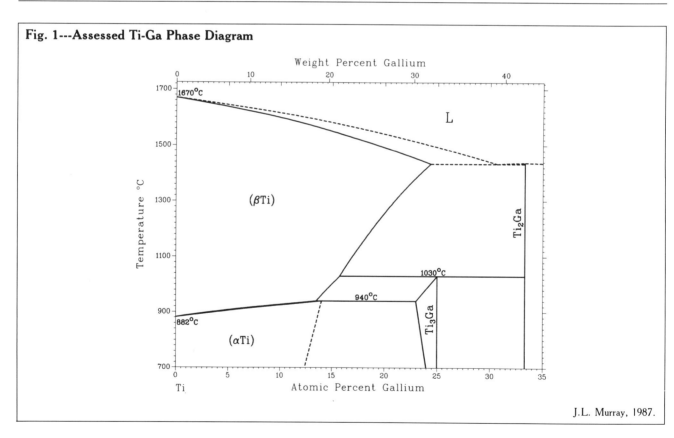

Fig. 1---Assessed Ti-Ga Phase Diagram

J.L. Murray, 1987.

---

*The diagram shows eight intermetallic compounds, all but one formed from the melt. According to [62Pot], three compounds melt congruently; two are formed by peritectic reactions; and one is formed in the solid state by a peritectoid reaction. Three eutectic reactions are postulated.

**For example, according to the diagram, in an as-cast 60 at.% Ga alloy, one expects primary Ti$_5$Ga$_4$, peritectic formation of TiGa, ending in the eutectic reaction L $\rightleftarrows$ TiGa$_2$ + TiGa. According to the diagram, Ti$_2$Ga$_3$ is formed in the solid state by a peritectoid reaction. The micrograph of the as-cast 60 at.% Ga alloy, however, shows primary TiGa, in a peritectic surround of Ti$_2$Ga$_3$ and a eutectic mixture of TiGa + TiGa$_2$ and TiGa$_2$ plates.

The Ga-Ti System

Table 2 Experimental Data on the (αTi)/Ti₃Ga Boundaries

Reference	Equilibrium composition, at.% Ga		Temperature, °C
	(αTi)	Ti₃Ga	
[71Sha]	5	21	660
	6	21	760
	8.4	19	810
	11.6	20.3	860
[58And]	11.6 to 13.1	...	900
	10	22 to 24	850
	10	...	800

- Finally, the manner in which [Shunk] completed the diagram to include the unexamined (Ti) liquidus was thermodynamically incorrect. The difficulty of constructing a consistent (βTi) liquidus suggests that the portion of the diagram based on experiment does not represent the binary equilibrium.

It is not possible to reconstruct a better phase diagram based on the available information. The present assessment (Fig. 1 and Table 1) is therefore limited to the Ti-rich part of the diagram, which has been more reliably established. The assessed diagram is constructed from the thermodynamic calculation by [73Kau], except for the homogeneity range of Ti₃Ga, which is drawn. The calculation reproduces both the experimental phase diagram and thermochemical data to within the experimental uncertainty. In the section on crystal structures, all the structures that have been found in Ti-Ga alloys are listed, some of which are probably not equilibrium phases.

The equilibrium phases of the Ti-Ga system include:

- The cph (αTi), bcc (βTi), and complex cubic (Ga) solution phases
- The ordered hexagonal phases Ti₃Ga and Ti₂Ga with the $D0_{19}$ and Ni₂In structures, respectively, which are well established as equilibrium phases
- The Ga-rich ordered fcc phase TiGa₃ and TiGa₂ with a related structure

Other structures that have been found only by [62Pot], [62Sch], or [64Sch] are listed in Table 2.

Equilibria in Ti-Rich Alloys.

The composition range 0 to 35 at.% Ga was examined by [57And1] and [58And] up to 1100 °C and in the as-cast condition. Equilibria were examined among (αTi), (βTi), Ti₃Ga, and Ti₂Ga. Alloys in composition increments of 2.5 at.% were equilibrated at temperatures of 50 °C increments and 20 °C increments for the (αTi)/(βTi) boundaries.

Addition of Ga to (Ti) increases the temperature of the (βTi) → (αTi) transformation, up to the peritectoid transformation (βTi) + Ti₃Ga ⇄ (αTi) at 940 ± 10 °C. Ti₃Ga decomposes by the peritectoid reaction (βTi) + Ti₂Ga ⇄ Ti₃Ga at 1030 ± 20 °C. From the microstructures of cast alloys, it is known that Ti₂Ga forms congruently from the melt and that there is a eutectic reaction L ⇄ (βTi) + Ti₂Ga [57And1, 58And].

[58And] located the (αTi)/(αTi) + Ti₃Ga boundary within 1 at.% at 900 °C and the Ti₃Ga/(αTi) + Ti₃Ga boundary within 1 at.% at 850 °C. [71Sha] studied the (αTi)/Ti₃Ga phase boundaries in the range 0 to 25 at.% Ga and 660 to 860 °C using microprobe analysis of annealed diffusion couples. These data conflict with the metallographic results of [58And], both boundaries lying to the Ti-rich side of the [58And] results, as summarized in Table 2.

It is expected that the precipitation of small coherent particles of Ti₃Ga would not be detected by optical microscopy and that the (αTi)/(αTi) + Ti₃Ga boundary would be shifted to more Ti-rich compositions. In the corresponding ordering reaction of the Ti-Al system, analysis of annealed diffusion couples, although a valuable tool for examining the phase diagram overall, did not give accurate (αTi)/Ti₃Al boundaries. The calculation tentatively used to construct the assessed diagram agrees somewhat better with the results of [58And] than with those of [71Sha]. The (αTi)/(αTi) + Ti₃Ga boundary should be studied by TEM to resolve this discrepancy.

[81Bel] suggested that another ordered hexagonal phase Ti₇Ga may exist based on electrical resistivity data. Attempts to discover the structure by X-ray analysis resulted in the observation of superlattice lines belonging

Table 3 Observed Structures in Ti-Ga Alloys

Phase	Pearson symbol	Space group	Strukturbericht designation	Prototype	Reference
(βTi)	cI2	Im3m	A2	W	[Pearson2]
(αTi)	hP2	P6₃/mmc	A3	Mg	[Pearson2]
Ti₃Ga	hP8	P6₃/mmc	$D0_{19}$	Ni₃Sn	[57And2, 78Bel]
Ti₂Ga	hP6	P6₃/mmc	$B8_2$	Ni₂In	[57And2, 62Pot]
Ti₅Ga₃	tI32	I4/mcm	$D8_m$	W₅Si₃	[62Sch, 62Pot]
Ti₅Ga₃O_x(a)	hP16	P6₃/mcm	$D8_8$	Mn₅Si₃	[63Bol, 62Pot]
Ti₅Ga₄	hP18	P6₃/mcm	...	Ti₅Ga₄	[62Sch]
TiGa	tP2	P4/mmm	$L1_0$	AuCu	[62Sch]
Ti₂Ga₃	tP10	P4/m	...	Ti₂Ga₃	[62Pot]
Ti₃Ga₅	tP32	P4/mbm	...	Al₅Ti₃	[82Mii]
TiGa₂	(a,b)	...	...	HfGa₂	[62Sch]
TiGa₃	tI8	I4/mmm	$D0_{22}$	Al₃Ti	[62Sch, 42Wal]
(Ga)	oC8	Cmca	A11	Ga	[Pearson2]

Note: Exact compositions are unknown. See text. (**a**) Superstructure of the $D0_{22}$ lattice. (**b**) Tetragonal.

Table 4   Lattice Parameters of Ti-Ga Alloys

Phase	Composition, at.%	Lattice parameters, nm			Reference
		a	b	c	
(αTi)	3	0.2944	...	0.4682	[74Bel](a)
	6	0.2937	...	0.4675	
	7	0.2930	...	0.4666	
	9	0.2927	...	0.4662	
	11.3	0.2919	...	0.4642	
	1.1	0.2947	...	0.4688	[55Den](a)
	2.5	0.2944	...	0.4683	
	3.1	0.2943	...	0.4680	
	3.7	0.2942	...	0.4677	
	4.9	0.2939	...	0.4677	[58And](a)
	11.6	0.2920	...	0.4668	
Ti₃Ga	25	0.576	...	0.464	[57And]
	25	0.57410	...	0.46350	[78Bel]
Ti₂Ga	33.3	0.4514	...	0.5501	[57And]
	33.3	0.4514	...	0.5501	[62Pot]
Ti₅Ga₃	37.5	1.0218	...	0.5054	[62Sch]
	37.5	1.022	...	0.5054	[62Pot]
Ti₅Ga₃Oₓ	37.5	0.7604	...	0.5288	[63Bol]
	37.5	0.7609	...	0.5308	[62Pot]
Ti₅Ga₄	44.4	0.7861	...	0.5452	[62Sch]
TiGa	50	0.3967	...	0.3967	[62Sch]
Ti₂Ga₃	60	0.6284	...	0.4010	[62Pot]
Ti₃Ga₅	62.6	1.117	...	0.399	[82Mii]
TiGa₂	66.7	0.3929	...	0.4061	[62Sch]
TiGa₃	75	0.3789	...	0.8734	[62Sch]
	75	0.5559	...	0.8109	[42Wal]
(Ga)	100	0.45197	0.45260	0.76633	[Pearson2]

(a) Digitized from graphical form.

Table 5   Gibbs Energies of the Ti-Ga System [73Kau]

Gibbs energies of the pure components

G^0(Ti,L)   = 0
G^0(Ga,L)   = 0
G^0(Ti,bcc) = − 16 234 + 8.368 T
G^0(Ga,bcc) =   8.368 T
G^0(Ti,cph) = − 20 585 + 12.134 T
G^0(Ga,cph) = − 2092 + 11.380 T

Solution phases

B(L) = − 41 840 B(bcc) = − 54 392 B(cph) = − 57 739

Intermetallic compounds

G^0(Ti₃Al) = − 29 298 + 8.4153 T
G^0(Ti₂Al) = − 31 994 + 7.9776 T

Note: Values are given in J/mol and J/mol · K.

to the Ti₃Ga structure. It is concluded that the most Ti-rich intermetallic phase is Ti₃Ga. Reports of Ti-rich compounds based on similar evidence have been shown incorrect in the Ti-Al system.

[73Kau] quoted unpublished data [72Bid] on the solidus; 1500 and 1425 °C are the solidus temperatures for 16 to 26 at.% Ga, respectively.

Crystal Structure and Lattice Parameters

All observed crystal structures of the Ti-Ga system are listed in Table 3, and lattice parameters are listed in Table 4. Homogeneity ranges are unknown for compound phases other than Ti₃Ga; neither is it known which of the ordered phases other than Ti₃Ga and Ti₂Ga are equilibrium compounds.

Thermodynamics

Activity measurements in the bcc phase suggest that a regular solution Gibbs energy can be used, with interaction parameter −54 400 J/mol [72Bid] or −51 900 J/mol [73Geg]. Based on these data and on the phase diagram data of [58And], [73Kau] modeled the thermodynamics of the Ti-rich phases and calculated a partial diagram.

The present evaluator has found no reason to modify the calculation; the [73Kau] Gibbs energies are listed in Table 5. The temperatures of the three-phase equilibria are reproduced, as well as the partial Gibbs energies of the bcc (βTi) phase. The regular solution model is suitable to represent the Ti-rich portion of the system, given the accuracy with which the phase diagram is presently known.

Gibbs energies of the solution phases are represented as:

$$G(i) = (1 - x)\, G^0(\text{Ti},i) + x\, G^0(\text{Ga},i) + RT[x \ln x + (1 - x)] + B(i)\, x(1 - x)$$

where i designates the phase; x is the atomic fraction of Ga; G^0 is the Gibbs energy of the pure metals; and $B(i)$ is the interaction parameter. These Gibbs energies are valid only in the composition range 0 to 35 at.% Ga. For a calculation of the entire diagram, it is probable that asymmetries in the interactions will have to be included.

Cited References

42Wal: H.J. Wallbaum, "The Isomorphy Among Al$_3$Ti, Ga$_3$Ti and Ga,Zr," *Z. Metallkd., 34,* 118-119 (1942) in German. (Crys Structure; Experimental)

55Den: J.M. Denney, "A Study of Electron Effects in Solid Solution Alloys of Titanium," dissertation, California Institute of Technology (1955). (Crys Structure; Experimental)

57And1: K. Anderko and U. Zwicker, "Constitution of the Titanium-Gallium System," *Naturwissenschaften, 44,* 510 (1957) in German. (Equi Diagram; Experimental)

57And2: K. Anderko, "Structural Investigation of the Systems Titanium-Gallium and Titanium-Indium," *Naturwissenschaften, 44,* 88 (1957) in German. (Crys Structure; Experimental)

***58And:** K. Anderko, "The Binary Systems of Titanium with Gallium, Indium, Germanium and of Zirconium with Gallium and Indium," *Z. Metallkd., 49,* 165-172 (1958) in German. (Equi Diagram; Experimental)

***62Pot:** M. Potzschke and K. Schubert, "Constitution of Several T4-B3 Homologous and Quasi Homologous Systems," *Z. Metallkd., 53,* 474-488 (1962) in German. (Equi Diagram; Experimental)

***62Sch:** K. Schubert, H.G. Meissner, and M. Potzschke, "Structure Data on Metallic Phases," *Naturwissenschaften, 49,* 57 (1962) in German. (Crys Structure; Experimental)

63Bol: H. Boller and E. Parthe, "Structure of $D8_8$ Phases Between 4a Metals and Al, Ga, In, and Sb," *Monatsh. Chem., 94,* 225-226 (1963) in German. (Crys Structure; Experimental)

64Sch: K. Schubert, H.G. Meissner, A. Rama, and W. Rossteutscher, "Structure Data on Metallic Phases," *Naturwissenschaften, 51*(12), 287 (1964) in German. (Crys Structure; Experimental)

***71Sha:** C.E. Shamblen and C.J. Rosa, "Ti$_3$Ga and αTi Interdiffusion Between 600 and 860 °C," *Metall. Trans., 2*(7), 1925-1931 (1971). (Equi Diagram; Experimental)

72Bid: L. Bidwell, private communication (1972). (Equi Diagram; Experimental)

73Geg: H.L. Gegel and M. Hoch, "Thermodynamics of α-Stabilized Ti-X-Y Systems," *Titanium Science and Technology,* Vol. 2, R.I. Jaffee and H.M. Burte, Ed., 923-933 (1973). (Thermo; Experimental)

73Kau: L. Kaufman and H. Nesor, "Phase Stability and Equilibria as Affected by the Physical Properties and Electronic Structure of Titanium Alloys," *Titanium Science and Technology,* Vol. 2, R.I. Jaffee and H.M. Burte, Ed., 773-789 (1973). (Equi Diagram; Thermo; Experimental)

74Bel: O.K. Belousov, "The Electron Structure and Some Physical Properties of Binary Titanium Alloys," *Izv. Akad. Nauk SSSR, Met.,* (2), 202-208 (1974) in Russian; TR: *Russ. Metall.,* (2), 119-123 (1974). (Crys Structure; Experimental)

78Bel: O.K. Belousov and I.I. Kornilov, "Lattice Parameters and Crystal Structure of the Compound Ti$_3$Ga," *Izv. Akad. Nauk SSSR, Met.,* (1), 195-197 (1978) in Russian; TR: *Russ. Metall.,* (1), 175-178 (1978). (Crys Structure; Experimental)

81Bel: O.K. Belousov, "Possibility of the Formation of an Ordered Ti$_7$Ga Phase," *Izv. Akad. Nauk SSSR, Met.,* (8), 195-200 (1981) in Russian; TR: *Russ. Metall.,* (3), 149-153 (1981). (Equi Diagram; Experimental)

82Mii: R. Miida, S. Hashimoto, and D. Watanabe, "New Type of A$_5$B$_3$ Structure in Al-Ti and Ga-Ti Systems; Al$_5$Ti$_3$ and Ga$_5$Ti$_3$," *Jpn. J. Appl. Phys., 21*(1), L59-L61 (1982). (Crys Structure; Experimental)

The Gd-Ti (Gadolinium-Titanium) System

157.25   47.88

By J.L. Murray

Equilibrium Diagram

There is general agreement that all Ti-rare earth phase diagrams are characterized by the absence of intermetallic compounds, limited mutual solubilities in the solid state, and, often, miscibility gaps in the liquid state. The assessed Ti-Gd phase diagram (Fig. 1) conforms to this pattern. However, there is a paucity of experimental data on this system, and where data exist, they conflict.

The two major experimental studies of the system were performed by [61Cro] and [62Sav]. The assessed diagram (Fig. 1 and Table 1) is based primarily on the data of [61Cro] because of the attention given to avoiding contamination during melting, the use of several experimental techniques to confirm results, and the detailed documentation of the experiments. The purity of Gd, however, was not good in either study and confirmatory work is still needed.

A calculation of the phase diagram from Gibbs energies was performed as part of this evaluation (see below). The experimental data of [61Cro] could be reproduced within estimated experimental uncertainties by the calculations, and Fig. 1 is drawn from these calculations. The

calculations confirmed the evaluator's judgment that the [61Cro] melting point data should be reinterpreted as evidence for a miscibility gap in the liquid phase and monotectic reaction (see below); these features are incorporated in Fig. 1.

[62Sav] used 99.9 wt.% pure Ti and 99 wt.% pure Gd and investigated the phase equilibria by thermal analysis, X-ray diffraction, metallography, and microhardness measurements. Based on the thermal analysis data, the reactions L $\rightleftarrows$ (βTi) + (αGd) and (βTi) + (αGd) $\rightleftarrows$ (αTi) were reported to occur at 1120 $\pm$ 5 and 880 $\pm$ 5 °C, respectively. Based on the microscopic and microhardness data, the solubility of Gd in (αTi) was reported to increase from 0.3 to 0.9 at.% between 600 and 850 °C. The solubility of Ti in (αGd) was reported to be less than 1.5 at.%.

[61Cro] determined solidus temperatures by the thermal gradient technique, using optical methods for the temperature measurement. Liquidus temperatures were estimated as the temperatures at which samples collapsed. If the data are interpreted in terms of a monotectic reaction L$_1$ $\rightleftarrows$ (βTi) + L$_2$, then the apparent liquidus temperature would be equal to that of the (βTi) liquidus

Fig. 1---Assessed Ti-Gd Phase Diagram

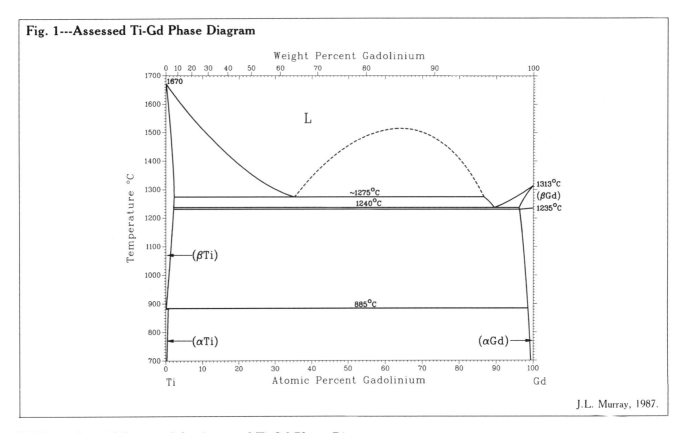

J.L. Murray, 1987.

Table 1 Special Points of the Assessed Ti-Gd Phase Diagram

Reaction	Compositions of the respective phases, at.% Gd			Temperature, °C	Reaction type
$L_1 \rightleftarrows L_2 + (\beta Ti)$	35	87	2	1275	Monotectic
$L \rightleftarrows (\beta Ti) + (\beta Gd)$	89	2	96	1240	Eutectic
$(\beta Gd) \rightleftarrows (\beta Ti) + (\alpha Gd)$	~ 96	2	~ 96	1230	Eutectoid
$(\beta Ti) + (\alpha Gd) \rightleftarrows (\alpha Ti)$	~ 0.5	~ 99	~ 0.6	885	Peritectoid
$L \rightleftarrows L_1 + L_2$		~ 63		~ 1515	Critical point
$L \rightleftarrows (\beta Ti)$		0		1670	Melting point
$(\beta Ti) \rightleftarrows (\alpha Ti)$		0		882	Allotropic transformation
$L \rightleftarrows (\beta Gd)$		100		1313	Melting point
$(\beta Gd) \rightleftarrows (\alpha Gd)$		100		1235	Allotropic transformation

temperature would be equal to that of the (βTi) liquidus for Ti content less than the monotectic composition and would be equal to the monotectic temperature for more Gd-rich alloys. Examination of the melting point data (Fig. 2) shows that this interpretation in terms of a monotectic reaction is a plausible treatment of the data. Metallographic data also suggest liquid immiscibility [60Bec]. Based on these observations and on thermochemical considerations, the monotectic reaction is shown on the equilibrium diagram. The estimated monotectic temperature, about 1275 °C, is based on thermodynamic calculations, and it is derived from the more accurately known eutectic temperature of 1240 °C.

[61Cro] estimated the eutectic composition to be between 73.3 and 85.3 at.% Gd, by metallographic examination of alloys that were slowly cooled from the melt by a technique designed to avoid further contamination during melting. The calculated eutectic composition is ~89 at.% Gd. In view of the experimental uncertainties due to

impurity of the Gd, the calculated value is preferred, because it is consistent with the overall melting data.

The invariant reaction associated with the βTi → αTi transformation was determined as 885 ± 5 °C by thermal analysis and resistivity measurements [61Cro] and as 880 ± 5 °C by thermal analysis [62Sav]. The value for pure Ti is 882 °C [Ti]. [61Cro] estimated the mutual solid solubilities metallographically:

Phase	Temperature, °C	Solubility limit, at.% Gd
(βTi)	1240	0.3
	1200	~ 0.6
(αTi)	850	< 0.8
	885	1.3
(αGd)	max sol	> 96.8
(βGd)	max sol	> 96.8

Fig. 2---Experimental Data of [61Cro]

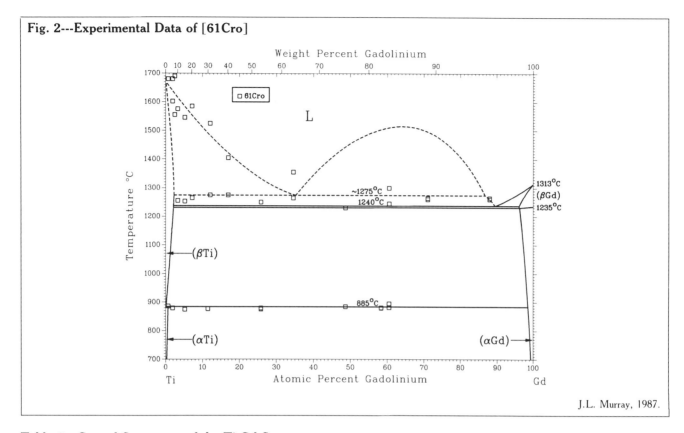

J.L. Murray, 1987.

Table 2   Crystal Structures of the Ti-Gd System

Phase	Homogeneity range, at.% Gd	Pearson symbol	Space group	Strukturbericht designation	Prototype	Reference
(βTi)	0 to ~ 2	cI2	Im3m	A2	W	[Pearson2]
(αTi)	0 to ~ 0.6	hP2	P6$_3$/mmc	A3	Mg	[Pearson2]
(βGd)	~ 96 to 100	cI2	Im3m	A2	W	[86Gsc]
(αGd)	~ 96 to 100	hP2	P6$_3$/mmc	A3	Mg	[86Gsc]

Table 3   Ti-Gd Thermodynamic Parameters

Properties of the pure components

G^0(Ti,L)   = 0
G^0(Gd,L)   = 0

G^0(Ti,bcc)   = − 16 234 +   8.368 T
G^0(Gd,bcc)   = − 10 054 +   6.339 T

G^0(Ti,cph)   = − 20 585 + 12.134 T
G^0(Gd,cph)   = − 13 996 +   8.933 T

Excess functions of solution phases

B(L)   =   26 800
C(L)   = −   6513

B(bcc)   =   47 280
C(bcc)   =   3560

B(cph)   =   46 523
C(cph)   =   3856

Note: Values are given in J/mol and J/mol · K.

In the assessed diagram, the solubility of Gd in (αTi) and (βTi) is shown as about 0.5 at.% at 885 °C. The maximum solubility of Ti in (βGd) is shown as 3.7 at.% at the eutectic temperature, decreasing with decreasing temperature.

The invariant reaction associated with the βGd → αGd transformation is too weak to be detected by thermal analysis. It is predicted to be eutectoid in character by the present thermodynamic calculations.

Metastable Phases

[76Wan] used TEM to examine splat-cooled Ti-Gd alloys for evidence of metastable extension of the solid solubility of Ti in (Gd). The observed maximum solubility was 4 to 5 at.% Ti, not significantly larger than the present estimate of the equilibrium solubility of Ti in (βGd).

Crystal Structure

Crystal structure data for Ti-Gd are listed in Table 2. Data on the lattice parameters as functions of composition are not available.

Thermodynamics

Experimental thermodynamic data are not available for the Ti-Gd system. The present calculations are based on the phase diagram only; the excess functions derived in this way are probably fairly realistic, first because the miscibility gaps occurring in all phases can be directly related to excess Gibbs energies, and second because the

117

derived Gibbs energies are consistent with the series of Ti-rare earth systems.

Parameters of the Gibbs energy expansions are listed in Table 3. For the solution phase i, the expansion is defined by:

$$G(i) = (1 - x) G^0(\text{Ti},i) + x G^0(\text{Gd},i) + RT [x \ln x + (1 - x) \ln (1 - x)] + x(1 - x) [B(i) + C(i) (1 - 2x)]$$

where x is the atomic fraction of Gd; G^0 is the Gibbs energy of the pure elements; and $B(i)$ and $C(i)$ are temperature-independent coefficients of expansion in Legendre polynomials. The Gibbs energies of Ti are from [70Kau]; those of Gd are based on heats of fusion and transformation from [Hultgren, E] and melting and transformation temperatures from [81Gsc].

Least-squares optimizations were based primarily on invariant temperature data and also on estimated solid solubilities. The calculated diagram is shown in Fig. 1 and 2. Assuming that the invariant temperatures are correct, it is judged that the Gibbs energies give a reasonably accurate picture of the thermodynamics of the system.

However, it should be understood that revisions of the phase diagram will also require revision of the Gibbs energies.

Cited References

60Bec: R. Beck, *U.S. Atom. Energy Comm. LAR-10, 93 p (1960). (Equi Diagram; Experimental)*

61Cro: J.G. Croeni, S.C. Rhoads, C.E. Armantrout, and H. Kato, "Titanium-Gadolinium Phase Diagram," *U.S. Bur. Mines, Rep. Invest. 5796, 14 p (1961). (Equi Diagram; Experimental)*

62Sav: E.M. Savitskii, "Equilibrium Diagram of the Gadolinium-Titanium System," *Russ. J. Inorg. Chem.*, 7(3), 358-359 (1962). *(Equi Diagram; Experimental)*

70Kau: L. Kaufman and H. Bernstein, *Computer Calculation of Phase Diagrams*, Academic Press, New York (1970). (Thermo; Theory)

76Wan: R. Wang, "Solubility of Liquid-Quenched Metastable cph Solid Solutions," *Mater. Sci. Eng.*, 23, 135-140 (1976). (Meta Phases; Experimental)

81Gsc: K.A. Gschneidner, Jr. and F.W. Calderwood, "Critical Evaluation of Binary Rare Earth Phase Diagrams," Rare-Earth Information Center, Iowa State University, Ames, IA (1981).

The Ge-Ti (Germanium-Titanium) System

72.59 47.88

By J.L. Murray

Equilibrium Diagram

The Ti-Ge phase diagram has been investigated only for Ti-rich alloys, and only a partial and unsatisfactory

diagram can be drawn at this time (Fig. 1). A number of additional equilibrium compounds have been identified, and enthalpies of mixing have been measured for the

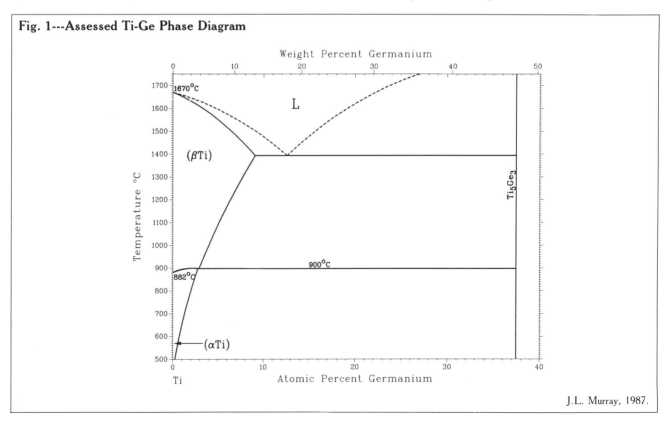

Fig. 1---Assessed Ti-Ge Phase Diagram

J.L. Murray, 1987.

Fig. 2---Experimental Data for the Ti-Ge Phase Diagram

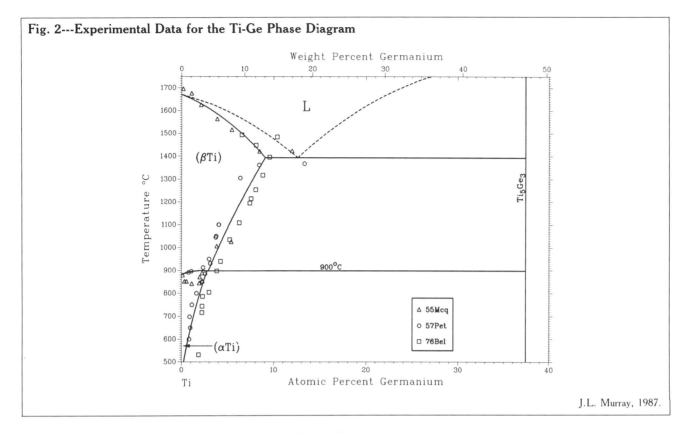

J.L. Murray, 1987.

Table 1   Special Points of the Assessed Ti-Ge Phase Diagram

Reaction	Compositions of the respective phases, at.% Ge			Temperature, °C	Reaction type
L ⇄ (βTi) + Ti$_5$Ge$_3$	12 to 15	~ 8 to 9	37.5	1360	Eutectic
(βTi) + Ti$_3$Ge ⇄ (αTi)	2.5	25	2.9	900	Peritectoid
L ⇄ (βTi)		0		1670	Melting point
(βTi) ⇄ (αTi)		0		882	Allotropic transformation
L ⇄ (Ge)		100		938.3	Melting point

Ge-rich liquid. The equilibrium phases and thermodynamic properties of the Ti-Ge system appear to be similar to those of the Ti-Si system.

An approximate phase diagram is shown in Fig. 1 and compared with experimental data in Fig. 2. The diagram was obtained from optimizations of Gibbs energies with respect to select experimental data. Special points of the diagram are summarized in Table 1.

The solid phases that have so far been identified in the Ti-Ge system are:

- The cph (αTi), bcc (βTi), and diamond cubic (Ge) solid solutions. The (αTi)/(βTi) phase equilibria have been examined experimentally and are discussed below. In addition, the maximum solubility of Ti in (Ge) is 1.8×10^{-6} at.% [64Vas], determined by composition analysis of single crystals of (Ge) grown from a liquid alloy.
- Ti$_5$Ge$_3$ is isomorphous with Ti$_5$Si$_3$. Its homogeneity range has not been investigated, but it probably has a range of several at.%, based on the crystal structure and by analogy with the Ti-Si system.

- Ti$_6$Ge$_5$ is isomorphous with Nb$_6$Sn$_5$, and TiGe is anti-isomorphous with TiSi.
- TiGe$_2$ is isomorphous with TiSi$_2$.
- A compound Ti$_3$Ge was found by [65Ros] in annealed powders, but was not found in phase equilibrium studies [55Mcq, 57Pet, 58And]. It is tentatively concluded that Ti$_3$Ge is a ternary phase stabilized by gaseous impurities.

Ti-Rich Alloys.

Experimental phase diagram data are shown in Fig. 2, and data on the three-phase equilibria are summarized in Table 2. The Ti-rich liquid solidifies by the eutectic reaction L ⇄ (βTi) + Ti$_5$Ge$_3$ [55Mcq, 57Pet, 70Yue, 76Bel], but reported eutectic temperatures range between 1360 and 1430 °C. [70Yue] determined the eutectic composition as 14.9 at.% Ge by chemical analysis of the eutectic region of a zone-melted alloy. Other values of the eutectic composition are less accurate, because they are based on metallographic examination of as-cast alloys [55Mcq, 57Pet, 76Bel].

Table 2 Experimental Data on Three-Phase Equilibria

Reference	Temperature, °C	Composition, at.% Ge		Experimental technique
		(βTi)	Eutectic	
Eutectic reaction				
[70Yue]	...	...	14.9	Chemical analysis of zone-melted alloy
[55Mcq]	1410 ± 10	~ 9.4	~ 16	Amount of eutectic and primary (βTi) in as-cast alloys
[57Pet]	1360 ± 10	8.6	13.4	Metallography, as-cast alloys
[76Bel]	1430	8.4	~ 12	Metallography

Reference	Temperature, °C	Composition, at.% Ge		Experimental technique
		(βTi)	Peritectic	
Peritectic reaction				
[55McQ]	897 ± 3	3.6	4.2	Metallography
[57Pet]	905 ± 10	2.1	2.7	Dilatometry, resistivity, X-ray diffraction, metallography
[76Bel]	860(a)	1.6	2.2	Resistivity

(a) [76Bel] also reported a congruent (βTi) → (αTi) transition at 830 °C and 1 at.% Ge.

Table 3 Crystal Structures of the Equilibrium Phases of the Ti-Ge System

Phase	Composition range, at.% Ge	Pearson symbol	Space group	Struktur-bericht designation	Prototype	Lattice parameters, nm			Reference
						a	b	c	
(αTi)	0 to 2	hP2	$P6_3/mmc$	A3	Mg	...	...	...	[Pearson2]
(βTi)	0 to 9	cI2	Im3m	A2	W	...	...	...	[Pearson2]
Ti$_3$Ge(a)	25	tI32	I4	...	Fe$_3$P	1.029	...	0.514	[65Ros]
Ti$_5$Ge$_3$	37.5	hP16	$P6_3/mcm$	$D8_8$-type	Mn$_5$Si$_3$	0.7537	...	0.5223	[51Pie]
TiGe	50	oP8	Pmm2	...	SiTi(b)	0.3809	0.6834	0.5235	[57Age, 59Age]
Ti$_6$Ge$_5$	45.5	oI44	Immm	...	Nb$_6$Sn$_5$	1.6915	0.7954	0.5233	[68Hal, 69Her]
TiGe$_2$	66.7	oF24	Fddd	C54	Si$_2$Ti	0.8594	0.5030	0.8864	[59Age, 44Wal]
(Ge)	100	cF8	Fd3m	A4	C(diamond)	0.56575	...	...	[Pearson2]

(a) The present evaluator believes Ti$_3$Ge to be a ternary phase. (b) Anti-isotypic with TiSi.

The eutectic reaction shown in Fig. 1 occurs at 1372 °C with (βTi) and (αTi) compositions 8.4 and 12.1 at.% Ge, respectively. Thus, Fig. 1 does not show the eutectic composition at the measured value of 14.9 at.% Gc, and it should only be used as a schematic of the phase equilibria. The discrepancies in the eutectic temperature and the (βTi) solvus rule out the possibility of drawing more quantitative phase boundaries at this time.

(αTi) and (βTi) form a peritectoid equilibrium with the most Ti-rich compound, which is probably Ti$_5$Ge$_3$. Reported versions of the diagram differ in three points:

- Ti$_5$Ge$_3$ was the most Ti-rich compound observed by [55Mcq, 57Pet, 58And], but there is also some evidence for Ti$_3$Ge as the first compound [65Ros, 76Bel].
- There are discrepancies of about 2 at.% in the position of the (βTi) solvus.
- [76Bel] claimed that there is a minimum in the (αTi)/(βTi) boundaries, but the combined data of [55Mcq] and [57Pet] indicate that the boundaries rise monotonically to the peritectoid isotherm.

[57Pet] used a combination of techniques to construct the diagram—resistivity (heating) for the (αTi)/(βTi) boundary and dilatometry, metallography, and X-ray diffraction for the (βTi) solvus. Solvus data obtained by the lattice parametric technique were consistent with metallographic examination of heat treated and quenched samples. Solvus data of [55Mcq] lie to the Ge-side of the

[57Pet] solvus, which may be interpreted as a failure to obtain equilibrium. The work of [76Bel] cannot be properly assessed because of insufficient information on the experimental techniques used. Therefore, greatest weight has been given to the [57Pet] data.

The proposal that there is an additional phase, Ti$_3$Ge, is based on the appearance of this phase in annealed powders [65Ros] and on a possible inflection in the (βTi) solvus [76Bel]. X-ray examination of two-phase alloys showed Ti$_5$Ge$_3$ in equilibrium with (αTi) [55Mcq, 57Pet, 58And], and the discrepancies in the (βTi) solvus data are inconsistent with metastable equilibrium between Ti$_5$Ge$_3$ and (αTi) or (βTi). Therefore, Ti$_3$Ge has not been included in the equilibrium diagram.

Crystal Structures

Crystal structures of the Ti-Ge phases, including those not shown on the phase diagram, are listed in Table 3.

Thermodynamics

Experimental Data.

Heats of mixing in the liquid phase were measured by [81Esi] and [78Shl], respectively, at 2000 K, and the data are shown in Fig. 3. The integral heats of mixing agree very well, but partial heats of mixing do not. The solid curve is from the phase diagram calculations described below.

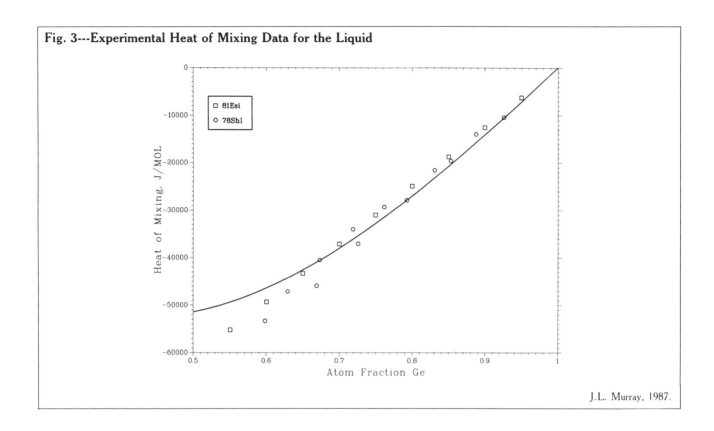

Fig. 3---Experimental Heat of Mixing Data for the Liquid

J.L. Murray, 1987.

Table 4 Thermodynamic Properties of the Ti-Ge System

Gibbs energies of the pure components and the compound

$G^0(\text{Ti,L})$ = 0
$G^0(\text{Ge,L})$ = 0

$G^0(\text{Ti,bcc})$ = $-16\,234 + 8.368\,T$
$G^0(\text{Ge,bcc})$ = $-6\,276 + 10.46\,T$

$G^0(\text{Ti,cph})$ = $-20\,585 + 12.134\,T$
$G^0(\text{Ge,cph})$ = $-418 + 12.13\,T$

$G^0(\text{Ge})$ = $36\,954 + 30.51\,T$

$G^0(\text{Ti}_5\text{Ge}_3)$ = $-127\,313 + 29.61\,T$

Excess Gibbs energies of the solution phases

$B(\text{L})$ = $-198\,308$
$C(\text{L})$ = $-42\,454$

$B(\text{bcc})$ = $-114\,681 - 25.72\,T$
$C(\text{bcc})$ = $-78\,997$

$B(\text{cph})$ = $-174\,978$
$C(\text{cph})$ = $-58\,918$

Note: Values are given in J/mol and J/mol · K.

Calculation of the Phase Diagram.

In view of the incompleteness of both phase diagram and thermochemical data, the present calculations are speculative and were undertaken to create a basis for future experiment and calculation.

Considering the phases L, (βTi), (αTi), and Ti$_5$Ge$_3$, the Gibbs energies of the solution phases are represented as:

$$G(i) = (1 - x) + x\,G^0(\text{Ge},i) + \text{RT}[x \ln x + (1 - x) \ln (1 - x)] + x(1 - x)\,[B(i) + C(i)\,(1 - 2x)]$$

where i designates the phase; x is the atomic fraction of Ge; G^0 is the Gibbs energies of the pure metals; and $B(i)$ and $C(i)$ are the interaction parameters. Gibbs energies of bcc and cph Ge were estimated from those used in the Ti-Si system.

An assumption concerning the excess entropy of the liquid is required, based on the small contribution of the excess entropy in the Ti-Si system, this term of the Gibbs energy was omitted. Similarly, because neither the heat of formation nor the melting point of Ti$_5$Ge$_3$ are known, these quantities had to be approximated and are again based on the Ti-Si system. It was also assumed that the (βTi) and (αTi) phases could be represented by subregular solutions. Gibbs energies for these phases, being based only on the phase diagram, can be used only for Ti-rich compositions. Despite the severity of these assumptions, a reasonable phase diagram can be calculated from the Gibbs energies given in Table 4. A measurement of the melting point and heat of formation of Ti$_5$Ge$_3$ would probably allow a quantitative calculation of the range 0 to 40 at.% Ge.

A further observation can be made about the liquid phase. A Gibbs energy function fitted only to the enthalpy of mixing data has an inflection that leads to a miscibility gap in the liquid near pure Ge. The present evaluators believe that there is probably no such gap and that melting occurs by a eutectic reaction L $\rightleftarrows$ TiGe$_2$ + (Ge). The Gibbs energy of the liquid was therefore constrained to have a positive definite second composition derivative.

121

Phase Diagrams of Binary Titanium Alloys

Cited References

44Wal: H.J. Wallbaum, "Intermetallic Compounds of Germanium," *Naturwissenschaften, 32*, 76 (1944) in German. (Crys Structure; Experimental)

51Pie: P. Pietrokowsky and P. Duwez, "Crystal Structure of Ti_5Si_3, Ti_5Ge_3, and Ti_5Sn_3," *Trans. AIME, 191*, 772-773 (1951). (Crys Structure; Experimental)

55Mcq: M.K. McQuillan, "A Study of the Titanium-Germanium System in the Region 0—11 At.% Germanium," *J. Inst. Met., 83*, 485-489 (1954-1955). (Equi Diagram; Experimental)

57Age: N. Ageev and V. Samsonov, "X-Ray Determination of Crystal Structure of TiSi and TiGe," *Dokl Akad. Nauk SSSR, 112*, 853-855 (1957) in Russian; TR: *Proc. Acad. Sci. USSR, Appl. Phys. Sect., 112*, 7-9 (1957). (Crys Structure; Experimental)

57Pet: V.C. Petersen and R.W. Huber, "The Titanium-Germanium System from 0 to 30% Germanium," U.S. Bur. Mines, Rep. Invest., 5365, 1-20 (2957). (Equi Diagram; Experimental)

58And: K. Anderko, "Binary Systems of Titanium with Gallium, Indium and Germanium, and of Zirconium with Gallium and Indium," *Z. metallkd., 49*, 165-172 (1958) in German. (Equi Diagram; Experimental)

59Age: N.V. Ageev and V.P. Samsonov, "An X-Ray Study of the Crystal Structures of Titanium Silicides and Germanides," *Zh. Neorg. Khim., 4*(7), 1590-1595 (1959) in Russian; TR: *Russ. J. Inorg. Chem., 4*(7), 716-719 (1959). (Crys Structure; Experimental)

64Vas: V.N. Vasilevskaya and E.G. Miselyuk, in *Radiatsionnaya Avtomatika Izotop i Yadernye Izlucheniya v Nauke i Tekhnike,* V.I. Stetsenko, Ed., Akad. Nauk Ukr. SSR. Kiev, 183-187 (1964). (Equi Diagram; Experimental)

65Ros: W. Rossteutscher and K. Schubert, "Structural Investigation of Several T4...5-B4...5 Systems," *Z. Metallkd., 56*(11), 813-822 (1965) in German. (Equi Diagram; Experimental)

68Hal: J. Hallais, P. Spinat, and R. Fruchart, "A New Ti Germanide, Ti_6Ge_5," *C.R. Acad. Sci. Paris, 267C*, 387-390 (1968). (Crys Structure; Experimental)

69Her: P. Herpin, P. Spinat, J. Hallais, R. Fruchart, J. Albrecht, and J. Ouvrard, "The Structure of Binary Compounds V_6Si_5 and Ti_6Ge_5," *C.R. Acad. Sci. Paris, 268*, 1750-1753 (1969) in German. (Crys Structure; Experimental)

70Yue: A.S. Yue and F.W. Crossman, "Diffusion-Phenomenon in Ti-Ti_5Ge_3 Eutectic," *Metall. Trans., 1*, 322-323 (1970). (Equi Diagram; Experimental)

76Bel: O.K. Belousov and I.I. Kornilov, "Solubility of Germanium in α-Ti," *Izv. Akad. Nauk, Met.,* (1), 168-169 (1976) in Russian; TR: *Russ. Metall.,* (1), 140-141 (1976). (Equi Diagram; Experimental)

81Esi: Yu.O. Esin, M.G. Valishev, A.F. Ermakov, O.V. Gel'd, and M.S. Petrushefskii, "The Enthalpies of Formation of Liquid Gemanium-Titanium and Nickel-Titanium Alloys," *Zh. Fiz. Khim., 55*, 753-754 (1981) in Russian; TR: *Russ. J. Phys. Chem., 55*(3), 421-422 (1981). (Thermo; Theory)

The H-Ti (Hydrogen-Titanium) System

1.00794   47.88

By A. San-Martin and F.D. Manchester

Equilibrium Diagram

The assessed Ti-H phase diagram is shown in Fig. 1 and 2, which are projections of a P-T-X surface on the T-X plane and the P-X plane, respectively, where P is pressure in Pa, T is temperature in K and °C, and X is the hydrogen concentration, expressed as $X = \mathrm{H/Ti}$ (the atomic ratio).

Fig. 1---Assessed Ti-H Phase Diagram

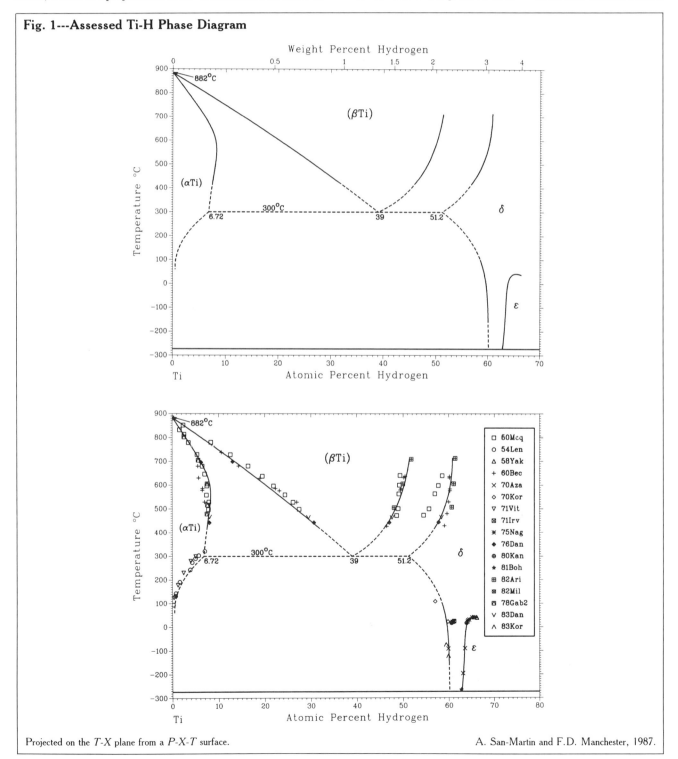

Projected on the T-X plane from a P-X-T surface.

A. San-Martin and F.D. Manchester, 1987.

Fig. 2---Assessed Ti-H Phase Diagram

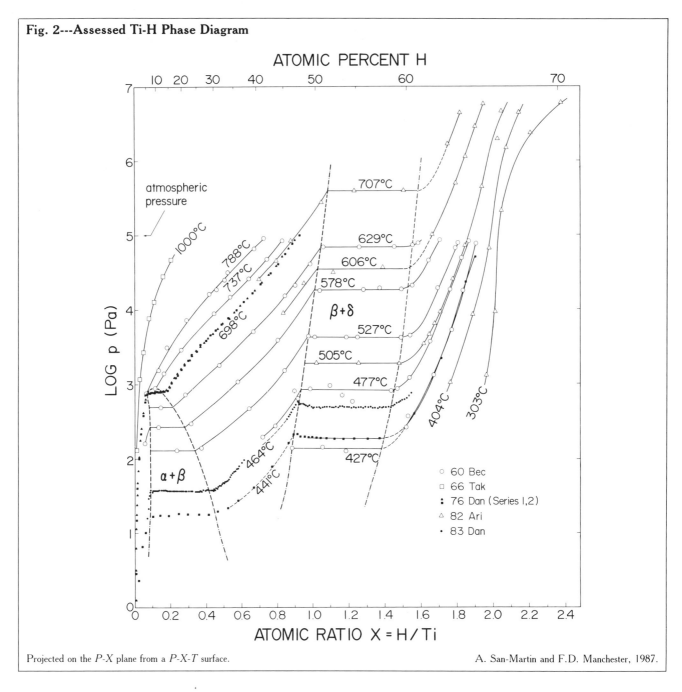

ATOMIC PERCENT H

LOG p (Pa)

atmospheric pressure

1000°C

788°C

737°C

698°C

707°C

629°C

606°C

578°C

β+δ

527°C

505°C

477°C

404°C

303°C

α+β

464°C

441°C

427°C

○ 60 Bec
□ 66 Tak
⦂ 76 Dan (Series 1,2)
△ 82 Ari
· 83 Dan

ATOMIC RATIO X = H/Ti

Projected on the *P-X* plane from a *P-X-T* surface.

A. San-Martin and F.D. Manchester, 1987.

Two presentations are necessary for a hydrogen–metal system, because the equilibrium pressure of the hydrogen surrounding the metal is always a significant thermodynamic variable, in contrast to most situations involving metallic alloys. The participation of hydrogen in the various phases of alloy systems is the best available example of hydrogen acting as a metal [71Gil]. The crystal structures and lattice parameters of the Ti-H phases are given in Tables 1 and 2, respectively. Figure 3 illustrates the existing phase relationships at high pressure (50 MPa) reported by [83Sha].

Recent work [84Num, 85Woo] evaluating evidence on the existence of a metastable hydride phase in the Ti-H system led both [84Num] and [85Woo] to propose relabeling the phases in the Ti-H system to correspond with the designations for the isostructural Zr-H system.

The composite phase diagram (Fig. 1) is of the eutectoid type, and consists of the following phases: (1) cph α; (2) bcc β; (3) two interstitial solid solutions of hydrogen based on the allotropic α and β forms of pure Ti; (4) δ, a fcc hydride; (5) ϵ, a tetragonally distorted fcc or fct hydride with axial ratio $c/a < 1$; and (6) γ, a metastable fct hydride with $c/a > 1$. Above the eutectoid temperature, the stability ranges of the α,β, and δ phases are delineated by temperature-composition values (see Tables 3 and 4) obtained from a composite set of isotherms. Figure 2 shows a simplified version.

Data previously reported by [50Mcq], [51Gib], [60Bec], [66Tak], and [82Ari] were compared with the results from high-precision work of [83Dan] (see also [76Dan]). The experiment measured the heat released when a small quantity of H reacts with the solid and

The H-Ti System

Table 1 Ti-H(D) Crystal Structures

Phase	System	Composition, at.% H(D)	Pearson symbol	Space group	Struktur-bericht designation	Prototype	Reference
(αTi)		0	$hP2$	$P6_3/mmc$	$A3$	Mg	[Pearson2]
(α)	Ti-H	0 to 8.38					
	Ti-D	0 to 8.34					
(βTi)		0	$cI2$	$Im3m$	$A2$	W	[Pearson2]
(β)	Ti-H	0 to 60.32					
	Ti-D	0 to ...					
δ	Ti-H	51.22 to 66.67	$cF12$	$Fm3m$	$C1$	CaF_2	[Pearson2]
	Ti-D						
ε	Ti-H	63.24 to 66.67	$tI6$	$I4/mmm$	$L'2_b$	ThH_2	[Pearson2]
	Ti-D						
Metastable phases(a)							
γ	Ti-H	1 to 2.9	$tP6$	$P4_2/n$	...	$\gamma H_{0.5}Zr$	[84Num]
	Ti-D						
?		39.67	...	$Pnnn$	...	...	[74Mir]

(a) Homogeneity ranges given in original works.

Table 2 TiH(D) Lattice Parameters

Phase	System	Composition, at.% H(D)	Lattice parameters, nm a	c	References
(αTi)		0	0.295111	0.468433(a)	[62Woo]
(α)	TiH	0 to 8.38	0.2951	0.4740(b)	[54Chr]
	TiD	0 to 8.34	...	...	[Pearson2]
(βTi)		0	0.33174(c)	...	[59Spr]
(β)	TiH	0 to 60.32	...	...	[Pearson2]
	TiD	0 to ...	0.336(d)	...	[74Mir]
	TiH	51.22 to 66.67	0.4397(e)	...	[31Hag]
			0.440	...	[54Len, 82Mil, 84Num]
			0.4403	...	[54Chr]
			0.4404	...	[70Aza]
			0.4405	...	[75Nag]
			0.4407	...	[71Irv]
	TiD		0.4440	...	[56Sid]
ε	TiH	63.24 to 66.67	0.4528(f)	0.4279	[58Yak]
	TiD		0.4516(f)	0.4267	[58Yak]
Metastable phases (g)					
γ	TiH	1.0 to 2.90	0.421	0.460	[84Num]
	TiD		0.420	0.459	[84Num]
		0.1 to 0.64	0.420	0.470	[85Woo]
?		39.76	0.434(h)	0.415	[74Mir]

(a) At 297 K. (b) For pure Ti, α = 0.2940 and c = 0.4680 nm [54Chr]. The listed value was measured at 31.51 at.% H. (c) At 1357 K. (d) At 653 K. (e) For dependence of lattice parameters on temperature and composition, see Fig. 4 and 5. (f) At 79 K. (g) Composition range given in original papers. (h) At 453 K.

the equilibrium pressure of the gas surrounding the metal. The composition of the TiH_x alloy was changed in steps of x ~0.004 at.% H, and the Ti sample used contained only about 30 ppm total impurities (oxygen 25 ppm, Fe 2 ppm, and <1 ppm other elements). In previous work, [82Dan] reported that the heat-flow calorimeter, operating at temperatures of 457 °C, was able to detect heat effects of 0.058 J, with a reproducibility of ~4%.

Below the eutectoid temperature, the α and δ phases were delineated (see Table 5) using X-ray diffraction measurements (XRD) [70Aza, 71Irv, 82Mil], electrical

resistivity (ER) [71Vit, 80Kan], and nuclear magnetic resonance (NMR) [70Kor, 83Kor].

The temperature of 300 °C for the eutectoid reaction β $\rightleftarrows$ α + δ was reported first by [50Mcq] and corroborated by [60Bec]. From thermal analysis, with heating and cooling rates of 1 °C/min, [54Len] reported that the average temperature of arrest was 319 °C on heating and 281 °C on cooling. However, 1 °C/min is a very large rate of temperature change for the kinetics of this experimental situation. This was demonstrated by [83Dan], who found that precipitation of the δ phase along the 464 °C isotherm

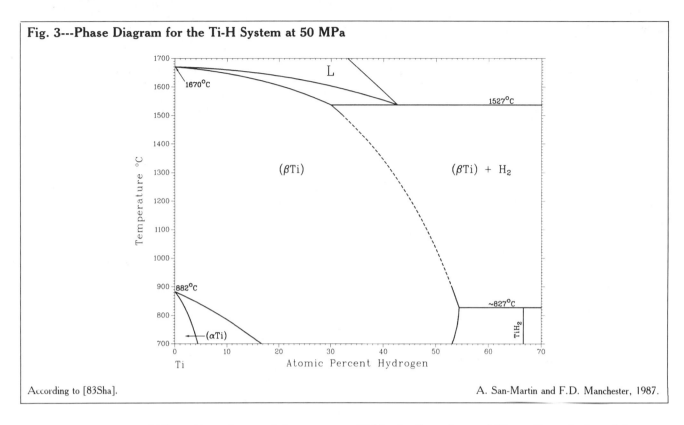

Fig. 3---Phase Diagram for the Ti-H System at 50 MPa

According to [83Sha].

A. San-Martin and F.D. Manchester, 1987.

Table 3 Experimental Phase Boundaries of the α + β Region

Temperature, °C	Composition, at.% H(D) α/(α ± β)	(α ± β)/β	Reference
882	0.00	0.00	[50Mcq](a)
850	...	2.63	[50Mcq](a)
832	2.06	...	[50Mcq](a)
815	2.63	...	[50Mcq](a)
802	2.44	...	[78Gab2](a)
779	...	8.09	[50Mcq](a)
737	...	10.55	[60Bec]
727	5.57	12.66	[50Mcq]
704	5.99	...	[78Gab2](a)
698	6.72	13.19	[76Dan](b)
679	6.37	16.60	[50Mcq]
	5.84	14.38	[60Bec]
636	7.06	19.36	[50Mcq]
629	6.28	19.09	[60Bec]
604	7.49	...	[78Gab2](a)
597	7.58	22.00	[50Mcq]
587	6.80	22.60	[60Bec]
578	6.81	23.31	[60Bec]
560	7.83	24.13	[50Mcq]
527	7.92	25.98	[50Mcq]
	7.24	27.11	[60Bec]
517	7.75	...	[78Gab2](a)
497	7.92	26.69	[50Mcq]
479	7.58	...	[78Gab2](a)
464	8.38	29.58	[83Dan](b)
441	7.92	30.65	[76Dan](b)
404	7.15	...	[78Gab2](a)
337	6.63	...	[78Gab2](a)

Note: All values extracted from (P-X) graphs in the original papers, except (a) and (b). (a) From the phase diagram in the original paper. (b) Data quoted by the authors.

Table 4 Experimental Phase Boundaries of the β + δ Region

Temperature, °C	Composition, at.% H(D) β/(β ± δ)	(β + δ)/δ	Reference
636	50.69	60.38	[60Bec]
...	49.55	58.81	[50Mcq]
629	50.45	60.38	[60Bec]
606	50.17	61.30	[82Ari]
596	49.42	57.93	[50Mcq]
585	49.77	60.44	[60Bec]
578	49.60	60.45	[60Bec]
560	49.34	56.75	[50Mcq]
527	48.80	60.25	[60Bec]
505	48.35	60.91	[82Ari]
496	49.19	55.87	[50Mcq]
468	48.77	54.69	[50Mcq]
464	47.78	58.68	[83Dan](a)
441	47.09	57.98	[76Dan](a)
427	46.75	59.28	[60Bec]

Note: Values read from graphs in the original papers, except (a). (a) Data quoted by authors.

occurred over time spans on the order of an hour. [83Dan] also commented that delays were longer for lower concentrations.

It is also anticipated that comparable precipitation times will be involved in locating the eutectoid temperature. Thus, the eutectoid limits quoted by [Hansen], which are based on [54Len], must be viewed as dependent on a particular rate of temperature change.

In general, hydride formation is associated with volume changes. Accommodation of this hydride-matrix misfit contributes significantly to the differences between the

Table 5 Experimental Phase Boundaries of the (α + δ) Region

Temperature, °C	Composition at.% H(D)	Reference
α/(α + δ)		
319	7.12	[54Len](a)
300	5.86	[54Len](a)
	4.83	[71Vit](b)
290	5.36	[54Len](a)
277	4.03	[71Vit](b)
271	4.78	[54Len](a)
242	3.83	[54Len](a)
227	2.60	[71Vit](b)
194	2.66	[54Len](a)
177	1.50	[71Vit](b)
174	2.12	[54Len](a)
145	1.08	[54Len](a)
136	0.45	[54Len](a)
127	0.75	[71Vit](b)
97	0.13	[54Len](a)
77	0.31	[71Vit](b)
59	...	[71Vit](b)
(α + δ)/δ		
109	57.08	[70Kor]
24	60.00	[70Kor]
20	60.00	[70Aza]
	60.94	[82Mil]
	61.24	[71Irv]
17	60.71	[80Kan]
− 73	59.51	[83Kor](c)
− 90	60.00	[70Aza]
−123	60.00	[83Kor](c)

(a) From Fig. 14 in the original paper. (b) Values calculated from $c = 8.6 \times 10^4 \exp(-2.52 \times 10^3/T)$; valid for $317 < T < 573$ K and where the H concentration is given in ppm by weight in the original work (here converted to at.%). (c) Data extracted from Fig. 6 of the original work.

Table 6 Experimental Boundary of the δ → ε Phase Transformation

Temperature, K	Composition at.% H(D)	Reference
315	65.87	[75Nag](a)
313	65.28	[75Nag](a)
310	66.44	[58Yak]
303	64.66	[75Nag](a)
295	64.29	[70Aza]
290	64.03	[80Kan](b)
280	64.29	[75Nag](a)
185	63.64	[70Aza]
79	63.24	[70Aza]
4 to 10	62.96	[81Boh](c)

Note: All values determined by XRD measurements, except (b) and (c). (a) Read from graph in the original work. (b) From electrical resistivity measurements. (c) From heat capacity measurements.

The retrograde character of the α/(α + β) phase boundary as reported by [78Gab2] leaves the 180° rule unviolated. Data from the very careful measurements of [76Dan] and [83Dan] shifted the α/(α + β) boundary to higher concentrations, $X_{max} = 0.0915$ (8.38 at.% H), at 464 °C. The data of [76Dan] and [83Dan], obtained with purer Ti than in other measurements, reduce the width of the α + β region, in agreement with previous findings [50Gib, 51Mcq] about the effect of impurities on the location of phase boundaries.

More reliable characterization of Ti-H alloys (concentration, impurities, etc.) is needed in future work. In the β + δ region, data from different authors for the solubility limits of the β phase (i.e.,β/(β + δ) boundary) are in good agreement, but differences exist for the (β + δ)/δ boundary (Fig. 1 and Table 4). The boundary drawn by [50Mcq] lies at much lower H concentration than those of [60Bec] and [82Ari]. However, the data points delineating the isotherm lines of [60Bec] and [82Ari] are rather sparse, leaving the composition at which the isotherms leave the two-phase region ill defined. The boundary points given by [76Dan] and [83Dan] (see Fig. 2 and Table 4) were obtained from considerably more experimental data, and they are probably more reliable. Using the [76Dan] and [83Dan] data, the β/(β + δ) and (β + δ)/δ boundaries were drawn to achieve maximum conformity with the data of [60Bec] and [82Ari]. Below the eutectoid temperature, the limit of H solubility in the α phase was delineated, using results from ER measurements [71Vit] and metallographic observation [54Len]. In XRD patterns taken at ~20 °C, [70Aza] and [82Mil] first detected the δ phase at $X = \sim0.17$ (~15 at.% H).

At the high concentration end of the α + δ region, the same type of disagreement exists as for the α/(α + δ) boundary. In this instance, the disagreement is between the results obtained by local measurements with the NMR method [70Kor, 83Kor], and those obtained from XRD [71Irv, 82Mil] and ER [80Kan]. Only the XRD data from [70Aza], which paid particular attention to the effects of concentration, impurities, and preparation methods, are in fairly good agreement with the NMR results.

Of all the boundaries of the δ phase, the boundary corresponding to the second-order transformation [58Yak] has been studied the most extensively (see "Crystal Structures and Lattice Parameters") and is the most well defined (Fig. 1). Data that this boundary is based on were obtained from heat capacity [81Boh], ER [80Kan], and

temperatures at which hydride forms (during cooling) and dissolves (during heating). In the Ti-H system, at low H concentration (less than 1000 ppm) and between 20 and 300 °C, [71Pat] found from ER measurements that the thermal hysteresis is ~10 °C for the location of the α/(α + δ) phase boundary. For a review of the Ti-H system, see [84Pul]

The extrapolated value for the composition of the β phase at the eutectoid temperature is $x = 0.64$ (39 at.% H). The assessed value lies between $x = 0.613$ (38 at.% H) and $x = 0.695$ (41 at.% H), reported by [50Mcq] and [60Bec], respectively. The assessed value differs greatly from the composition of $x = 0.786$ (44 at.% H) determined by [54Len] from metallographic examination. The compositions of the α and β phases at the eutectoid temperature were determined to be 0.072 (6.7 at.% H) and 1.05 (51.2 at.% H), respectively.

The α/(α + β) boundary is retrograde, based on P-T-X measurements of [78Gab2] (Fig. 1 and Table 3). These results were presented by [78Gab2] in a small T-X diagram; however, the purity of the materials used was not reported. Below 500 °C, the [78Gab2] results modified the straight-line extrapolation of this boundary to $x = 0.086$ (7.9 at.% H) at 300 °C, which was proposed by [50Mcq] and is generally accepted [Hansen, 60Lib, 68Mue, 76Kip, 83Kub2]. The uncertainty about impurity levels in the work of [78Gab2] leaves the difference between these two sets of results unresolved at the present time.

Fig. 4---Concentration Dependence of Ti-H Lattice Parameters of the α + δ, δ, and ε Phase Regions at Room Temperature

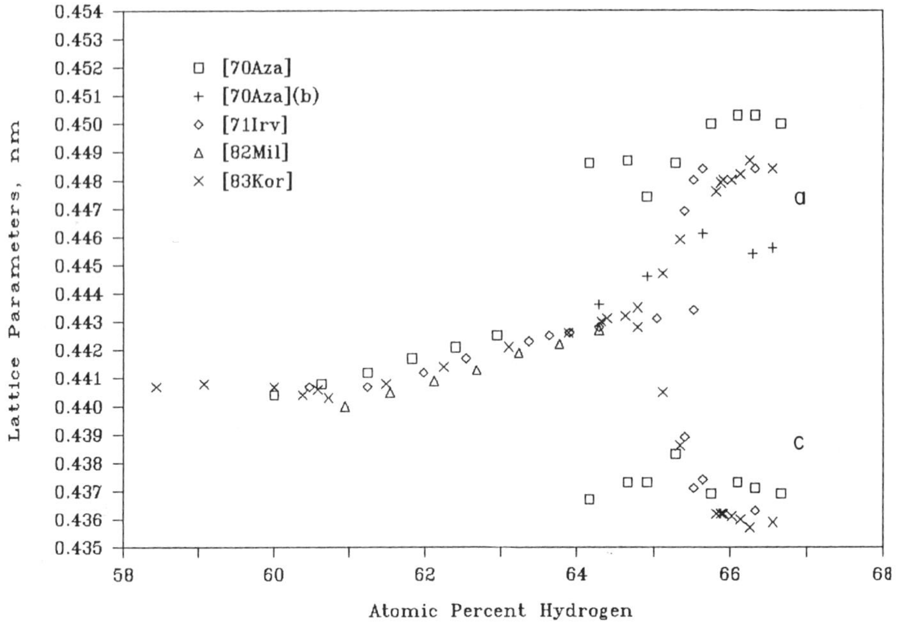

(a) Taken from graphs in the original papers. (b) Fast-cooled samples after charging with hydrogen.   A. San-Martin and F.D. Manchester, 1987.

XRD measurements [58Yak, 70Aza, 75Nag] (see Table 6).

[76Kip] and [78Gab2] analyzed the effect of pressure on the phase relationships of the Ti-H system, using data from previous compilations [62Liv, 66Kol, 67Gal, 68Mue]. Postulated phase diagrams are available for pressures of 0.5 and 1 kPa [78Gab2] and for pressures of 0.1 and 100 kPa and 10 MPa [76Kip]. However, the results of [76Kip] for 10 MPa should be considered with caution, because they were based on data from [51Gib], which deviate considerably from Fig. 2.

[83Sha] experimentally investigated the effects of high pressure (up to 50 MPa) on the solubility of H in Ti between 900 and ~1670 °C. [83Sha] used the value 1667 °C for the melting temperature of Ti (T_{fus}), which according to [Melt] is 1670 °C. [83Sha] reported that at 50 MPa, H lowers the melting point of Ti. The most significant change in T_{fus} was observed between 10 and 25 MPa (around 100 °C), and the total effect at the maximum H concentration amounted to ~130 °C. At 50 MPa, solid Ti dissolves about 30 at.% H, whereas the solubility in liquid Ti exceeds 40 at.%. These results were used by [83Sha] to construct the upper portion of the phase diagram shown in Fig. 3. This diagram includes a eutectoid reaction at 1537 °C, L ⇄ β + H₂, and a peritectoid reaction of the type postulated by [83Sha], β + H₂ ⇄ TiH₂, at 827 °C.

Crystal Structures and Lattice Parameters

In the Ti-H and Ti-D systems, the metal sublattice transforms, depending on the H(D) concentration, temperature, and cooling rate. It changes from a cph structure for pure Ti (α phase), through bcc (β phase), to fcc (δ phase), or fct with $c/a < 1$ (ε phase). In all of these phases, the tetrahedral or octahedral interstitial sites are occupied at random by H(D) atoms. Only for a metastable fct phase with $c/a > 1$ have deviations from random distribution of D atoms [74Mir] and ordered arrangements of H and D atoms [84Num, 85Woo] been reported. Crystal structure and lattice parameter data are presented in Tables 1 and 2, respectively, and Fig. 4 and 5.

α Phase.

In the cph α-phase structure, there are four tetrahedral sites per unit cell located at (0, 0, 3/4), (0, 0, 5/8), (1/3, 2/3, 1/8), and (1/3, 2/3, 7/8) and two octahedral sites at (2/3, 2/3, 1/4) and (2/3, 2/3, 3/4). The location of the atoms of the light interstitial elements in the α phase remains unresolved. From inelastic neutron scattering (INS) measurements at 599 K and $x = 0.07$, [82Hem] deduced that the majority of H atoms occupy octahedral interstitial sites and that the balance of H atoms occupy tetrahedral sites. [82Kho], from INS results obtained at 588 K and $x = 0.05$, concluded that the H atoms are situated in the tetrahedral rather than the octahedral sites. This type of occupancy is consistent with earlier NMR results between 433 and 593 K for $x = 0.08$ [70Kor]. An analogous situation applies for the locations of the D atoms. Thus, [82Hem], from well-defined INS results for TiD_x ($x = 0.09$) at 601 K, concluded that D atoms occupy tetrahedral sites. In contrast, results of neutron diffraction (ND) experiments at 648 K on TiD_x ($x = 0.075$) assigned around 70% of D atoms to the octahedral sites, with the remainder on the tetrahedral interstices [76Alp].

β Phase.

The β phase bcc structure has 12 tetrahedral interstices at (1/2, 1/4, 0) plus permutations and six octahedral interstices at (1/2, 1/2, 0) and (1/2, 0, 0) plus permutations. From ND measurements on deuterated samples of $x = $

Fig. 5---Variations of the Lattice Parameters of Ti-H (D) with Absolute Temperature for Various H(D)Concentrations

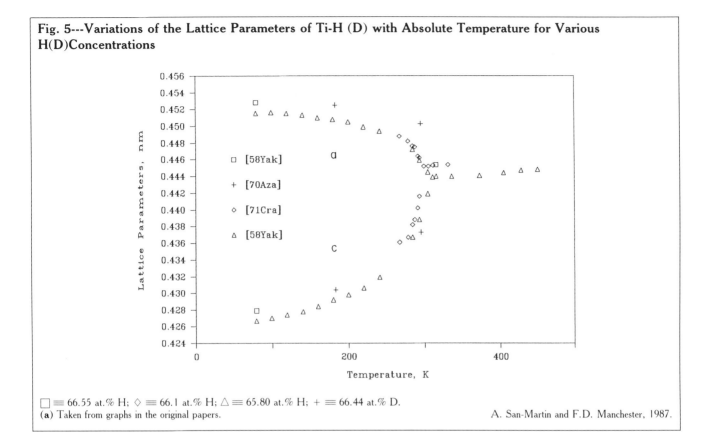

□ ≡ 66.55 at.% H; ◇ ≡ 66.1 at.% H; △ ≡ 65.80 at.% H; + ≡ 66.44 at.% D.
(**a**) Taken from graphs in the original papers.

A. San-Martin and F.D. Manchester, 1987.

0.66 and 0.67 at 653 and 673 K, respectively, [74Mir] and [76Alp] concluded that D atoms are distributed over the tetrahedral sites, although [76Alp] found a better fit to their data if between 15 and 30% of D atoms were assumed to be located in the octahedral sites. The bcc lattice is expanded by absorption of H(D) on these interstitial sites. [74Mir] reported a lattice parameter of $a_0 = 0.336$ nm, which represents an increase of around 6% compared with the value of a_0 for pure βTi, extrapolated from 1156 K (Table 2).

δ Phase.

The δ phase exists over a concentration range of $1.50 < x < 1.94$ (60 to 66 at.% H or D) [56Sid, 58Sof, 66Sam, 70Aza, 71Irv, 75Nag, 82Mil]. X-ray diffraction and ND experiments at room temperature [56Sid] revealed a C1 crystal structure (CaF$_2$ type). The unit cell contains four Ti atoms located at (0, 0, 0) and (1/2, 1/2, 0) plus permutations and eight H(D) atoms situated in the tetrahedral interstices at (1/4, 1/4, 1/4) and (3/4, 1/4, 1/4) plus permutations, (3/4, 3/4, 1/4) plus permutations, and (3/4, 3/4, 3/4). In this arrangement, each Ti atom could be surrounded by eight H(D) atoms located at a distance of $a_0\sqrt{3}/4$ and by 12 Ti atoms at the much longer distance of $a_0/\sqrt{2}$ (i.e., the nearest neighbors of each atom, Ti or H(D), are the unlike ones). In this phase, the lattice parameter a_0 increases linearly (see Fig. 4) with increasing H(D) concentration (approximately 0.15 at.% H or D [70Aza, 71Irv, 75Nag, 82Mil]. Approaching $x = 1.94$ from below, a broadening of diffraction lines from planes {200}, {220}, {311}, and {400} was observed [56Sid, 58Yak, 58Sof, 66Sam, 70Aza, 71Cra, 72Bal]. Below a critical temperature (T_c) (see ε phase), each of these lines splits into a pair, whereas the {111} and {222} planes remain sharp and

single [58Yak]. The fcc structure transforms to an fct structure and apparently never completely occupies the available tetrahedral sites; therefore, the δ phase typically is a nonstoichiometric or defect structure.

ε Phase.

The reversible fcc ⇄ fct transformation that produces the ε phase is higher than first order in character and occurs at a critical temperature T_c [58Yak]. In samples prepared with high-purity Ti, [58Yak] measured a T_c value of 310 ± 4 K for H and D concentrations of $X = 1.99$ and 1.98, respectively. T_c depends on concentration [70Aza] and is affected by the purity of the starting materials and the preparation methods [70Aza, 71Irv] (see Fig. 4). [70Aza] reported that the transition was not observed for $X < 1.70$, even at 79 K.

[75Nag] corroborated the concentration dependence of T_c reported by [70Aza] and showed, by adding controlled impurities to the Ti-H system, that as the impurity content rises the critical H concentration (X_c) for the onset of the fcc ⇄ fct transformation decreases. Different values of X_c have been reported for "room temperature" in the literature ($X = 1.80$ [70Aza], $X = 1.85$ [75Nag], $X = 1.90$ [71Irv], $X = 1.924$ [71Cra], and $X = 1.97$ [70 Duc]). However, a critical concentration specification requires a more specific temperature reference than "room temperature". [58Yak] established that the fcc ⇄ fct transformation consists of a continuous contraction of the cubic cell along a [001] axis and expansion along the other two ([100] and [010]). Accordingly, the axial ratio c/a increases continuously (see Fig. 5) as a function of temperature up to $T_c = 310 \pm 4$ K, with a minimum value of 0.945 at 79 K for concentrations of H and D of $X = 1.99$ and 1.98, respectively. The volume of the unit cell increases contin-

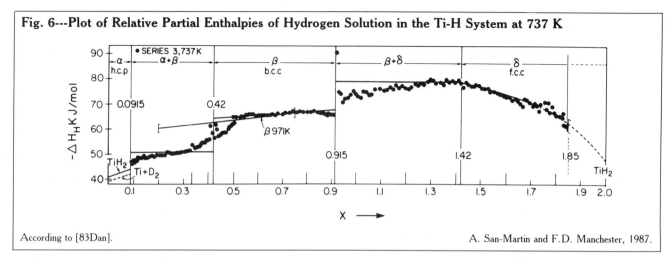

Fig. 6---Plot of Relative Partial Enthalpies of Hydrogen Solution in the Ti-H System at 737 K

According to [83Dan].

A. San-Martin and F.D. Manchester, 1987.

uously with temperature throughout the transition region, even through and beyond T_c.

The tetragonality produces shorter Ti-Ti distances for eight of the 12 Ti atoms nearest metal neighbors of Ti in the original cubic structure of the δ phase. For the eight atoms, the distance of nearest approach decreases (in the case of Ti-D) from 0.3140 to 0.3107 nm (approximately 1%), and for the other four atoms, it increases to 0.3193 nm. The eight H(D) atoms move slightly nearer to 0.1920 nm from the original distance of 0.1923 nm (around 0.2%). All distances are quoted to within 0.0001 nm, as given by [58Yak].

γ Phase.

The γ phase precipitates in the form of thin platelets in the αTi matrix, at low H(D) concentrations ($0.01 < X < 0.03$ H or D [84 Num] and $0.001 < X < 0.0064$ D [85Woo]). From electron diffraction patterns, [84Num] identified the crystal structure as fct, with $c/a = 1.093 \pm 0.001$ and with no difference between H and D precipitates. The reported lattice parameters were $a_0 = 0.421$ and $c = 0.460$ nm for H and $a_0 = 0.420$ and $c = 0.459$ nm for D. In addition, a (110)-type superlattice reflection was detected. The D precipitates studied by [85Woo] also revealed an fct structure with $a_0 = 0.420$ and $c = 0.470$ nm and $c/a = 1.12$. [84Num] assumed the fct hydride to be isostructural with the γ phase of zirconium hydride ($c/a = 1.08$ and H atoms located in the tetrahedral sites on alternate {110} planes). Consequently, [85Num] assigned the space group $P4_2/n$ to the precipitated phase and defined the unit cell as containing four Ti atoms located at (0, 0, 0) and (1/2, 1/2, 0) plus permutations and four H(D) atoms at (1/4, 1/4, 1/4), (3/4, 3/4, 3/4), (1/4, 1/4, 3/4), and (3/4, 3/4, 1/4). According to this model, the composition should be TiH.

Metastable Phases

[84Num] showed that an fct ($c/a = 1.09$) phase* precipitates when αTi samples are charged with H or D by the gas-equilibrium method ($T = 464\,°C$, $0.01 < X < 0.03$) and are kept at 500 °C for about 10 h and cooled at a rate of approximately 1 °C/min. Using TEM, [84Num] observed that thin-plane and slightly bent platelets precipitate in prism habit planes {01$\bar{1}$0} and near-basal habit planes {02$\bar{2}$5}, respectively. [85Woo] confirmed the existence of γTiD in specimens charged with D ($T = 397\,°C$, $0.001 \leq X$

*The existence of an fct phase with $c/a < 1$ was previously reported by [52Cra], [67Gol], and [74Mir], but unfortunately was not always supported by sufficiently convincing experimental data.

≤ 0.0064) and subjected to different cooling regimes. Some specimens were water quenched; others were given a further anneal, followed by a slow cooling. TEM experiments showed that γTiD precipitates on {10$\bar{1}$0} prism habit planes and on {20$\bar{7}$2} near-basal habit planes. This is in good agreement with the habit planes calculated by [85Woo], using a dislocation mechanism for the formation of the γ phase by martensitic transformation (previously proposed by [81Wea] for γZrH).

Thermodynamics

The available thermodynamic data up to 1968 were reviewed by [68Mue]. [83Kub1] presented a critical evaluation of thermodynamic properties in the Ti-H and Ti-D systems up to 1976. As a criterion for data evaluation, [83Kub1] stated that "only those that have used the purest possible Ti metal and the corresponding clear environmental conditions may be accepted." Thus, the data selected for review in [83Kub1] were restricted mainly to the results of [76Dan]. Other experimental data for Ti-H and Ti-D evaluations were extracted from the following sources—calorimetric measurements were obtained from [31Sie], [60Sta], [62Sta], and [66Wed], and pressure-composition-temperature (P-X-T) equilibrium relationships were obtained from [50Mcq], [56Haa], [60Mor], [66Lak], [66Tak], [67Gio], [69Mcq], [69Yav], and [76Nag].

Other publications concerning the thermodynamics of the Ti-H (Ti-D) system were reviewed by the present evaluators [58Mel, 60Sof, 74Sch1, 74Sch2, 75Mal, 75Wal, 77Sor, 78Gab1, 79Aga, 79Lyn, 80Hof, 80She, 81Sar, 84Bou]. However, the present evaluation benefited most from a recent and precise thermodynamic study of the Ti-H system by [83Dan] at 736 to 745 K. The investigation (extending over the wide composition range of $0 < X < 1.85$) was carried out using the combined calorimetric-equilibrium pressure method previously described in [76Dan]. Furthermore, important improvements in the experimental technique [82Dan], along with the use of very pure Ti samples, provided more reliable results.

Figure 6 shows the observed partial molar enthalpy of solution of H ($\Delta_H H$) plotted against H concentration (X). [83Dan] combined his calorimetric results (i.e., $\Delta_H H$) with equilibrium measurements (i.e., the relative partial molar Gibbs energy, $\Delta_H G$) on the same material and calculated the partial molar entropy of hydrogen (S_H). [83Dan] first calculated the corresponding values

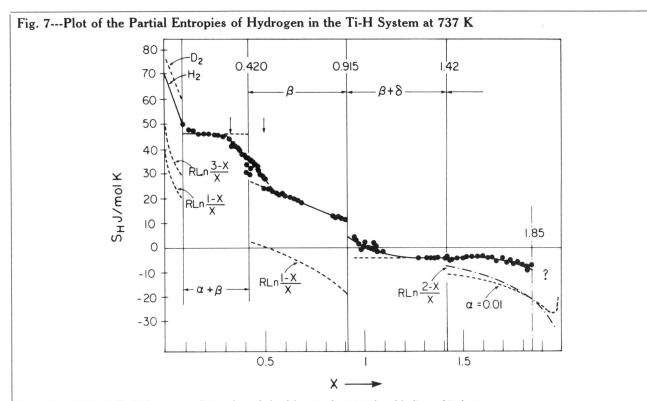

Fig. 7---Plot of the Partial Entropies of Hydrogen in the Ti-H System at 737 K

According to [83Dan]. The broken curves refer to values calculated from simple statistical models discussed in the text.

A. San-Martin and F.D. Manchester, 1987.

Table 7 Relative Partial Molar Enthalpies of Solution of H(D) in Ti [83Dan]

Phase	Ti impurities, ppm	Activation temperature, K	Composition range (a), X	Temperature, K	Coefficients −A	Coefficients −B
Ti-H						
α	(b)	737	0 to 0.0915	737 to 746	40.17	34.30
	(c)	800	0 to 0.0915	737	41.00	38.20
	(c)	1200	0 to 0.0915	737	41.84	45.56
β	(b)	737	0.51 to 0.915	746	60.67	6.99
	(d)	1123(d)	0.55 to 0.92	706 to 721	58.79	5.27
	(d)	1123(e)	0.2 to 0.750	971	57.57	10.71
Ti D						
α	(b)	737	0 to 0.0910	746	38.70	33.47
	(d)	1123(d)	0 to 0.0917	707	42.46	25.10
β	(b)	1123(d)	0.613 to 0.786	707	53.97	31.46

(a) X is the H concentration, expressed as $X = $ H/Ti, the atomic ratio. (b) O_2, 25; Fe, 2; others <1. (c) O_2, 980; N_2, 240; C, 100; Fe, 400. (d) O_2, 360; S, 10; C, 15; Al, 10; Cu, 18; Fe, 22; Ni, 20; Zr, 27; others <10. (e) From [76Dan].

of the relative partial molar entropy ($\Delta_H S$). These entropies are referred to the standard state of 1/2 mole of hydrogen gas at 1 bar pressure and at the considered temperature. Adopting values of $1/2\ S_{H_2}$ from [71Stu], [83Dan] calculated the partial molar entropies of H from $S_H = \Delta_H S + 1/2\ S^0_{H_2}(T)$. Figure 7 shows a plot of the dependence of S_H on X at the experimental temperature of 737 K.

A summary of the most accurate values of $\Delta_H H$ and $\Delta_D H$ in αTi and βTi is given in Table 7. The data from [76Dan] and [83Dan] are fitted to the straight-line relationships $\Delta_H H = A + BX$ kJ/mol, where $X = $ H/Ti (or $X = $ D/Ti).

In the α phase, the limiting values of $\Delta_H H$ at infinite dilution $\Delta_H H\infty$ vary between −40.17 and −41.84 kJ/mol. They appear to correlate with activation treatment and Ti purity. The less exothermic value corresponds to the most pure Ti metal. Activated at the temperature of the experiment, this yields $\Delta_H H = $ −40.17 kJ/mol as the preferred value. [76Dan] previously reported values of −44.35 kJ/mol at 773 K and −45.60 kJ/mol at 971 K. Agreement is quite good, if impurity levels and activation temperatures are taken into account.

The experimental value of H_H closest to the preferred value is −45.19 kJ/mol [50Mcq] ($753 < T < 1140$ K

Table 8 Relative Partial Molar Enthalpies and Entropies of H(D) Solution in βTi at Infinite Dilution

Temperature, K		Enthalpy $(-\Delta H^\infty_{H(D)})$, kJ/mol	Entropy $(-\Delta^\infty_{H(D)}S)$, J/mol · K	Reference
Hydrogen				
753 to 1223	....	58.2	...	[50Mcq]
1173 to 1773	....	54.1	...	[66Tak]
1073 to 1273	....	67.4	61.6	[74Sch1]
1173 to 1423	....	59.4	67.4	[76Nag]
1160 to 1260	....	54.5	66.1	[79Lyn]
Deuterium				
1173 to 1423	....	59.4	...	[76Nag]
1160 to 1260	....	52.1	64.9	[79Lyn]

From [79Lyn].

Table 9 Relative Partial Enthalpies and Entropies of Solution of H in Liquid Ti

Reference	Concentration, at.% H(D)	Temperature, K	Enthalpy $(\Delta_H H)$, kJ/mol	Entropy (ΔS^{ex}), J/mol · k
[66Lak] ...	<9.0	1873 to 2573	− 45.19	− 44.98(a)
[69Yav] ...	<9.0	1928 to 2073	− 47.07	− 48.95(a)

(a) As quoted by [83Kub1], using experimental values of [66Lak] and [69Yav].

Table 10 Standard Gibbs Energies of Formation of TiH₂

Temperature, K	Gibbs energy $(\Delta_f G^0)$,		
	Calorimetric [83Dan]	Equilibrium [82Ari]	Data analysis [71Stu](a)
677	....	...	− 42.6	− 54.435
737	....	− 32.49(b)	− 34.5(a)	− 44.058
778	....	...	− 28.6	− 38.104
879	....	...	− 13.3	− 23.797

Note: Standard states are Ti and H₂. (a) Interpolated value. (b) Calculated from experimental values of [83Dan].

Table 11 Standard Enthalpies and Entropies of Formation of TiH₂

Temperature, K	Enthalpy $(\Delta_f H^0)$ kJ/mol	Entropy $(\Delta_f S^0)$, J/mol · K	Reference
298.15	....	− 123.4	− 125.5	[60Sta]
298.15	....	− 144.348(a)	− 131.512(b)	[71Stu]
677 to 980	....	− 179	− 145	[82Ari]
737	....	− 136.95	...	[83Dan]

Note: Standard states are Ti and H₂. (a) From results of [31Sie], extrapolated using data from [51Gib]. (b) From the data of [60 Sta].

with Ti impurities of Si, Fe, Cr, Mg, Sb, and Cu <700 ppm). [67Gio] and [76Nag] found no difference between $\Delta_H H$ and $\Delta_D H$ values. Using u.h.v. techniques, [67Gio] reported values of −50.63 kJ/mol ($0.001 < X < 0.015$; 773 $< T < 1073$ K; Ti impurities 900 ppm Fe, 400 ppm Mg, 400 ppm Mn, <100 ppm U, and <100 ppm Sn). [75Nag] reported values of −52.72 kJ/mol ($X < 0.001$; $623 < T < 1073$ K; Ti of 99.9% purity). For the partial entropy, S_H (S_D) converted into an excess entropy, S_H^{ex} ($S_H^{ex} = S_H - S_{conf}$, where S_{conf} is the ideal configurational entropy**), [83Dan] found no dependence on composition. The average values of S_H^{ex} and S_D^{ex} are 35.27 and 42.68 J/mol·K, respectively, if S_{conf} is defined on the basis of one interstitial site per Ti atom (Fig. 7, see curve $R \ln (1 - X)/X$. [83Dan] also explored the possibility of a site degeneracy, which would appear if both types of interstitial sites (tetrahedral or octahedral) were occupied. Calculations for those conditions (Fig. 7, see curve $R \ln (3 - X)/X$) provide a minimum value of 25.94 J/mol·K for S_H^{ex}. All of these values greatly exceed the expected contribution of the vibrational entropy of H in Ti, which was calculated by [83Dan] to be 12.13 J/mol·K (19.25 J/mol·K for D) at 773 K. From these values of S^{ex}, [83Dan] was only able to predict a rather large positive contribution from other entropy terms, due to lack of appropriate experimental data.

The phase rule requires that $\Delta_H H$ and P_{equi} retain constant values throughout a mixed phase region. It can be seen from Fig. 6 that $\Delta_H H$ does not satisfy this condition for the α + β two-phase region, whereas P_{equi} (Fig. 2) does satisfy it for the same two-phase region. Above $X = 0.33$, the dependence of $\Delta_H H$ on composition is even more pronounced, and [83Dan] hypothesized on the origin of these discrepancies.

For the β phase (see Table 7), [83Dan] suggested a limiting value for $\Delta_H H$ of −58.99 ± 2 kJ/mol between 746 and 771 K. For S^{ex} at $X = 0.5$, [83Dan] calculated a value of 24.27 J/mol·K. Table 8 presents values for the relative partial molar enthalpy (ΔH) and entropy (ΔS) of H (D) solution at infinite dilution. [79Lyn] determined $\Delta_{H(D)} H^\infty$ and $S_{H(D)}^\infty$ from P-X-T data following a method described by [72Fla], which requires that Sievert's law be satisfied by the experimental data. For liquid Ti, the thermody-

namic conditions ($\Delta_H H$ and $\Delta_H S^{ex}$) of H solution were reported by [66Lak] and [69Yav] and are given in Table 9.

In the two-phase (β + δ) region (Fig. 6), [83Dan] attributed the observed concentration dependence of $\Delta_H H$ (X) to underestimation of heat effects, because nucleation of the δ phase altered the nucleation kinetics. The δ phase was investigated by [83Dan] up to $X = 1.85$ ($P_{equi} = 90$ 650 Pa). The $\Delta_H H$ values changed from −79.50 kJ/mol at $X = 1.42$ to −62.76 kJ/mol at $X = 1.85$.

Figure 7 shows results for S_H (experimental curve) and calculated values of S_{conf}. [83Dan] calculated the curve $R \ln (2 - X)/X$, assuming that the vacancies were randomly distributed among the tetrahedral sites for compositions $X < 2$. For the other curve, [83Dan] assumed that S_{conf} was a function of the deviation from stoichiometry and of α, the intrinsic disorder in the completely stoichiometric alloy. The value of 0.01, assigned to α, is typical at these temperatures [83Dan]. This curve (α = 0.01) appears to follow the trend of the experiments. Although [83Dan] assigned 8.70 J/mol·K to the vibrational entropy and −2.51 J/mol·K to the electronic entropy term, the difference in S_H^{ex} between the experimental and calculated values amounted to 6.28 J/mol·K. Tables 10 and 11 give the standard Gibbs energies, enthalpies, and entropies of formation for the hydride of composition $X = 2$. [83Dan] determined $\Delta_f H^0$ (TiH₂ at 737 K) as −136.94

**The terminology adopted here is that of [83Dan], because his calculations are used.

Table 12 Lattice and Electronic Contribution to the Heat Capacity for the Ti-H System

Composition, at.% H(D)	Electronic specific heat coefficient (γ), mJ/mol $\cdot$ K^2	Debye temperature (θ_D), K
0.00	3.383 ± 0.002	426.2 ± 0.6
29.08	3.246 ± 0.003	415.0 ± 0.8
44.13	3.147 ± 0.003	406.6 ± 0.6
54.55	3.108 ± 0.003	395.2 ± 0.5
59.84	2.841 ± 0.002	385.0 ± 0.5
60.78	2.812 ± 0.002	384.8 ± 0.5
61.54	2.793 ± 0.002	383.1 ± 0.4
63.77	4.041 ± 0.008	293.1 ± 1.0
64.29	4.750 ± 0.009	280.5 ± 0.8
64.91	5.44 ± 0.02	268.2 ± 1.6
65.57	5.50 ± 0.02	251.6 ± 1.4
65.88	5.24 ± 0.02	246.4 ± 1.5
66.22	5.10 ± 0.02	241.3 ± 0.9
66.56	4.46 ± 0.01	238.4 ± 0.8

From [81Boh].

kJ/mol by summing all the measured heat effects up to $X = 1.85$ (-130.25 kJ/mol) and by adding the extrapolated contribution from 1.85 to 2. The generally accepted value of $\Delta_f S^0$ [71Stu, 83Kub1] was obtained from the heat capacity measurements ($X = 1.971$, $24 < T < 363$ K) of [60Sta].

[82Ari], using the same method of a previous investigation [79Ari], determined $\Delta_f G^0$ of TiH$_x$ ($1.6 < x < 2.0$) from the dependence of Ti activity, a_{Ti}, on H$_2$ fugacity. The integration of a_{Ti} (based on the Gibbs-Duhem relation [79Ari]) started from the α phase and extended up to the δ phase. Data from [50Mcq] ($0 < X < 0.8P$, < 0.1 MPa) were used by [79Ari] and [82Ari] to complement his own (P-X-T) equilibrium measurements ($0.8 < X < 2.3$, 1 kPa $< P < 6.5$ MPa, $576 < T < 980$ K).

Measurements of heat capacity (C_p) at low temperature ($1.8 < T < 10$ K) and its dependence on composition over the range $0 < X < 2$ were reported by [81Boh]. Earlier, [70Duc] reported results for just two concentrations in the same temperature range. Numerical results from [81Boh] for the electronic specific heat coefficient, γ, and the Debye temperature, θ_D, are listed in Table 12. In the two-phase $\alpha + \delta$ region ($0 < X < 1.5$), both θ_D and γ decrease linearly with increasing H content. For $X > 1.6$, θ_D shows a strong decrease. Totally different is the behavior of the electronic contribution from $X = 1.5$ and above. In the cubic δ region, it shows a small decrease, followed by a sharp maximum ($X = 1.90$) in the fct phase. These results are consistent with theoretical predictions of an "electronically driven" fcc $\rightleftarrows$ fct transformation [76Swi, 78Kul, 79Gup].

γ coefficient is connected [76Swi] to the electronic density of states at the Fermi level N(E$_F$) by $\gamma = 2\pi k^2 N(E_F) (1 + \lambda)/3$, where k is the Boltzmann's constant and λ is the electron-phonon interaction parameter. Magnetic contributions were discounted by the experiments of [81Boh].

Cited References

31Hag: G. Hagg, "X-Ray Investigation on the Hydrides of Titanium, Zirconium, Vanadium and Tantalum," *Z. Phys. Chem. B*, *11*, 433-454 (1931) in German. (Crys Structure; Experimental)

31Sie: A. Sieverts and A. Gotta, "Properties of Metal Hydrides. III. Titanium Hydride," *Z. Anorg. Chem.*, *199*, 384-386 (1931) in German. (Thermo; Experimental)

50Gib: T.R.P. Gibb, Jr. and H.W. Kruschwitz, Jr., "The Titanium-Hydrogen System and Titanium Hydride. I. Low Pressure Studies," *J. Am. Chem. Soc.*, *72*, 5365-5369 (1950). (Crys Structure; Experimental)

***50Mcq:** A.D. McQuillan, "An Experimental and Thermodynamic Investigation of the Hydrogen-Titanium System," *Proc. R. Soc. (London)., Ser. A*, *204*, 309-322 (1950). (Equi Diagram, Thermo; Experimental; #)

51Gib: T.R.P. Gibb, Jr., J.J. McSharry, and R.W. Bragdon, "The Titanium-Hydrogen System and Titanium Hydride. II. Studies at High Pressure," *J. Am. Chem. Soc.*, *73*, 1751-1755 (1951). (Thermo; Experimental)

51Mcq: A.D. McQuillan, "The Titanium-Hydrogen System for Magnesium-Reduced Titanium," *J. Inst. Met.*, *79*, 371-378 (1951). (Equi Diagram; Experimental; #)

52Cra: C.M. Craighead, G.A. Lenning, and R.I. Jaffee, "Nature of the Line Markings in Titanium and Alpha Titanium-Alloys," *Trans. Metall. AIME*, *194*, 1317-1319 (1952). (Meta Phases, Experimental)

54Chr: A. Chretien, W. Freundlich, and M. Bichara, "Study of the Titanium-Hydrogen System: Preparation of the Titanium Hydride, TiH$_2$," *C. R. Acad. Sci. (Paris)*, *238*, 1423-1424 (1954) in French. (Crys Structure; Experimental)

54Len: G.A. Lenning, C.M. Craighead, and R.I. Jaffee, "Constitution and Mechanical Properties of Titanium-Hydrogen Alloys," *Trans. Metall. AIME*, *200*, 367-376 (1954). (Equi Diagram; Experimental; #)

56Haa: R.M. Haag and F.J. Shipko, "The Titanium-Hydrogen System," *J. Am. Chem. Soc.*, *78*, 5155-5159 (1956). (Equi Diagram; Thermo; Experimental; #)

56Sid: S.S. Sidhu, L.R. Heaton, and D.D. Zauberis, "Neutron Diffraction Studies of Hafnium-Hydrogen and Titanium-Hydrogen Systems," *Acta Crystallogr.*, *9*, 607-614 (1954). (Crys Structure; Experimental)

58Mel: G.A. Melkonian, "Contribution to the Titanium-Hydrogen System," *Z. Phys. Chem. Neue Folge*, *17*, 120-124 (1958) in German. (Thermo; Experimental)

58Sof: V.V. Sofina, Z.M. Aza and N.N. Orlova, "X-Ray Analysis of the Phases in the Zr-H and Ti-H Systems," *Kristallografiya*, *3*, 539-544 (1958) in Russian. (Crys Structure; Experimental)

***58Yak:** H.L. Yakel, Jr., "Thermocrystallography of Higher Hydrides of Titanium and Zirconium," *Acta Crystallogr.*, *11*, 46-51 (1958). (Crys Structure; Experimental)

59Spr: J. Spreadborough and J.W. Christian, "The Measurement of the Lattice Expansions and Debye Temperatures of Titanium and Silver by X-Ray Methods," *Proc. Phys. Soc. London*, *74*, 609-615 (1959). (Crys Structure; Experimental)

***60Bec:** R.L. Beck, "Research and Development of Metal Hydrides," Summary Rep., USAEC Rep. LAR-10, Denver Research Institute, 60-65.77-80 (Nov 1960). (Equi Diagram, Thermo; Experimental; #)

60Lib: G.G. Libowitz, "The Nature and Properties of Transition Metal Hydrides," *J. Nucl. Mater.*, *2*, 1-22 (1960). (Equi Diagram; Review; #)

60Mor: J.R. Morton and D.S. Stark, "The Dissociation Pressures of Titanium and Zirconium Deuterides as Functions of Composition and Temperature," *Trans. Faraday Soc.*, *56*, 351-356 (1960). (Thermo; Experimental)

60Sof: V.V. Sofina and N.G. Pavlovskaya, "Equilibria in the Titanium-Hydrogen and Zirconium-Hydrogen Systems at Low Pressures," *Zh. Fiz. Khim.*, *34*, 1104-1109 (1960) in Russian; TR: *Russ. J. Phys. Chem.*, *34*, 525-528 (1960). (Thermo; Experimental)

60Sta: B. Stalinski and Z. Bieganski, "Heat Capacity and Thermodynamic Functions of Titanium Hydride, TiH$_2$, Within the Range 24 to 363 K," *Bull. Acad. Pol. Sci. Ser. Sci. Chim.*, *8*, 243 (1960). (Thermo; Experimental)

62Liv: V.A. Livanov, A.A. Buhanova, and B.A. Kolachev, *Hydrogen in Titanium*, Daniel Davey and Co., Inc., New York,(1965); cited in [78Gab2]. (Equi Diagram)

62Sta: B. Stalinski and Z. Bieganski, "Thermodynamic Properties of Nonstoichiometric Titanium Hydrides Within the Range 24 to 360 K," *Bull. Acad. Pol. Sci., Ser. Sci. Chim.*, *10*, 247 (1962). (Thermo; Experimental)

62Woo: R.M. Wood, "The Lattice Constants of High Purity α Titanium," *Proc. Phys. Soc. London, 80*, 783-786 (1962). (Crys Structure; Experimental)

66Kol: B.A. Kolachev, "Hydrogen Embrittlement of Non-Ferrous Metals," Moscow (1966), Israel Program for Scientific Translation, Jerusalem (1968); cited in [78Gab2]. (Equi Diagram; Compilation)

66Lak: V.I. Lakomsky and N.N. Kalinyuk, "Solubility of Hydrogen in Liquid Titanium and Nickel," *Izv. Akad. Nauk SSSR, Metall.*, (2), 149-155 (1966) in Russian; TR: *Russian Metall.* (2), 80-85 (1966). (Thermo; Experimental)

66Sam: G.V. Samsonov and M.M. Antonova, "Phase Diagrams of the Systems Formed by Hydrogen with Titanium, Zirconium, Vanadium and Niobium," *Ukr. Khim. Zh., 32*, 555-559 (1966) in Russian; TR: *Sov. Prog. Chem., 32*, 421-424 (1966). (Crys Structure; Experimental)

66Tak: S. Takeuchi, T. Honma, and S. Ikeda, "Solubility of Hydrogen in Titanium at 900 °C to 1500 °C," *Sci. Rep. Res. Inst. Tohoku Univ., Ser A, 18*, 161-170 (1966). (Thermo; Experimental)

66Wed: G. Wedler and H. Strothenk, "Electric and Calorimetric Measurements in the Titanium-Hydrogen System at 273 K," *Z. Phys. Chem. Neue Folge, 48*, 86-101 (1966) in German. (Thermo; Experimental; #)

67Gal: N.A. Galaktinova, "Hydrogen in Metals," *Metallurgiya, Moscow* (1967) in Russian, as cited in [78Gab2]. (Equi Diagram; Compilation)

67Gio: T.A. Giorgi and F. Ricca, "Thermodynamic Properties of Hydrogen and Deuterium in α-Titanium," *Al Nuovo Cimento, Suppl.*, *5*(2), 472-482 (1967). (Thermo; Experimental)

67Gol: H.J. Goldschmidt, *Interstitial Alloys*, Ch. 9, Butterworths, London (1967). (Equi Diagram, Meta Phases; Compilation; #)

68Mue: W.M. Mueller, J.P. Blackledge, and G.G. Libowitz, "Titanium Hydrides," in *Metal Hydrides,* Academic Press, New York and London, 336-383 (1968). (Equi Diagram; Compilation)

69Mcq: A.D. McQuillan and A.D. Wallbank, "Thermodynamic Behaviour of Dilute Solutions of Hydrogen and Deuterium in Titanium and Zirconium," *J. Chem. Phys., 51*, 1026-1031 (1969). (Thermo; Experimental)

69Yav: V.I. Yavoiskii, L.B. Kosterov, A.D. Chuchuriukin, M.I. Musatov, A.F. Bushkariev, and N.G. Vilyaeva, "Solubility of Hydrogen in Solid and Liquid Titanium," *Izv. V.U.Z. Tsvetn. Metall.*, (1), 106-112 (1969) in Russian. (Thermo; Experimental)

70Aza: Z.M. Azarkh and P.I. Gavrilov, "Structural Changes in Titanium Hydride at Large Hydrogen Concentrations," *Kristallografiya, 15*, 275-279 (1970) in Russian; TR: *Sov. Phys. Crystallogr., 15*, 231-234 (1970). (Crys Structure; Experimental)

70Duc: F. Ducastelle, R. Caudron, and P. Costa, "Electronic Properties of Hydrides of the Ti-H and Zr-H Systems," *J. Phys. (Paris), 31*, 57-64 (1970) in French. (Crys Structure; Experimental)

70Kor: C. Korn and D. Zamir, "NMR Study of Hydrogen Diffusion in the Three Different Phases of the Titanium-Hydrogen System," *J. Phys. Chem. Solids, 31*, 489-502 (1970). (Equi Diagram, Thermo; Experimental; #)

71Cra: R.L. Crane and S.C. Chattoraj, "A Room-Temperature Polymorphic Transition of Titanium Hydride," *J. Less-Common Met., 25*, 225-227 (1971). (Crys Structure; Experimental)

71Gil: J.J. Gilman, "Lithium Dihydrogen Fluoride—An Approach to Metallic Hydrogen," *Phys. Rev. Lett., 26*, 546-548 (1971). (Crys Structure; Theory)

71Irv: P.E. Irving and C.J. Beevers, "Some Metallographic and Lattice Observations on Titanium Hydride," *Metall. Trans., 2*, 613-615 (1971). (Crys Structure; Experimental)

71Pat: N.E. Paton, B.S. Hickman, and D.H. Leslie, "Behavior of Hydrogen in α-Phase Ti-Al Alloys," *Metall. Trans., 2*, 2791-2796 (1971). (Equi Diagram; Experimental)

71Stu: D.R. Stull and H. Prophet, "Janaf Thermochemical Tables," 2nd ed., NSRDS-NBS 37, U.S. Govt. Printing Office, Washington, DC (1971). (Thermo; Compilation)

71Vit: R.S. Vitt and K. Ono, "Hydrogen Solubility in α Titanium," *Metall. Trans., 2*, 608-609 (1971). (Equi Diagram; Experimental)

72Bal: H.D. Bale and S.B. Peterson, "X-Ray Diffraction Study of the Structural Transformation in TiH_2," *Solid State Commun., 11*, 1143-1145 (1972). (Crys Structure; Experimental)

72Fla: T.B. Flanagan and W.A. Oates, "Thermodynamics of Metal/Hydrogen Systems," *Ber. Bunsenges. Phys. Chem., 76*, 706-714 (1972). (Thermo; Theory)

74Mir: N.F. Miron, V.I. Shcherbak, V.N. Bykov, and V.A. Levdik, "Neutron Diffraction Study of the Structures of the Metastable γ and the High-Temperature β Phases in the Ti-D System," *Kristallografiya, 19*, 754-758 (1974) in Russian; TR: *Sov. Phys. Crystallogr., 19*, 468-470 (1975). (Crys Structure; Meta Phases; Experimental)

74Sch1: E. Schurmann, T. Kootz, H. Preisendanz, P. Schuler, and G. Kauder, "On the Solubility of Hydrogen in the Systems Titanium-Aluminium-Hydrogen, Titanium-Vanadium-Hydrogen in the Temperature Range 800 to 1000 °C and at Hydrogen Pressures Between 0.1 and 520 mbar. Part I: Theory and Experimental Results," *Z. Metallkd., 65*, 167-172 (1974) in German. (Thermo; Experimental)

74Sch2: E. Schurmann, T. Kootz, H. Preisendanz, P. Schuler, and G. Kauder, "On the Solubility of Hydrogen in the Systems Titanium-Aluminium-Hydrogen, Titanium-Vanadium-Hydrogen in the Temperature Range 800 to 1000 °C and at Hydrogen Pressures Between 0.1 and 520 mbar. Part II: Thermodynamic Evaluation," *Z. Metallkd., 65*, 249-255 (1974) in German. (Thermo; Experimental)

75Mal: N.I. Malyavskii and V.S. Parbuzin, "Characteristics of the Isotopic Effect in Equilibrium Pressure for Hydrogen Absorbed by Metals," deposited document of the Chemistry Faculty of Lomonosov State University, Moscow, 13 p (1975) in Russian. (Thermo; Theory)

75Nag: H. Nagel and R.S. Perkins, "Crystallographic Investigation of Ternary Titanium Vanadium Hydrides," *Z. Metallkd., 66*, 362-366 (1975). (Crys Structure; Experimental)

75Wal: A.D. Wallbank and A.D. McQuillan, "Thermal Transpiration Correction of Hydrogen Equilibrium Pressure Measurements in Metal/Hydrogen Solution," *Trans. Faraday Soc., 71*, 685-689 (1975). (Thermo; Experimental)

76Alp: H.A. Alperin, H. Flotow, J.J. Rush, and J.J. Rhyne, "Deuterium-Site Occupancy in the α and β Phases of $TiDx$," Proc. Conf. on Neutron Scattering, Vol. I, R.M. Moon, Ed., National Technical Information Service, U.S. Department of Commerce, Springfield, VA, 517-521 (1976). (Crys Structure; Experimental)

76Dan: P. Dantzer, O.J. Kleppa, and M.E. Melnichak, "High-Temperature Thermodynamics of the $Ti\text{-}H_2$ and $Ti\text{-}D_2$ Systems," *J. Chem. Phys., 64*, 139-147 (1976). (Thermo; Experimental)

76Kip: C.C. Kiparisov, Yu.V. Levinskii, and V.P. Lukyanov, "Phase Equilibrium in the Ti-H System at Various Temperatures and Pressures," *Izv. V.U.Z. Tsvetn. Metall.*, (1), 100-103 (1976) in Russian. (Equi Diagram; Theory; #)

76Nag: M. Nagasaka and T. Yamashina, "Solubility of Hydrogen and Deuterium in Titanium and Zirconium Under Very Low Pressure," *J. Less-Common Met., 45*, 53-62 (1976) (Thermo; Experimental)

76Swi: A.C. Switendick, "Influence of the Electronic Structure on the Titanium-Vanadium-Hydrogen Phase Diagram," *J. Less-Common Met., 49*, 283-290 (1976). (Thermo; Theory)

77Sor: V.P. Sorokin, E.V. Levakov, and A.Ya. Malyshev, "Heat of Sorption of Hydrogen by Titanium at 295 K," *Zh. Fiz. Khim., 51*, 2804-2806 (1977) in Russian; TR: *Russ. J. Phys. Chem., 51*, 1635-1636 (1977). (Thermo, Experimental)

78Gab1: R.M. Gabidulin, B.A. Kolachev, and E.V. Krasnova, "Thermodynamic Analysis of Dissolution and Interaction Between Hydrogen Atoms in Metals," *Izv. V.U.Z. Tsvetn. Metall.*, (6), 98-102 (1978) in Russian. (Thermo; Theory)

*78Gab2: R.M. Gabidulin, B.A. Kolachov, A.A. Bukhanova, and E.V. Shchekoturova, "A Thermodynamic Investigation of the Hydrogen-Titanium System," Titanium and Titanium Alloys: Scientific and Technological Aspects, Proc. 3rd Int. Conf. Moscow, 1976, Vol. 2, A. Belov, Ed., 419-428 (1978) in Russian; TR: Vol. 2, W.J. Case, Ed., 1365-1375, Plenum Press, New York, (1982). (Equi Diagram, Thermo; Experimental; #)

78Kul: N.I. Kulikov and V.N. Borzunov, "Band Model of Martensitic Phase Transition in Titanium Dihydride," *Izv. Akad. Nauk SSSR Neorg. Mater.*, *14*, 1659-1663 (1978) in Russian; TR: *Inorg. Mater.(USSR)*, *19*, 1292-1296 (1978). (Thermo; Theory)

79Aga: E.V. Agababyan, S.L. Kharatyan, M.D. Nersesyan, and A.G. Merzhanov, "Combustion Mechanism of Transition Metals in Conditions of Strong Dissociation (Illustrated by the Titanium-Hydrogen System)," *Fiz., Goren. Vzry.*, *15*, 3-9 (1979). in Russian. (Thermo; Experimental)

79Ari: M. Arita, R. Kinaka, and M. Someno, "Application of the Metal-Hydrogen Equilibration for Determining Thermodynamic Properties in the Ti-Cu System," *Metall. Trans. A*, *10*, 529-534 (1979). (Thermo; Experimental)

79Gup: M. Gupta, "Electronically Driven Tetragonal Distortion in TiH_2," *Solid State Commun.*, *29*, 47-51 (1979). (Thermo; Theory)

79Lyn: J.F. Lynch and J. Tanaka, "The Dilute Solution of Hydrogen and Deuterium in β-Titanium," *Scr. Metall.*, *13*, 599-604 (1979). (Thermo; Experimental)

80Hof: F. Hofmann and W. Auer, "Kinetic Studies on the Hydrogen Absorption in α-Titanium," *Ber. Bunsenges. Phys. Chem.*, *84*, 1168-1174 (1980) in German. (Thermo; Experimental)

80Kan: K. Kandasamy and N.A. Surplice, "The Effects of Hydrogen on the Resistivity of Some Transition Metals," Physics of Transition Metals, 1980, Proc. Int. Conf., University of Leeds, P. Rhodes, Ed., Conference Series No. 55, The Institute of Physics, Bristol and London, 587-590 (1980). (Equi Diagram; Experimental)

80She: M.N. Shetty and K.P. Singh, "Strain-Energy Model for Solid Solubility Limits in Zr-H, Ti-H and Zr-Nb-H Systems," Proc. Interdiscip. Meet. Hydrogen Met., Bhabba At. Res. Cent., Bombay, India, 201-213 (1980). (Crys Structure, Thermo; Theory)

***81Boh:** K. Bohmhammel, G. Wolf, G. Gross, and H. Madge, "Investigations of the Molar Heat Capacity at Low Temperatures in the TiH_x System," *J. Low Temp. Phys.*, *43*, 521-532 (1981). (Equi Diagram; Experimental)

81Sar: Yu.S. Sardanyan, S.L. Kharatyan, Yu.M. Grigorev, and A.G. Merzhanov, "Kinetics of the High-Temperature Interaction of Titanium with Hydrogen," *Izv. Akad. Nauk SSSR Met.*, (2), 216-222 (1981) in Russian. (Thermo; Experimental)

81Wea: G.C. Weatherly, "The Precipitation of γ-Hydride Plates in Zirconium," *Acta Metall.*, *29*, 501-512 (1981). (Meta Phases; Experimental)

82Ari: M. Arita, K. Shimizu, and Y. Ichinose, "Thermodynamics of the Ti-H System," *Metall. Trans. A*, *13*, 1329-1336 (1982). (Thermo; Experimental)

82Dan: P. Dantzer and A. Guillot, "A New High-Temperature Heat-Flow Calorimeter," *J. Phys. E Sci. Instr.*, *15*, 1373-1375 (1982). (Thermo; Experimental)

82Hem: R. Hempelmann, D. Richter, and B. Stritzker, "Optic Phonon Modes and Superconductivity in α Phase (Ti, Zr)-(H, D) Alloys," *J. Phys. F Met. Phys.*, *12*, 79-86 (1982). (Crys Structure; Experimental)

82Kho: R. Khoda-Bakhsh and D.K. Ross, "Determination of the Hydrogen Site Occupation in the α Phase of Zirconium Hydride and in the α and β Phases of Titanium Hydride by Inelastic Neutron Scattering," *J. Phys. F Met. Phys.*, *12*, 15-24 (1982). (Crys Structure; Experimental)

82Mil: P. Millenbach and M. Givon, "The Electrochemical Formation of Titanium Hydride," *J. Less-Common Met.*, *87*, 179-184 (1982). (Crys Structure; Experimental)

***83Dan:** P. Dantzer, "High Temperature Thermodynamics of H_2 and D_2 in Titanium, and in Dilute Titanium Oxygen Solid Solutions," *J. Phys. Chem. Solids*, *44*, 913-923 (1983). (Thermo; Experimental)

83Kor: C. Korn, "NMR Study Comparing the Electronic Structures of ZrH_x and TiH_x," *Phys. Rev. B*, *28*, 95-111 (1983). (Equi Diagram, Crys Structure; Experimental)

***83Kub1:** O. Kubaschewski, "Thermochemical Properties," in *Titanium: Physico-Chemical Properties of its Compounds and Alloys*, Special Issue No. 9, K.L. Komarek, Ed., International Atomic Energy Agency, Vienna, 3-71 (1983). (Thermo; Compilation)

83Kub2: O. Kubaschewski-von Goldbeck, "Phase Diagrams," in *Titanium: Physico-Chemical Properties of its Compounds and Alloys*, Special Issue No. 9, K.L. Komarek, Ed., International Atomic Energy Agency, Vienna, 75-197 (1983). (Equi Diagram; Compilation; #)

83Sha: V.I. Shapovalov, N.P. Serdyuk, and A.L. Titkov, "Titanium-Hydrogen Phase Diagram," *Izv. V.U.Z. Tsvetn. Metall.*, (6), 74-78 (1983) in Russian. (Thermo, Pressure; Experimental; #)

84Bou: G. Boureau, "The Configurational Entropy of Hydrogen in Body Centered Metals," *J. Phys. Chem. Solids*, *45*, 973-974 (1984). (Thermo; Theory)

***84Num:** H. Numakura and M. Koiwa, "Hydride Precipitation in Titanium," *Acta Metall.*, *32*, 1799-1807 (1984). (Meta Phases; Experimental; #)

84Pul: M.P. Puls, "Elastic and Plastic Accommodation Effects on Metal-Hydride Solubility," *Acta Metall.*, *32*, 1259-1269 (1984). (Equi Diagram; Theory)

***85Woo:** O.T. Woo, G.C. Weatherly, C.E. Colean, and R.W. Gilbert, "The Precipitation of γ-Deuterides (Hydrides) in Titanium," *Acta Metall.*, *33*, 1897-1906 (1985). (Meta Phases; Experimental)

The Hf-Ti (Hafnium-Titanium) System

178.49   47.88

By J.L. Murray

Equilibrium Diagram

Both Ti and Hf have low-temperature cph (α) and high-temperature bcc (β) modifications. Ti and Hf are completely miscible in both (αTi,αHf) and (βTi,βHf) phases. For brevity, the phases will be referred to as α and β in the following discussion.

The assessed diagram is given in Fig. 1, and special points of the diagram are summarized in Table 1. The present assessment does not differ significantly from that of [Hansen]. The assessed diagram is calculated from Gibbs energies obtained by least-squares optimization with respect to experimental phase diagram data. Calculation of the diagram has improved the accuracy of the width of the two-phase regions.

Liquidus and Solidus.

The liquidus and solidus have a minimum (congruent) point at 15 ± 2 at.% Hf and 1650 ± 10 °C [57Hay, 60Tho, 66Cha, 69Rud]. Experimental determinations of the liquidus and solidus are summarized in Table 2, and experimental data are compared with the assessed diagram in Fig. 2.

The solidus and liquidus of [59Tyl] agree qualitatively with the data of [57Hay], [66Cha], and [69Rud]. The congruent melting temperature of 1600 °C found by [59Tyl] disagrees with the data of other authors. The low minimum temperature found by [59Tyl] seems to reflect scatter in their data, because the melting points at other compositions lie more in line with other work. [57Hay],

Table 1 Special Points of the Ti-Hf System

Reaction	Compositions, at.% Hf	Temperature, °C	Reaction type
L ⇌ (βTi,βHf)	15	1650	Congruent
(βTi,βHf) ⇌ (αTi,αHf)	20	800	Congruent
L ⇌ (βTi)	0	1670	Melting point
L ⇌ (βHf)	100	2231	Melting point
(βTi) ⇌ (αTi)	0	882	Allotropic transformation
(βHf) ⇌ (αHf)	100	1740	Allotropic transformation

Fig. 1---Assessed Ti-Hf Phase Diagram

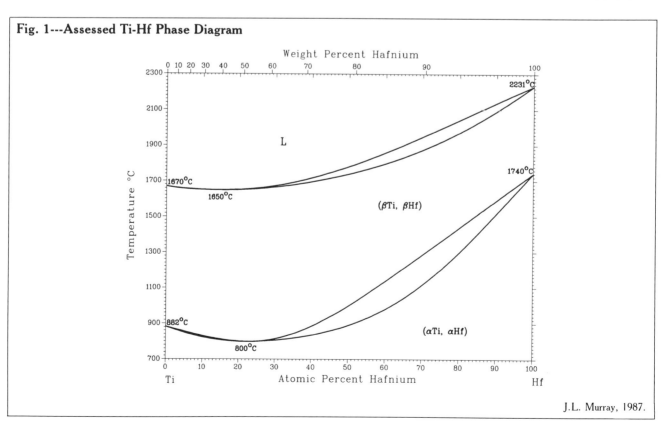

J.L. Murray, 1987.

Fig. 2---Experimental Data for the β and Liquid Phases

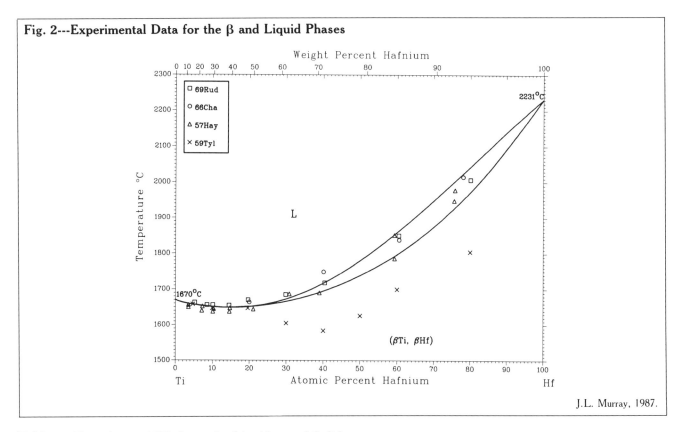

J.L. Murray, 1987.

Table 2   Experimental Work on the Liquidus and Solidus

| | | Congruent melt | | |
Reference	Phase boundary	Composition, at.% Hf	Temperature, °C	Experimental technique
[57Hay]	Solidus/liquidus	12.6	1640	Resistivity
[59Tyl]	Solidus	15.2	1600	Thermal analysis
[66Cha]	Solidus	...	...	Optical observation of melting
[69Rud]	Solidus/liquidus	15	1656	Incipient melting

[66Cha], and [69Rud] agree within the experimental uncertainty. [69Rud] reported congruent melting over a large composition range (0 to 40 at.% Hf) and a narrow two-phase region at higher Hf concentrations. The data of [57Hay] imply a wider two-phase region.

α/β Boundaries.

The α/β boundaries have a minimum (congruent) point at 20 ± 2 at.% Hf and 800 ± 10 °C [57Hay, 60Tho, 66Cha, 69Rud]. Experimental determinations of the α/β boundaries are summarized in Table 3, and experimental data are compared with the assessed diagram in Fig. 3.

At the lowest temperatures (for Ti-rich alloys), the results of the various experiments are in good agreement. The discrepancies in reported compositions of the minima are not indicative of serious discrepancies in the data, because the phase boundaries are very flat in this region.

At 90 at.% Hf, [66Cha] found the transformation temperature to be higher than one would expect from the transformation temperature of pure Hf. The data at 90 at.% Hf have not been used to construct the diagram.

[62Img] found that 15 at.% Hf alloys heat treated at temperatures below 815 °C and quenched were two phase.

indicates nonequilibrium conditions. This effect and the narrowness of the two-phase field make metallographic examination a difficult technique for phase equilibrium determinations in the Ti-Hf system.

[72Rud] showed that one finds different temperatures for the beginning and end of the α/β transformation depending on the experimental probe. The differential thermal analysis (DTA) data of these authors have been chosen over resistivity and dilatometric measurements as most likely to represent equilibrium, and they agree well with [57Hay], [62Img], and [66Cha].

Metastable Phases

No experimental study of metastable phase formation in Ti-Hf alloys is presently available. The behavior of Ti-Hf alloys is likely to be similar to that of Ti-Zr alloys. In the Ti-Zr system, metastable phase transformations take place as nonequilibrium transformations of the β phase. Over a wide composition range near the minimum of the α/β phase boundaries, α phase can form martensitically from β. In slightly contaminated Ti-Zr alloys, the metastable ω-phase can also form as an intermediate phase in the decomposition of β to α.

Table 3   Experimental Work on α/β Boundaries

| | Congruent transformation | | |
Reference	Composition, at.% Hf	Temperature, °C	Experimental technique
[57Hay]	~ 20	~ 790	DTA (heating)
[59Tyl]	19.2	770	DTA, metallography, dilatometry
[62Img]	(a)	(a)	Metallography
[66Cha]	~ 25	~ 800	DTA
[75Rud]	~ 30	795	DTA (heating), dilatometry, resistivity

(a) Not determined.

Fig. 3---Experimental Data for the α and β Phases

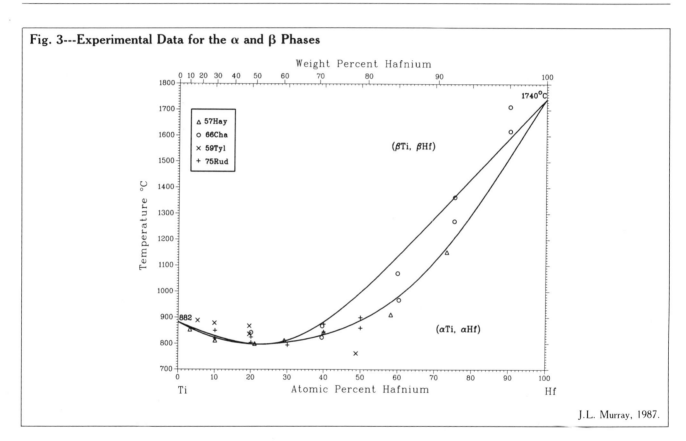

J.L. Murray, 1987.

Crystal Structures and Lattice Parameters

Crystal structure data for the Ti-Hf system are given in Table 4. The lattice parameters of α phase alloys [66Cha, 69Rud] are plotted in Fig. 4. Lattice parameters for the β phase have not been measured.

Thermodynamics

No experimental determination of excess Gibbs energies has been made for the Ti-Hf system. The assessed phase diagram (Fig. 1) is the result of a thermodynamic calculation based on phase diagram data only. From the complete miscibility of Ti and Hf in all equilibrium phases, it can be deduced that the system is nearly ideal and that the excess Gibbs energies can be represented by the regular solution approximation.

The Gibbs energies of the solution phases are represented as:

$$G(i) = (1 - x)\, G^0(\text{Ti},i) + G^0(\text{Hf},i)\, x + RT[x \ln x + (1 - x) \ln (1 - x)] + B(i)\, x(1 - x)$$

where i designates the phase; x is the atomic fraction of Hf; G^0 is the Gibbs energies of the pure components; and $B(i)$ is the regular solution interaction parameter. The pure component Gibbs energies were taken from [70Kau] and slightly adjusted to reproduce the assessed pure metal transformation points.

Starting values of the regular solution parameters were taken from [70Kau]. The positions of the congruent transformations are determined by differences between excess Gibbs energies of the appropriate phases. Therefore, based on the available data, only differences among excess Gibbs energies can be determined. The regular solution parameter for the bcc phase was taken from [70Kau], and excess Gibbs energies of the cph and liquid phases were determined with respect to the bcc phase. For the calculation of the liquidus and solidus, the congruent point was fixed at 1650 °C, and data of [57Hay], [66Cha], and [69Rud] were used. The small excess entropy contribution to the liquid phase Gibbs energy was included to better reproduce the curvature of the phase boundaries at high Hf concentrations. The calculated solidus agrees

Table 4   Crystal Structures of the Ti-Hf System [Pearson2]

Phase	Homogeneity range, at.% Hf	Pearson symbol	Space group	Strukturbericht designation	Prototype
(βTi,βHf)	0 to 100	cI2	Im3m	A2	W
(αTi,αHf)	0 to 100	hP2	P6₃ mmc	A3	Mg

Fig. 4---Lattice Parameters *a* and *c* for (αTi,αHf) Alloys

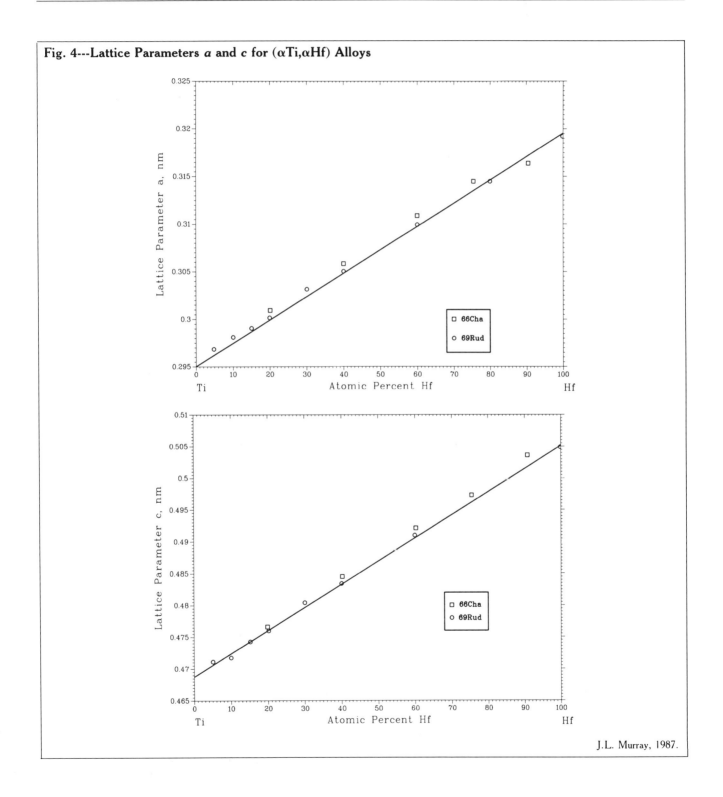

J.L. Murray, 1987.

with the data of [57Hay] and the liquidus with the data of [66Cha] and [69Rud].

Similarly, for the calculation of the bcc/cph boundaries, the congruent point was fixed at 800 °C, and data of [57Hay], [66Cha], and [75Rud] were used. The calculated diagram appears to be an acceptable resolution of the discrepancies in the experimental data.

The interaction parameters used to calculate the assessed diagram are compared with the starting values [70Kau] in Table 5. It must be emphasized that only the differences in Gibbs energies are regarded as accurate.

Cited References

57Hay: T. Hayes and D.K. Deardorff, "The Phase Diagram of the Ti-Hf System," U.S. Atom. Energy Comm. USBM-V-345 (1957). (Equi Diagram; Experimental)

59Tyl: M.A. Tylkina, A.I. Pekarev, and E.M. Savitskii, "Phase Diagram of the Titanium-Hafnium System," *Zh. Neorg. Khim.,* 4(10), 2320-2322 (1959) in Russian; TR: *Russ. J. Inorg. Chem.,* 4(10), 1059-1060 (1959). (Equi Diagram; Experimental)

60Tho: D.E. Thomas and E.T. Hayes, "The Metallurgy of Hafnium," USAEC Tech. Rep., 205-206 (1960). (Equi Diagram; Experimental)

62Img: A.G. Imgram, D.N. Williams, and H.R. Ogden, "Tensile Properties of Binary Titanium-Zirconium and Titanium-Hafnium Alloys," *J. Less-Common Met.,* 4, 217-225 (1962). (Equi Diagram; Experimental)

66Cha: Y.A. Chang, USAF Tech. Rep. AFML-TR-65-2, Part II, Vol. V (1966). (Equi Diagram; Experimental)

69Rud: E. Rudy, "Compilation of Phase Diagram Data," USAF Tech. Rep. AFML-TR-65-2, Part V (1969). (Equi Diagram; Experimental)

Table 5 Thermodynamic Functions for the Ti-Hf System

Properties of the pure components

$G^0(Ti,L)$ = 0
$G^0(Hf,L)$ = 0

$G^0(Ti,bcc)$ = $-16\,234 + 8.368\,T$
$G^0(Hf,bcc)$ = $-20\,980 + 8.368\,T$

$G^0(Ti,cph)$ = $-20\,585 + 12.134\,T$
$G^0(Hf,cph)$ = $-28\,635 + 12.172\,T$

Excess quantities of the solution phases

	[70Kau]	This evaluation
B(bcc)	2038	2 038
B(cph)	6473	$10\,750 - 2.5\,T$
B(L)	-5209	$-9\,500 + 2.5\,T$

Note: Values are given in J/mol; T in K.

70Kau: L. Kaufman and H. Bernstein, *Computer Calculations of Phase Diagrams,* Academic Press, New York (1970). (Thermo; Experimental)

72Rud: E. Rudy, USAF Tech. Rep. AFML-TR-65-2, Part VIII (1972). (Equi Diagram; Experimental)

75Rud: G.I. Ruda, I.I. Kornilov, and V.V. Vavilova, "The Influence of Hafnium on the Polymorphic Transformation Temperature of Titanium," *Izv. Akad. Nauk SSSR, Met.,* (5), 203-205 (1975) in Russian; TR: *Russ. Metall.,* (5), 160-162 (1975). (Equi Diagram; Experimental)

The Hg-Ti (Mercury-Titanium) System

200.59   47.88

By J.L. Murray

Equilibrium Diagram

The boiling point of Hg is 327 °C, and the melting points of Hg and Ti are −39 and 1670 °C, respectively [Melt]. Therefore, only limited portions of the phase diagram can be studied, and these represent phase boundaries in equilibrium with the vapor.

The phase diagram (Fig. 1) is based on the vapor pressure and X-ray study of [73Lug]. The detail of the Hg-rich side (Fig. 2) is based on [63Jan] and shows additional data from [67Wee]. Only the range 0 to 700 °C has been investigated, and the high-temperature bcc (βTi) phase therefore does not appear.

Structures occurring in the Ti-Hg system are summarized in Table 1. The equilibrium solid phases of the system are:

- The cph (αTi) solution. The solubility of Hg in (αTi) is sufficiently low that it cannot be determined by lattice parameter measurements [73Jan].
- Ti_3Hg, with the $A15$ structure [68Kur]. [54Pie] identified two allotropes of Ti_3Hg, the high-tem-

perature form having the $L1_2$ structure. [68Kur] showed that this phase is actually an oxide phase with the perovskite structure.

- TiHg with the $L1_0$ structure. Structural data on TiHg are from [54Pie] and [70Pus], and the phase equilibria are from [73Lug]. Both [54Pie] and [70Pus] stated that the alloys showed a pronounced tendency to decompose at room temperature.
- Two modifications of a near-equiatomic phase, designated αX and βX. αX, with a wide, temperature-dependent homogeneity range, was found in the vapor pressure studies of [73Lug]. The crystal structure could not be determined. According to [73Lug], there is a narrow vertical two-phase region (TiHg + X), and TiHg has only a narrow range of homogeneity. The present evaluator finds the existence of a stoichiometric $L1_0$ phase difficult to accept and the shapes of the phase boundaries difficult to explain. Further work will probably suggest a better interpretation of the vapor pres-

The Hg-Ti System

Fig. 1---Assessed Ti-Hg Phase Diagram

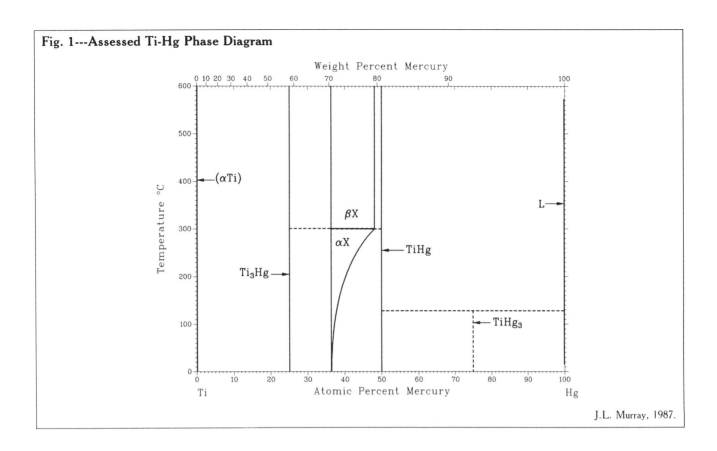

J.L. Murray, 1987.

Fig. 2---Detail of the Hg-Rich Portion of the Ti-Hg System

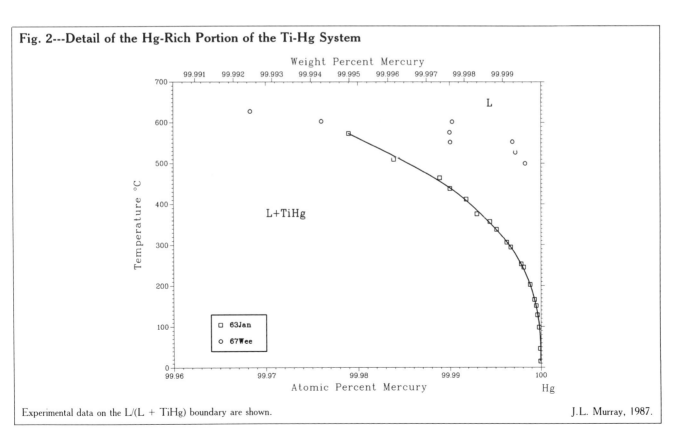

Experimental data on the L/(L + TiHg) boundary are shown.

J.L. Murray, 1987.

Table 1 Crystal Structures of the Ti-Hg system

Phase	Homogenity Range, at.% Hg	Pearson symbol	Space group	Struktur-bericht designation	Proto-type	Lattice parameters, nm		Reference
						a	c	
(βTi)	0 – ?	$cI2$	$Im3m$	$A2$	W	0.33065	...	(a)
(αTi)	~ 0	$hP2$	$P6_3/mmc$	$A3$	Mg	0.29511	0.46843	(a)
Ti_3Hg	25	$cP8$	$Pm3n$	$A15$	Cr_3Si	0.5189	...	[68Kur]
						0.51888	...	[54Pie]
αX	36.6 to 47.8	(a)	...	...	...	...	...	[73Lug]
βX	36.6 to 47.8	(a)	...	...	...	...	...	[73Lug]
TiHg	50	tP_2	$P4/mmm$	$L1_0$	AuCuI	0.3009	0.4041	[54Pie]
						0.301	0.404	[70Pus]
$TiHg_3$	75	(a) ...	...	...	...	...	...	...
(Hg)	100	$hR1$	$R\bar{3}m$	$A10$	Hg	0.3005(c)	...	(a)

(a) Unknown. (b) Lattice parameters for the pure elements are from [Pearson], that for pure βTi is for 900 °C. (c) $\alpha = 70° 32'$.

sure measurements. Meanwhile, Fig. 1 essentially reproduces the diagram of [73Lug].

$TiHg_3$ also has an unknown structure. [73Lug] conjectured its presence from the decomposition of amalgams stored for several months, and determined its decomposition temperature to be 128 °C by differential thermal analysis. If $TiHg_3$ is an equilibrium phase, the (Hg) liquidus must represent a metastable equilibrium of the liquid with TiHg.

The solubility of Ti in liquid Hg is determined by chemically analyzing the equilibrated, filtered liquid [32Irv, 63Jan, 67Wee]. The results of [32Irv] are not quantitative, and the results of [63Jan] and [67Wee] differ. The measurements of [63Jan] extend over the range 15 to 575 °C. Different filtration techniques were used above and below the boiling point of Hg, and the two sets of data lie on a single straight line in a log x versus $1/T$ plot. [67Wee] worked in a smaller temperature range, and their data show more scatter. [67Wee] indicated a larger temperature dependence of the solubility than [63Jan]. In a survey of several alloy systems, this discrepancy occurs consistently, not only with [63Jan] but also with work of other investigators. The assessed liquidus (Fig. 2) is based on [63Jan].

Cited References

32Irv: N.M. Irvin and A.S. Russell, "The Solubilities of Copper, Manganese, and Some Sparingly Soluble Metals in Mercury," *J. Chem. Soc.,* 891-898 (1932). (Equi Diagram; Experimental)

54Pie: P. Pietrokowsky, "A Cursory Investigation of Intermediate Phases in the Systems Ti-Zn, Ti-Hg, Zr-Zn, Zr-Cd, and Zr-Hg by X-Ray Powder Diffraction Methods," *Trans. AIME, 200,* 219-226 (1954). (Equi Diagram; Experimental)

63Jan: G. Jangg and H. Palman, "The Solubility of Several Metals in Mercury," *Z. Metallkd., 54*(6), 364-369 (1963) in German. (Equi Diagram; Experimental)

67Wee: J.R. Weeks, "Liquidus Curves and Corrosion of Fe, Cr, Ni, Co, V, Cb, Ta, Ti, Zr, in 500-750 C Mercury," *Corrosion, 23,* 98-106 (1967). (Equi Diagram; Experimental)

68Kur: F. Kurka and P. Ettmayer, "The Perovskite Oxide Ti_3HgO," *Monatsh. Chem., 99,* 1836-1838 (1968) in German. (Equi Diagram; Experimental)

70Pus: M. Puselj and Z. Ban, "X-Ray Studies in the System Ti-Hg-Zn," *Z. Naturforsch., 25*(2), 315 (1970). (Equi Diagram; Experimental)

73Jan: G. Jangg and E. Lugscheider, "The Solubility of Mercury in Several Metals," *Monatsh. Chem., 104,* 1269-1275 (1973) in German. (Equi Diagram; Experimental)

73Lug: E. Lugscheider and G. Jangg, "Ti-Hg, Zr-Hg and Hf-Hg Systems," *Z. Metallkd., 64*(10), 711-715 (1973) in German. (Equi Diagram; Experimental)

The In-Ti (Indium-Titanium) System

114.82   47.88

By J.L. Murray

The assessed Ti-In phase diagram (Fig. 1) is based on (1) metallographic and X-ray diffraction work [54Lev, 58And] on the Ti-side; (2) thermal analysis, X-ray diffraction, and optical observation of melting [62Joh] on the

Fig. 1---Assessed Ti-In Diagram

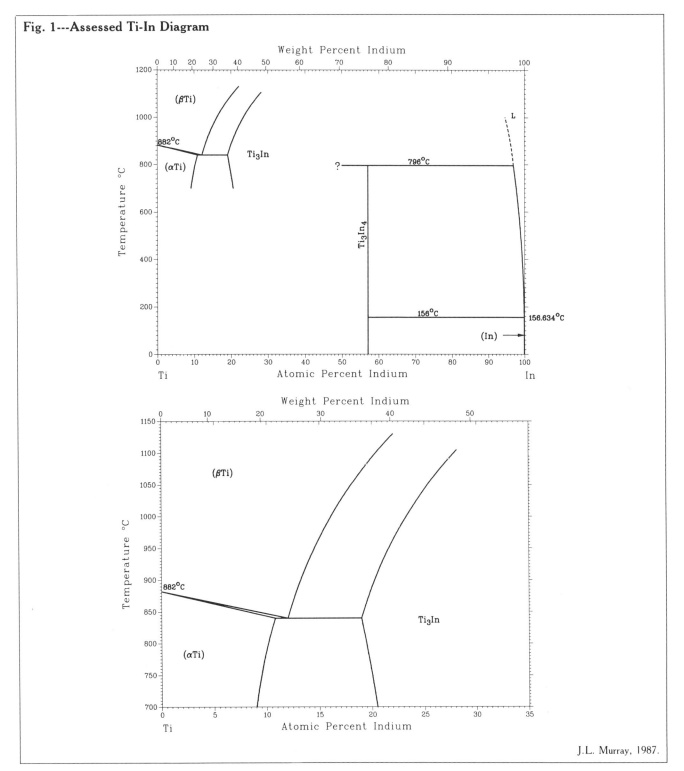

J.L. Murray, 1987.

Phase Diagrams of Binary Titanium Alloys

Table 1 Special Points of the Assessed Ti-In Phase Diagram

Reaction	Compositions of the respective phases, at.% In			Temperature, °C	Reaction type
$(\beta Ti) \rightleftarrows (\alpha Ti) + Ti_3In$	~ 12	~ 10	~ 21	~ 850	Eutectoid
$L + X(a) \rightleftarrows Ti_3In_4$	97.1	...	57.2	796	Peritectic
$L + Ti_3In_4 \rightleftarrows (In)$	~ 100	57.2	~ 100	156	Peritectic
$L \rightleftarrows (\beta Ti)$		0		1670	Melting point
$(\beta Ti) \rightleftarrows (\alpha Ti)$		0		882	Allotropic transformation
$L \rightleftarrows (In)$		100		156.634	Melting point

(a) Unidentified phase.

Table 2 Crystal Structures of the Ti-In System

Phase	Homogeneity range, at.% In	Pearson symbol	Space group	Struktur- bericht designation	Prototype	Reference
(αTi)	0 to ~ 10	$hP2$	$P6_3/mmc$	$A3$	Mg	[Pearson2]
(βTi)	0 to ~ 12	$cI2$	$Im3m$	$A2$	W	[Pearson2]
Ti_3In	21 to ?	$hP8$	$P6_3/mmc$	$D0_{19}$	Ni_3Sn	[57And]
Ti_3In_2	(a)	$tP4$	$P4/mmm$	$L1_0$	AuCu-type	[64Sch, 65Ram]
Ti_3In_4	57.1	$tP14$	$P4/mbm$	...	Ti_3In_4	[62Joh, 65Ram]
(In)	~ 100	$tI2$	$I4/mmm$	$A6$	In	[Pearson2]

(a) Unknown.

Table 3 Room-Temperature Lattice Parameters of Ti-In Intermetallic Phases

Phase	Lattice parameters, nm		Reference
	a	c	
Ti_3In	0.589	0.476	[58And]
	0.592	0.478	[65Ram]
Ti_3In_2	0.4203	0.4238	[65Ram]
TiIn	0.423	...	[58And]
Ti_3In_4	0.998	0.298	[65Ram]
	1.0094	0.3052	[62Joh]

In-rich side; and (3) X-ray diffraction of three samples of intermediate composition [65Rau]. Additional structure data were reported by [57And2], [64Sch], and [64Sch]. The diagram is only very incompletely known.

Special points of the assessed diagram are listed in Table 1. Crystal structure and lattice parameter data are listed in Tables 2 and 3, respectively.

The equilibrium solid phases are:

- The solid solutions, (αTi), (βTi), and (In), based on the cph, bcc, and tetragonal forms of the pure metals
- The ordered cph phase Ti_3In, which is established as the most Ti-rich intermetallic phase
- Ti_3In_4, with an ordered fcc structure, which is established as the phase in equilibrium with liquid In
- Phases of ordered fcc structure, designated TiIn(h) and Ti_3In_2 by [65Ram]. Neither the homogeneity range of these phases nor their liquidus and solidus curves have been investigated.

Ti-Rich Alloys.

Ti-rich alloys were studied by [54Lev] and [58And]. [54Lev] examined four alloys containing 1.8 to 15.2 at.

In the range 800 to 1000 °C by metallography. [58And] examined ten alloys in the range 4 to 31 at.% In at 750 to 1050 °C by metallography and X-ray diffraction. In is soluble in both (αTi) and (βTi). [54Lev] and [58And] agreed that the eutectoid reaction $(\beta Ti) \rightleftarrows (\alpha Ti) + Ti_3In$ occurs slightly below 850 °C and with a (βTi) composition of about 12 at.% In. [54Lev] found the maximum solubility of In in (αTi) to be less than 4 at.% In; [57And1] and [58And] did not corroborate this finding, but reported the maximum solubility to be over 10 at.% In.

The results of [58And] are tentatively preferred because of the larger number of alloys examined, the probability that purer alloys were used, and the verification of metallographic results by X-ray diffraction. At 800 °C, Ti_3In has a homogeneity range of at least 21 to 26 at.% In [58And]. [58And] determined the $(\beta Ti)/Ti_3In$ boundaries to within 2 at.% up to 1050 °C. The $(\beta Ti)/[(\beta Ti) + Ti_3In]$ boundary was drawn with a kink [58And, Elliott] to account for a single one-phase (βTi) alloy (16.4 at.% In, 950 °C). [54Lev], however, found 15.2 at.% In alloys to be two-phase up to 1000°C. The (βTi) boundary has been smoothed to conform more plausibly with the combined findings of [54Lev] and [58And].

In-Rich Alloys.

[58And] reported that a 30 at.% In alloy annealed at 950 °C contained a two-phase assemblage of Ti_3In and a phase that probably had the $L1_2$ structure. They noted that composition changes of as much as 10 at.% were caused by In loss during arc-melting. [65Ram] identified the following phases by X-ray diffraction: TiIn, with an $L1_2$ structure, identified as a high-temperature phase; Ti_3In_2 with the $L1_0$ structure; and Ti_3In_4 with a crystal structure of tetragonal symmetry.

The most In-rich compound, Ti_3In_4, forms by a peritectic reaction between the liquid and an unidentified phase at 796 °C [62Joh]. [62Joh] determined the composition of Ti_3In_4 by X-ray diffraction and density measurements on crystals removed from the In-rich liquid. Based on the possible space groups attributed at Ti$_3$In$_4$ by [62Joh], this phase is identified with the Ti$_3$In$_4$ phase observed by [63Sch] and [65Ram]. [Joh] estimated the composition of the liquidus at the peritectic temperature as 97.1 at.% In, based on the disappearance of the peritectic arrest and visual observation of the beginning of solidification.

Cited References

54Lev: D.W. Levinson, D.J. McPherson, and W. Rostoker, "Constitution of Titanium Alloys Systems—Supplement 1," WADC Tech. Rep., 53-41, 23-24 (1954). (Equi Diagram; Experimental)

57And1: K. Anderko, K. Sagel, and U. Zwicker, "Ordered Hexagonal Phases in the Titanium-Aluminum and Titanium-Indium Systems," *Z. Metallkd.,* 48 57-58 (1957) in German. (Crys Structure; Experimental)

57And2: K.Anderko, "Structure Determinations in the Ti-Ga and Ti-In Systems," *Naturwissenschaften, 44*(4), 88 (1957) in German. (Crys Structure; Experimental)

*58And: K. Anderko, "Binary Systems of Titanium with Gallium, Indium and Germanium, and of Zirconium with Gallium and Indium," *Z. Metallkd., 49,* 165-172 (1958) in German. (Equi Diagram; Experimental)

*62Joh: R.G. Johnson and R.J. Prosen, "The Indium Rich Side of the Indium-Titanium Systems," *Trans. AIME, 224,* 397-398 (1962). (Equi Diagram; Experimental)

63Sch: K. Schubert, K. Frank, R. Gohle, A. Maldonado, H.G. Meissner, A. Raman, and W. Rossteutscher, "Structure Data on Metallic Phases," *Naturwissenschaften, 50,* 41 (1963) in German. (Crys Structure; Experimental)

64Sch: K. Schubert, H.G. Meissner, A. Raman, and W. Rossteutscher, "Structure Data on Metallic Phases," *Naturwissenschaften, 51*(12), 287 (1964) in German. (Crys Structure; Experimental)

65Ram: A. Raman and K. Schubert, "The Constitution of Some Alloy Series Related to TiAl$_3$," *Z. Metallkd., 56*(1), 44-52 (1965) in German. (Equi Diagram; Experimental)

The Ir-Ti (Iridium-Titanium) System

192.2   47.88

By J.L. Murray

Equilibrium Diagram

The equilibrium solid phases of the Ti-Ir system are:

- The cph solid solution, (αTi), which is the equilibrium form of pure Ti below 882 °C
- The bcc (βTi) solid solution, stable above 882 °C. The maximum solubility of Ir in (βTi) is approximately 15 at.%
- The fcc (Ir) solid solution. The maximum solubility of Ti in (Ir) is approximately 11 at.%
- Ti$_3$Ir, a compound with the βW (W$_3$O) structure. The homogeneity range of Ti$_3$Ir is 25 to 27 at.% Ir. It is formed from the liquid by the peritectic reaction L + βTiIr $\rightleftarrows$ Ti$_3$Ir

- Two compounds each with a wide homogeneity range about the stoichiometry TiIr. The high-temperature form, designated βTiIr, has the CsCl structure. For Ti-rich alloys, βTiIr is stable between the melt and room temperature; on the Ir-rich side, it transforms at about 1750 °C to the low-temperature form, αTiIr. αTiIr has been assigned almost as many structures as there have been investigators of the system. The structure of αTiIr has tentatively been listed in this evaluation as monoclinic.
- The compound TiIr$_3$ with the ordered fcc Cu$_3$Au structure. Its homogeneity range is 73 to 77 at.% Ir.

Table 1   Special Points of the Assessed Ti-Ir Phase Diagram

Reaction	Compositions of the respective Phases, at.% Ir			Temperature, °C	Reaction type
L $\rightleftarrows$ (βTi) + Ti$_3$Ir	17	15	26	1468	Eutectic
(βTi) $\rightleftarrows$ (αTi) + Ti$_3$Ir	5	1	25	720	Eutectoid
L $\rightleftarrows$ αTiIr + TiIr$_3$	60	58	75	2000	Eutectic
L + αTiIr $\rightleftarrows$ Ti$_3$Ir	23	35	27	1515	Peritectic
L + (Ir) $\rightleftarrows$ TiIr$_3$	74	89	75	...	Peritectic
L $\rightleftarrows$ αTiIr		50		2130	Congruent
L $\rightleftarrows$ (βTi)		0		1670	Melting point
(βTi) $\rightleftarrows$ (αTi)		0		882	Allotropic transformation
L $\rightleftarrows$ (Ir)		100		2447	Melting point

Fig. 1---Assessed Ti-Ir Phase Diagram

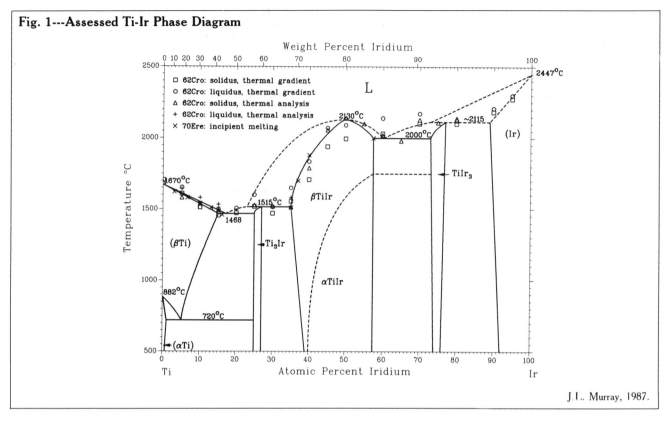

J.L. Murray, 1987.

A Ti$_2$Ni-type phase is often found in Ti transition metal alloys. It is an equilibrium phase of several Ti systems, but it is more usually a ternary phase stabilized by oxygen. [60Nev] showed that this structure does not appear in binary Ti-Ir alloys, but that it does appear in ternary Ti-Ir-O alloys.

The major investigations of this system are [62Cro], [70Ere], and [77Kan]. The results of [70Ere] were also published in [66Ere], [72Ere], and [75Sht]. Table 1 summarizes the special points of the Ti-Ir phase diagram, which is shown in Fig. 1. For the high-temperature Ir-rich portion of the diagram, further experimental work is necessary before a reasonably accurate diagram can be drawn.

Ti-Rich Liquidus and Solidus.

Addition of Ir lowers the melting point of Ti, so that the liquidus and solidus terminate at the eutectic isotherm where L $\rightleftarrows$ (βTi) + Ti$_3$Ir. Melting point data were obtained by [62Cro] using both thermal analysis and the gradient technique and by [70Ere] and [72Ere] using the Pirani-Alterthum (optical) method to determine temperatures of incipient melting. [62Cro] roughly estimated liquidus temperatures as the temperatures of sample collapse in thermal gradient experiments. From the solidus and liquidus data (Fig. 2), the eutectic temperature and maximum solubility of Ir in (βTi) are known approximately. An average of the thermal analysis data [62Cro, 70Ere], which agree within experimental error, was used to locate the eutectic isotherm at 1468 ± 10 °C. Solidus data bracket the maximum solubility between 15 and 16 at.% Ir; the (βTi)/(βTi) + Ti$_3$Ir boundary was used to place the solubility more precisely at 15 at.% Ir. The liquidus data are less accurate than the solidus data; they permit one to conclude that the melting range is narrow, but do not strongly constrain the eutectic composition.

(αTi) and (βTi).

[70Ere] placed an upper bound of 1 at.% on the maximum solubility of Ir in (αTi), based on metallographic examination of annealed alloys.

The eutectoid reaction (βTi) $\rightleftarrows$ (αTi) + Ti$_3$Ir occurs at approximately 720 °C and 5 at.% Ir, based primarily on the work of [77Kan]. [62Cro] predicted a eutectoid reaction at approximately 500 °C, based on X-ray, dilatometric, and metallographic work on a 5 at.% Ir alloy. [70Ere] concluded that the eutectoid temperature must be less than 500 °C, based on the failure to see thermal arrests corresponding to an invariant reaction above that temperature. [77Kan], however, made detailed metallographic and X-ray diffraction studies of Ti-rich alloys and demonstrated the existence of a eutectoid reaction at a temperature between 700 and 750 °C. The reaction occurred sluggishly, which explains the failure to detect it by differential thermal analysis.

Above 700 °C, the data of [72Ere] and [77Kan] for the (βTi) transus are mutually consistent; the data given by [72Ere] below 700 °C may represent a metastable extension of the (βTi) transus.

Experimental data on the boundaries of the single-phase (βTi) field are summarized in Table 2. The (βTi)/(βTi) + Ti$_3$Ir boundary below about 900 °C, as observed by [62Cro] and [70Ere], is inconsistent with the accepted eutectoid reaction. [64Rau] reported the maximum solubility of Ir in (βTi) as 15 at.%, but did not describe the experiments on which the estimate was based.

Ti$_3$Ir.

Ti$_3$Ir has the A15 structure [58Nev, 68Reu]. A homogeneity range about stoichiometry was not observed (or investigated) by [66Ere], [70Ere], or [62Cro]. [76Jun], in an

The Ir-Ti System

Table 2   Experimental Data on the (βTi) Phase Boundaries

| Reference | (βTi) transus | | (βTi) /βTi) + Ti₃Ir | | Experimetal technique |
	Composition, at.% Ir	Temperature, °C	Composition, at.% Ir	Temperature, °C	
[70 Ere]	~ 5	~ 600	> 10	800	Metallography, X-ray diffraction
	...	...	~ 10 to 13	1000	
	...	...	~ 10 to 13	1200	
	...	...	~ 14	1400	
[76Jun]	...	...	15 to 18	1400	X-ray diffraction, superconducting T_C
[77Kan]	5 to 8	700 to 500	...	...	X-ray diffraction, optical metallography

Fig. 2---Comparison of Liquidus and Solidus Data with the Assessed Phase Diagram

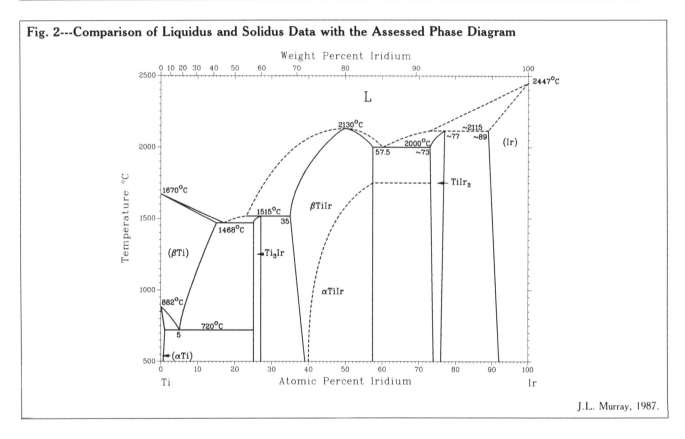

J.L. Murray, 1987.

X-ray study of alloys in the range 23 to 30 at.% Ir, found that the single-phase Ti₃Ir region at 800 °C extends from 25 to 27 at.% Ir. This range is also consistent with changes in slope of the superconducting critical temperatures as a function of composition.

Ti₃Ir forms from the liquid by the peritectic reaction L + Ti₃Ir ⇌ βTiIr. [62Cro] and [70Ere] agreed in their placement of the peritectic temperature within the limits of experimental error (1510 and 1520 °C, respectively), and therefore, the average of the two values is used in the assessed diagram.

Composition Range 40 to 60 at.% Ir.

There are two equilibrium near-equiatomic phases— a high-temperature, Ti-rich phase, βTiIr, with the CsCl structure and a low-temperature, Ir-rich form, αTiIr [62Cro, 70Ere]. For alloys containing less than 40 at.% Ir, βTiIr remains stable down to room temperature [70Ere]. The transformation between βTiIr and αTiIr at the high Ir side occurs at approximately 1750 °C on cooling [70Ere,

62Cro]. Inconsistent reports of the structure of αTiIr have not yet been satisfactorily resolved (see Crystal Structure and Lattice Parameters).

βTiIr melts congruently at 2130 ± 20 °C, based on solidus data of [62Cro] and [70Ere]. [62Cro] and [70Ere] agreed within the limits of experimental error about the temperature of congruent melting (2140 and 2120 °C, respectively), and therefore, 2130 °C is used in the assessed diagram. At 2000 °C, the eutectic reaction L ⇌ βTiIr + TiIr₃ occurs; again [62Cro] and [70Ere] agreed concerning the eutectic temperature. Solidus and liquidus data are shown in Fig. 2. There is reasonable agreement between [62Cro] and [70Ere] on the solidus, but liquidus data are very scattered.

The single-phase βTiIr field extends to 35 at.% Ir at 1515 °C. The maximum Ir content of single-phase βTiIr is 57.5 at.% at 2000 °C, and the phase boundary is nearly vertical through the transformation to αTiIr down to 1200 °C. In Fig. 1, the extent of the combined βTiIr/αTiIr region is 39 to 57.5 at.% Ir at low temperature (600 °C).

Table 3 Crystal Structures of the Ti-Ir System

Phase	Homogeneity range, at.% Ir	Pearson symbol	Space group	Strukturbericht designation	Prototype	Reference
(αTi)	0 to 1	$hP2$	$P6_3/mmc$	$A3$	Mg	[Pearson2]
(βTi)	0 to 14.5	$cI2$	$Im3m$	$A2$	W	[Pearson2]
Ti$_3$Ir	25 to 27	$cP8$	$Pm3n$	$A15$	Cr$_3$Si	[68Reu]
αTiIr	39 to 55	$cP2$	$Pm3m$	$B2$	CsCl	[62Dwi]
TiIr	41 to 55	(a)	...	...	...	[62Cro]
TiIr$_3$	73 to 77	cP_4	$Pm3m$	$L1_2$	AuCu$_3$	[59Dwi]
(α'Ti)	(b)	$hP2$	$P6_3/mmc$	A_3	Mg	(c)
ω	(b)	$hP3$	$P6/mmm$	ωMnTi	...	(c)
(Ir)	89 to 100	$cF4$	$Fm3m$	$A1$	Cu	[Pearson2]

(a) Monoclinic. (b) Metastable. (c) These are the generally accepted structures of the martensite in dilute Ti-alloy and the "ideal" ω phase. However, distortion of these structures have also been observed in some systems.

Table 4 Lattice Parameters of Cubic Phases

Phase	Composition, at.% Ir	Lattice parameter, nm	Reference
Ti$_3$Ir	25	0.5007	[56Gel]
	25	0.50101 ± 0.00004	[58Nev]
	25	0.5000	[62Cro]
	(a)	0.5009	[64Ram]
	25	0.50082 ± 0.00001	[66Ere]
	25	0.50087	[68Reu]
	25	0.5012 ± 0.0002	[76Jun]
	26	0.5008	
	27	0.5002	
(βTi)	0	0.332	[62Cro]
	15	0.322	
TiIr$_3$	75	0.3845	[59Dwi]
	75	0.385	[62Cro]
	~ 75	0.3858	[64Ram]
	73 to 77	0.3845	[66Ere]
(Ir)	100	0.384	[62Cro]
	90	0.385	

(a) Two-phase.

The homogeneity range of βTiIr was determined by metallographic examination of equilibrated alloys [62Cro, 70Ere]. On the Ti-rich side, [62Cro] placed the minimum Ti content of βTiIr between 35 and 40 at.% Ir. [70Ere] noted that determination of this point is difficult; they identified a 35 at.% Ir alloy as single phase at 1515 °C. There is slight disagreement about the homogeneity range at 800 °C. [62Cro] identified a 40 at.% alloy as two phase, whereas [70Ere] determined it to be single phase.

At the Ir-rich side, the solidus and metallographic measurements of [62Cro] and [70Ere] agree.

[64Ram] reported four distinct phases in this composition range—Ti$_{65}$Ir$_{35}$, Ti$_{55}$Ir$_{45}$, αTiIr, and Ti$_3$Ir$_7$. They attributed a NbRu-type structure to αTiIr, a CsCl-type structure to Ti$_{65}$Ir$_{35}$, and the tetragonal AuCuI structure to the other phases. These phases are tabulated as distinct equilibrium phases in data compilations [Pearson2, Landoldt], and this is almost certainly erroneous. The phase designated Ti$_{65}$Ir$_{35}$ is clearly βTiIr. Its composition, deduced from the volume fraction in a 33 at.% Ir alloy annealed at 1100 °C, agrees with the assessed phase diagram (Fig. 1). The 40 at.% Ir alloys, as-cast and heat treated at 820 °C, showed sharp CsCl lines and faint indications of tetragonal distortion. The alloys containing 60 at.% Ir or more showed the presence of TiIr$_3$, again consistent with the assessed phase diagram.

Ir-Rich Alloys—TiIr$_3$, (Ir), and the Liquid.

TiIr$_3$ has the ordered Cu$_3$Au structure [59Dwi, 64Ram, 64Sch, 70Ere]; this structure is common to binary systems of Ti with an fcc transition metal, either as a stable or a metastable phase (e.g., Ti-Ni, Ti-Pd). [62Cro] drew TiIr$_3$ as a line compound, based on examination of alloys in 5 at.% intervals. [70Ere], however, found that TiIr$_3$ has a homogeneity range of 73 to 77 at.% Ir at all temperatures investigated (1500 to 2000 °C), and this homogeneity range is shown on the assessed diagram.

[62Cro] bracketed the solubility of Ti in (Ir) between 10 and 15 at.% in the temperature range 1200 to 1850 °C. [70Ere] further bracketed the solubility between 7.5 and 12.5 at.% in the temperature range 1500 to 2000 °C. In Fig. 1, the solubility of Ti in (Ir) varies between 11 at.% at 2150 °C and 8 at.% at 600 °C.

Table 5 Structures and Lattice Parameters of Alloys near 50 at.% Ir

Composition, at.% Ir	Structure	Lattice parameters, nm a	b	c	Angle	Reference
50	Tetragonal (distorted CsCl)	0.2925	...	0.3446	...	[59Dwi2] [62Dwi]
50	Monoclinic	0.2926	0.3463	...	$\alpha = 90.92°$	[62Cro]
50	Monoclinic	0.2990	0.2883	0.3525	$\alpha = 90°52'$	[66Ere]
40	CsCl	0.3106	...	...	...	
50	NbRu-type	0.417	0.411	0.346	...	[64Ram]
45	CuAuI	0.420	...	0.337	...	
55	CuAuI	0.409	...	0.352	...	

Table 6 Experimental Thermodynamic Data for the Ti-Ir System

Reference	Composition, at.% Ir	Temperature, °C	Activities a_{Ti}	a_{Ir}	Gibbs energy of mixing, kJ/mol
[76Cho]	50	2000	3.32×20^{-4}	0.135	-83.0 ± 6.3
	60(a)	2000	1.67×10^{-4}	0.187	...
	75	2000	0.176×10^{-4}	0.35	-58.6 ± 4.2
[79Pel]	7	1623	0.54	1.95×10^{-5}	-18.0 ± 1.2
	11.5	1623	0.32	2.01×10^{-5}	-30.1 ± 2.5
	20.4(a)	1623	2.59×10^{-2}	3.54×10^{-5}	...

Note: Standard states, βTi and fcc Ir. (a) Two-phase alloys.

The melting temperature of Ir is 2447 °C [Melt]; addition of Ti to Ir depresses the melting point. Based on clear microstructural evidence, both [62Cro] and [70Ere] reported that TiIr$_3$ is formed from the liquid by a peritectic reaction. However, there is a large discrepancy in their reported peritectic temperatures—2115 °C [62Cro] and 2315 °C [70Ere].

The Ir-rich portion of the diagram requires further experimental study. The peritectic temperature should be located to within at least 50 °C, and at least one tie-line should be determined between the peritectic temperature and the melting point of Ir.

Metastable Phases

Metastable phases are formed from (βTi) during quenching. In Ti-rich alloys, the cph phase (α″Ti) can form martensitically, and at slightly higher Ir contents, ω phase forms as an intermediate phase in the decomposition of (βTi) into (αTi) at low temperatures. Both [70Ere] and [77Kan] placed the maximum Ir content for the martensitic transformation between 2 and 5 at.% Ir.

ω Phase.

ω appears as a transitory phase during the transformation of metastable single-phase β alloys to two-phase (αTi) + (βTi). ω phase does not form if (βTi) transforms martensitically, and if (βTi) alloys are aged at too high a temperature, equilibrium (αTi) precipitates directly, and ω does not appear.

ω phase is of two types. Athermal (as-quenched) ω forms during rapid quenching of (βTi) alloys and has the same composition as the (βTi) matrix; aged (isothermal) ω forms during aging of metastable (βTi) alloys at temperatures usually between 200 and 450 °C, accompanied by composition changes.

[77Kan] observed ω phase in a 5 at.% Ir alloy, quenched from the (βTi) field and aged at 570 °C. ω was not observed in a 2 at.% Ir alloy similarly heat treated.

Crystal Structure and Lattice Parameters

The structures of equilibrium and metastable phases of the Ti-Ir system are summarized in Table 3. Lattice parameter measurements are presented in Table 4.

The structure of αTiIr has been the subject of controversy. It has been variously identified as orthorhombic NbRu [64Sch, 64Ram]; tetragonal AuCuI [64Sch, 64Ram]; tetragonal distortion of CsCl [62Dwi]; monoclinic [62Cro, 70Ere]; and monoclinic distortion of AuCuI [Landoldt]. [70Ere] verified the monoclinic structure tentatively proposed by [62Cro], and they found no evidence that the various structures are associated with distinct phases rather than the degree of ordering in a single phase. Various c/a ratios agree very well, despite different struc-

ture identifications. Structure and lattice parameter data are summarized in Table 5.

Experimental Thermodynamic Data

Using Knudsen cell mass spectrometry, [76Cho] measured vapor pressures of Ti over Ir-rich alloys in the temperature range 1850 to 2100 K. [79Pel] studied Ti-rich alloys in the temperature range 1480 to 1770 K by the same experimental technique. The results are summarized in Table 6. Using Gibbs energies of the pure components from [70Kau], a (βTi) regular solution parameter of $-220\,000$ J/mol can be derived from [79Pel]. Both results [76Cho, 79Pel] indicate that the phase equilibria are dominated by very strong attractive interaction Gibbs energies.

[70Kau] calculated the Ti-Ir phase diagram using the regular solution approximation for Gibbs energies of the solid solutions and TiIr. Ti$_2$Ir was not considered. Except for the (βTi) liquidus and solidus, there was no agreement, even qualitative, between calculated and observed phase diagrams. Further regular solution calculations were performed as part of the present assessment; better agreement could be obtained for Ti-rich compositions. However, a eutectic rather than a peritectic reaction was predicted between TiIr$_3$ and (Ir). This means that some qualitative features of the excess Gibbs energies fail to be modeled as regular solutions, as indeed is to be expected for a strongly interacting system. In the absence of heat of formation or heat of mixing data, it is not possible at present to make a realistic estimate of Gibbs energies, and the calculated diagrams are not shown.

Cited References

56Gel: S. Geller, "A Set of Effective Coordination Number (12) Radii for the beta-Wolfram Structure Elements," *Acta Crystallogr., 9*, 885 (1956). (Crys Structure; Experimental)

58Nev: M.V. Nevitt, "Atomic Size Effects in Cr$_3$O-Type Structure," *Trans. AIME, 212*, 350-355 (1958). (Crys Structure; Experimental)

59Dwi: A.E. Dwight and P.A. Beck, "Close-Packed Ordered Structures in Binary AB_3 Alloys of Transition Elements," *Trans. AIME, 215*, 976-979 (1959). (Crys Structure; Experimental)

60Nev: M.V. Nevitt, J.W. Downey, and R.A. Morris, "A Further Study of Ti$_2$Ni-Type Phases Containing Titanium, Zirconium, or Hafnium," *Trans. AIME, 218*, 1019-1023 (1960). (Equi Diagram; Experimental)

*62Cro: J.G. Croeni, C.E. Armantrout, and H. Kato, "Titanium-Iridium Phase Diagram," U.S. Bureau of Mines, Report of Investigation 6079 (1962). (Equi Diagram; Experimental)

62Dwi: A.E. Dwight, "Equiatomic Phases," USAEC Report ANL-6677, 258-259 (1962). (Crys Structure; Experimental)

64Rau: Ch. Raub and J. Hull, "Superconductivity of Solid Solutions of Ti and Zr with Co, Rh and Ir," *Phys. Rev., 133*(4A), A932-A934 (1964). (Equi Diagram; Experimental)

*64Ram: A. Raman and K. Schubert, "Structural Investigations on Some Alloy Systems Homologous and Quasi-Homologous to

T^4-T^9," *Z. Metallkd.*, 55(11), 704-710 (1964) in German. (Crys Structure; Experimental)

64Sch: K. Schubert, H.G. Meissner, A. Raman, and W. Rossteutscher, "Structure Data on Intermetallic Phases (9)," *Naturwissenschaften, 51*(1), 287 (1964) in German. (Crys Structure; Experimental)

*66Ere: V.N. Eremenko, T.D. Shtepa, and V.G. Sirotenko, "Intermediate Phases in Alloys of Titanium with Iridium, Rhodium, and Osmium," *Poroshk. Metall., 42*(6), 68-72 (1966) in Russian; TR: *Sov. Powder Metall., 42*(6), 487-490 (1966). (Equi Diagram, Crys Structure; Experimental)

68Reu: E.C. van Reuth and R.M. Waterstrat, "Atomic Ordering in Binary A15-Type Phases," *Acta Crystallogr., B, 24,* 186-196 (1968). (Crys Structure; Experimental)

*70Ere: V.N. Eremenko and T.D. Shtepa, "Phase Diagram of the Ti-Ir System," *Izv. Akad. Nauk SSSR, Met.,* (6), 198-203 (1970) in Russian; TR: *Russ. Metall.,* (6), 127-130 (1970). (Equi Diagram; Experimental)

70Kau: L. Kaufman and H. Bernstein, *Computer Calculation of Phase Diagrams,* Academic Press, New York (1970). (Thermo; Review)

*72Ere: V.N. Eremenko and R.D. Shtepa, "Phase Equilibria in the Binary Systems of Titanium with Ruthenium, Osmium, Rhodium, Iridium, and Palladium," *Colloq. Int. CNRS,* (205),

403-413 (1972) in German. (Equi Diagram, Crys Structure; Experimental)

*75Sht: T.D. Shtepa, "Interaction of Titanium with Platinum-Group Metals," *Fiz. Khim. Kondens. Faz. Sverkhtverd. Mater.,* 175-191 (1975) in Russian. (Equi Diagram, Crys Structure; Experimental)

*76Cho: U.V. Choudary, K.A. Gingerich, and L.R. Cornwell, "High Temperature Thermodynamic Investigation of the Iridium-Rich Portion of the Titanium-Iridium System by Knudsen Cell Mass Spectrometry and Predicted Enthalpies of Formation for Some Transition Metal-Iridium Intermetallic Compounds," *J. Less-Common Met., 50*(2), 201-211 (1976). (Thermo; Theory)

76Jun: A. Junod, R. Flukiger, and J. Muller, "Superconductivity and Specific Heat in Titanium-Base A15 Compounds," *J. Phys. Chem. Solids, 37,* 27-31 (1976) in French. (Equi Diagram, Crys Structure; Experimental)

*77Kan: V. Kandarpa, L.R. Cornwell, and K.A. Gingerich, "An X-Ray and Metallographic Study of Titanium-Iridium Alloys in the Titanium-Iridium System," *Microstructure Sci., 5,* 383-393 (1977). (Equi Diagram; Experimental)

*79Pel: M. Pelino, S.K. Gupta, L.R. Cornwell, and K.A. Gingerich, "High Temperature Thermodynamic Investigation of the Titanium-Rich Portion of the Titanium-Iridium System by Knudsen Cell Mass Spectrometry," *J. Less-Common Met., 68,* P31-P28 (1979). (Thermo; Theory)

The K-Ti (Potassium-Titanium) System

39.0983   47.88

By C.W. Bale

Equilibrium Diagram

The assessed K-Ti phase diagram is shown in Fig. 1. The system is almost completely immiscible in both the solid and liquid states. There is no evidence of intermetallic formation in this or in any other alkali metal–Group IVA system. The extent of solubility is very limited, and it follows from thermodynamic calculations that the un-

Fig. 1---Assessed K-Ti Phase Diagram

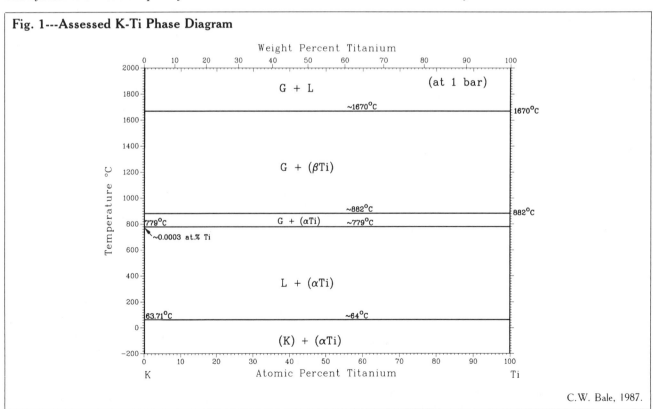

C.W. Bale, 1987.

Table 1 Crystal Structures and Lattice Parameters of the K-Ti System

Phase	Composition, at.% Ti	Pearson symbol	Space group	Struktur-bericht designation	Prot-type	Lattice parameters, nm		Comment	Reference
						a	*c*		
(K)	0	cI2	Im3m	A2	W	0.5321	...	AtRT	[King3]
(αTi)	100	hP2	P6₃/mmc	A3	Mg	0.29503	0.46836	AtRT	[King2]
(βTi)	100	cI2	Im3m	A2	W	0.33065	...	Above 900 °C	[King2]

ivariant temperatures in Fig. 1 are virtually identical with phase transitions of the pure components. Crystal structure data for the elements are summarized in Table 1.

[70Ste] attempted to measure the solubility of Ti in liquid K in the temperature range 747 to 1068 °C. The study was part of a program to measure the solubilities in liquid K of five transition elements (Mo, W, V, Ti, and Zr). K containing less than 20 wt. ppm of oxygen was equilibrated in sealed Ti capsules (approximately 3 ml) for 24 h. The liquid was drained and quenched. The Ti content was measured by optical spectrographic analysis, and it was found to be less than the detection limit of 4 wt. ppm Ti

(0.0003 at.% Ti) for the apparatus. There were no other reported solubility measurements for this system.

By analogy with other alkali metal–Group IVA systems, it is probable that the solubility of K in solid and liquid Ti is very limited. From vapor-pressure calculations, it is estimated that the composition of the gas phase at 2000 °C and 1 atm is approximately 0.04 at.% Ti, in equilibrium with almost pure liquid Ti.

Cited Reference

70Ste: S. Stecura, "Solubilities of Molybdenum, Tungsten, Vanadium, Titanium and Zirconium in Liquid Potassium," *Corrosion by Liquid Metals*, Plenum Press, New York, 601-611 (1970). (Equi Diagram; Experimental)

The La-Ti (Lanthanum-Titanium) System

138.9055   47.88

By J.L. Murray

Equilibrium Diagram

Like all Ti-rare earth systems, Ti-La contains no intermetallic phases and has a miscibility gap in the liquid phase. The equilibrium solid phases are those based on the pure elements:

- bcc (βTi) and cph (αTi), the high- and low-temperature forms, respectively.
- bcc (γLa) and fcc (βLa), high- and low-temperature forms of La in the temperature range under consideration above 600 °C; dcph (αLa) is stable below 310 °C [81Gsc].

The only published experimental study of the system [57Sav, 62Sav] is limited both in the range examined and in the purity of the alloys. Those Ti-rare earth phase diagrams for which more accurate and extensive data are available can be calculated from Gibbs energies, which nearly approximate regular or subregular solutions The present evaluator has therefore used a thermodynamic calculation to test the self-consistency of the diagram

Table 1 Special Points of the Ti-La Phase Diagram

Reaction	Compositions of the respective phases, at.% La			Temperature, °C	Reaction type
(LTi) ⇌ (βTi) + (LTi)	~ 9	~ 3	~ 98	~ 1550	Monotectic
(βTi) + (LLa) ⇌ (La)	~ 0.5	~ 99.9	~ 98	~ 960	Peritectic
(βTi) + (γLa) ⇌ (αTi)	~ 0.4	~ 98.4	~ 1	~ 900	Eutectoid
(γLa) ⇌ (βLa) + (αTi)	~ 100	~ 98.8	~ 0.8	~ 825(a)	Eutectoid
L ⇌ (βTi)		0		1670	Melting point
(βTi) ⇌ (αTi)		0		882	Allotropic transformation
L ⇌ (γLa)		100		917	Melting point
(γLa) ⇌ (βLa)		100		865	Allotropic transformation
β(La) ⇌ (αLa)		100		310	Allotropic transformation

(**a**) [62Sav] reported 800 °C, but the calculated value is preferred (see text).

Phase Diagrams of Binary Titanium Alloys

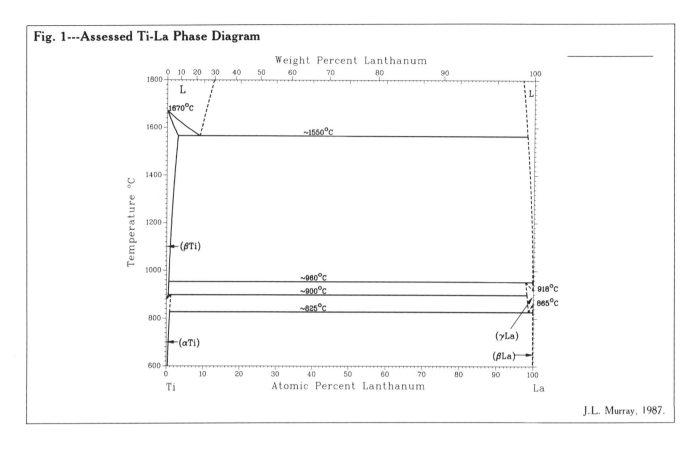

Fig. 1---Assessed Ti-La Phase Diagram

Weight Percent Lanthanum

Temperature °C

Atomic Percent Lanthanum

J.L. Murray, 1987.

Table 2 Crystal Structures of the Ti-La System

Phase		Homogeneity range, at.% La	Pearson symbol	Space group	Struktur-bericht designation	Proto-type	Lattice parameters(a), nm		Comment
							a	c	
(αTi)	......	0 to ~ 1	hP2	P6₃/mmc	A3	Mg	0.29511	0.46843	...
(βTi)	......	0 to ~ 3	cI2	Im3m	A2	W	0.33065	...	At 900 °C
(γLa)	......	~ 98 to 100	cI2	Im3m	A2	W	0.426	...	At 887 °C
(βLa)	......	~ 100	cF4	Fm3m	A1	Cu	0.5303	...	At 325 °C
(αLa)	......	~ 100	hP4	P6₃/mmc	A3′	La	0.37740	1.2171	...

(a) Lattice parameters are for the pure elements. Ti parameters are from [Pearson]; La parameters are from [86Gsc].

reported by [57Sav] and [62Sav]. The assessed diagram (Fig. 1) is the result of the calculation.

Special points of the assessed diagram are summarized in Table 1, and crystal structure data are given in Table 2. Because the diagram is only roughly known and the La-rich side was constructed by thermodynamic extrapolation, Table 1 and Fig. 1 should be assumed to contain substantial inaccuracies.

Experimental Data.

[57Sav] and [62Sav] studied the diagram in the range 0 to 7 at.% La by microscopy, thermal analysis, and dilatometry. The La used contained 1.6 wt.% impurities. The temperatures of invariant reactions at 1550 and 900 °C were determined by thermal analysis and dilatometry, and the measurements were probably accurate to about 30 and 10 °C, respectively. The other invariant temperatures appear to have been determined by microscopic methods, and they have an unknown but larger uncertainty. The solubility of La in (αTi) or (βTi) was found to be about 1 at.% over the entire range examined, 600 to 1600 °C.

Thermodynamic Calculations

Gibbs energies of the solution phases are represented as:

$$G(i) = (1 - x) G^0(Ti,i) + x G^0(La,i) + RT[x \ln x + (1 - x) \ln (1 - x)] + x (1 - x) [B^i + C^i (1 - 2x)]$$

where i designates the phase; x is the atomic fraction of La; G^0 is the Gibbs energies of the pure components; and B^i and C^i are coefficients describing the excess Gibbs energies. The regular solution approximation ($C = 0$) was used for the hexagonal and fcc phases. The lattice stability parameters of La are based on data in [Hultgren, E]. Input experimental data were the temperatures of the three-phase reactions and the compositions of (αTi) and (βTi) at the three-phase equilibria [62Sav]. Based on such limited experimental data, optimization calculations give only inadequate fits to the data if too few parameters are simultaneously varied, but do not converge at all if too many parameters are varied. Trial and error had to be used to weight the input data and determine the number of coefficients that could be meaningfully determined.

Table 3 Gibbs Energies of the Ti-La System

Properties of the pure components

$G^0(\text{Ti,L}) = 0$
$G^0(\text{La,L}) = 0$

$G^0(\text{Ti,bcc}) = -16\,234 + 8.368\ T$
$G^0(\text{La,bcc}) = -6\,196 + 5.203\ T$

$G^0(\text{Ti,hex}) = -20\,585 + 12.134\ T$
$G^0(\text{La,hex}) = -9\,682 + 8.570\ T$

$G^0(\text{Ti,fcc}) = -17\,238 + 12.134\ T$
$G^0(\text{La,fcc}) = -9\,318 + 7.9456\ T$

Interaction parameters of the solution phases

$B(\text{L})$ = 51 891	$B(\text{bcc})$ = 47 697
$C(\text{L})$ = -11 625	$C(\text{bcc})$ = 6 765
$B(\text{hex})$ = 44 647	$B(\text{fcc})$ = 70 557

Note: Values are given in J/mol and J/mol · K.

Results of the optimizations are given in Table 3. Further calculations on the Ti-La system will require experimental determination of the solubility of Ti in (La).

Cited References

57Sav: E.M. Savitskii and G.S. Burkhanov, "Phase Diagrams of Titanium-Lanthanum and Titanium-Cerium Alloys," *Zh. Neorg. Khim.*, 2(11), 2609-2616 (1957) in Russian; TR: *J. Inorg. Chem.*, 2, 199-219 (1957). (Equilibrium Diagram; Experimental)

62Sav: E.M. Savitskii and G.S. Burkhanov, "Phase Diagrams of Titanium Alloys with the Rare Earth Metals," *J. Less-Common Met.*, 4, 301-314 (1962). (Equilibrium Diagram; Experimental)

81Gsc: K.A. Gschneidner, Jr. and F.W. Calderwood, "Critical Evaluation of Binary Rare Earth Phase Diagrams," IS-RIC-PR-1, Rare-Earth Information Center, Iowa State University, Ames, IA (1981). (Crys Structures; Review)

The Li-Ti (Lithium-Titanium) System

6.941 47.88

By C.W. Bale

Equilibrium Diagram

The assessed Li-Ti phase diagram is shown in Fig. 1. The system is virtually completely immiscible in both the solid and liquid states. There is no evidence of intermetallic formation in this or in any other alkali metal–Group IVA system. Crystal structure data for the elements are summarized in Table 1.

Solubility of Ti in Liquid Li.

[61Lea] held liquid Li in Ti crucibles for up to 24 h in the temperature range 717 to 917 °C. Tests showed that

---

Fig. 1---Assessed Li-Ti Phase Diagram

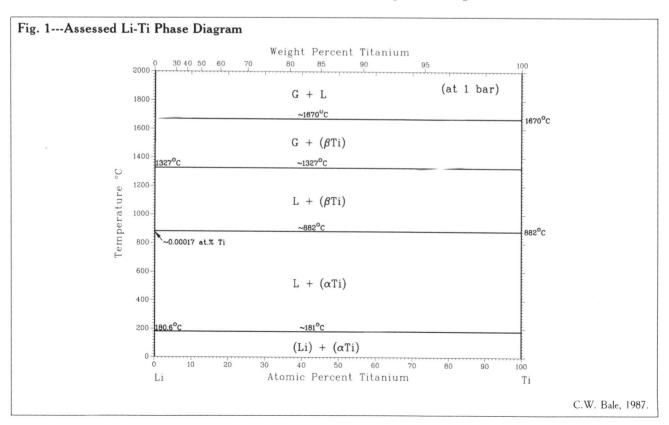

C.W. Bale, 1987.

Table 1 Crystal Structures and Lattice Parameters of the Li-Ti System [King2]

Phase	Composition, at.% Ti	Pearson symbol	Space group	Struktur-bericht designation	Proto-type	Lattice parameters, nm		Comment
						a	*c*	
(αLi)	 0	hP2	P6₃/mmc	A3	Mg	0.3111	0.5093	Below 72 K
(βLi)	 0	cI2	Im3m	A2	W	0.35093	...	At RT
(αTi)	 100	hP2	P6₃/mmc	A3	Mg	0.29503	0.46836	At RT
(βTi)	 100	cI2	Im3m	A2	W	0.33065	...	Above 900 °C

Fig. 2---Solubility of Ti in Liquid Li

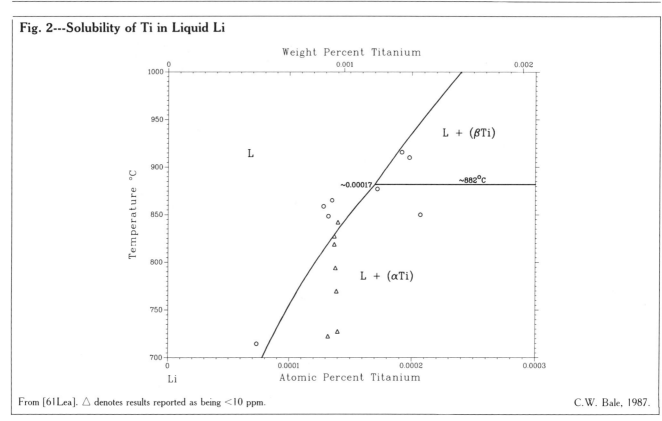

From [61Lea]. △ denotes results reported as being <10 ppm.

C.W. Bale, 1987.

equilibrium was attained after approximately 6 h. Half-gram samples were withdrawn, and Ti content was determined by colorimetry. The results were presented in a small diagram (redrawn in Fig. 2). It was reported that the measurements were reproducible after both increasing and decreasing temperatures. The solubilities of Ni, Cr, Fe, and Mo were also determined. The nitrogen contamination level was reported as 50 to 100 ppm. There were no other details of impurities, gaseous atmosphere, or equipment given in the report. The strong affinity of both Ti and Li for oxygen and the volatility of Li would require strict experimental conditions. It was reported that the temperature dependence of the solubilities agreed well with theoretical slopes obtained from size factors of the liquid solution [58Sta].

The solid line in Fig. 2 is given by the equation:

$$\log (\text{at.\% Ti}) = -2.07 - 1975/T \quad (\text{Eq 1})$$

where T is in K.

The solubility of Ti is so small that it does not appear in Fig. 1.

By a similar technique, [61Byc] determined one solubility data point, 0.014 at.% Ti at 900 °C. This value is higher than the Ti 0.002 at.% value calculated from Eq 1, but it still indicates that the solubility is very limited.

Earlier data of [50Jes] (0.005 to 0.54 at.% Ti at 732 to 1016 °C) and [50And] (0.5 to 1.4 at.% Li at 732 to 1016 °C) indicated a more extensive solubility. These results are not considered accurate in the present assessment.

Solubility of Li in Solid Ti.

[63Pek] reported solubilities of Li in (αTi) and (βTi) in the range 300 to 1350 °C (Fig. 3). Ti chips were heated in an excess of Li in closed Ti containers. Unreacted Li was removed with boiling distilled water, and Li contents in the Ti were determined by flame photometry. Figure 3 shows that the solubility of Li in solid Ti increases from 0.0007 at.% Li at 300 °C to approximately 0.029 at.% Li at 1050 °C. A discontinuity was observed between 1050 and 1150 °C, and the solubility was reduced to approximately 0.00015 at.% Li at the higher temperatures.

[Shunk] attributed the discontinuity to the peritectic reaction L + (βTi) ⇄ (αTi). [79Mof] incorporated this peritectic reaction in his diagram and showed that the transition temperature is increased by approximately 200 °C with the dissolution of only 0.029 at.% Li in (αTi) and little or no dissolution of Li in (βTi). Thermodynamic considerations indicate that this is only possible if the αTi → βTi enthalpy of transformation is a few joules (~16 J). According to [83Cha], the αTi → βTi transition occurs at

Li-Ti System

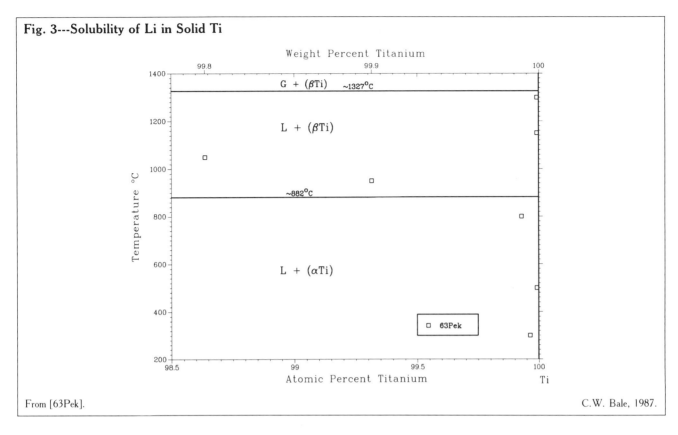

Fig. 3---Solubility of Li in Solid Ti

From [63Pek].

C.W. Bale, 1987.

893 °C, with a corresponding enthalpy of transition of 4.17 kJ. Consequently, the data in Fig. 3 are not consistent with thermodynamic calculations. The peritectic temperature could be increased by 200 °C if the solubility of Li in (αTi) were extensive (~7.5 at.% Li), but this is unlikely. According to [83Pet], there was extensive intercrystalline penetration of Li. This would create artificially high Li concentrations, and consequently, the data of [63Pek] for the dissolution of Li in (αTi) are not considered accurate.

Vapor Phase.

At 1 atm pressure and above the temperature of fusion of Ti (1668 °C), the vapor phase, which is almost pure Li, is in equilibrium with virtually pure liquid Ti. From vapor-pressure calculations, it is estimated that the composition of the gas phase—Li(g) + Li$_2$(g) + Ti(g)—at 2000 °C is approximately 0.04 at.% Ti in equilibrium with almost pure liquid Ti.

Cited References

50And: R.C. Anderson and H.R. Stephan, "Progress Report on Materials Tested in Lithium," U.S. Atomic Energy Comm. NEPA Report No. 1652 (1950). (Equi Diagram; Experimental)

50Jes: O.S. Jesseman, W.S. Fleshman, G.D. Roben, R. Anderson, A.L. Grunewald, and V.P. Calkins, "Preliminary Investigation of Metallic Elements in Molten Lithium," U.S. Atomic Energy Comm. NEPA Report No. 1465 (1950). (Equi Diagram; Experimental)**58Sta:** S.W. Strauss, J.L. White, and B.F. Brown, "The Atomic Size Effect and Alloying Behavior in Liquid Metals," *Acta Metall.*, *6*, 604 (1958). (Equi Diagram; Theory)

61Byc: Yu.F. Bychov, A.N. Rozanov, and V.B. Yakovleva, "Determination of the Solubility of Metals in Lithium," *Sov. J. Atom. Energy*, *7*, 987 (1961). (Equi Diagram; Experimental)

61Lea: H.W. Leavenworth and R.E. Clearly, "The Solubility of Ni, Cr, Fe, Ti and Mo in Liquid Lithium," *Acta Metall.*, *9*, 519 (1961). (Equi Diagram; Experimental)

63Pek: A.I. Pekarev, E.M. Savitskii, and M.A. Tylkina, "Interaction of Lithium with Titanium at Elevated Temperatures," *Tr. Inst. Met. Baikova, Akad. Nauk SSSR*, *12*, 198 (1963) in Russian. (Equi Diagram; Experimental)

79Mof: W.G. Moffatt, *The Handbook of Binary Phase Diagrams*, and suppl., General Electric Co., Schenectady, NY (1979). (Equi Diagram; Compilation)

83Cha: M.W. Chase, "Heats of Transition of the Elements," *Bull. Alloy Phase Diagrams*, *4*(1), 124 (1983). (Equi Diagram; Compilation)

The Mg-Ti (Magnesium-Titanium) System

24.305   47.88

By J.L. Murray

Equilibrium Diagram

There is very little mutual solubility of Mg and Ti in any phase, and no intermetallic compounds occur in the Ti-Mg system. Thus, the equilibrium solid phases are the low-temperature cph (αTi) and (Mg) solid solutions and the bcc solid solution, (βTi), based on the high-temperature form of pure Ti.

Most of the experimental work on this system has been concerned with the solubility of Ti in liquid and solid Mg [49Aus, 53Eis, 59Obi, 68Fin]. The solubilities of Mg in (βTi) and (αTi) below 1200 °C were investigated by [55Fre] and [59Obi] and solid solubility estimates were given by [68Fin].

The assessed diagram is shown in Fig. 1; only the Mg-rich portion should be considered quantitatively reliable. The assessed diagram was drawn from thermodynamic calculations described below. Gibbs energies for the Mg-rich liquid, (Mg), and (αTi), were estimated by least-squares optimization with respect to select phase diagram data (see Fig. 2). The calculated phase boundaries and original data coincide within experimental uncertainty; in the absence of experimental thermodynamic data, however, extrapolation to high temperatures or Ti-rich compositions is not recommended.

Salient points of the system are summarized in Table 1, and crystal structures are listed in Table 2.

Mg-Rich Alloys.

There is agreement [49Aus, 53Eis, 68Fin] that the solubility of Ti in solid (Mg) is greater than in the liquid at the Mg melting point and, therefore, that there is a peritectic reaction L + (αTi) ⇌ (Mg). The peritectic temperature was reported to differ from the Mg melting point by no more than 0.5 °C [49Aus] and 1 °C [68Fin]; both determinations were made by thermal analysis.

For determination of the liquidus, only analysis of the composition of the saturated liquid has been found to be a suitable method [49Aus, 53Eis, 59Obi, 68Fin]. The work of [53Eis], which relied on the assumption that only the Ti that had dissolved in Mg was soluble in H_2SO_4, is not used here. Between [49Aus] and [68Fin], there are discrepancies of about 0.01 at.% Ti, but the solubilities reported by [59Obi] are consistently larger by 0.1 at.% than those reported by [49Aus] and [68Fin]. The work of [68Fin] is preferred, because equilibrium at each temperature was approached from above and below, and no difference in solubilities was found.

The (Mg) solvus is based on electrical resistivity data from [68Fin]. Solubilities could not be determined by electron or light microscopy, because of the presence of insoluble impurities in the microstructures. At the higher temperature, the resistivity data were very scattered. According to [68Fin], this was most likely due to the

Fig. 1---Assessed Ti-Mg Phase Diagram

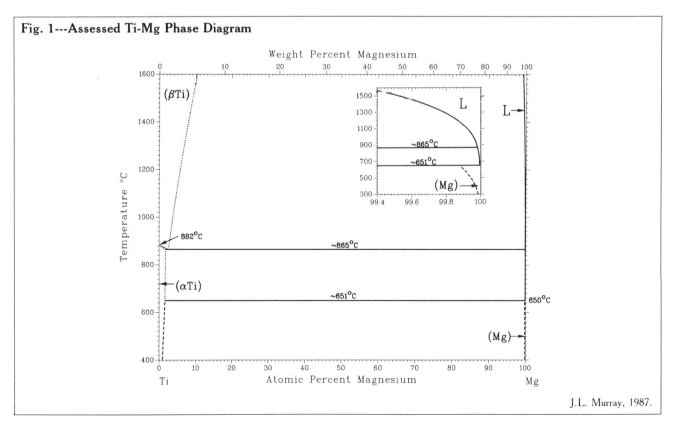

J.L. Murray, 1987.

Fig. 2---Experimental Data on the Mg-Rich Portion of the Ti-Mg Phase Diagram vs the Assessed Phase Diagram

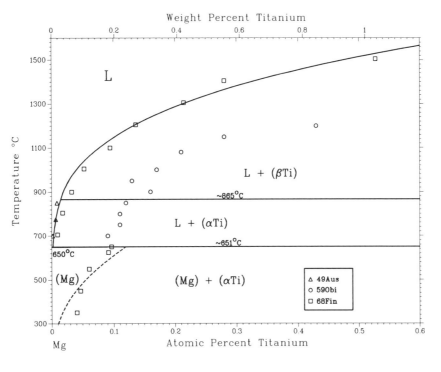

J.L. Murray, 1987.

Table 1 Special Points of the Assessed Ti-Mg Phase Diagram

Reaction	Compositions of the respective phases, at.% Mg			Temperature, °C	Reaction type
L + (αTi) ⇌ (Mg)	99.999	(1.5)	99.88	~ 651	Peritectic
(βTi) ⇌ (αTi) + (LMg)	(2.4)	(1.6)	99.986	~ 865	Catatectic(a)
L ⇌ (βTi)		0		1670	Melting point
(βTi) ⇌ (αTi)		0		882	Allotropic transformation
L ⇌ (Mg)		100		650	Melting point

Note: Values in parentheses are obtained by extrapolation and should be regarded as very uncertain. (a) Also sometimes designated "inverted peritectic".

Table 2 Crystal Structures of the Ti-Mg System [Pearson2]

Phase	Homogeneity range, at.% Mg	Pearson symbol	Space group	Struktur-bericht designation	Prototype
(αTi)	0 to 1.6(a)	hP2	P6₃/mmc	A3	Mg
(βTi)	0 to 2.4(a)	cI2	Im3m	A2	W
(Mg)	99.88 to 100	hP2	P6₃/mmc	A3	Mg

(a) Approximately, at 865 °C.

vaporization of Mg at the heat treatment temperature. The reported solubility at 350 °C appears to be too high, based on the failure of this one point to lie on a straight line in a ln X vs $1/T$ plot. It may be that equilibrium was not reached at the lowest temperature. In view of these difficulties with the experimental data, the calculated solvus is considered an adequate fit of the data and is shown in the assessed diagram (Fig. 1 and 2).

Ti-Rich Alloys.

On the Ti-rich side of the diagram, estimates of the solubility of Mg were made by [55Fre], [59Obi], and [68Fin]. [68Fin], by chemical analysis of particles extracted from liquid Mg, gave 2.6 at.% Mg as the solubility of Mg in (βTi) at 1400 °C. [55Fre], based on metallographic work, found at least 2.9 at.% Mg to be soluble in both (βTi) and (αTi) near 882 °C. According to [59Obi], the maximum observed solubility is 1.2 at.% Mg in (βTi) at 1200 °C.

Based on a discontinuity in the slope of the Mg-rich liquidus, [68Fin] gave 865 °C as the temperature of the invariant reaction associated with the (βTi) to (αTi) transformation. [55Fre] used alloys containing considerable oxygen, so that the two-phase (αTi) + (βTi) field apparently violates the phase rule and the invariant temperature could not be determined. However, based on the tendency of oxygen to stabilize (αTi), it was thought that Mg stabilizes (βTi), in agreement with [68Fin]. [59Obi] reported a peritectic-type reaction at about 890 °C and reported solubilities of Mg in (βTi) between 0.6 and 1.2 at.% in the range 900 to 1200 °C. The higher estimates [55Fre, 68Fin] are preferred, but they must be considered rough estimates only of unknown accuracy.

The catatectic (inverted peritectic) reaction (βTi) $\rightleftarrows$ (αTi) + (LMg) is shown in the assessed diagram at 865 °C. The (βTi) and (αTi) compositions are roughly estimated as 2.6 and 1.5 at.% Mg, respectively, at 865 °C.

Thermodynamics

There are no experimental thermochemical data available on the Ti-Mg system; however, rough thermodynamic calculations can be based on phase diagram data alone. The assessed phase diagram (Fig. 1 and 2) is calculated from Gibbs energies optimized with respect to select experimental data [49Aus, 55Fre, 68Fin].

The Gibbs energies are approximated by the subregular solution model:

$$G^i = (1 - x) G^0(\text{Ti},i) + x G^0(\text{Mg},i) + RT[x \ln x + (1 - x) \ln (1 - x)] + x (1 - x) [B(i) + C(i) (1 - 2x)]$$

where i is the phase; x is the atomic fraction of Mg; G^0 the Gibbs energies of the pure components; and $B(i)$ and $C(i)$ are the expansion coefficients of the excess Gibbs energies. The pure component Gibbs energies are from [70Kau]. Optimized Gibbs energy parameters are listed in Table 3.

For optimization calculations, the data of [68Fin] were used for the Mg-rich phase boundaries. The temperature of the catatectic reaction (βTi) $\rightleftarrows$ (αTi) + (LMg) was deduced from the liquidus data [68Fin], and rough estimates of the solubility of Mg in (βTi) and (αTi) also were taken from that work.

The miscibility gap of the cph phase was most susceptible to a thermodynamic treatment. An excess entropy was included in the cph Gibbs energy to reproduce the

Table 3   Gibbs Energies of the Ti-Mg System

Properties of the pure elements

$G^0(\text{Ti},\text{L})$ = 0
$G^0(\text{Mg},\text{L})$ = 0

$G^0(\text{Ti},\text{bcc})$ = − 16 234 + 8.368 T
$G^0(\text{Mg},\text{bcc})$ = − 4 351 + 6.694 T

$G^0(\text{Ti},\text{cph})$ = − 20 585 + 12.134 T
$G^0(\text{Mg},\text{cph})$ = − 8 954 + 9.707 T

Estimated interaction parameters of the solution phases

$B(\text{L})$ = 77 020
$B(\text{bcc})$ = 33 608
$B(\text{cph})$ = 21 779 + 22.165 T
$C(\text{cph})$ = − 9 467

Note: Values are given in J/mol and J/mol · K.

(Mg) solvus. For the bcc and liquid phases, the data are sufficient only to determine regular solution Gibbs energies. The Gibbs energy of the liquid was estimated from the (Mg) liquidus; that of the bcc phase by the invariant reaction at 865 °C.

A syntectic reaction (LTi) + (LMg) $\rightleftarrows$ (βTi) is predicted at about 2000 °C. However, because the excess Gibbs energies are very large in magnitude, there is no physical justification for the regular solution approximation. Therefore, the details of the complete calculated phase diagram should not be used, except as a qualitative topological completion of the partial diagram.

Cited References

*49Aus: K.T. Aust and L.M. Pidgeon, "Solubility of Titanium in Liquid Magnesium," *Metall. Trans., 185*, 585-587 (1949). (Equi Diagram; Experimental)

53Eis: H. Eisenreich, "Investigation of the Solubility of Titanium in Magnesium," *Metall, 7*, 1003-1006 (1953) in German. (Equi Diagram; Experimental)

55Fre: J.W. Fredrickson, "Preliminary Investigation of the System Ti-Mg," *Trans. AIME, 203*, 368 (1955). (Equi Diagram; Experimental)

*59Obi: I. Obinata, Y. Takeuchi, and R. Kawanishi, "The Titanium-Magnesium System," *Metall, 13*, 392-397 (1959) in German. (Equi Diagram; Experimental)

*68Fin: L.C. Fincher and D.H. Desy, "The Magnesium-Titanium Phase Diagram to 1.0 pct Titanium," *Trans. AIME, 242*, 2069-2073 (1968). (Equi Diagram; Experimental)

70Kau: L. Kaufman and H. Bernstein, *Computer Calculation of Phase Diagrams*, Academic Press, New York, (1970). (Thermo; Review)

The Mn-Ti (Manganese-Titanium) System

54.9380   47.88

By J.L. Murray

Equilibrium Diagram

The assessed Ti-Mn phase diagram is shown in Fig. 1, and its salient features are summarized in Table 1. The crystal structures of the compounds TiMn$_3$ and βTiMn, the positions of several phase boundaries, and the temperatures of several of the proposed three-phase equilibria have not yet been determined experimentally or fully confirmed. The assessed phase diagram is drawn from a calculation of the diagram from Gibbs energies optimized with respect to select phase diagram data. In some instances, thermodynamic considerations were used as a critical tool in the data assessment. The calculations are discussed below.

The equilibrium solid phases of the Ti-Mn system are:

- The bcc (βTi) and (δMn) solid solutions, stable in pure Ti above 882 °C and in pure Mn between 1244 and 1136 °C. The maximum solubility of Mn in (βTi) is 30 at.% at 1174 °C, and that of Ti in (δMn) is about 6 at.% at 1204 °C.
- The Ti-rich low-temperature cph (αTi) solid solution. The maximum solubility of Mn in (αTi) is 0.4 at.%.
- The fcc (γMn) solid solution, stable in pure Mn between 1100 and 1136 °C. Approximately 0.6 at.% Ti is soluble in (γMn).

- The complex cubic (βMn) solid solution, stable in pure Mn between 720 and 1100 °C. The solubility of Ti in (βMn) is approximately 5 at.% at 1145 °C.
- The complex cubic low-temperature (αMn) solid solution. The presence of Ti strongly stabilizes this phase, which exists in Ti-Mn alloys up to about 1148 °C.
- Near-equiatomic compounds αTiMn and βTiMn, with compositions of 50.5 and 52 at.% Mn, respectively.
- TiMn$_2$, a Laves phase of the MgZn$_2$ type with a broad homogeneity range. TiMn$_2$ melts congruently at 1325 °C.
- The compound TiMn$_3$, stable between about 950 and 1250 °C.
- The compound TiMn$_4$ of composition 81.5 at.% Mn, stable between approximately 930 and 1230 °C.

(βTi) Liquidus and Solidus.

Experimental data on the Ti-rich region of the phase diagram are shown in Fig. 2. A eutectic reaction L ⇌ (βTi) + βTiMn occurs at 1180 ± 2 °C and 39 at.% Mn. The (βTi) liquidus and solidus are based on thermal analysis, microstructural, and incipient melting studies [53May]. An uncertainty of about ± 25 °C is estimated for these boundaries. The eutectic temperature is based on thermal analysis studies of [57Hel], (1181 ± 2 °C) and [53May]

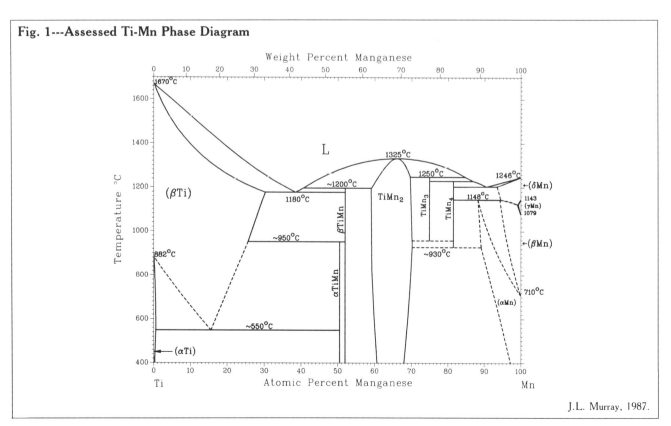

Fig. 1---Assessed Ti-Mn Phase Diagram

J.L. Murray, 1987.

(1175 °C). The maximum solubility of Mn in (βTi) is 30 at.% Mn [53May], based primarily on extrapolation of the (βTi)/(βTi) + βTiMn boundary to the eutectic temperature.

(βTi)/(αTi) Equilibria.

The (βTi)/[(αTi) + (βTi)] boundary, (βTi) transus, was determined by [53May] and [51Mcq] by metallography and the hydrogen pressure technique, respectively. Addi-

Table 1 Special Points of the Assessed Ti-Mn Phase Diagram

Reaction	Compositions of the respective phases, at.% Mn			Temperature, °C	Reaction type
(δMn) ⇌ (βMn) + (γMn)	99.3	99.1	99.4	1120 ± 2	Eutectoid
(δMn) + (αMn) ⇌ (βMn)	97	88.6	~ 95	~ 1145	Peritectoid
(δMn) + TiMn$_4$ ⇌ (αMn)	96	81.5	89.5	5	Peritectoid
TiMn$_4$ ⇌ (αMn) + TiMn$_2$	81.5	90	70	930	Eutectoid
TiMn$_3$ ⇌ TiMn$_4$ + TiMn$_2$	75	81.5	70	950	Eutectoid
L + TiMn$_2$ ⇌ TiMn$_3$	85	70	75	1250 ± 2	Peritectic
L + TiMn$_3$ ⇌ TiMn$_4$	90	75	81.5	1230 ± 2	Peritectic
L ⇌ (δMn) + TiMn$_4$	90.5	~ 94	81.5	1204 ± 2	Eutectic
L ⇌ TiMn$_2$		66.7		1325 ± 2	Congruent
L + TiMn$_2$ ⇌ βTiMn	59	43	52	1200	Peritectic
L ⇌ βTiMn + (βTi)	39.2	52	30	1180 ± 2	Eutectic
βTiMn + (βTi) ⇌ αTiMn	52	25	50.5	950	Peritectoid
(βTi) ⇌ (αTi) + αTiMn	15	0.4	50.5	~ 550	Eutectoid
L ⇌ (βTi)		0		1670	Melting point
(βTi) ⇌ (αTi)		0		882	Allotropic transformation
L ⇌ (δMn)		100		1246	Melting point
(δMn) ⇌ (γMn)		100		1143	Allotropic transformation
(γMn) ⇌ (βMn)		100		1079	Allotropic transformation
(βMn) ⇌ (αMn)		100		710	Allotropic transformation

Fig. 2---Experimental Data for the Ti-Rich Portion of the Phase Diagram

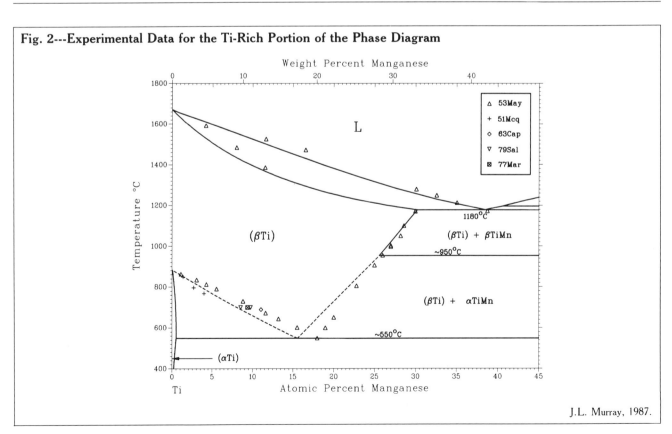

J.L. Murray, 1987.

tional phase boundary data at 700 °C were reported by [63Cap], [77Mar], and [79Sal], based on volume fractions of (αTi) and (βTi) in equilibrated two-phase alloys [77Mar, 79Sal] and on analysis of superconducting critical temperature data [63Cap].

As shown in Fig. 2, there are some discrepancies among the data. The hydrogen pressure technique consistently appears to underestimate the (βTi) transus temperature in a number of Ti-systems; the effect of oxygen or of an insufficiently rapid quench is to overestimate the (βTi) transus. Another complication in assessment of the data is that the eutectoid reaction is very sluggish; the eutectoid composition and temperature could have been estimated by [53May] only from the intersection of appropriate phase boundaries. The technique of annealing and quenching from successively higher temperatures was used by [54Fro] to determine a eutectoid temperature of 675 °C, but alloys based on Mg-reduced (less pure) Ti were used.

Verification of phase equilibria is therefore needed in this portion of the diagram. The tentative assessed (βTi) transus lies between determinations of [51Mcq] and [53May] and is based on optimization of Gibbs energies with respect to the data of [53May]. That the observed transus temperatures [53May] are consistently too high is supported by the observation by [79Sal] that a 9 at.% Mn alloy was single phase (βTi) at 700 °C. Thermodynamic calculations support the judgment that the equilibrium phase boundary lies between the two data sets. Thermodynamic calculations also allow imposition of the constraint that the slope of the (βTi) transus must be consistent with the thermochemical properties of pure Ti.

Experimental data on the (αTi) solvus are given in Table 2. [53May], [62Bor], and [63Luz] found the solubility of Mn in (αTi) to be 0.4 ± 0.1 at.% at the eutectoid temperature, using metallography supplemented by hardness measurements [62Bor, 63Luz] and resistivity measurements [63Luz]. [63Cap] derived a solubility of 0.5 at.% from superconductivity data. The calculated maximum solubility of Mn in (αTi) is 0.45 at.%.

Near-Equiatomic Compounds αTiMn and βTiMn.

The two phases near the stoichiometry TiMn (αTiMn and βTiMn) were designated φ and ρ, respectively, by [62Wat]. The phase nomenclature is changed in the present assessment to indicate the approximate stoichiometry and relative stability. The high-temperature, higher Mn phase, βTiMn, is formed by a peritectic reaction at 1200 °C. The lower temperature phase, αTiMn, is formed by the peritectoid reaction (βTi) + βTiMn ⇌ TiMn.

There has been controversy about this region of the phase diagram, primarily because it was originally assumed that only one phase existed between the (Ti) solutions and TiMn₂. Thus, evidence for the peritectic reaction [53May] and evidence for a peritectoid reaction [53Ell, 53Ros] were viewed as contradictory. [54Mar] first suggested that there may be several distinct compounds in this region. Although reports of several proposed phases were due to misleading etching effects, there are two phases of near equiatomic composition—one stable only to about 950 °C, the other stable to at least 1150 °C and, therefore, probably to the melt [62Wat].

This assessment is based on [62Wat]. The peritectic reaction is placed at 1200 °C based on thermal analysis on cooling [53May]. The peritectoid reaction occurs between 930 and 960 °C [62Wat]. The compositions of βTiMn and αTiMn are 52 and 50.5 at.% Mn, respectively, based on the X-ray and microstructural work of [62Wat].

Laves Phase TiMn₂.

The Laves phase TiMn₂ was first identified as an equilibrium phase by [39Lav]. It melts congruently at 1325 °C [57Hel] and has a homogeneity range of 59 to 70 at.% Mn [58Mur, 62Wat]. Experimental data on the melting point and other invariant temperatures are given in Table 3.

The TiMn₂ liquidus was determined by thermal analysis [53May, 57Hel, 58Mur, 76Sve]. Experimental data on TiMn₂ are shown in Fig. 3. The data of [57Hel] are preferred because of the care taken to avoid contamination and the attention to the possibility of composition changes during melting.

The thermal analysis data of [53May] and [57Hel] are in good agreement. Data of [76Sve] cannot be quantitatively compared because of inaccuracy in reading data from the published graph. The discrepancy between data of [58Mur] and the assessed liquidus was probably caused by excessive Mn loss during melting.

The homogeneity range of TiMn₂ was examined by [58Mur] and [62Wat] using microstructural and X-ray diffraction methods. Both lattice parameters and microstructure indicated that the Mn-rich boundary of the TiMn₂ field lies at 69 ± 1 at.% Mn, essentially independent of temperature. On the Ti-rich side, reported values of the phase boundary of TiMn₂ are 55 at.% Mn [62Wat], 59 at.% Mn [58Mur], ~62 at.% Mn [53May], and 64 at.% Mn [76Sve]. The lattice parameter data are ambiguous concerning the solubility limit. The interpretation of [58Mur] is accepted here, which does not conflict with the microstructural work of [62Wat]. Lattice parameters of [58Mur] and [62Wat] also agree over most of the compo-

Table 2 Experimental Determinations of Mn Solubility in (αTi)

Reference	Composition, at.% Mn	Temperature, °C	Method/comment
[53May]	0.4	575	Metallography, cold rolled alloys
	0.4	675	
	0.2	800	
[62Bor]	0.3 to 0.7	550	Short anneals
[63Luz]	0.3 to 0.5	400	Resistivity, hardness, metallography
	0.4 to 0.6	530	
[63Cap]	0.54	690	Magnetic properties(a)

(a) [63Cap] compared the Curie constants of equilibrated two-phase alloys with those of supersaturated (αTi) solid solutions. The curves intersect at the (αTi) solvus.

Table 3 Experimental Data on Invariant Reactions

Reaction	Reference	Temperature, °C	Comment
L $\rightleftarrows$ (βTi) + βTiMn	[57Hel]	1181	Thermal analysis; heating, cooling
	[53May]	1175	Thermal analysis; cooling
(βTi) $\rightleftarrows$ (αTi) + αTiMn	[53Ell]	> 900	Metallography and X-ray diffraction
	[62Wat]	930 to 960	Metallography and X-ray diffraction
L + TiMn$_2$ $\rightleftarrows$ βTiMn	[53May]	~ 1200	Thermal analysis; cooling
L $\rightleftarrows$ TiMn$_2$	[57Hel]	1325	Thermal analysis; heating, cooling
	[53May]	1335	Thermal analysis; cooling
	[76Sve]	1310	Thermal analysis; cooling
	[60Sav]	1330	Thermal analysis; heating
	[58Mur]	1335	Thermal analysis; heating
L + TiMn$_2$ $\rightleftarrows$ TiMn$_3$	[57Hel]	1250	Thermal analysis; heating, cooling
	[60Sav]	1230	Thermal analysis; heating
L + TiMn$_3$ $\rightleftarrows$ TiMn$_4$	[57Hel]	1230	Thermal analysis; heating, cooling
L $\rightleftarrows$ TiMn$_4$ + (δMn)	[57Hel]	1204	Thermal analysis; heating, cooling
	[58Mur]	1185	Interpreted as L → TiMn$_2$
	[60Sav]	1195	Interpreted as L → TiMn$_4$
(δMn) + (αMn) $\rightleftarrows$ (βMn)	[57Hel]	~ 1145	Metallography; this reaction was not distinguished from (δMn) + TiMn$_4$ → (αMn)
	[61Wat]	~ 1160	Metallography, X-ray diffraction
(δMn) + TiMn$_4$ $\rightleftarrows$ (αMn)	[61Wat]	~ 1140	Metallography, X-ray diffraction

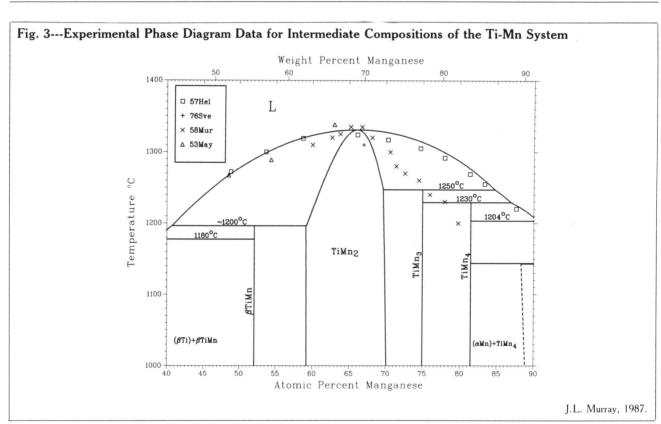

Fig. 3---Experimental Phase Diagram Data for Intermediate Compositions of the Ti-Mn System

J.L. Murray, 1987.

sition range. [58Mur], however, examined more alloys annealed at several temperatures, thus providing a somewhat stronger basis for determining the homogeneity range.

TiMn$_3$ and TiMn$_4$.

The existence of one or more intermetallic compounds between TiMn$_2$ and pure Mn was deduced from thermal

analysis results [57Hel]. Thermal arrests at 1250 and 1230 °C were interpreted as peritectic reactions involving unknown phases.

X-ray and metallographic work [61Wat] identified the two phases as TiMn$_3$ (75 at.% Mn) and TiMn$_4$ (about 81.5 at.% Mn). The reactions were interpreted as L + TiMn$_2$ $\rightleftarrows$ TiMn$_3$ and L + TiMn$_3$ $\rightleftarrows$ TiMn$_4$. At low temperature, (αMn) is in equilibrium with TiMn$_2$ [61Wat]. The esti-

Fig. 4---Experimental Data for the Mn-Rich Portion of the Phase Diagram

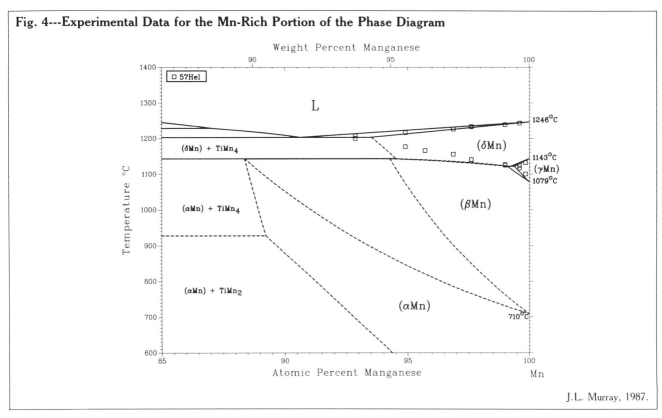

J.L. Murray, 1987.

mated temperatures of the solid-state reactions involving TiMn$_3$ and TiMn$_4$, 930 and 950 °C, are taken from [61Wat]. [58Mur] did not report TiMn$_3$ or TiMn$_4$, but their thermal analysis data suggest that one or more invariant reactions occur between 1325 and 1183 °C.

Mn-Rich Phases: (αMn), (βMn), (γMn), and (δMn).

The (δMn) liquidus and solidus were investigated by [57Hel] and [58Mur]. Thermal analysis data are shown in Fig. 4. Both authors showed the formation of (δMn) by a eutectic reaction.

The effect of Ti additions on the allotropic transformations (δMn) $\rightleftarrows$ (γMn) $\rightleftarrows$ (βMn) was investigated by [57Hel] and [60Sav], both using thermal analysis. There are substantial disagreements between the two investigations. The major disagreement concerns the stability of (γMn) relative to (δMn); [60Sav] showed (γMn) as strongly stabilized up to a peritectic reaction L + (δMn) $\rightleftarrows$ (γMn), whereas [57Hel] found clearer evidence for a eutectoid reaction (δMn) $\rightleftarrows$ (γMn) + (βMn). According to [57Hel], this reaction occurs at 1120 °C, and less than 1 at.% of Ti is soluble in (γMn). [57Hel] further indicated that the (δMn) and (βMn) phases extend to approximately 94 at.% Mn. The work of [57Hel] is preferred because of the care taken to maintain and to document sample purity.

On the extent of the (αMn) field, there is only limited information:

- A 95.5 at.% Mn alloy was two-phase, (βMn) + TiMn$_4$ at 1110 °C [57Hel].
- At 1000 °C, alloys containing 88.7 and 91.5 at.% Mn were two-phase (αMn) + (βMn) [58Mur].
- A 91 at.% Mn alloy was single-phase (αMn) at 1140 °C, and (αMn) is stable up to about 1150 °C [61Wat].
- [60Sav] reported, on the basis of thermal analysis, that the solubility of Ti in (αMn) is not more than

about 1 at.% and that (αMn) is stable only up to a peritectoid reaction (αMn) + TiMn$_4$ $\rightleftarrows$ (βMn) at 730 °C.

Based on the first three items, the reaction (αMn) + (δMn) $\rightleftarrows$ (βMn) probably occurs near 1150 °C. Based primarily on invariant temperatures, differences among Gibbs energies of the (Mn) solution phases were determined. The calculated phase boundaries lie within about 5 °C and 1 at.% of the input data and were used to draw the assessed diagram. The (αMn)/[(αMn) + TiMn$_2$] boundary is extrapolated to low temperature based on thermodynamic calculations. The boundaries in this composition and temperature region may require significant modifications as further experimental data on the extent of the (αMn) region become available.

Magnetic Properties

Pure (αMn) is antiferromagnetic with a Nèel temperature, T_N, of 95 K; T_N is lowered by the addition of Ti. [76Wil] used electrical resistivity measurements to determine the effect of Ti additions on T_N. T_N was taken to be the temperature where the derivative of the resistivity dp/dT is a minimum; experimental errors were estimated to be ± 2 °C. Using alloys with Ti contents of 0.7, 1.8, and 2.9 at.%, the change in T_N was found to be linear, with a slope of −10 °C per at.% Ti.

Metastable Phases

Metastable phases are formed from the (βTi) phase on cooling. In Ti-rich alloys, the cph phase (α'Ti) can form martensitically during quenching. At sufficiently high Mn content, metastable (βTi) phase can be retained as a metastable phase after quenching. The ω phase also

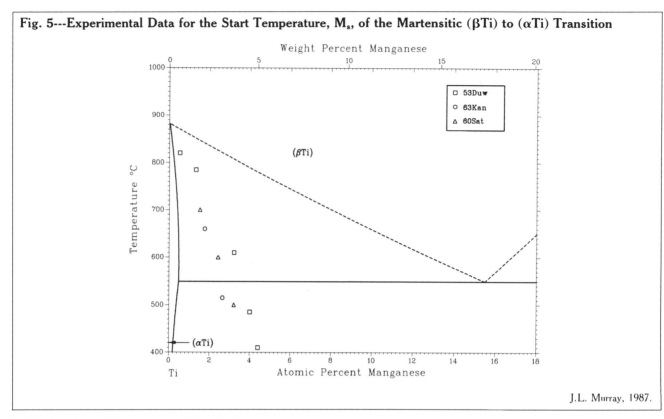

Fig. 5---Experimental Data for the Start Temperature, M_s, of the Martensitic (βTi) to (αTi) Transition

J.L. Murray, 1987.

Table 4 Crystal Structures of the Ti-Mn System

Phase	Homogeneity range, at.% Mn	Pearson symbol	Space group	Struktur-bericht designation	Prototype	Reference
(αTi)	0 to 0.4	$hP2$	$P6_3/mmc$	$A3$	Mg	[Pearson2]
(βTi)	0 to 30	$cI2$	$Im3m$	$A2$	W	[Pearson2]
βTiMn	52	(a)	...	...	...	...
αTiMn	50.5	t^*58	...	...	...	[62Wat]
TiMn$_2$	60 to 70	$hP12$	$P6_3/mmc$	$C14$	MgZn$_2$	[39Lav, 58Mur, 62Wat]
TiMn$_3$	75	(b)	...	...	...	[61Wat]
TiMn$_4$	81.5	$hR53$	$R\bar{3}m$	...	~ δ (Mo,Ni)	[61Wat, 62Wat]
(αMn)	88 to 100	$cI58$	$I\bar{4}3m$	$A12$	αMn	[Pearson2]
(δMn)	91 to 100	$cI2$	$Im3m$	$A2$	W	[Pearson2]
(βMn)	94 to 100	$cP20$	$P4_132$	$A13$	βMn	[Pearson2]
(γMn)	99.4 to 100	$cF4$	$Fm3m$	$A1$	Cu	[Pearson2]
(α'Ti)	(c)	$hP2$	$P6_3/mmc$	$A3$	Mg	[53Duw]
ω	(c)	$hP3$	$P6/mmm$	...	ωMnTi	[68Miy]

(a) Undetermined. (b) Orthorhombic. (c) Metastable.

appears as an intermediate phase in the decomposition of (βTi) into equilibrium (αTi) + (βTi).

Martensite Formation.

In some Ti systems, depending on alloy composition, either a cph (α'Ti), or an orthorhombic (α''Ti) structure forms martensitically during quenching of dilute (βTi) alloys. In Ti-Mn, only the (α'Ti) martensite has been reported.

[53Duw], [60Sat], and [70Hua] measured start temperatures, M_s, for the martensite transformation; the data are shown in Fig. 5. The values of M_s found by [60Sat] and [63Kan] are lower than those of [53Duw]; this is consistently true for several Ti systems studied by these investigators. According to [53Duw] and [62Age], (βTi) is

retained to room temperature when the Mn content exceeds about 4 to 5 at.%.

Metastable ω Phase.

In Ti-Mn alloys, ω phase can form either during quenching from the (βTi) region (athermal or as-quenched ω), or during aging of metastable (βTi) alloys below approximately 400 °C (isothermal or aged ω) [54Fro, 69Hic]. [62Age] reported athermal ω phase between 2.6 and 4.8 at.% Mn.

When ω phase forms during aging, the compositions of ω and (βTi) approach a metastable equilibrium at 5.1 and 19.1 at.% Mn; after longer aging, equilibrium (αTi) appears [69Hic]. Aging above 420 °C [60Dya] produces

Fig. 6---Lattice Parameters of TiMn₂

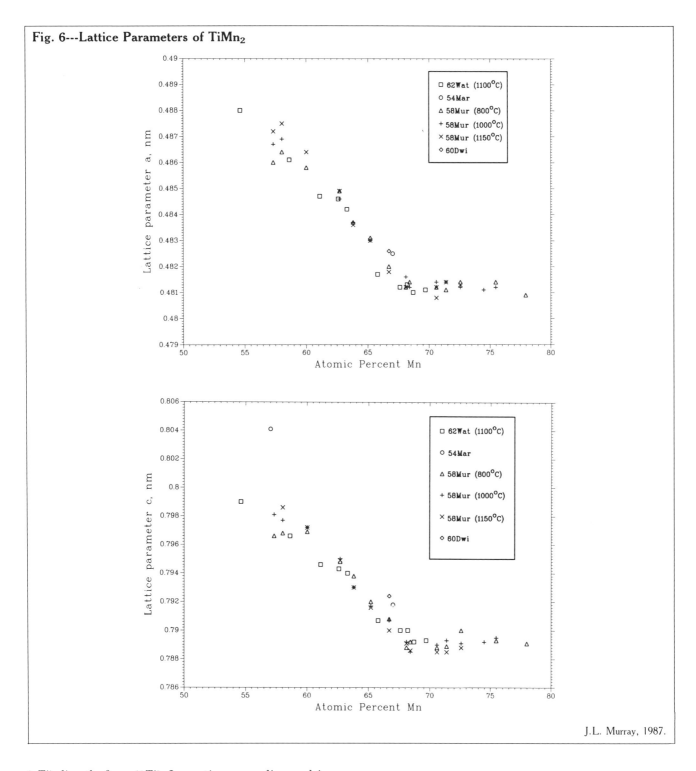

J.L. Murray, 1987.

(αTi) directly from (βTi). In continuous cooling and isothermal aging experiments, [60Dya], [68Dya], [71Miy], and [72Ika] found ω phase to precipitate at approximately 370 and 420 °C, respectively. On heating, the transformation (βTi) + ω → (βTi) + (αTi) occurs at 450 to 500 °C [60Dya, 68Dya, 72Ika].

ω phase is coherent with the bcc matrix. The ω precipitates have a cubic morphology, as in other alloy systems with relatively large misfit between the ω precipitate and the bcc matrix [69Wil]. [75Apa] studied short-range ω-like order in 13 and 24 at.% Mn alloys.

Crystal Structures and Lattice Parameters

The crystal structures of the equilibrium and metastable phases of the Ti-Mn system are given in Table 4. TiMn₂ is the only intermetallic phase whose structure is definitively established [39Lav, 58Mur, 62Wat]. That of βTiMn has not been examined at all. Structural information on αTiMn, TiMn₃, and TiMn₄ is due to [61Wat, 62Wat]. The structure of TiMn₄ was compared to the ternary rhombohedral phase of the Mo-Cr-Co system, that of TiMn₃ to a δ(Mo,Ni) phase [61Wat]. TiMn₃ was tenta-

165

Table 5 Lattice Parameters of Ti-Mn Phases

Phase	Reference	Composition, at.% Mn	Lattice parameters, nm	
			a	c
(βTi)	[53May]	5.4	0.3242	...
		6.1	0.3244	...
		8.9	0.3232	...
		11.6	0.3212	...
		13.3	0.3203	...
		18.0	0.3200	...
(βMn)	[60Sav]	98.9	0.632	...
		95.8	0.622	...
(αMn)	[60Sav]	98.9	0.8922	...
		95.8	0.8868	...
		92.6	0.8944	..
αTiMn	[62Wat]	50.5	0.819	1.281
TiMn$_4$	[62Wat]	81.5	1.1003	1.9446

tively assigned an orthorhombic unit cell, but a tetragonal cell was also considered. The powder pattern of βTiMn corresponds to that reported by [54Mar], but no further structural analysis has been made. αTiMn was first reported to be a σ phase [53Ell, 53Ros], but [62Wat] argued that a large tetragonal cell and structure corresponding to a φ phase of the Hf-Re system allows the diffraction pattern to be more satisfactorily indexed.

The structure of ω phase is essentially one of hexagonal symmetry. However, in some alloy systems, the structure of ω phase is intermediate between the "true" ω structure and bcc, in which case the symmetry is trigonal. For Ti-Mn alloys, [68Miy] determined that ω phase has hexagonal symmetry.

Lattice parameters of TiMn$_2$ quenched from various equilibration temperatures are shown as a function of composition in Fig. 6. The lattice parameters of other phases of the system are listed in Table 5. The lattice parameters of (αTi) are not changed by small additions of Mn [53May].

Thermodynamics

Experimental high-temperature thermodynamic data are not available for Ti-Mn alloys. The present calculations were undertaken as an aid in assessing the phase diagram. The starting point of the calculation was a set of theoretical estimates of excess quantities made by [78Kau]. Because no direct experimental thermodynamic data are available, only Gibbs energy differences are determined with any accuracy. Assuming that there is no seriously anomalous temperature or composition dependence of the Gibbs energies, the calculation of the phase diagram can be used to assess those features of the phase diagram that depend only on properties of the pure elements or excess Gibbs energy differences. Thermodynamic parameters used in this evaluation are listed in Table 6, with the parameters of [78Kau].

The Gibbs energies of the solution phases are represented as:

$$G(i) = G^0(\text{Ti},i)(1-x) + (\text{Mn},i)x + RT[x \ln x + (1-x)\ln(1-x)] + B(i)x(1-x) + C(i)x(1-x)(1-2x)$$

where i designates the phase; x is the atomic fraction of Mn; G^0 is the Gibbs energies of the pure components; and $B(i)$ and $C(i)$ are parameters in the expansion of the excess quantities. Pure element Gibbs energies are taken from

[70Kau] and were adjusted slightly to match the transformation temperatures of pure Mn. The calculated transitions L → δMn → γMn → βMn → αMn occur at 1244, 1136, 1100, and 720 °C, respectively. These temperatures are updated to conform with more recent tables in the assessed diagram, but not in the thermodynamic functions.

For TiMn$_2$, a Wagner-Schottky Gibbs energy function was used:

$$G(i) = G^0(i) + RT[x_{\text{Ti}}\ln x_{\text{Ti}} + x_{\text{Mn}}\ln x_{\text{Mn}} + v\ln v + s\ln s - (1+v)\ln(1+v)] + C(i)s + D(i)v$$

The Wagner-Schottky compound is conceptually resolved into Ti and Mn sublattices; v is the concentration of vacancies on the Ti sublattice; s is the concentration of substitutional Ti atoms on the Mn sublattice; x_{Ti} and x_{Mn} are the concentrations of Ti and Mn on their respective sublattices; and x is the overall composition. All concentrations are referred to the total number of atoms. The concentrations of vacancies and substitutions are determined by minimizing the Wagner-Schottky Gibbs energy with respect to the variables s and v. $G^0(i)$ can be thought of roughly as the Gibbs energy of a corresponding line compound.

The other compounds of the system were assumed to be line compounds, with compositions as listed in Table 6. All Gibbs energies are referenced to the same standard state, the liquid.

Initial values of the thermodynamic parameters were taken from [78Kau]; for the several compounds not considered in [78Kau], rough estimates were made. Experimental data used as input to the optimization calculations were the liquidus [57Hel, 53May], the (βTi) boundaries [53May], the homogeneity range of TiMn$_2$ [58Mur], and the temperatures and compositions of three-phase reactions. The latter were weighted most heavily in optimizations. Three-phase equilibria could be reproduced within 3 °C of the experimental or estimated values.

The first notable feature of the phase diagram calculation is the discrepancy between the calculated (βTi) eutectoid composition and the data of [53May] (Fig. 2). The calculation tends to support our judgment that the equilibrium boundary lies between the data of [51Mcq] and those of [53May]. However, experimental verification is needed for the (βTi) transus and eutectoid temperature, using high-purity alloys and a rapid quench from the equilibration temperature.

Table 6 Thermodynamic Properties of the Ti-Mn System

Properties of the pure elements

$G^0(\text{Ti,L}) = 0$ $G^0(\text{Ti,bcc}) = -16\,234 + 8.368\,T$

$G^0(\text{Mn,L}) = 0$ $G^0(\text{Mn,bcc}) = -14\,644 + 9.653\,T$

$G^0(\text{Ti,cph}) = -20\,585 + 12.134\,T$ $G^0(\text{Ti},\alpha\text{Mn}) = -15\,983 + 12.134\,T$

$G^0(\text{Mn,cph}) = -9\,205 + 7.113\,T$ $G^0(\text{Mn},\alpha\text{Mn}) = -20\,878 + 14.790\,T$

$G^0(\text{Ti},\beta\text{Mn}) = -13\,054 + 12.134\,T$ $G^0(\text{Ti,fcc}) = -17\,238 + 12.134\,T$

$G^0(\text{Mn},\beta\text{Mn}) = -18\,619 + 12.515\,T$ $G^0(\text{Mn,fcc}) = -16\,401 + 10.900\,T$

Properties of solution phases

	Present evaluation	[78Kau]
$B(\text{L})$	$-12\,302$	$-12\,552$
$C(\text{L})$	$-5\,250$	0
$B(\text{bcc})$	-467	0
$C(\text{bcc})$	$-2\,122$	0
$B(\text{cph})$	$40\,000$	$23\,012$
$B(\alpha\text{Mn})$	$-3\,000 - 10\,T$	$-1\,569 - 9.205\,T$
$B(\beta\text{Mn})$	$-9\,000$	$-10\,042$
$B(\text{fcc})$	0	$-2\,720$

Intermetallic phases

αTiMn	$-15\,464 + 5.494\,T$	$-19\,309 + 12.432\,T$
βTiMn	$-23\,724 + 5.271\,T$	
TiMn_4	$-20\,705 + 8.798\,T$	
TiMn_3	$-22\,525 + 8.675\,T$	
TiMn_2	$G^0(\text{TiMn}_2) = -26\,522 + 10.234\,T$	$-19\,872 + 15.918\,T$
	$C(\text{TiMn}_2) = 42\,969 - 9.697\,T$	
	$D(\text{TiMn}_2) = 52\,308$	

Note: Values are given in J/mol; T in K.

A second noteworthy feature is the equilibria among TiMn_2, αTiMn, and βTiMn. A three-phase equilibrium occurs among αTiMn, βTiMn, and TiMn_2, but is ignored in the calculations because it reflects very small Gibbs energy differences between αTiMn and βTiMn, which cannot be determined from presently available phase diagram data.

Finally, note the slight discrepancy between the calculation and the experimental data on the $(\beta\text{Mn})/(\beta\text{Mn}) + \text{TiMn}_4$ boundary (Fig. 4). Again, the three-phase equilibria near 1145 °C are known only roughly and the calculation appears to present the best compromise among various invariant reactions. In this composition region, very small changes in Gibbs energies can alter the topology of the phase diagram.

In summary, the Gibbs energy functions listed in Table 6 are thermodynamically plausible and reproduce the phase diagram within the present estimate of the uncertainties. A definitive calculation will require experimental determination of heats of mixing and partial Gibbs energies of the liquid phase and heats of formation of the intermetallic compounds.

Cited References

39Lav: F. Laves and H.J. Wallbaum, "Crystal Chemistry of Titanium Alloys," *Naturwissenschaften, 27,* 674-675 (1939) in German. (Crys Structure; Experimental)

51Mcq: A.D. McQuillan, "The Effect of the Elements of the First Long Period on the α-β Transformation in Ti," *J. Inst. Met., 80,* 363-368 (1951). (Equi Diagram; Experimental)

53Duw: P. Duwez, "The Martensite Transformation Temperature in Titanium Binary Alloys," *Trans. ASM, 45,* 934-940 (1953). (Meta Phases; Experimental)

53Ell: R.P. Elliott and W. Rostoker, "Structure of the Phase TiMn and the Indexing of Powder Patterns of σ-Type Phases," *Trans. AIME, 197,* 1203-1204 (1953). (Crys Structure; Experimental)

53May: D.J. Maykuth, H.R. Ogden, and R.I. Jaffee, "Titanium-Manganese System," *Trans. AIME, 197,* 225-231 (1953). (Equi Diagram; Experimental)

53Ros: W. Rostoker, R.P. Elliot, and D.J. McPherson, "Comments on Titanium-Manganese System," *Trans. AIME, 197,* 1566-1568 (1953). (Crys Diagram; Experimental)

54Fro: P.D. Frost, W.M. Parris, L.L. Hirsch, J.R. Doig, and C.M. Schwartz, "Isothermal Transformation of Titanium-Manganese Alloys," *Trans. ASM, 46,* 1056-1074 (1954). (Equi Diagram, Meta Phases; Experimental)

54Mar: H. Margolin and E. Ence, "Titanium-Manganese Phases," *Trans. AIME, 200,* 1268-1269 (1954). (Equi Diagram, Crys Structure; Experimental)

*57Hel: A. Hellawell and W. Hume-Rothery, "The Constitution of Alloys of Iron and Manganese with Transition Elements of the First Long Period," *Phil. Trans. Roy. Soc., London, 249,* 417-454 (1957). (Equi Diagram; Experimental)

58Mur: Y. Murakami and T. Enjyo, "On the TiMn_2-Mn Range of the Binary Ti-Mn System," *J. Jpn. Inst. Met., 22,* 261-265 (1958) in Japanese. (Equi Diagram; Experimental)

60Dwi: A.E. Dwight, "Laves-Type Phases," AEC Report ANL-330 156-158 (1960). (Crys Structures; Experimental)

60Dya: M.A. D'Yakaova and I.N. Bogachev, "Decomposition of the β Solid Solution in the Alloy Titanium-Manganese," *Fiz. Met. Metalloved., 10,* 896-902 (1960) in Russian; TR: *Phys. Met. Metallogr., 10*(6), 103-108 (1960). (Meta Phases; Experimental)

60Sat: T. Sato, S. Hukai, and Y.C. Huang, "The M_s Points of Binary Titanium Alloys," *J. Aust. Inst. Met., 5*(2), 149-153 (1960). (Meta Phases; Experimental)

60Sav: E.M. Savitskii and Ch.V. Kopetskii, "Phase Diagram of the Manganese-Titanium and Manganese-Zirconium Systems," *Zh. Neorg. Khim., 5,* 2422-2434 (1960); TR: *Russ. J.*

Inorg. Chem., 5, 1173-1179 (1960). (Equi Diagram; Experimental)

*61Wat: R.M. Waterstrat, "Identification of Intermediate Phases in the Manganese-Titanium System," *Trans. AIME, 221,* 687-690 (1961). (Equi Diagram, Crys Structure; Experimental)

62Age: N.V. Ageev and Z.M. Smirnova, "Stability of the β-Phase in Titanium-Manganese Alloys," *Titanium and Its Alloys, Akad. Nauk SSSR, 1,* 17-24 (1962). (Meta Phases; Experimental)

62Bor: N.G. Boriskina and K.P. Myasnikova, "Solubility of Iron, Manganese, and Copper in α-Titanium," *Titanium and its Alloys, Akad. Nauk SSSR, 7,* 61-67 (1962). (Equi Diagram; Experimental)

*62Wat: R. Waterstrat, B.N. Das, and P.A. Beck, "Phase Relationships in the Titanium-Manganese System," *Trans. AIME, 224,* 512-518 (1962). (Equi Diagram; Experimental)

63Cap: J.A. Cape, "Superconductivity and Localized Magnetic States in Ti-Mn Alloys," *Phys. Rev., 132*(4), 1486-1492 (1963). (Equi Diagram; Experimental)

63Kan: H. Kaneko and Y.C. Huang, "Continuous Cooling Transformation Characteristics of Titanium Alloys of Eutectoidal Type (I)," *J. Jpn. Inst. Met.,* 27, 1393-1397 (1963). (Equi Diagram, Meta Phases; Experimental)

63Luz: L.P. Luzhnikov, V.M. Novikova, and A.P. Mareev, "Solubility of β-Stabilizers in α-Titanium," *Metalloved. Term. Obrab. Met.,* (2), 13-16 (1963) in Russian; TR: *Met. Sci. Heat Treat.,* (2), 78-81 (1963). (Equi Diagram; Experimental)

68Dya: M.A. D'Yakova and I.N. Bogachev, "Phase Transformations in Titanium Alloys Under Nonequilibrium Conditions," NASA Tech. Trans., F-596 153-159 (1968). (Meta Phases; Experimental)

68Miy: M. Miyagi, M. Morikawa, and S. Shin, "On the Crystal Structure of ω Phase in Ti-Mn and Ti-Mo Alloys," *J. Jpn. Inst. Met.,* 32, 756-761 (1968) in Japanese. (Meta Phases; Experimental)

69Hic: B.S. Hickman, "ω Phase Precipitation in Alloys of Titanium with Transition Metals," *Trans. AIME, 245,* 1329-1336 (1968). (Meta Phases; Experimental)

69Wil: J.C. Williams and M.J. Blackburn, "The Influence of Misfit on the Morphology and Stability of the ω Phase in Titanium-Transition Metal Alloys," *Trans. AIME, 245,* 2352-2355 (1969). (Meta Phases; Experimental)

70Hua: Y.C. Huang, S. Suzuki, H. Kaneko, and T. Sato, "Thermodynamics of the M_s Points of Binary Titanium Alloys," Sci. Technol. Appl. Titanium, Proc. Int. Conf., R. I. Jaffee, Ed., 691-698 (1970). (Meta Phases; Experimental)

70Kau: L. Kaufman and H. Bernstein, *Computer Calculation of Phase Diagrams,* Academic Press, New York (1970). (Equi Diagram, Thermo; Theory)

71Miy: M. Miyagi and S. Shin, "Isothermal Transformation Characteristics of Metastable β-type Titanium Alloys," *J. Jpn. Inst. Met.,* 35, 716-722 (1971) in Japanese. (Meta Phases; Experimental)

72Ika: H. Ikawa, S. Shin, and M. Masaki, "Continuous Cooling Transformation in Binary α + β and Metastable β Titanium Alloys," *Yosetsu Gakkai-Shi, 41*(4), 394 (1972) in Japanese. (Meta Phases; Experimental)

75Apa: N.N. Aparov, L.V. Lyasotskiy, and Yu.D. Tyapkin, "On the Structure of Metastable Titanium-Manganese Solid Solutions," *Fiz. Metal. Metalloved., 40,* 1107-1110 (1975); TR: *Phys. Met. Metallogr., 40,* 187-190 (1975). (Meta Phases; Experimental)

76Sve: V.V. Svechnikov and V.V. Petkov, "Laves Phases in Alloys of Mn with Transition Metals of Groups IVA-VA," *Izv. Akad. Nauk Ukr SSR, Metallofiz., 64,* 24-29 (1976). (Equi Diagram; Experimental)

76Wil: W. Williams and J.L. Stranform, "Antiferromagnetism of the α-Mn System," *J. Magn. Magn. Mater., 1,* 271-285 (1976). (Equi Diagram; Experimental)

77Mar: H. Margolin, E. Levine, and M. Young, "The Interface Phase in α-β Titanium Alloys," *Metall. Trans. A, 8,* 373-377 (1977). (Meta Phases; Experimental)

78Kau: L. Kaufman, "Coupled Phase Diagrams and Thermochemical Data for Transition Metal Binary Systems-III," *Calphad, 2*(2), 117-146 (1978). (Thermo; Theory)

79Sal: Y. Saleh and H. Margolin, "Bauschinger Effect During Cyclic Straining of Two Ductile Phase Alloys," *Acta Metall., 27,* 535-544 (1979). (Equi Diagram; Experimental)

The Mo-Ti (Molybdenum-Titanium) System

95.94   47.88

By J.L. Murray

Equilibrium Diagram

The equilibrium solid phases of the Ti-Mo system are: (1) the bcc (βTi,Mo) solid solution, in which Ti and Mo are completely miscible above the transformation temperature of pure Ti (882 °C); and (2) the cph (αTi) solid solution with restricted solubility of Mo.

There have been two conflicting descriptions of the phase stability of (βTi,Mo). [51Duw], [51Han], [70Ron], and [72Ron] reported that the temperature of the (βTi) transus decreases monotonically with Mo content and discovered no evidence for a miscibility gap in (βTi,Mo). [77Ter] and [78Ter], however, found evidence for a monotectoid reaction (βTi) ⇄ (αTi) + (Mo). The critical composition and temperature of the miscibility gap were reported as 20 at.% Mo and 795 °C; the temperature of the monotectoid reaction was reported as 675 °C. The experimental evidence for the miscibility gap was consistent results of resistivity, X-ray, metallographic, and TEM studies. Differential thermal analysis results from this laboratory support the monotectoid reaction, but at a temperature of about 695 °C [84Mca].

The monotectoid reaction is incorporated in the assessed diagram (Fig. 1 and Table 1). Because there is considerable scatter in the experimental phase boundary data, nonlinear least-squares optimization of Gibbs energies with respect to select data was relied upon as a tool in the data assessment. Several phase boundaries of the assessed diagram are the result of thermodynamic calculations to be described below.

Solidus and Liquidus.

The melting temperature of pure Mo is 2623 °C [Melt]. The solidus was investigated by [51Han] and [69Rud]. [69Rud] used the Pirani technique to determine incipient melting temperatures and estimated the liquidus to lie slightly above the temperature at which samples collapsed. [51Han] determined temperatures of incipient melting by metallographic examination of quenched specimens, with an uncertainty of ±25 °C. The data of [51Han] agree with those of [69Rud], except at Ti-rich compositions. [51Han] gave 1720 °C for the melting point of pure Ti, in disagreement with the presently accepted value of 1670 °C. Data are shown in Fig. 2.

The solidus data of [51Han] and [69Rud] are accurately fitted by thermochemical calculations. Because the temperature of sample collapse cannot be used as a quantitative measure of the liquidus, the present calculation is used to predict the liquidus.

Solubility of Mo in (αTi).

The maximum solubility of Mo in (αTi) is approximately 0.4 at.% Mo at ~695 °C [51Han, 74Flo]. The heat treatments used by [51Han] were chosen to ensure that attainment of equilibrium; a 0.375 at.% Mo alloy was very close to the (αTi) solvus. In a study of transformations as

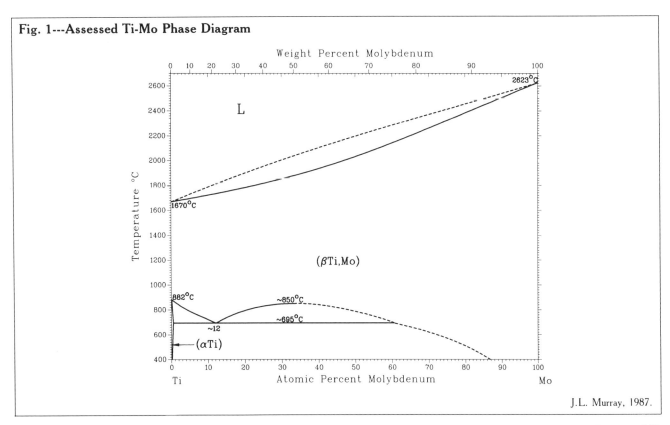

Fig. 1---Assessed Ti-Mo Phase Diagram

J.L. Murray, 1987.

Table 1 Special Points of the Assessed Ti-Mo Phase Diagram

Reaction	Compositions of the respective phases, at.% Mo			Temperature, °C	Reaction type
(βTi) ⇄ (αTi) + (Mo)	12	0.4	~ 60	~ 695	Monotectoid
(βTi,Mo) ⇄ (βTi) + (Mo)		~ 33		~ 850	Critical point
L ⇄ (βTi))		0		1670	Melting point
L ⇄ (Mo)		100		2623	Melting point
(βTi) ⇄ (αTi)		0		882	Allotropic transformation

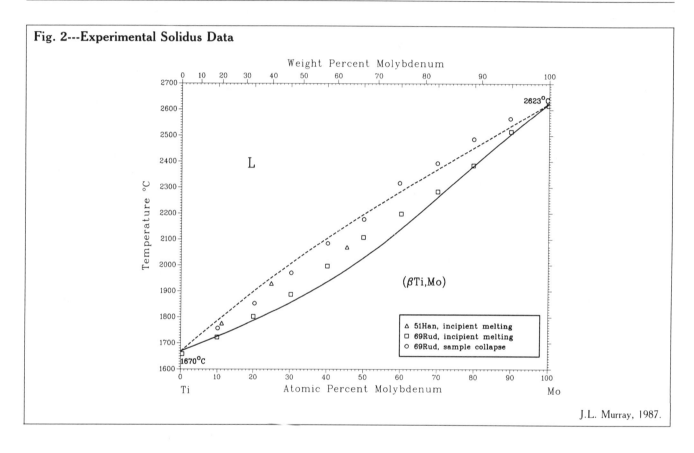

Fig. 2---Experimental Solidus Data

J.L. Murray, 1987.

a function of heat treatment and quench method, [74Flo] found a 0.25 at.% Mo specimen to be within the two-phase field at 750 °C. By X-ray microanalysis, [70Ron] determined the composition of (αTi) in equilibrium with (βTi) in annealed diffusion couples as 0.5 at.% Mo. Solubility data reported by [63Luz] are not used, because they chose heat treatments to correspond to those used in industry, and they did not necessarily produce equilibrium.

(βTi) Transus and Miscibility Gap.

Data on the (βTi) transus are shown in Fig. 3. Determinations were made by the following experimental techniques:

References	Technique
[51Duw, 51Han, 64Wil]	Metallography
[61Bun, 77Ter, 78Ter]	Electrical resistivity
[70Ron, 72Ron]	X-ray microanalysis of annealed diffusion couples

The effect of oxygen contamination is to raise the apparent temperature of the transus. Therefore, the high transus temperatures observed by [51Duw], for example, are not used to establish the boundary.

The work of [76Ter], [77Ter], and [78Ter] established the miscibility gap and monotectoid reaction in the equilibrium diagram. The results of metallography, TEM, electrical resistivity, and lattice parameter measurements were in agreement, not only about the existence of the two bcc phases, but also about the positions of the phase boundaries. [77Ter] reported the monotectoid reaction at 675 °C based on electrical resistivity data. Recent differential thermal analysis results [84Mca] indicate that the reaction occurs at somewhat higher temperature, about 695 °C. Because of the greater purity of the samples [84Mca], this value is accepted for the equilibrium diagram. The monotectoid composition is placed at 12 at.% Mo.

[77Ter, 78Ter] also estimated points on the Mo-rich side of the miscibility gap, based on metallographic work. These values are implausible from a thermodynamic

Fig. 3---Experimental Data on the (βTi,Mo)/(αTi) Phase Equilibria

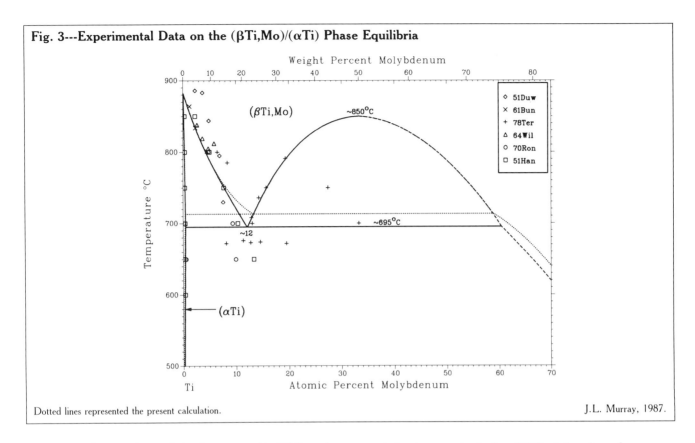

Dotted lines represented the present calculation.

J.L. Murray, 1987.

standpoint, but can be explained as a result of failure to reach equilibrium. Also, the boundary at which coherent precipitates begin to appear may differ substantially from the equilibrium (incoherent) phase boundary.

The miscibility gap has been calculated by optimization of the bcc excess Gibbs energy with respect to miscibility gap data on the Ti-rich side, and the calculated miscibility gap is used to draw the assessed diagram. The (βTi) transus data reported by [51Han] and [70Ron] below the monotectoid temperature are tentatively interpreted as observations of the metastable extension of this phase boundary, which could, in principle, be observed for compositions not lying within the spinodal.

Clustering in Retained bcc Alloys.

There has been controversy about whether or not (βTi) retained during quenching exhibits a tendency toward short-range order [61Dup, 73Mor] or toward clustering [72Cha].

Short-range order was found by [61Dup] in five alloys and by [73Mor] in an 8 at.% Mo alloy, using diffuse X-ray scattering. [61Dup] reported the short-range order parameter as constant with concentration, rather than with a maximum at some stoichiometry, as expected. Anomalies were also found in the interatomic spacings. The results of diffuse X-ray scattering experiments conflict with more direct experiments, and therefore, the existence of short-range order is not accepted in this evaluation.

The evidence for clustering comes from thermodynamic measurements (see below) and phase equilibrium studies. [72Cha] used X-ray and metallographic studies to examine an 11 at.% Mo sample heat treated above the temperature of ω phase formation (~550 °C), and they found two bcc phases. They proposed that the phase separation was the effect of a metastable miscibility gap

and short-range clustering. [70Kou] found no phase separation after prolonged aging of a 15 at.% Mo alloy at 350 °C.

Metastable Phases

Metastable phases are formed from the (βTi,Mo) solution retained during quenching. In Ti-rich alloys, the cph phase (α'Ti) can form martensitically during quenching; in alloys of slightly higher Mo content, an orthorhombic distortion of the cph structure (α''Ti) is formed. The ω phase is formed as an intermediate phase in the decomposition of metastable (βTi,Mo) to equilibrium (αTi). At successively higher Mo concentrations, (α'Ti), (α'''Ti), (αTi'') + (βTi), (βTi) + (ω), and (βTi) are found after quenching from the single phase (βTi,Mo) region. A summary of reported composition ranges of the various metastable phases is given in Table 2. Discrepancies are not large and can be attributed to differences in oxygen contents and different sensitivity of the experimental probes to the presence of fine precipitates. The effect of oxygen impurities is to widen slightly the stability range of (α'Ti) and (α''Ti) [75Age].

Martensitic Transformations.

The start temperature, M_s, for the martensitic transformation was studied as a function of concentration by [51Duw], [52Del], [60Sat], [61Bun], and [73Fed] (see also [70Hua]). Data are shown in Fig. 4. [51Duw] and [60Sat] used thermal analysis, [52Del] used a metallographic technique, and [73Fed] used Young's modulus and other elastic property measurements. The M_s values of [51Duw] and [61Bun] agree and are considerably higher than the M_s temperatures detected by [52Del] and [73Fed]. From a

Table 2 Observed Composition Ranges of Metastable Phases

Reference	Composition, at.% Mo		
	(α'Ti)	(α"Ti)	ω
[58Bag]	0 to 2	2 to 4	4 to ?
[70Col]	...	...	~ 6.5 to ~ 14
[72Fed]	0 to 2.5	2.5 to 4.5	4.5 to ?
[73Mor]	...	4.7 to 6.9	6.4 to 11
[74Gus]	0 to 2	2 to 6	5 to 10
[59Age, 75Age] ...	0 to 2	3.1 to 6	6 to 10
[75Kol]	0 to 2	2 to 4.5	4.3 to 6
[79Dav]	0 to 2	2 to 5.3	...
[80Lei2]	0 to 3	2 to 5	5 to ~ 14

TEM study of martensite structures in dilute alloys, [74Flo] concluded that M_s is greater than 750 °C in 0.5 at.% Mo, in agreement with the results of [51Duw] and [61Bun]. In alloys containing more than 5 to 6 at.% Mo, the bcc phase, or (βTi) + ω, is retained completely during quenching, that is, the M_s temperature falls below room temperature [52Del, 51Duw].

Martensitic (α'Ti) forms between 0 and 2 at.% Mo. The morphology of the observed product in the most dilute alloys is referred to as "packet," "lath," or "colony". The transition from the colony microstructure to an acicular microstructure takes place over the composition range 0.25 to 2.0 at.% Mo [58Bag, 79Dav].

In alloys containing 2 to 4 at.% Mo, the (α"Ti) martensite has an orthorhombic structure. Martensitic (α"Ti) can be produced by deformation of metastable (βTi) or (βTi) + ω alloys containing 5 to 8 at.% Mo [63Woo, 68Bla, 76Whi].

ω Phase.

The ω phase can form either during quenching from the (βTi) region (athermally), or after low-temperature aging of quenched alloys. Data on the composition range in which athermal ω can be found are summarized in Table 2. ω phase formation starts at approximately 4 at.% Mo, but there is no good agreement about the maximum Mo content at which ω is formed. [69Hic] found no ω phase in as-quenched alloys containing between 8 and 14 at.% Mo. [72Flo] found no evidence of ω phase in the diffraction patterns of an as-quenched 11 at.% Mo alloy, but only diffuse streaking indicative of ω-like defects.

The aged form of ω occurs in alloys of composition between approximately 4 and 15 at.% Mo [69Hic, 73Col, 75Bag]. In alloys of higher Mo content, ω is discernible as diffuse streaking in the diffraction patterns. The optimal aging temperature is approximately 350 °C. If an aging temperature of 500 °C is used, equilibrium (αTi) is formed directly from (βTi) [69Hic]. During aging, the compositions of ω and (βTi) tend toward saturation values, and then equilibrium (αTi) begins to precipitate. The saturation composition of ω was reported as: 4 to 4.5 at.% Mo [69Hic], ~4 at.% Mo [73Col], and 3.4 at.% Mo [74Gys]. (βTi) tends toward compositions of 14 to 16 at.% Mo [69Hic].

The ω phase can also be found in dilute Ti-Mo (3 at.% Mo) alloys after quenching from the (βTi) region and high-pressure soaking at room temperature [79Lei, 80Lei1, 80Lei2].

Crystal Structures and Lattice Parameters

Crystal structure data are summarized in Table 3, and lattice parameter data are given in Fig. 5 and Table 4.

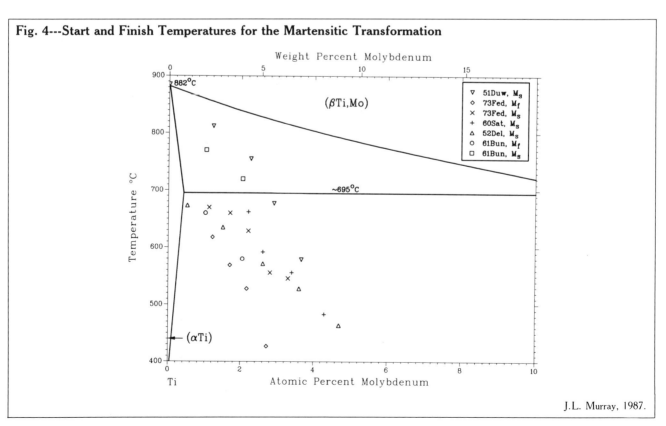

Fig. 4---Start and Finish Temperatures for the Martensitic Transformation

J.L. Murray, 1987.

Mo-Ti System

Table 3 Crystal Structures and Lattice Parameters in the Ti-Mo System

Phase	Homogeneity range, at.% Mo	Pearson symbol	Space group	Struktur- bericht desigation	Proto- type	Reference
α	0 to 0.4	hP2	P6₃/mmc	A3	Mg	[Pearson2]
β	0 to 100	cI2	Im3m	A2	W	[Pearson2]
α′	(a)	hP2	P6₃/mmc	A3	Mg	[Pearson2]
α″	(a)	oC4	Cmcm	A20	αU	[79Dav]
ω	(a)	hP3	P6/mmm	...	ωMnTi	[80Lei]

(a) Metastable.

Table 4 Lattice Parameters of Metastable Phases

| Reference | Phase | Composition, at.% Mo | Lattice parameters, nm | | |
			a	b/√3	c
[79Dav]	(α′Ti)	1.0	0.2945	...	0.4681
	(α′Ti)	2.0	0.2943	...	0.4676
	(α″Ti)	3.1	0.2965	0.2913	0.4662
	(α″Ti)	4.2	0.2994	0.2881	0.4644
[70Wil]	(α″Ti)	4.0	0.3001	0.2886	0.4657
[80Lei2](a)	(α′Ti)	3	0.2954	...	0.4650
	(α″Ti)	3	0.3007	0.2899	0.4650
	ω	~ 3	0.4612	...	0.2821
		~ 5	0.4602	...	0.2820
		~ 11.5	0.4591	...	0.2815
		~ 15	0.4587	...	0.2808
[75Age]	(α″Ti)	3.1	0.2980	0.2880	0.4649

(a) High-pressure data.

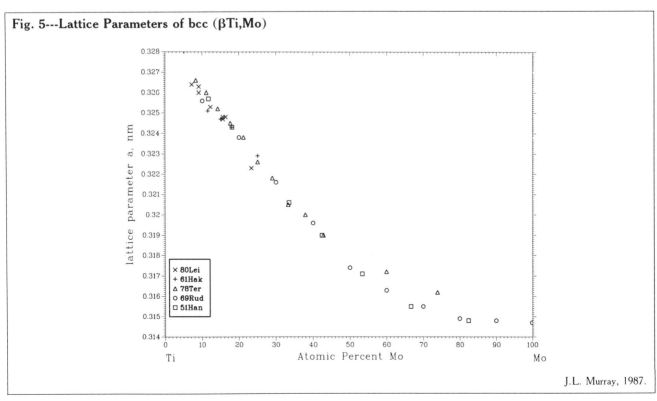

Fig. 5---Lattice Parameters of bcc (βTi,Mo)

J.L. Murray, 1987.

Table 5 Thermodynamic Properties of the Ti-Mo System

Properties of the pure components

$G^0(\text{Ti,bcc}) = 0$	$G^0(\text{Mo,bcc}) = 0$
$G^0(\text{Ti,L}) = 16\,234 - 8.368\,T$	$G^0(\text{Mo,L}) = 24\,267 - 8.368\,T$
$G^0(\text{Ti,cph}) = -4\,351 + 3.77\,T$	$G^0(\text{Mo,cph}) = 8\,368$

Excess Gibbs energies

Reference	B(L)	C(L)	B(bcc)	C(bcc)	B(α)
[80Bre]	$18\,125 - 13.6\,T$	...	$36\,500 - 18.7\,T$	...	$42\,404$
[78Kau]	$6\,537$	...	$5\,192$	...	$15\,359$
This work	$14\,943$	...	$15\,657$	5238	$36\,647$

Note: Values are given in J/mol and J/mol · K.

Lattice parameters of martensitic (α'Ti) do not vary with composition, as do lattice parameters of martensitic (α''Ti). The data of [80Lei2] were obtained after quenching from the (βTi) region and high-pressure soaking. Error estimates were ±0.0006 nm.

[80Lei1] represented atomic volumes, V, as a function of atomic fraction of Mo, x, as:

$$V_\omega(x) = 10.47\,(1 - 0.152\,x)\ \text{cm}^3/\text{mol}$$

$$V_\beta(x) = 10.60\,(1 - 0.180\,x)\ \text{cm}^3/\text{mol}$$

$$V_\alpha(x) = 10.66\,(1 - 0.302\,x)\ \text{cm}^3/\text{mol}$$

Thermodynamics

Vapor pressure measurements were made by [63Luz] using the Langmuir method. [71Hoc] determined the activity of Ti in the bcc phase using the Knudsen cell technique. The Knudsen cell experiment is preferred over the Langmuir cell experiment, because it is an equilibrium method, but the uncertainties in the partial Gibbs energy of Ti become very large at high Mo concentrations. Fitting the experimental data in the regular solution approximation, [71Hoc] found the bcc excess Gibbs energy to be $x(1 - x)\,(12\,134 \pm 2\,900)$ J/mol, where x is the atomic fraction of Mo. This value implies a metastable miscibility gap with a critical temperature of 459 °C.

The regular solution approximation requires an average between a higher value for Ti-rich alloys and a lower value for Mo-rich compositions. The observed asymmetry of the excess Gibbs energy is qualitatively in agreement with the asymmetry of the bcc phase miscibility gap and the calculated Gibbs energies to be described below.

Calculation of the Diagram.

The Gibbs energies of the solution phases are represented as:

$$G(i) = G^0(\text{Ti},i)\,(1 - x) + G^0(\text{Mo},i)\,x + RT(x \ln x +$$
$$(1 - x) \ln (1 - x)) + x(1 - x)\,[B(i) + C(i)\,(1 - 2x)]$$

where i designates the phase; x is the atomic fraction of Mo; G^0 is the Gibbs energies of the pure metals; and $B(i)$ and $C(i)$ are parameters in the polynomial expansion of the excess Gibbs energies. Gibbs energies of the pure elements are from [70Kau].

Subregular solution parameters for the L and bcc phases and a regular solution parameter for the cph phase were optimized using the miscibility gap and monotectoid reaction data [77Ter, 84Mca], (αTi) solvus and (βTi) transus data [51Han], and solidus data [69Rud]. Miscibility gap data were taken only from the Ti-rich side of the critical point. Gibbs energy parameters are listed in Table 5. The calculated liquidus and solidus were used to draw

the assessed diagram Fig. 1. The calculated cph/bcc boundaries are compared with the assessed diagram and the experimental data in Fig. 3. The calculated monotectic temperature is high compared to the observed temperature, because the slope of the calculated (βTi) transus changes significantly as it nears the miscibility gap. This feature is common to many calculations of systems with miscibility gaps in the solid state.

Previous calculations of the phase diagrams are those of [75Che], [78Kau], and [80Bre]. [75Che], [78Kau], and [80Bre] assumed that the miscibility gap is metastable and therefore that excess Gibbs energies are quite low. [80Bre] recognized the bcc miscibility gap [77Ter], but used the regular solution approximation; his calculation therefore deviates from the observed diagram more than the present calculation. Excess Gibbs energies reported by [80Bre] are listed in Table 5 to illustrate the variation in Gibbs energies, depending on assumptions about the equilibrium diagram.

Cited References

51Duw: P. Duwez, "Effect of Rate of Cooling on the α-β Transformation in Titanium and Titanium-Molybdenum Alloys," *Trans. AIME, 191,* 765-771 (1951). (Meta Phases; Experimental)

51Han: M. Hansen, E.L. Kamen, and H.D. Kessler, "Systems Titanium-Molybdenum and Titanium-Columbium," *Trans. AIME, 191,* 881-888 (1951). (Equi Diagram; Experimental)

52Del: D.J. Delazaro, M. Hansen, R.E. Riley, and W. Rostoker, "Time-Temperature-Transformation Characteristics of Titanium Molybdenum Alloys," *Trans. AIME, 194,* 265-268 (1952). (Meta Phases; Experimental)

58Bag: Yu.A. Bagariatskii, G.I. Nosova, and T.V. Tagunova, "Factors in the Formation of Metastable Phases in Titanium-Base Alloys," *Dokl. Akad. Nauk SSSR, 122,* 593-596 (1958) in Russian; TR: *Soviet Phys. Dokl., 3,* 1014-1018 (1958). (Meta Phases; Experimental)

59Age: N.V. Ageev and L.A. Petrova, "Breakdown of the β-Solid Solution in Titanium-Molybdenum Alloys," *Zh. Neorg. Khim., 4,* 1924 (1959) in Russian; TR: *Russ. J. Inorg. Chem., 4,* 871-872 (1959). (Meta Phases; Experimental)

60Sat: T. Sato, S. Hukai, and Y.C. Huang, "The M_s Points of Binary Titanium Alloys," *J. Aust. Inst. Met., 5(2),* 149-153 (1960). (Meta Phases; Experimental)

61Bun: K. Bungardt and K. Ruedinger, "Phase Transformations in Titanium-Molybdenum Alloys," *Z. Metallkd., 52,* 120-135 (1961) in German. (Meta Phases; Experimental)

61Dup: J.M. Dupouy and B.L. Averbach, "Atomic Arrangements in Titanium-Molybdenum Solid Solutions," *Acta Metall., 9,* 755-763 (1961). (Equi Diagram; Experimental)

63Luz: L.P. Luzhnikov, W.M. Novikova, and A.P. Mareev, "Solubility of β-Stabilizers in α-Titanium," *Metalloved. Term. Obrab. Met.,* (2), 13-16 (1963) in Russian; TR: *Met. Sci. Heat Treat.,* (2), 78-81 (1963). (Equi Diagram; Experimental)

63Woo: R.M. Wood, "Martensitic α and ω Phases as Deformation Products in a Titanium-15% Molybdenum Alloy," *Acta Metall.*, *11*, 907-914 (1963). (Meta Phases; Experimental)

64Wil: A.J. Williams, "A Study of the Constitution of the Titanium-Rich Corner of the Titanium-Aluminum-Molybdenum System," Dept. Mines, Ottawa, Canada, Research Report R132 (1964). (Experimental)

68Bla: M.J. Blackburn and J.C. Williams, "Phase Transformations in Ti-Mo and Ti-V Alloys," *Trans. AIME*, *242*, 2461-2469 (1968). (Meta Phases; Experimental)

69Hic: B.S. Hickman, "ω Phase Precipitation in Alloys of Titanium with Transition Metals," *Trans. AIME*, *245*, 1329-1335 (1969). (Meta Phases; Experimental)

69Rud: E. Rudy, "Compilation of Phase Diagram Data," Tech. Rep. AFML-TR-65-2, Part V, Wright Patterson Air Force Base (1969). (Experimental)

70Col: E.W. Collings and J.C. Ho, "Physical Properties of Titanium Alloys," Sci. Technol. Appl. Titanium, Proc. Int. Conf., R.I. Jaffee, Ed., 331-347 (1970). (Meta Phases; Experimental)

70Hua: Y.C. Huang, H. Kaneko, T. Sato, and S. Suzuki, "Continuous Cooling-Transformation of β-Phase in Binary Titanium Alloys," Sci. Technol. Appl. Titanium, Proc. Int. Conf., R.I. Jaffee, Ed., 695 (1970). (Meta Phases; Experimental)

70Kau: L. Kaufman and H. Bernstein, *Computer Calculation of Phase Diagrams*, Academic Press, New York (1970). (Thermo; Theory)

70Kou: M.K. Koul and J.F. Breedis, "Phase Transformations in β Isomorphous Titanium Alloys," *Acta Metall.*, *18*, 579-588 (1970). (Meta Phases; Experimental)

70Ron: G.N. Ronami, S.M. Kuznetsova, S.G. Fedotov, and K.M. Konstantinov, "Determination of the Phase Boundaries in Ti Systems with V, Nb, and Mo by the Diffusion-Layer Method," *Vestn. Moskov. Univ., Fiz.*, *25*(2), 186-189 (1970) in Russian; TR: *J. Moscow Univ. Phys.*, *25*(2), 55-57 (1970). (Equi Diagram; Experimental)

70Wil: J.C. Williams and B.S. Hickman, "Tempering Behavior of Orthorhombic Martensite in Titanium Alloys," *Metall. Trans.*, *1*, 2648-2650 (1970). (Meta Phases; Experimental)

71Hoc: M. Hoch and R. Wiswanathan, "Thermodynamics of Titanium Alloys, III. The Ti-Mo System," *Metall. Trans.*, *2*, 2765-2767 (1971). (Thermo; Experimental)

72Cha: V. Chandrasekaran, R. Taggart, and D.H. Polonis, "Phase Separation Processes in the β Phase of Ti-Mo Binary Alloys," *Metallogr.*, *5*, 393-398 (1972). (Meta Phases; Experimental)

72Fed: S.G. Fedotov, V.S. Lyasotskaya, K.M. Konstantinov, A.A. Kutsenko, and E.P. Sinodova, "Phase Transformations in Alloys of the System Titanium-Molybdenum," Nov. Konstr. Mater.-Titan, Tr. Nauch.-Teckh. Soveshch. Met. Metalloved. Primen. Titana, 8th, 37-41 (1972) in Russian. (Meta Phases; Experimental)

72Flo: H.M. Flower and P.R. Swann, "In Situ Observation of a Radiation-Induced Phase Transformation," Electron Microsc. Proc. 5th Eur. Congr. Electron Microsc., 570-571 (1972). (Meta Phases; Experimental)

72Ron: G.N. Ronami, "Determination of Phase Equilibria in Superconducting Alloys Using Electron Beam Microanalysis," *Kristall. Technik.*, *7*(6), 615-638 (1972) in German. (Equi Diagram; Experimental)

73Col: E.W. Collings, J.C. Ho, and R.I. Jaffee, "Physics of Titanium Alloys*—II, Fermi Density-of-States Properties, and Phase Stability of Ti-Al and Ti-Mo," Sci. Technol. Appl. Titanium, Proc. Int. Conf., R.I. Jaffee, Ed., Vol. 2, 831-842 (1973). (Meta Phases; Experimental)

73Fed: S.G. Fedotov, K.M. Konstantinov, E.F. Sidorova, and N.F. Kvasova, "Transformation of Titanium-Molybdenum Martensite," *Izv. Akad. Nauk SSSR, Met.*, (5), 225-230 (1973) in Russian; TR: *Russ. Metall.*, (5), 155-160 (1973). (Meta Phases; Experimental)

73Mor: J.P. Morniroli and M. Gantois, "Formation Conditions for the Omega Phase of Titanium-Niobium and Titanium-Molyb-

denum Alloys," *Mem. Sci. Rev. Met.*, *70*(11), 831- 842 (1973) in French. (Meta Phases; Experimental)

74Flo: H.M. Flower, S.D. Henry, and D.R.F. West, "The α-β Transformation in Dilute Ti-Mo Alloys," *J. Mater. Sci.*, *9*(1), 57-64 (1974). (Equi Diagram; Experimental)

74Gus: L.N. Guseva and I.V. Egiz, "Metastable Phase Diagram of High-Purity Titanium-Molybdenum Alloys," *Metalloved. Term. Orab. Met.*, (4), 71-72 (1974); TR: *Met. Sci. Heat Treat.*, (2), 355-356 (1974). (Meta Phases; Experimental)

74Gys: A. Gysler, W. Bunk, and V. Gerold, "Precipitation of the ω Phase in Titanium-Molybdenum Alloys," *Z. Metallkd.*, 65, 411-417 (1974) in German. (Meta Phases; Experimental)

75Age: N.V. Ageev, L.N. Gukseva, and L.K. Dolinskaya, "Metastable Phases in Quenched Titanium-Molybdenum and Titanium-Vanadium Alloys and the Influence of Small Quantities of Oxygen on Them," *Izv. Akad. Nauk SSSR, Met.*, (4), 151-156 (1975) in Russian; TR: *Russ. Metall.*, (4), 113-117 (1975). (Meta Phases; Experimental)

75Bag: R.G. Baggerly, "Determination of ω Phase Volume Fraction in Single Crystals of Beta Titanium Alloys," *Metallogr.*, 8, 361-373 (1975). (Meta Phases; Experimental)

75Che: D.B. Chernov and A.Y. Shinyayev, "Calculation of the Interaction Parameters of Titanium with Vanadium, Niobium and Molybdenum," *Izv. Akad. Nauk SSSR, Met.*, (5), 212-219 (1975) in Russian; TR: *Russ. Metall.*, (5), 167 (1975). (Thermo; Theory)

75Kol: B.A. Kolachev, F.S. Mamonova, and V.S. Lyasotskaya, "Structure of Quenched Alloys of the Titanium-Molybdenum System," *Izv. Tsvetn. Metall.*, (6), 130-133 (1975) in Russian. (Crys Structure; Experimental)

76Ter: S. Terauchi, H. Matsumoto, T. Sugimoto, and K. Kamei, "Investigation of the Titanium-Molybdenum Binary Phase Diagram," Titanium and Titanium Alloys, Scientific and Technological Aspects, Moscow, USSR, 1335-1349 (1976). (Equi Diagram; Experimental)

76Whi: J.J. White and E.W. Collings, "Analysis of Calorimetrically Observed Superconducting Transition-Temperature Enhancement in Titanium-Molybdenum (5 at.%)-Based Alloys," AIP Conf. Proc. Issue,*J. Magn. Magn. Mater.*, Jt. MMM-Intermag. Conf., *34*, 75-77 (1976). (Equi Diagram; Experimental)

77Ter: S. Terauchi, H. Matsumoto, T. Sugimoto, and K. Kamei, "Investigation of the Titanium-Molybdenum Binary Phase Diagram," *J. Jpn. Inst. Met.*, *41*(6), 632-637 (1977) in Japanese. (Equi Diagram; Experimental)

78Kau: L. Kaufman and H. Nesor, "Coupled Phase Diagrams and Thermochemical Data for Transition Metal Binary Systems—II," *Calphad, 2* 81-108 (1978). (Thermo; Theory)

78Ter: S. Terauchi, H. Matsumoto, T. Sugimoto, and K. Kamei, "Solid Phase Transformation of the Ti-Mo Binary System," *Technol. Rept. Kansai Univ.*, *19*, 61-71 (1978). (Equi Diagram; Experimental)

79Dav: R. Davis, H.M. Flower, and D.R.F. West, "Martensitic Transformations in Ti-Mo Alloys," *J. Mater. Sci.*, *14*(3), 712-722 (1979). (Meta Phases; Experimental)

79Lei: Ch. Leibovitch, A. Rabinkin, M. Ron and E. Gartstein, "Mössbauer Study of Metastable Phases in Ti-3 at.% Mo Alloy," *J. Phys. Colloq.*, *40*, C2-604-C2-607 (1979). (Meta Phases; Experimental)

80Bre: L. Brewer, *Molybdenum, Physico-Chemical Properties of Its Compounds and Alloys*, Special Issue No. 7, UNIPUB, New York (1980). (Equi Diagram; Compilation)

80Lei1: Ch. Leibovitch and A. Rabinkin, "Metastable Diffusionless Equilibria Ti-Mo and Ti-V Systems Under High Pressure Conditions," *Calphad*, *14*(1), 13-26 (1980). (Equi Diagram, Meta Phases; Experimental)

80Lei2: Ch. Leibovitch, E. Gartstein, and A. Rabinkin, "The Structural Stability and Superconductivity of Ti-Mo Alloys Under Pressure. Part I. Structural Stability," *Z. Metallkd.*, *71*, 438-447 (1980). (Crys Structure; Experimental)

84Mca: A.J. McAlister, private communication (1984). (Equi Diagram; Experimental)

The N-Ti (Nitrogen-Titanium) System

14.0067   47.88

By H.A. Wriedt and J.L. Murray

Equilibrium Diagram

The equilibrium solid phases of the Ti-N system (Fig. 1) are: (1) the terminal cph solid solution, (αTi), based on Ti below 882 °C, with a wide range of compositions; (2) the terminal bcc solid solution, (βTi), based on Ti above 882 °C, also with a wide range of compositions; (3) the tetragonal Ti_2N phase, also called $Ti_2N(\beta)$, ϵ, or γ phase, with a small range of composition near 33.3 at.% N; (4) the

fcc TiN phase (also called δ, TiN_{1-x}, or TiN_x), with a wide range of compositions; and (5) the bct δ' phase, also called $Ti_2N(\alpha)$, with a small range of compositions near 37.5 at.% N.

The structures of the crystalline phases reported in the literature are listed in Table 1. In addition to the alternative designations for the Ti_2N, TiN, and δ' phases noted above, somewhat confusing designations are used in describing other Ti-N phases. The term "α'" either may

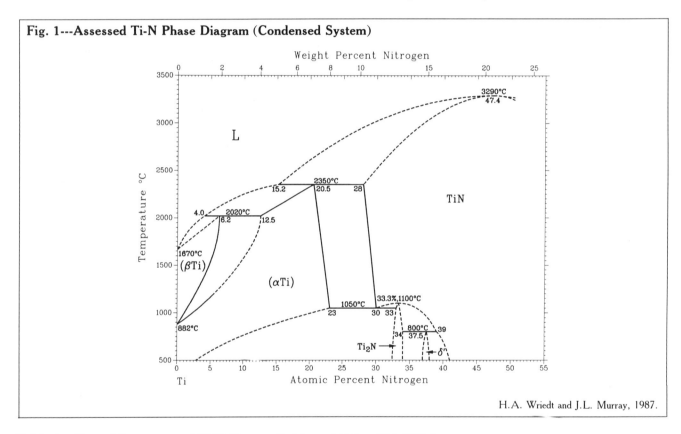

Fig. 1---Assessed Ti-N Phase Diagram (Condensed System)

H.A. Wriedt and J.L. Murray, 1987.

Table 1   Crystal Structures of Ti-N Phases–Stable and Other (0.1 MPa)

Phase	Composition, at.% N	Pearson symbol	Space group	Strukturbericht designation	Prototype	Temperature, °C	Lattice parameters, nm a	b	c	Reference
Stable phases										
(αTi) ...	0 to 22	hP2	$P6_3/mmc$	A3	Mg	25	0.29511(a)	...	0.46843(a)	[62Woo]
(βTi) ...	0 to 6	cI2	Im3m	A2	W	886	0.3306(a)	...	...	[58Sku]
Ti_2N	33	tP6	$P4_2/mnm$	C4	Anti-O_2Ti (rutile)	RT(b)	0.4943	...	0.3036	[62Hol]
TiN ...	30 to 55	cF8	Fm3m	B1	NaCl	25	0.4241 ± 0.0002(c)	...	...	[49Ehr, 67 Gri, 75Nag, 77Arb1]
δ'	38	tI12	$I4_1/amd$	C_c	Si_2Th	RT(b)	0.4198	...	0.8591	[77Nag]
Other phases										
α'	?	(d)		C4	Anti-O_2Ti (rutile)					[83Sun]

(a) Pure Ti (0 at.% N). (b) Exact temperature unspecified. (c) 50.0 at.% N. (d) Tetragonal.

denote (αTi) when it is formed martensitically from (βTi) during fast cooling [59Kau, 70Kou], or it may denote a tetragonal phase formed by ordering of the N atoms in supersaturated cph martensite with 1.6 at.% N [83Sun]. The term "α″" has been used [83Sun] to denote cph martensite also called α′.

The phase diagram derived experimentally by [54Pal] has been the basis for subsequent comprehensive diagrams [65Mcc, 71Tot, 74Woo] (Fig. 2). In most qualitative features, except the omission of the δ′ phase and, probably, the nature of the (αTi)-TiN-Ti₂N equilibrium, it remains current. Work subsequent to that of [54Pal] has often, however, disagreed with it quantitatively, especially with respect to the locations of boundaries of the (αTi), TiN, and Ti₂N phases. These later studies, most of which have dealt only with portions of the diagram, often

differ significantly among themselves, as well as from [54Pal]. In some instances, this can be attributed to variations in the impurity contents of the Ti-N specimens. The sensitivity of some boundary locations to oxygen impurity is well documented [53Sto]. The assessed phase diagram is shown in Fig. 1, and invariants of the condensed system at 0.1 MPa are listed in Table 2. Figure 3 presents the experimental data on which Fig. 1 was based.

Terminal Solid Solutions, (αTi) and (βTi).

(αTi) and (βTi) exist stably only at low nitrogen fugacities. (βTi) saturates with respect to (αTi) for sufficiently high nitrogen fugacities. With increasing nitrogen fugacity, at lower temperatures, (αTi) eventually saturates with respect to solid Ti₂N.

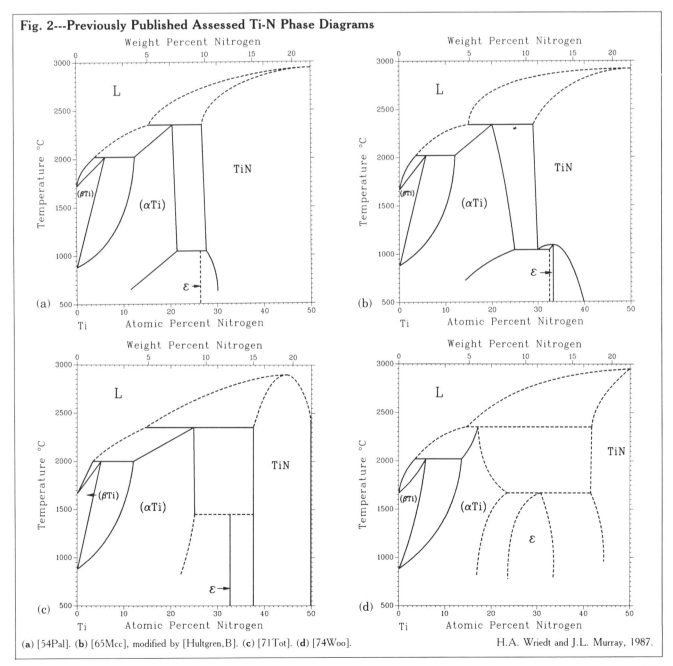

Fig. 2---Previously Published Assessed Ti-N Phase Diagrams

(a) [54Pal]. (b) [65Mcc], modified by [Hultgren,B]. (c) [71Tot]. (d) [74Woo].

H.A. Wriedt and J.L. Murray, 1987.

Fig. 3---Experimental Data for the Ti-N Phase Diagram

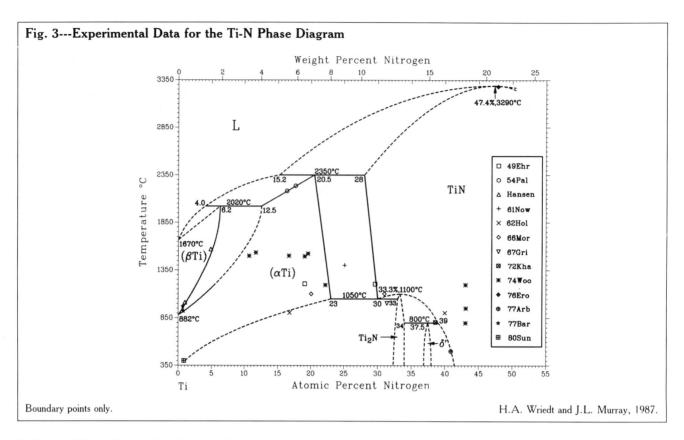

Boundary points only.

H.A. Wriedt and J.L. Murray, 1987.

Table 2   Three-Phase Equilibria and Other Transformations at 0.1 MPa Pressure (Condensed System)

Reaction(a)	Compositions of the respective phases, at.% N			Temperature, °C	Reaction type
L + (βTi) + (αTi)	4.0	6.2	12.5	2020 ± 25	Peritectic
L + (αTi) + TiN	15.2	20.5	28	2350 ± 25	Peritectic
(αTi) + TiN + Ti₂N	23	30	33	1050 ± 60	Probably eutectoid, possibly peritectoid
Ti₂N + δ' + TiN	34	37.5	39	800 ± 100	Probably peritectoid
Congruent transformations: pure component					
αTi ⇌ βTi	0			883 ± 3	Polymorphic
βTi ⇌ L	0			1670 ± 6	Fusion
Congruent transformations: other					
TiN ⇌ L(b)	47.4			∼ 3290	Fusion
Ti₂N ⇌ TiN(c)	33.3			∼ 1100	Polymorphic

(a) Stable equilibria involving ω are possible at elevated hydrostatic pressure. (b) Only actual observation under pressures greater than or equal to ∼ 1MPa. (c) Occurrence if (αTi) + TiN + Ti₂N equilibrium is eutectoid.

At higher temperatures, saturation is with respect to TiN. The range of (βTi) in the equilibrium diagram for the condensed system is delimited by two boundaries with two-phase regions—the solidus and the N-rich compositions where it coexists with (αTi). For the (αTi) range, the four boundaries correspond to coexistence with (βTi), with liquid (the solidus), with TiN, and with Ti₂N.

The compositions of coexisting (αTi) and (βTi) were determined metallographically by [50Jaf] up to 950 °C and by [53Sto] and [54Pal] up to 1400 °C. Their adopted values agree well. [54Pal] compared the results obtained with sponge Ti and with a purer iodide Ti. He found that the two-phase field is broadened by impurities. [77Bar] examined a single alloy (0.7 at.% N) by dilatometry; he claimed

that the purity of his sample was higher than those of [54Pal]. Consistent with this claim, the two-phase field was observed to be narrower. Compositions on the boundaries of the (αTi) + (βTi) field were also reported by [74Woo] for (αTi) up to 1525 °C and by Wasilewski [Hansen] for (βTi) at 1000 and 1560 °C, both from analysis of diffusion couples. The Wasilewski data, like those of [77Bar], indicate that the (βTi) boundary is at higher N concentrations than the [54Pal] data indicated.

The boundaries of the (αTi) + (βTi) field end at a peritectic equilibrium L + (αTi) + (βTi). This peritectic reaction and the (αTi) and (βTi) solidus curves were examined only by [54Pal], who used metallographic and incipient melting techniques. The temperature of this peritectic was placed at 2020 ± 25 °C; the compositions of (βTi) and (αTi) coexisting with liquid were estimated as 6.2 and 12.5 at.% N, respectively, from an extrapolation of the boundaries of the (αTi) + (βTi) field to intersect the solidus curves.

Because of the known tendency of impurities to broaden the two-phase field, the boundaries of the (βTi) + (αTi) field are drawn in Fig. 1 to yield the narrowest field consistent with the available data. The temperature and compositions shown for the peritectic reaction at their high-temperature ends are those of [54Pal]. In view of the difficulty of maintaining sample purity in melting-point determinations near 2000 °C, an uncertainty of ± 25 °C attributed by [54Pal] to the value 202 °C for this peritectic temperature may be an underestimate.

The only determination of the (βTi) liquidus is also by [54Pal]. Because the 1725 °C value of [54Pal] for the melting point of pure βTi is 55 °C higher than the value adopted here, the original locations shown for the liquidus and solidus are unacceptable. The forms of these curves in Fig. 1 between the correct melting point for pure Ti (1670 °C) and their upper ends at the 2020 °C peritectic equilibrium are schematic. The peritectic liquid is shown with 4.0 at.% N, as indicated by [54Pal].

According to [54Pal], the (αTi) solidus runs from the peritectic equilibrium at 2020 °C to a second peritectic equilibrium, L + (αTi) + TiN, located at 2350 °C with 20.5 at.% N in (αTi). Although [71Tot] and [74Woo] have since suggested higher (25 at.% N) or lower (17.5 at.% N) values, respectively, for (αTi) at the upper peritectic equilibrium, the existence and attributed temperature of this equilibrium have not been disputed. The suggested relocations arose from dissatisfaction with the location of the (αTi)/(αTi) + TiN boundary as determined by [54Pal] (see discussion later).

No experimental data for the (αTi) liquidus were found, except for the location of the low-temperature end; that is, 4.0 at./% N at the 2020 °C peritectic equilibrium [54Pal]. In Fig. 1, this curve is depicted schematically in the position suggested by [54Pal].

The (αTi)/(αTi) + TiN boundary runs from its high-temperature end at the upper peritectic equilibrium to its low-temperature end where (αTi), TiN, and Ti₂N are in equilibrium. Controversy has existed not only over the boundary location, but over the temperature and the type of three-phase equilibrium (peritectoid or eutectoid) at the lower terminus. [49Ehr] located the upper limit of (αTi) at 19 at.% N with specimens prepared at 1200 °C, although this limit is not explicitly related to this temperature. This value is lower than the limits shown by [54Pal], which are 20.5 at.% at 2350 °C increasing to about 21.2 at.% at a 1050 °C peritectoid equilibrium. [61Now] found a higher limit at 1400 °C of about 25 at.%. The (αTi)/(αTi) + TiN boundary in the diagram of [65Mcc] agrees with that of [54Pal] at the upper peritectic equilibrium, but is closer to the datum of [61Now] at lower temperature, where the boundary is shown terminating at 24.5 at.% N, 1050 °C. [66Mor] located the (αTi) boundary at 20 at.% N in specimens quenched from 1100 °C. [74Woo] represented the (αTi) boundary as running from 17.5 at.% at 2350 °C to about 23 at.% at a peritectoid equilibrium near 1675 °C, although the evidence for this location is unclear.

The temperature of the (αTi)-TiN-Ti₂N equilibrium was reported to be 1050 °C [54Pal], 1450 °C [71Tot] (unknown basis), and 1600 to 1700 °C [74Woo]. The [71Tot] value may be based on the statement of [61Now] that Ti₂N decomposes above 1400 °C, but this value may

have been affected by the presence of B. Other values were >1593 °C [68Ren], 1110 °C [70Mcd], > 1500 °C [74Bar], and about 2000 °C [77Bar]. The specimens that indicated that the (αTi)-TiN-Ti₂N equilibrium is above 1110 °C, except for those of [61Now] with B possibly present, were from diffusion experiments, where Ti₂N was observed at room temperature between (αTi) and TiN layers produced in the nitrogen gradient. In the experiments of [68Ren] and [74Woo], cooling of the specimens after the diffusion anneal was not severe; [74Bar] and [77Bar] did not describe their cooling technique, although [77Bar] called it "very rapid". This allows speculation that Ti₂N may have precipitated during cooling. [70Mcd], who used more severe quenching that [68Ren] or [74Woo], observed no Ti₂N for anneals above 1110 °C. In common with [62Hol] and [77Arb1], he noted that Ti₂N can nucleate very quickly, in this instance between (αTi) and TiN. [73Arb] found no Ti₂N in specimens Ga-quenched from 1400 °C. Because it is more difficult to believe that Ti₂N, if stable above 1100 °C, could be absent from rapidly quenched specimens [54Pal, 70Mcd, 73Arb] than that it could be present through inadequate quenching, the value of 1100 °C has been adopted for the upper stability limit of Ti₂N, in essential agreement with [54Pal], [65Mcc], [70Mcd], [74Lev], and [77Arb1].

With the exception of [65Mcc], all investigators have depicted the (αTi)-TiN-Ti₂N equilibrium as peritectoid in type. [65Mcc] depicted it as a eutectoid type at 1050 °C, with Ti₂N decomposing congruently at 1100 °C. This could be justified if it were shown that TiN exists at compositions below 33.3 at.% N at 1100 °C. The preponderance of such evidence (discussed later) favors the existence of the eutectoid type. The equilibrium is therefore depicted in Fig. 1 as eutectoid at 1050 °C.

Data on the (αTi)/(αTi) + Ti₂N boundary are relatively sparse. The first measurements were made by [54Pal], who reported values from 21.2 at.% at 1050 °C down to slightly below 13 at.% at 700 °C. [62Hol] located the boundary at 900 °C at 17 at.%, near the value of 18 at.% of [54Pal]. [74Woo] tabulated their experimental values for the high N limit of the (αTi) range between 750 and 1525 °C and identified them as compositions of (αTi) coexisting with Ti₂N. The values vary unsystematically with temperature from 16.5 to 22 at.%, with the highest value at 1185 °C. They are not clearly related to the boundary drawn by [74Woo]. [80Sun] found that the limit of (αTi) was less than 1.6 at.% N at 400 °C. The diagrams of [65Mcc] and [71Tot] depict the N content of (αTi) at the 1050 °C invariant as 24.5 and 25 at.%. The location of the line in Fig. 1 is a compromise at 1050 °C, below which it is fitted to the [62Hol] and [80Sun] data.

Ti₂N (ε).

The existence of this tetragonal phase was first recorded by [54Pal]; however, the unit cell dimensions and composition were reported incorrectly. [61Now] confirmed the existence of Ti₂N, but found that its composition range was narrow and near 33 at.% N. [62Hol] also indicated that the composition range was narrow, with a probable upper limit of 33.3 at.% N at 900 °C. From the small spread of lattice parameters for Ti₂N in specimens prepared at 845, 996, and 1092 °C and from their proximity to the 900 °C values of [62Hol], [70Mcd] concluded that the Ti₂N range was narrow at 845 to 1100 °C. [74Woo] reported compositions between 24 to 28.5 at.% N (low limit) and 31 to 35 at.% N (high limit) at 795 to 1525 °C. The values at each composition limit varied unsystematically with temperature; they were discordant from those of all other investigators. [77Arb1] found that specimens with 31 at.% N, annealed at 1000 or 500 °C, were single-phase Ti_2N, whereas 29 or 34 at.% N specimens annealed at 500 °C were duplex. They concluded that the probable range of Ti_2N is between 31 and 33 at.% N. N-rich and N-poor Ti_2N coexist with δ' and with (αTi), respectively. The phase relationships and homogeneity field of Ti_2N in Fig. 1 with an upper limit of 1100 °C are adapted from the findings of [62Hol], [65Mcc], and [77Arb1].

TiN (δ).

This phase includes the stoichiometric composition TiN in a broad range extending from about 30 to over 54 at.% N and from low temperatures to a solidus which runs from 2350 °C up. The upper limits of N concentration are undetermined.

In early determinations of the solidus, melting points of phases termed TiN were measured by [25Fri] (2930 °C)* and by [31Agt] (2950 °C).** Using a similar procedure of melting the specimen by direct electrical heating, [74Ett] obtained the value 2945 ± 30 °C for the melting point of specimens with 48.4 to 48.8 at.% N, under N_2 pressures up to 5 MPa. [54Pal] used the [31Agt] value and sketched, without experimental points, the TiN solidus from this 2950 °C melting point at 50 at.% N, shown as congruent, to its lower temperature end at the upper peritectic equilibrium (2350 °C), where the indicated composition was 27 at.% N. Although the position of the solidus apparently has not been determined over its whole length, nor has the 2350 °C value for the peritectic terminal point been superseded by later determination, several modified values have been proposed for the N concentration in peritectic TiN: 29 at.% N [65Mcc], 38 at.% N [71Tot], and 42 at.% N [74Woo]. In all such proposed diagrams, this point was shown as the lower limit of N concentration in TiN for any temperature. The validity of the 2950 °C value for the melting point at 50 at.% N, which has been adopted in all published diagrams, is rendered doubtful by observations of the onset of melting in TiN as a function of N_2 pressure by [76Ero]. Above 1 MPa N_2 pressure, congruent melting was found at 3287 ± 60 °C, with 47.4 at.% N in the phases. Below 1 MPa N_2 pressure, it was found that melting was incongruent, with less N in liquid, and that the solidus temperature declined with decreasing N_2 pressure. A graphical relation [76Ero], based on the experimental data down to about 2430 °C, corresponds approximately to:

$$\log p = 8.68 - 31\,100/T$$

where p is the N_2 pressure in MPa, and T is in K. However, the N concentrations in TiN were not reported for the lower N_2 pressures. These data contradict the finding of [74Ett] with respect to the effect of N_2 pressure on the solidus temperature, but indicate a solidus temperature of 2940 °C at 0.1 MPa N_2, in good agreement with [74Ett] and the earlier determinations. Inconsistency remains regarding the N concentration at which the solidus temperature is 2940 °C; it would be significantly lower than 50 at.% according to [76Ero]. The apparently discontinuous change in melting point with variation of N_2 pressure observed by [76Ero] near 1 MPa is initially surprising. However, [77Hoj] has shown that, with increasing N content of the TiN phase at or very near 50 at.% N, the equilibrium pressure of N_2 gas at 1100 °C

suddenly starts to increase radically, an effect associated with the approach to complete filling of octahedral sites for N. The [76Ero] value of 3290 °C for the congruent melting point of TiN with 47.4 at.% N is adopted.

The only measurements related to the TiN liquidus above 2350 °C are the "melting points" of "TiN" near 2950 °C of [25Fri], [31Agt], and [74Ett] and the datum of [76Ero] indicating a congruent melting composition of 47.4 at.% N at 3290 °C. As noted in the discussion of the solidus, the older "melting point" determinations probably were not for congruent melting and therefore are not points on the liquidus. The [76Ero] congruent melting composition and temperature are adopted as the coordinates of the maximum on the TiN liquidus; the coordinates indicated by [54Pal] are used for the low-temperature end at the 2350 °C peritectic equilibrium.

The TiN/TiN + δ' boundary at 500 °C was located by [77Arb1] at 41 at.% N. [72Kha] indicated that TiN at this boundary contains about 38.5 at.% N at 800 °C, which is the peritectoid temperature suggested by [77Arb1] and adopted herein.

Definitive placement of the TiN/TiN + Ti_2N boundary is impossible because of uncertainties in the form of the diagram as well as an insufficiency of data. The present authors prefer the form suggested by [65Mcc], with an (αTi) + TiN + Ti_2N eutectoid equilibrium at 1050 °C and congruent decomposition of Ti_2N with about 33 at.% N at 1100 °C, over the form suggested by [54Pal], with an (αTi) + Ti_2N + TiN peritectoid equilibrium.

In the adopted form, the TiN/TiN + Ti_2N boundary runs from the eutectoid composition of 30 at.% N through a maximum at the congruent decomposition temperature of Ti_2N to the peritectoid Ti_2N + δ' + TiN equilibrium at 38.5 at.% N and 800 °C. The boundary as sketched in Fig. 1 is mainly schematic, but is compatible with the data of 40 ± 2 at.% N and 900 °C [62Hol] and 38.5 at.% N and 800 °C [72Kha]. [66Mor] observed no Ti_2N in specimens with more than 31 at.% N quenched from 1100 °C. The [74Woo] points on the TiN/TiN + Ti_2N boundary at 43 at.% N from 800 to 1200 °C disagree with all of the foregoing. The boundary as drawn is compatible at its termini with the TiN/(αTi) + TiN and TiN/δ' + TiN boundaries.

Definitive placement of the TiN/(αTi) + TiN boundary is also hampered by the uncertainty regarding the type of three-phase equilibrium occurring at its lower temperature terminus and by the sparseness and lack of agreement of the data. A primary difficulty in placement of the boundary is the absence of usable experimental data above 1600 °C. [83And] indicated a lower limit of 28 at.% N at 1727 to 2227 °C, but the basis is unclear. However, consistent with the adoption of the eutectoid form, no investigation has located any part of the boundary above 33 at.% N, the established composition of Ti_2N. The data of [54Pal] at 1000 to 1400 °C indicated a location 2 and 3 at.% N lower than the points 30 and 31 at.% N of [49Ehr] and of [66Mor] at 1200 and 1100 °C, respectively. The point of [67Gri] at 31 at.% and 1000 °C is in agreement with [66Mor], but tends to place the eutectoid equilibrium lower than the 1050 °C value adopted. The boundary location suggested by the work of [73Sam] at 1000 °C is rejected because of the use of hydrogen as a probing agent. The data of [77Arb1] indicate that the boundary lies below 29 at.% at 1600 °C and above 27 at.% at 1400 °C. The boundary location in Fig. 1 is very close to that adopted by [65Mcc]. Upward adjustment in the N concentration rela-

---

*Not adjusted to IPTS-68 because of uncertainty of the correction.
**English translation.

tive to the [54Pal] value was made at the higher temperature terminus, consistent with the choice of a higher concentration at the eutectoid end.

δ'.

The existence of the δ' phase, under the designation "Ti$_2$N(α)", was apparently first discovered by [69Lob], who obtained a bct phase by aging at 500 °C a single-phase 33 at.% N alloy quenched from 1400 °C. The "Ti$_2$N(α)" transformed to Ti$_2$N, termed "Ti$_2$N(β)" by [69Lob], at 900 °C and was not observed to re-form during subsequent lower temperature annealing. [77Nag], apparently independently, found that aging an alloy with 38 at.% N at 500 °C produced a bct structure with long-range order of the N atoms. It was classified in the same space group as that of the [69Lob] phase. [77Arb1, 77Arb2] termed "δ'" the tetragonal phase that they observed in alloys with 34 to 40.5 at.% N annealed at 500 °C and identified it with the "Ti$_2$N(α)" phase of [69Lob]. δ' was associated with Ti$_2$N below about 37 at.% and with TiN above 38 at.%, consistent with δ' containing about 37.5 at.% N. This suggests that the actual N content of the alloy used by [69Lob] was greater than the nominal 33 at.%. The upper limit of its stability was placed at 800 °C, where it coexisted with Ti$_2$N and TiN in a peritectoid equilibrium. Figure 1 follows the diagram suggested by [77Arb1].

Gas Phase.

Except for pure Ti, TiN is the only phase of the binary system for which vaporization has been studied. The earliest experimental investigation of TiN vaporization was by [55Hoc], who stated that the phase vaporized as Ti and N$_2$ molecules. [62Aki] examined mass spectrometrically the vapor formed at 1690 °C, but found no nitride species. [69Ste] and [70Koh] detected TiN molecules in the vapor over solid TiN with 43.8 at.% N at 2006 to 2183 °C. The amounts were less than 0.002% of the total gas mixture. The molecular ratio N$_2$/Ti varies with the composition of the condensed TiN. In at least three studies [55Hoc, 69Ryk, 70Ryk, and 83And], congruent vaporizing compositions were determined, and in the latter two, these compositions were determined as functions of temperature. They agree that the proportion of N$_2$ in the vapor over TiN with any particular composition increases with temperature above the temperature of congruency (see Thermodynamics). [83And] presented analytical expressions for the individual N$_2$ and Ti pressures over solid TiN at four compositions from 35.6 to 48.0 at.% in various ranges of temperatures.

Metastable Phases

Metastable extension of the (αTi) solid-solution range to N concentrations much above the stable limit was achieved by [78Mit] by deposition of Ti sputtered in N$_2$ + Ar gas mixtures. X-ray diffraction of the films 0.3μ thick indicated that the solutions contained as much as 31.5 at.% N.

[83Sun] reported that martensitic (αTi) with 1.6 at.% N, after quenching from 1300 °C and aging at 50 to 100 °C, precipitated an order transition phase (α') with tetragonal symmetry, which they distinguished from stable Ti$_2$N.

Crystal Structures and Lattice Parameters

Details of the crystal structures at 0.1 MPa pressure are summarized in Table 1. With the exception of the

Table 3 Lattice Parameters of (αTi)

Investigator	Composition, at.% N	Lattice parameters, nm	
		a	c
[49Cla](a)	0	0.29504	0.46833
	0.34	0.29503	0.46860
	1.02	0.29506	0.46905
	1.35	0.29511	0.46923
[49Ehr](a)	0	0.2949	0.4727
	9.1	0.2958	0.4742
	16.7	0.2963	0.4772
[54Pal](a)	0.5	0.2953	0.4726
	5.7	0.2956	0.4772
	10.1	0.2962	0.4758
	17.5	0.2963	0.4775
[62Hol]	6.5	0.2959	0.4739
	10.7	0.29621	0.47547
	13.8	0.29681	0.47732
[77Bar]	0	0.2950	0.4682
	0.70	0.2951	0.4685
	0.82	0.2951	0.4684
	3.50	0.2953	0.4708
	4.72	0.2956	0.4715
	5.90	0.2960	0.4733
	7.94	0.2964	0.4750
	9.08	0.2966	0.4760

Note: Parameters obtained by X-ray diffraction at room temperature. (a) Points read from plots.

[61Now] data, measurements of the dilation of the (αTi) lattice by N have been for alloys with less than 18 at.% N (Table 3). Linearity of the dilation at low concentration (0 to 1.4 at.% N) was indicated by [49Cla]; the data of [49Ehr], [54Pal], and [62Hol] for higher concentrations were presented to indicate nonlinearity. However, there are inconsistencies in the degree and even the direction of convexity of a and c curves among these studies, perhaps partly because of use in the latter two sets of lattice parameters for pure Ti from other sources [49Cla, 55Sza]. The data of [77Bar], about which the earlier data scatter, indicate linearity up to 9 at.% N, with da/dx = 1.8 × 10^{-4} nm and dc/dx = 8.9 × 10^{-4} nm, where x denotes concentrations, at.% N (Fig. 4). [61Now] reported lattice parameters of a = 0.3043 nm and c = 0.5120 nm for (αTi) in Ti-N-B alloys with 17 to 27 at.% N, but these data are not used because of the possible presence of B. The high N solutions were said to undergo ordering of the N atoms with formation of an anti-AsNi or anti-CdI$_2$ type lattice. These ordered structures apparently have not been investigated further; however, [83Sun] reported the occurrence at 1.6 at.% N of the "tetragonally-ordered" phase called α' (see Metastable Phases.)

The effect of dissolved N on the lattice parameter of (βTi) is apparently undetermined.

Although [54Pal] correctly identified the tetragonal nature of Ti$_2$N, their lattice parameter values were incorrect. [61Now], in confirming the finding of tetragonality, obtained the essentially correct a value of 0.4945 nm, but was undecided between c values of 0.304 and 0.607 nm. [62Hol] identified the structure as antirutile.

The ranges of the lattice parameters found by [62Hol] and subsequent investigators [69Lob], [70Mcd], [77Arb1], and [77Bar] are given in Table 4. The overall range (a = 0.4942 ± 0.0004 nm, c = 0.3035 ± 0.0005 nm) is consistent with the phase having a very limited range of compositions.

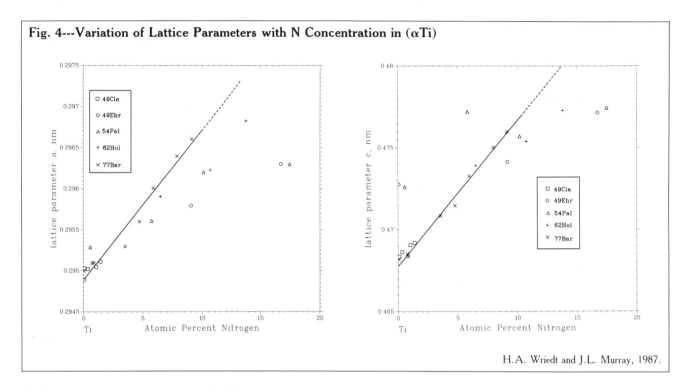

Fig. 4---Variation of Lattice Parameters with N Concentration in (αTi)

H.A. Wriedt and J.L. Murray, 1987.

Table 4 Lattice Parameters of Ti₂N

Investigator	Composition, at.% N	Boundary Temperature, °C	Boundary Coexisting phase	Lattice parameters, nm a	Lattice parameters, nm c
[62Hol]	24.2(a)	900	(αTi)	0.49414	0.30375
	32.9	...	...	0.49428	0.30357
	34.2(a)	900	TiN	0.49452	0.30342
[69Lob]	33.3	...	...	0.4946	0.3030
[70Mcd]	(a)	845	TiN	0.4946	0.30357
	(a)	996	TiN	0.4938	0.3034
	(a)	1092	TiN	...	0.3039
[77Arb1]	27.0, 29.0(a)	500	(αTi)(b)	0.4942	0.3034
	31.0	...	...	0.4942	0.3034
	34.0(a)	500	δ'(b)	0.4942	0.3034
[77Bar]	...	?	?	0.4939	0.3030

Note: Parameters obtained by X-ray diffraction at room temperature. (**a**) Composition outside single-phase range. (**b**) Not quenched.

According to [39Bra1], Van Arkel established in 1924 that TiN is of NaCl-type. Lattice parameters have been determined by X-ray diffraction for compositions from about 30 to 55 at.% N. The derivative of the lattice parameter with respect to N content is discontinuous at the stoichiometric TiN composition [49Ehr]. The study of [49Ehr] apparently was the first to determine the variation of the lattice parameter at N contents below 50 at.%. He showed that the derivative was positive between 30.6 and 50.0 at.% N. Earlier, [39Bra2] had measured lattice parameters at 50.0 to 53.7 at.% and had shown that the derivative is negative in this range. Numerous investigations have been made of the lattice parameter (Table 5).

Experimental data are present in Fig. 5, where considerable scatter and, in several instances, large systematic differences are discernible among data from different sources. For stoichiometric TiN, values range from 0.4239 to 0.4246 nm, and the [67Gri] data extrapolate to a value outside this range. The systematic variations are probably attributable to impurities, or to analytical error. Selection of the best data was hampered by lack of information supplied in some reports regarding purity. Lattice parameters reported by [62Hol], [69Lob], [75Nag], [77Aiv], [77Arb1], [77Arb2], and [77Chr] were used in establishing the least-squares relation:

$$a \pm 0.0002 = 0.4159 + 0.000164x$$

between lattice parameter a (nm) and composition x (at.% N) at $x < 50$. This relation yields a value of 0.4241 nm for the lattice parameter at 50 at.% N, as does the least-squares relation:

$$a \pm 0.0002 = 0.4415 - 0.000348x$$

based on data of [77Hoj] and [82Sae]. The data selected were the more recent and older concordant values that yielded convergence of the equations at 50 at.% N. Both the [75Nag] and [77Chr] data were retained, although not in close agreement. The [49Ehr], [67Gri], [72Kha], and

Table 5 Lattice Parameters of TiN

Investigator	Composition, at.% N	Lattice parameter, nm
Nitrogen concentration ≤ 50 at.%		
[39Bra2]	50.0	0.4244
[49Ehr]	50.0	0.4243
[50Duw]	50.0	0.4246
[62Hol]	41.52	0.42259
[64Hou]	48.7	0.4242
[67Gri]	31.2	0.4223
	33.6	0.4228
	36.4	0.4232
	39.5	0.4236
	42.5	0.4241
	45.7	0.4240
	48.0	0.4243
	50.0	0.4243
[69Lob]	33.3	0.42152
[72Kha]	35.0	0.4212
	38.4	0.4231
	38.7	0.4232
	39.8	0.4231
	40.7	0.4231
	42.5	0.4235
	44.2	0.4237
	48.5	0.4240
	49.7	0.4244
	49.8	0.4244
[73Arb]	43.0	0.4236
	45.0	0.4236
[75Nag]	37.9	0.42209
	39.4	0.42228
	40.5	0.42242
	42.4	0.42272
	44.2	0.42294
	47.1	0.42339
	48.6	0.42370
	50.0	0.42389
[77Aiv]	49.5	0.4240
[77Arb1, 77Arb2]	29.0	0.4209
	35.0	0.4215
	37.0	0.4218
	41.5	0.4224
	42.0	0.4225
	43.0	0.4228
	45.0	0.4233
	50.0	0.4242
[77Chr]	37.1	0.4219
	37.9	0.4221
	39.8	0.4223
	43.5	0.4231
	45.1	0.4234
	45.2	0.4235
	45.4	0.4236
	46.5	0.4236
	46.8	0.4237
	47.4	0.4239
	50.0	0.4241

[73Arb] data were omitted because of their large displacements from each other and most other work or because they were superseded [72Kha, 73Arb] by later work with common authorship. The difference in slope of the [39Bra2] data from that of [82Sae] was large, and the

Table 5 (Continued)

Investigator	Composition, at.% N	Lattice parameter, nm
Nitrogen concentration > 50 at.%		
[39Bra2]	50.7	0.4242
	51.2	0.4237
	51.9	0.4234
	52.6	0.4229
	53.7	0.4222
[61Gor]	54.1	0.4210
[67Gri]	51.5	0.4237
[67Str]	52.6	0.4240
[77Hoj]	50.25	0.4240
[82Sae]	50.50	0.4239
	51.69	0.4237
	52.38	0.4233
	52.61	0.4233
	53.05	0.4230
	53.70	0.4228
	54.13	0.4228
	54.75	0.4224

Note: Parameters obtained by X-ray diffraction at room temperature.

older data were also incompatible with the selected values below 50 at.% N.

[72Bil] examined TiN specimens for ordering. No long-range order was found, but short-range order was discerned by electron diffraction at compositions from about 38 to 43 at.% N at temperatures up to about 1500 °C. However, [77Arb1] suggested that δ' is related to TiN, an implication that is interpreted as referring to long-range ordering.

The second tetragonal nitride, δ', was originally identified by [69Lob] as a low-temperature, ordered form of cubic TiN at the Ti_2N composition, but with a bct structure. [77Arb1], in confirming the relationship to TiN, noted that it formed from quenched TiN by a first-order reaction. They also showed that the narrow range of equilibrium δ' compositions was near 37 not 33.3 at.% N. The investigation of [77Nag] agreed with these findings regarding structure. Contrary to the findings of [69Lob] and [77Nag] that δ' is bct, the report* of [77Arb2] described the structure as fct. The report of an fct structure is apparently incorrect. There is good agreement among the various investigations on the lattice parameters except for [77Nag] (Table 6). This difference may be related to the very long aging times used by [77Nag].

Thermodynamics

Apart from a determination of the heats of formation, there are no thermodynamic data for the terminal phases, (αTi), and (βTi), except at the pure Ti extremity. For the liquid phase, there are data only at 0 and 50 at.% N. Among Ti_2N, δ', and TiN, there are thermodynamic data only for TiN.

[66Mor] determined by combustion calorimetry heats of formation at 25 °C of Ti-N alloys with up to 49.5 at.% N prepared at 1100 °C and quenched. Alloys so prepared were (αTi) when they contained less than 20 at.% N and TiN when they contained more than 31 at.% N; alloys

*English translation.

Fig. 5---Variation of Lattice Parameters with N Concentration in TiN

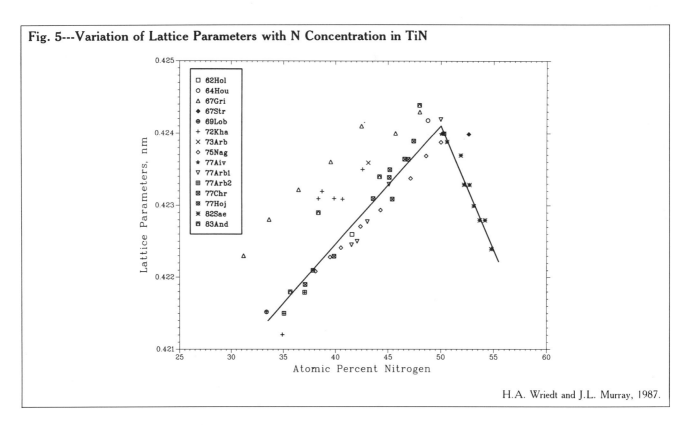

H.A. Wriedt and J.L. Murray, 1987.

Table 6 Lattice Parameters of δ′

Investigator	Composition, at.% N	Boundary Temperature, °C	Boundary Coexisting phase	Lattice parameters, nm a	Lattice parameters, nm c
[69Lob]	33.3(a)	500	?	0.4140	0.8805
[77Arb1]	~ 34 to 37(a)	500	Ti₂N	0.4144	0.8781
	~ 38 to 40(a)	500	TiN	0.4144	0.8781
[77Arb2]	35 to 37(a)	500	?	0.4144	0.8787
[77Nag]	37.9(b)	...	...	0.4198	0.8591

Note: Parameters obtained by X-ray or electron diffraction at room temperature. (**a**) Reported compositions outside the range for single-phase δ′. (**b**) Single phase.

with 20 to 31 at.% N were duplex with Ti₂N present. The data were included in the [Hultgren, B] review of the Ti-N system.

Thermodynamic data for TiN may be divided for descriptive purposes into those for the special composition, stoichiometric TiN, and those for all other compositions. The low-temperature and high-temperature thermophysical and thermochemical properties for stoichiometric TiN were reviewed and tabulated by [68Stu] and by [Hultgren, B] in 1969. The breadth of literature coverage in the latter review, which included newer determinations, was significantly greater. In addition to the [66Mor] data, [Hultgren, B] also tabulated heats of formation at 1657 °C for alloys with 44 to 50.3 at.% N. For the latter alloys at 1657 °C, integral and partial enthalpies, entropies, and Gibbs energies of formation were also tabulated.

No calorimetric measurements of the thermodynamic properties of TiN were found for the period subsequent to that surveyed by [Hultgren, B]. However, the vapor over various compositions of the phase were investigated experimentally. [69Ryk] determined Ti and N₂ pressures by a Knudsen effusion method over TiN with 46 at.% N at

1870 °C, where it was considered to evaporate congruently. They calculated a value for the standard heat of formation of this solid alloy (279 kJ/g per formula weight of TiN₀.₈₆) that was in fair agreement with the calorimetric value (−290 kJ) of [66Mor]. [69Ste] (and in modified form [70Koh]) reported the heat of formation of solid stoichiometric TiN as 337.7 kJ/mol (compared to 326 kJ/mol [66Mor]), based on mass spectrometric examination of vapor from a Knudsen celll containing TiN with 43.8 at.% N at temperatures above 1977 °C. The heat of formation of gaseous TiN from elemental Ti and N₂ gas were given as +7.5 ± 29 kJ/mol TiN [69Ste] and −3.3 kJ/mol TiN [70Koh]. Unlike [69Ryk], [69Ste], and [70Koh], who studied only one composition, [83And] studied the vapor over four single-phase compositions between 35.6 and 48.0 at.% N at 1227 to 2227 °C. Component partial pressures and activities were evaluated, and the concentration dependence of activity was discussed. The standard heat of formation of solid stoichiometric TiN was reported as 345.4 kJ/mol. Congruently vaporizing compositions were determined in two investigations, with quite different results:

$x/(100 - x) = 1.46 - 2.8 \times 10^{-4} \, T$ [70Ryk]

or

$x/(100 - x) = 0.93 - 7.3 \times 10^{-5} \, T$ [83And]

where T is temperature in K, and x is N concentration in at.%. These equations agree only at about 2300 °C.

Suggestions for Further Experimental Research

Many features of the Ti-N phase diagram are placed in accordance with their original and only determination by [54Pal]. These include the two peritectic equilibria and most of the univariant curves of the condensed system. Except for individual points, the liquidus segments are mostly unknown and the (βTi) liquidus as determined by [54Pal] terminates at a melting value for pure Ti that is incorrect by 55 °C. Uncertainty still attaches to the nature of the (αTi)-TiN-Ti$_2$N equilibrium and to the integral question of whether Ti$_2$N decomposes congruently. Together, the questions raised indicate the scientific desirability of redetermining much of the diagram with newer methods and higher purity materials.

Cited References

25Fri: E. Friederich and L. Sittig, "Production and Properties of Nitrides," *Z. Anorg. Chem.*, 143, 293-320 (1925) in German. (Equi Diagram; Experimental)

31Agt: C. Agte and K. Moers, "Methods for Producing Pure High-Melting Carbides, Nitrides and Borides and Description of Some of Their Properties," *Z. Anorg. Chem.*, 198, 233-275 (1931) in German. (Equi Diagram; Experimental)

39Bra1: A. Brager, "An X-Ray Examination of Titanium Nitride I. Single Crystal Investigation," *Acta Physico. URSS*, 10(4), 593-600 (1939). (Crys Structure; Experimental)

39Bra2: A. Brager, "An X-Ray Examination of Titanium Nitride III. Investigation by the Powder Method," *Acta Physico. URSS*, 11(4), 617-632 (1939). (Crys Structure; Experimental)

49Cla: H.T. Clark, Jr., "The Lattice Parameters of High Purity α Titanium; The Effects of Oxygen and Nitrogen on Them," *Trans. AIME*, 185, 588-589 (1949). (Crys Structure; Experimental)

49Ehr: P. Ehrlich, "On the Binary Systems of Titanium with the Elements Nitrogen, Carbon, Boron and Beryllium," *Z. Anorg. Chem.*, 259, 1-41 (1949). (Crys Structure, Equi Diagram; Experimental)

50Duw: P. Duwez and F. Odell, "Phase Relationships in the Binary Systems of Nitrides and Carbides of Zirconium, Columbium, Titanium, and Vanadium," *J. Electrochem. Soc.*, 97(10), 299-304 (1950). (Crys Structure; Experimental)

50Jaf: R.I. Jaffee, H.R. Ogden, and D.J. Maykuth, "Alloys of Titanium with Carbon, Oxygen, and Nitrogen," *Trans. AIME*, 188, 1261-1266 (1950). (Equi Diagram; Experimental)

53Sto: L. Stone and H. Margolin, "Titanium-Rich Regions of the Ti-C-N, Ti-C-O, and Ti-N-O Phase Diagrams," *Trans. AIME*, 197, 1498-1503 (1953). (Equi Diagram; Experimental)

*54Pal: A.E. Palty, H. Margolin, and J.P. Nielsen, "Titanium-Nitrogen and Titanium-Boron Systems," *Trans. ASM*, 46, 312-328 (1954). (Equi Diagram; Experimental)

55Hoc: M. Hoch, D.P. Dingledy, and H.L. Johnston, "The Vaporization of TiN and ZrN," *J. Am. Chem. Soc.*, 77, 304-306 (1955). (Thermo; Experimental)

55Sza: I. Szántó, "On the Determination of High-Purity α-Titanium Lattice Parameters," *Acta Tech. Acad. Sci. Hung.*, 13, 363-372 (1955). (Crys Structure; Experimental)

58Sku: P. Skulari and L. Chvátalová, "X-Ray Examination of Titanium," *Hutnicke Listy*, 13, 899-908 (1958) in Czechoslovakian. (Crys Structure; Review)

59Kau: L. Kaufman, "The Lattice Stability of Metals I. Titanium and Zirconium," *Acta Metall.*, 7(8), 575-587 (1959). (Thermo; Theoretical)

61Gor: N.S. Gorbunov, N.A. Shishakov, G.G. Sadikov, and A.A. Babad-Zakhryapin, "Neutronographic Study of Carbides and Nitrides of Titanium," *Izv. Akad. Nauk SSSR, Otdel. Khim. Nauk*, (11), 2093-2095 (1961) in Russian; TR: *Acad. Nauk SSSR, Bull. Div. Chem. Sci.*, (11), 1953-1955 (1961). (Crys Structure; Experimental)

61Now: H. Nowotny, F. Benesovsky, C. Brukl, and O. Schob, "The Ternaries: Titanium-Boron-Carbon and Titanium-Boron-Nitrogen," *Monatsh. Chem.*, 92, 403-414 (1961). (Crys Structure, Equi Diagram; Experimental)

62Aki: P.A. Akishin and Yu.S. Klodeev, "Mass Spectrometric Investigation of the Composition of the Vapour Phase Above the Nitrides of Zirconium, Titanium, and Boron," *Zh. Neorg. Khim.*, 7(4), 941-942 (1962); TR: *Russ. J. Inorg. Chem.*, 7(4), 486 (1962). (Thermo; Experimental)

62Hol: B. Holmberg, "Structural Studies on the Titanium-Nitrogen System," *Acta Chem. Scand.*, 16(5), 1255-1261 (1962). (Crys Structure, Equi Diagram; Experimental)

62Woo: R.M. Wood, "The Lattice Constants of High Purity α Titanium," *Proc. Phys. Soc.*, 80, 783-786 (1962). (Crys Structure; Experimental)

64Hou: C.R. Houska, "Thermal Expansion and Atomic Vibration Amplitudes for TiC, TiN, ZrC, ZrN and Pure Titanium," *J. Phys. Chem. Sol.*, 25, 359-366 (1964). (Crys Structure; Experimental)

*65Mcc: L.A. McClaine and C.P. Coppel, "Equilibrium Studies of Refractory Nitrides, Part I. Details of the Apparatus and Studies of the Ti-N System," U.S. Air Force Systems Command, Res. Technol. Div., Tech. Rept. AFML-TR-65-299 (1965) (AD-474087). (Equi Diagram; Experimental, Review)

66Mor: M.P. Morozova and M.M. Khernburg, "Enthalpy of Formation of Titanium Nitrides as a Function of Their Composition," *Zhur. Fiz. Khim.*, 40(5), 1125-1128 (1966) in Russian; TR: *Russ. J. Phys. Chem.*, 40(5), 604-606 (1966). (Thermo; Experimental)

67Gri: P. Grieveson, "An Investigation of the Ti-C-N System," *Proc. Brit. Ceram. Soc.*, (8), 137-153 (1967). (Crys Structure, Equi Diagram, Thermo; Experimental)

67Str: M.E. Straumanis, C.A. Faunce, and W.J. James, "Bonding, Lattice Parameter, Density and Defect Structure of TiN Containing an Excess of N," *Acta Metall.*, 15(1), 65-71 (1967). (Crys Structure; Experimental)

68Ren: E.H. Rennhack, W.C. Coons, and R.A. Perkins, "High-Temperature Stability of Epsilon Ti$_x$N," *Trans. TMS-AIME*, 242, 343-345 (1968). (Equi Diagram; Experimental)

68Stu: D.R. Stull and H. Prophet, "JANAF Thermochemical Tables," 2nd ed., NSRDS-NBS37, U.S. Gov. Printing Office, Washington, DC (1968).

69Lob: G. Lobier and J.-P. Marcon, "Study and Structure of a New Phase of the Sub-Nitride of Titanium Ti$_2$N," *Compt. Rend. Acad. Sci. Paris (Ser. C)*, 268, 1132-1135 (1969) in French. (Crys Structure, Equi Diagram; Experimental)

69Ryk: E.A. Ryklis, A.S. Bolgar, and V.V. Fesenko, "Vaporization Rate and Thermodynamic Properties of Titanium Nitride," *Porosh. Met.*, 9(6), 62-64 (1969); TR: *Sov. Powder Met. Met. Ceram.*, 9(6), 478-480 (1969). (Thermo; Experimental)

69Ste: C.A. Stearns and F.J. Kohl, "The Dissociation Energy of Gaseous Titanium Mononitride," Nat. Aeronautics Space Admin. Tech. Note NASA-TN-D-5027 (1969). (Thermo; Experimental)

70Koh: F.J. Kohl and C.A. Stearns, "Mass Spectrometric Studies of the Vaporization of Refractory Carbides and Nitrides," Nat. Aeronaut. Space Admin. Report NASA SP-227, Aerospace Structural Materials, 173-185 (1970). (Thermo; Experimental)

70Kou: M.K. Koul and J.F. Breedis, "Phase Transformations in β Isomorphous Titanium Alloys," *Acta Metall.*, 18, 579-588 (1970). (Meta Phases; Experimental)

70Mcd: N.R. McDonald and G.R. Wallwork, "The Reaction of Nitrogen with Titanium Between 800 and 1200 °C," *Oxid. Met.*, 2(3), 263-283 (1970). (Equi Diagram; Experimental)

Phase Diagrams of Binary Titanium Alloys

70Ryk: E.A. Ryklis, A.S. Bolgar, O.P. Kulik, S.A. Shvab, and V.V. Fesenko, "Evaporation of Titanium Nitride at High Temperatures," *Zhur. Fiz. Khim., 44*(5), 1292-1297 (1970) in Russian; TR: *Russ. J. Phys. Chem., 44*(5), 720-722 (1970). (Thermo; Experimental)

71Tot: L.E. Toth, *Transition Metal Carbides and Nitrides,* Academic Press, New York (1971). (Equi Diagram; Experimental)

72Bil: J. Billingham, P.S. Bell, and M.H. Lewis, "Vacancy Short-Range Order in Substoichiometric Transition Metal Carbides and Nitrides with the NaCl Structure. I. Electron Diffraction Studies of Short-Range Ordered Compounds," *Acta Crystallogr., A28,* 602-608 (1972). (Crys Structure; Experimental)

72Kha: V.V. Khaenko, E.T. Kachkovskaya, and O.A. Frenkel, "Condition of Titanium and Vanadium Mononitride Particles after Nitriding and Heat Treatment," *Porosh. Met., 11*(7), 34-39 (1972) in Russian; TR: *Sov. Powder Met. Met. Ceram., 11*(7), 541-545 (1972). (Equi Diagram; Experimental)

73Arb: M.P. Arbuzov, B.V. Khaenko, and E.T. Kachkovskaya, "The Real Structure of Titanium Mononitride in Its Homogeneity Range," *Porosh. Met., 12*(6), 69-74 (1973) in Russian; TR: *Sov. Powder Met. Met. Ceram., 12*(6), 490-493 (1973). (Crys Structure; Experimental)

73Sam: G.V. Samsonov and M.M. Antonova, "Formation of Compounds in Systems of Titanium and Zirconium Nitride Alloys with Hydrogen," *Zhur. Prikl. Khim., 46*(1), 13-17 (1973) in Russian; TR: *Russ. J. Appl. Chem., 46*(1), 11-15 (1973). (Equi Diagram; Experimental)

74Bar: J.-P. Bars, E. Etchessahar, J. Debuigne, and A. Lerous, "On the Nitriding of Titanium by Nitrogen at High Temperature," *Compt. Rend. Acad. Sci. Paris (Ser. C), 278,* 581-584 (1974) in French. (Equi Diagram; Experimental)

74Ett: P. Ettmayer, R. Kieffer, and F. Hattinger, "Determination of Melting Points of Metal Nitrides Under Nitrogen Pressure," *Metall, 28*(12), 1151-1156 (1974) in German. (Equi Diagram; Experimental)

74Lev: Yu.V. Levinskii, "*P-T* Projection of the Phase Diagram of the System Ti-N," *Izv. Akad. Nauk SSSR Neorg. Mater, 10*(9), 1628-1631 (1974) in Russian; TR: *Inorg. Mater., 10*(9), 1403-1405 (1974). (Equi Diagram; Review)

74Woo: F.W. Wood, P.A. Romans, R.A. McCune, and O.G. Paasche, "Phases and Interdiffusion Between Titanium and Its Mononitride," U.S. Bur. Mines, Rept. Inv 7943 (1974). (Equi Diagram; Experimental)

75Nag: S. Nagakura, T. Kusunoki, F. Kakimoto, and Y. Hirotsu, "Lattice Parameter of the Non-Stoichiometric TiN_x," *J. Appl. Crystallogr., 8,* 65-66 (1975). (Crys Structure; Experimental)

76Ero: M.A. Eron'yan, R.G. Avarbé, and T.N. Danisina, "Effects of Equilibrium Nitrogen Pressure on the Melting Points of TiN_n and HfN_n," *Teplofiz. Vys. Temp., 14*(2), 398-399 (1976) in Russian; TR: *High Temp., 14*(2), 359-360 (1976). (Equi Diagram; Experimental)

77Aiv: M.I. Aivazov and T.V. Rezchikova, "Nature of the Formation of the Phases in the M-Sc-N and M-Mn-N Systems," *Zhur. Neorg. Khim., 22*(2), 458-463 (1977) in Russian; TR: *Russ. J.*

Inorg. Chem., 22(2), 250-253 (1977). (Crys Structure; Experimental)

***77Arb1:** M.P. Arbuzov, S.Ya. Golub, and B.V. Khaenko, "X-Ray Investigation of Titanium Nitrides," *Izv. Akad. Nauk SSSR, Neorg. Mater., 13*(10), 1779-1789 (1977) in Russian; TR: *Inorg. Mater., 13*(10), 1434-1437 (1977). (Equi Diagram; Experimental)

77Arb2: M.P. Arbuzov, S.Ya. Golub, B.V. Khayenko, "Study of the Precipitation of an Ordered δ' Phase from Titanium Mononitrides," *Fiz. Met. Metalloved, 44*(3), 666-668 (1977) in Russian; TR: *Phys. Met. Metallogr., 44*(3), 194-197 (1977). (Crys Structure, Equi Diagram; Experimental)

77Bar: J.-P. Bars, E. Etchessahar, and J. Debuigne, "Kinetic Diffusional and Morphological Study of the Nitriding of Titanium by Nitrogen at High Temperature: Mechanical and Structural Properties of Alpha Solid Solution Ti-Nitrogen," *J. Less-Common Met., 52,* 51-76 (1977). (Crys Structure, Equi Diagram; Experimental)

77Chr: A.N. Christensen and S. Fregerslev, "Preparation, Composition, and Solid State Investigations of TiN, ZrN, NbN and Compounds from the Pseudo-Binary Systems NbN-NbC, NbN-TiC and NbN-TiN," *Acta Chem. Scand., A31*(10), 861-868 (1977). (Crys Structure; Experimental)

77Hoj: J. Hojo, O. Iwamoto, Y. Maruyama, and A. Kato, "Defect Structure, Thermal and Electrical Properties of Ti Nitride and V Nitride Powders," *J. Less-Common Met., 53,* 277-286 (1977). (Crys Structure; Experimental)

77Nag: S. Nagakura and T. Kusunoki, "Structure of TiN_x Studied by Electron Diffraction and Microscopy," *J. Appl. Crystallogr., 10,* 52-56 (1977). (Crys Structure, Equi Diagram; Experimental)

78Mit: H. Mitsuhashi, Y. Igasaki, and M. Kanenko, "Properties of Titanium Solid Solutions Prepared by Reactive Sputtering," *J. Crystallogr. Growth, 45,* 350-354 (1978). (Meta Phases; Experimental)

80Sun: D. Sundararaman, V. Seetharaman, and V.S. Raghunathan, "Phase Transformations in a Ti-1.6 at.% N Alloy," *Titanium '80,* Vol. 2, 1521-1532, TMS-AIME, Warrendale, PA (1980). (Meta Phases, Equi Diagram; Experimental)

82Sae: Y. Saeki, R. Matsuzaki, A. Yajima, and M. Akiyama, "Reaction Process of Titanium Tetrachloride with Ammonia in the Vapor Phase and Properties of the Titanium Nitride Formed," *Bull. Chem. Soc. Jpn., 55*(10), 3193-3196 (1982). (Crys Structure; Experimental)

83And: R.A. Andrievskii, Yu.F. Khromov, D.E. Svistunov, and R.S. Yurkova, "Partial Thermodynamic Characteristics of Titanium Nitride," *Zhur. Fiz. Khim., 57*(7), 1641-1644 (1983) in Russian; TR: *Russ. J. Phys. Chem., 57*(7), 996-998 (1983). (Thermo; Experimental)

83Sun: D. Sundararaman, A.L.E. Terrance, V. Seetharaman, and V.S. Raghunathan, "Electron Microscopy Study of Microstructural Changes in a Ti–1.6 at.% N Alloy," *Trans. Jpn. Inst. Met., 24*(7), 510-513 (1983). (Crys Structure, Meta Phases; Experimental)

The Na-Ti (Sodium-Titanium) System

22.98977   47.88

By C.W. Bale

Equilibrium Diagram

There are no published experimental data for the Na-Ti phase diagram (Fig. 1). By analogy with the Na-Zr and other alkali metal–Group IVA systems, it is probable that the Na-Ti system is almost completely immiscible in both the solid and liquid states.

If the solubilities are very limited, it follows from thermodynamic considerations that the univariant temperatures in the phase diagram are virtually identical with the transition temperatures of the pure components.

From vapor-pressure calculations, it is estimated that the composition of the gas phase at 2000 °C and 1 atm is approximately 0.04 at.% Ti, in equilibrium with almost pure liquid Ti.

Crystal structure data for the elements are summarized in Table 1.

Table 1   Crystal Structures and Lattice Parameters of the Na-Ti System [King2]

Phase	Composition, at.% Ti	Pearson symbol	Space group	Strukturbericht designation	Prototype	Lattice parameters, nm		Comment
						a	c	
(αNa)	0	$hP2$	$P6_3/mmc$	$A3$	Mg	0.3767	0.6154	Below 36 K
(βNa)	0	$cI2$	$Im3m$	$A2$	W	0.42096	...	At RT
(αTi)	100	$hP2$	$P6_3mmc$	$A3$	Mg	0.29503	0.46836	At RT
(βTi)	100	$cI2$	$Im3m$	$A2$	W	0.33065	...	Above 900 °C

Fig. 1---Assessed Na-Ti Phase Diagram

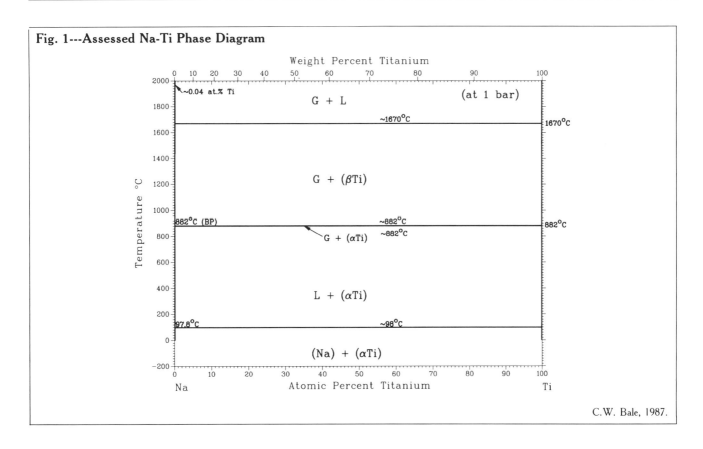

C.W. Bale, 1987.

The Nb-Ti (Niobium-Titanium) System

92.9064   47.88

By J.L. Murray

Equilibrium Diagram

The equilibrium solid phases of the Ti-Nb system are:

- The bcc (βTi,Nb) solid solution, with a complete range of solubility above 882 °C. For brevity, the phase (βTi,Nb) will be referred to as β. The β/[β + (αTi)] boundary will be referred to as the β transus.
- The low-temperature cph (αTi) solid solution, with restricted solubility of Nb.

The assessed phase diagram (Fig. 1) does not differ qualitatively from previous evaluations [Hansen, Elliott, Shunk]. The stable equilibrium diagram is without invariant reactions, congruent transformations, or critical points. The transformation points of the pure components are:

Reaction	Composition, at.% Nb	Temperature, °C
L ⇌ (βTi)	0	1670
L ⇌ (Nb)	100	2469
(βTi) ⇌ (αTi)	0	882

The assessed phase diagram is drawn from the present thermodynamic calculations, which reproduce ex-perimental data within the uncertainties arrived at in the present assessement. Thermodynamic calculations will be discussed in detail below.

Solidus and Liquidus.

The solidus was investigated by [51Han], [69Rud], and [69Zak]. [51Han] microscopically examined heat treated for signs of incipient melting; an uncertainty of ± 25 °C was assigned to the data based on the temperature interval examined. For Ti-rich alloys, the reported melting point of 1720 °C (vs the assessed value of 1670 °C) indicates that ± 50 °C is a more realistic uncertainty. [69Rud] and [69Zak] made optical observations of incipient melting; an uncertainty of ± 20 °C is estimated for Ti-rich alloys and somewhat more for the Nb-rich alloys. The data are in good agreement and are compared with the assessed phase diagram in Fig. 2.

Experimental data are not available for the liquidus, because of the narrow melting range. The liquidus shown in Fig. 1 and 2 was determined by thermodynamic calculations. Based on the assumption that regular solution parameters cannot differ from those chosen by more than ± 10 kJ/mol, the uncertainty in the liquidus is no more than ± 2 at.%.

Solubility of Nb in (αTi).

The maximum solubility of Nb in (αTi) is 2.2 ± 0.5 at.% at about 600 to 650 °C. Experimental data on the

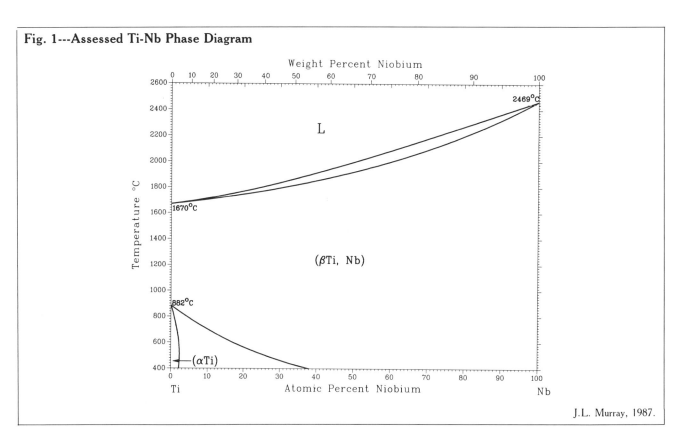

Fig. 1---Assessed Ti-Nb Phase Diagram

J.L. Murray, 1987.

Fig. 2---Experimental Solidus Data vs the Assessed Phase Diagram

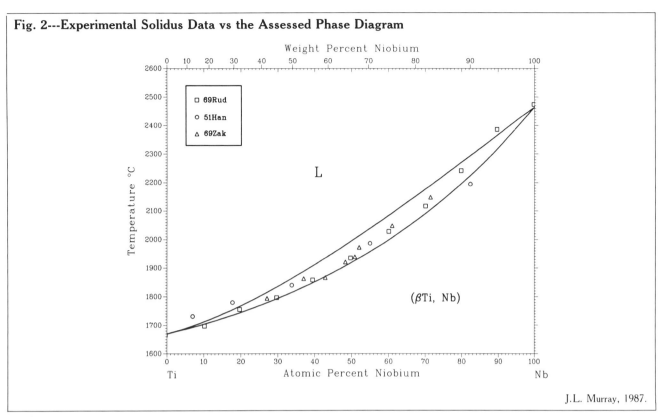

J.L. Murray, 1987.

Fig. 3---Experimental Data on the (αTi)/β Phase Boundaries vs the Assessed Diagram Based Primarily on [66Bro]

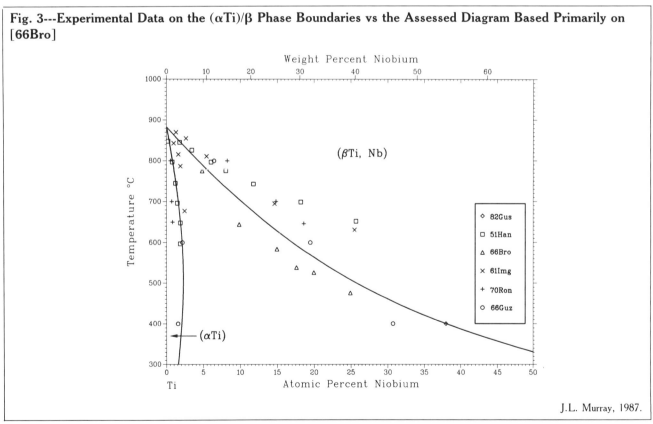

J.L. Murray, 1987.

(αTi) solvus are compared with the assessed diagram in Fig. 3. An additional value of about 2.5 at.% at 600 °C was cited by [65Rau].

The highest solubilities were found by investigators who cold-worked their samples prior to annealing to attain equilibrium; the same correlation between high solubility and cold-work has been found in other analogous titanium systems, e.g., Ti-Ta.

β Transus.

Experimental data on the β transus [51Han, 61Img, 64Bro, 66Bro, 66Guz, 70Ron, 82Gus] show considerable scatter, as shown in Fig. 3. Determinations by [51Han], [61Img], [61Sha], and [66Guz] using classical metallographic techniques and by [70Ron] using microprobe analysis of annealed diffusion couples are essentially in agreement, except at the lowest temperatures. Below about 600 °C, reaching equilibrium becomes difficult even in very long anneals.

[64Bro] determined the β transus by resistivity measurements on cooling and verified the results by rapid quenching experiments [66Bro]. The results of [64Bro, 66Bro] showed the β transus to be lower in temperature than the determinations cited above, by as much as 150 °C at 25 at.% Nb. In other Ti-based systems, for example Ti-Cr, rapid quenching methods have been shown to be necessary to determine the β transus. The effect of sample contamination is also to raise the apparent temperature of the β transus.

The present assessment is therefore based primarily on [64Bro] and [66Bro]. The assessed phase boundary is placed somewhat above the resistivity (cooling) results. The recent determination of the volume fractions of (αTi) and β in equilibrium at 400 °C [82Gus] supports placing the equilibrium β transus somewhat above the resistivity results.

Nonexistence of Equilibrium Intermetallic Compounds.

Several intermetallic compounds have been proposed as equilibrium phases of the Ti-Nb system—TiNb [61Sha], Ti_3Nb [57Gru, 68Kal], $TiNb_2$, Ti_2Nb, and Ti_5Nb [58Gru]. The reports of compound phases were based on anomalies in physical properties, for example, the Hall constant [58Gru] and electrical resistivity and thermal conductivity [68Kal]. [66Guz] and [69Zak] attempted to verify the existence of these compounds by X-ray diffraction, but superlattice lines or metallographic evidence for the phases were not found at any compositions. Anomalies in physical properties can probably be adequately explained in terms of the metastable phases that form during quenching and low-temperature aging.

Metastable Phases

Hexagonal and Orthorhombic Martensites.

The β phase can transform martensitically during quenching to the cph (α'Ti) form for low Nb contents, or to an orthorhombic distortion of the cph form (α''Ti) for higher Nb contents. At low Nb concentrations, (α'Ti) appears regardless of quenching rate. The (α'Ti) martensite is hexagonal in structure and above 0.25 at.% Nb displays an acicular morphology. The concentration range in which (α'Ti) is found was reported as:

Reference	Concentration range, at.% Nb
[63Bor]	0 to > 5.4
[64Bro]	0 to ~ 7.5
[58Bag]	0 to ~ 5
[72Mor]	0 to ~ 11

The product of the transformation in this compilation range is independent of quenching rate [72Mor].

At higher Nb concentrations, the proportions of (α''Ti), β, and ω depend on cooling rate [70Jep, 72Mor, 73Mor, 75Ple]. For example, [58Bag] found that the orthorhombic martensite (α''Ti) appears between 8 and 15 at.% Nb and that ω appears between 17 and 18 at.% Nb. By using higher quenching rates, [72Mor] could produce (α''Ti) at 28 at.% Nb, but they also found ω after slow quenches.

Start temperatures, M_s, of the martensite transformation [53Duw, 60Sat, 64Bro, 66Bro, 70Jep, 75Ple, 80Obs] are shown superimposed on the equilibrium diagram in Fig. 4.

The shape memory effect is obtained in this system; alloys deformed below M_s recover their shape upon heating above the start temperature of the reverse martensitic transformation [71Bak, 80Obs]. The shape memory effect does not occur if ω phase precipitation intervenes [80Obs].

Tetragonal Martensite τ.

A tetragonal martensite, τ, closely related to (α''Ti) in structure was reported by [68Hat, 74Kad, 79Lya, 81Pat]. The τ phase is formed only as a result of cold working. According to [79Lya] and consistent with the other observations, τ forms in the composition range 15 to 26 at.% Nb and disappears during reheating to 200 °C.

Metastable (βTi,Nb) Miscibility Gap.

There is some evidence to suggest that the unmixing of bcc alloys within a metastable miscibility gap has been observed experimentally. For a 34 at.% Nb alloy, [78Men] interpreted a mottled contrast in the matrix of aged specimens as evidence of phase separation by a spinodal mechanism. In a quenched and aged 31.4 at.% Nb alloy, [80Lyo] reported a new tetragonal phase β' with a morphology similar to that of ordered domains. It can be tentatively suggested that β' may be coherent phase-separated β. [71Men] reported coherent, ellipsoidal precipitates of a second bcc phase in a 21.7 at.% Nb alloy aged at 375 °C; however, in these observations, there is some conflict with other work on the ω phase.

ω Phase.

The ω phase can form either during quenching from the β field (athermally) [58Bag, 69Hic, 72Bal, 81Lyo], or during aging at temperatures below about 450 °C [67Bra, 69Hic, 80Osa, 81Lyo]. ω phase is not formed in alloys that have undergone the martensitic transformation [69Hic, 73Mor]. Above about 500 °C, (αTi) precipitates directly from β [69Hic, 72Bal, 73Mor].

The ideal ω structure is hexagonal, with symmetry $P6/mmm$. In some alloys, the ω phase is distorted to a trigonal structure of symmetry $P\bar{3}m1$, but in Ti-Nb alloys, the structure appears to be hexagonal [69Hic]. [72Bal] deduced from the appearance of forbidden reflections that the aged form of ω is ordered. These lines, however, may also be due to distortion of the hexagonal lattice [69Sas].

ω phase precipitates as ellipsoidal particles coherent with the bcc matrix [67Bra, 69Hic, 72Bal, 73Mor], indicative of low misfit with the matrix. The growth kinetics of the particles was investigated by [80Osa] and [81Lyo].

As-quenched ω is found in the approximate composition range 13 to 18 at.% Nb [58Bag, 69Hic]. [69Hic] found ω in the range 9 to 30 at.% Nb after aging at 450 °C. [80Osa] proposed a wider range on the Nb-side; ω was

Fig. 4---Start Temperatures, M$_s$, of the Martensitic β to (α'Ti) or β to (α"Ti) Transformation Superimposed on the Equilibrium Diagram

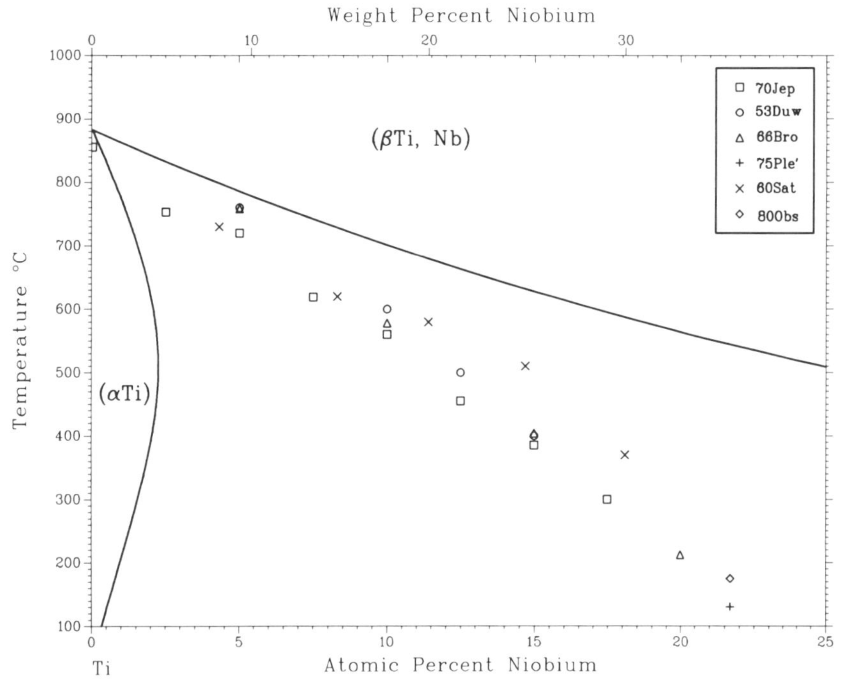

J.L. Murray, 1987.

Table 1 Crystal Structures of the Ti-Nb System

Phase	Homogeneity range, at.% Nb	Pearson symbol	Space group	Struktur-bericht designation	Prototype	Reference
(αTi)	0 to 2.5	hP2	P6$_3$/mmc	A3	Mg	[Pearson2]
(α'Ti)	0 to ~5	hP2	P6$_3$/mmc	A3	Mg	[58Bag]
(βTi,Nb)	0 to 100	cI2	Im3m	A2	W	[Pearson2]
(α"Ti)	~ 8 to 28	oC4	Cmcm	A20	αU	[58Bag]
ω	9 to 30	hP3	P6/mmm	...	ωMnTi	[69Hic]
τ	15 to 26	(a)	C4/mmm	...	...	[68Hat]

(a) Body-centered tetragonal.

found in a 36 at.% Nb alloy, but the precipitation was very slow.

During aging, the β matrix is depleted in Nb, and a metastable two-phase equilibrium between ω and β is approached. Based on lattice parameter data, [69Hic] found compositions of 9 and 30 at.% Nb for ω and β, respectively, at 450 °C. Based on analysis of SAXS data, [81Lyo] found a ratio x_ω:x_β of 0.86 for alloys aged at the same temperature.

The ω phase is an equilibrium phase of pure Ti at high pressure [Ti]. [73Afo] studied ω phase formation in 10 and 20 at.% Nb alloys under pressure; ω phase was formed at 30 and 50 kbar, respectively. The structure and lattice parameters of ω formed under pressure were found to be the same as those of ω produced by appropriate heat treatment. Using a shock wave of amplitude 320 ± 20 kbar, [79Sik] also produced ω in 10

and 20 at.% Nb alloys previously annealed in the two-phase β + (αTi) region.

Crystal Structures and Lattice Parameters

Crystal structure data on the equilibrium and meta-stable phases of the Ti-Nb system are summarized in Table 1.

Lattice parameters for the β phase retained during quenching and for martensitic (α"Ti) are shown in Fig. 5 and 6, respectively. For (α"Ti), c is not plotted, because it is independent of composition [58Bag, 0.468 nm; 72Mor, 0.470 nm]. Also, b/√3 rather than b is plotted so that the degree of distortion of the lattice can be easily seen. Lattice parameters of the (α'Ti) martensite do not differ measurably from those of pure αTi.

Additional lattice parameter data for the ω and τ phases are given in Table 2.

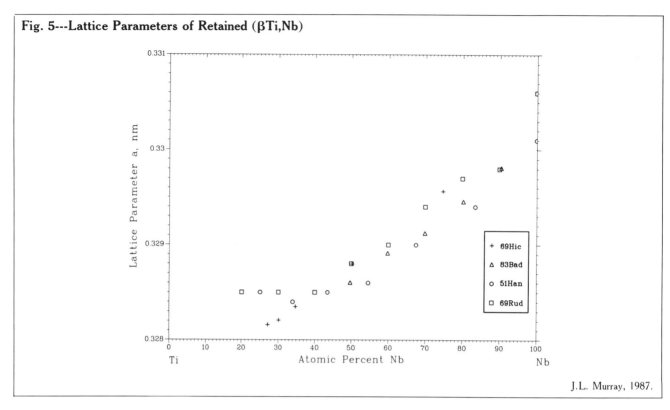

Fig. 5---Lattice Parameters of Retained (βTi,Nb)

J.L. Murray, 1987.

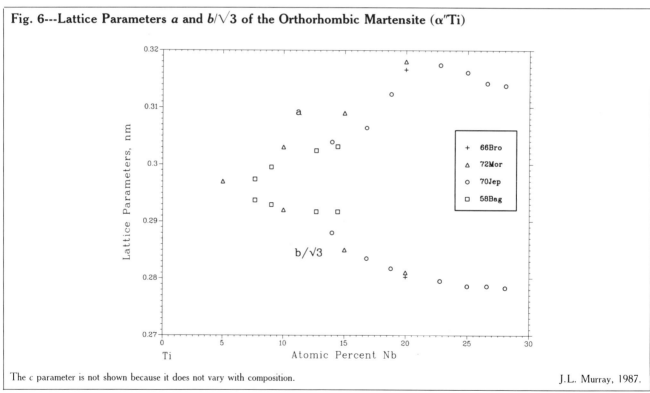

Fig. 6---Lattice Parameters *a* and *b*/√3 of the Orthorhombic Martensite (α″Ti)

The *c* parameter is not shown because it does not vary with composition.

J.L. Murray, 1987.

Thermodynamics

Experimental Data.

[64Rud] performed diffuse X-ray scattering experiments and found that there is a tendency to clustering rather than short-range order in this system. Thus, they concluded that the deviations from ideality are positive. Experimental determinations of thermodynamic quantities have not been made for the Ti-Nb system.

Calculations of the Phase Diagram.

Previous calculations of the phase diagram were

Table 2   Lattice Parameters of Metastable Phases

Reference	Structure	Composition, at.% Nb	Lattice parameters, nm	
			a	c
[69Hic] ... ω (aged)		...	0.4627	0.2836
[68Hat] ... τ		25	0.328	0.318
[81Pat] ... τ		21.7	0.328	0.3225
[74Kad] .. τ		14.7	0.312	0.355
		18.1	0.317	0.353
		21.7	0.324	0.352
		25.6	0.324	0.349

made by [64Rud], [75Che], and [78Kau], all based on phase diagram data. [64Rud] used phase diagram data given in [Hansen] and assumed zero solubility of Nb in (αTi). The assumption of zero solubility is not valid for this system, and these results they are not reproduced here. [75Che] used an average of the data of [66Bro] and the diagram of [Hansen]. Gibbs energies of [78Kau] reproduce the [Hansen] diagram.

The present calculations were undertaken because of the re-evaluation of the experimental data, resulting in a significant change in the β transus. The assessed phase diagram (Fig. 1) is the result of the present calculations using the regular solution approximation for the excess Gibbs energies.

The Gibbs energies of the three equilibrium phases are represented as:

$$G(i) = G^0(Ti,i) (1 - x) + G^0(Nb,i) x + RT [x \ln x + (1 - x) \ln (1 - x) + B(i) x(1 - x)$$

where i designates the phase; x is the atomic fraction of Nb; G^0 are the properties of the pure components; and $B(i)$ is the regular solution interaction parameter. The properties of the pure components are from [70Kau]. Numerical values of the thermodynamic parameters are summarized in Table 3.

For any nearly (± 10 kJ/mol) ideal solution model for the excess Gibbs energy of the bcc phase, corresponding functions can be found for the cph and liquid phases that reproduce the (αTi) solvus and the solidus. Thus, only the β transus depends on the magnitude of the exess functions, rather than differences between the various phases. Because of the presence of the metastable miscibility gap, the β transus is lower, the lower the bcc phase regular solution parameter. In order to exactly reproduce the resistivity results of [66Bro], the bcc solution parameter must be negative, contrary to observation [64Rud]. However, there is justification for believing that the resistivity data underestimate the temperature of the β transus. In the present calculations, the lowest excess Gibbs energies are chosen subject to the constraint that all the excess quantities must be positive, and the assessed phase diagram is reproduced within the assessed uncertainty.

The present results do not yet succeed in producing a completely satisfactory picture of the thermodynamics of the system. First, the critical point of the calculated bcc miscibility gap lies near room temperature (compared to 513 °C, [78Kau]), and there is some evidence that the metastable miscibility gap has been observed experimentally. Second, in the present calculation, the T_0 curve, the locus of equal Gibbs energies of the cph and bcc phases, lies below the experimental observed start temperatures for the martensite transformation; T_0, however, should lie above M_s.

In conclusion, the present calculation matches the phase diagram data more accurately than previous calcu-

Table 3   Thermodynamic Properties of the Ti-Nb System

Properties of the pure components

$G^0(Ti,L) = 0$
$G^0(Nb,L) = 0$
$G^0(Ti,bcc) = -16\,234 + 8.368\,T$
$G^0(Nb,bcc) = -22\,928 + 8.37\,T$
$G^0(Ti,cph) = -20\,585 + 12.134\,T$
$G^0(Nb,cph) = -16\,652 + 11.72\,T$

Excess Gibbs energies

	$B(L)$	$B(bcc)$	$B(cph)$
[78Kau]	13 075	13 075	13 075
[75Che]	...	13 389	13 451
Present work	1 500	5 200	10 200

Note: Values are given in J/mol and J/mol · K.

lations, but the previous calculations present a more consistent overall picture of the system thermodynamics, including the miscibility gap and martensitic transformation. In addition to measurements of thermodynamic properties of the bcc phase, additional critical experiments are needed to redetermine the temperature of the β transus at a few compositions in the range 10 to 20 at.% Nb.

Cited References

*51Han: M. Hansen, E.L. Kamen, H.D. Kessler, and D.J. McPherson, "Systems Titanium-Molybdenum and Titanium-Columbium," *Trans. AIME, 191*, 881-888 (1951). (Equi Diagram; Experimental)

53Duw: P. Duwez, "The Martensite Transformation Temperature in Titanium Binary Alloys," *Trans. ASM, 45*, 934-940 (1953). (Meta Phases; Experimental)

57Gru: N.V. Grum-Grzhimailo, "Diffusion in Titanium-Niobium Alloys," *Izv. Akad. Nauk SSSR Ofdel, Tech. Nauk*, (7), 24-28 (1957) in Russian. (Equi Diagram; Experimental)

58Bag: Yu.A. Bagariatskii, G.I. Nosova, and T.V. Tagunova, "Factors in the Formation of Metastable Phases in Titanium-Base Alloys," *Dokl. Akad. Nauk SSSR, 122*, 593-596 (1958) in Russian; TR: *Sov. Phys. Dokl., 3*, 1014-1018 (1958). (Equi Diagram; Experimental)

58Gru: N.V. Grum-Grzhimailo, "Hall's Coefficient for Binary Alloys of Titanium with Niobium," *Zh. Neorg. Khim., 3*(7), 1715-1716 (1958) in Russian; TR: *Russ. J. Inorg. Chem. (USSR), 3*(7), 328-330 (1958). (Equi Diagram; Experimental)

60Sat: T. Sato, S. Hukai, and Y.C. Huang, "The M_s Points of Binary Titanium Alloys," *J. Aust. Inst., 5*(2), 149-153 (1960). (Meta Phases; Experimental)

*61Img: A.G. Imgram, D.N. Williams, R.A. Wood, H.R. Ogden, and R.I. Jaffee, "Metallurgical and Mechanical Characteristics of High-Purity Titanium-Base Alloys," WADC Tech. Rep 59-595, Battelle Memorial Inst. (1961). (Equi Diagram; Experimental)

61Sha: K.I. Shakhova and P.B. Budberg, "Constitution Diagram of the Titanium-Niobium System," *Izv. Akad. Nauk SSSR Otd. Tekh. Metall. Top.*, (4), 56-68 (1961) in Russian. (Equi Diagram; Experimental)

63Bor: B.A. Borok, E.K. Novikova, L.S. Golubeva, R.P. Shchegoleva, and N.A. Rucheva, "Dilatometric Investigation of Binary Alloys of Titanium," *Metall. Term. Obrab. Met.*, (2), 32-36 (1963) in Russian; TR: *Met. Sci. Heat Treat.*, (2), 94-98 (1963). (Equi Diagram; Experimental)

64Bro: A.R.G. Brown, D. Clark, J. Eastabrook, and J.S. Jepson, "The Titanium-Niobium System," *Nature, 201*, 914-915 (1964). (Equi Diagram, Meta Phases; Experimental)

64Rud: P.S. Rudman, "An X-ray Diffuse-Scattering Study of the Nb-Ti B.C.C. Solution," *Acta Metall., 12*, 1381-1388 (1964). (Equi Diagram; Experimental)

Phase Diagrams of Binary Titanium Alloys

65Rau: Ch.J. Raub and U. Zwicker, "Superconductivity of α-Titanium Solid Solutions with Vanadium, Niobium, and Tantalum," *Phys. Rev., 137*(1A), A142-A143 (1965). (Equi Diagram; Experimental)

*66Bro: A.R.G. Brown and K.S. Jepson, "Physical Metallurgy and Mechanical Properties of Titanium-Niobium Alloys," *Mem. Sci. Rev. Metall., 63*(6), 575-584 (1966) in French. (Equi Diagram, Meta Phases; Experimental)

*66Guz: L.S. Guzei, E.M. Sokoloyskaya, and A.T. Grigor'ev, "Phase Diagram of the Niobium-Titanium System," *Vestn. Mosk. Univ., Chem., 21*(5), 79-82 (1966) in Russian; TR: *J. Moscow Univ., Chem., 21*(5), 406-409 (1966). (Equi Diagram; Experimental)

67Bra: W.G. Brammer and C.G. Rhodes, "Determination of ω Phase Morphology in Ti−35% Nb by Transmission Electron Microscopy," *Philos. Mag., 16*, 477-486 (1967). (Meta Phases; Experimental)

68Hat: B.A. Hatt and V.G. Rivlin, "Phase Transformation in Superconducting Ti-Nb Alloys," *Brit. J. Appl. Phys. (J. Phys. D) Ser., 2, 1*, 1145 (1968). (Equi Diagram; Experimental)

68Kal: G.P. Kalinin and O.P. Elyutin, "The Anomalies of the Properties of Titanium Alloys with β Stabilizing Elements," *Metalloved. Term. Obrab. Met.*, (4), 52-54 (1968) in Russian; TR: *Met. Sci. Heat Treat.*, (4), 301-302 (1968). (Equi Diagram; Experimental)

69Hic: B.S. Hickman, "ω Phase Precipitation in Alloy of Titanium with Transition Metals," *Trans. AIME, 245*, 1329-1335 (1969). (Meta Phases; Experimental)

*69Rud: E. Rudy, "Compendium of Phase Diagram Data," Tech. Rep. AFML-TR-65-2, Part V, Wright Patterson Air Force Base (1969). (Equi Diagram; Experimental)

69Sas: S.L. Sass, "The ω Phase in a Zr−25 at.% Ti Alloy," *Acta Metall., 17*, 813-820 (1969). (Experimental)

*69Zak: A.M. Zakharov, V.P. Pshokin, and A.I. Baikov, "On the Existence of the Compound Niobium Titanide in the Niobium-Titanium System," *Izv. Tsvetn. Metall.*, (6),104-108 (1969) in Russian. (Equi Diagram; Experimental)

70Jep: K.S. Jepson, A.R.G. Brown, and J.A. Gray, "The Effect of Cooling Rate on the β Transformation in Titanium-Niobium and Titanium-Aluminum Alloys," Sci. Technol. Appl. Titanium, Proc. Int. Conf., R.I. Jaffee, Ed., 677-690 (1970). (Equi Diagram; Experimental)

70Kau: L. Kaufman and H. Bernstein, *Computer Calculation of Phase Diagrams*, Academic Press, New York (1970). (Thermo; Theory)

*70Ron: G.N. Ronami, S.M. Kuznetsova, S.G. Fedotov, and K.M. Konstantinov, "Determination of the Phase Boundaries in Ti Systems with V, Nb, and Mo by the Diffusion-Layer Method," *Vest. Mosk. Univ., Fiz, 25*(2), 186-189 (1970) in Russian; TR: *J. Moscow Univ. Phys., 25*(2), 55-57 (1970). (Equi Diagram; Experimental)

71Bak: C. Baker, "The Shape-Memory Effect in a Titanium−35 wt.% Niobium Alloy," *Met. Sci., 5*, 92-100 (1971). (Meta Phases; Experimental)

71Men: M.G. Mendiratta, G. Lutjering, and S. Weissman, "Strength Increase in Ti−35 wt.% Nb Through Step-Aging," *Metall. Trans., 2*, 2599-2605 (1971). (Equi Diagram; Experimental)

72Bal: A.T. Balcerzak and S.L. Sass, "The Formation of the ω Phase in Ti-Nb Alloys," *Metall. Trans., 3*, 1601-1605 (1972). (Meta Phases; Experimental)

72Mor: J.P. Morniroli and M. Gantois, "Study of the Martensites α′ and α″ in Ti-Nb Alloys," *C.R. Acad. Sci., Paris, 275C*, 869-871 (1972) in French. (Meta Phases; Experimental)

73Afo: N.S. Afonikova, V.F. Degtyareva, Yu.A. Litvin, A.G. Rabin'kin, and Yu.A. Skakov, "Superconducting Properties and Structure of Ti-Nb Alloys Subjected to Hydrostatic Pressures up to 120 kbar," *Fiz. Tverd. Tela, 15*, 1096-1101 (1972) in Russian; TR: *Sov. Phys. Solid State, 15*(4), 746-749 (1973). (Meta Phases; Experimental)

73Mor: J.P. Morniroli and M. Gantois, "Formation Conditions for the ω Phase in Titanium-Niobium and Titanium-Molybdenum Alloys," *Mem. Sci. Rev. Metall., 70*(11), 831-842 (1973) in French. (Meta Phases; Experimental)

74Kad: G.N. Kadykova, M.M. Gadzoyeva, and T.V. Obkhodova, "Effect of Cold Deformation on Phase Transformations in Titanium-Niobium Alloys," *Izv. Akad. Nauk SSSR, Met.*, (3), 165-170 (1974) in Russian; TR: *Russ. Metall.*, (3), 102-107 (1974). (Meta Phases; Experimental)

75Che: D.B. Chernov and A.Ya. Shinyayev, "Calculation of the Interaction Parameters of Titanium with Vanadium, Niobium, and Molybdenum," *Izv. Akad. Nauk SSSR, Met.*, (5), 212-219 (1975) in Russian; TR: *Russ. Metall.*, (5), 167-172 (1975). (Equi Diagram; Experimental)

75Ple: J. Plejewski, C. Texler, and G. Cizeron, "Analysis of the Structural Changes in a Titanium−35 wt.% Niobium Alloy During Quenching and Tempering," *C.R. Acad. Sci., Paris, 280*, 173-176 (1975) in French. (Equi Diagram; Experimental)

78Kau: L. Kaufman and H. Nesor, "Coupled Phase Diagrams and Thermochemical Data for Transition Metal Binary Systems—II," *Calphad, 2*(1), 81-108 (1978). (Thermo; Theory)

78Men: E.S.K. Menon, S. Banerjee, and R. Krishnan, "Phase Separation in Ti−34 at.% Nb Alloy," *Trans. Ind. Inst. Met., 31*(5), 305-307 (1978). (Meta Phases; Experimental)

79Lya: I.V. Lyasotskiy, G. Kadykova, and Yu.D. Tyapkin, "Structure of Solid Solutions of Titanium-Based Alloys with a b.c.c. Lattice and Formation of Tetragonal Phase During Cold Deformation," *Fiz. Met. Metalloved., 46*(1), 144-150 (1979) in Russian; TR: *Phys. Met. Metallogr., 46*(1), 120-126 (1979). (Meta Phases; Experimental)

79Sik: V.N. Sikorov and V.F. Degtyareva, "Phase Transformation in Ti-Nb Alloys Exposed to Shock Waves," *Fiz. Metal. Metalloved., 48*(1), 211-213 (1979) in Russian; TR: *Phys. Met. Metallogr., 48*(1), 181-182 (1979). (Meta Phases; Experimental)

80Lyo: O. Lyon, "Study of Structures in Titanium-Niobium Alloys (40 and 47 wt.%) After Quenching and Aging Between 250 and 350 °C," *J. Microsc. Spectrosc. Electron, 5*, 303-308 (1980) in French. (Meta Phases; Experimental)

80Obs: B. Obst and D. Pattanayak, "Structural Effects in the Superconductor $NbTi_{65}$," *J. Low Temp. Phys., 41*(5/6), 595-609 (1980). (Experimental)

80Osa: K. Osamura, E. Matsubara, T. Miyatani, Y. Murakami, T. Horiuchi, and Y. Monju, "Effects of Cold-Working and Oxygen Addition on Precipitation Behavior in Superconducting Ti-Nb Alloy," Titanium '80, Sci. and Tech., Proc. 4th Int. Conf. Ti., Kyoto, Japan, 1369-1377 (1980). (Meta Phases; Experimental)

81Lyo: O. Lyon, "Isothermal Formation of ω Phase in Ti−35 wt.% Alloy Studied by Electron Microscopy and Small Angle X-ray Scattering," *J. Less-Common Met., 81*, 103-113 (1981) in French. (Equi Diagram; Experimental)

81Pat: D. Pattanayak, B. Obst, and U. Wolfstieg, "X-Ray Phase Determination in Niobium-Titanium Superconductors," *Z. Metallkd., 72*(7), 481-486 (1981) in German. (Meta Phases; Experimental)

82Gus: L.N. Guseva and L.K. Dolinskaya, "Metastable Phase Equilibria in the Systems Ti-V and Ti-Nb," *Dokl. Akad. Nauk SSSR, 266*(3), 634-637 (1982) in Russian; TR: *Dokl. Chem., 334-337* (1982). (Equi Diagram; Experimental)

83Bad: W. Badan and A. Weiss, "X-Ray and 1H-NMR Studies on Niobium-Titanium-Hydrides," *Z. Metallkd., 74*(2), 89-93 (1983) in German. (Crys Structure; Experimental)

The Nd-Ti (Neodymium-Titanium) System

144.24   47.88

By J.L. Murray

Equilibrium Diagram

Like all Ti-rare earth systems, Ti-Nd contains no intermetallic phases and has a miscibility gap in the liquid phase. The equilibrium solid phases are those based on the pure elements: (1) bcc (Ti) and cph (αTi), the high- and low-temperature forms, respectively; and (2) bcc (βNd) and cph (αNd), high- and low-temperature forms, respectively.

The only published experimental study of the system is that of [57Sav, 62Sav], limited both in the range examined and in the purity of the alloys. Those Ti-rare earth phase diagrams for which more accurate and extensive data are available can be calculated from regular or Gibbs energy equations. The present evaluator has therefore used a thermodynamic calculation to test the self-consistency of the diagram reported by [57Sav, 62Sav]. The assessed phase diagram (Fig. 1) is the result of the calculation.

Special points of the assessed diagram are summarized in Table 1, and crystal structure data are given in

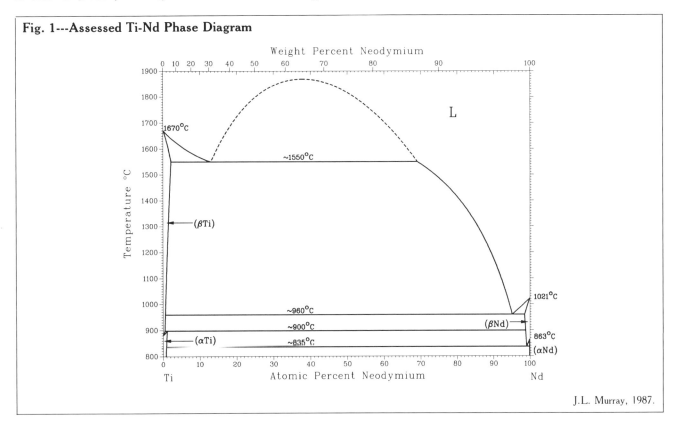

Fig. 1---Assessed Ti-Nd Phase Diagram

J.L. Murray, 1987.

Table 1 Special Points of the Ti-Nd Phase Diagram

Reaction	Compositions of the respective phases, at.% Nd			Temperature, °C	Reaction type
(LTi) $\rightleftarrows$ (βTi) + (LNd)	~ 13	~ 2	~ 69	~ 1550	Monotectic
L $\rightleftarrows$ (βTi) + (βNd)	~ 95	~ 0.5	~ 99	~ 960	Eutectic
(βTi) + (γNd) $\rightleftarrows$ (αTi)	~ 0.4	~ 99	~ 1	~ 900	Peritectoid
(βNd) $\rightleftarrows$ (αNd) + (αTi)	~ 99	~ 98.8	~ 0.8	~ 835(a)	Eutectoid
L $\rightleftarrows$ (βTi)		0		1670	Melting point
(βTi) $\rightleftarrows$ (αTi)		0		882	Allotropic transformation
L $\rightleftarrows$ (βNd)		100		1021	Melting point
β(Nd) $\rightleftarrows$ αNd)		100		863	Allotropic transformation

(a) [62Sav] reported 800 °C, but the calculated value is preferred (see text).

195

Table 2 Crystal Structures of the Ti-Nd System

Phase	Homogeneity range, at.% Nd	Pearson symbol	Space group	Struktur-bericht designation	Proto-type	Reference
(αTi)	0 to $\sim$ 1	$hP2$	$P6_3/mmc$	$A3$	Mg	[Pearson2]
(βTi)	0 to $\sim$ 3	$cI2$	$Im3m$	$A2$	W	[Pearson2]
(βNd)	$\sim$ 100	$cI2$	$Im3m$	$A2$	W	[86Gsc]
(αNd)	$\sim$ 100	$hP4$	$P6_3/mmc$	$A3'$	La	[86Gsc]

Table 2. Because the diagram is only roughly known and because the Nd-rich side was constructed by thermodynamic extrapolation, Table 1 and Fig. 1 should be assumed to contain substantial inaccuracies.

Experimental Data.

[57Sav] and [62Sav] studied the diagram in the range 0 to 6 at.% Nd by microscopy, thermal analysis, and dilatometry. Their Nd contained 0.2 wt.% other rare earths. The temperatures of the invariant reactions at 1550 and 900 °C were determined by thermal analysis and dilatometry, and the measurements were probably accurate to about ±30 and ±10 °C, respectively. The other invariant reaction temperature, 800 °C, appears to have been determined by microscopic methods, and it has an unknown but larger uncertainty. The solubility of Nd in (Ti) was found to be about 0.6 at.% at 600 °C.

Table 3 Gibbs Energies of the Ti-Nd System

Properties of the pure components

G^0(Ti,L) = 0
G^0(Nd,L) = 0

G^0(Ti,bcc) = $-$ 16 234 + 8.368 T
G^0(Nd,bcc) = $-$ 7 142 + 5.519 T

G^0(Ti,cph) = $-$ 20 585 + 12.134 T
G^0(Nd,cph) = $-$ 10 171 + 8.186 T

Excess enthalpies of the solution phases

B(L) = 33 064	B(bcc) = 50 000
C(L) = $-$ 6 574	C(bcc) = 5 000
B(cph) = 65 000	
C(cph) = $-$ 20 000	

Note: Values are given in J/mol and J/mol $\cdot$ K.

Thermodynamic Calculations

Gibbs energies of the solution phases are represented as:

$$G(i) = (1 - x) G^0(\text{Ti},i) + x G^0(\text{Nd},i) + RT[x \ln x + (1 - x) \ln (1 - x)] + x (1 - x) [B(i) + C(i) (1 - 2x)]$$

where i designates the phase; x is the atomic fraction of Nd; G^0 is the Gibbs energies of the pure components; and $B(i)$ and $C(i)$ are coefficients for the excess Gibbs energies. The regular solution approximation ($C = 0$) was used for the hexagonal and fcc phases. The Gibbs energies for Nd are based on enthalpy of fusion and transformation data [Hultgren, E] and the accepted temperatures of melting and transformation [86Gsc].

Input experimental data were the temperatures of the three-phase reactions and the compositions of (αTi) and (βTi) at the three-phase equilibria [62Sav]. Based on such limited experimental data, optimization calculations give only inadequate fits to the data if too few parameters are simultaneously varied, but do not converge at all if too many parameters are varied. Trial and error had to be used to weight the input data and determine the number of coefficients that could be meaningfully determined. The results of the optimizations are given in Table 3. Further calculations on the Ti-Nd system will require experimental determination of the solubility of Ti in (βNd) and (αNd).

Cited References

57Sav: E.M. Savitskii and G.S. Burkhanov, "Phase Diagrams of Titanium-Lanthanum and Titanium-Cerium Alloys," *Zh. Neorg. Khim.*, 2(11), 2609-2616 (1957) in Russian; TR: *J. Inorg. Chem.*, 2, 199-219 (1957). (Equi Diagram; Experimental)

62Sav: E.M. Savitskii and G.S. Burkhanov, "Phase Diagrams of Titanium Alloys with the Rare Earth Metals," *J. Less-Common Met.*, 4, 301-314 (1962). (Equi Diagram; Experimental)

86Gsc: K.A. Gschneidner, Jr. and F.W. Calderwood, in *Handbook of the Physics and Chemistry of Rare Earths*, Vol. 8, K.A. Gschneidner, Jr. and L. Eyring, Ed., North-Holland Physics Publishing, Amsterdam (1986). (Crys Structure; Review)

The Ni-Ti (Nickel-Titanium) System

58.69   47.88

By J.L. Murray

Equilibrium Diagram

The Ti-Ni system is unique among Ti systems in that the liquidus and solidus are precisely determined over their whole extent. The assessed equilibrium diagram is shown in Fig. 1, and its salient points are summarized in Table 1. The system is of particular interest because of the shape memory alloys based on TiNi.

The equilibrium solid phases of the Ti-Ni system are:

- The low-temperature cph (αTi) and high-temperature bcc (βTi) solid solutions. The maximum solubility of Ni in (αTi) and (βTi) is about 0.2 and 10 at.%, respectively.
- The fcc (Ni) solid solution. The maximum solubility of Ti in (Ni) is 13.8 at.% Ti.
- Ti_2Ni, an fcc ordered structure with 96 atoms per unit cell. It is formed by a peritectic reaction at 984 °C.

Fig. 1---Assessed Ti-Ni Phase Diagram

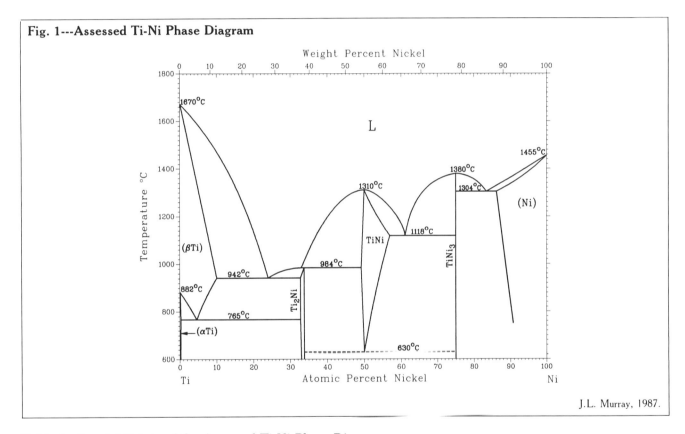

J.L. Murray, 1987.

Table 1   Special Points of the Assessed Ti-Ni Phase Diagram

Reaction	Compositions of the respective phases, at.% Ni			Temperature, °C	Reaction type
L $\rightleftarrows$ (βTi) + Ti$_2$Ni	24	10	33.3	942	Eutectic
L + TiNi $\rightleftarrows$ Ti$_2$Ni	32	49.5	33.3	984	Peritectic
L $\rightleftarrows$ TiNi + TiNi$_3$	61	57	75	1118	Eutectic
L $\rightleftarrows$ TiNi$_3$ + (Ni)	83.5	75	86.3	1304	Eutectic
(βTi) $\rightleftarrows$ (αTi) + Ti$_2$Ni	4.5	0.2	33.3	765	Eutectoid
TiNi $\rightleftarrows$ Ti$_2$Ni + TiNi$_3$	49.5	33.3	75	630	Eutectoid
L $\rightleftarrows$ TiNi		50		1310	Congruent
L $\rightleftarrows$ TiNi$_3$		75		1380	Congruent
L $\rightleftarrows$ (Ni)		100		1455	Melting point
L $\rightleftarrows$ (βTi)		0		1670	Melting point
(βTi) $\rightleftarrows$ (αTi)		0		882	Allotropic transformation

- TiNi, with the ordered bcc CsCl structure. TiNi melts congruently at 1310 °C and may not be an equilibrium phase below approximately 630 °C. The homogeneity range of TiNi is 49.5 to 57 at.% Ni.
- Stoichiometric TiNi$_3$, a four-layer cph ordered structure. TiNi$_3$ is formed from the liquid by a eutectic reaction at 1304 °C and melts congruently at 1390 °C.

(βTi) Liquidus and Solidus.

The Ti-rich liquidus descends to a very deep eutectic at 942 °C, with a eutectic composition of 24 at.% Ni. The maximum solubility of Ni in (βTi) is 10 at.% at 942 °C. In addition to the eutectic temperature and composition, accurate data are available for the whole extent of the liquidus and solidus [54Poo]. The (βTi) liquidus measurements were part of a complete determination of the diagram above 850 °C by [54Poo]. The liquidus was measured by a microscopic technique developed for treating reactive, high-melting metals [54Hum]. Microscopic and X-ray methods were used to determine the solidus. Experimental data are shown in Fig. 2; experimental determinations of the invariant temperatures are summarized in Table 2. The determinations by [49Lon] and [53Mar] agree qualitatively with [54Poo], but because they are less accurate, only the data of [54Poo] are used to determine the assessed phase boundaries.

(αTi)/(βTi) Boundaries and Eutectoid Reaction.

[53Mar] estimated the solubility of Ni in (αTi) to be less than 0.2 at.%, based on microscopic examination of equilibrated alloys, and this is the only information presently available on the (αTi) solvus. In the following discussion, the (βTi)/(βTi) + (αTi) and the (βTi)/(βTi) + Ti$_2$Ni boundaries are designated the (βTi) transus and

(βTi) solvus respectively. The (βTi) transus and solvus intersect at a eutectoid point at 4.5 at.% Ni and 765 °C.

The (βTi) phase equilibria were examined by [53Mar], [54Mcq], and [54Poo] using metallography. [51Mcq] used the hydrogen pressure technique, and [74Bas] used microprobe analysis of equilibrated two-phase alloys and annealed diffusion couples. [51Mcq] and [54Mcq] examined the (βTi) transus only, [54Poo] the solvus only, and [53Mar] and [74Bas] both boundaries. The discrepancies are striking. Figure 2 includes a detail of the eutectoid region.

For a given composition, the (βTi) transus of [51Mcq] lies at the lowest temperature, and that of [53Mar] lies at the highest temperature, with [74Bas] in between. The [53Mar] results can be discounted as inconsistent with the thermodynamic properties of pure Ti. [54Mcq] attributed the discrepancy to the decomposition of single-phase (βTi) during an insufficiently rapid quench, and this explanation can be accepted as consistent with observations on many Ti-based eutectoid systems and with studies of the (βTi)/(αTi) martensite transformation. Hydrogen pressure results, however, consistently lie below the best metallographic work for several Ti systems. The (βTi) transus data of [74Bas] come from diffusion couple experiments only, and it cannot be accurately judged how closely these experiments represent equilibrium. For the (βTi) solvus, [74Bas] also examined equilibrated bulk alloys. The EPMA determinations lie in a band about 2 at.% wide about 1 at.% to the Ni-rich side of the metallographic work [53Mar, 54Poo].

The assessed solvus is based primarily on the work of [54Poo]; the assessed (βTi) transus lies above that of [51Mcq] and [54Mcq], but slightly below that of [74Bas].

Assessment of this part of the diagram is hampered by the absence of a reliable direct determination of the

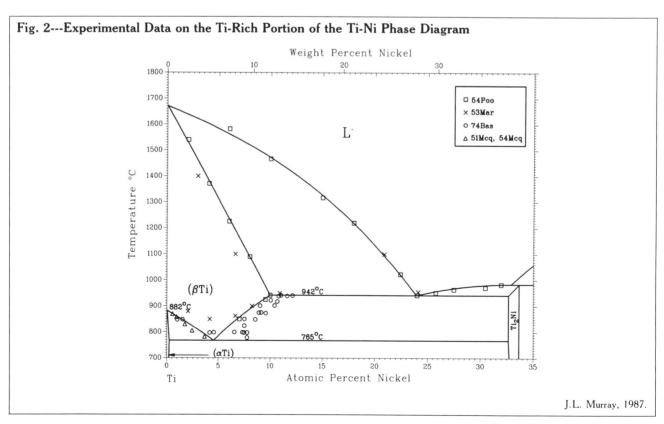

Fig. 2---Experimental Data on the Ti-Rich Portion of the Ti-Ni Phase Diagram

J.L. Murray, 1987.

Table 2   Experimental Determinations of Invariant Temperatures

Reaction	Reference	Temperature, °C	Note
L $\rightleftarrows$ (Ti) + Ti$_2$Ni	[54Poo]	942	Thermal analysis, cooling
	[53Mar]	955	Incipient melting
	[49Lon]	960	As quoted by [53Mar]
(βTi) $\rightleftarrows$ (αTi) + Ti$_2$Ni	[53Mar]	770	Metallography
	[74Bas]	770	EPMA annealed diffusion couples
	[49Lon]	765	As quoted by [53Mar]
L + TiNi $\rightleftarrows$ Ti$_2$Ni	[54Poo]	984	Microscopic and quenching methods
	[53Mar]	1015	Incipient melting
	[71Was]	1025	X-ray diffraction, metallography
TiNi $\rightleftarrows$ Ti$_2$Ni + TiNi$_3$	[54Poo]	>600	X-ray diffraction
	[69Kos]	625 to 650	TEM
	[73Gup]	650 ± 10	TEM, SAD
L $\rightleftarrows$ TiNi	[54Poo]	1310	Thermal analysis, metallography
	[53Mar]	1240	Optical observation of melting
L $\rightleftarrows$ TiNi + TiNi$_3$	[54Poo]	1118	Thermal analysis, quenching experiments
	[41Wal]	1110	Thermal analysis
	[38Vog]	1102	Thermal analysis
	[53Mar]	1110	Incipient melting
L $\rightleftarrows$ TiNi$_3$	[54Poo]	1380	Thermal analysis
	[41Wal]	1378	Thermal analysis
	[38Vog]	1378	Thermal analysis
L $\rightleftarrows$ (Ni) + TiNi$_3$	[54Poo]	1304	Thermal analysis, microscopy
	[38Vog]	1287	Thermal analysis, microscopy

eutectoid temperature and composition. Therefore, an uncertainty of as much as 2 at.% and 10 °C should be assumed for the eutectoid point.

Ti$_2$Ni.

Ti$_2$Ni has an ordered fcc structure with 96 atoms per unit cell [39Lav2, 50Duw, 59Yur, 63Mue]. This structure usually occurs as a ternary phase stabilized by oxygen in Ti-based systems, but it has been verified to be a stable binary phase in Ti-Ni [52Ros, 54Poo, 58Yur, 59Yur].

There is no detectable variation of the lattice parameter of Ti$_2$Ni with overall alloy composition. Microstructures indicate that the maximum homogeneity range is less than 2 at.% [54Poo, 58Yur] and suggest about a 1 at.% range at 700 °C.

[41Wal] found Ti$_2$Ni to melt at a temperature above 1600 °C, indicating that alloys were contaminated through the starting materials or in melting. In a study of the ternary Ti-Ni-C system, [59Sto] pointed out that use of carbon crucibles raises the apparent melting point of Ti$_2$Ni. In the binary system, Ti$_2$Ni forms by the peritectic reaction L + TiNi $\rightleftarrows$ Ti$_2$Ni [53May, 54Poo]. [53May] reported the peritectic temperature as 1015 °C using metallographic methods; [54Poo] reported it as 984 °C based on thermal analysis. The work of [54Poo] is accepted because of their greater precaution to avoid contamination of the alloys. [54Poo] also verified thermal analysis results by a microscopic study of the liquidus.

TiNi.

In the assessed phase diagram, a single equiatomic phase with the ordered bcc CsCl structure is shown. There is no major disagreement concerning the extent of the single-phase region above 900 °C [54Poo, 61Pur, 71Was, 74Bas]. The phase boundary on the Ti-rich side is essentially vertical, but on the Ni-rich side the homogeneity range decreases sharply with temperature. The assessed

diagram is based on the metallographic and X-ray work of [54Poo] and the diffusion work of [74Bas]. A detail of the equiatomic region of the diagram, including experimental data, is shown in Fig. 3.

The assessed phase diagram shows a very uncertain eutectoid decomposition of TiNi to Ti$_2$Ni + TiNi$_3$ at 630 ± 15 °C. Considerable controversy surrounds the question of whether the eutectoid reaction properly belongs in the equilibrium diagram. Arguments in favor of the eutectoid reaction near 630 °C were given by [54Poo], [63Gil], [69Kos], and [73Gup]; arguments against were made by [65Wan], [70Was], [71Was], and [75Hon].

Experimental evidence and arguments concerning on the eutectoid reaction are summarized as:

* [50Duw] found X-ray evidence of decomposition during cooling of a 50 at.% alloy and postulated a eutectoid reaction at about 800 °C. [53Mar], finding no difference between the structures produced by annealing at 750 °C and 900 °C, concluded that the eutectoid decomposition does not occur at any temperature.
* [54Poo] found both Ti$_2$Ni and TiNi$_3$, together with TiNi, in X-ray diffraction patterns of powder samples annealed at 600 °C. [54Poo] took precautions to avoid contamination of the samples, so that oxygen contamination is less likely a factor in this work than in some subsequent work. [63Gil] verified the result of [54Poo] for filings, but found that the decomposition did not take place in bulk samples and concluded that TiNi can exist in either a stable or a metastable state at low temperatures.
* [69Kos], using TEM, found different structures in alloys quenched from 625 and 650 °C. From 650 °C, Ti$_2$Ni or TiNi$_3$ were found in a matrix of TiNi; from 625 °C, Ti$_2$Ni was found in a matrix of a "transition

Fig. 3---Experimental Data on the Central Portion of the Ti-Ni Phase Diagram

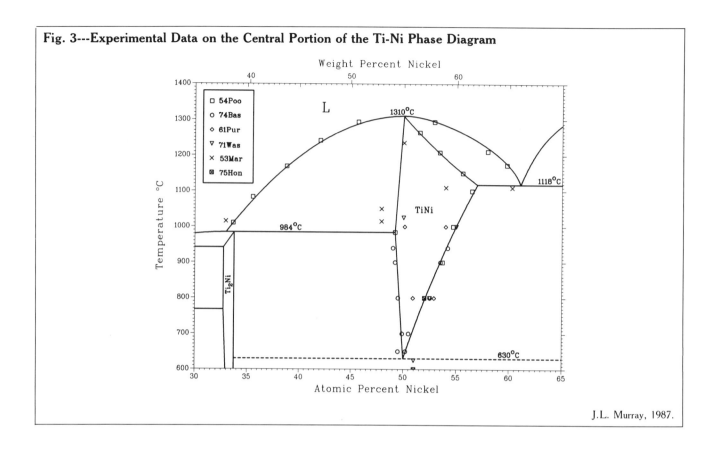

J.L. Murray, 1987.

phase". After a 1-month anneal at 600 °C, only Ti_2Ni and $TiNi_3$ were present. Using TEM, selected area diffraction (SAD), and X-ray diffraction, [73Gup] verified the results of [69Kos]. They tentatively identified the transition phase as an fcc phase.

- [65Wan] claimed that TiNi has the disordered bcc structure at high temperature and undergoes a transformation to a complex cubic structure—the "9 Å" phase, or TiNi(II)—at approximately 700 °C. They concluded that previous investigators had confused powder patterns of the martensitic phase with patterns of Ti_2Ni + $TiNi_3$.

- Using X-ray diffraction, [71Was] found an isothermal transformation at 625 °C, but interpreted it as a peritectoid reaction with a new stable compound of stoichiometry Ti_2Ni_3. They concluded that TiNi is present in the equilibrium diagram to low temperatures, although in a very restricted composition range. Ti_2Ni_3 was identified as a 10H stacking fault variant of $TiNi_3$ that had previously been found by [68Pfe] in compositions near $TiNi_3$. Ti_2Ni_3 was not, however, demonstrated to be in equilibrium with $TiNi_3$ at appropriate compositions and temperatures.

- The extension of the TiNi phase range below 400 °C by [75Hon] was based on the effect of heat treatment on the start temperature of the martensite transformation.

- From heat of formation data (see Thermodynamics), one can determine which phases are stable at zero temperature, if one assumes that the compounds have nearly the same heat capacities. Based on the data of [56Kub], TiNi is stable to

low temperature; based on the data of [81Gac], it is not.

In any case, the CsCl structure is not stable with respect to the martensite at low temperature. The eutectoid reaction is very tentatively shown in Fig. 1. It is clear that attainment of equilibrium is slow. Cold working aids decomposition, and decomposition would thus be recognized more easily in filings than in bulk samples. TEM also appears to be a more suitable tool for determining this portion of the equilibrium diagram. The presence of Ti_2Ni_3 in equiatomic alloys [71Was] has been verified [73Gup], but there is no evidence to suggest that it is an equilibrium phase.

$TiNi_3$ and (Ni) Equilibria.

$TiNi_3$ was first detected in the liquidus determination of [38Vog], and its structure was first examined by [39Lav1]. $TiNi_3$ forms congruently from the melt at 1380 °C and melts by the eutectic reaction L $\rightleftarrows$ $TiNi_3$ + (Ni) at 1304 °C. The assessed Ni-rich liquidus is based on the thermal analysis data of [54Poo]. The previous determination [38Vog] is in reasonable agreement. Experimental data on Ni-rich alloys are compared with the assessed boundaries in Fig. 4.

Based on lattice parameters of $TiNi_3$ in two-phase alloys, the homogeneity range of $TiNi_3$ is very small [54Poo]. In the assessed diagram, $TiNi_3$ is represented as a line compound.

The (Ni) solvus was determined above 750 °C by [38Vog, 51Tay, 54Poo, 58Bag, 62Koe]. [62Koe] used microscopic and electrical resistivity measurements. [38Vog], [51Tay], [54Poo] and [58Bag] used the lattice parametric method, and [54Poo] used metallography as well. The [38Vog] data can probably be discounted due to

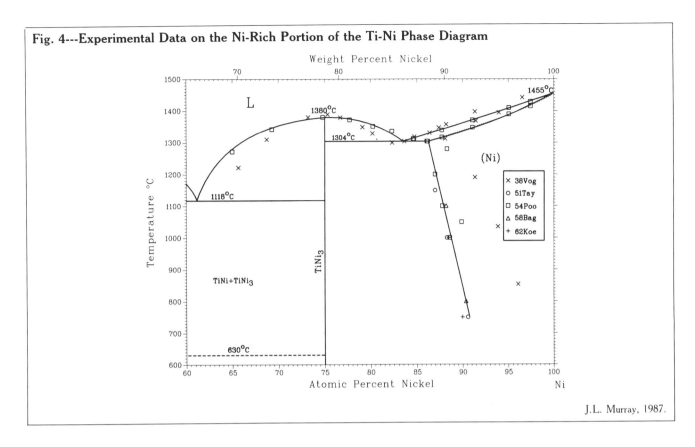

Fig. 4---Experimental Data on the Ni-Rich Portion of the Ti-Ni Phase Diagram

J.L. Murray, 1987.

sample impurity. The lattice parameter measurements of [54Poo] led to solubilities about 1 at.% lower than indicated by the other work. The assessed solvus is based on [51Tay], [54Poo], and [58Bag].

Magnetic Transitions

Pure Ni is ferromagnetic below 358 °C, and the effect of Ti additions is to lower the Curie temperature. Experimental data on the Curie temperature [37Mar, 51Tay, 59Sto, 70Ard, 63Ben, 78Yao] are shown in Fig. 5. At higher Ti contents, the lower transition temperatures are preferred because segregation during quenching would tend to raise the Curie temperature. TiNi, Ti_2Ni, and $TiNi_3$ are paramagnetic [57Phi, 60Nev, 62Bue].

Metastable Phases

Ti-Based Alloys.

(βTi) transforms martensitically to (α'Ti) in alloys containing less than 2 to 3.5 at.% Ni [56Pol, 60Bar]. Start temperatures, M_s, for the martensite transformation are:

Reference	Composition, at.% Ni	Temperature, °C
[60Sat]	2.5	660
	4.0	555
	6.0	415
[73Hua]	2.5	665
[60Bar]	2.0	760
	3.0	650
	3.5	525

In alloys containing more than about 3.5 at.% Ni, the (βTi) $\rightarrow$ (α'Ti) transformation occurs by diffusion and not by a martensitic mechanism [56Pol, 60Bar, 73Fed, 74Gus, 78Gus]. This substantiates the claim by [54Mcq] that phase separation during quenching introduces error into the determination of the equilibrium phase boundaries by metallographic techniques. [60Sat] reported M_s temperatures for alloys containing up to 6 at.% Ni, which conflict with the above reports. The reason for the discrepancy is not known.

Formation of (αTi) can be suppressed in alloys containing over about 6 to 8 at.% Ni [56Pol, 60Bar, 59Age, 74Gus]. Retained (βTi) is accompanied by ω phase [73Fed, 74Gus, 78Gus].

The ω phase forms from metastable (βTi) either at fixed composition during quenching from the region (athermal ω), or during aging below about 400 °C [59Age]. During aging, the ω phase Ti content approaches a saturation value, and the (βTi) phase becomes enriched in Ni. Quantitative determinations of these compositions for Ti-Ni are not available. Particles of aged ω are cubic in shape and align in rows parallel to <100> directions [73Ika].

Shape Memory Alloys.

Considerable attention has been given to the martensitic and other metastable transformations of TiNi. The unusual mechanical properties of these alloys were first discussed by [62Bue] and [63Bue]; [75Was] and other workers [75Per] reviewed the extensive literature on the shape memory effect. The scope of this review does not permit extended coverage of physical and mechanical properties of TiNi; the present discussion is limited to metastable phase equilibria, the martensite transition, and the structures of the metastable phases.

Phase Diagrams of Binary Titanium Alloys

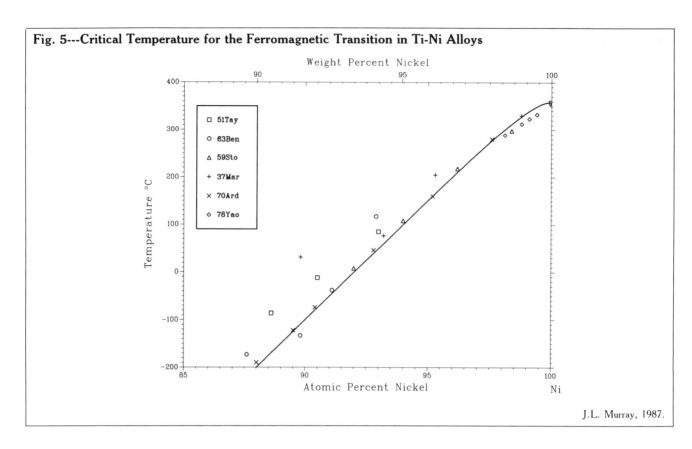

Fig. 5---Critical Temperature for the Ferromagnetic Transition in Ti-Ni Alloys

J.L. Murray, 1987.

The structure changes in TiNi can be roughly separated into: (1) the martensite transformation proper, in which the high-temperature $B2$ structure is transformed to a monoclinic structure, and (2) "premartensitic instabilities" or "precursor phenomena" occurring above the martensite start temperature in the parent bcc phase. Although some consensus has been reached on the approximate structure and stability range of the martensite, the term "premartensitic instabilities" embraces a variety of imprecisely characterized phenomena, not all necessarily connected with the martensitic transformation.

The martensite structure is a monoclinic distortion of the AuCd ($B19$) structure [68Mar, 68Sas, 71Heh, 71Ots, 71San, 74Zij, 78Kha, 79Che, 81Mic, 83Buh]. The most recent structure determination is that of [83Buh] by neutron powder diffraction; they identified the symmetry as $P2_1/m$. Lattice parameters and other structural information are given under "Crystal Structures and Lattice Parameters." [68Mar] and [68Sas] proposed that there are two structurally distinct monoclinic martensite phases. [65Dau] proposed a triclinic structure that is a slight distortion of the monoclinic lattice. [79Che] observed a transition from a monoclinic to a triclinic structure at -20 °C for Ti-rich alloys. Comparison of compositions and temperatures at which triclinic and monoclinic martensites are seen [e.g., 65Dau, 79Che, 83Buh] does not at present suggest a resolution of the discrepancies in terms of an additional monoclinic to triclinic transition.

The martensite has also been described as a mixture of a hexagonal and two triclinic phases [72Wan], as a mixture of several ordered stacking variants of close-packed planes [71Gup], [71Nag1], and [71Nag2] of hexagonal and rhombohedral symmetry.

Measurements of start and finish temperatures for the martensitic transformation on cooling and heating

(M_s, M_f, A_s, and A_f) were made by [68Wan, 71Kor, 79Sha, 80Mil, 81Mel], and these data are given in Table 3. These and other data [67Han, 71San, 74San, 78Kha, 83Buh] are compared graphically in Fig. 6. [78Kha] and [83Buh] used X-ray and neutron powder diffraction, respectively; others used electrical resistivity and magnetic properties. M_s can be influenced by cooling rate [71Ots], thermal cycling [68Wan], and other aspects of sample history; consequently, scatter is thus expected in the measurements.

Above M_s (at about 40 to 50 °C), a reversible and diffusionless, but non-martensitic, transition is seen by calorimetry [66Dau, 67Ber, 68Ber], differential thermal analysis [68Wan], internal friction [65Wan, 78Kha], resistivity, and other physical properties [66Spi, 68Wan, 74San]. [65Dau] concluded that the transition is a higher order displacive transition based on the continuous distortion of the parent lattice as the temperature was lowered; [67Ber] and [68Ber] found the shape in the heat capacity characteristic of a higher order transition. The underlying structural change has been described in terms of instabilities in lattice displacement waves (soft phonons) [71San, 82Mic, 82Moi, 68Cha] and also in terms of an additional phase transition to a rhombohedral phase quite independent of the martensite transition [65Dau, 69Kos, 72Wan, 78Kha, 79Che, 80Lin]. The two views are not contradictory. In terms of lattice displacement waves, [82Mic] and [82Moi] showed that most but not all of the displacements could be associated with the martensite transformation. Other waves may lead to the rhombohedral phase. The rhombohedral structure is not well characterized, but can be connected with many reports of a tripling of the $B2$ cubic cell [65Dau, 68Cha, 68Wan, 71Ots, 72Wan, 74San, 74Zij]. There is agreement between the temperature of the displacive transition and that at which the rhombohedral phase appears in diffraction experiments.

Table 3 TiNi Martensitic Transformation Temperatures

Reference	Composition, at.% Ni	M_s	M_f	Temperature, °C A_s	A_f	Technique
[71Kor]	46.6	57	12	81	117	Dilatometry
	47.6	37	18	79	134	
	49.6	33	13	75	114	
	50.2	− 51	30	33	32	
	51	− 136	− 178	0	− 94	
	51.5	− 4	− 38	− 12	46	
	52.8	28	− 14	44	278	
[81Mel]	49.4	57	5	63	106	Dilatometry
	49.7	20	− 20	39	77	
	50.4	− 30	− 53	− 12	0	
[79Sha]	50	78	45	100	120	X-ray diffraction
[80Mil]	49.7	45	...	67	...	DTA (as-received material)
	50	44	...	120	...	
	50.1	10	...	52	...	
	50.5	− 9, − 29	...	21	...	
[68Wan]	51	20 to 25	...	60	...	Electrical, magnetic properties
[79Che]	48.1	100	60	123	140	Electrical resistivity
	48.6	101	74	178	153	
	49.0	66	16	56	93	
	49.5	47	19	53	80	
	50.5	5	− 31	8	44	
	51.0	− 52	− 85	− 39	− 34	

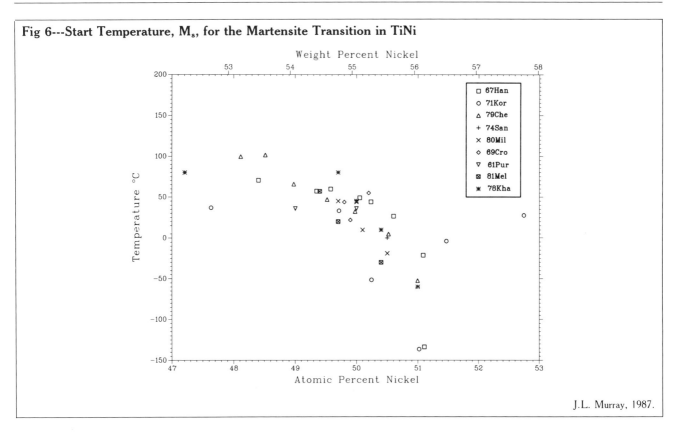

Fig 6---Start Temperature, M_s, for the Martensite Transition in TiNi

J.L. Murray, 1987.

Based on symmetries, neither a tripling of the cubic cell nor a rhombohedral distortion of the cell can be understood as leading to the monoclinic martensite lattice. Therefore, the present judgment is that the premartensitic instability and the rhombohedral distortion are distinct phenomena. It is therefore suggested that the term "premartensitic instability" should, in this system, be used only to describe softening of phonons associated with the transition to the monoclinic B19-type structure.

Age-Hardening of Ni-Rich Alloys (γ'TiNi$_3$).

The precipitation of equilibrium TiNi$_3$ from supersaturated (Ni) at low temperatures is preceded by a metastable phase, usually called γ' designated γ'TiNi$_3$ in the present assessment [57Bag, 57Buc, 59Buc, 60Mih, 61Bag, 69Sai]. $\gamma'TiNi_3$ has the ordered Cu_3Au structure. Because of the coherency strains, $\gamma'TiNi_3$ has a tetragonal rather than cubic symmetry [61Bag]. $\gamma'TiNi_3$ is coherent with the (Ni) matrix; it forms as a modulated structure of cuboidal particles aligned in <100> directions with faces parallel to (100) planes [59Buc, 60Mih, 69Sai, 70Ard, 76Lau].

Experimental work has been done on (1) the position of the coherent solvus, (2) the composition of $\gamma'TiNi_3$, and (3) the mechanism of the transformation and position of the coherent spinodal. Magnetic properties [63Ben, 68Ard, 68Coh, 69Ras, 70Ard, 81Has], X-ray diffraction [61Bag, 78Has], and field ion microscopy [74Sin] were used as probes of the compositon of the matrix and precipitate. Experimental data are shown in Fig. 7. By the same techniques, the precipitate composition was estimated as 88 at.% Ni [74Sin], 83 to 85 at.% Ni [78Has], and 87 at.% Ni [63Ben, 70Ard]. The precipitate in the initial stages of aging is thus often called Ni_6Ti.

A modulated microstructure and satellites in diffraction patterns suggest a spinodal decomposition mechanism. However, the formation of a modulated structure from the fcc solid solution could proceed by several mechanisms: (1) ordered particles formed by a nucleation and growth mechanism could become aligned during aging because of the elastic properties of the matrix, (2) partitioning of the disordered fcc solid solution could occur by a spinodal mechanism, with the appearance of a modulated structure from the beginning of the precipitation, or (3) a continuous variation of the order parameter (spinodal ordering) could occur.

For an 86 at.% Ni alloy, [74Sin] showed that the precipitation of $\gamma'TiNi_3$ does occur by spinodal decompo-

sition, with simultaneous composition partitioning and ordering. [76Lau] examined early stages of decomposition of a 88 at.% Ni alloy in order to distinguish between spinodal decomposition that begins as partitioning into solute-rich and solute-poor regions followed by ordering, on the one hand, and continuous ordering on the other hand. [76Lau] concluded that the former process occurs. This does not necessarily contradict [74Sin] because of the composition difference in the alloys used in the two studies. The sequence of metastable phase boundaries is (from the Ni-rich side to the Ti-rich side) equilibrium solvus, coherent solvus, spinodal clustering, and spinodal ordering.

Metastable transition phases leading to the formation of equilibrium $TiNi_3$ are also found on the Ti-rich side of stoichiometry. [69Kos], [70Nag], [70Pos], and [73Gup] described a transition structure as an fcc phase with a lattice parameter of 1.52 nm. [68Pfe] identified several stacking fault variants of $TiNi_3$ as 9-, 10-, and 21-plane sequences of close-packed planes.

Rapid Solidification.

[80Ino] rapidly solidified alloys by melt spinning in the composition range 77 to 87 at.% Ni and found that amorphous alloys were not formed. [73Pol], using cooling rates of 10^7 to 10^8 K/s (splat cooling), examined several compositions near the three eutectic minima in the liquidus and found that an amorphous phase could be produced in the composition range 61 to 65.5 at.% Ni, but not near the very deep eutectic at 942 °C and 24 at.% Ni. In addition, [73Pol] observed the extended solubility of Ti in (Ni) to be 22.3 at.% compared to the equilibrium solubility of 13.9 at.% Ti.

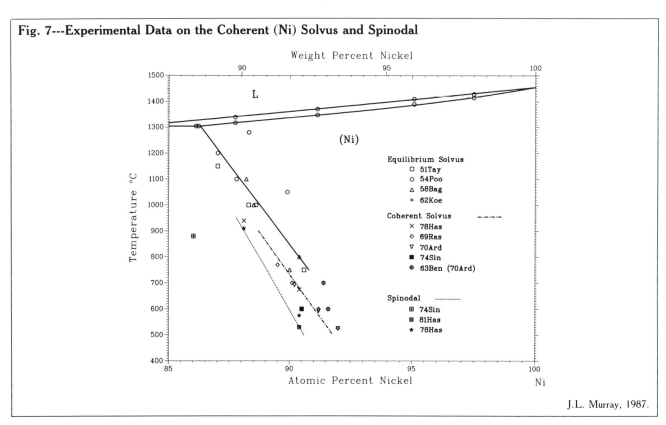

Fig. 7---Experimental Data on the Coherent (Ni) Solvus and Spinodal

J.L. Murray, 1987.

Crystal Structures and Lattice Parameters

The crystal structures of the equilibrium and metastable phases of the Ti-Ni system are summarized in Table 4. Crystal structures and lattice parameters of the TiNi martensites are given in Table 5. The crystal structure of Ti_2Ni [39Lav2, 50Duw, 52Ros, 59Yur, 62Yur, 63Mue] is closely related to that of Fe_3W_3C ($E9_3$, $cF112$). Ti_2Ni has 96 atoms per unit cell: 48 Ti in (f), 32 Ni in (e), and 16 Ti in (c). The most precise lattice parameters are those of [54Poo] and [59Yur] and of [63Mue], determined by X-ray and neutron powder diffraction, respectively:

Reference	Ti_2Ni lattice parameter, nm
[54Poo]	1.13198 ± 0.0002
[59Yur]	1.1278 ± 0.0001 (1.301)
[63Mue].....................	1.13193 ± 0.00002

The large discrepancy between [59Yur] and the other work is not explicable.

Despite the similar scattering power of Ti and Ni, sufficient evidence has accumulated to show that equilibrium TiNi has the ordered bcc CsCl structure [57Phi,

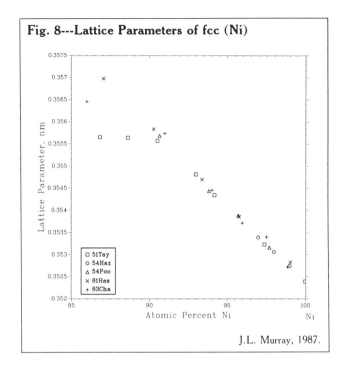

Fig. 8---Lattice Parameters of fcc (Ni)

J.L. Murray, 1987.

Table 4 Crystal Structures of the Ti-Ni System

Phase	Homogeneity range, at.% Ni	Pearson symbol	Space group	Stukturbericht designation	Prototype	Reference
(αTi)	0 to 0.2	$hP2$	$P6_3/mmc$	$A3$	Mg	[Pearson2]
$(\alpha'Ti)$	0 to 3.5	$hP2$	$P6_3/mmc$	$A3$	Mg	[Pearson2]
(βTi)	0 to 10	$cI2$	$Im3m$	$A2$	W	[Pearson2]
ω	~ 8	$hP3$	$P6/mmm$ or $P3m1$	... ...	ωMnTi ωCrTi	[Pearson2]
Ti_2Ni	33.3	$cF96$	$Fd3m$	$E9_3$	Ti_2Ni	[63Mue]
TiNi' ....~ 49 to 53		$mP4$	$P2_1/m$	...	...	[83Bun]
TiNi	49.5 to 57	$cP2$	$Pm3m$	$B2$	CsCl	[75Hon]
$\gamma''TiNi_3$ ....~ 73		$hR21$	$R3m$	...	...	[71Bha]
$TiNi_3$	75	$hP16$	$P6_3/mmc$	$D0_{24}$	$TiNi_3$	[50Tay]
$\gamma'TiNi_3$ ....~ 83 to 88		$cP4$	$Pm3m$	$L1_2$	$AuCu_3$	[57Buc]
(Ni)...............	86.1 to 100	$cF4$	$Fm3m$	$A1$	Cu	[Pearson2]

Table 5 Crystal Structure and Lattice Parameters of TiNi Martensite

Reference	Composition, at.% Ni	Symmetry	Lattice parameters, nm a	b	c	Angle
[83Buh]	49.2	$P2_1/m$	0.2884(2)	0.4665(3)	0.4110(2)	γ = 98.1°
[81Mic]	...	$P2_1/m$	0.2885(4)	0.4622(5)	0.4120(5)	γ = 96.8°
[71Ots]	50	$P2/c$	0.2889	0.4622	0.4120	γ = 96.8°
[71San, 71Heh]	50.5	Monoclinic	0.288	0.463	0.414	γ = 97°
[73Gup]	48	Monoclinic	0.29	0.463	0.411	γ = 96.7°
[65Dau]	48	Triclinic	0.460	0.286	0.411	α = 90.1° γ = 90.9° γ = 96.7°
[68Mar]	50	Monoclinic	0.519	0.496	0.425	γ = 99°
	50	Monoclinic	0.519	0.552	0.425	γ = 116°
[72Wan]	51	Triclinic	0.424	0.300	0.6701	β = 108.4° γ = 96.1°

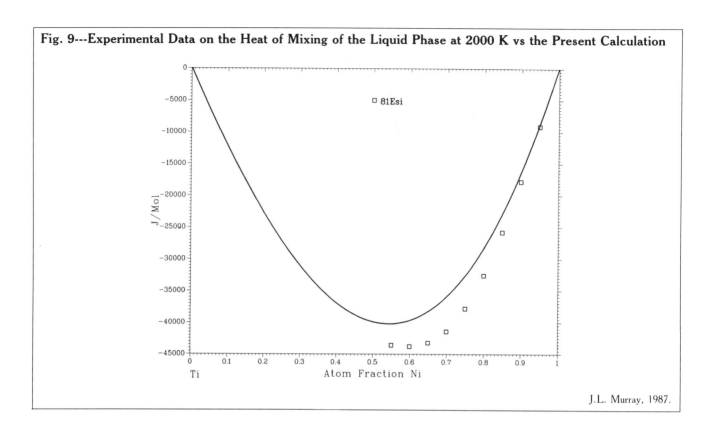

Fig. 9---Experimental Data on the Heat of Mixing of the Liquid Phase at 2000 K vs the Present Calculation

J.L. Murray, 1987.

59Dwi, 60Stu, 61Pur, 63Gil, 65Dau, 71Ots, 75Hon]. The suggestion [72Wan] that TiNi has a more complex ordered structure appears to pertain to a metastable rather than a stable phase. Lattice parameters were reported as functions of composition and of temperature [60Stu, 61Pur], but the data are scattered and cannot be used to determine the homogeneity range. At stoichiometry, the lattice parameter is generally agreed to be 0.3015 ± 0.0005 nm.

[39Lav1], [50Duw] and [50Tay] identified TiNi$_3$ as a four-layer ordered cph structure with the stacking sequence ABAC. The structure can also be described as a 4H modulation of the Cu$_3$Au structure [73Gup]. Lattice parameters were measured by [39Lav], [50Tay], [54Poo], [58Bag] and [69Sai]. The most precise lattice parameters are those measured by [54Poo]:

x = 74.77 at.% Ni

a = 0.510885 ± 0.000005 nm

c = 0.831874 ± 0.000005 nm

Lattice parameters of supersaturated (αTi) and (βTi) as a function of composition are not available, because phase separation invariably occurs during quenching for all but the most dilute alloys. Lattice parameters of fcc (Ni) [50Tay, 54Haz, 54Poo, 80Has, 83Cha] are shown in Fig. 8.

The ω phase is a hexagonal phase of symmetry $P6/mmm$, which in some systems is distorted to a trigonal structure of symmetry $P\bar{3}m1$. The present assignment of a hexagonal symmetry is not based on detailed structural studies.

Lattice parameters for single-phase γ'TiNi$_3$ cannot be defined, because γ'TiNi$_3$ is coherent with the fcc matrix and because the phase has variable composition. Lattice parameters as a function of heat treatment temperature and overall alloy composition can be found in [61Bag].

[71Bha] determined the structure of a metastable phase occurring in a 73 at.% Ni alloy to be a 21-layer stacking of close-packed layers. The phase is designated γ''TiNi$_3$ in Table 5. It has lattice parameters of: a_h = 0.2549 and c_h = 4.3648.

Thermodynamics

Experimental Data.

Experimental data are available for the integral heat of mixing [81Esi] and partial Gibbs energy [72Ger] of the liquid and for heats and Gibbs energies [56Kub, 79Lev, 81Gac, 83Cha] of formation of the compounds.

Heats of mixing [81Esi] were obtained from partial heats of solution by high-temperature calorimetry at 2000 K. The data are shown in Fig. 9. Partial Gibbs energies [72Ger] were determined by Knudsen cell/mass spectrometry between 1475 and 1725 °C. The data at 1700 °C, where all alloys are single-phase liquid, are shown in Fig. 10 with the results of the present calculations. Heats and Gibbs energies of formation are compiled in Table 6.

Calculation of the Phase Diagram.

The phase diagram was previously calculated by [78Kau]. The calculations by the present evaluator and [85Sau] incorporate more recent thermochemical and phase diagram data. Gibbs energies of the solution phases are represented as:

$$G(i) = (1 - x) G^0(\text{Ti},i) + x G^0(\text{Ni},i) + RT [x \ln x + (1 - x) \ln (1 - x)] + x(1 - x) [B(i) + C(i) (1 - 2x)]$$

where i designates the phase; x is the atomic fraction of Ni; G^0 is the Gibbs energies of the pure metals; and $B(i)$ and $C(i)$ are parameters describing the excess functions.

Fig. 10---Experimental Data on the Partial Gibbs Energies of the Liquid Phase at 1700 °C, vs the Present Calculation

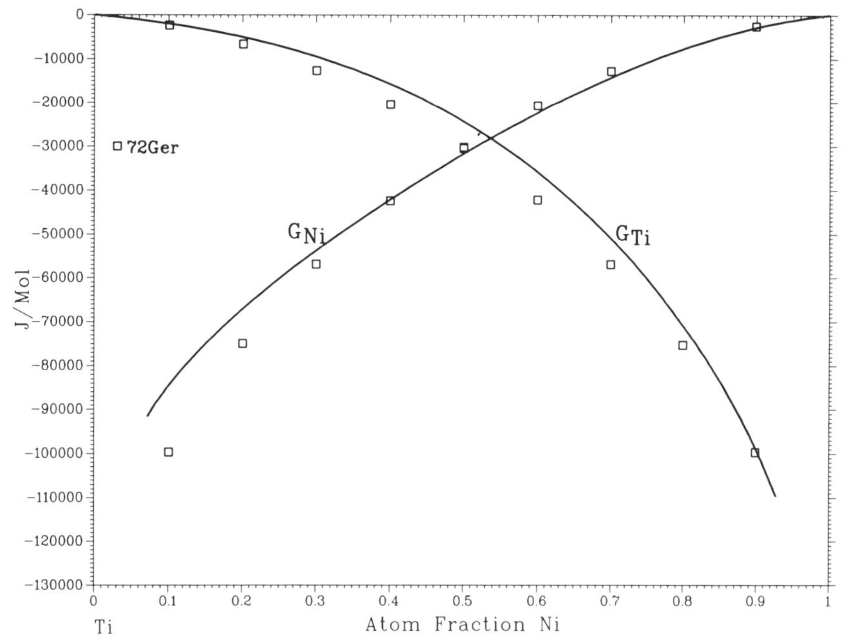

J.L. Murray, 1987.

Table 6 Heats and Gibbs Energies of Formation of the Compounds

Reference	Phase	Temperature, °C	Enthalpy of formation, J/mol	Gibbs energy of formation, J/mol
[56Kub]	Ti_2Ni	25	− 26 800	...
	TiNi	25	− 33 900	...
	$TiNi_3$	25	− 34 700	...
[81Gac]	Ti_2Ni	929	− 29 300	...
	TiNi	1187	− 34 000	...
	$TiNi_3$	1240	− 42 900	...
[79Lcv]	$TiNi_3$	1027	...	− 43 600
	TiNi	1027	...	− 31 600
[83Cha]	$TiNi_3$	...	...	− 54 000 + 11.5 T

TiNi, Ti_2Ni, and $TiNi_3$ were represented as line compounds.

Results.

The Gibbs energies resulting from the present calculations are listed in Table 7. Calculated thermodynamic quantities are compared with experimental data in Fig. 9 and 10; the calculated diagram is shown in Fig. 11. The Gibbs energy of the liquid is in better accord with partial Gibbs energies [72Ger] than with the heat of mixing data [81Esi]. The agreement with phase diagram data is within experimental uncertainty for the Ti-rich portion of the diagram.

The enthalpies of the compounds are based on [56Kub]; the Gibbs energies of the solid phases are based on the phase diagram. Because accurate liquidus and solidus data are available over a wide temperature range,

the bcc excess entropy could be determined. The fcc excess entropy was determined primarily by the solvus; the excess entropy of the cph phase should be considered uncertain.

The calculation deviates from the experimental diagram in the range 50 to 70 at.% Ni. The liquidus is reproduced by the present calculations. The use of a Wagner-Schottky defect model for TiNi does not succeed in reproducing the homogeneity range and leads to similar discrepancies between the calculated and assessed liquidus.

Cited References

37Mar: V. Marian, "Ferromagnetic Curie Points and the Absolute Saturation of Some Nickel Alloys," *Ann. Phys. (Paris), 7,* 459-527 (1937) in French. (Equi Diagram, Magnetism; Experimental)

Table 7 Gibbs Energies for the Ti-Ni System

Gibbs energies of the pure components

G^0(Ti,L)   = 0
G^0(Ni,L)   = 0

G^0(Ti,bcc) = − 16 234 + 8.368 T
G^0(Ni,bcc) = − 11 966 + 13.556 T

G^0(Ti,cph) = − 20 585 + 12.134 T
G^0(Ni,cph) = − 16 567 + 11.464 T

G^0(Ti,fcc) = − 17 238 + 12.134 T
G^0(Ni,fcc) = − 17 614 + 10.209 T

Interaction parameters for the solution phases

B(L)   = − 159 492 + 47.195 T
C(L)   =   23 788 − 7.195 T

B(cph) =   − 44 350 + 29.29 T

B(bcc) =   − 94 000 + 12.5 T
C(bcc) =   35 000

B(fcc) = − 101 508 + 19.66 T
C(fcc) =   − 28 632 − 1.49 T

Compound phases

G(Ti$_2$Ni) = − 49 120 + 17.208 T
G(TiNi$_3$) = − 55 585 + 15.962 T
G(TiNi)   = − 54 600 + 18.133 T

Note: Values are given in J/mol and J/mol · K.

38Vog: R. Vogel and H.J. Wallbaum, "The System Iron-Nickel-Ni$_3$Ti-Fe$_2$Ti," *Arch. Eisenhuetten., 12*, 299-305 (1938) in German. (Equi Diagram; Experimental)

39Lav1: F. Laves and H.J. Wallbaum, "The Crystal Structure of Ni$_3$Ti and Si$_2$Ti (Two New Types)," *Z. Kristallogr., 101*, 78-93 (1939) in German. (Crys Structure; Experimental)

39Lav2: F. Laves and H.J. Wallbaum, "On the Crystal Chemistry of Titanium Alloys," *Naturwissenschaften, 49*, 674-675 (1939) in German. (Crys Structure; Experimental)

41Wal: H.J. Wallbaum, "The Systems of Iron with Titanium, Zirconium, Niobium, and Tantalum," *Arch. Eisenhuetten., 10*, 521-526 (1941) in German. (Equi Diagram; Experimental)

49Lon: J.R. Long, E.T. Hayes, D.C. Root, and C.E. Armentrout, "A Tentative Titanium-Nickel Diagram," *Bureau of Mines RI 4463* (1949). (Equi Diagram; Experimental)

50Duw: P. Duwez and J.L. Taylor, "The Structure of Intermediate Phases in Alloys of Titanium with Iron, Cobalt, and Nickel," *Trans. AIME, 188*(9), 1173-1176 (1950). (Crys Structure; Experimental)

50Tay: A. Taylor and R.W. Floyd, "Precision Measurements of Lattice Parameters of Non-Cubic Crystals," *Acta Crystallogr., 3*, 285-289 (1950). (Crys Structure; Experimental)

51Mcq: A.D. McQuillan, "The Effect of the Elements of the First Long Period on the α-β Transformation in Ti," *J. Inst. Met., 80*, 363-368 (1951). (Equi Diagram; Experimental)

***51Tay:** A. Taylor and R.W. Floyd, "The Constitution of Nickel-Rich Alloys of the Nickel-Chromium-Titanium System," *J. Inst. Met., 80, 577-587 (1951-52). (Equi Diagram; Experimental)*

52Ros: W. Rostoker, "Observations on the Occurrence of Ti$_2$X Phases," *Trans. AIME, 194*, 209-210 (1952). (Equi Diagram, Crys Structure; Experimental)

***53Mar:** H. Margolin, E. Ence, and J.P. Nielsen, "Titanium-Nickel Phase Diagram," *Trans. AIME, 197*, 243-247 (1953). (Equi Diagram; Experimental)

53May: D.J. Maykuth, H.R. Ogden, and R.I. Jaffee, "Titanium-Manganese System," *Trans. AIME, 197*, 225-230 (1953). (Equi Diagram; Experimental)

54Haz: T.H. Hazlett and E.R. Parker, "Effect of Some Solid Solution Alloying Elements on the Creep Parameters of Nickel," *Trans. ASM, 46*, 701-715 (1954). (Crys Structure; Experimental)

54Hum: W. Hume-Rothery and D.M. Poole, "Methods for Determining the Liquidus Points of Titanium-Rich Alloys," *J. Inst. Met., 82*, 490-492 (1953-54). (Equi Diagram; Experimental)

Fig. 11---Calculated Ti-Ni Phase Diagram

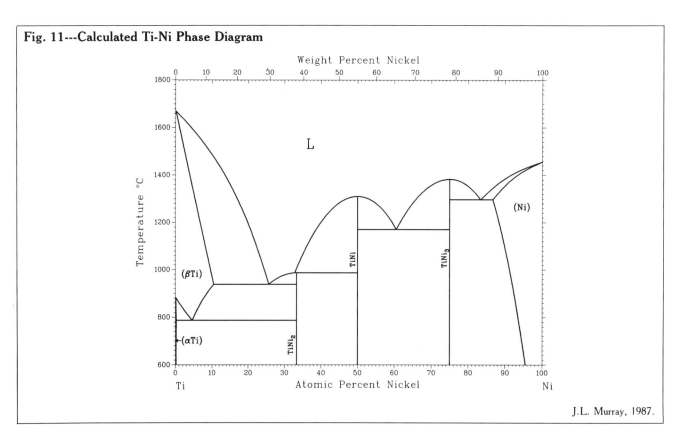

J.L. Murray, 1987.

Ni-Ti System

*54Mcq: A.D. McQuillan, "The Re-Investigation on a Nickel-Titanium Alloy and Observations on β/(α + β) Boundaries in Titanium Systems," *J. Inst. Met., 82*, 47-48 (1953-54). (Equi Diagram, Meta Phases; Experimental)

*54Poo: D.M. Poole and W. Hume-Rothery, "The Equilibrium Diagram of the System Nickel-Titanium," *J. Inst. Met., 83*, 473-480 (1954-55). (Equi Diagram; Experimental)

56Kub: O. Kubaschewski, H. Villa, and W.A. Dench, "The Reaction of Titanium Tetrachloride with Hydrogen in Contact with Various Refractories," *Trans. Faraday Soc., 52*, 214-222 (1956). (Thermo; Experimental)

56Pol: D.H. Polonis and J.G. Parr, "Phase Transformations in Titanium-Rich Alloys of Nickel and Titanium," *Trans. AIME, 206*, 531-536 (1956). (Meta Phases; Experimental)

57Bag: Yu.A. Bagariatskii and Yu.D. Tiapkin, "The Mechanism of Structure Transformations in Age Hardening Alloys Based on Nickel," *Sov. Phys. Crystallogr., 2*, 414-421 (1957). (Meta Phases; Experimental)

57Buc: C. Buckle and J. Manenc, "Pre-Precipitation and Transition Phase in a Nickel-Titanium Alloy During Isothermal Aging," *C.R. Acad. Sci., 224*, 1643-1646 (1957) in French. (Meta Phases; Experimental)

57Phi: T.V. Philip and P.A. Beck, "CsCl-Type Ordered Structures in Binary Alloys of Transition Elements," *Trans. AIME, 209*, 1269-1270 (1957). (Crys Structure; Experimental)

58Bag: Yu.A. Bagariatskii and Yu.D. Tyapkin, "X-ray Diffraction Determination of the Solubility Limits in Coarse-Grained Specimens by the Oscillation Method," *Zh. Neorg. Khim., 3*(4), 934-935 (1958) in Russian; TR: *Russ. J. Inorg. Chem., 3*(4), 151-158 (1958). (Equi Diagram; Experimental)

58Yur: G.A. Yurko, J.W. Barton, and J.G. Parr, "The Composition Range of Ti$_2$Ni," *Trans. AIME, 212*, 698-700 (1958). (Equi Diagram; Experimental)

59Age: N.V. Ageev and L.A. Petrova, "Stability of the β-Phase in Alloys of Titanium with Iron and Nickel," *Zh. Neorg. Khim., 4*(5), 1092-1099 (1959) in Russian; TR: *Russ. J. Inorg. Chem., 4*(5), 496-499 (1959). (Meta Phases; Experimental)

59Buc: C. Buckle, B. Genty, and J. Manenc, "Features of Precipitation in a Nickel-Titanium Alloy," *Rev. Metall., 56*, 247-259 (1959) in French. (Meta Phases; Experimental)

59Dwi: A.E. Dwight, "CsCl-Type Equiatomic Phases in Binary Alloys of Transition Elements," *Trans. AIME, 215*, 283-286 (1959). (Crys Structure; Experimental)

59Sto: E.R. Stover and J. Wulff, "The Nickel-Titanium-Carbon System," *Trans. AIME, 215*, 127-136 (1959). (Equi Diagram; Experimental)

59Yur: G.A. Yurko, J.W. Barton and J.G. Parr, "The Crystal Structure of Ti$_2$Ni," *Acta Crystallogr., 12*, 909-911 (1959). (Crys Structure; Experimental)

60Bar: J.W. Barton, G.R. Purdy, R. Taggart, and J.G. Parr, "Structure and Properties of Titanium-Rich Titanium-Nickel Alloys," *Trans. AIME, 218*, 844-849 (1960). (Meta Phases; Experimental)

60Mih: J.R. Mihalisin and R.F. Decker, "Phase Transformations in Nickel-Rich Nickel-Titanium Aluminum Alloys," *Trans. AIME, 218*, 507-515 (1960). (Meta Phases; Experimental)

60Nev: M.V. Nevitt, "Magnetization of the Compound TiFe," *J. Appl. Phys., 31*(1), 155-157 (1960). (Equi Diagram; Experimental)

60Sat: T. Sato, S. Hukai, and Y. Huang, "The M$_s$ Points of Binary Titanium Alloys," *J. Aust. Inst. Met., 5*(2), 149-153 (1960). (Meta Phases; Experimental)

60Stu: H.P. Stuwe and Y. Shimomura, "Lattice Constants of Cubic Phases FeTi, CoTi, NiTi," *Z. Metallkd., 51*(3), 180-181 (1960) in German. (Crys Structure; Experimental)

61Bag: Yu.A. Bagariatskii and Yu.D. Tiapkin, "Fresh Structural Data on the Decomposition of Supersaturated Solid Solutions of Titanium in Nickel and Nichrome," *Kristallografiya, 5*,(6), 882-890 (1961) in Russian; TR: *Sov. Phys. Crystallogr., 5*(6), 841-847 (1961). (Meta Phases; Experimental)

61Pur: G.R. Purdy and J.G. Parr, "A Study of the Titanium-Nickel System Between Ti$_2$Ni and TiNi," *Trans. AIME, 221*(6), 636-639 (1961). (Equi Diagram; Experimental)

62Bue: W.J. Buehler and R.C. Wiley, "TiNi—Ductile Intermetallic Compound," *Trans. ASM, 55*, 269-276 (1962). (Equi Diagram, Meta Phases; Experimental)

62Koe: W. Koester and R. Christ, "Hardness of Alloys Based on Nickel-Chromium and Nickel-Chromium-Cobalt with Additions of Aluminum and Titanium," *Arch. Eisenhuetten., 11*, 791-904 (1962). (Equi Diagram; Experimental)

62Yur: G.A. Yurko, J.W. Barton, and J.G. Parr, "The Crystal Structure of Ti$_2$Ni (A Correction)," *Acta Crystallogr., 15*, 1309 (1962). (Crys Structure; Experimental)

63Ben: D.H. Ben Israel and M.E. Fine, "Precipitation Studies in Ni$_{10}$ at.% Ti," *Acta Met., 11*, 1051-1059 (1963). (Equi Diagram; Experimental)

63Bue: W.J. Buehler, J.V. Gilfrich, and R.C. Wiley, "Effect of Low-Temperature Phase Changes on the Mechanical Properties of Alloys near Composition TiNi," *J. Appl. Phys., 34*(5), 1475-1477 (1963). (Equi Diagram, Meta Phases; Experimental)

63Gil: J.V. Gilfrich, "X-Ray Diffraction Studies on the Titanium-Nickel System," *Adv. X-Ray Anal., 6*, 74-84 (1963). (Equi Diagram, Crys Structure; Experimental)

63Mue: M.H. Mueller and H.W. Knott, "The Crystall Structure of Ti$_2$Cu, Ti$_2$Ni, Ti$_4$Ni$_2$O, and Ti$_4$Cu$_2$O," *Trans. AIME, 227*, 674-678 (1963). (Crys Structure; Experimental)

65Dau: D.P. Dautovich and G.R. Purdy, "Phase Transformations in TiNi," *Can. Met. Quart., 4*(2), 129-143 (1965). (Meta Phases, Crys Structure; Experimental)

65Wan: F.E. Wang, W.J. Buehler, and S.J. Pickart, "Crystal Structure and a Unique 'Martensitic' Transition of TiNi," *J. Appl. Phys., 36*(10), 3232-3239 (1965). (Meta Phases, Crys Structure; Experimental)

66Dau: D.F. Dautovich, Z. Melkvi, G.R. Purdy, and C.V. Stager, "Calorimetric Study of a Diffusionless Phase Transition in TiNi," *J. Appl. Phys., 37*, 2513-2514 (1966). (Meta Phases; Experimental)

66Spi: S. Spinner and A.G. Rozner, "Elastic Properties of NiTi as a Function of Temperature," *J. Acoust. Soc. Am., 40*(5), 1009-1015 (1966). (Meta Phases; Experimental)

67Ber: H.A. Berman and E.D. West, "Anomalous Heat Capacity of TiNi," *J. Appl. Phys., 38*(11), 4473-4476 (1967). (Meta Phases; Experimental)

67Han: J.E. Hanlon, S.R. Butler, and R.J. Wasilewski, "Effect on Martensitic Transformation on the Electrical and Magnetic Properties of NiTi," *Trans. AIME, 239*(9), 1323-1326 (1967). (Meta Phases; Experimental)

68Ard: A.J. Ardell, "Reply to Comments on 'Further Applications of the Theory of Particle Coarsening'," *Scr. Metall., 2*, 173-176 (1968). (Meta Phases; Experimental)

68Ber: H.A. Berman, E.D. West, and A.G. Rozner, "Engineering Data Obtained for Titanium-Nickel Alloy," *NBS Tech. News Bull., 52*, 75-76 (1968). (Thermo; Experimental)

68Cha: K. Chandra and G.R. Purdy, "Observations of Thin Crystals of TiNi in Premartensitic States," *J. Appl. Phys , 39*(5), 2176-2181 (1968). (Meta Phases; Experimental)

68Coh: J.B. Cohen and M.E. Fine, "Comments on 'Further Applications of the Theory of Particle Coarsening'," *Scrip. Metall., 2*, 153-154 (1968). (Meta Phases; Experimental)

68Mar: M.J. Marcinkowski, A.S. Sastri, and D. Koskimaki, "Martensitic Behaviour in the Equi-atomic Ni-Ti Alloy," *Philos. Mag., 18*, 945-958 (1968). (Meta Phases; Experimental)

68Pfe: H.U. Pfeifer, S. Bhan, and K. Schubert, "On the Ti-Ni-Cu System and Some Quasi-Homologous Alloys," *J. Less-Common Met., 14*, 291-302 (1968) in German. (Equi Diagram, Crys Structure; Experimental)

68Sas: A.S. Sastri, M.J. Marcinkowski, and D. Koskimaki, "Nature of the Ni-Ti Martensite Transformation," *Phys. Status Solidi, 25*, K67-K69 (1968). (Meta Phases; Experimental)

68Wan: F.E. Wang, B.F. DeSavage, and W.J. Buehler, "The Irreversible Critical Range in the TiNi Transition," *J. Appl. Phys., 39*(5), 2166-2175 (1968). (Meta Phases, Crys Structure; Experimental)

69Cro: W.B. Cross, A.H. Kariotis, and F.J. Stimler, "Nitinol Characterization Study," *NASA CR 1433* Sep (1969). (Meta Phases; Experimental)

69Kos: D. Koskimaki, M.J. Marcinkowski, and A.S. Sastri, "Solid State Diffusional Transformations in the Near-Equiatomic Ni-Ti Alloys," *Trans. AIME, 245*(9), 1883-1890 (1969). (Equi Diagram, Meta Phases; Experimental)

69Ras: P.K. Rastogi and A.J. Ardell, "The Coherent Solubilities of γ' in Ni-Al, Ni-Si and Ni-Ti Alloys," *Acta Metall., 17*, 595-602 (1969). (Meta Phases; Experimental)

69Sai: K. Saito and R. Watamabe, "Precipitation in Ni_{12} at.% Ti Alloy," *Jpn. J. Appl. Phys., 8*(1), 27-35 (1969); TR: *Trans. Nat. Res. Inst. Met., 11*(3), 27-35 (1969). (Equi Diagram; Experimental)

70Ard: A.J. Ardell, "The Growth of γ' Precipitates in Aged Ni-Ti Alloys," *Met. Trans., 1*, 525-534 (1970). (Meta Phases; Experimental)

70Nag: A. Nagasawa, "A New Phase Transformation in the NiTi Alloy," *J. Phys. Soc. Jpn., 29* 1386 (1970). (Meta Phases, Crys Structure; Experimental)

70Pos: V.S. Postnikov, V.S. Lebedinskiy, V.A. Yevsyukov, I.M. Sharshakov, and M.S. Pesin, "Phase Transformations in the Intermetallic Compound TiNi," *Fiz. Met. Metalloved., 29*(2), 364-369 (1970) in Russian; TR: *Phys. Met. Metallogr., 29*(2), 139-144 (1970). (Meta Phases; Experimental)

70Was: R.J. Wasilewski, S.R. Butler, and J.E. Hanlon, "Discussion of Solid State Diffusional Transformations in the Near-Equiatomic Ni-Ti Alloys," *Metall. Trans., 1*, 1459- 1461 (1970). (Equi Diagram; Experimental)

71Bha: S. Bhan, "Structure of High Temperature Ti $(Ti_{0.11}Ni_{0.89})_3$ Phase," *J. Less-Common Met., 25*, 215-220 (1971). (Crys Structure; Experimental)

71Gup: S.P. Gupts, A.A. Johnson, and K. Mukherjee, "Comments on the Phase Transformation in the NiTi Alloy," *J. Phys. Soc. Jpn., 31*, 605-606 (1971). (Meta Phases, Crys Structure; Experimental)

71Heh: R.F. Hehemann and G.D. Sandrock, "Relations Between the Premartensitic Instability and the Martensite Structure in TiNi," *Scr. Metall., 5*, 801-806 (1971). (Crys Structure; Experimental)

71Kor: I.I. Kornilov, Ye.V. Kachur, and O.K. Belousov, "Dilatation Analysis of Transformation in the Compound TiNi," *Fiz. Met. Metalloved., 32*(2), 420-422 (1971) in Russian; TR: *Phys. Met. Metallogr., 32*(2), 190-193 (1971). (Meta Phases; Experimental)

71Nag1: A. Nagasawa, "Martensite Transformation and Memory Effect in the NiTi Alloy," *J. Phys. Soc. Jpn., 31*(1), 136-147 (1971). (Meta Phases; Experimental)

71Nag2: A. Nagasawa, "Characteristics of Stacking Faults in the NiTi Martensite," *J. Phys. Soc. Jpn., 30*, 587 (1971). (Meta Phases; Experimental)

71Ots: K. Otsuka, T. Sawamura, and K. Shimizu, "Crystal Structure and Internal Defects of Equiatomic TiNi Martensite," *Phys. Status Solidi (a), 5*, 457-470 (1971). (Meta Phases, Crys Structure; Experimental)

71San: G.D. Sandrock, A.J. Perkins, and R.F. Hehemann, "The Premartensitic Instability in Near-Equiatomic TiNi," *Metall. Trans., 2*(10), 2769-2781 (1971). (Meta Phases; Experimental)

71Was: R.J. Wasilewski, S.R. Butler, J.E. Hanlon, and D. Worden, "Homogeneity Range and the Martensitic Transformation in TiNi," *Metall. Trans., 2*(1), 229-238 (1971). (Equi Diagram, Meta Phases; Experimental)

72Ger: R.M. German and G.R. St. Pierre, "The High Temperature Thermodynamic Properties of Ni-Ti Alloys," *Metall. Trans., 3*, 2819-2823 (1972). (Thermo; Experimental)

72Wan: F.E. Wang, S.J. Pickart, and H.A. Alperin, "Mechanism of the TiNi Martensitic Transformation and the Crystal Structures of TiNi-II and TiNi-III Phases," *J. Appl. Phys., 43*(1), 97-112 (1972). (Meta Phases; Experimental)

73Fed: S.G. Fedotov and N.F. Kvasova, "Phase Relations in Metastable Titanium-Nickel Alloys," *Dokl. Akad. Nauk SSSR, 209*(5), 1084-1087 (1973) in Russian; TR: *Sov. Phys. Dokl., 18*(4), 270-272 (1973). (Meta Phases; Experimental)

73Gup: S.P. Gupta and A.A. Johnson, "Diffusion Controlled Solid State Transformation in the Near-Equiatomic TiNi," *Mater. Sci. Eng., 11*, 283-297 (1973). (Equi Diagram, Meta Phases; Experimental)

73Hua: Y.C. Huang, S. Suzuki, H. Kaneko, and T. Sato, "Continuous Cooling-Transformation of β-Phase in Binary Titanium Alloys," Sci. Technol. Appl. Titanium, Proc. Int. Conf., R.I. Jaffee, Ed., 695-698 (1973). (Meta Phases; Experimental)

73Ika: H. Ikawa, S. Shin, M. Miyagi, and M. Morikawa, "Some Fundamental Studies on the Phase Transformation from β Phase to α Phase in Titanium Alloys," Sci. Technol. Appl. Titanium, Proc. Int. Conf., R.I. Jaffee, Ed., 1545 (1973). (Meta Phases; Experimental)

73Pol: A.F. Polesya and L.S. Sluipchenko, "Formation of Amorphous Phases and Metastable Solid Solutions in Binary Ti and Zr Alloys with Fe, Ni, and Cu," *Izv. Akad. Nauk SSSR, Met.*, (6), 173-178 (1973) in Russian; TR: *Russ. Metall.*, (6), 103-107 (1973). (Meta Phases; Experimental)

74Bas: G.F. Bastin and G.D. Rieck, "Diffusion in the Titanium-Nickel System. I. Occurrence and Growth of the Various Intermetallic Compounds," *Metall. Trans., 5*(8), 1817-1826 (1974). (Equi Diagram; Experimental)

74Gus: L.N. Guseva and L.K. Dolinskaya, "Metastable Phases in Titanium Alloys with Group VIII Elements Quenched from the β-Region," *Izv. Akad. Nauk SSSR, Met.*, (6), 195-202 (1974) in Russian; TR: *Russ. Metall.*, (6), 155-159 (1974). (Meta Phases; Experimental)

74San: G.D. Sandrock, "Premartensitic Behavior of the Electrical Resistivity of NiTi," *Metall. Trans., 5*(1), 299-301 (1974). (Meta Phases; Experimental)

74Sin: R. Sinclair, J.A. Leake, and B. Ralph, "Spinodal Decomposition of a Nickel-Titanium Alloy," *Phys. Status Solidi (a), 26*, 285-298 (1974). (Meta Phases; Experimental)

74Zij: S.R. Zijlstra, J. Beijer, and J.A. Klostermann, "An Electron Microscopical Investigation on the Martensitic Transformation in TiNi," *J. Mater. Sci., 9*, 145-154 (1974). (Meta Phases; Experimental)

75Hon: T. Honma and H. Takei, "Effect of Heat Treatment on the Martensitic Transformation in TiNi Compound," *J. Jpn. Inst. Met., 39*, 175-182 (1975) in Japanese. (Equi Diagram; Experimental)

75Per: J. Perkins, *Shape Memory Effects in Alloys*, Plenum Press, New York (1975). (Meta Phases; Experimental)

75Was: R.J. Wasilewski, "The Shape Memory Effect in TiNi: One Aspect of Stress-Assisted Martensitic Transformation," in *Shape Memory Effects in Alloys*, Plenum Press, New York, 245-271. (Meta Phases; Review)

76Lau: D.E. Laughlin, "Spinodal Decomposition in Nickel Based Nickel-Titanium Alloys," *Acta Metall., 24*, 63-58 (1976). (Meta Phases; Experimental)

78Gus: L.N. Guseva and L.K. Dolinskaya, "Formation Conditions of Athermal ω Phase in Alloys of Titanium with Transition Elements," *Krist. Strukt. Svoistva. Met. Splavov*, 59-63 (1978) in Russian. (Meta Phases; Experimental)

78Has: K. Hashimoto and T. Tsujimoto, "X-ray Diffraction Patterns and Microstructures of Aged Ni-Ti Alloys," *Trans. Jpn. Inst. Met, 19*, 77-84 (1978). (Meta Phases; Experimental)

78Kau: L. Kaufman and H. Nesor, "Coupled Phase Diagrams and Thermochemical Data for Transition Metal Binary Systems—III," *Calphad, 2*(2), 81-108 (1978). (Thermo; Theory)

78Kha: V.N. Khachin, Yu.I. Paskal, V.E. Gunter, A.A. Monasevich, and V.P. Sivokha, "Structural Transformations, Physical Properties and Memory Effects in the Nickel-Titanium and Titanium-Based Alloys," *Fiz. Met. Metalloved., 46*(3), 511-520 (1978) in Russian; TR: *Phys. Met. Metallogr., 46*(3), 49-57 (1978). (Meta Phases; Experimental)

78Yao: Y.D. Yao and S. Arajs, "Determination of the Curie Temperature of Ferromagnetic Alloys Using the Deviation from Matthiessen's Rule," *Phys. Status Solidi (b), 89*, K201-K205 (1978). (Equi Diagram; Experimental)

79Che: D.B. Chernov, Yu.I. Paskal, V.E. Gyunter, L.A. Monasevich, and E.M. Savitskii, "The Multiplicity of Structural Transitions in Alloys Based on TiNi," *Dokl. Akad. Nauk SSSR, 247*, 854-857 (1979) in Russian; TR: *Sov. Phys. Dokl., 24*(8), 664-666 (1979). (Meta Phases; Experimental)

79Lev: G.A. Levshin and V.I. Alekseev, "Thermodynamic Properties of Nickel-Titanium Alloys," *Zh. Fiz. Khim., 53*, 769-772

(1979) in Russian; TR: *Russ. J. Phys. Chem., 53*(3), 437-439 (1979). (Thermo; Experimental)

80Ino: A. Inoue, K. Kobayashi, C. Suryanarayana, and T. Masumoto, "An Amorphous Phase in Co-Rich Co-Ti Alloys," *Scr. Metall., 14,* 119-123 (1980). (Meta Phases; Experimental)

80Lin: H.C. Ling and R. Kaplow, "Phase Transitions and Shape Memory in NiTi," *Metall. Trans. A, 11,* 77-83 (1980). (Meta Phases; Experimental)

80Mil: R.V. Milligan, "Determination of Phase Transformation Temperatures of TiNi Using Differential Thermal Analysis," Titanium '80, Ti Sci. Tech., Proc. Int. Conf. Kyoto, Japan, May 18-22, T. Kimuzi, Ed., 1461-1467 (1980). (Meta Phases; Experimental)

81Esi: Yu.O. Esin, M.G. Valishev, A.F. Ermakov, O.V. Gel'd, and M.S. Petrushevskii, "The Enthalpies of Formation of Liquid Germanium-Titanium and Nickel-Titanium Alloys," *Zh. Fiz. Khim., 55,* 753-754 (1981) in Russian; TR: *Russ. J. Phys. Chem., 55*(3), 421-422 (1981). (Thermo; Experimental)

81Gac: J.C. Gachon, M. Notin, and J. Hertz, "The Enthalpy of Mixing of the Intermediate Phases in the Systems FeTi, CoTi, and NiTi by Direct Reaction Calorimetry," *Thermochim. Acta, 48,* 155-164 (1981). (Thermo; Experimental)

81Has: K. Hashimoto, T. Tsujimoto, and K. Saito, "Magnetic Study on Phase-Decomposition in a Ni–9.6 at.% Ti Alloy,"

Trans. Jpn. Inst. Met., 22(11), 798-806 (1981). (Meta Phases; Experimental)

81Mel: K.N. Melton and O. Mercier, "The Mechanical Properties of NiTi-Based Shape Memory Alloys," *Acta Metall., 29,* 393-398 (1981). (Meta Phases; Experimental)

81Mic: G.M. Michal and R. Sinclair, "The Structure of TiNi Martensite," *Acta Crystallogr., B37,* 1803-1807 (1981). (Meta Phases, Crys Structure; Experimental)

82Mic: G.M. Michal, P. Moine, and R. Sinclair, "Characterization of the Lattice Displacement Waves in Premartensitic TiNi," *Acta Metall., 30,* 125-138 (1982). (Meta Phases; Experimental)

82Moi: P. Moine, G.M. Michal, and R. Sinclair, "A Morphological Study of "Premartensitic" Effects in TiNi," *Acta Metall., 30,* 109-121 (1982). (Meta Phases; Experimental)

83Buh: W. Buhrer, R. Gotthardt, A. Kulik, O. Mercier, and F. Staub,"Powder Neutron Diffraction Study of Nickel-Titanium Martensite," *J. Phys. F: Met. Phys., 13,* L77-L81 (1983). (Meta Phases; Experimental)

83Cha: G. Chattopadhyay and H. Kleykamp, "Phase Equilibria and Thermodynamic Studies in the Titanium-Nickel and Titanium-Nickel-Oxygen Systems," *Z. Metallkd., 74,* 182-197 (1983). (Thermo; Experimental)

85Sau: N. Saunders, University of Surrey, private communication (1985).

The O-Ti (Oxygen-Titanium) System

15.9994   47.88

By J.L. Murray and H.A. Wriedt

Introduction

This assessment of the Ti-O system covers the phase equilibria and crystal structures of the condensed phases in the composition range between pure Ti and TiO_2. The thermodynamic properties of the Ti oxides have been studied and assessed extensively [75Cha]. The present assessment does not duplicate [75Cha]; coverage of thermochemical properties is limited to recent work. So far as possible, melting points and solid-state transition temperatures in the present diagram agree with [75Cha], so that the two assessments may be used together.

The assessed Ti-O phase diagram is shown in Fig. 1, and its important features have been summarized in Table 1. The temperature range in which a reasonable equilibrium diagram can be constructed excludes some phase transitions of the higher oxides, and it has not been possible to include all of the observed higher oxide phases in a diagram. However, Tables 1, 2, and 3 contain complete listings of the phases and phase transitions.

O has a large solubility in low-temperature cph (αTi), and it stabilizes (αTi) with respect to the high-temperature bcc form, (βTi). At low temperature, the ordered cph phases Ti_2O, Ti_3O, and, possibly, Ti_6O are formed with some homogeneity range.

Structures of the monoxides are based on the NaCl structure of the high-temperature γTiO form. Four additional structural modifications were identified, which here are designated βTiO, αTiO, $βTi_{1-x}O$, and $αTi_{1-x}O$. In this assessment, "TiO" refers to the monoxides without

restriction to a particular variety. The phase boundaries separating these phases, except for the disordering of αTiO, were not determined; the phase boundaries of the monoxides in equilibrium with (αTi) and with $βTi_2O_3$ were determined, but without distinguishing the various monoxide modifications.

The stable condensed phase richest in O is rutile (TiO_2). In addition to rutile, TiO_2 has two nonequilibrium low-pressure forms (anatase and brookite) and two nonequilibrium high-pressure forms (TiO_2-II and TiO_2-III).

Between the monoxides and TiO_2 is a series of discrete phases with stoichiometry Ti_nO_{2n-1}, where $n \geq 2$, which are called Magneli phases. It was suggested that discrete equilibrium phases exist for $n \leq 99$ [72Roy]. Ti_nO_{2n-1} ($4 \leq n \leq 10$) phases have crystal structures derived from the rutile structure by crystallographic shear. Closely related structures were observed and described as coherent intergrowths of Magneli phases [70And] or as families of phases based on different crystallographic shear operations [69Bur, 71Bur1, 71Bur2]. Magneli phases undergo one or more structural, electrical, or magnetic transitions at low temperature.

Equilibrium Diagram

Terminal Solid Solutions and Suboxides

(αTi) + (βTi) Phase Field.

Experimental data on the (αTi)/(βTi) equilibria are compared with the assessed diagram in Fig. 2. The following experimental techniques were used:

Phase Diagrams of Binary Titanium Alloys

J.L. Murray and H.A. Wriedt, 1987.

Fig. 1---Assessed Ti-O Phase Diagram

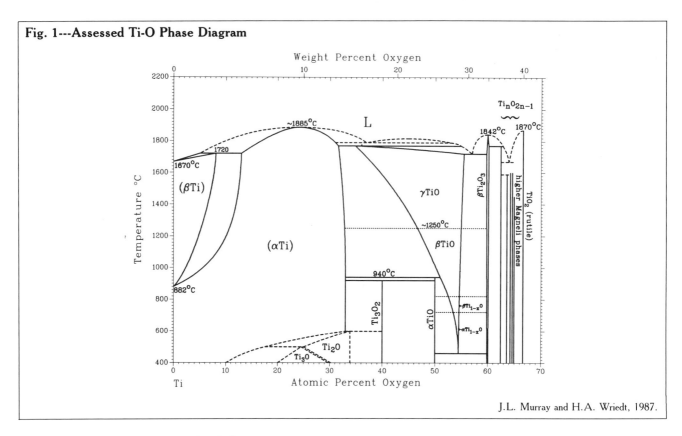

Table 1 Special Points of the Ti-O System

Reaction	Compositions of the respective phases, at.% O			Temperature, °C	Reaction type
L + (αTi) ⇌ (βTi)	5	13	8	1720 ± 25	Peritectic
L ⇌ (αTi)		~ 24		1885 ± 25	Congruent
(αTi) + Ti_3O_2 ⇌ Ti_2O	33.3	40	33.9	~ 600	Peritectoid
(αTi) + Ti_2O ⇌ Ti_3O	~ 17	~ 25	~ 24.5	~ 500	Peritectoid
L ⇌ (αTi) + L	~ 37	~ 31	~ 53	~ 1800	Monotectic (?)
L + (αTi) ⇌ γTiO	~ 55	31.4	34.5	1770	Peritectic
γTiO ⇌ βTiO		...		~ 1250	Unknown
$βTiO ⇌ βTi_{1-x}O$		...		...	Unknown
$βTi_{1-x}O ⇌ αTi_{1-x}O$		...		...	Unknown
(αTi) + βTiO ⇌ αTiO	33.3	51	50	940	Peritectoid
(αTi) + αTiO ⇌ Ti_3O_2	32.4	50	40	920	Peritectoid
$αTi_{1-x}O ⇌ αTiO + βTi_2O_3$	54.5	50	60	460	Eutectoid
$L ⇌ γTiO + βTi_2O_3$	~ 57	54.5	59.8	1720	Eutectic
$L ⇌ βTi_2O_3$		60		1842	Congruent
$L + βTi_2O_3 ⇌ βTi_3O_5$	63	60.2	62.5	1770	Peritectic
$βTi_2O_3 ⇌ αTi_2O_3$		60		~ 180	Unknown
$βTi_3O_5 ⇌ αTi_3O_5$		62.5		187	Unknown
$γTi_4O_7 ⇌ βTi_4O_7$		63.64		~ 123	Unknown
$βTi_4O_7 ⇌ αTi_4O_7$		63.64		~ 148	Unknown
$L ⇌ βTi_3O_5 + ?$	~ 64	62.5	...	~ 1670	Eutectic
$βTi_3O_5 + βTi_5O_9 ⇌ γTi_4O_7$	62.5	64.29	63.64	~ 1500	Peritectoid
$L ⇌ TiO_2$		66.7		1870	Congruent
L ⇌ (βTi)		0		1670	Melting point
(βTi) ⇌ (αTi)		0		882	Allotropic transformation

- Metallography [50Jaf, 53Bum, 56Sch]
- Thermoelectric power measurements and metallography [51Jen]
- Thermodynamic measurements [54Kub, 57Mah, 78Tet]

- Analysis of diffusion specimens [54Was]

[51Jen] showed that the impurities present in commercial-grade Ti were sufficient to cause a marked (>1 at.%) broadening of the (αTi) + (βTi) field. Thermoelectric

Table 2 Experimental Congruent and Incongruent Melting Temperatures

Phase	Reference	Temperature, °C	Comment
γTiO	[37Daw]	1750	Incongruent
	[54Dev]	1830	
	[56Nis]	1750	
	[75Cha]	1750	
βTi$_2$O$_3$	[54Dev]	1920	Congruent
	[56Sch]	1800	
	[56Nis]	1830	
	[60Bra]	1820 ± 15	
	[66Wah, 67Gil]	1842 ± 10	
	[73Sly]	1770	
	[75Cha]	1842	
βTi$_3$O$_5$	[66Wah, 67Gil]	1774 ± 10	Incongruent
	[54Dev]	1900	
	[73Sly]	1810	
	[75Cha]	1777	
γTi$_4$O$_7$	[71Ham]	<1670	Probably incongruent
	[73Sly]	1720	
	[75Cha]	1677 ± 20	
Rutile	[73Sly]	1741	Congruent
	[60Bra]	1870	
	[54Dev]	1830 ± 10	
	[56Nis]	1825	
	[75Cha]	1870	

power data [51Jen] for high-purity alloys showed well defined discontinuities in slope for the (αTi) transus, but were difficult to interpret for the (βTi) transus above 1050 °C.

For the (βTi) transus, the data of [56Sch] are preferred, because they were corroborated by the diffusion experiments of [54Was]. The data of [51Jen] are preferred for the low-temperature part of the (αTi) transus. For the (αTi) transus above 1300 °C, where the data of [53Bum] and [56Sch] were scattered and disagreed with each other by as much as 2 at.%, there is no clear basis for preferring the data of either. In constructing the assessed transus, the liquidus and solidus have been considered. The (αTi) boundaries must be drawn so that the liquidus and solidus can be extrapolated to a metastable melting point of pure cph Ti. The roughly extrapolated melting point falls at about 1425 °C, in good agreement with the theoretical prediction of [70Kau].

Ti-Rich Liquidus and Solidus.

Optical pyrometry was used by [53Bum] and [56Sch] to determine the temperature of the peritectic reaction L + (αTi) ⇄ (βTi). [53Bum] found 1740 ± 25 °C (relative to 1720 °C for "pure" Ti); [56Sch] found 1720 ± 25 °C (relative to 1660 °C for "pure" Ti). The value 1720 ± 25 °C is used in this evaluation.

(αTi) with 24 at.% O melts congruently at 1885 ± 25 °C, according to optical pyrometric melting point data [53Bum, 56Sch, 65Kor] (Fig. 3). The schematic (αTi)

Table 3 Low-Temperature Transitions of the Higher Oxides

Transition	Temperature, K	Experimental technique/comment	Reference
βTi$_2$O$_3$ → αTi$_2$O$_3$	473 ± 20	Heat content	[46Nay]
	453	Heat capacity	[73Bar]
	433 to 473	Lattice parameters	[58Pea]
	390 to 450	Lattice parameters	[68Rao]
	660 ± 30	Néel temperature (neutron diffraction, electrical resistivity)	[63Abr]
	~ 450	Magnetic susceptibility	[67Kcy1]
βTi$_3$O$_5$ → αTi$_3$O$_5$	460	Metallic → semiconductor	[69Bar]
	462	Magnetic susceptibility (heating)	[70Mul, 72Dan]
	432	Magnetic susceptibility (cooling)	
	450	Heat content	[46Nay]
	373 to 393	X-ray diffraction (corrected by [71Ash])	[59Asb, 61Mag]
	460		[71Asb]
	448	DTA, X-ray diffraction, electrical conductivity	[71Rao]
	450	Magnetic susceptibility	[65Por]
α'Ti$_3$O$_5$ → αTi$_3$O$_5$	250	Metastable transformation	[71Asb]
γTi$_4$O$_7$ → βTi$_4$O$_7$	149 ± 2	Metallic → semiconductor	[69Bar]
	150	Metallic → semiconductor	[70Mul, 72Dan]
	150	Discontinuity in lattice parameters	[70Mar]
βTi$_4$O$_7$ → αTi$_4$O$_7$	125	Structural transition	[70Mul]
	125	8 to 10 K hysteresis	[69Bar]
γTi$_5$O$_9$ → βTi$_5$O$_9$	130	Magnetic susceptibility	[69Bar, 70Dan]
βTi$_5$O$_9$ → αTi$_5$O$_9$	125	Magnetic susceptibility	[69Bar]
βTi$_6$O$_{11}$ → αTi$_6$O$_{11}$	130	Electrical conductivity (DTA places transition lower)	[69Bar]
	122	Magnetic susceptibility	[70Mul, 67Key1, 67Key2]
Ti$_7$O$_{13}$ through Ti$_{10}$O$_{19}$	~ 150	Magnetic susceptibility	[67Key1, 67Key2]

Fig. 2---Experimental Data for the (αTi)/(βTi) Phase Boundaries

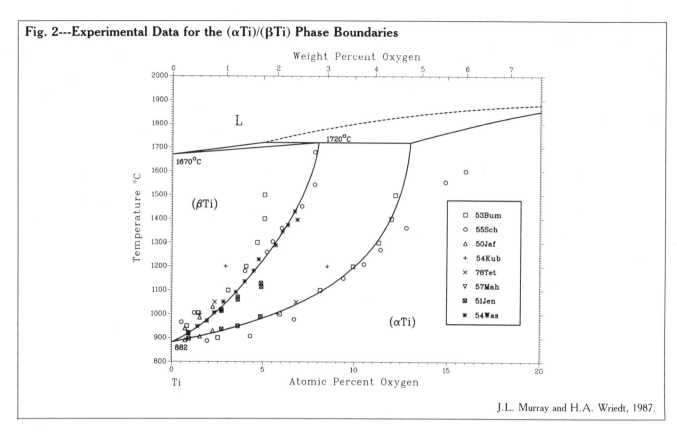

J.L. Murray and H.A. Wriedt, 1987.

Fig. 3---Experimental Liquidus Data for the Ti-O System

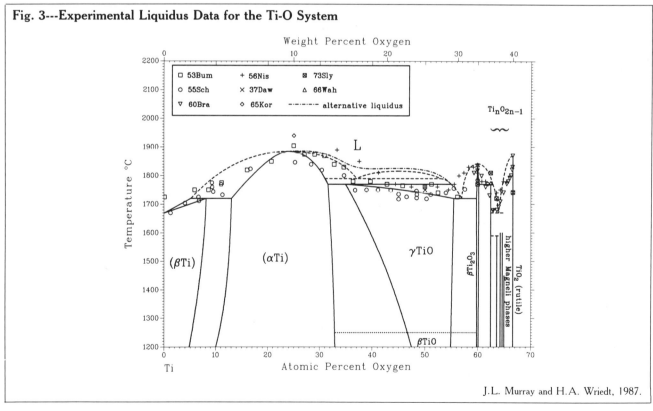

J.L. Murray and H.A. Wriedt, 1987.

liquidus (Fig. 1), is based on the endpoint compositions at the invariant reactions. Experimental liquidus data for the whole system are shown in Fig. 3.

(αTi) Homogeneity Range.

[53Bum], [56Sch], and [70Jos] examined the (αTi)/ (αTi) + TiO boundary from 600 °C to the melting point,

Fig. 4---Experimental Data for Phase Boundaries Involving the Hexagonal Phase

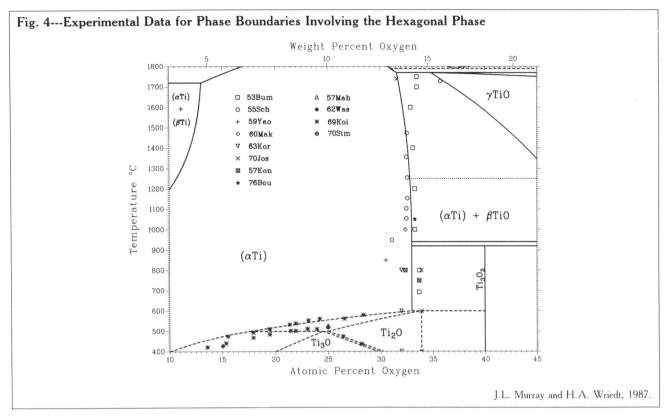

J.L. Murray and H.A. Wriedt, 1987.

using X-ray diffraction (XRD) and metallography on quenched specimens. Additional data were provided over limited temperature ranges by [60Mak] (lattice parameter); [59Yao] and [63Vas] (magnetic susceptibility); [66Dub] (hardness); [76Bou] [78Tet] and (partial Gibbs energy); and [57Kon], [63Kor], [71Dub1], [71Dub2], and [71Dub3] (microstructure and lattice parameter). Selected experimental data are shown in Fig. 4.

Below 1000 °C, there is a discrepancy between the vertical phase boundary reported by [53Bum] and the decreasing O solubility at lower temperature reported by [56Sch]. The magnetic susceptibility results of [59Yao] agreed with those of [56Sch] but are not sufficiently accurate to be considered definitive. [70Jos] suggested that [56Sch] may have misinterpreted banded microstructures associated with etching effects on faulted (αTi) alloys as evidence for TiO precipitation. The work of [53Bum] is supported by lattice parameter data [70Jos] and further metallographic work [57Kon, 63Kor]; therefore, it has been used in the present assessment.

Above 1400 °C, the assessed diagram is based on [70Jos], who showed that single-phase (αTi) alloys transformed on heating to (αTi) + TiO structures. [70Jos] placed the solubility of O in (αTi) between 30.9 and 31.6 at.% O at the L + (αTi) + γTiO + L peritectic temperature.

Ordered Hexagonal Phases (Suboxides).

Several qualitatively different pictures were given for the phase equilibria involving the ordered hexagonal phases. Ti$_2$O (also designated α' [70Jos, 69Yam] or Ti$_2$O$_{1-y}$ [57And1, 62Hol]) and Ti$_3$O are well established phases. Ti$_6$O [63Kor, 70Kor, 79Dav] and Ti$_{12}$O also were proposed as equilibrium phases. However, a single-ordered phase with compositions ranging from near Ti$_2$O to quite low O content may encompass the phases distin-

guished previously as Ti$_3$O and Ti$_6$O, and possibly Ti$_{12}$O, as well. In the present assessment, this phase has been designated Ti$_3$O to avoid proliferation of nomenclature.

In all of the ordered hexagonal phases, O resides in octahedral sites in layers alternating with Ti. Ti$_2$O has the anti-CdI$_2$ structure (space group $P\bar{3}m1$) with alternate O layers vacant, and additional vacancies distributed randomly in the occupied layer [57And1, 59Now, 62Hol, 69Yam, 70Yam, 71Dub1, 71Dub2, 71Dub3]. For alloys containing up to 33.3 at.% O (at which all sites in the plane are occupied) ordering can occur within the occupied plane, lowering the symmetry to $P\bar{3}1c$ [62Hol, 69Yam, 70Kor, 70Yam, 71Dub1, 71Dub2, 71Dub3]. Three equivalent sites are available in the plane, which are filled successively at Ti$_6$O, Ti$_3$O and Ti$_2$O.

Thus, the special compositions Ti$_6$O and Ti$_3$O may have distinct structures; alternatively, they may belong to a single phase in which the sites are filled successively but in completely continuous fashion. $P\bar{3}m1$ is a halving subgroup of $P6_3/mmc$ (of the completely disordered phase); therefore, a second-order transition, (αTi) ⇄ Ti$_2$O, is not ruled out by symmetry considerations. $P\bar{3}1c$, however, is not a halving subgroup of $P\bar{3}m1$; consequently, a second-order transition is ruled out for the Ti$_2$O ⇄ Ti$_3$O transition, by the Landau criterion.

Ti$_2$O was discovered by [57And1], [57And2], [59Now], and [60Mak]. Single-phase Ti$_2$O was found within the range 22.5 to 33.3 at.% O in as-solidified alloys [71Dub1]. [62Hol] found Ti$_2$O in stoichiometric alloys quenched from 1800 °C; in other alloys, it was found only after low-temperature annealing. Ti$_3$O was found only in 25 at.% O alloys heat treated at 400°C. These results were verified in all of their qualitative features by [66Dub], [68Mod], and [70Jos].

Based primarily on physical and mechanical properties of quenched alloys, it was proposed that Ti$_2$O

[57And1, 62Hol, 68Mod] or Ti_3O [63Kor, 65Kor, 66Dub] is stable to the melt. [70Jos] presented indirect microstructural evidence that Ti_2O forms in the solid state. Work at temperature [57Mah, 62Was, 66Yam, 69Hir, 69Koi, 70Sim2, 70Yam] showed that both Ti_2O and Ti_3O are formed only below about 600 °C. Observations of the ordering reactions by resistivity [62Was], heat capacity [56Mah, 69Koi], and structural analysis [70Sim1, 70Sim2, 70Yam] are in good agreement, as shown in Fig. 4. Two-phase structures were not found by optical microscopy [68Mod]. This area of the assessed diagram is a tentative, schematic construction based on the data shown in Fig. 4.

From C_p-vs-temperature and order parameter-vs-temperature measurements, [69Koi] and [69Hir] proposed that both ordering reactions, $(\alpha Ti) \rightleftarrows Ti_2O \rightleftarrows Ti_3O$, proceed as second-order phase transitions. As pointed out above, this assertion for the second reaction is ruled out by the Landau criterion. The present authors believe that the heat capacity data of [69Koi] are interpreted most plausibly as evidence for a three-phase reaction of (αTi), Ti_2O, and Ti_3O, as shown schematically in the assessed diagram.

The diagram proposed by [63Kor], [65Kor], [70Kor], and [73Kor] differed from all others. Ti_6O was shown as stable to 820 °C with a wide two-phase $Ti_6O + (\alpha Ti)$ field. Ti_3O was shown as a line compound in the single-phase (αTi) field (sic) and stable to the melt. Ti_2O was not shown. None of these features was verified; consequently, this diagram has not been considered further. Ti_6O was also thought to be a distinct phase by [66Yam], [71Dub1], [71Dub2], [71Dub3], and [79Dav]. Because concrete evidence for a phase transition between Ti_3O and Ti_6O was lacking, and also from structural considerations, the present authors do not consider Ti_6O to be a distinct phase. In summary, the ordered cph phases were shown in various compilations either as discrete compounds or as phases of wide homogeneity range produced from the disordered phase by second-order transitions. They were shown as persisting to the melting point, or as disordering above about 600 °C. According to crystal structure data, ordering on the O sublattice occurs over a wide composition range, and it is erroneous to show these phases as discrete compounds. Evidence for the persistence of the ordered phases to the melting point was indirect; direct observations of disordering at lower temperatures have been preferred as the basis for the assessed diagram. Detailed placement of phase boundaries, however, is still very uncertain.

Ti_3O_2.

Ti_3O_2 (usually designated δ in the experimental literature) was observed [53Bum, 56Sch, 57Kon] to occur only below 800 to 900 °C. Originally, the crystal structure was thought incorrectly to be tetragonal. Ti_3O_2 was not obtained as a single phase [57And1, 59And]. Microstructures of alloys containing Ti_3O_2 are not typically peritectoid, but the conditions of temperature and composition range in which Ti_3O_2 forms suggested that it is produced by the reaction $(\alpha Ti) + \alpha TiO \rightleftarrows Ti_3O_2$ [53Bum, 56Sch, 57And1, 57Kon, 59And]. The crystal structure was determined by [59And] to be hexagonal, an ordered structure with ideal stoichiometry Ti_3O_2. In the assessed diagram, the peritectoid reaction has been set at 920 °C, on the basis of the values 925 and 910 °C of [53Bum] and [56Sch], respectively. The location of the composition at 40 at.% O is from [57And1] and [57Kon].

The Monoxides

Liquidus and Solidus.

Over the composition region 34.5 to 55.6 at.% O, the homogeneity range of γTiO, alloys begin to melt in the narrow range of 1770 to 1720 °C. These data were interpreted previously in three ways:

- The reaction of L, γTiO, and Ti_2O_3 is peritectic. This implies that γTiO melts congruently with a minimum melting point [56Nis, 70Jos, Elliott].
- The reaction of L, γTiO, and Ti_2O_3 is eutectic and occurs at a temperature below 1770 °C. The melting of γTiO is then incongruent. If the peritectic composition for the reaction $L + (\alpha Ti) \rightleftarrows \gamma TiO$ is placed near 35 at.% O, then the two-phase $L + \gamma TiO$ field is a narrow lens [53Bum, 66Wah].
- The reaction of L, γTiO, and Ti_2O_3 is eutectic and it occurs below 1770 °C; γTiO melts incongruently, as above. However, if the peritectic composition is placed nearer 55 at.%, both the $L + (\alpha Ti)$ and $L + \gamma TiO$ two-phase fields broaden as much as 20 at.% over a narrow (50 °C) temperature interval, as the peritectic temperature is approached from above and below (see Fig. 3).

These constructions are very unusual, and require care to draw a diagram consistent with the rules of phase diagram construction. In this assessment, the observed melting points have been interpreted in terms of a liquid miscibility gap (see Fig. 3).

In addition to the determinations of incipient melting temperatures shown in Fig. 3 and Table 2, two observations were reported about the melting in this region:

- By metallographic examination of alloys quenched from temperatures near the melting point, [56Sch] found evidence for $(\alpha Ti) + L$ and $\gamma TiO + L$ phase fields over a wide composition range, but a small temperature range.
- [51Jen], in a discussion of [53Bum], noted that there appeared to be a pronounced "immiscibility in the liquid melts between TiO and the higher oxides." [53Bum] corroborated the gravity segregation during preparation of compositions between TiO and TiO_2.

Based on these data, a liquid miscibility gap is shown in Fig. 1. Other investigators who prepared monoxides by arc melting did not mention such segregation [e.g., 64Den, 72Hul], and it is possible that in the binary system, the miscibility gap is metastable. Therefore, the critical temperature has been placed close to the monotectic temperature. If the miscibility gap were metastable, the liquidus would assume approximately the shape proposed by [56Sch]. The two possible liquidus curves are compared in Fig. 3.

The presence of a liquid miscibility gap, as shown in Fig. 1, would explain the sudden broadening of the $L + (\alpha Ti)$ phase field, whether the gap appears in equilibrium or is metastable. The miscibility gap as drawn in the present diagram is entirely qualitative.

High- and Low-Temperature Monoxides.

Five structural modifications of the monoxide were reported:

- γTiO, the high-temperature form, has the NaCl structure and a wide homogeneity range. The γTiO

phase was recognized by [39Ehr] and since then has been established firmly as an equilibrium phase.

- βTiO, a cubic superstructure of ideal stoichiometry TiO, was observed by [68Hil1] in alloys heat treated below 1250 °C.
- αTiO has a narrow homogeneity range about the equiatomic composition and a monoclinic structure involving vacancy ordering [67Wat, 68Hil2]. It was established as an equilibrium phase by a variety of experimental techniques.
- βTi$_{1-x}$O has an orthorhombic structure related to that of αTiO, also based on vacancy ordering. [65Wat] and [67Wat] identified it as a metastable transition phase formed to the Ti-side of stoichiometry; [68Hil1], on the other hand, identified it as an intermediate-temperature phase (720 to 820 °C) formed in O-rich alloys (54.5 at.% O). According to [69Ver], the same basic characteristics were found in the diffraction patterns of alloys in the entire 41.2 to 54.5 at.% O range.
- αTi$_{1-x}$O has a bct structure and is the low-temperature form at O-rich compositions (55.6 at.% O) [68Hil1, 68Wat]. The high-O monoxide structures also were designated Ti$_4$O$_5$ [68Hil1], TiO$_{1.2}$ [60Ari], or γ'' [70Jos].
- Originally, only a single, NaCl-type monoxide with a wide homogeneity range was identified [39Ehr, 53Bum]. The first evidence for the low-temperature stoichiometric phase αTiO was found by [56Sch], [56Kuy], and [56Wan]; the two-phase field between αTiO and the high-temperature form was determined primarily by [65Por], [66Wah], [72Roy], and [72Suz]. However, these investigations did not take into account the other structural

modifications. There is no definitive basis for judging which modifications are equilibrium binary bulk phases or where the phase boundaries and invariant reactions should be placed. The assessed diagram indicates by dotted horizontal lines the temperature ranges in which the phases were observed to form [68Hil1]; these lines do not represent quantitative phase boundaries. The following discussion of the boundaries involving (αTi), βTi$_2$O$_3$, or αTiO refers simply to the "high-temperature monoxide".

- In the assessed diagram, αTiO is shown to form from the high-temperature phase by a peritectoid reaction at 940 °C. According to [53Bum], [56Sch], and [70Jos], the homogeneity range of the high-temperature form extends to about 48 at.% O down to 700 °C, and αTiO forms by a congruent transformation. The peritectoid construction of Fig. 1 is based on O activity measurements [72Suz, 78Tet] and on XRD and metallography [65Por, 66Wah, 72Roy]. It is consistent with additional data of [56Kuy], [57And1], [59Vol], and [60Ari] as shown in Fig. 5. The peritectoid construction is based on the observation by [66Wah] that a 48.7 at.% O alloy equilibrated at 1000 °C was two phase; the alloy would have been expected to be single phase for a congruent transformation anywhere below 1000 °C. The γTiO field extends to 34.9 at.% O at the peritectic temperature, according to [70Jos]. This conclusion is consistent with the data of [72Suz] in the range 1000 to 1600 °C.
- There is good agreement on the temperature of the (αTi) + βTiO $\rightleftharpoons$ αTiO peritectoid reaction. Metallography as well as X-ray and electron diffraction

---

Fig. 5---Experimental Data for Phase Boundaries Involving the Monoxide Phases

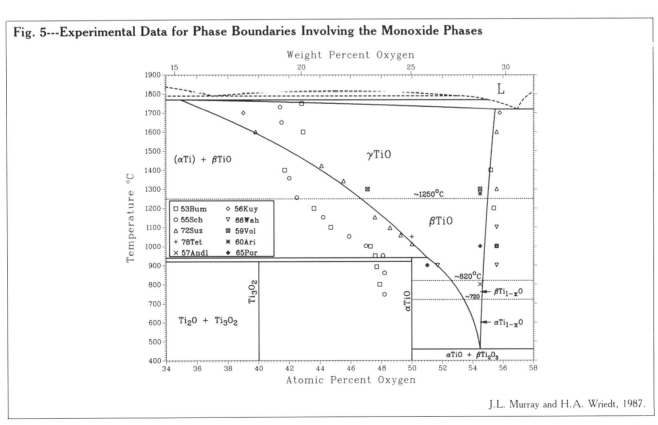

J.L. Murray and H.A. Wriedt, 1987.

placed the peritectoid reaction between 900 and 950 °C [56Wan, 57And1, 58Pea, 66Wah, 67Wat]. The assessed value of 940 °C is based on DTA [65Por, 72Roy]; the thermal analysis (heating) of [56Sch] can also be reinterpreted to give a peritectoid temperature of about 945 °C. A transformation temperature of 990 °C, based on heat content [46Nay] is too high. The narrow homogeneity range of αTiO shown in the present diagram is based on metallographic work [65Por, 66Wah]. [70Jos] proposed a somewhat larger homogeneity range (48 to 50 at.% O).

- Location of the eutectoid reaction αTi$_{1-x}$O $\rightleftarrows$ αTiO + βTi$_2$O$_3$ at 460 °C is based on thermal analysis and microscopy [65Por, 72Roy].

- Experimental data on the O-rich limit of the monoxide phases, obtained by XRD and metallography, located this boundary between 54.5 and 55.5 at.% O [53Bum, 56Kuy, 57And1, 59Vol, 60Ari, 65Por, 66Wah, 70Jos, 72Roy]. Experimental uncertainty is sufficient to explain the spread; consequently, the assessed phase boundary is drawn between 54.5 at.% O at low temperature and 55.5 at.% O at the eutectic temperature.

- [60Bri] proposed that some phase separation occurs within the γTiO phase field, because single-phase specimens could be prepared away from stoichiometry, but 50 at.% alloys were found to be disproportionate. An explanation does not suggest itself at present.

Higher Oxides

Liquidus and Solidus.

βTi$_2$O$_3$ and TiO$_2$ (rutile) melt congruently. The melting points are 1842 °C [66Wah] and 1870 ± 15 °C [60Bra], respectively. [60Bra] showed that the observed melting point of rutile varied between 1800 and 1870 °C with the partial pressure of O. The value 1870 °C corresponded to stoichiometric rutile [72Roy]. Other optical pyrometry on the melting point gave values ranging between 1825 and 1850 °C, (see Table 2 [75Cha]). According to [60Bra] and [72Roy], these measurements pertained to slightly reduced rutile.

The liquidus between Ti$_2$O$_3$ and TiO$_2$ was determined under argon at 10^5 Pa [60Bra]. Between Ti$_2$O$_3$ and TiO$_2$, at least one eutectic reaction must occur; [60Bra] and [56Nis] agreed that the reaction is L $\rightleftarrows$ βTi$_2$O$_3$ + TiO$_2$ at 1670 ± 15 °C.

If the above interpretation is correct, the intermediate Magneli phases decompose by solid-state reactions below 1670 °C. Incongruent melting points were reported for βTi$_3$O$_5$ and γTi$_4$O$_7$ (Table 2), but [65Por] claimed to have shown that the Magneli phases are stable until they melt. On the basis of their O-potential measurements, [74Gre] and [83Zad] presented a more complicated picture. According to [74Gre], the Magneli phases with n odd are less stable than those with n even. [83Zad] verified this for Ti$_7$O$_{13}$ and Ti$_9$O$_{17}$, which were not observed as single phase above 1500 °C. From O-potential data and XRD, [83Zad] concluded tentatively that γTi$_4$O$_7$ is unstable with respect to βTi$_3$O$_5$ and Ti$_5$O$_9$ above 1500 °C.

This assessment accepts the temperature value 1670 ± 15 °C for a eutectic reaction involving unknown solid phases. On the basis of liquidus data [60Bra], a peritectic reaction βTi$_2$O$_3$ + L $\rightleftarrows$ βTi$_3$O$_5$, is shown in the assessed diagram. The fact that 1670 °C is the best value reported for the melting temperature of γTi$_4$O$_7$ [71Ham] is consistent with the hypothesis that, on heating, γTi$_4$O$_7$ first decomposes in the solid state and then melts by a eutectic reaction.

Magneli Phases.

In early work, the region including the Magneli phases was identified as a single phase of wide homogeneity range [39Ehr]. The Magneli phases, or crystallographic shear structures, were established later as discrete, stoichiometric equilibrium phases, Ti$_n$O$_{2n-1}$, both by structural studies [57And1] and by oxygen-potential measurements [64Roy, 73Mer, 83Zad]. The contrary assertion by [63Bog] was regarded by [70And] as a misinterpretation of the tensimetric data. In addition to the series ($4 \leq n \leq 10$), other families of structures based on other shear planes were established [69Bur, 70And, 71Bur1, 71Bur2]. Some estimates of the largest value of n were 36 [72Por] and 40 [70And]. [65Por] and [72Roy] suggested that the largest value may be as high as 99. The lower estimates, based on structural data, are probably more realistic.

The assessed diagram, Fig. 1, displays the Magneli phases with stoichiometries Ti$_4$O$_7$ through Ti$_7$O$_{13}$. Additional members of this family and other families, produced by different crystallographic shear operations, were shown to be possible theoretically and demonstrated experimentally [69Bur, 70And, 71Bur1, 71Bur2]. There is some evidence that higher index crystallographic shear structures are stable only at lower temperatures [71Bur2].

Modifications of Ti$_2$O$_3$.

The homogeneity range of βTi$_2$O$_3$ (59 to 61 at.% O), reported by [39Ehr], was widely quoted; a narrower range (59.8 to 60.2 at.% O), based on lattice parameter data [57And1], is preferred.

Ti$_2$O$_3$ undergoes a transition from semiconducting αTi$_2$O$_3$ at low temperature to metallic βTi$_2$O$_3$ at about 180 °C. There has been some disagreement concerning the mechanism and temperature of the transition (see Table 3). [58Pea] and [68Rao] found rapid changes in the dimensions of the unit cell between 160 and 200 and between 117 and 177 °C, respectively, but no symmetry change. [63Abr] reported that antiferromagnetic αTi$_2$O$_3$ with monoclinic symmetry changed to rhombohedral βTi$_2$O$_3$; other investigators reported only a distension of the rhombohedral unit cell [58Pea, 68Rao]. [46Nay] found an effect in the heat content at 200 ± 20 °C and [73Bar] found a λ-type peak in the heat capacity at 180 °C. In approximate agreement, [67Key1] and [67Key2] found a change in the magnetic susceptibility at about 130 °C. On the other hand, [63Abr], by neutron diffraction, found that antiferromagnetic ordering occurred below 390 ± 30 °C and connected it with a change in electrical resistivity occurring over the range 180 to 630 °C.

In the present assessment, the thermochemical and lattice parameter data are preferred tentatively as indicators of the phase transition, which is placed at 180 °C.

Modifications of Ti$_3$O$_5$.

This phase has at least two equilibrium modifications: αTi$_3$O$_5$, stable below about 190 °C [59Asb, 71Rao]; and the high-temperature form, βTi$_3$O$_5$, (anosovite), probably stable to the melting point [51Rus, 59Asb]. The βTi$_3$O$_5$ $\rightleftarrows$ αTi$_3$O$_5$ transition occurs with a hysteresis of about 30 °C [71Rao]. Experimental data on the transition temperatures are listed in Table 3. The transition is from a metallic conductor to a semiconductor. βTi_3O_5 was stabilized to room temperature by Fe impurities [59Asb, 71Rao].

[72Roy] reported an additional phase transition at about 1200 °C. The basis for its inclusion was not specified; consequently it has been omitted from the assessed diagram. Another structural modification, $\alpha'Ti_3O_5$, formed at 600 to 925 °C [71Asb], but because of the irreversibility of the $\alpha'Ti_3O_5 > \alpha Ti_3O_5$ transformation, $\alpha'Ti_3O_5$ has been considered metastable. [69Iwa] identified three forms of Ti_3O_5 that they designated D, D', and M. Structurally, the D and D' probably could be identified as βTi_3O_5 and M as αTi_3O_5. [69Iwa] distinguished the three forms according to the products of oxidation under different conditions (rutile or a mixture of anatase and rutile). The D type was thought to be metastable.

Modifications of Ti_4O_7.

This phase undergoes two transitions: $\gamma Ti_4O_7 \rightleftarrows \beta Ti_4O_7$ at 150 K and $\beta Ti_4O_7 \rightleftarrows \alpha Ti_4O_7$ at 125 K, which bring it from a metallic conductor at room temperature to an insulator below 125 K [69Bar, 70Mar, 73Mar]. [70Mar] observed the upper transition only from a discontinuity in the lattice parameters; at the lower transition, new X-ray reflections were observed [69Bar, 70Mar]. [73Mar] reported that when purer samples were used, both transitions involved only the rearrangement of atoms within the unit cell. [84Lep] verified the structures of γTi_4O_7 and αTi_4O_7, but attributed to βTi_4O_7 a five-fold increase in the cell size.

Modifications of Ti_5O_9 and Higher Magneli Phases.

Ti_5O_9, Ti_6O_{11}, Ti_7O_{13}, Ti_8O_{15}, Ti_9O_{17}, and $Ti_{10}O_{19}$ were shown similarly to undergo electrical transitions. In Ti_5O_9, two transitions occurred very close together [69Bar]. The higher Magneli phases were studied more cursorily, but they appear to have transitions at approximately 150 K [67Key1, 67Key2, 70Mul]. Data are summarized in Table 3. In magnetic susceptibility studies, measurements extended to a sharp change, suggestive of another phase transition [72Dan].

TiO_2 (Rutile).

Five polymorphs of TiO_2 exist: anatase and brookite, which are low-temperature, low-pressure forms; TiO_2-II and TiO_2-III, which are formed from anatase or brookite under pressure; and rutile, the stable phase at all temperatures and ambient pressure. The polymorphic transformations anatase $\rightarrow$ rutile and brookite $>$ rutile do not occur reversibly [66Vah, 68Dac]. This fact and the heat of transformation data [67Nav, 79Mit] showed that anatase and brookite are not stable at any temperature. Therefore,

in the assessed phase diagram (Fig. 1), only rutile is shown. Temperature-pressure data for the polymorphic transformations are discussed in the section "Metastable Phases."

Experimental data on the homogeneity range of rutile are summarized in Table 4. The tabulated compositions vary between 66.6 at.% O at 1000 °C to about 66.3 at.% O near the melting point. In Fig. 1, TiO_2 is represented simply as a stoichiometric line compound.

Metastable Phases

Terminal Solid Solution, (Ti).

Above a critical cooling rate of 3000 °C/s, the $(\beta Ti) \rightleftarrows (\alpha Ti)$ transition is diffusionless, i.e., either martensitic or massive in mechanism. [74Cor] measured the start temperature, M_s, of the (assumed) martensitic transformation as:

Composition, at.% O	M_s, °C
0	802 ± 10
0.6	836
0.9	861
1.0	901
1.2	849
1.5	832

The maximum in M_s at about 1 at.% O was stated to be real, but its origin was not explained.

Anatase, Brookite, TiO_2-II, and TiO_2-III.

As already noted, heat of transformation data and irreversibility of the anatase $\rightarrow$ rutile and brookite $\rightarrow$ rutile transformations indicate that rutile is the stable phase at all temperatures. Similarly, the transformations of anatase and brookite to the high-pressure forms, TiO_2-II and TiO_2-III, occur irreversibly.

The nonequilibrium transformations have been examined experimentally:

Transformation	References
Anatase $\rightarrow$ rutile	[58Cza, 61Rao, 66Vah, 68Dac]
Brookite $\rightarrow$ rutile	[68Dac]
TiO_2-II $\rightarrow$ rutile	[68Dac, 66Ben]
TiO_2-II $\rightarrow TiO_2$-III	[67Mcq, 78Liu, 81Mam]
Anatase $\rightarrow TiO_2$-III	[68Dac, 79Ohs]
Rutile $\rightarrow TiO_2$-II	[71Nic, 78Liu, 80Mam, 66Ben]

Table 4   Homogeneity Range of Rutile (TiO_x to TiO_2)

Reference	Atomic ratio of TiO_x, x	Temperature, °C	Experimental method
[39Ehr]	~ 1.9	...	Thermogravimetry
[57Bra]	~ 1.969	~ 1800	...
[61Str]	1.983	1400	Lattice parameters, density
[63Blu]	1.992	1100	Isopiestic, emf
[75Pic]	~ 1.99	1050	Partial Gibbs energy
[77Ban]	1.992	1000	Electrical conductivity
[78Koz]	1.990	850 to 950	Solid galvanic cell
[83Zad]	1.988	1300	Oxygen potentials
	1.985	1400	
	1.982	1500	

Fig. 6---Observed Conditions for Persistence or Induced Transformation of Unstable Anatase (or Brookite) and TiO$_2$-II

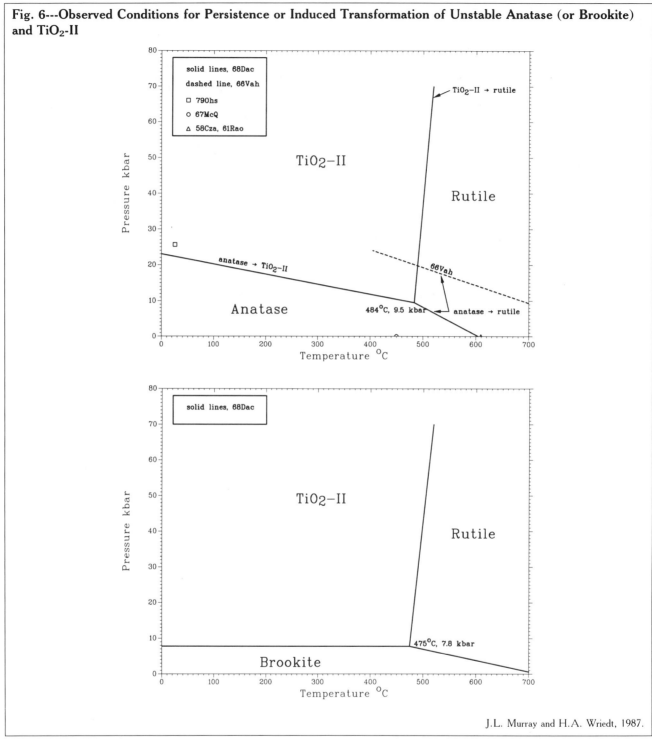

J.L. Murray and H.A. Wriedt, 1987.

The transformations anatase → rutile, anatase → TiO$_2$-II, and TiO$_2$-II → rutile can be represented in a nonequilibrium temperature-pressure diagram, as can the corresponding transformations with brookite replacing anatase. Figure 6 shows the diagrams determined by [68Dac] with quenching and X-ray techniques in the range 0 to 80 kbar, 200 to 700 °C. In the figure, the data of [58Cza], [61Rao], [66Vah] for the anatase → rutile transformation and of [79Ohs] for the anatase to TiO$_2$-II transformation are compared with the results of [68Dac].

The diagrams in Fig. 6 are neither stable nor metastable equilibrium diagrams from which heats of transformation could be calculated. They indicate conditions for which the rate of transformation in the appropriate direction is appreciable. Of the high-pressure phases, TiO$_2$-II is formed more easily and the only one that can be retained during quenching.

The kinetics of the anatase → rutile transformation were studied by [58Cza], who used X-ray spectroscopy as a probe of the volume fractions of the phases and found that the transformation occurred infinitely slowly at 610 °C.

Table 5 Crystal Structures of the Ti-O System

Phase	Homogeneity range, at.% O			Pearson symbol	Space group	Struktur-bericht designation	Prototype	Reference
(βTi)	0	to	8	$cI2$	$Im3m$	$A2$	W	[Pearson2]
(αTi)	0	to	31.9	$hP2$	$P6_3/mmc$	$A3$	Mg	[Pearson2]
Ti$_3$O	~ 20	to	~ 30	$hP{\sim}16$	$P\bar{3}1c$	...	...	[71Dub1,71Dub2]
Ti$_2$O	~ 25	to	33.4	$hP3$	$P\bar{3}m1$	...	Anti-CdI$_2$	[61Mag, 62Hol]
Ti$_3$O$_2$	~ 40			$hP{\sim}5$	$P6/mmm$	...	...	[59And]
γTiO	34.9	to	55.5	$cF8$	$Fm3m$	$B1$	NaCl	[Pearson2]
βTiO	...			(a)	...	...	...	[68Hil1]
αTiO	~ 50			$mC16$	$A2/m$	...	...	[67Wat]
	...			...	$B*/*$	...	...	[68Hil2]
βTi$_{1-x}$O	~ 55.5			$oI12$	$I222$	...	...	[65Wat, 66Yam, 68Hil1]
αTi$_{1-x}$O	~ 55.5			$tI18$	$I4/m$	...	...	[68Wat, 68Hil1]
βTi$_2$O$_3$	59.8	to	60.2	$hR30$	$R\bar{3}c$	$D5_1$	αAl$_2$O$_3$	[58Pea, 68Rao, 62New]
αTi$_2$O$_3$	59.8	to	60.2	$hR30$	$R\bar{3}c$	$D5_1$	αAl$_2$O$_3$	[57And1, 58Pea, 68Rao]
βTi$_3$O$_5$	62.5			(b)	...	...	Anosovite	[59Asb, 51Rus]
αTi$_3$O$_5$	62.5			$mC32$	$C2/m$	...	...	[57Asb, 59Asb, 61Mag]
α'Ti$_3$O$_5$(c)	...			$mC32$	Cc	...	V$_3$O$_5$	[71Asb]
γTi$_4$O$_7$	63.6			$aP44$	$P\bar{1}$	...	...	[63And, 73Mer, 82Lep1]
βTi$_4$O$_7$	63.6			$aP44$	$P\bar{1}$	...	...	[73Mer, 84Lep]
αTi$_4$O$_7$	63.6			$aP44$	$P\bar{1}$	...	...	[73Mer, 63And, 84Lep]
γTi$_5$O$_9$	64.3			$aP28$	$P\bar{1}$	...	...	[60And, 77Mar]
βTi$_6$O$_{11}$	64.7			$aC68$	$A\bar{1}$	...	...	[63And, 82Lep1]
Ti$_7$O$_{13}$	65.0			$aP40$	$P\bar{1}$	...	...	[63And, 82Lep1]
Ti$_8$O$_{15}$	65.2			$aC92$	$A\bar{1}$	...	...	[63And, 82Lep1]
Ti$_9$O$_{17}$	65.4			$aI52$	$P\bar{1}$	...	...	[63And, 82Lep1]
Anatase(c)	...			$tI12$	$I4_1/amd$	$C5$	Anatase	[57And, 55Cro]
Rutile	~ 66.7			$tP6$	$P4_2/mnm$	$C4$	Rutile	[56Bau, 57And, 55Cro, 61Str]
Brookite(c)	...			$oP24$	$Pbca$	$C21$	Brookite	[59Wey, 61Yog]
TiO$_2$-II(d)	...			$oP12$	$Pbcn$	...	αPbO$_2$	[66Ben, 67Sim, 67Mcq]
TiO$_2$-III(d)	...			${\sim}hP48$	(e)	...	...	[78Liu]

(a) Cubic. (b) Monoclinic. (c) Metastable phase. (d) High-pressure phase. (e) Hexagonal.

[61Rao] extrapolated thermal analysis data to zero heating rate to obtain a transformation temperature of 610 ± 10 °C. [66Vah] examined the transition by electrical resistivity measurements in samples quenched from 3.8 to 24 kbar, and 20 to 1000 °C. Transformation of rutile to TiO$_2$-II occurs at pressures above 7 GPa [71Nic, 78Liu, 80Mam] and temperatures above about 700 °C [66Ben]. The transformation of TiO$_2$-II to TiO$_2$-III at about 20 to 30 GPa was observed with XRD and Raman spectroscopy [67Mcq, 78Liu, 81Mam]. The crystallographic mechanisms of these transformations were examined by [64Sha] and [70Sim2].

Crystal Structures and Lattice Parameters

Tables 5 and 6 contain crystal structure and room temperature lattice parameter data on the observed phases of the Ti-O system; those phases that are probably not equilibrium binary phases at low pressure are distinguished by footnotes. For the Magneli phases, comparison of lattice parameters from various authors has been made by application of the cell transformations, described below.

Measurements of the lattice parameter of (αTi) as a function of composition are shown in Fig. 7. The values for very dilute alloys are listed in Table 7. Because (βTi) could not be retained by quenching, its lattice parameters were not measured. The structures of Ti$_2$O and Ti$_3$O and their

relationships are discussed above. Note that the early structure determinations of Ti$_6$O [66Yam] and Ti$_3$O [68Jos] are not listed, because they were superseded by later work, including neutron diffraction [69Yam, 70Kor, 70Yam, 71Dub1, 71Dub2].

γTiO has the NaCl structure [39Ehr]; lattice parameter values are shown in Fig. 8. Other forms of the monoxide—βTiO, αTiO, βTi$_{1-x}$O, and αT$_{1-x}$O—have ordered structures based on γTiO. The intermediate-temperature phase βTiO identified by [68Hil1] is cubic with a lattice parameter three times that of the γTiO parameter. The structure of the equilibrium low-temperature form αTiO was determined by [67Wat] and [68Hil1]. In every third (110)γTiO plane, half the O and half the Ti atoms were missing, in ordered fashion. The ordered orthorhombic structure βTi$_{1-x}$O was identified in thin foils for Ti-rich compositions [65Wat, 69Ver] and for O-rich compositions [68Hil1]. The structure differs from that of αTiO in that only the O vacancies were ordered.

The high-temperature βTi$_3$O$_5$ phase was reported originally to be orthorhombic [51Rus]; the accepted monoclinic structure is a slight distortion of the orthorhombic lattice. The effect of impurities is to lessen the distortion [59Asb]. The names "anosovite" and "pseudobrookite" refer to the βTi$_3$O$_5$ structure.

Several choices were made for the cell upon which to index the structures of the Magneli phases, Ti$_n$O$_{2n-1}$ for $4 \le n \le 9$ [63And, 82Lep1]. One choice was the rutile cell and another was the following: for odd n, the cell is

Table 6 Room-Temperature Lattice Parameters of the Ti-O System

Phase	Composition, at.% O	Lattice parameters						Reference
		a, nm	b, nm	c, nm	α	β	γ	
Ti_3O	25	0.51411	...	0.95334	...	...	...	[71Dub1, 71Dub2]
		0.506	...	0.956	...	...	...	[70Kor]
		0.506	...	0.948	...	...	...	[70Kor]
Ti_2O	33.3	0.29593	...	0.48454	...	...	...	[71Dub1, 71Dub2]
		0.29593	...	0.48454	...	...	...	[62Hol]
		0.296	...	0.483	...	...	...	[70Kor]
Ti_3O_2	~ 40	0.49915	...	0.28794	...	...	...	[59And]
βTiO	50	1.254	...	...	...	...	...	[68Hil1]
αTiO	50	0.5855	0.9340	0.4142	...	107.53°	...	[67Wat]
		0.9355	0.5868	0.4135	...	107.53°	...	[68Hil2]
$\beta Ti_{1-x}O$	...	0.2981	0.9086	0.3986	...	...	...	[66Wat, 68Hil1]
$\alpha Ti_{1-x}O$	...	0.6594	...	0.4171	...	...	...	[68Wat]
		0.6632	...	0.4156	...	...	...	[68Hil1]
αTi_2O_3(a) ...	60	0.5431	...	...	56.58°	...	...	[62Str]
	59.8	0.5160	...	...	56.80°	...	...	[57And2]
	60.0	0.5425	...	...	56.73°	...	...	
	60.2	0.5432	...	...	56.56°	...	...	
	60	0.542	...	...	56.9°	...	...	[61Mag]
	60	0.5428	...	...	56.65°	...	...	[58Pea]
	60	0.543258	...	...	56.75°	...	...	[74Rob]
βTi_3O_5	62.5	0.982	0.378	0.997	...	91.0°	...	[59Asb]
		0.9484	0.3755	0.9735	...	90.0°	...	[51Rus]
		0.9828	0.3776	0.9898	...	91.32°	...	[69Iwa]
		0.990	0.378	1.002	...	90.75°	...	[71Rao]
αTi_3O_5	62.5	0.9757	0.3802	0.9452	...	93.11°	...	[57Asb, 57And2]
		0.97524	0.38020	0.94419	...	91.547°	...	[61Mag]
		0.976	0.380	0.943	...	91.58°	...	[69Iwa]
		0.9752	0.38020	0.9442	...	91.55°	...	[59Asb]
		0.980	0.379	0.945	...	91.75°	...	[71Rao]
$\alpha' Ti_3O_5$	62.5	1.0120	0.5074	0.9970	...	138.15°	...	[71Asb]
γTi_4O_7	63.6	0.5604	0.7137	1.2478	95.072°	95.16°	108.77°	[63And]
		0.5593	0.7125	1.2456	95.02°	95.21°	108.73°	[73Mer]
		0.5600	0.7133	1.2466	95.05°	95.17°	108.71°	[71Mar]
		0.5593	0.7125	2.043	67.63°	57.17°	108.73°	[82Lep1](a)
βTi_4O_7	63.6	0.5590	0.7128	1.2483	95.03°	95.34°	108.89°	[73Mer]
		0.6918	1.1142	1.5127	90.64°	92.79°	91.45°	[84Lep](b)
αTi_4O_7	63.6	0.5591	0.7131	1.2487	95.00°	98.33°	108.88°	[73Mer]
		0.5626	0.7202	2.0260	67.90°	57.69°	109.68°	[84Lep](b)
βTi_5O_9	64.3	0.5569	0.7120	0.8865	97.55°	112.34°	108.50°	[60And]
		0.5577	0.7117	2.632	67.24°	57.04°	108.51°	[77Mar](b)
γTi_6O_{11}	64.7	0.5566	0.7144	2.407	98.5°	120.8°	108.5°	[61And]
		0.5552	0.7126	3.2234	66.94°	57.08°	108.50°	[82Lep1](b)
βTi_7O_{13}	65.0	0.554	0.713	1.536	98.9°	125.5°	108.5°	[61And]
		0.5537	0.7132	3.8152	66.70°	57.12°	108.50°	[82Lep1](b)
βTi_8O_{15}	65.2	0.557	0.710	3.746	97.2°	128.8°	109.6°	[61And]
		0.55261	0.7133	4.4060	66.54°	57.18°	108.51°	[82Lep1](b)
βTi_9O_{17}	65.4	0.55272	0.71413	2.22788	99.26°	130.34°	108.50°	[61And]
		0.55241	0.71421	5.0031	66.41°	57.20°	108.53°	[82Lep1](b)
Rutile	66.667	0.4594	...	0.2959	...	...	...	[56Bau]
		0.4593	...	0.2959	...	...	...	[57And2]
		0.45929	...	0.29591	...	...	...	[55Cro]
		0.459373	...	0.295812	...	...	...	[61Str]
Anatase	66.7	0.3786	...	0.9517	...	...	...	[57And2]
		0.3785	...	0.9514	...	...	...	[55Cro]
Brookite ...	66.7	0.925	0.546	0.516	...	...	...	[61Yog]
		0.9184	0.5447	0.5145	...	...	...	[59Wey]
TiO_2-II	66.7	0.4515	0.5497	0.4939	...	...	...	[67Sim]
		0.4529	0.5464	0.4905	...	...	...	[67Mcq]
		0.4531	0.5498	0.4900	...	...	...	[66Ben]
TiO_2-III	66.7	0.922	...	0.5685	...	...	...	[78Liu]

(a) Temperature-dependent lattice parameters were determined by [68Rao] and [58Pea]. (b) See text for a description of the cell.

Fig. 7---Experimental Measurements of the Variation of the Lattice Parameters of cph (αTi) with O Concentration

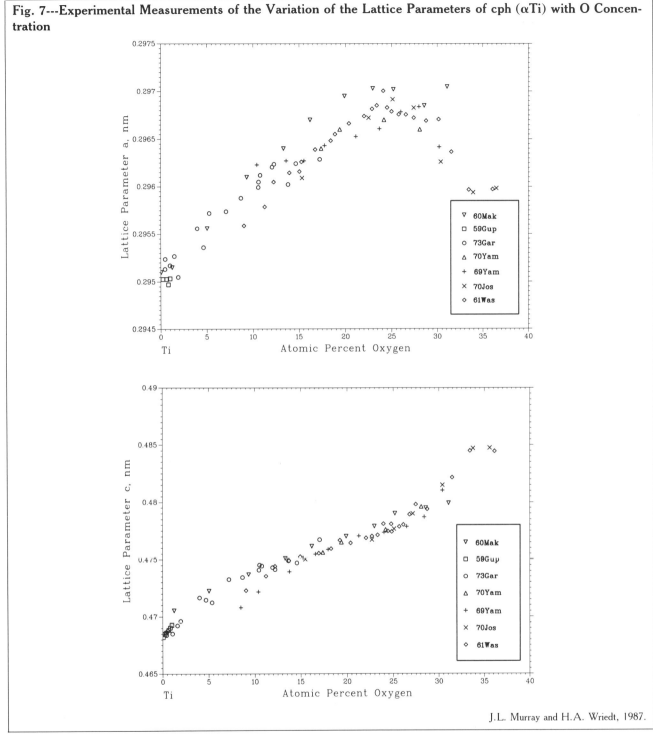

J.L. Murray and H.A. Wriedt, 1987.

primitive, for even n, the cell is A face centered [61And, 73Mar]. [82Lep1] chose a nonconventional body-centered cell, which allows a more convenient comparison of the various structures. The relationship among the rutile (R), [61And] (A), and [82Lep1] (L) cells was given by [82Lep1]:

$$a_L = a_A = a_R - c_R$$

$$b_L = b_A = -a_R - b_R - c_R$$

$$c_L = (2n - 5)a_A + (n - 2)b_A + pc_A = -(2n - 1)c_R$$

where $p = 1$ for n even and $p = 2$ for n odd.

In Tables 5 and 6, symmetry and lattice parameter data are as listed originally by the authors.

[63And] and [67And] described the structures of the Magneli phases as derivatives of the rutile structure by the following crystallographic operations: every $2n^{\text{th}}$ O plane parallel to $(121)_{\text{rutile}}$ was removed to adjust the composition, and adjacent rutile slabs were displaced by $1/2(011)_{\text{rutile}}$ to collapse the structure.

More complex structures appeared in higher oxides. [69Bur] and [71Bur1] described another family of oxides with the same generic formula Ti_nO_{2n-1} that were

Table 7   Lattice Parameters of Dilute (αTi) Alloys at 21 °C [77Dec]

Composition, at. ppm O		Lattice parameters, nm	
Activation	Gravimetric	a	c
90	...	0.295112	0.468272
...	126	0.295135	0.468288
...	192	0.295126	0.468243
300	430	0.295129	0.468299
1026	1085	0.295121	0.468315
1336	1561	0.2956162	0.468369
1900	2130	0.295142	0.468435
2332	2586	0.295138	0.468434
...	2950	0.295182	0.468462
3104	3200	0.295149	0.468458
...	3380	0.295133	0.468465
...	4160	0.295226	0.468515

Fig. 8---Experimental Measurements of the Variation of the Lattice Parameter of γTiO with O Concentration

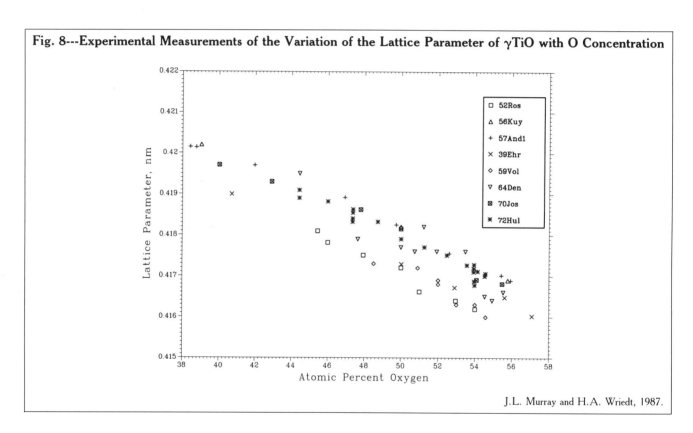

J.L. Murray and H.A. Wriedt, 1987.

generated by the removal of the (132)$_{rutile}$ O planes. Phases were identified for $16 \leq n$ (even) ≤ 36 [71Bur1, 71Bur2]. [71Bur2] resolved the crystallographic operations into combinations of a (121) operation (which altered the composition) and the introduction of a (011) antiphase boundary. The (132) structure was described as an ordered coherent intergrowth structure of crystallographic shear (CS) and antiphase boundary (APB) structures:

$$Ti_{16}O_{31} (132) = Ti_9O_{17}(CS) + Ti_7O_{14} (APB)$$

In this way, new families of generic shear structures were derived. The existence of (253)-type crystallographic shear structure was verified by electron diffraction [71Bur2].

The temperature dependence of the lattice parameters of γTi$_4$O$_7$ and βTi$_4$O$_7$ was examined by [70Mar].

Thermodynamics

Because the thermodynamic data on the Ti-O system were reviewed extensively [75Cha], the present coverage is restricted to a summary of selected previous assessments and a literature survey of experimental work since 1975.

[75Cha] updated [71Stu] for the Ti-O system, covering the solid phases βTiO, αTiO, βTi$_2$O$_3$, Ti$_4$O$_7$, βTi$_3$O$_5$, αTi$_3$O$_5$, TiO$_2$ (anatase), TiO$_2$ (rutile), and the liquids at corresponding compositions. Heats and entropies of formation of the compounds from the pure elements in their standard states are given in Table 8.

[64Kau] analyzed thermochemical data for the interstitial solutions (αTi) and (βTi) and the monoxide γTiO. Data for nonstoichiometric compositions of the compound

Table 8 Thermodynamic Properties of Intermetallic Phases [75Cha]

Phase	Entropy of formation(a) $(\Delta_f S^0)$, kJ/mol·K	Enthalpy of formation(a) $(\Delta_f H^0)$, kJ/mol	Enthalpy of fusion $(\Delta_{fus} H)$, kJ/mol	Enthalpy of transition $(\Delta_{trs} H)$, kJ/mol
αTiO	34.77 ± 2	− 542.7 ± 12	...	...
βTiO	38.1	− 538.5 ± 12	41.8	...
βTi$_2$O$_3$	77.24 ± 0.2	− 1520.9 ± 8	104.6	...
αTi$_3$O$_5$	129.37	− 2459.15 ± 4	...	13.26 ± 6.3
βTi$_3$O$_5$	157.620	− 2446.188	171	...
βTi$_4$O$_7$	198.7 ± 12	− 3404.5 ± 6.3	226	...
Anatase	49.907 ± 0.3	− 938.72 ± 2	57.99	...
Rutile	50.29 ± 0.17	− 939.89 ± 1.2	66.9 ± 17	...

(a) At 25 °C.

were analyzed in terms of the Wagner-Schottky model; a general model was presented for interstitial solid solutions, with a variable number of available interstitial sites. Comparison was made between observed and calculated values of the enthalpy of formation and vacancy concentration of γTiO as a function of composition. It was stated that the calculated (αTi)/(βTi) phase boundaries were within 2 at.% of the assessed boundaries of [Hansen].

The following is recent experimental work:

- High-temperature O activities on the O-rich side of the diagram were measured by [78Koz], [82Gra], and [83Zad]. Heats of formation derived by [83Zad] are listed in Table 9.
- High-temperature partial heats of formation were measured for compositions near TiO$_2$ and TiO [75Pic, 76Bou, 78Tet, 81Tet]. [78Tet] corrected previous work [76Bou]. The results differed from previous work, and appeared also to differ from the present assessed phase diagram.
- Relative partial Gibbs energies for dilute (αTi) solutions were given by [73Rez]. Expressions for relative partial Gibbs energies of O were given for six alloys containing 0.15 to 0.9 at.% O over the range 1000 to 1150 °C.

[79Mit] determined the enthalpy changes for the transformations anatase → rutile and brookite → rutile as 698 °C by solution calorimetry, supplemented by DSC and DTA:

For $\Delta_{trs} H$(anatase → rutile): 698 °C = −3.26 ± 0.84 kJ/mol

For $\Delta_{trs} H$(brookite → rutile): 698 °C = −0.71 ± 0.4 kJ/mol

For $\Delta_{trs} H$(anatase → rutile): 698 °C = −2.93 ± 1.3 kJ/mol (DSC)

Temperature-pressure transformation data could not be used to determine $\Delta_{trs} H$, because rutile is the stable phase at all temperatures. [79Mit] attributed the discrepancy between their results and previous solution calorimetry data [67Nav] to failure to dissolve the samples completely in the solvent in the earlier work.

Table 9 Heats of Formation of Oxides at 1400 °C [83Zad]

Phase	Heat of formation $(\Delta_f H)$, kJ/mol	Phase	Heat of formation $(\Delta_f H)$, kJ/mol
Ti$_3$O$_5$	− 2414	Ti$_7$O$_{13}$	− 6130
Ti$_4$O$_7$	− 3351	Ti$_8$O$_{15}$	− 7054
Ti$_5$O$_9$	− 4276	Ti$_9$O$_{17}$	− 7979
Ti$_6$O$_{11}$	− 5212	TiO$_{10}$O$_{19}$	− 8912

Note: Mol is mole of atoms.

Cited References

37Daw: W. Dawhill and K. Schroter, "Preparation and Properties of Titanium Monoxide," *Z. Anorg. Chem., 233*, 178-183 (1937). (Equi Diagram; Experimental)

39Ehr: P. Ehrlich, "Phase Relations and Magnetic Properties in the Titanium-Oxygen System," *Z. Electrochem., 45*, 362-370 (1939) in German. (Equi Diagram; Experimental)

46Nay: B.F. Naylor, "High-Temperature Heat Contents of TiO, Ti$_2$O$_3$, Ti$_3$O$_5$, and TiO$_2$," *J. Amer. Chem. Soc., 68*, 1077-1068 (1946). (Equi Diagram; Experimental)

50Jaf: R.I. Jaffee, H.R. Ogden, and D.J. Maykuth, "Alloys of Titanium with Carbon, Oxygen, and Nitrogen," *Trans. Metall. AIME, 188*, 1261-1266 (1950). (Equi Diagram; Experimental)

*51Jen: A.E. Jenkins and H.W. Worner, "The Structure and Some Properties of Titanium-Oxygen Alloys Containing 0-5 at. Percent Oxygen," *J. Inst. Met., 80*, 157-166 (1951). (Equi Diagram, Experimental)

51Rus: A.A. Rusakov and G.S. Ladanov, "Crystal Structure and Stoichiometry of Ti$_3$O$_5$ (Anosovite)," *Dokl. Akad. Nauk SSSR, 77*, 411-414 (1951) in Russian. (Equi Diagram, Crys Structure; Experimental)

52Ros: W. Rostoker, "Observations on the Lattice Parameters of the Alpha and TiO Phases in the Titanium-Oxygen System," *Trans. Metall. AIME, 194*, 981-982 (1952). (Crys Structure; Experimental)

*53Bum: E.S. Bumps, H.D. Kessler, and M. Hansen, "The Titanium-Oxygen System," *Trans. Metall. ASM, 45*, 1008-1028 (1953). (Equi Diagram; Experimental)

54Dev: R.C. DeVries, R.Roy, and E.F. Osborn, "The System TiO$_2$-SiO$_2$," *Trans. J. Brit. Ceram. Soc., 53*, 525-540 (1954). (Equi Diagram; Experimental)

54Kub: O. Kubaschewski and W.A. Dench, "The Free-Energy Diagram of the System Titanium-Oxygen," *J. Inst. Met., 82*, 87-91 (1954). (Equi Diagram, Thermo; Experimental)

54Was: R.J. Wasilewski and G.L. Kehl, "Diffusion of Nitrogen and Oxygen in Titanium," *J. Inst. Met., 83*, 94-104 (1954). (Equi Diagram; Experimental)

55Cro: D.T. Cromer and K. Herrington, "The Structures of Anatase and Rutile," *J. Am. Chem. Soc., 77*, 4708-4709 (1955). (Cry Structure; Experimental)

56Bau: W.H. Baur, "Determination of Refined Crystal Structure for Several Compounds of the Rutile Type TiO$_2$, SnO$_2$, GeO$_2$ and MgF$_2$," *Acta Crystallogr., 9*, 515-520 (1956). (Crys Structure; Experimental)

56Kuy: U. Kuylenstierna and A. Magneli, "A New Modification of Titanium Monoxide," *Acta Chem. Scand., 10*(7), 1195-1196 (1956). (Equi Diagram, Crys Structure; Experimental)

56Nis: H. Nishimura and H. Kimura, "On the Equilibrium Diagram of Titanium-Oxygen-Carbon System (II). The Titanium-Oxygen System," *J. Jpn. Inst. Met. 20*, 524-527 (1956) in Japanese. (Equi Diagram; Experimental)

Phase Diagrams of Binary Titanium Alloys

*56Sch: T.H. Schofield, "The Constitution of the Titanium-Oxygen Alloys in the Range 0.35 Weight Per Cent Oxygen," *J. Inst. Met.*, *84*, 47-53 (1956). (Equi Diagram; Experimental)

56Wan: C.C. Wang and N.J. Grant, "Transformation of the TiO Phase," *J. Met.*, *8*, 184-185 (1956). (Equi Diagram; Experimental)

*57And1: S. Andersson, B. Collen, U. Kuylenstierna, and A. Magneli, "Phase Analysis Studies on the Titanium-Oxygen System," *Acta Chem. Scand.*, *11*, 1641-1652 (1957). (Equi Diagram, Crys Structure; Experimental)

57And2: S. Andersson, B. Collen, G. Kruuse, U. Kuylenstierna, A. Magneli, H. Pestmalis, and S. Asbrink, "Identification of Titanium Oxides by X-Ray Powder Patterns," *Acta Chem. Scand.*, *11*, 1653-1657 (1957). (Crys Structure; Experimental)

57Asb: S. Asbrink and A. Magneli, "Note on the Crystal Structure of Trititanium Pentoxide," *Acta Chem. Scand.*, *11*(9), 1606-1607 (1957). (Crys Structure; Experimental)

57Kon: I. Koncz and M. Koncz-Deri, "Formation of the Delta-Phase by Oxidation of Alpha-Titanium," *Period. Polytech.*, *1*, 67-87 (1957). (Equi Diagram; Experimental)

57Mah: A.D. Mah, K.K. Kelley, N.L. Gellert, E.G. King, and C.J. O'Brien, "Thermodynamic Properties of Titanium-Oxygen Solutions and Compounds," *Bur. Mines Rep. Invest.* 5316 (1957). (Equi Diagram, Thermo; Experimental)

58Cza: A.W. Czanderna, C.N.R. Rao, and J.M. Honig, "The Anatase-Rutile Transition Part 1. – Kinetics of the Transformation of Pure Anatase," *Trans. Faraday Soc.*, *54*, Part 7, 1067-1073 (1958). (Meta Phases; Experimental)

58Pea: A.D. Pearson, "Studies on the Lower Oxides of Titanium," *J. Phys. Chem. Solids*, *5*, 316-327 (1958). (Equi Diagram, Crys Structure; Experimental)

59And: S. Anderson, "The Crystal Structure of the So-Called Alpha-Titanium Oxide and Its Structural Relation to the Omega-Phases of Some Binary Alloy Systems of Titanium," *Acta Chem. Scand.*, *13*, 415-419 (1959). (Equi Diagram, Crys Structure; Experimental)

59Asb: S. Asbrink and A. Magneli, "Crystal Structure Studies on Trititanium Pentoxide, Ti_3O_5," *Acta Crystallogr.*, *12*, 575-581 (1959). (Equi Diagram, Crys Structure; Experimental)

59Gup: D. Gupta and S. Weinig, "The Dislocation-Oxygen Interaction in Alpha Titanium and Its Effect on the Ductile-to-Brittle Transition," *Trans. Metall. AIME*, *215*, 209-216 (1959). (Crys Structure; Experimental)

59Now: H. Nowotny and E. Dimakopoulou, "The Ti_2O Phase," *Monatsh. Chem.*, *90*, 620-622 (1959) in German. (Crys Structure; Experimental)

59Vol: E. Volf, S.S. Tolkachev, and I.I. Kozhina, "X-Ray Study of the Lower Oxides of Titanium and Vanadium," *Vestn. Leningrad Univ.*, *14*(10), Ser. Fiz. Khim., (2), 87-92 (1959) in Russian. (Equi Diagram, Crys Structure; Experimental)

59Wey: B. Weyl, "Precision Determination of the Crystal Structure of Brookite, TiO_2," *Z. Krist.*, *111*, 401-420 (1959) in German. (Crys Structure; Experimental)

59Yao: Y.L. Yao, "Magnetic Susceptibilities of Titanium-Rich Titanium-Oxygen Alloys," *Trans. Metall. AIME*, *215*, 851-854 (1959). (Equi Diagram; Experimental)

60And: S. Andersson, "The Crystal Structure of Ti_5O_9," *Acta Chem. Scand.*, *14*, 1161-1172 (1960). (Crys Structure; Experimental)

60Ari: S.M. Ariya and N.I. Bogdanova, "Electrical Conductivity of Certain Titanium and Vanadium Oxides," *Sov. Phys. Solid State*, (1), 936-939 (1960). (Equi Diagram; Experimental)

*60Bra: G. Brauer and W. Littke, "Melting Point and Thermal Dissociation of Titanium Dioxide," *J. Inorg. Nucl. Chem.*, *16*, 67-76 (1960) in German. (Equi Diagram; Experimental)

60Bri: N.F.H. Bright, "X-Ray Studies in the Ti-O System," *Adv. X-Ray Anal.*, *4*, 175-193 (1960). (Equi Diagram; Experimental)

60Mak: E.S. Makarov and L.M. Kuzneov, *Zh. Strukt. Khim.*, *1*(2), 170-177 (1960) in Russian. (Crys Structure; Experimental)

61Mag: A Magneli, S. Andersson, S. Asbrink, S. Westman, and B. Holmberg, "Crystal Chemistry of Titanium, Vanadium, and Zirconium Oxides at Elevated Temperatures," *U.S. Dept. Comm., Office Tech. Serv.*, PB Rep. 145-923, 82 (1961). (Crys Structure; Experimental)

61Rao: C.N.R. Rao, "Kinetics and Thermodynamics of the Crystal Structure Transformation of Spectroscopically Pure Anatase to Rutile," *Can. J. Chem.*, *39*, 498-500 (1961). (Meta Phases; Experimental)

61Str: M.E. Straumanis, T. Ejima, and W.J. James, "The TiO_2 Phase Explored by the Lattice Constant and Density Method," *Acta Crysallogr.*, *14*, 493-497 (1961). (Equi Diagram, Crys Structure; Experimental)

61Was: R.J. Wasilewski, "Thermal Expansion of Titanium and Some Ti-O Alloys," *Trans. Metall. AIME*, *221*, 1231-1235 (1961). (Crys Structure; Experimental)

61Yog: S.R. Yoganarasimhan and C.N.R. Rao, "Titanium Dioxide (Brookite), TiO_2," *Anal. Chem.*, *33*(1), 155 (1961). (Crys Structure; Experimental)

62Hol: B. Holmberg, "Disorder and Order in Solid Solutions of Oxygen in Alpha-Titanium," *Acta Chem. Scand.*, *16*, 1245-1250 (1962). (Experimental)

62New: R.E. Newnham and Y.M. De Haan, "Refinement of the Alpha Al_2O_3, V_2O_3 and Cr_2O_3 Structures," *Z. Krist.*, *117*, 235-237 (1962). (Crys Structure; Experimental)

62Str: M.E. Straumanis and T. Ejima, "Imperfections within the Phase Ti_2O_3 and its Structure Found by the Lattice Parameter and Density Method," *Acta Crystallogr.*, *15*, 404-409 (1962). (Crys Structure; Experimental)

62Was: R.J. Wasilewski, "Electrical Resistivity of Titanium-Oxygen Alloys," *Trans. Metall. AIME*, *224*, 8-12 (1962). (Experimental)

63Abr: S.C. Abrahams, "Magnetic and Crystal Structure of Titanium Sesquioxide," *Phys. Rev. Lett.*, *130*(6), 2230-2237 (1963). (Equi Diagram, Crys Structure; Experimental)

63And: S. Andersson and L. Jahnberg, "Crystal Structure on the Homologous Series Ti_nO_{2n-1}, V_nO_{2n-1} and Ti_nO_{2n-1}, Cr_2O_{2n-1}," *Ark. Kem.*, *21*, 413-426 (1963). (Crys Structure; Experimental)

63Blu: R.N. Blumenthal and D.H. Whitmore, "Thermodynamic Study of Phase Equilibria in the Titanium-Oxygen System within the $TiO_{1.95}$-TiO_2 Region," *J. Electrochem. Soc.*, *110*(1), 92-93 (1963). (Equi Diagram, Thermo; Experimental)

63Bog: N.I. Bogdanova, G.P. Pirogovskaya, and S.M. Ariya, "Higher Oxides of Titanium," *Zh. Neorg. Khim.*, *8*, 785-787 (1963) TR: *Russ. J. Inorg. Chem.*, *8*, 401-402 (1963). (Equi Diagram; Experimental)

63Kor: I.I. Kornilov and V.V. Glazova, "Phase Diagrams of Ti_6O and Ti_3O in the Titanium-Oxygen System," *Dokl. Akad. Nauk SSSR*, *150*(2), 313-316 (1963) in Russian. (Equi Diagram; Experimental)

63Vas: Y.V. Vasilev, D.D. Khrycheva, and S.M. Ariya, "Magnetic Susceptibility of the Lower Oxides of Titanium," *Russ. J. Inorg. Chem.*, *8*, 402-404 (1963) in Russian. (Equi Diagram; Experimental)

64Den: S.P. Denker, "Relation of Bonding and Electronic Band Structure to the Creation of Lattice Vacancies in TiO," *J. Phys. Chem. Solids*, *25*, 1397-1405 (1964). (Crys Structure; Experimental)

64Kau: L. Kaufman and E.V. Clougherty, "Thermodynamic Factors Controlling the Stability of Solid Phases at High Temperatures and Pressures," *AIME Metall. Soc. Conf., Metallurgy at High Pressures and High Temperatures*, Vol. 22, Gordon and Breach, Science Publishers, Inc., NY 322-380 (1964).

64Roy: R. Roy, "Controlled O_2 Including High Oxygen Pressure Studies in Several Transition Metal-Oxygen Systems," *CNRS Colloq. Int.*, (149), 27-33 (1964). (Equi Diagram; Experimental)

64Sha: R.D. Shannon and J.A. Pask, "Topotaxy in the Anatase-Rutile Transformation," *Am. Min.*, *49*, 1707-1717 (1964). (Meta Phases; Experimental)

65Kor: I.I. Kornilov and V.V. Glazova, "Thermal Stability of the Compound Ti_3O in the Oxygen-Titanium System," *Zh. Neorg. Chem.*, *10*(7), 1660-1662 (1965) in Russian; TR: *Russ. J. Inorg. Chem.*, *10*(7), 905-907. (Equi Diagram; Experimental)

65Por: V.R. Porter, "Studies in the Titanium-Oxygen System and the Defect Nature of Rutile," *Diss. Abst.*, *27*(11), 6809 (1965). (Equi Diagram; Experimental)

O-Ti System

65Wat: D. Watanabe, "Electron Diffraction Study on the Titanium-Oxygen Alloy System," Int. Conf. Electron Diffraction and Crystal Defects, Melbourne, C3 (1965). (Equi Diagram, Crys Structure; Experimental)

66Ben: N.A. Bendeliani, S.V. Popova, and L.F. Vereshchagin, "New Modification of Titanium Dioxide Obtained at High Pressures," *Geochem. Int.*, 3(3), 387-390 (1966). (Meta Phases, Crys Structure; Experimental)

66Dub: A. Dubertret and P. Lehr, "Hardness Study of Titanium-Oxygen Alloys," *C.R. Acad. Hebd. Sèances Sci.*, 263, 591-594 (1966) in French. (Experimental)

66Vah: F.W. Vahldiek, "Phase Transition of Titanium Dioxide Under Various Pressures," *J. Less-Common Met.*, 11, 99-110 (1966). (Equi Diagram, Meta Phases; Experimental)

66Wah: P.G. Wahlbeck and P.W. Gilles, "Reinvestigation of the Phase Diagram for the System Titanium-Oxygen," *J. Am. Ceram. Soc.*, 49(1), 180-183 (1966). (Equi Diagram; Experimental)

66Yam: S. Yamaguchi, M. Koiwa, and M. Hirabayashi, "Interstitial Superlattice of Ti_6O and Its Transformation," *J. Phys. Soc. Jpn.*, 21, 2096 (1966). (Crys Structure; Experimental)

67And: J.S. Anderson and B.G. Hyde, "On the Possible Role of Dislocations in Generating Ordered and Disordered Shear Structures," *J. Phys. Chem. Solids*, 28, 1393-1408 (1967). (Crys Structure; Experimental)

67Gil: P.W. Gilles, K.D. Carlson, J.F. Franzen, and P.G. Wahlbeck, "High-Temperature Vaporization and Thermodynamics of the Titanium Oxides. I. Vaporization Characteristics of the Crystalline Phases," *J. Chem. Phys.*, 46(7), 2461-2465 (1967). (Equi Diagram; Experimental)

67Key1: L.K. Keys and L.N. Mulay, "Magnetism of the Titanium-Oxygen System," *Jpn. J. Appl. Phys.*, 6, 122-123 (1967). (Equi Diagram; Experimental)

67Key2: L.K. Keys and L.N. Mulay, "Magnetic-Susceptibility Studies on the Magneli Phases of the Titanium-Oxygen System," *J. Appl. Phys.*, 38(1), 1466-1404 (1967). (Equi Diagram; Experimental)

67Mcq: R.G. McQueen, J.C. Jamieson, and S.P. Marsh, "Shock-Wave Compression and X-Ray Studies of Titanium Dioxide," *Science*, 155, 1401-1407 (1967). (Meta Phases; Experimental)

67Nav: A. Navrotsky and O.J. Kleppa, "Enthalpy of the Anatase-Rutile Transformation," *J. Am. Ceram. Soc.*, 50(11), 626 (1967). (Equi Diagram, Thermo; Experimental)

67Sim: P.Y. Simons and F. Dachille, "The Structure of TiO_2 II, a High Pressure Phase of TiO_2," Part 2, *Acta Crystallog.*, 23, 334-335 (1967). (Crys Structure; Experimental)

67Wat: D. Watanabe and J.R. Castles, "The Ordered Structure of TiO," *Acta Crystallogr.*, 23, 307-313 (1967). (Equi Diagram, Crys Structure; Experimental)

68Dac: F. Dachille, P.Y. Simons, and R. Roy, "Pressure-Temperature Studies of Anatase, Brookite, Rutile and TiO_2 II," *Am. Min.*, 53, 1929-1939 (1968). (Meta Phases; Experimental)

68Hil1: E. Hilti, "New Phases in the Titanium-Oxygen System," *Naturwissenschaften*, 55, 130-131 (1968) in German. (Equi Diagram, Crys Structure; Experimental)

68Hil2: E. Hilti and F. Laves, "X-Ray Investigation of the Low-Temperature Modification of Titanium Monoxide," *Naturwissenschaften*, 55, 131 (1968) in German. (Equi Diagram, Crys Structure; Experimental)

68Jos: A. Jostsons and A.S. Malin, "The Ordered Structure of Ti_3O," *Acta Crystallogr.*, B24, 211-213 (1968). (Crys Structure; Experimental)

68Mod: M.S. Model and G.Y. Shubina, "Investigations of the Structure and Properties of Solid Solutions of Oxygen in Titanium and Zirconium," *Izv. Akad. Nauk SSSR, Met.*, (6), 143-147 (1968) in Russian; TR: *Russ. Met.*, (6), 97-108 (1968). (Equi Diagram; Experimental)

68Rao: C.N.R. Rao, R.E. Loehman, and J.M. Honig, "Crystallographic Study of the Transitions in Ti_2O_3," *Phys. Lett. (A)*, 27(5), 271-272 (1968). (Equi Diagram; Experimental)

68Wat: D. Watanabe, O. Terasaki, A. Jostsons, and J.R. Castles, "The Ordered Structure of $TiO_{1.25}$," *J. Phys. Soc. Jpn.*, 25, 292 (1968). (Crys Structure; Experimental)

69Bar: R.F. Bartholomew and D.R. Frankel, "Electrical Properties of Some Titanium Oxides," *Phys. Rev. Lett.*, 187(3), 828-833 (1969). (Equi Diagram; Experimental)

69Bur: L.A. Bursill, B.G. Hyde, O. Terasaki, and D. Watanabe, "On a New Family of Titanium Oxides and the Nature of Slightly-Reduced Rutile," *Philos. Mag.*, 347-359 (1969). (Crys Structure; Experimental)

69Hir: M. Hirabayashi, M. Koiwa, and S. Yamaguchi, "Interstitial Order-Disorder Transformation in the Titanium-Oxygen System," *Mechanism of Phase Transformations in Crystalline Solids*, Inst. Met., London, 207-211 (1969). (Equi Diagram; Experimental)

69Iwa: H. Iwasaki, N.F.H. Bright, and J.F. Rowland, "The Polymorphism of the Oxide Ti_3O_5," *J. Less-Common Met.*, 17, 99-110 (1969). (Equi Diagram, Crys Structure; Experimental)

69Koi: M. Koiwa and M. Hirabayashi, "Interstitial Order-Disorder Transformation in the Ti-O Solid Solutions. II. A Calorimetric Study," *J. Phys. Soc. Jpn.*, 27(4), 801-806 (1969). (Equi Diagram; Experimental)

69Ver: A.W. Vere and R.E. Smallman, "The Structure of the Low-Temperature Modification of Titanium Monoxide," *Mechanism of Phase Transformation in Crystalline Solids*, Inst. Met., London, 212-219 (1969). (Equi Diagram, Crys Structure; Experimental)

69Yam: S. Yamaguchi, "Interstitial Order-Disorder Transformation in the Ti-O Solid Solution. I. Ordered Arrangement of Oxygen," *J. Phys. Soc. Jpn.*, 27(1), 155-163 (1969). (Equi Diagram, Crys Structure; Experimental)

70And: J.S. Anderson and A.S. Khan, "Equilibria of Intermediate Oxides in the Titanium-Oxygen System," *J. Less-Common Met.*, 22, 219-223 (1970). (Equi Diagram; Experimental)

70Jos: A. Jostsons and R. McDougall, "Phase Relationships in Titanium-Oxygen Alloys," Sci. Technol. Appl. Titanium, Proc. Int. Conf., R.I. Jaffee, Ed., 745-763 (1970). (Equi Diagram; Experimental)

70Kor: I.I. Kornilov, V.V. Vavilova, L.E. Fykin, R.P. Ozerov, S.P. Solowiev, and V.P. Smirnov, "Neutron Diffraction Investigation of Ordered Structures in the Titanium-Oxygen System," *Metall. Trans.*, 1, 2569-2571 (1970). (Equi Diagram; Experimental)

70Mar: M. Marezio, P.D. Dernier, D.B. McWhan, and J.P. Remeika, "X-Ray Diffraction Studies of the Metal Insulator Transitions in Ti_4O_7, V_4O_7, and VO_2," *Mater. Res. Bull.*, 5, 1015-1024 (1970). (Equi Diagram, Crys Structure; Experimental)

70Mul: L.N. Mulay and W.J. Danley, "Cooperative Magnetic Transitions in the Titanium-Oxygen System A New Approach," *J. Appl. Phys.*, 41(3), 877-879 (1970). (Equi Diagram; Experimental)

70Sim1: G.W. Simmons and E.J. Scheibner, "Order-Disorder Phenomena at the Surface of Alpha-Titanium-Oxygen Solid Solutions," *J. Mater.*, 933-949 (1970). (Equi Diagram; Experimental)

70Sim2: P.Y. Simons and F. Dachille, "Possible Topotaxy in the TiO_2 System," *Am. Min.*, 55, 403-415 (1970). (Meta Phases, Experimental)

*70Yam: S. Yamaguchi, K. Hiraga, and M. Hirabayashi, "Interstitial Order-Disorder Transformation in the Ti-O Solid Solution. IV. A Neutron Diffraction Study," *J. Phys. Soc. Jpn.*, 28(4), 1014-1023 (1970). (Equi Diagram; Experimental)

71Asb: G. Asbrink, S. Asbrink, A. Magneli, H. Okinaka, K. Kosuge, and S. Kachi, "A Ti_3O_5 Modification of V_3O_5-Type Structure," *Acta Chem. Scand.*, 25(10), 3889-3890 (1971). (Equi Diagram, Crys Structure; Experimental)

71Bur1: L.A. Bursill and B.G. Hyde, "Crystal Structures in the {132} *CS Family of Higher Titanium Oxides Ti_nO_{2n-1}*," *Acta Crystallogr. B*, 27, 210-215 (1971). (Crys Structure; Experimental)

71Bur2: L.A. Bursill, B.G. Hyde, and D.K. Philip, "New Crystallographic Shear Families Derived from the Rutile Structure, and the Possibility of Continuous Ordered Solid Solution," *Philos. Mag.*, 23, 1501-1513 (1971). (Crys Structure; Experimental)

71Dub1: A. Dubertret, "The Solid Solutions Alpha-Titanium-Oxygen and Alpha-Zirconium-Oxygen, Part I," *Mètaux-Corros.-Ind., 46(545)*, 1-24 (1971) in French. (Crys Structure; Experimental)

71Dub2: A. Dubertret, "The Solid Solution Alpha-Titanium-Oxygen and Alpha-Zirconium-Oxygen, Part II," *Mètaux-Corros.-Ind., 46(546)*, 69-83 (1971) in French. (Crys Structure; Experimental)

71Dub3: A Dubertret and P. Lehr, "Zirconium-Oxygen and Titanium-Oxygen Solid Solutions," *Metalloved., Mater. Simp.*, N.V. Ageev, Ed., 381-388 (1971) in Russian. (Crys Structure; Experimental)

71Ham: P.J. Hampson and P.W. Gilles, "High-Temperature Vaporization and Thermodynamics of the Titanium Oxides. VII. Mass Spectrometry and Dissociation Energies of TiO(g) and TiO_2(g)," *J. Chem. Phys., 55(8)*, 3712-3728 (1971). (Thermo; Experimental)

71Mar: M. Marezio and P.D. Dernier, "The Crystal Structure of Ti_4O_7, A Member of the Homologous Series Ti_nO_{2n-1}," *J. Solid State Chem., 3*, 340-348 (1971). (Crys Structure; Experimental)

71Nic: M. Nicol and M.Y. Fong, "Raman Spectrum and Polymorphism of Titanium Dioxide at High Pressures," *J. Chem. Phys., 54(7)*, 3167-3170 (1971).

71Rao: C.N.R. Rao, S. Ramdas, R.E. Loehman, and J.M. Honig, "Semiconductor-Metal Transition in Ti_3O_5," *J. Solid State Chem., 3*, 83-88 (1971). (Equi Diagram, Crys Structure; Experimental)

71Stu: D. Stull and H. Prophet, "JANAF Thermochemical Tables," NBS STP No. 37, Washington, D.C. (1971). (Thermo; Experimental)

72Dan: W.J. Danley and L.N. Mulay, "Magnetic Studies on the Ti-O System: Experimental Aspects and Magnetic Parameters," *Mater. Res. Bull., 7*, 739-748 (1972). (Equi Diagram; Experimental)

72Hul: J.K. Hulm, C.K. Jones, R.A. Hein, and J.W. Gibson, "Superconductivity in the TiO and NbO Systems," *J. Low Temp. Phys., 7(3/4)*, 291-307 (1972). (Crys Structure; Experimental)

72Por: V.N. Porter, W.B. White, and R. Roy, "Optical Spectra of the Intermediate Oxides of Titanium, Vanadium, Molybdenum, and Tungsten," *J. Solid State Chem., 4*, 250-254 (1972). (Equi Diagram; Experimental)

72Roy: R. Roy and W.B. White, "Growth of Titanium Oxide Crystals of Controlled Stoichiometry and Order," *J. Cryst. Growth, 13/14*, 78-83 (1972). (Equi Diagram; Experimental)

72Suz: K. Suzuki and K. Sambongi, "High-Temperature Thermodynamic Properties in Ti-O System," *Tetsu-to-Haganè, (J. Iron Steel Inst. Jpn.), 58(12)*, 1579-1593 (1972) in Japanese. (Equi Diagram; Experimental)

73Bar: H.L. Barros, G.V. Chandrashekhar, T.C. Chi, J.M. Honig, and R.J. Sladek, "Specific Heat of Single-Crystal Undoped and V-Doped Ti_2O_3," *Phys. Rev. B, Condens. Matter, 7(12)*, 5147-5152 (1973). (Equi Diagram; Experimental)

73Gar: E.A. Garcia, J. Com-Nougue, X. Lucas, G. Beranger, and P. Lacombe, "Variation of the Lattice Parameter of the Interstitial Ti-O Solid Solution as a Function of Oxygen Concentration," *C.R. Hebd. Sèances Acad. Sci., 277*, 1291-1293 (1973) in French. (Crys Structure; Experimental)

73Kor: I.I. Kornilov, "Relation of an Anomaly of Titanium Oxidation to a New Phase Diagram of the Titanium-Oxygen System with Suboxides," *Dokl. Akad. Nauk SSSR, 208(1-3)*, 356-359 (1973) in Russian. (Equi Diagram; Experimental)

73Mar: M. Marezio, D.B. McWhan, P.D. Dernier, and J.P. Remeika, "Structural Aspects of the Metal-Insulator Transitions in Ti_4O_7," *J. Solid State Chem., 6*, 213-221 (1973). (Crys Structure; Experimental)

73Mer: R.R. Merritt, B.G. Hyde, L.A. Bursill, and D.K. Philp, "The Thermodynamics of the Titanium-Oxygen System: An Isothermal Gravimetric Study of the Composition Range Ti_3O_5 to TiO_2 at 1304 K," *Philos. Mag., 274*, 628-661 (1973). (Equi Diagram, Thermo; Experimental)

73Rez: V.A. Reznichenko and F.V. Khalimov, "Determination of a Variation in Free Energy in the Titanium Oxygen System Using an emf Method," *Protsessy Proizvod. Titana Ego Dvuokisi*, N.V. Ageev, Ed., 193-197 (1973) in Russian. (Thermo; Experimental)

73Sly: N.P. Slyusar, A.D. Krivorotenko, E.N. Fomichev, A.A. Kalashnik, and V.P. Bondarenko, "Experimental Investigation of the Enthalpy of Titanium Oxide in the Temperature Range 500-2000 Degree K," *Teplofiz. Vys. Temp., 11(1)*, 213-215 (1973) in Russian; TR: *High Temp., 11(1)*, 190-192 (1973). (Equi Diagram; Experimental)

74Cor: M. Cormier and F. Claisse, "Beta-Alpha Phase Transformation in Ti and Ti-O Alloys," *J. Less-Common Met., 34*, 181-189 (1974). (Meta Phases; Experimental)

74Gre: I.E. Grey, C. Li, and A.F. Reid, "A Thermodynamic Study of Iron in Reduced Rutile," *J. Solid State Chem., 11*, 120-127 (1974). (Thermo; Theory)

74Rob: W.R. Robinson, "The Crystal Structures of Ti_2O_3, A Semiconductor, and $(Ti_{0.900}V_{0.100})_2O_3$, A Semimetal," *J. Solid State Chem., 9*, 255-260 (1974). (Crys Structure; Experimental)

75Cha: M.W. Chase, J.L. Curnutt, H. Prophet, and R.A. McDonald, "JANAF Thermochemical Tables Supplement," *J. Phys. Chem. Data, (4)* (1975). (Thermo; Experimental)

75Pic: C. Picard and P. Gerdanian, "Thermodynamics Study of Oxides TiO_{2-x} at 1050 °C," *J. Solid State Chem., 14*, 66-77 (1975) in German. (Thermo; Experimental)

76Bou: G. Boureau and P. Gerdanian, "Thermodynamic Study of Interstitial Solid Solutions of Oxygen in Titanium at 1050 °C," *Acta Metall., 24*, 717-723 (1976). (Thermo; Experimental)

77Bau: J.F. Baumard, D. Panis, and M. Anthony, "A Study of Ti-O System Between Ti_3O_5 and TiO_2 at High Temperature by Means of Electrical Resistivity," *J. Solid State Chem., 20*, 43-51 (1977). (Thermo; Experimental)

77Dec: M. Dechamps, A. Quivy, G. Baur, and P. Lehr, "Influence of the Distribution of the Interstitial Oxygen Atoms on the Lattice Parameters in Dilute cph Titanium-Oxygen Solid Solutions (90-4000 ppm at)," *Scr. Metall., 11*, 941-945 (1977). (Crys Structure; Experimental)

78Koz: A. Kozlowska-Rog and G. Rog, "Study on Phase Equilibria in the Titanium-Oxygen System within the Composition Range $TiO_{1.960}$-TiO_2 by the Solid Galvanic Cell Technique," *Pol. J. Chem., 52*, 607-611 (1978). (Equi Diagram, Thermo; Experimental)

78Liu: L.G. Liu, "A Fluorite Isotype of SnO_2 and A New Modification of TiO_2 Implications for the Earth's Lower Mantle," *Science, 199*, 422-424 (1978). (Meta Phases; Experimental)

78Tet: R. Tetot, C. Picad, G. Boureau, and P. Gerdanian, "High Temperature Thermodynamics of the Titanium-Oxygen System for O/Ti 1," *J. Chem. Phys., 69(1)*, 326-331 (1978). (Thermo; Experimental)

79Dav: D. David, E.A. Garcia, X. Lucas, and G. Beranger, "Study of the Diffusion of Oxygen in Alpha-Titanium, Oxidized Between 700 and 950 °C," *J. Less-Common Met., 65*, 51-69 (1979). (Equi Diagram; Experimental)

79Mit: T. Mitsuhashi and O.J. Kleppa, "Transformation Enthalpies of the TiO_2 Polymorphs," *J. Am. Chem. Soc., 62,(7-8)*, 356-357 (1979). (Equi Diagram, Thermo; Experimental)

79Ohs: T. Ohsaka, S. Yamaoka, and O. Shimomura, "Effect of Hydrostatic Pressure on the Raman Spectrum of Anatase (TiO_2)," *Solid State Commun., 30*, 345-347 (1979). (Meta Phases; Experimental)

80Mam: J.F. Mammone, S.K. Sharma, and M. Nicol, "Raman Study of Rutile (TiO_2) at High Pressures," *Solid State Commun., 34*, 799-802 (1980). (Meta Phases; Experimental)

81Mam: J.F. Mammone, M. Nicole, and S.K. Sharma, "Raman Spectra of TiO_2-II, TiO_2-III, SnO_2, and GeO_2 at High Pressure," *J. Phys. Chem. Solids, 42*, 379-384 (1981). (Meta Phases; Experimental)

81Tet: R. Tetot, C. Picard, and P. Gerdanian, "Direct Measurements of $\delta H(O_2)$ at 1323 K for Nonstoichiometric TiO," *J. Chem. Phys., 75(3)*, 1365-1367 (1981). (Thermo; Experimental)

82Gra: B. Granier and P.W. Gilles, "High-Temperature Vaporization and Thermodynamics of the Titanium Oxides XVII. Approximate Oxygen Potentials in Metal-Rich Solutions," *High Temp.-High Pressures, 14*, 383-386 (1982). (Thermo; Experimental)

82Lep1: Y. Le Page and P. Strobel, "Structural Chemistry of the Magneli Phases Ti_nO_{2n-1} ($4 \leq n: \leq 9$). I. Cell and Structure Comparisons," *J. Solid State Chem., 43,* 314-319 (1982). (Crys Structure; Experimental)

82Lep2: Y. Le Page and P. Strobel, "Structural Chemistry of the Magneli Phases Ti_nO_{n-1} ($4 \leq n \leq 9$). II. Refinements and Structural Discussion," *J. Solid State Chem., 4,* 273-281 (1982). (Crys Structure; Experimental)

83Zad: S. Zador and C.B. Alcock, "A Thermodynamic Study of Magneli and Point Defect Phases in the Ti-O System," *High Temp. Sci., 16,* 187-207 (1983). (Thermo; Experimental)

84Lep: Y. Le Page and M. Marezio, "Structural Chemistry of Magneli Phases Ti_nO_{2n-1} ($4 \leq n \leq 9$). IV. Superstructure in Ti_4O_7 at 140 K," *J. Solid State Chem., 53,* 13-21 (1984). (Crys Structure; Experimental)

The Os-Ti (Osmium-Titanium) System

190.2   47.88

By J.L. Murray

Equilibrium Diagram

The equilibrium solid phases of the Ti-Os system are

- The bcc (βTi) solid solution, based on the equilibrium solid phase of pure Ti above 882 °C. The maximum solubility of Os in (βTi) is approximately 23 at.% at 1710 °C.
- The ordered bcc (CsCl) compound TiOs. TiOs melts congruently at approximately 2160 °C. It has a wide homogeneity range, 38 to 51 at.% Os at 1710 °C.

- The cph solid solutions, (αTi) and (Os). (αTi) is based on the equilibrium phase of Ti below 882 °C. The solubility of Os in (αTi) is probably less than 1 at.% at 600 °C; the solubility of Ti in (Os) is approximately 21 at.% at 2100 °C.

The important points of the assessed Ti-Os phase diagram (Fig. 1) are summarized in Table 1. The assessed phase diagram is based on the results of a single experimental study [71Ere, 72Ere, 75Sht]; experimental data are shown in Fig. 2. [71Ere] clearly established the topology of the phase diagram. However, they noted that

Fig. 1---Assessed Ti-Os Phase Diagram

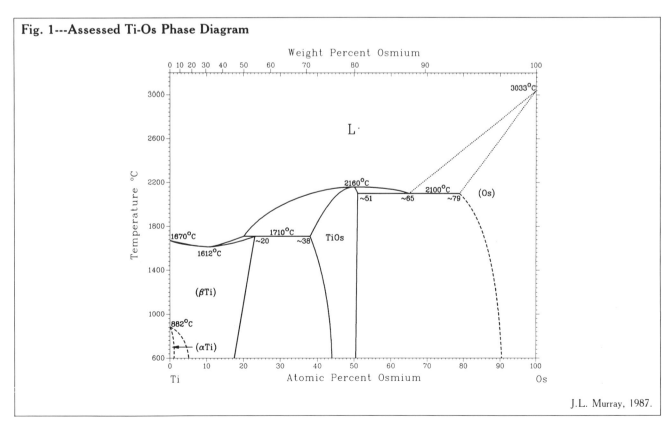

J.L. Murray, 1987.

229

Table 1 Special Points of the Assessed Ti-Os Phase Diagram

Reaction	Compositions of the respective phases, at.% Os			Temperature, °C	Reaction type
L + TiOs ⇌ (βTi)	20	38	23	1710	Peritectic
L ⇌ TiOs + (Os)	65	51	79	2100	Eutectic
L ⇌ TiOs		50		2160	Congruent
L ⇌ (βTi)		10		1612	Congruent
L ⇌ (βTi)		0		1670	Melting point
L ⇌ (Os)		100		3033	Melting point
(βTi) ⇌ (αTi)		0		882	Allotropic transformation

Fig. 2---Experimental Data on the Ti-Os Phase Diagram

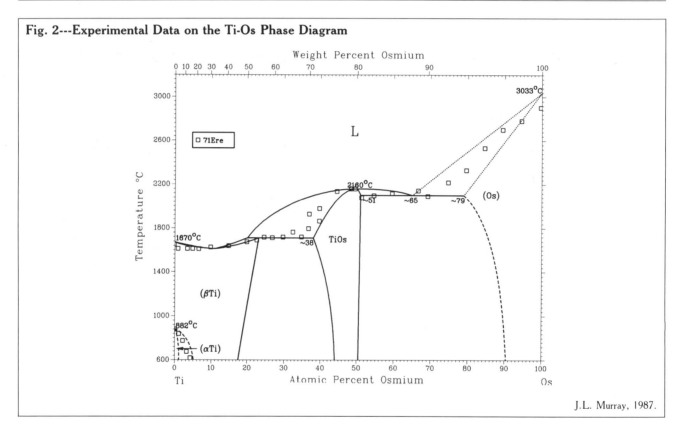

J.L. Murray, 1987.

equilibrium was very difficult to achieve over the composition ranges 20 to 45 and 67 to 95 at.% Os. This, together with some internal inconsistencies in the data and the implausibility of several phase boundaries from the point of view of thermodynamics, suggests that further experimental work is needed to define the phase boundaries. In particular, the (αTi)/(βTi) boundaries and the (Os) liquidus, solidus and solvus require further examination.

Liquidus and Solidus, (Os) Solvus.

The melting point of pure Os is 3033 °C [Melt]. Various determinations of this melting point reported in the literature differ by as much as 100 °C [Hultgren,E]. Oxygen contamination lowers the melting point of Os [Hultgren,E].

The solidus was studied by [71Ere] using the Pirani (optical) method. [71Ere] observed pure Os to melt at 2913 °C; this discrepancy with the assessed value of 3033 °C suggests that contamination of the Os probably affected the experiments over a wide composition range. Thus, uncertainties in the equilibrium binary phase

boundaries are somewhat larger than would be expected from the scatter in the data and the errors intrinsic to the experimental method.

The eutectic reaction L ⇌ TiOs + (Os) was observed at 2100 °C, and the eutectic composition was estimated, from microstructural data, as 65 at.% Os. [71Ere] reported the maximum solubility of Ti in (Os) to be approximately 21 at.% (79 at.% Os), based on microstructural data. This conflicts with solidus data near the eutectic temperature. [71Ere] drew the solidus to agree exactly with their experimental points. The resulting solidus is slightly retrograde in character; a retrograde would be a very unusual feature for this system. In Fig. 1, both liquidus and solidus are sketched as straight lines between the melting point of Os and the appropriate compositions at the eutectic isotherm.

Based on microstructural observations of annealed samples, the solubility of Ti in (Os) decreases to between 5 and 10 at.% Ti at 1000 °C. [71Ere] showed the solubility as decreasing significantly between the eutectic temperature and 1800 °C, then remaining essentially constant and

Fig. 3---Assessed Ti-Os Phase Diagram vs Thermodynamic Calculations

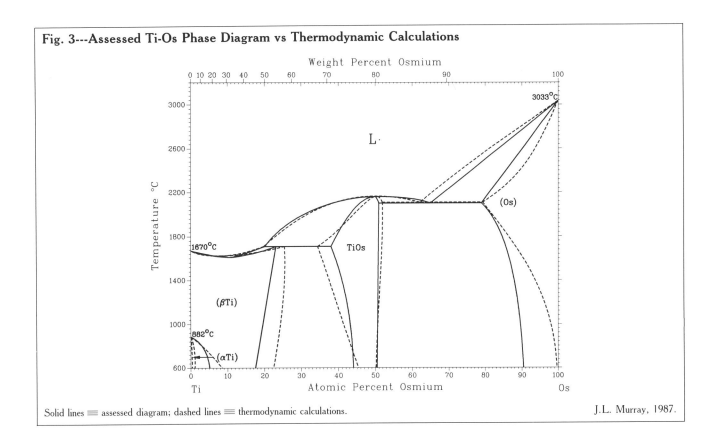

Solid lines ≡ assessed diagram; dashed lines ≡ thermodynamic calculations.

J.L. Murray, 1987.

large down to 1000 °C. The assessed solvus is redrawn from those authors' phase diagram with a smoother curvature, but it is not inconsistent with the microstructural data. Moreover, the very large solubilities observed at relatively low temperatures suggest that the solid solution may be metastable. Although the experimental data [71Ere] are used to draw Fig. 1, the solvus is shown as a dashed line to indicate that smaller solubilities are more likely. The calculated solvus, Fig. 3, can also be considered as an approximation to the equilibrium curve.

Addition of Os to Ti decreases the melting point, until a congruent melting minimum is reached at approximately 10 at.% Os and 1612 °C. The (βTi) solidus and liquidus are flat and narrow, and the solidus data are somewhat scattered, so that an accurate determination of the composition of the minimum is not possible. Liquidus and solidus rise from the minimum to a peritectic reaction L + TiOs ⇄ (βTi) at 1710 °C. The solubility of Os in (βTi) increases from approximately 20 at.% at 1000 °C to approximately 23 at.% Os at the peritectic equilibrium at 1710 °C, determined by thermal analysis and metallographic examination of alloys equilibrated at temperatures between 1000 °C and the peritectic temperature.

(αTi)/(βTi) Boundaries.

[71Ere] determined the (βTi)/(βTi) boundaries using DTA and metallographic examination of alloys annealed at 600 and 800 °C. Alloys containing 1 at.% Os were single-phase at both 800 and 600 °C. The (βTi) transus was placed between 3 and 4 at.% Os at 800 °C and between 4 and 5 at.% Os at 600 °C. DTA data for the (βTi) transus lie within the two-phase region as determined by metallography. Details of the experimental procedure were not

given. No other determination of these boundaries has yet been made.

The observation of a single-phase (αTi) structure alloy at 1 at.% Os and 800 °C (a very large solubility at that temperature) suggests that difficulties, either in equilibrating alloys or in identifying the phases, may have caused errors in both the (αTi) solvus and the (βTi) transus. In the systems Ti-Ru, Ti-Rh, and Ti-Ir, the work of [72Ere] can be compared with the work of others. In each instance, additional experiments on the (αTi)/(βTi) boundaries tended to place the (βTi) transus at higher Ti contents. For example, in Ti-Ir, work subsequent to [72Ere] demonstrated the existence of a previously unobserved eutectoid reaction of (αTi), (βTi), and the Ti-rich intermetallic compound.

Therefore, the equilibrium maximum solubility of Os in (αTi) is probably lower than 1 at.%, and the two-phase region is probably somewhat wider than that drawn in Fig. 1. In thermodynamic calculations described below, the Gibbs energy of (βTi) was determined by the phase relations of (βTi) at temperatures above 1000 °C. Using this Gibbs energy, the calculated (βTi) transus reaches 8.4 at.% Os at 600 °C.

TiOs.

TiOs, with the ordered bcc (CsCl) structure, was first reported by [39Lav], and [55Jor] verified the structure. [71Ere] found the composition range of TiOs to be 38 to 51 at.% Os at 1710 °C by metallographic studies and observed TiOs to melt congruently at 2160 °C by DTA. The experimental TiOs solidus was also determined by DTA. The solidus data are very scattered; however, they define the general shape of the solidus and are consistent with the observed composition range of single-phase TiOs.

231

Table 2 Crystal Structures of the Ti-Os System

Phase	Homogeneity range, at.% Os	Pearson symbol	Space group	Struktur-bericht designation	Prototype	Reference
(αTi)	0 to 1	hP2	P6₃/mmc	A3	Mg	[Pearson2]
(βTi)	0 to 23	cI2	Im3m	A2	W	[Pearson2]
TiOs	38 to 51	cP2	Pm3m	B2	CsCl	[55Jor]
(Os)	79 to 100	hP2	P6₃/mmc	A3	Mg	[Pearson2]

Table 3 Ti-Os Lattice Parameter Data

Reference	Phase	Composition, at.% Os	Lattice parameters, nm	
			a	c
[55Jor]	TiOs	50	0.307	...
[66Ere]	TiOs	50	0.3081	...
[71Ere]	TiOs	50	0.308	...
		45	0.309	...
		4	0.325	...
		10	0.324	...
		15	0.321	...
		20	0.319	...
	(Os)	90	0.274	0.435
		95	0.274	0.433
		100	0.274	0.432

Table 4 Thermodynamic Parameters of the Ti-Os System

Properties of the pure components [70Kau]

G^0(Ti,bcc) = $- 16\,234 + 8.368\,T$
G^0(Ti,L) = 0
G^0(Ti,cph) = $- 20\,585 + 12.134\,T$
G^0(Os,bcc) = $- 31\,757 + 9.623\,T$
G^0(Os,L) = 0
G^0(Os,cph) = $- 27\,614 + 8.368\,T$

Excess Gibbs energies of solution phases

Present evaluation		Regular solution approximation [70Kau]	
B(L)	= $- 46\,000$	B(L)	= $- 60\,969$
C(L)	= $- 20\,000$		
B(bcc)	= $- 82\,660 + 20\,T$	B(bcc)	= $- 68\,224$
C(bcc)	= $1\,000$		
B(cph,Os)	= $- 32\,000$	B(cph)	= $- 32\,451$
C(cph,Os)	= $- 20\,000$		
B(cph,Ti)	= $- 10\,000$		

TiOs Gibbs energy (this evaluation)

G(TiOs) = $- 53\,811 + 12\,T$
C(TiOs) = $84\,830 - 10\,T$
D(TiOs) = $60\,000$

Note: Values are given in J/mol and J/mol · K.

Metastable Phases

[71Ere] found that metastable (βTi) can be retained during quenching in alloys of Os content greater than approximately 3 at.%. For lower Os content, (βTi) transforms martensitically to the cph (α'Ti) form during quenching. This is the only experimental data now available on the metastable phases of the Ti-Os system. Alloys of Ti with transition metals have many properties in common, and it is possible to predict some qualitative

features of the metastable equilibria. In particular, the Ti-Os system is analogous to the Ti-Ru system. In that system, cph (α'Ti) can form martensitically in Ti-rich alloys, and at slightly higher Ru contents an orthorhombic distortion of the cph structure (α''Ti) is formed. The ω phase forms as an intermediate phase in the decomposition of (βTi) into two-phase (αTi) + (βTi) at low temperatures.

Crystal Structures and Lattice Parameters

The crystal structures of the established equilibrium phases are summarized in Table 2. Lattice parameters are separately given in Table 3.

Thermodynamics

Experimental determinations of thermodynamic properties of Ti-Os alloys are not available. The present thermodynamic calculations were performed in order to predict a more probable shape for the (Os) solvus than the presently available experimental data and to provide a basis for the calculation of ternary diagrams. The thermodynamic parameters used in this calculation are compared in Table 4 to the parameters previously calculated by [70Kau].

Gibbs energies of the solution phases are represented as:

$$G(i) = G^0(\text{Ti},i)\,(1 - x) + G^0(\text{Os},i)\,x + RT\,[x \ln x + (1 - x) \ln (1 - x)] + B^i\,x(1 - x) + C^i\,x(1 - x)\,(1 - 2x)$$

where i designates the phase, x is the atomic fraction of Os, G^0 is the Gibbs energies of the pure metals, $B(i)$ and $C(i)$ are coefficients in the subregular expansion of the excess Gibbs energies. The Gibbs energies of the pure elements are from [70Kau].

TiOs was modeled with a Wagner-Schottky Gibbs energy:

$$G(\text{TiOs}) = G^0(\text{TiOs}) + RT[x_{\text{Ti}} \ln x_{\text{Ti}} + x_{\text{Os}} \ln x_{\text{Os}} +$$
$$v \ln v + s \ln s - (1 + v) \ln (1 + v)] +$$
$$C s + D v$$

The Wagner-Schottky compound is conceptually resolved into Ti and Os sublattices; v is the concentration of vacancies on the Ti sublattice and s is the concentration of substitutional Ti atoms on the Os sublattice. x_{Ti} and x_{Os} are the concentrations of Ti and Os on their respective sublattices, and x is the overall composition. All concentrations are referred to the total number of atoms. The concentrations of vacancies and substitutions are determined by minimizing the Wagner-Schottky Gibbs energy with respect to the variables s and v.

The regular solution approximation [70Kau] was used as the starting point of the present calculations. The (Os)/L phase boundaries, for example, can be adequately reproduced by regular solution Gibbs energies. The regular solution model, however, is intrinsically unable to model the peritectic equilibrium of L, (βTi), and TiOs. (As estimated by [70Kau], the melting point of metastable bcc Os is quite low compared to that of the equilibrium cph Os, between 1200 and 1300 °C. Therefore, the bcc liquidus and solidus, having reached a minimum at about 10 at.% Os, must at yet higher Os content find a maximum and then descend to the metastable melting point. The regular solution model does not give rise to double extrema.)

Gibbs energies were arrived at by first determining Gibbs energies of the solution phases by trial and error matching of the phase diagram. The difference between the subregular contributions to the (βTi) and liquid phases was determined by the double extrema for the T_0 curve between those phases. The subregular term of the liquid phase was determined by matching the liquidus involving TiOs.

The Gibbs energy of TiOs was next estimated in the same way, starting with a rough estimate of the heat of fusion. The parameters $C(\text{TiOs})$ and $D(\text{TiOs})$ were determined from the composition width of TiOs.

The basic features of the phase diagram are reproduced by these Gibbs energies. The calculated eutectic temperature is 2108 °C (compared to 2100 °C); the peritectic temperature is 1710 °C (compared to 1710 °C); the (Os) solvus descends from 85 at.% Os at the eutectic temperature to 97.5 at.% Os at 1000 °C; the solubility of Os in (βTi) decreases from 25 to 22 at.% Os at 600 °C (compared to approximately 20 at.%); the congruent melt of (βTi) occurs at 1626 °C and 8 at.% Os. The calculated phase diagram is compared to the assessed experimental diagram in Fig. 3. It should be noted that, in the absence of experimental thermodynamic data, these Gibbs energies can only be considered rough estimates.

Cited References

39Lav: F. Laves and H.J. Wallbaum, "Crystal Chemistry of Titanium Alloys," *Naturwissenschaften, 27*, 674 (1939) in German. (Crys Structure; Experimental)

55Jor: C.B. Jordon, "Crystal Structure of TiRu and TiOs," *Trans. AIME, 188*, 832-833 (1955). (Crys Structure; Experimental)

70Kau: L. Kaufman and H. Bernstein, *Computer Calculations of Phase Diagrams*, Academic Press, NY (1970). (Thermo; Theory)

71Ere: V.N. Eremenko, R.D. Shtepa, and E.L. Semenova, "Phase Diagram of Ti-Os," *Izv. Akad. Nauk SSSR, Met.*, (4), 210-213 (1971) in Russian; TR: *Russ. Metall.*, (4), 147-149 (1971).

72Ere: V.N. Eremenko and R.D. Shtepa, "Phase Equilibria in the Binary Systems of Titanium with Ruthenium, Osmium, Rhodium, Iridium, and Palladium," *Colloq. Int. CNRS*, (205), 403-413 (1972). (Equi Diagram; Experimental)

75Sht: T.D. Shtepa, "Interactions of Titanium with Platinum-Group Metals," *Fiz. Khim. Kondens. Faz. Sverkhtverd. Mater.*, 175-191 (1975) in Russian.

The P-Ti (Phosphorus-Titanium) System

30.97376 47.88

By J.L. Murray

Equilibrium Diagram

The assessed Ti-P phase diagram (Fig. 1) is based primarily on work by [65Ere] and [68Sne]. There is some disagreement in the range between Ti_2P and Ti_4P_3, which is tentatively attributed to solid-state transformations occurring during quenching. The diagram is thus uncertain and incomplete. Other work on the system has primarily been on preparation and crystal structure analysis of the compounds. The special points of the diagram are summarized in Table 1, and a listing of the observed phases and their crystal structures is given in Table 2.

The equilibrium solid phases of the Ti-P system are:

- The solid solutions, cph (αTi) and bcc (βTi). The maximum solubility of P in (βTi) is shown as 0.3 at.% P; the solubility of P in (αTi) has not been determined, but based on the (βTi) solvus, it must be very low. Phases based on pure P are not considered in the present assessment.

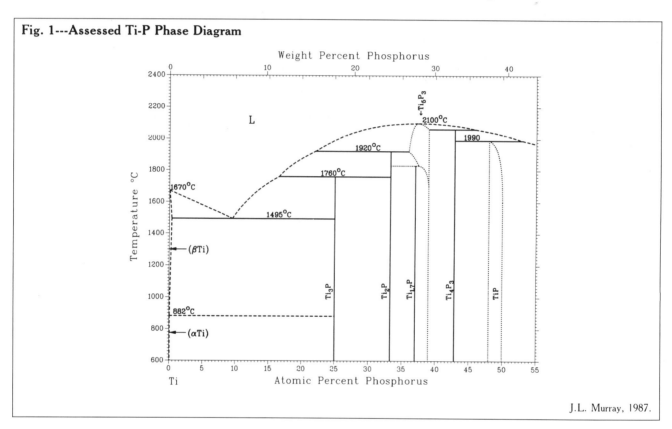

Fig. 1---Assessed Ti-P Phase Diagram

J.L. Murray, 1987.

Table 1 Special Points of the Assessed Ti-P Phase Diagram

Reaction	Compositions of the respective phases, at.% P			Temperature, °C	Reaction type
$L \rightleftarrows (\beta Ti) + Ti_3P$	~ 0.3	9.5	25	1495	Eutectic
$(\beta Ti) \rightleftarrows (\alpha Ti) + Ti_3P$	~ 0	~ 0	25	~ 882	Eutectoid (?)
$L + Ti_2P \rightleftarrows Ti_3P$	~ 17	33.3	25	1760	Peritectic
$L + Ti_5P_3 \rightleftarrows Ti_2P$	~ 22	~ 36	33.3	1920	Peritectic
$L \rightleftarrows Ti_5P_3$		37.5		2100	Congruent
$Ti_2P + Ti_5P_3 \rightleftarrows Ti_{1.7}P$	33.3	37.5	37	1830	Peritectoid (hypothetical)
$Ti_5P_3 + L \rightleftarrows Ti_4P_3$	~ 39	~ 45	42.9	2060	Peritectic
$Ti_5P_3 + L \rightleftarrows TiP$	42.9	~ 53	48	1990	Peritectic
$L \rightleftarrows (\beta Ti)$		0		1670	Melting point
$(\beta Ti) \rightleftarrows (\alpha Ti)$		0		882	Allotropic transformation

234

Table 2 Crystal Structures of the Equilibrium Phases of the Ti-P System

Phase	Homogeneity range, at.% P	Pearson symbol	Space group	Strukturbericht designation	Proto-type	Lattice parameters, nm			Reference
						a	b	c	
(βTi)	0 to 0.3	$cI2$	$Im3m$	A2	W	...	...	...	[Pearson2]
(αTi)	0 to ?	$hP2$	$P6_3/mmc$	A3	Mg	...	...	...	[Pearson2]
Ti_3P	25	$tP32$	$P4_2/n$	...	Ti_3P	0.99592	...	0.49869	[67Lun]
						0.9956	...	0.4988	[63Lun]
						0.928	...	0.499	[65Ere]
Ti_2P	33.3	(a)	...	...		1.15314 to 1.15276	...	0.34575 to 0.34580	[67Lun]
		(b)	...	...	Fe_2P	0.6715	...	0.3462	[65Ere]
Ti_5P_3	~ 36 to ~ 39	$hP16$	$P6_3/mcm$	$D8_8$	Mn_5Si_3	0.72381	...	0.4088	[65Bar]
						0.72226	...	0.50936	[67Lun](c)
						0.72297	...	0.50950	[67Lun](d)
						0.7234	...	0.4090	[64Bra]
$Ti_{1.7}P$ ...	37.0	$oP?$	$P2_12_12_1$	...	...	0.74376	0.97522	0.65047	[68Sne]
Ti_4P_3 ...	42.9	(e)	...	...	...	0.74298	...	...	[68Sne]
						0.7425	...	...	[65Ere]
Ti_3P_2	38	(f)	...	...	...	0.7483	...	1.0495	[65Ere]
TiP	48 to 50	$hP8$	$P6_3/mmc$	B_i	AsTi	0.3499	...	1.1700	[67Sne]
						0.34991 to 0.34988	...	1.17025 to 1.17000	[67Lun]
						0.3487	...	1.165	[54Sch]
						0.348	...	1.162	[59Shl]
TiP_2	66.7	$tI12$	$I4/mcm$	C16	Al_2Cu	0.61812	0.33455	0.82578	[68Sne, 67Lun]
						0.6654	0.3346	0.8256	[64Hul]

(a) Hexagonal. (b) Trigonal. (c) Single crystal. (d) Powders. (e) Cubic. (f) Tetragonal.

- TiP and TiP_2 with the AsTi and Al_2Cu structures, respectively. The former is a superstructure of NiAs.
- The subphosphides—Ti_3P, Ti_2P, and Ti_4P_3—appear in both the diagrams of [65Ere] and [68Sne]. In addition, several phases have been observed in the range 36 to 40 at.% P. [68Sne] reported two phases, $Ti_{1.7}P$ and Ti_5P_3. [65Ere] reported a single compound, of yet a different structure, which they designated Ti_3P_2. In the assessed diagram, the phases $Ti_{1.7}P$ and Ti_5P_3 have been included. This region requires further experimental work.

Assessment of the Diagram.

[65Ere] provided a phase diagram covering the range 0 to 50 at.% P. They used thermal analysis, X-ray diffraction, and microscopic examination for its determination. Thermal analysis data were plotted for the three-phase equilibria, and uncertainties were assigned to the invariant temperatures. Information on experimental techniques, heat treatments, purity of the alloys, or microstructures was not given, however, and it is therefore not possible to assess the accuracy of the two-phase boundaries or to resolve discrepancies, except hypothetically. [68Sne] used visual determinations of melting points (± 50 °C), X-ray diffraction of single-crystal and powder samples, and metallography. [68Sne] stated that they verified most of the diagram of [65Ere], but they did not report sufficient details of their experiments to permit a quantitative data assessment.

Any determination of the phase relations for Ti-rich alloys must take into account that, in most alloy preparation techniques, TiP is the first phase to form [38Bil, 54Sch, 63Lun, 67Lun] and is probably metastable in all Ti-rich alloys [67Lun, 68Sne].

It was originally believed that the compounds of the system were TiP and perhaps one subphosphide. [38Bil] and [54Sch] found evidence for a subphosphide of proposed composition Ti_2P coexisting with TiP; however, neither the composition nor the structure were determined. [60Ver] prepared alloys by reacting phosphine with Ti powder and identified Ti_2P and possibly Ti_3P by chemical analysis of products formed after different reaction times.

The existence of the most Ti-rich compound, Ti_3P, was established by crystal structure analysis [63Lun, 67Lun]. Concerning the phase equilibria involving Ti_3P, [65Ere] and [68Sne] are in agreement. [65Ere] placed a eutectic reaction, L $\rightleftarrows$ (βTi) + Ti_3P, at 1495 ± 5 °C with eutectic composition 9.5 at.% P; their experimental basis for assigning this eutectic composition is unclear. At 1450 and 1300 °C, the solubility of P in (βTi) was given as 0.2 [65Ere] and 0.08 at.% [68Sne]. [68Sne] verified the eutectic temperature and composition by optical pyrometry and metallographic examination of as-cast alloys.

Ti_2P was prepared by [65Ere], [67Lun], and [68Sne]. According to [65Ere], Ti_3P and Ti_2P melt peritectically at 1760 ± 10 °C and 1920 ± 10 °C, respectively. [68Sne] verified these reactions and also noted that a metastable eutectic between Ti_2P and (βTi) was observed in arc-melted alloys. Further study of this reaction might be of use in constructing the liquidus. No direct measurements of the liquidus in the range 10 to 40 at.% P have been made, and this curve is therefore purely hypothetical at present. Finally, from lattice parameter measurements, the homogeneity range of Ti_3P is narrow. The range of Ti_2P has not been investigated, and it is assumed to be a strictly stoichiometric phase.

[65Ere] found two reactions involving (αTi) at 905 and 880 °C. Based on the low solubility of P in (βTi), the three-phase reaction of (αTi), (βTi), and Ti_3P probably lies close to the pure metal transformation point. A eutectoid reaction is therefore shown at 882 °C. A change in the kinetics of formation of TiP was observed near 900 °C by [77Uga]. Our present hypothesis is that the reaction observed by [65Ere] at 905 °C is connected with metastable TiP.

Beyond Ti_2P, the phase diagram becomes complicated. [65Ere] showed a tetragonal phase Ti_3P_2 (but with composition 38 at.% P), which melts congruently at 2098 °C, has a broad homogeneity range near the melting point, and is stoichiometric below 1800 °C. A stoichiometric cubic phase Ti_4P_3 melts by a peritectic reaction at 2060 ± 15 °C [65Ere, 68Sne]. [65Ere] did not observe Ti_5P_3, reported by [64Bra], [65Bar], and [67Lun], or $Ti_{1.7}P$, reported by [68Sne]. Ti_5P_3 was stable up to at least 1500 °C [68Sne]. [68Sne] verified the stability and symmetry of Ti_4P_3 and also confirmed that Ti_4P_3 forms by a peritectic reaction. They observed that powder patterns of as-cast alloys in the range 35 to 45 at.% P are very complex.

Ti_5P_3 has some homogeneity range [65Bar, 67Lun], but the lattice parameter variation with composition is not reproducible. Because of its structure, Ti_5P_3 is easily contaminated, and contamination may also influence its range of stability. The present hypothesis is that Ti_5P_3 is stable to the liquidus and has a homogeneity range of several percent at high temperatures. $Ti_{1.7}P$ and possibly other structures could be formed during quenching of high-temperature single-phase Ti_5P_3. Complex powder patterns and discrepancies among various structure determinations may be explicable in terms of two-phase or transformed structures. This region of the diagram requires careful microstructural characterization of alloys and would probably require high-temperature structural determinations.

TiP was assigned the homogeneity range 47.9 to 48.5 at.% P by [38Bil], using X-ray diffraction, and 48.6 to 49.4 at.% P by [66Gin], using the Knudsen cell technique. The latter work established only the Ti-rich boundary and cited unpublished work that placed the P-rich side nearer 50 at.% P. Lattice parameters of TiP do not vary significantly with composition [67Lun]. The approximate range 48 to 50 at.% P is shown in the assessed diagram. TiP melts by a peritectic reaction at 1990 ± 20 °C [65Ere, 68Sne]. The enthalpy of formation of TiP from the pure components is −132.6 J/mol, based on the difference between heats of combustion of the compound and unalloyed mixtures [59Shc].

TiP_2 has been examined by [64Hul], [67Lun], and [68Sne]. It has metallic conductivity [64Hul]. In this composition range, inclusion of the gaseous phase and an account of the effect of pressure are needed to describe the phase diagram. Accordingly, the phase diagram is drawn only in the range 0 to 55 at.% P.

Cited References

38Bil: W. Blitz, A. Rink, and F. Wiechmann, "Reaction of Titanium with Phosphorus," *Z. Anorg. Chem., 238,* 395-405 (1938) in German

54Sch: N. Schonberg, "An X-ray Investigation of Transition Metal Phosphides," *Acta Chem. Scand., 8,* 226-239 (1954). (Equi Diagram; Experimental)

59Shc: S.A. Shchukarev, M.P. Morozova, and Li Miao-hsiu, "Enthalphy of Formation of Compounds of Titanium with Elements of the Main Subgroup of Group V," *Zh. Obshch. Khim., 29,* 2465 (1959) in Russian; TR: *J. Gen. Chem. USSR, 29,* 2427-2429 (1959). (Equi Diagram; Experimental)

60Ver: L.L. Vereikina and G.V. Samsonov, "A Simple Preparation of Titanium Phosphides," *Zh. Neorg. Khim., 5,* 1888-1889 (1960) in Russian; TR: *Russ. J. Inorg. Chem., 5*(8), 916-917 (1960). (Equi Diagram; Experimental)

63Lun: T. Lundstrom, "A Note on the Crystal Structure of Ti_3P and V_3P," *Acta Chem. Scand., 17*(4), 1166-1167 (1963). (Equi Diagram; Experimental)

64Bra: G. Brauer and K. Gingerich, "Crystal Structure of the Titanium Phosphide Ti_5P_3," *Angew. Chem., 76,* 187-188 (1964) in German. (Equi Diagram; Experimental)

64Hul: F. Hulliger, "New Representatives of the $NbAs_2$ and $ZrAs_2$ Structures," *Nature, 204,* 775 (1964). (Equi Diagram; Experimental)

65Bar: H. Barnighausen, M. Knausenberger, and G. Brauer, "The Crystal Structure of $TiP_{0.63}$," *Acta Crystallogr., 19*(1), 1-6 (1965) in German. (Equi Diagram; Experimental)

***65Ere:** V.M. Eremenko and V.E. Listovnichii, "Phase Diagram of Titanium-Phosphorus System," *Dopovidi Akad. Nauk UKR SSR,* (9), 1176-1179 (1965) in Russian. (Equi Diagram; Experimental)

66Gin: K.A. Gingerich, "The Application of High Temperature Mass Spectrometry to the Study of the Composition Function of Thermodynamic Properties in Non-Stoichiometric Compounds," Advan. Mass. Spectrom., Proc. Conf. 3, 1009-1016 (1966). (Equi Diagram; Experimental)

67Lun: T. Lundstrom and P. Snell, "Studies of Crystal Structures and Phase Relationships in the TiP System," *Acta Chem. Scand., 21,* 1343-1352 (1967). (Equi Diagram; Experimental)

67Sne: P. Snell, "The Crystal Structure of TiP," *Acta Chem. Scand., 27,* 1773-1776 (1967). (Equi Diagram; Experimental)

68Sne: P. Snell, "Phase Relationships in the Ti-P System with some Notes on the Crystal Structures of TiP_2 and ZrP_2", *Acta Chem. Scand., 22,* 1942-1952 (1968). (Equi Diagram; Experimental)

77Uga: Ya.A. Ugai, A.A. Illarinov, O. Gukov, and S.M. Dorokhin, "Reaction of Titanium with Phosphorus at High Temperatures," *Izv. Akad. Nauk SSSR, Neorg. Mater., 13*(6), 1086-1087 (1977) in Russian; TR: *Inorg. Mater. USSR, 13,* 885-886 (1977). (Equi Diagram; Experimental)

The Pb-Ti (Lead-Titanium) System

207.2   47.88

By J.L. Murray

Equilibrium Diagram

The Ti-Pb phase diagram is known only approximately and only in the range 0 to 40 at.% Pb. A preliminary X-ray and microscopic study of the range 12 to 100 at.% Pb was made by [51Now], using sintered compacts. One intermetallic phase, Ti$_4$Pb, was found. [55Far] examined the region 0 to 35 at.% Pb using microscopy, X-ray diffraction, and visual observation of melting. The Ti$_4$Pb phase was verified, and evidence was found for an additional intermetallic phase. The assessed phase diagram (Fig. 1) is based entirely on the work of [55Far] and is essentially identical to that of [Hansen]. Dashed lines

indicate that quantitative phase boundary data are not available, or that the boundary requires further experimental work. Special points of the assessed diagram are given in Table 1, and crystal structures of the equilibrium phases are given in Table 2.

The equilibrium solid phases of the Ti-Pb system are:

- The bcc (βTi), cph (αTi), and (Pb) solid solutions. The maximum solubilities of Pb in (βTi) and (αTi), respectively, are 16 and 4.2 ± 0.2 at.% [55Far].
- Ti$_4$Pb has the ordered hexagonal $D0_{19}$ structure [51Now] with ideal stoichiometry Ti$_3$Pb. The com-

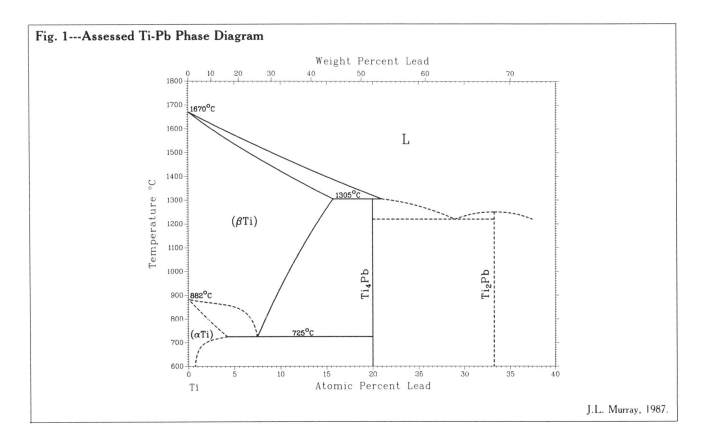

Fig. 1---Assessed Ti-Pb Phase Diagram

J.L. Murray, 1987.

Table 1 Special Points of the Assessed Ti-Pb Phase Diagram

Reaction	Compositions of the respective phases, at.% Pb			Temperature, °C	Reaction type
L + (βTi) $\rightleftarrows$ Ti$_4$Pb	~ 21	~ 16	~ 20	1305	Peritectic
(βTi) $\rightleftarrows$ (αTi) + Ti$_4$Pb	7.5	4.2	~ 20	725	Eutectoid
L $\rightleftarrows$ Ti$_4$Pb + Ti$_2$Pb		~ 20	~ 33	1200 to 1300	Eutectic
L + Ti$_2$Pb + (Pb)		~ 33			Unknown
L $\rightleftarrows$ (βTi)		0		1670	Melting point
(βTi) $\rightleftarrows$ (αTi)		0		882	Allotropic transformation
L $\rightleftarrows$ (Pb)		100		327.502	Melting point

237

Phase Diagrams of Binary Titanium Alloys

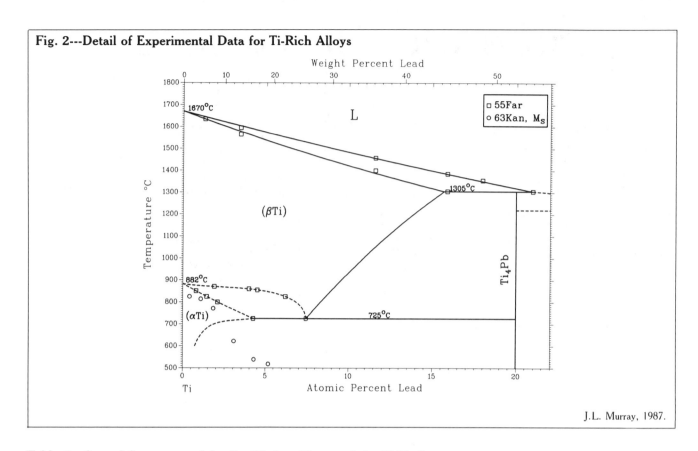

Fig. 2---Detail of Experimental Data for Ti-Rich Alloys

J.L. Murray, 1987.

Table 2 Crystal Structures of the Equilibrium Phases of the Ti-Pb System

Phase	Homogeneity range, at.% Pb	Pearson symbol	Space group	Struktur-bericht designation	Proto-type	Lattice parameters, nm		Reference
						a	c	
(αTi)	0 to 4.2	hP2	$P6_3/mmc$	A3	Mg	...	...	...
(βTi)	0 to ~16	cI2	Im3m	A2	W	...	...	[Pearson2]
Ti$_4$Pb	~20	hP8	$P6_3/mmc$	$D0_{19}$	Ni$_3$Sn	0.5985	0.4846	[51Now]
TiPb$_2$	~33	(a)	...	...	...	...	...	...
(Pb)	100	cF4	Fm3m	A1	Cu	0.49502	...	[Pearson2]

(a) Unknown.

position of Ti$_4$Pb was determined by density and lattice parameter measurements [51Now] and verified by [55Far] and [56Rab]. Its homogeneity range is small.

- An additional intermetallic phase was shown by X-ray diffraction to be neither Ti$_4$Pb nor (Pb), but its composition and structure are otherwise unknown. [55Far] found the stoichiometry Ti$_2$Pb to be the most likely, based on microstructures of as-cast alloys. Analogy with the Ti-Bi system also points to Ti$_2$Pb as a likely stoichiometry. In this assessment, the phase is designated Ti$_2$Pb, but further investigation is clearly needed.

Experimental data of [55Far] are shown in Fig. 2. The liquidus and solidus temperatures were taken to be the temperature at which rounding of the corners of the specimen occurred and the temperature at which the alloy flowed freely, respectively. Liquidus and solidus temperatures were assigned an accuracy of ±10 °C. The temperature of the peritectic reaction L + (βTi) ⇌ Ti$_4$Pb was reported as 1305 ±10 °C, based on melting point data and

the microstructure of a 15.8 at.% Pb alloy equilibrated at 1300 °C. The reaction of L, Ti$_4$Pb, and Ti$_2$Pb was tentatively identified as a eutectic reaction, based on microstructures of as-cast alloys, but its temperature was not determined. The (βTi)/(βTi) + Ti$_4$Pb boundary was based on microstructures of samples equilibrated at temperatures between 950 and 1200 °C.

Figure 2 also shows data on the (αTi)/(βTi) phase boundaries. The eutectoid reaction (βTi) ⇌ (αTi) + Ti$_4$Pb was placed at 725 ± 10 °C and 7.5 ± 0.5 at.% Pb. The temperature was judged to be below 725 °C because of the appearance of [(αTi) + (βTi)] microstructures at that temperature and to be above 700 °C because of the complete transformation of (βTi) retained from a heat treatment at 1100 °C. Placement of the eutectic composition at 7.5 at.% Pb was based on appearance of equal fractions of [(αTi) + (βTi)] and [(βTi) + Ti$_4$Pb] at 7.5 at.% Pb and 775 °C.

[78Fra] studied the kinetics of the eutectoid decomposition of a 5.3 at.% Pb alloy; the results are consistent with a eutectoid temperature of 725 °C. [78Fra] also found the reaction to be very sluggish, requiring two orders of magnitude more time than the analogous reaction at the same temperature in the Ti-Bi system.

The (αTi)/(βTi) phase boundaries are drawn with dotted lines, because thermodynamic considerations suggest these boundaries are in need of modification. First, the sharp bend in the (βTi)/(βTi) + (αTi) boundary is symptomatic of some anomalous behavior of the bcc Gibbs energy function, which, if real, would also be reflected in an anomaly of the (βTi)/(βTi) + Ti$_4$Pb boundary. Second, the (βTi) and (αTi) phase boundaries are shown by [55Far] with opposite curvatures, which is implausible near a retrograde point of the (βTi)/(βTi) + (αTi) boundary. Uncertainty in these boundaries can be attributed to the unknown effect of impurities on the phase equilibria, sluggishness of the eutectoid reaction, and possible difficulties in identifying transformed (βTi) metallographically. Moreover, [55Far] noted that the (αTi) range is particularly ill-defined because varying amounts of Ti$_4$Pb were found in (αTi) alloys.

[60Sat] and [63Kan] measured the start temperature, M$_s$, of the martensitic (βTi) → (αTi) transformation, and these nonequilibrium data are also shown in Fig. 2.

Cited References

*51Now: H. Nowotny and J. Pesl, "Investigation of the Titanium-Lead System," Monatsh. Chem., 82, 344-347 (1951) in German. (Equi Diagram; Experimental)

*55Far: P. Farrar and H. Margolin, "Titanium-Lead System," Trans. AIME, 203, 101-104 (1955). (Equi Diagram; Experimental)

56Rab: B.N. Rabinovich and D.M. Chizhikov, "Study of Mutual Solubility of Titanium and Lead at Temperatures 600, 700, and 800 °C," Izv. Akad. Nauk SSSR, Otd. Tekh. Nauk, (7), 114-117 (1956) in Russian. (Equi Diagram; Experimental)

60Sat: T. Sato, S. Hukai, and Y.C. Huang, "The M$_s$ Points of Binary Titanium Alloys," J. Aust. Inst. Met., 5(2), 149-153 (1960). (Equi Diagram; Experimental)

63Kan: H. Kaneko and Y.C. Huang, "Some Considerations on the Continuous Cooling Diagrams and M$_s$ Points of Titanium Base Alloys," J. Jpn. Inst. Met., 27(8), 403-407 (1963) in Japanese. (Equi Diagram; Experimental)

78Fra: G.W. Franti, J.C. Williams, and H.I. Aaronson, "A Survey of Eutectoid Decomposition in Ten Ti-X Systems," Metall. Trans. A, 9, 1641-1649 (1978). (Equi Diagram; Experimental)

The Pd-Ti (Palladium-Titanium) System

106.42   47.88

By J.L. Murray

Equilibrium Diagram

The major phase diagram investigations of the complete composition range of the Ti-Pd system were conducted by [58Nis], [60Rud], [68Rau], and [72Ere2]. In addition, [65Ros], [68Kra], [73Wil2], and [79Eva] studied restricted composition and temperature regimes.

There is no feature of this system that all investigators agree upon, except that all alloys melt at some temperature. The number of observed intermetallic compounds has ranged from one [58Nis, 60Rud] to eight [68Rau]. The early investigations of [58Nis] and [60Rud] are considered obsolete because of the impurity of the Ti used.

In spite of many apparent difficulties in the literature, the assessed phase diagram (Fig. 1, Table 1) represents a resolution of most of the conflicting reports. The partial phase diagram determinations of [65Ros], [73Wil2] and [79Eva] resolved several discrepancies between [68Rau] and [72Ere2]. Most of the conclusions of [72Ere2] (see also [72Ere1] or [75Sht]) were corroborated by independent work. However, some differences were irreconcilable, particularly for Pd-rich alloys, possibly because of the appearance of metastable precipitates. Discrepancies that call for further experimental work are discussed below.

The following solid phases have been observed in the Ti-Pd system:

- The cph (αTi) solid solution. The cph structure is the equilibrium form of Ti below 882 °C. The solubility of Pd in (αTi) is less than 1 at.%.
- The bcc (βTi) solid solution, based on the equilibrium structure of Ti above 882 °C. The maximum solubility of Pd in (βTi) is about 45 at.%.
- The fcc (Pd) solid solution. The maximum solubility of Ti in (Pd) is about 22 at.%.
- Ti$_2$Pd, a stoichiometric compound that probably has the MoSi$_2$ structure and is stable only below 960 °C.
- Two compounds with a homogeneity range of 47 to 53 at.% Pd about the stoichiometry TiPd. The high-temperature form, βTiPd, has the CsCl structure. βTiPd transforms to αTiPd with the B19 structure at temperatures ranging between 420 and 515 °C.
- The essentially stoichiometric phase Ti$_2$Pd$_3$, with an orthorhombic structure similar to that of VAu$_2$.
- Three variants of the MoSi$_2$ structure, TiPd$_2$, TiPd$_{2-}$, and Ti$_3$Pd$_5$. TiPd$_2$ has a homogeneity range of 65 to 67 at.% Pd, and the polymorphic transformation between its two structural variants occurs at approximately 1280 °C. Ti$_3$Pd$_5$ appears to be a distinct stoichiometric phase.
- The stoichiometric compound TiPd$_3$ with the hexagonal TiNi$_3$ structure.
- The compound γ with the ordered fcc Cu$_3$Au structure. If γ is an equilibrium binary phase, its phase

239

Fig. 1---Assessed Ti-Pd Phase Diagram

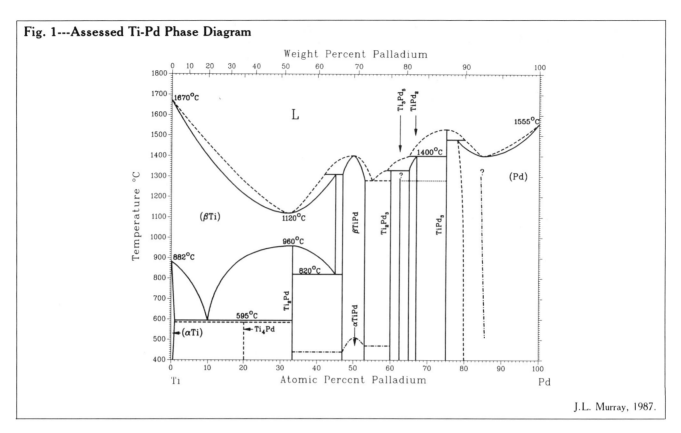

J.L. Murray, 1987.

Table 1 Special Points of the Assessed Ti-Pd Phase Diagram

Reaction	Compositions of the respective phases, at.% Pd			Temperature, °C	Reaction type
$(\beta Ti) \rightleftarrows (\alpha Ti) + Ti_2Pd$	9.8	<1	33.3	595	Eutectoid
$L \rightleftarrows \alpha TiRh + Ti_2Pd_3$	~ 58	~ 53	60	1280	Eutectic
$L + TiPd_2 \rightleftarrows Ti_2Pd_3$	~ 59	65	60	1330	Peritectic
$L + TiPd_3 \rightleftarrows TiPd_2$	~ 65	75	67	1400	Peritectic
$(\beta Ti) \rightleftarrows Ti_2Pd + \alpha TiPd$	~ 45	33.3	~ 47	820	Eutectoid
$L + \alpha TiPd \rightleftarrows (\beta Ti)$	~ 42	~ 47	~ 45	1310	Peritectic
$TiPd_3 + L \rightleftarrows (Pd)$	75	80	~ 78	1480	Peritectic
$(\alpha Ti) + Ti_2Pd \rightleftarrows Ti_4Pd$	1	33.3	20	~ 550 to 585	Peritectoid
$L \rightleftarrows (\beta Ti)$		32		1120	Congruent
$L \rightleftarrows \alpha TiPd$		50		1400	Congruent
$L \rightleftarrows TiPd_3$		75		1530	Congruent
$L \rightleftarrows (Pd)$		~ 85		1400	Congruent
$(\beta Ti) \rightleftarrows Ti_2Pd$		33.3		960	Congruent
$L \rightleftarrows (Pd)$		100		1555	Melting point
$L \rightleftarrows (\beta Ti)$		0		1940	Melting point
$(\beta Ti) \rightleftarrows (\alpha Ti)$		0		882	Allotropic transformation

equilibria with the liquid and with Ti_3Pd are unknown. It may be a metastable phase that forms in the supersaturated (Pd) solid solution because nucleation of equilibrium $TiPd_3$ is prevented. The latter possibility, however, is speculation.

- The stoichiometric compound Ti_4Pd with the $A15$ structure. Ti_4Pd may be a ternary phase that is stabilized by oxygen.
- Similarly, a phase with the Ti_2Ni structure is sometimes found, but is certainly a ternary phase.

- A phase Ti_4Pd_3 of unknown structure has been proposed as a high-temperature phase, but it is not accepted in the assessed diagram.

(βTi)/Liquid Boundaries and Extent of the (βTi) Field.

[68Rau] and [72Ere2] measured the (βTi) solidus by thermal analysis and by the Pirani-Alterthum method, respectively (see Fig. 2). [72Ere2] found a minimum in the solidus at 1120 °C and between 30 and 35 at.% Pd. Liquidus and solidus were shown as terminating at a

Pd-Ti System

Fig. 2---Experimental Liquidus and Solidus Data

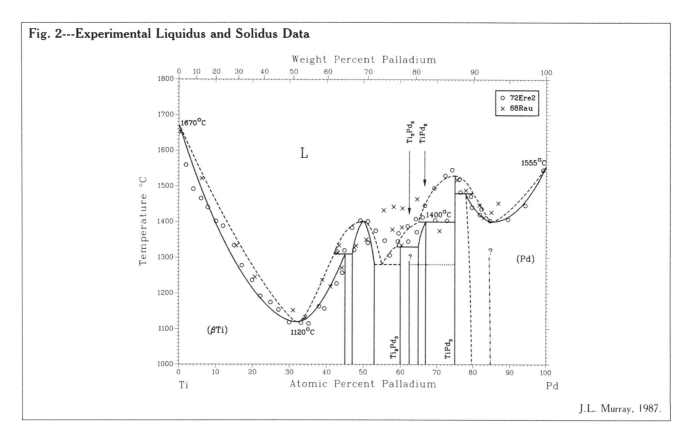

J.L. Murray, 1987.

peritectic reaction, (βTi) + L ⇄ TiPd at approximately 1310 °C. The maximum solubility of Pd in (βTi) was reported to be between 45 and 46 at.% Pd at the peritectic temperature. [68Rau] proposed a different phase diagram; the solidus terminated at a eutectic point at 33 at.% Pd and 1140 °C, with a maximum solubility of Pd in (βTi) of 31 at.% Pd. According to [68Rau], a peritectic reaction occurs (L + TiPd ⇄ Ti₄Pd₃) at 1280 °C. They did not determine the structure of the proposed phase Ti₄Pd₃.

The (βTi) solidus and liquidus are based on the work of [72Ere2]. The congruent melting point is placed at 32 at.% Pd and 1120 °C. The congruent transformation of (βTi) to Ti₂Pd is accepted, and the compound Ti₄Pd₃ is not included in the assessed diagram. The microstructures reproduced by [68Rau] and [72Ere2] are not inconsistent with each other; they do not show evidence of a eutectic reaction and they indicate that reactions occurred in solid single-phase alloys during quenching or annealing. Furthermore, additional information about the extent of the (βTi) field is implicit in the relatively well-known (βTi)/Ti₂Pd boundary. The congruent transformation of (βTi) to Ti₂Pd implies that the (βTi) field extends to compositions larger than 33 at.% Pd and therefore that Ti₄Pd₃ is probably not an equilibrium compound (see below).

[72Ere2] did not observe two-phase (βTi) + βTiPd alloys. The (βTi)/(βTi) + βTiPd boundary was estimated from the composition of (βTi) in equilibrium at the peritectic temperature and of βTiPd in equilibrium at the eutectoid temperature, 820 °C. Additional experiments to locate the (βTi) + βTiPd field would be of theoretical interest because it is possible for a bcc to CsCl ordering transition to be of higher order than first. It is thus possible that [72Ere2] did not find a two-phase field because none exists.

(αTi)/(βTi) Boundaries and Ti-Rich Compounds.

The (βTi) transus was precisely determined by [73Wil2], but further experimental work is needed to identify the compound or compounds in equilibrium with (αTi) and (βTi). That Ti₂Pd occurs as an equilibrium phase was established by X-ray diffraction [60Nev, 62Nev, 65Dwi] and metallography and X-ray diffraction [65Ros, 73Wil2].

Single-phase alloys have been produced at 20 at.% Pd [66Rau, 72Ere2, 68Rau] after long anneals. [66Rau] attributed the A15 structure to Ti₄Pd and stated that it is not stabilized by impurities. [68Rau] claimed that Ti₄Pd is stable to 780 °C, where it decomposes by the peritectoid reaction (βTi) + Ti₂Pd ⇄ Ti₄Pd. [72Ere2] found two thermal arrests near the eutectoid temperature and proposed that both a peritectoid reaction, (αTi) + Ti₂Pd → Ti₄Pd, at 585 °C and a eutectoid reaction, (βTi) ⇄ Ti₂Pd + (αTi), at 615°C occur. [82Wat], on the other hand, did not find Ti₄Pd after prolonged annealing of bulk samples. [71Tis] produced an A15 phase in thin-film diffusion couples after long anneals, and they attributed the stoichiometry Ti₃Pd to this phase rather than Ti₄Pd, as dictated by the ordered structure. Ti₄Pd is provisionally included in the assessed diagram; the uncertainty about its stability as a binary phase is indicated by the dashed lines. The major problem yet to be experimentally resolved is to reconcile the stoichiometry of the observed phase Ti₄Pd with that of the structure attributed to it, "Ti₃Pd".

The (αTi)/(βTi) boundaries terminate in a eutectoid reaction (βTi) ⇄ Ti₂Pd + (αTi) at 595 °C and 9.8 at.% Pd [73Wil2]. [73Wil2] showed that optical microscopy is virtually useless for determining the composition and temperature of the eutectoid point, because in the optical

241

Fig. 3---Experimental Phase Boundary Data for Ti-Rich Alloys

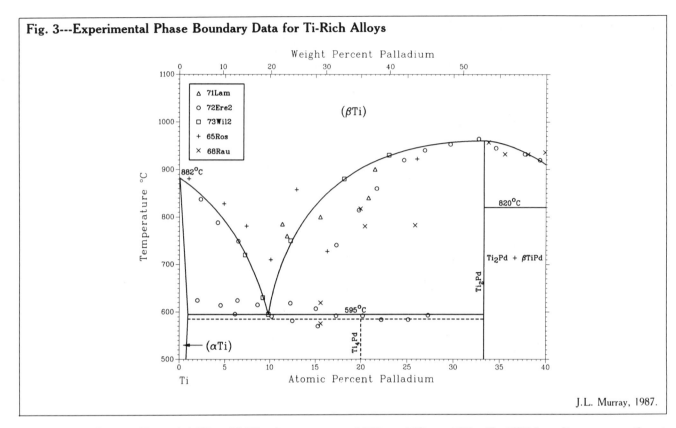

J.L. Murray, 1987.

microscope, (αTi) + (βTi) and (βTi) + Ti₂Pd microstructures appear to be identical. They relied instead on X-ray diffraction and TEM; they also demonstrated the high purity of their alloys. The assessed (αTi)/(βTi) and (βTi)/Ti₂Pd boundaries are also taken from [73Wil2].

Experimental data involving (αTi), (βTi), and Ti₂Pd are shown in Fig. 3. The diffusion data of [71Lam, 73Lam] agree well with the assessed (βTi)/[(βTi) + Ti₂Pd] boundary [73Wil2], except that the microprobe data show more scatter. The thermal analysis data on the eutectoid temperature and the metallographic data on the (βTi) transus [72Ere2] are also consistent with the assessed diagram. The earlier metallographic and X-ray observations of [65Ros] and [68Rau] disagree about not only the stability of the compounds, but also about the (βTi) transus.

The congruent transformation of Ti₂Pd to (βTi) as reported by [72Ere2] is supported by the earlier work of [65Ros], who noted that 25 and 33.3 at.% Pd alloys quenched from 927 and 1038 °C showed only diffraction lines of Ti₂Pd and that the microstructures showed evidence that (βTi) has a tendency to decompose to Ti₂Pd during quenching. The strong curvature of the (βTi)/(βTi) + Ti₂Pd phase boundary as determined by [73Wil2] and [71Lam] is also indirect evidence for the congruent transformation of (βTi) to Ti₂Pd. Finally, the congruent transformation of (βTi) to Ti₂Pd is not inconsistent with the thermal analysis data of [68Rau]; thermal arrests were observed in the composition range 33.3 to about 46 at.% Pd at a maximum temperature of 950 °C. [68Rau], however, interpreted the arrests as evidence of a peritectoid isotherm and a phase boundary (Ti₂Pd + Ti₄Pd₃)/Ti₄Pd₃.

Equiatomic Compounds.

The high-temperature equiatomic compound βTiPd has the ordered CsCl structure [70Don, 72Ere2], the low-temperature form αTiPd has the AuCd B19 structure

[68Rau, 70Don, 72Ere2]. βTiPd melts congruently at 1400 °C [72Ere2, 80Bor]; its homogeneity range is 47 ± 1 to 53 ± 1 at.% Pd and displays no significant variation with temperature [70Don, 72Ere2]. The following values of the homogeneity range have been reported:

Reference	Homogeneity range, at.% Pd	Temperature, °C	Experimental technique
[68Rau]	~ 49 to 51	600 to 1200	Metallography
[72Ere2]	~ 47 to 53	900 to 1200	Metallography, X-ray diffraction
[71Lam]	42.9 to 49	900	Microprobe analysis of annealed diagram couples
[70 Don]	46 to 54	...	X-ray diffraction

Other versions of the phase diagram in this composition range have also been presented. [68Rau] showed a peritectic reaction (L + Ti₃Pd₅ ⇌ αTiPd) at 1350 °C. The diagrams of [58Nis] and [65Ros] showed no equiatomic phase at all. However, re-examination of the as-cast microstructures illustrated in [58Nis] and [68Rau] supports the congruent melt [72Ere2]. That is, there is no discrepancy in the experimental data on this portion of the diagram, only in the interpretation of the data.

[68Rau] and [72Ere2] observed the βTiPd → αTiPd transformation using differential thermal analysis. [68Rau] placed the transformation between 510 and 520°C, and [72Ere2] placed it at approximately 540 °C. It is not known whether these temperatures represent the transformation on heating or on cooling, or an average of the two. [70Don] used high-temperature X-ray diffraction to locate M_s, M_f, A_s, and A_f temperatures (beginning and

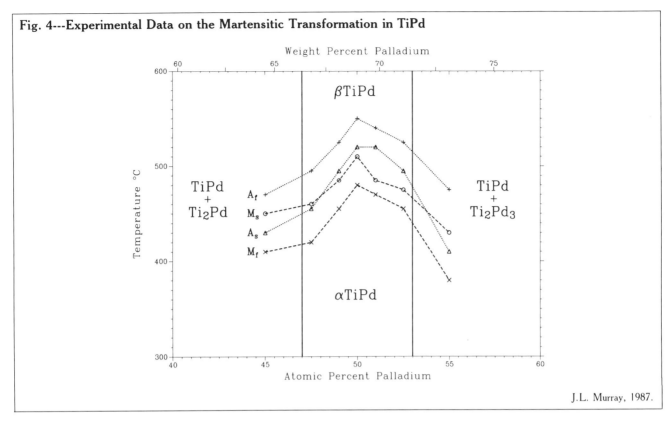

Fig. 4---Experimental Data on the Martensitic Transformation in TiPd

Weight Percent Palladium

βTiPd

TiPd
+
Ti₂Pd

A_f
M_s
A_s
M_f

TiPd
+
Ti₂Pd₃

αTiPd

Temperature °C

Atomic Percent Palladium

J.L. Murray, 1987.

end of the transformation on cooling and heating). The microstructures were typically martensitic, but in a martensitic transformation, M_s must be less than A_s, whereas in TiPd the reverse is true. The average, $(M_s + A_s)/2$, is the best presently available approximation to the locus, T_0, of composition and temperature at which βTiPd and αTiPd have equal Gibbs energies. The equilibrium phase boundaries between βTiPd and αTiPd are not known. M_s, M_f, A_s, and A_f temperatures are plotted in Fig. 4.

Composition Range 60 to 70 at.% Pd.

For this region of the diagram, conflicting reports have been made about both the number and structures of equilibrium compounds and also the types of the three-phase equilibria. All investigators agreed that TiPd₂ is a stable phase with the MoSi₂ structure. [72Ere2] found a stoichiometric compound Ti₂Pd₃ in addition to TiPd₂, and they observed TiPd₂ over the composition range 65 to 67 at.% Pd. Microprobe analysis [71Lam] of annealed diffusion couples also revealed two compounds: TiPd₂ and another compound with a homogeneity range of approximately 57.5 to 59.5 at.% Pd (Ti₂Pd₃). Occurrence as an equilibrium phase is a necessary, but not sufficient, condition for its formation in diffusion couples; however, the appearance of these two phases corroborates the X-ray and metallographic findings of [72Ere2].

[68Rau] claimed that Ti₃Pd₅ rather than Ti₂Pd₃ should appear in the equilibrium diagram. [68Kra] made structure determinations of both Ti₂Pd₃ and Ti₃Pd₅, as well as two forms of TiPd₂. The two TiPd₂ phases and Ti₃Pd₅ are related by having the MoSi₂ structure and its variants. A thermal effect near 1280 °C and a transformation microstructure demonstrated the existence of a polymorphic transformation within the TiPd₂ phase [72Ere2], and it is possible that the phases designated TiPd₂ and TiPd₂· may be low- and high-temperature forms. A single

TiPd₂ phase field is shown in Fig. 1, with a dotted line at the temperature of the polymorphic transformation. In the assessed diagram, Ti₂Pd₃ and TiPd₂ are shown in equilibrium with the liquid. Ti₃Pd₅ is also included in Fig. 1, but equilibria with the other phases of the system are not known and therefore not indicated. Further experimental work should be done to confirm the existence of Ti₃Pd₅ as an equilibrium binary phase.

The disagreements concerning the three-phase equilibria are as follows. [68Rau] proposed a cascade of peritectic reactions between the proposed (βTi) eutectic isotherm at 1280 °C and a peritectic isotherm at 1390 °C (L + TiPd₃ → Ti₃Pd₅). They proposed a peritectoid decomposition of TiPd₂. [72Ere2], on the other hand, proposed a eutectic reaction of Ti₂Pd₃ and αTiPd at 1280 °C. The eutectic isotherm was followed by a cascade of peritectic reactions up to the formation of TiPd₃ at 1400 °C (L + TiPd₂ ⇌ TiPd₃). As-cast microstructures of alloys in this composition range from both [68Rau] and [72Ere2] show evidence of eutectic, which is not compatible with the diagram of [68Rau]. Therefore, the diagram of [72Ere2] is tentatively accepted, except that the Ti₃Pd₅ phase has also been included. In the thermal analysis data, there is some correspondence between the invariant temperatures in the two studies (1280 vs 1310 °C, 1350 vs 1330 °C, 1390 vs 1400 °C, according to [68Rau] and [72Ere2], respectively).

Pd-Rich Alloys (Liquidus, Solidus, Solvus, and TiPd₃).

The melting point of pure Pd is 1555 °C [Melt]. In Pd-rich alloys, contradictions exist in the literature analogous to those discussed in connection with Ti-rich alloys. [72Ere2] reported that (Pd) melts congruently at 1400 °C and that the solubility of Ti in (Pd) is 22 at.%. They included in this composition range one compound, TiPd₃, with the Ti₃Ni structure. [68Rau], on the other hand,

Phase Diagrams of Binary Titanium Alloys

Table 2 Crystal Structures of the Ti-Pd System

Phase	Homogeneity range, at.% Pd		Pearson symbol	Space group	Struktur-bericht designation	Prototype	Reference
(αTi)	0 to	~ 1	hP2	P6₃/mmc	A3	Mg	[Pearson2]
(βTi)	0 to	45	cI2	Im3m	A2	W	[Pearson2]
Ti₄Pd	20		cP8	Pm3n	A15	βW	[66Rau]
Ti₂Pd	33.3		tI6	I4/mmm	C11_b	MoSi₂	[65Dwi]
βTiPd	47 to	53	cP2	Pm3m	B2	CsCl	[70Don]
αTiPd	47 to	53	oP4	Pmma	B19	AuCd	[70Don]
Ti₂Pd₃	60		oC20	Cmcm	...	~Au₂V	[68Kra]
Ti₃Pd₅	62.5		tP8	P4/mmm	~ C11_b	...	[68Kra]
TiPd₂₋	65 to	67	tI6	I4/mmm	C11_b	MoSi₂	[68Kra]
TiPd₂	65 to	67	(a)	...	...	...	[68Kra]
TiPd₃	75		hP16	P6₃/mmc	DO₂₄	TiNi₃	[68Rau]
γ	75 to	84	cP4	P4/mmm	L1₂	AuCu₃	[68Rau]
(Pd)	100 to	85	cF4	Fm3m	A1	Cu	[Pearson2]

(a) Orthorhombic distortion of MoSi₂.

Table 3 Lattice Parameters of Ti-Pd Compounds

Phase	Reference	Composition, at.% Pd	Lattice parameters, nm		
			a	b	c
(Pd)	[79Eva]	100	0.38908	...	...
		98.2	0.38882	...	...
		97.1	0.38875	...	...
		95.2	0.38870	...	...
		91.8	0.38863	...	...
		91.3	0.38861	...	...
		80	0.38846	...	...
Ti₄Pd	[66Rau]	20	0.5005	...	...
Ti₂Pd	[60Nev, 62Nev]	33.3	0.3090	...	1.0054
	[72Ere2]	33.3	0.3095	...	1.005
βTiPd	[65Dwi]	50	0.456	0.281	0.489
	[70Don]	50	0.455	0.278	0.486
αTiPd	[70Don]	50	0.3180(a)	...	...
Ti₂Pd₃	[68Kra]	60	0.461	1.433	0.464
Ti₃Pd₅	[68Kra]	62	0.3263	...	1.1436
TiPd₂₋	[68Kra]	66.7	0.324	...	0.848
TiPd₂	[68Kra]	66.7	0.341	0.307	0.856
TiPd₃	[58Nis]	75	0.5489	...	0.8964
	[79Eva]	75	0.54895	...	0.89739
γ	[68Rau]	80	0.3985	...	...
	[79Eva]	80	0.3885	...	...

(a) Measured at 700 °C.

showed a eutectic reaction at 1450 °C with a more restricted solubility of 14 at.% Ti in (Pd); they included two compounds, Ti₃Pd with the Ti₃Ni structure and TiPd4 (γ) with the Cu₃Au structure.

[72Ere2] determined points on the solidus by the Pirani-Alterthum method; the liquidus has not been determined. According to [72Ere2], (1) the solidus has a minimum at approximately 85 at.% Pd, (2) the maximum solubility of Ti in (Pd) is at least 22 at.%, and (3) TiPd₃ melts congruently at 1450 °C and enters the peritectic reaction TiPd₃ + L ⇄ (Pd). The congruent melting of (Pd) is accepted, on the basis of as-cast microstructures [58Nis, 72Ere2, 68Rau].

[72Ere2] and [58Nis] found that alloys crystallized as solid solutions and did not detect superlattice lines in any of their annealed fcc alloys. [68Rau] observed superlattice lines of the ordered fcc L1₂ structure in an 80 at.% Pd alloy and therefore postulated the existence of a stoichiometric compound TiPd₄.

[79Eva, 80Eva] verified the existence of the ordered fcc structure. [79Eva] examined lattice parameters, microhardness, and magnetic properties of fcc alloys between 80 and 100 at.% Pd. Alloys containing less than 85 at.% Pd exhibited long-range order; the lattice parameters of the ordered alloys lay on a continuous straight line with parameters of the disordered alloys. Alloys containing less than 95 at.% Pd exhibited short-range order. Hardness and magnetic measurements supported the absence of two-phase alloys. Evidence of nonequilibrium behavior was given by [79Eva], who noted that strong fcc lines appeared in the diffraction patterns of 74 at.% alloys. [68Kra] also noted the appearance of fcc lines in alloys

Table 4 Thermodynamic Properties of the Liquid

Composition, at.% Pd	[77Cho]			Calculated, this evaluation		
	G, J/mol	ln activity (Pd)	(Ti)	G_{mix}, J/mol	ln activity (Pd)	(Ti)
85	− 39 200	− 3.27	− 8.99	− 24 624	− 0.37	− 8.46
75	− 56 400	− 3.97	− 7.20	− 35 288	− 0.86	− 6.50
60	− 69 000	− 6.47	− 4.83	− 44 440	− 1.96	− 4.19
50	− 66 900	− 7.62	− 3.23	− 46 169	− 2.96	− 2.96

containing as little as 50 at.% Pd that had not been stress-relieved.

It is clear that further experimental investigation is required before the equilibrium diagram can be drawn with any certainty. The ordered γ phase may be a metastable phase that appears in supersaturated fcc (Pd) during quenching. If that is so, then the equilibrium solubility of Ti in (Pd) is much more restricted than has been observed. The observed solubility, however, is shown in Fig. 1. The dot-dash line represents the composition beyond which ordering was observed in (Pd) [79Eva, 80Eva].

Crystal Structures and Lattice Parameters

The structures of the equilibrium and metastable phases of the Ti-Pd system are summarized in Table 2. Lattice parameter data are listed in Table 3.

[39Lav] identified the structure of Ti₂Pd as Ti₂Ni-type. Ti₂Ni-type structures are often stabilized in Ti-transition metal systems by oxygen, and this structure is almost certainly not correct. [60Nev], [62Nev], and [65Dwi] reported a MoSi₂-type structure. [65Ros] observed that their diffraction patterns did not agree with those of [65Dwi], but they also noted the difficulty of making an unambiguous structure determination and did not rule out the MoSi₂ structure. [73Wil2] found diffraction patterns similar to those of [65Ros], but did not pursue the problem further. [80Bor] verified the MoSi₂ structure of Ti₂Pd, but did not describe details of the experiments. Ti₂Pd is provisionally listed as a MoSi₂-type structure.

The conflict between the observed stoichiometry and structure of Ti₄Pd has been discussed above and is yet to be resolved. The structure of αTiPd was determined by [65Dwi], and structures of both αTiPd and βTiPd were determined by high-temperature X-ray measurements of [70Don]. The X-ray work of [68Kra] is the source for the structure data on Ti₂Pd₃, Ti₃Pd₅, TiPd₂, and TiPd₂.; complete descriptions of the structures of all but the last were given.

Thermodynamics

Experimental Thermodynamic Data.

[77Cho] used the Knudsen cell mass spectrometric technique to measure activities of Ti and Pd in Pd-rich liquid alloys at 1600 °C. Gibbs energies of mixing were derived from the activity data. The results are summarized in Table 4. The excess Gibbs energies implied by the Gibbs energies of mixing are extremely large, even compared to other Ti-platinum metal systems. Based on the mixing Gibbs energy at 50 at.% Pd, a regular solution parameter of −224 400 J/mol is obtained. The experimental data also imply, however, that the excess Gibbs energy

Table 5 Ti-Pd Thermodynamic Properties

Gibbs energy of the pure components [70Kau]

G^0(Ti,L) $= 0$
G^0(Ti,bcc) $= − 16 234 + 8.368 T$
G^0(Ti,cph) $= − 20 585 + 12.134 T$
G^0(Ti,fcc) $= − 17 238 + 12.134 T$

G^0(Pd,L) $= 0$
G^0(Pd,bcc) $= − 9 581 + 11.715 T$
G^0(Pd,cph) $= − 14 184 + 9.623 T$
G^0(Pd,fcc) $= − 15 230 + 8.368 T$

Interaction parameters of solution phases

This evaluation	[70Kau] (calculated)
B(L) $= − 141 500$ C(L) $= 0$	B(L) $= − 142 210$
B(bcc) $= − 115 500$ C(bcc) $= 43 000$	B(bcc) $= − 129 574$
B(fcc) $= − 182 500$ C(fcc) $= − 62 000$	B(bcc) $= − 150 160$
B(cph) $= − 77 500$ C(cph) $= 0$	B(cph) $= − 108 110$

Note: Values are given in J/mol and J/mol · K.

does not follow a regular solution model; the excess functions are even larger in magnitude in Pd-rich alloys.

Calculations of the Diagram.

An incomplete diagram containing the solution phases and TiPd₃ was calculated by [70Kau] in the regular solution approximation, and some more detailed calculations were performed as part of this evaluation. The regular solution approximation is intrinsically unable to reproduce the double extrema observed in the (βTi) and (Pd) liquidus and solidus curves, and therefore, comparison of that calculation to the observed diagram cannot really be made. The present calculations in the subregular solution approximation succeed in reproducing the liquidus and solidus of the solution phases. However, as mentioned above, the observed diagram contains many features that are very implausible from the standpoint of thermodynamic modeling, and if one attempts to calculate the liquidus, these implausibilities are reflected in the Gibbs energies of the compounds. Both the direct thermodynamic quantities and the phase diagram require further experimental work before a reasonable calculation can be performed. Therefore, a calculated diagram is not included in this assessment. The solution phase Gibbs energies according to [70Kau] are compared with the present values in Table 5.

The Gibbs energies of the solution phases are represented as:

$$G(i) = G^0(\text{Ti},i)\,(1 − x) + G^0(\text{Pd},i)\,x + RT[x \ln x + (1 − x) \ln (1 − x)] + B(i)\,x\,(1 − x) + C(i)\,x\,(1 − x)(1 − 2x)$$

where i designates the phase; x is the atomic fraction of Pd, $G^0(i)$ is the Gibbs energies of the pure metals; and $B(i)$ and $C(i)$ are the interaction parameters. The pure component Gibbs energies are taken from [70Kau].

Cited References

39Lav: F. Laves and H.J. Wallbaum, "Crystal Chemistry of Titanium Alloys," *Naturwissenschaften, 27,* 674-675 (1939) in German. (Crys Structure; Experimental)

*58Nis: H. Hishimura and T. Hiramatsu, "On the Corrosion Resistance of Titanium Alloys. The Equilibrium Diagram of the Titanium-Palladium System," *Nip. Kinzoku Gakkaishi, 22,* 88-91 (1958) in Japanese. (Equi Diagram; Experimental)

60Nev: M.V. Nevitt, "A_2B Phases," USAEC Report ANL-6330, 164-165 (1960). (Crys Structure; Experimental)

60Rud: A.A. Rudnitskii and N.A. Birun, "Phase Diagram of the Titanium-Palladium System," *Zh. Neorg. Khim., 5*(11), 2414-2421 (1960) in Russian; TR: *Russ. J. Inorg. Chem., 5*(11), 1169-1173 (1960). (Equi Diagram; Experimental)

62Nev: M.V. Nevitt and J.W. Downey, "A Family of Intermediate Phases Having the Si₂Mo-Type Structure," *Trans. AIME, 224*(2), 195-196 (1962). (Crys Structure; Experimental)

65Dwi: A.E. Dwight, R.A. Conner, and J.W. Downey, "Equiatomic Compounds of the Transition and Lanthanide Elements with Rh, Ir, Ni and Pt," *Acta Crystallogr., 18,* 835-839 (1965). (Crys Structure; Experimental)

*65Ros: H.W. Rosenberg and D.B. Hunter, "The Titanium-Rich Portion of the Ti-Pd Phase Diagram," *Trans. AIME, 233*(4), 681-685 (1965). (Equi Diagram; Experimental)

66Rau: E. Raub and E. Roeschel, "On Some New A15 Phases," *Naturwissenschaften, 53*(1), 17 (1966) in German. (Crys Structure; Experimental)

*68Kra: P. Krautwasser, S. Bhan, and K. Schubert, "Structural Investigations in the Systems Ti-Pd and Ti-Pt," *Z. Metallkd., 59*(9), 724-729 (1968) in German. (Equi Diagram, Crys Structure; Experimental)

68Rau: E. Raub and E. Roeschel, "The Titanium-Palladium System," *Z. Metallkd., 59*(2), 112-114 (1968) in German. (Equi Diagram, Crys Structure; Experimental)

*70Don: H.C. Donkersloot and J.H.N. van Vucht, "Martensitic Transformations in Gold-Titanium, Palladium-Titanium and Platium-Titanium Alloys near the Equiatomic Composition," *J. Less-Common Met., 20,* 83-91 (1970). (Equi Diagram, Crys Structure; Experimental)

70Kau: L. Kaufman and H. Bernstein, *Computer Calculation of Phase Diagrams,* Academic Press, New York (1970). (Thermo; Review)

71Lam: P. Lamparter, S. Steeb, and A. Gukelberger, "Microprobe Measurements of Diffusion in Pd-V, Pd-Ti, and Metal-Ceramic Systems," *High Temp. High Pressures, 3,* 727-740 (1971). (Equi Diagram; Experimental)

71Tis: T.C. Tisone and J. Brobek, "Diffusion in Thin Films Ti-Au, Ti-Pd, and Ti-Pt Couples," *J. Vacuum Sci. Technol., 9*(1), 271-275 (1971). (Equi Diagram; Experimental)

72Ere1: V.N. Eremenko and T.D. Shtepa, "Phase Equilibria in the Binary Systems of Titanium with Ruthenium, Osmium, Rhodium, Iridium, and Palladium," *Colloq. Int. CNRS,* (205), 403-413 (1972) in German. (Equi Diagram, Crys Structure; Experimental)

72Ere2: V.N. Eremenko and T.D. Shtepa, "Phase Diagram of the System Titanium-Palladium," *Porosh. Metall.,* (3), 75-81 (1972) in Russian; TR: *Sov. Powder Metall.,* (3), 228-233 (1972). (Equi Diagram; Experimental)

73Lam: P. Lamparter, T. Krabichler, and S. Steeb, "Diffusion Experiments in the System Palladium-Titanium Using Microprobe Analysis," *Z. Metallkd., 64*(10), 720-724 (1973) in German. (Equi Diagram; Experimental)

73Wil1: J.C. Williams, "Kinetics and Phase Transformations," Sci. Technol. Appl. Titanium, Proc. Int. Conf., R.I. Jaffee, Ed., 1433 (1973). (Meta Phases; Review)

*73Wil2: J.C. Williams, H.I. Aaronson, and B.S. Hickman, "The Eutectoid Region of the Ti-Pd System," *Metall. Trans., 4*(4), 1181-1183 (1973). (Equi Diagram; Experimental)

75Sht: T.D. Shtepa, "Interaction of Titanium with Platinum-Group Metals," *Fiz. Khim. Kondens. Faz. Sverkhtverd. Mater.,* 175-191 (1975) in Russian. (Equi Diagram, Crys Structure; Experimental)

*77Cho: U.V. Choudary, K.A. Gingerich, and L.R. Cornwell, "Mass-Spectrometric Investigation of the High Temperature Thermodynamic Stabilities of Pd-Ti Alloys," *Metall. Trans. A,* 8, 1487-1491 (1977). (Thermo; Theory)

*79Eva: J. Evans and I.R. Harris, "An Investigation of Some Palladium-Titanium and Some-Palladium-Titanium-Hydrogen Alloys," *J. Less-Common Met., 64,* P39-P57 (1979). (Equi Diagram, Crys Structure; Experimental)

80Bor: N.G. Boriskina and E.M. Kenina, "Phase Equilibria in the Ti-TiPd-TiNi System Alloys," *Titanium '80, Sci. and Techn., Proc. Fourth Int. Conf. on Titanium,* H. Kimura and O. Izumi, Ed., 2917-2927 (1980). (Equi Diagram; Experimental)

80Eva: J. Evans, I.R. Harris, and P.F. Martin, "Some Further Observations on the System Palladium-Titanium and Palladium-Titanium-Hydrogen," *J. Less-Common Met., 75,* P49-P53 (1980). (Equi Diagram, Crys Structure; Experimental)

82Wat: R. Waterstrat, private communication (1982).

The Pt-Ti (Platinum-Titanium) System

195.08   47.88

By J.L. Murray

Equilibrium Diagram

[57Nis] performed the only comprehensive phase diagram study of this system. All the information concerning the liquid phase relations is from [57Nis]. Additional work has been done on the structures and composition ranges of the intermetallic compounds, and four compounds were not accounted for by [57Nis]. Thus, significant modifications of the diagram of [57Nis] are required, but the additional information provided by subsequent studies is not sufficient to determine a complete new phase diagram. The assessed phase diagram is shown in Fig. 1, and special points are given in Table 1.

The observed solid phases of the Ti-Pt system are:

- The bcc (βTi) solid solution. Pure Ti has the bcc structure above 882 °C. The maximum solubility of Pt in (βTi) is approximately 10 at.%.
- The cph (αTi) solid solution. Pure Ti has the cph structure below 882 °C. The maximum solubility of Pt in (αTi) is approximately 0.5 at.%.
- The fcc (Pt) solid solution. According to [57Nis], the solubility of Ti in (Pt) is between 18 and 20 at.% at 1000 °C. However, subsequent observations of a phase TiPt$_8$ [65Pie] and extended composition range of γ revealed that the [57Nis] solvus was incorrect.

- The compound Ti$_3$Pt, with the A15 structure. Ti$_3$Pt has a homogeneity range of 22 ± 2 to 29 ± 2 at.% Pt at 500 °C.
- Two equiatomic compounds—the high temperature form, βTiPt, has the CsCl structure, and the low-temperature form, αTiPt, has the B19 (AuCd) structure. The homogeneity range of βTiPt at 600 °C is estimated to be 46 to 54 at.% Pt.
- The stoichiometric compound Ti$_3$Pt$_5$. The phase relations among Ti$_3$Pt$_5$, the liquid, and the neighboring compounds have not been determined.
- The compound TiPt$_{3-}$ with the TiNi$_3$ structure. The composition of TiPt$_{3-}$ is slightly to the Ti-rich side of stoichiometry.
- The compound γ with the ordered fcc Cu$_3$Au structure. γ occurs over a composition range of at least 75 to 80 at.% Pt. Its equilibrium relations with TiPt$_{3-}$, TiPt$_8$, and (Pt) require further experimental study.
- TiPt$_8$ with the ordered fcc MoNi$_4$ structure. TiPt$_8$ has the same structure as TiCu$_4$ and TiAu$_4$ and disorders at approximately 1080 ± 20 °C.

L/(βTi) Boundaries.

[57Nis] reported a eutectic reaction L $\rightleftarrows$ (βTi) + Ti$_3$Pt, based on optical metallography and differential thermal analysis. [57Nis] estimated the maximum solubility of Pt in (βTi) at the eutectic temperature to be 10

Fig. 1---Assessed Ti-Pt Phase Diagram

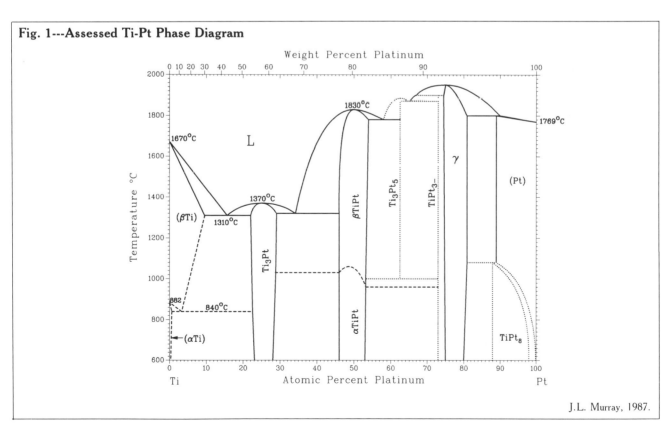

J.L. Murray, 1987.

Table 1 Special Points of the Assessed Ti-Pt Phase Diagram

Reaction	Compositions of the respective phases, at.% Pt			Temperature, °C	Reaction type
Observed reactions [57Nis]					
L ⇄ (βTi) + Ti₃Pt	~ 16	~ 10	25	1310	Eutectic
(βTi) ⇄ (αTi) + Ti₃Pt	~ 3	~ 0.5	25	840	Eutectoid
L ⇄ Ti₃Pt + βTiPt	~ 34	25	45	1320	Eutectic
L ⇄ βTiPt + γ	~ 58	~ 54	75	1780(a)	Eutectic
γ ⇄ L + (Pt)	75	~ 90	~ 80	~ 1800(b)	Peritectic
L ⇄ Ti₃Pt		25		1370	Congruent
L ⇄ βTiPt		50		1830	Congruent
L ⇄ γ		75		1950	Congruent
L ⇄ (βTi)		0		1670	Melting point
(βTi) ⇄ (αTi)		0		882	Allotropic transformation
L ⇄ (Pt)		100		1769	Melting point
Reactions not yet investigated					
L ⇄ Ti₃Pt₅ + αTiPt	...	...	...	...	Eutectic?
L ⇄ Ti₃Pt₅ + TiPt₃₋	...	...	...	...	Eutectic?
γ + L ⇄ TiPt₃₋	...	...	...	...	Peritectic?
Ti₃Pt₅ ⇄ βTiPt + TiPt₃₋	...	...	...	...	Eutectoid?
γ—(Pt)—TiPt₈	...	...	...	...	Metastable?

(a) If Ti₃Pt₅ is formed by an interaction with the liquid phase, then this reaction must be replaced, possibly by the two eutectic reactions listed below.
(b) If γ is to be included as an equilibrium phase, then this must be replaced, possibly by the peritectic reaction listed below.

at.%. From the dependence of the superconducting transition temperature on composition, [76Jun] also concluded that the maximum solubility is between 10 and 15 at.%.

The extent of the (βTi) field has been the subject of controversy in several related systems (Ti-Rh and Ti-Pd, for example). For Ti-Pd and Ti-Rh, evidence has been given, on the one hand, for a eutectic reaction and restricted solubility of Pt-group metals in (βTi) and, on the other hand, for a congruent melting point and extended composition range of (βTi). The micrographs of [57Nis] suggest that re-examination of the Ti-Pt system may also bring similar contradictions to light; the eutectic was absent from a 7.6 at.% Pt as-cast alloy, which is expected to contain eutectic, based on the phase diagram of Fig. 1.

(αTi)/(βTi) Phase Boundaries.

The only information available on the (αTi)/(βTi) equilibria comes from [57Nis], who proposed a eutectoid reaction (βTi) ⇄ (αTi) + Ti₃Pt at 840 °C. The proposed eutectoid composition and temperature were based on optical metallography and differential thermal analysis studies.

In the Ti-Pd system, the eutectoid point proposed by the same authors deviated significantly from the evaluated diagram (735 °C, compared to 595 °C, respectively). Several possible sources of error can be proposed: (1) (αTi) + (βTi) microstructures were misinterpreted as eutectoid microstructures (a TEM study may be required to distinguish between (αTi) + Ti₃Pt and (βTi) + Ti₃Pt microstructures); and (2) the impurity content of the alloys may have a significant effect on the homogeneity range of (βTi) and the eutectoid temperature.

The evaluated diagram (Fig. 1) shows a eutectoid reaction as proposed by [57Nis]. The dashed lines indicate that further experimental work is suggested.

Ti₃Pt.

[57Nis] presented differential thermal analysis and optical metallographic evidence that Ti₃Pt forms from the melt by a congruent transformation. [68Kra] believed Ti₃Pt to be a ternary compound stabilized by oxygen, but the existence of Ti₃Pt as a binary equilibrium phase was verified by [68Reu], who probably obtained purer alloys. [76Jun] investigated the homogeneity range of Ti₃Pt as part of a superconductivity study.

Lattice parameters of alloys annealed at 500 °C implied a range of 22 ± 2 to 29 ± 2 at.% Pt for the single-phase Ti₃Pt field [76Jun]. In the assessed diagram, this estimate is used for the maximum composition range.

Equiatomic Compounds.

The fact that the equiatomic compound has two allotropes was not originally recognized [57Nis, 64Ram, 65Dwi, 68Kra]; the low-temperature B19 form, αTiPt, was designated TiPt. [68Kra] noted a maximum in the magnetic susceptibility of a 60 at.% Pt alloy at 1040 °C; this probably corresponded to the transformation of βTiPt to αTiPt.

[70Don] studied the CsCl to B19 structural transformation using high-temperature X-ray diffraction. The equilibrium phase relations were not determined. They measured the start and finish temperatures for the transformations on heating and cooling (A$_s$, A$_f$, M$_s$, and M$_f$, respectively, see Table 2). The average, (A$_s$ + M$_s$)/2, is the best available estimate of the temperature at which the Gibbs energies of αTiPt and βTiPt are equal. This quantity is plotted in Fig. 1.

On the Pt-rich side, [70Don] also noted that above 600 °C a 55 at.% Pt alloy showed precipitation of TiPt₃₋, and a 52.5 at.% Pt alloy did not. Thus, in Fig. 1, the single-phase region is shown as terminating at 54 at.% Pt. Constitution data were not given on the Ti-rich side, and the [57Nis] estimate of 46 at.% Pt for the beginning of the single-phase region has been used.

Composition Range 60 to 75 at.% Pt (Ti₃Pt₅ and TiPt₃₋)

[39Lav] and [64Ram] mentioned a phase Ti₂Pt₃, but did not determine its structure. [68Kra] found that the

Table 2   Martensitic βTiPt/αTiPt Transformation [70Don]

Composition, at.% Pt	Start temperatures, °C		$T_0 \cong (M_s + A_s)/2$, °C
	M_s	A_s	
45.0	1035	1025	1030
47.5	1060	1035	1052
50.0	1070	1040	1055
52.5	1000	995	998

stoichiometry of this phase is more precisely Ti_3Pt_5 and determined its structure. Both [39Lav] and [57Nis] recognized the existence of γ, calling it a stoichiometric compound $TiPt_3$. [64Ram], [64Sch] and [68Kra] also found the hexagonal $TiNi_3$ structure in slightly Ti-rich off-stoichiometric alloys. This phase was also observed by [73Rau] in ternary Ti-Pt-C alloys. The composition of $TiPt_{3-}$ has not been investigated further. A rough estimate of 73 at.% Pt was used to draw the assessed diagram.

[69Sin] found three phases present in a 73 at.% Pt alloy annealed at 900 °C—$TiPt_{3-}$, γ, and αTiPt. The presence of αTiPt rather than Ti_3Pt_5 suggests the possibility of a eutectoid reaction $Ti_3Pt_5 \rightarrow \alpha TiPt + TiPt_{3-}$ above 900 °C.

[57Nis], accounting only for the phases TiPt and γ, proposed that there is a eutectic reaction, $L \rightleftarrows TiPt + \gamma$, and that γ melts congruently at 1950 °C. The phase relations involving Ti_3Pt_5 and $TiPt_{3-}$ have not been investigated experimentally. In the published micrographs [57Nis], a eutectic was present in as-cast alloys containing less than 66 at.% Pt, but a 73 at.% Pt alloy showed evidence of a peritectic reaction. Therefore, in Fig. 1, the Pt-rich compounds are sketched in the following manner: the eutectic reaction at 1780 °C with eutectic composition 58 at.% Pt is identified as the reaction $L \rightleftarrows \alpha TiPt + Ti_3Pt_5$ rather than $L \rightleftarrows \alpha TiPt + \gamma$. A congruent melt is shown for Ti_3Pt_5 and another eutectic reaction $L \rightleftarrows Ti_3Pt_5 + TiPt_{3-}$. The temperatures of the hypothetical reactions were chosen arbitrarily, and this region of Fig. 1 should be viewed as a sketch of the possible topology of the phase diagram.

Pt-Rich Alloys (γ, TiPt$_8$, Liquid).

The melting point of pure Pt is 1769 °C [Melt]. The addition of Ti raises the liquidus and solidus, which terminate at a peritectic equilibrium with γ at about

1800 °C [57Nis]. [57Nis] estimated the composition of the liquidus as 90 at.% Pt at the peritectic temperature based on their thermal analysis data. They estimated the composition of (Pt) at the peritectic temperature to be 81 ± 2 at.% Pt based on the solvus data at 1000 °C.

According to [57Nis], γ is a stoichiometric compound. According to [73Rau], [76Jac], and [76Mes], however, γ has a wide composition range to the Pt-rich side of stoichiometry, and the maximum solubility of Ti in the disordered (Pt) solid solution is only between 11 and 13 at.%.

Moreover, [65Pie] discovered an additional ordered phase, $TiPt_8$, with a structure similar to that of $TiAu_4$. $TiPt_8$ was also said to occur over an (unspecified) homogeneity range. [68Kra], [73Rau], and [76Mes] verified the existence of $TiPt_8$. [68Kra] observed the superlattice lines of $TiPt_8$ to disappear above 1100 °C.

[76Mes] made X-ray measurements in support of an emf study of solid alloys and combined his and other data [65Dwi, 73Rau, 76Jac] to propose a phase diagram for Pt-rich alloys. Alloys used by [76Mes] and [76Jac] were saturated with oxygen. Alloys used by [73Rau] were saturated with carbon; the carbon contents were as high as 5 to 10 at.%. [76Mes] argued that because the oxygen content was low and the carbon content did not alter lattice parameters neither oxygen nor carbon significantly altered the binary phase equilibria. In agreement with [68Kra], [76Mes] found that $TiPt_8$ disordered at 1080 ± 20 °C. Single-phase $TiPt_8$ was observed in the composition range 89 to 98 at.% Pt. Single-phase γ was observed between 75 and 84 at.% Pt. [76Mes] drew a peritectoid isotherm, $(Pt) + \gamma \rightleftarrows TiPt_8$, at 1080 °C, between approximately 86 and 89 at.% Pt. The two-phase regions (Pt) + γ and γ + $TiPt_8$ were shown as nearly vertical. The two-phase region (Pt) + $TiPt_8$ was shown as narrow and approximately horizontal and extending from 89 to 98 at.% Pt; therefore it extrapolates to an order/disorder transformation in pure Pt.

Because there is no transformation in pure Pt, it is considered that the two-phase field (Pt) + $TiPt_8$ proposed by [76Mes] is evidence that the phase diagram is a ternary section that shows significant influence of impurities and does not represent binary equilibrium. Therefore, only the following information has been used to sketch the approximate phase equilibria in Fig. 1.

The liquid end of the peritectic isotherm is placed at 90 at.% Pt, in agreement with the [57Nis] liquidus data.

Table 3   Crystal Structures of Observed Phases of the Ti-Pt System

Phase	Homogeneity range, at.% Pt		Pearson symbol	Space group	Strukturbericht designation	Prototype	Reference
(αTi)	0	to 0.5	$hP2$	$P6_3/mmc$	$A3$	Mg	[Pearson2]
(βTi)	0	to 10	$cI2$	$Im3m$	$A2$	W	[Pearson2]
Ti_3Pt	22	to 29	$cP8$	$Pm3n$	$A15$	βW	[68Reu]
Ti_2Pt	(a)		$cF96$	$Fd3m$	$E9_3$	$NiTi_2$	[39Lav]
βTiPt	46	to 54	$oP2$	$Pmma$	$B2$	CsCl	[70Don]
αTiPt	46	to 54	$oP4$	$Ibam$	$B19$	AuCd	[70Don]
Ti_3Pt_5	62.5		$oI32$	$Ibam$	...	Au_5Zn_2Ga	[68Kra]
$TiPt_{3-}$	<75		$hP16$	$P6_3/mmc$	$D0_{24}$	Ni_3Ti	[68Kra]
γ	75	to 81	$tP4$	$Pm3m$	$L1_2$	$AuCu_3$	[76Mes]
$TiPt_8$	89	to 98	$tI18$	$I4/m$	$D1_a$	$MoNi_4$	[76Mes]
(Pt)	81	to 100	$cF4$	$Fm3m$	$A1$	Cu	[Pearson2]

(a) Observed by [39Lav], but it is a ternary phase of the Ti-Pt-O system according to [60Nev].

Phase Diagrams of Binary Titanium Alloys

Table 4 Lattice Parameters of Intermetallic Compounds

Phase	Reference	Composition, at.% Pt	Lattice parameters, nm a	b	c
Ti₃Pt	[52Duw]	25	0.5033	...	...
	[57Nis]	25	0.5024(a)	...	...
	[68Pie]	25	0.50311	...	...
		25	0.50308	...	...
	[68Reu]	25	0.50309	...	...
		25	0.50327	...	...
	[76Jun]	23	0.5028	...	...
		25	0.50335	...	...
		27	0.5034	...	...
βTiPt	[70Don]	50	0.3172(a)	...	...
αTiPt	[65Dwi]	50	0.4592	0.2761	0.4838
	[64Sch, 64Ram]	50	0.459	0.276	0.482
	[70Don]	50	0.455	0.273	0.479
Ti₃Pt₅	[68Kra]	62.5	0.8312	...	0.3897
TiPt₃₋	[64Sch, 64Ram]	75	0.552	...	0.9029
	[69Sin]	73(b)	0.5520	...	0.9019
γ	[57Nis]	75	0.3916(c)	...	...
	[43Wal]	75	0.3898	...	...
TiPt₈	[65Pie]	89	0.8312	...	0.3897

(a) Measured at 1100 °C. (b) Three phases present. (c) The unit for these lattice parameters was given as Angstroms, but may actually be kX.

According to subsequent work, the two-phase (Pt) + L region should be considerably narrower than the 9 at.% proposed by [57Nis]. This qualitative feature is supported by thermodynamic calculation and does not conflict with [57Nis]. Therefore, the peritectic composition is placed at 89 at.% Pt. The single-phase γ field is shown as extending to 81 at.% Pt; the two-phase γ + (Pt) field is drawn as vertical. The order-disorder transformation at 1080 °C is accepted, also observed in binary alloys by [68Kra]. The phase boundaries of TiPt₈ are sketched to conform to the peritectoid reaction proposed by [76Mes], but with the shape of the two-phase region (Pt) + TiPt₈ in better accord with thermodynamic principles. The Pt-rich region should be viewed as a sketch rather than a quantitative representation of the phase relations. A congruent order-disorder transformation TiPt₈ ⇄ (Pt) and a eutectoid reaction with γ are also consistent with the experimental data.

Metastable Phases

In Ti-rich alloys, the cph phase (α'Ti) can form martensitically from (βTi) during quenching. [60Sat] measured the temperature of the martensite transformation on cooling (M_s). M_s decreased from about 882 °C in pure Ti to about 750 °C at 2.5 at.% Pt. Between 2.5 and 5 at.% Pt, M_s remained constant at about 750 °C. M_s is monotonically decreasing with concentration for almost all additions to Ti; the flattening of the M_s curve is an unusual feature that the Ti-Pt system has in common with Ti-Al, Ti-Sn, and Ti-Ag.

Crystal Structures and Lattice Parameters

The crystal structures of all the observed phases in the Ti-Pt system are given in Table 3. Lattice parameters are listed in Table 4.

Thermodynamics

Experimental Thermodynamic Data.

[76Mes] and [81Wor] made Ti activity measurements in Pt-rich alloys, in the temperature range 1150 to 1400 K, and also reported unpublished work of [76Jac]. They reported Gibbs energies of formation for the compounds TiPt₈ and γ at 1300 K as −34.3 and −74.6 kJ/mol, respectively. Thermodynamic data were supplemented by X-ray identification of phases present in the alloys.

The alloys used by both [76Mes] and [76Jac] were saturated with oxygen. Arguments were presented that, because the solubility of oxygen in the Ti-Pt alloys is less than 0.4 at.%, the results represent the equilibrium binary Ti-Pt system. However, as discussed above, the Pt-rich region of the phase diagram proposed by [76Mes] suggests that even this small oxygen content has a significant influence on the Ti-Pt phase equilibria. Therefore, data on oxygen-saturated alloys, strictly speaking, pertain to the ternary Ti-Pt-O system and only approximately to the binary Ti-Pt system.

Thermodynamic Calculations.

In this system, the scarcity of both phase diagram and thermochemical data precludes the use of thermodynamic calculations as a tool for evaluating the phase diagram. However, the present thermodynamic calculations comprise a rough approximation to the excess Gibbs energies, primarily of the solution phases (see Fig. 2 and Table 5). The agreement among the calculated diagram, the rough estimate of the phase diagram, and the thermodynamic measurements of [76Mes] is good. Considering the gaps in the present knowledge of the phase diagram, this agreement is partly fortuitous; however, it does indicate that the parameters of the present assessment, or those of [70Kau], can be used for qualitative, predictive extrapolation of the binary diagram to ternary systems.

250

Pt-Ti System

Fig. 2---Regular Solution Calculation of Ti-Pt Phase Diagram

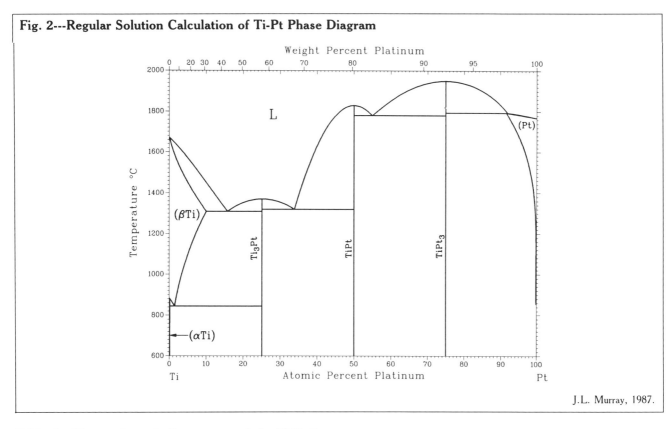

J.L. Murray, 1987.

Table 5 Thermodynamic Parameters of the Ti-Pt System

Lattice stability parameters of Ti and Pd [70Kau]

$G^0(\text{Ti,L}) = 0$
$G^0(\text{Ti,bcc}) = -16\ 234 + 8.368\ T$
$G^0(\text{Ti,cph}) = -20\ 585 + 12.134\ T$
$G(\text{Ti,fcc}) = -17\ 238 + 12.134\ T$

$G^0(\text{Pt,L}) = 0$
$G^0(\text{Pt,fcc}) = -17\ 071 + 8.368\ T$
$G^0(\text{Pt,bcc}) = -11\ 422 + 11.715\ T$
$G^0(\text{Pt,cph}) = -16\ 025 + 9.623\ T$

Regular solution parameters of the Ti-Pt solution phases

This evaluation	[70Kau]
$B(\text{L}) = -133\ 150$	$B(\text{L}) = -133\ 150$
$B(\text{bcc}) = -118\ 000$	$B(\text{bcc}) = -120\ 897$
$B(\text{cph}) = -80\ 000$	$B(\text{cph}) = -99\ 433$
$B(\text{fcc}) = -144\ 500$	$B(\text{fcc}) = -141\ 482$

Compound Gibbs energies

This evaluation	[76Mes]
$G(\text{Ti}_3\text{Pt}) = -66\ 849 + 20.82\ T$	...
$G(\text{TiPt}) = -50\ 747 + 2.54\ T$	...
$G(\gamma) = -113\ 988 + 35.37\ T$	$G(\gamma) = -101\ 535 + 17.69\ T$

Note: Values are given in J/mol and J/mol · K.

Because it is not known how to draw the equilibria involving Ti₃Pt₅, TiPt₃., and the liquid, or TiPt₃., γ, and (Pt), the data reported by [57Nis] was used as the experimental input for the phase diagram calculations. Phases that were not accounted for by [57Nis] are also not accounted for in the thermodynamic calculations.

In the present calculations the solution phases—(αTi), (βTi), (Pt), and the liquid—are represented as regular solutions with Gibbs energy functions:

$$G(\text{i}) = G^0(\text{Ti},i)(1-x) + G^0(\text{Pt},i)x + RT[x \ln x + (1-x)\ln(1-x)] + B(i)x(1-x)$$

251

where i designates the phase; x is the atomic fraction of Pt; G^0 is the Gibbs energies of the pure metals; and $B(i)$ is the regular solution parameters. Pure component Gibbs energies are from [70Kau]. The compounds are modelled as stoichiometric compounds. The line compound approximation allows comparison of the calculated and observed liquidus and heats of formation (which do not depend significantly on the compound width) using the fewest parameters.

When small adjustments were made to the [70Kau] regular solution parameters, the calculated (βTi)/liquid and (αTi)/(βTi) phase boundaries agreed with the data of [57Nis]. On the Pt-rich side, [57Nis] proposed that at the peritectic temperature the (Pt)/liquid two-phase field is 10 to 11 at.% wide. The calculated two-phase field, on the other hand, is less than 1 at.% wide and was allowed to deviate from the estimate of [57Nis].

After the phase boundaries of the solution phases were fitted to the [57Nis] diagram, the Gibbs energies of the compounds were calculated. The Gibbs energy of Ti_3Pt was determined by the (βTi) eutectic temperature and the congruent melting point. The calculated eutectoid temperature was then 845 °C, compared to the observed 840 °C. The remaining calculated three-phase equilibrium temperatures agree with the experimental values. The calculated composition of the eutectic involving TiPt and γ is 54 at.% compared to the observed 58 at.%. The calculated composition of the eutectic involving Ti_3Pt and TiPt is 34 at.%, in agreement with the observed value.

The calculated (Pt) solvus disagrees with that observed by [57Nis]; the calculated solubility of Ti in (Pt) decreases rapidly from its maximum value at the peritectic temperature. This discrepancy occurs in other Ti-Pt-metal systems. It is in accord with the hypothesis that Pt-rich compounds may be metastable phases formed from the supersaturated solid solution.

Cited References

39Lav: F. Laves and H.J. Wallbaum, "Crystal Chemistry of Titanium Alloys," *Naturwissenschaften, 27*, 674-675 (1939) in German. (Crys Structure; Experimental)

43Wal: H.J. Wallbaum, "On the Alloy Chemistry of Transition Metals," *Naturwissenschaften, 31*, 91-92 (1943) in German. (Crys Structure; Experimental)

52Duw: P. Duwez and C.B. Jordan, "The Crystal Structure of Ti_3Au and Ti_3Pt," *Acta Crystallogr., 5*, 213-214 (1952) in German. (Crys Structure; Experimental)

57Nis: H. Nishimura and T. Hiramatsu, "On the Corrosion Resistance of Titanium Alloys. The Equilibrium Diagram of the Titanium-Platinum System," *Nippon Kinzoku. Gakkaishi, 21*, 469-473 (1957). (Equi Diagram; Experimental)

58Nev: M.V. Nevitt, "Atomic Size Effects in Cr_3O-Type Structure," *Trans. AIME, 212*, 350-355 (1958). (Crys Structure; Experimental)

60Nev: M.V. Nevitt, J.W. Downey, and R.A. Morris, "A Further Study of Ti_2Ni-Type Phases Containing Titanium, Zirconium or Hafnium," *Trans. AIME, 218*, 1019-1023 (1960). (Equi Diagram; Experimental)

60Sat: T. Sato, S. Hukai, and Y.C. Huang, "The M_s Points of Binary Titanium Alloys," *J. Aust. Inst. Met., 5*(2), 149-153 (1960). (Meta Phases; Experimental)

64Ram: A. Raman and K. Schubert, "Structural Investigations on Some Alloy Systems Homologous and Quasi-homologous to T^4-T^9," *Z. Metallkd., 55*(11), 704-710 (1964) in German. (Crys Structure;Experimental)

64Sch: K. Schubert, A. Raman, and W. Rossteutscher, "Structure Data on Intermetallic Phases (10)," *Naturwissenschaften, 51*(2), 506-507 (1964) in German. (Crys Structure; Experimental)

65Dwi: A.E. Dwight, R.A. Conner, and J.W. Downey, "Equiatomic Compounds of the Transition and Lanthanide Elements with Rh, Ir, Ni and Pt," *Acta Crystallogr., 18*, 835-839 (1965). (Crys Structure; Experimental)

65Pie: P. Pietrokowsky, "Novel Ordered Phase, Pt_8Ti," *Nature, 206*, 291 (1965). (Crys Structure; Experimental)

68Kra: P. Krautwasser, S. Bhan, and K. Schubert, "Structural Investigations in the Systems Ti-Pd and Ti-Pt," *Z. Metallkd., 59*(9), 724-729 (1968). (Equi Diagram, Crys Structure; Experimental)

68Reu: E.C. Van Reuth and R.M. Waterstrat, "Atomic Ordering in Binary $A15$-Type Phases," *Acta Crystallogr., B23*, 186-196 (1968). (Crys Structure; Experimental)

69Sin: A.K. Sinha, "Close-Packed Ordered AB_3 Structures in Binary Transition Metal Alloys," *Trans. AIME, 245*, 237-240 (1969). (Crys Structure; Experimental)

70Don: H.C. Donkersloot and J.H.N. Van Vucht, "Transformations in Gold-Titanium, Palladium-Titanium and Platinum-Titanium Alloys near the Equiatomic Composition," *J. Less-Common Met., 20*, 83-91 (1970). (Equi Diagram, Crys Structure; Experimental)

70Kau: L. Kaufman and H. Bernstein, *Computer Calculation of Phase Diagram*, Academic Press, New York (1970). (Thermo; Review)

73Rau: E. Raub and G. Falkenburg, "Reactions of the Platum Metals with the Carbides of the Group IV and V Elements of the Periodic Table," *Metall, 7*, 669-679 (1973) in German. (Equi Diagram; Experimental)

76Jac: T. Jacob and C.B. Alcock, unpublished data (1976); cited by [76Mes]. (Thermo; Theory)

76Jun: A. Junod, R. Flukiger, and J. Muller, "Superconductivity and Specific Heat in Titanium-Base $A15$ Compounds," *J. Phys. Chem. Solids, 37*, 27-31 (1976) in French. (Equi Diagram; Crys Structure; Experimental)

76Mes: P.J. Meschter and W.L. Worrell, "An Investigation of High-Temperature Thermodynamic Properties in the Pt-Ti Systems," *Metall. Trans. A, 7*, 299-305 (1976). (Thermo; Theory)

81Wor: W.L. Worrell and T.A. Ramanarayanan, "Electrochemical Cell Investigations of Platum Binary Systems at Elevated Temperatures," *Chemical Metallurgy—A Tribute to Carl Wagner*, Proc. Conf. TMS/AIME, 15086, 69-81 (1981). (Thermo; Experimental)

The Pu-Ti (Plutonium-Titanium) System

244   47.88

By J.L. Murray

The assessed Ti-Pu phase diagram is shown in Fig. 1, and its special points are summarized in Table 1. There are no intermetallic compounds; the equilibrium solid phases are the solid solutions based on the various allotropic forms of the pure components. The phases and their structures are listed in Table 2. All investigators agree that (1) melting occurs by the peritectic reaction L + (βTi) ⇌ (εPu), (2) there is a miscibility gap in the bcc (βTi,εPu) phase whose critical point lies above the peritectic temperature, and (3) there is a eutectoid reaction (βTi) ⇌ (αTi) + (εPu). Concerning the detailed placement of the phase boundaries and the invariant temperatures, the disagreements are large. The discrepancies are exacerbated by the lack of detailed documentation of experimental procedures, which would allow a reasoned assessment to be made. The phase diagram must therefore be considered extremely uncertain.

Experimental work was reported by [61Poo], [61Ell], [65Kut], and [71Lan]. [61Ell] dealt only with the (δPu)/(δ'Pu) equilibria. The others covered the entire composition range. The effect of pressure on the transformation points of Pu and dilute Ti alloys was examined by [67Lip]: unfortunately, the two-phase field was not taken into account, and only averages of transformation points on heating and cooling were reported.

[61Poo] investigated the diagram by the techniques of thermal analysis, dilatometry, X-ray diffraction, and metallography. Alloys were arc-melted. No further information

tion was given on details of the experiments, or on the purity of the metals used, or composition control of the alloys.

[65Kut] reported data from thermal analysis and X-ray diffraction on Pu-rich alloys in the form of a figure, but otherwise experimental details were undocumented.

[71Lan] used dilatometry, metallography, and chemical analysis of equilibrated two-phase alloys and of heat treated and quenched diffusion couples. Dilatometric arrests on heating and cooling were tabulated. Impurity levels in the Ti and Pu used to make alloys were 320 ppm and 400 ppm, respectively.

Based on lattice parameter measurements, [57Wal] reported that a maximum of about 30 at.% Ti is soluble in (εPu) and that the low solubility of Pu in (αTi) could not be determined by this method.

The reported values of invariant temperatures are compared in Table 3. The first arrest temperatures on heating and cooling [71Lan] are also listed separately.

Data points on the phase boundaries are shown in Fig. 2. Note that two points of [71Lan] that lie above the peritectic temperature on the (βTi) solidus were reported as points on the miscibility gap. The datum of [61Poo] on the (εPu)/(εPu) + (αTi) boundary denotes a two-phase alloy. [65Kut] placed the (βTi) eutectoid point at 6 at.% Pu, in significant disagreement with the other work.

Based on the hypothesis that undercooling is a greater likelihood than superheating and on the agree-

Fig. 1---Assessed Ti-Pu Phase Diagram

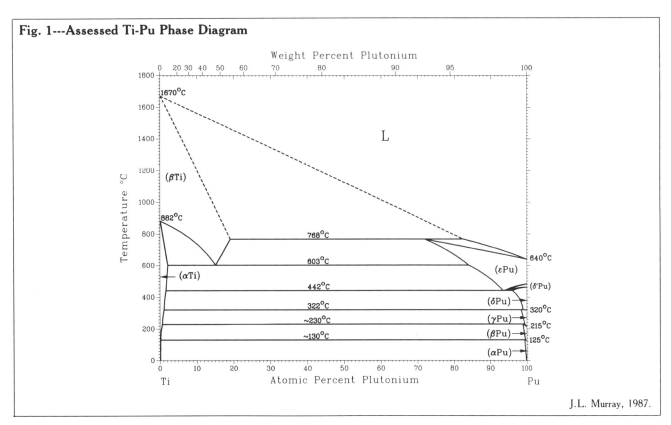

J.L. Murray, 1987.

253

Phase Diagrams of Binary Titanium Alloys

Table 1 Special Points of the Assessed Ti-Pu Phase Diagram

Reaction	Compositions of the respective phases, at.% Pu			Temperature, °C	Reaction type
L + (βti) ⇌ (εPu)	82.5	19	72	768	Peritectic
(βTi) ⇌ (αTi) + (εPu)	15	2	84	603	Eutectoid
(δ'Pu) ⇌ (δPu) + (εPu)	95	96.5	94.2	446	Eutectoid
(εPu) ⇌ (δPu) + (αTi)	93.5	96	(~ 1.5)	442	Eutectoid
(αTi) + (δPu) ⇌ (γPu)	(1)	(~ 99)	(~ 98.5)	322	Peritectoid
(αTi) + (γPu) ⇌ (βPu)	(0.5)	(~ 99.5)	(~ 99)	230	Peritectoid
(αTi) + (βPu) ⇌ (αPu)	(0.1)	(~ 99.7)	(~ 99.5)	~ 130	Peritectoid
L ⇌ (βTi)		0		1670	Melting point
(βTi) ⇌ (αTi)		0		882	Allotropic transformation
L ⇌(εPu)		100		640	Allotropic transformation
(εPu) ⇌ (δ'Pu)		100		483	Allotropic transformation
(δ'Pu) ⇌ (δPu)		100		463	Allotropic transformation
(δPu) ⇌ (γPu)		100		320	Allotropic tranformation
(γPu) ⇌ (βPu)		100		215	Allotropic transformation
(βPu) ⇌ (αPu)		100		125	Allotropic transformation

Note: Compositions in parentheses are the values used to draw Fig. 1, but they have not been determined experimentally.

Table 2 Crystal Structures of the Ti-Pu System

Phase	Homogeneity range, at.% Pu		Pearson symbol	Space group	Struktur-bericht designation	Proto-type	Reference
(αTi)	0 to	~ 2	hP2	P6₃/mmc	A3	Mg	[Pearson2]
(βTi,εPu)	0 to	19	cI2	Im3m	A2	W	[Pearson2]
	72 to	100					
(δ'Pu)	~ 95 to	100	tI2	...	...	...	[Pearson2]
(δPu)	~ 96 to	100	cF4	Fm3m	A1	Cu	[Pearson2]
(αPu)	~ 98.5 to	100	oF8	Fddd	...	...	[Pearson2]
(βPu)	~ 99 to	100	mC34	C2/m	...	...	[Pearson2]
(γPu)	~ 99.5 to	100	mP16	P2₁/m	...	...	[Pearson2]

ment between reported peritectic temperatures [61Poo, 71Lan], the present evaluator notes that the heating data of [71Lan] are comparable to the [61Poo] values. [61Poo] is tentatively chosen as the basis for the assessed invariant temperatures.

The (δPu)/(δ'Pu) boundaries are based on [61Ell]. High-temperature X-ray diffraction was used to locate the invariant temperatures to about ±2 °C and the eutectoid composition to ±0.25 at.%. The work of [61Ell] is preferred to [65Kut], whose (εPu) ⇌ (δ'Pu) + (αTi) reaction has not been verified by any other work.

A detail of the (δ'Pu) region is shown in Fig. 3. [61Ell] noted a "self-plating" effect, apparently unique to Ti-Pu alloys; reaction with oxygen causes Ti to be depleted in the bulk and to form an oxide layer.

Metastable Phases

A metastable bcc (εPu) solid solution can be retained to room temperature by rapid solidification (10^6 to 10^8 °C/s) [73Ell]. Alloys containing 55 to 80 at.% Pu had a single-phase bcc structure; alloys containing 80 to 100 at.% Pu had the fcc (δPu) structure and had probably transformed from (εPu) in the solid state. [65Kut] also reported (δPu) phase retained in alloys that contained less

than 73 at.% Pu and were quenched from the (εPu) region. [65Kut] also reported (βPu), (αPu), and a complex metastable phase in more dilute Pu alloys.

The transformations (βTi) → (αTi) and (βPu) → (αPu) can occur by a martensitic mechanism [64Dav, 70Lan]. [64Dav] examined the kinetics of the (βPu) → (αPu) transition; [70Lan] examined the effect of oxygen on solute segregation during the (βTi) → (αTi) transition during quenching.

Crystal Structures and Lattice Parameters

Crystal structures of the equilibrium phases are summarized in Table 2. Lattice parameters of the extended (βTi,εPu) and (δPu) solid solutions [73Ell] are listed in Table 4.

Thermodynamics

A thermodynamic calculation of the diagram was attempted, and some preliminary regular solution results are reported in Table 5. Gibbs energies of liquid, bcc, cph, fcc, and (δ'Pu) phases were represented as:

$$G(i) = G^0(\text{Ti},i)(1 - x) + x\, G^0(\text{Pu},i) + RT\,[x \ln x + (1 - x) \ln (1 - x)] + x(1 - x)\, B(i)$$

Table 3   Experimental Data on the Three-Phase Equilibria

| | | | Invariant temperatures, °C | | |
| | | | | [71Lan] | |
Reaction	[60Poo]	[65Kut]	Heating	Average	Cooling
L + (βTi) ⇄ (εPu)	770	730	...	765	...
(βTi) ⇄ (αTi) + (εPu)	603	580	588	572	565
(εPu) ⇄ (δPu) + (αTi)	442(a)	430(b)	441	430	420
(αTi) + (δPu) ⇄ (γPu)	322	305	...	280	...
(αTi) + (γPu) ⇄ (βPu)	...	235	...	225	...
(αTi) + (βPu) ⇄ (αPu)	...	130	...	155	...

(a) This value due to [61Ell]. (b) [65Kur] interpreted the reaction as (δ'Pu) ⇄ (δPu) + (αTi).

Fig. 2---Experimental Data vs the Assessed Phase Diagram

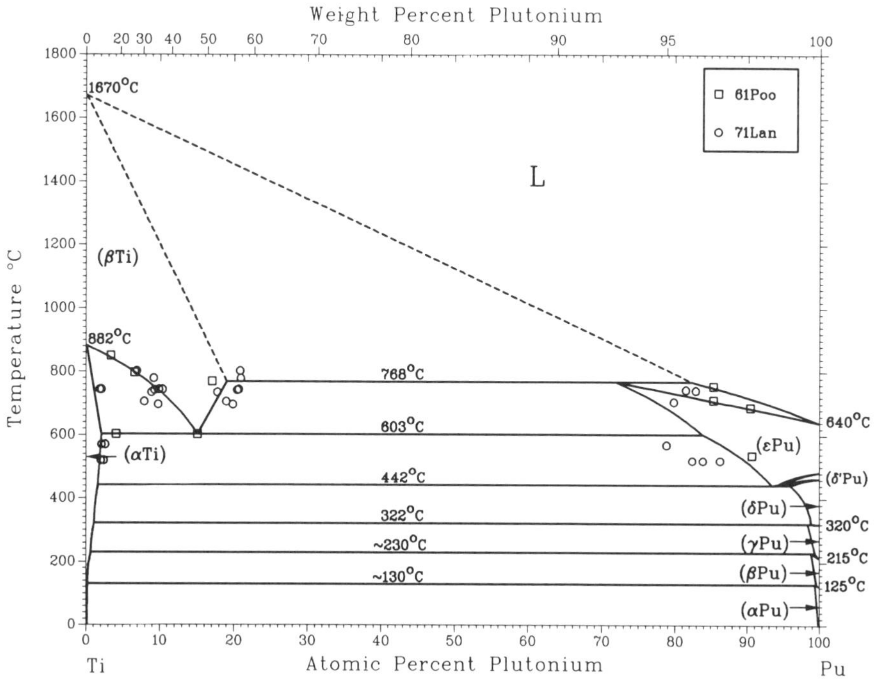

J.L. Murray, 1987.

Table 4   Lattice Parameters of (βTi,εPu) and (δPu) [73Ell]

Phase	Composition, at.% Pu	Lattice parameter, nm
(δPu)	95	0.4610
	90	0.4559
	85	0.4588
	85	0.4572
	80	0.4505
	80	0.4515
	75	0.4502
(βTi),εPu)	80	0.3473
	75	0.3463
	70	0.3458
	65	0.3445
	55	0.3420

where i designates the phase; x is the atomic fraction of Pu; G^0 is the Gibbs energies of the pure metals; and $B(i)$ is the regular solution parameters describing the excess Gibbs energies. (γPu), (βPu), and (αPu) were assumed to have no homogeneity range.

Optimizations with respect to select phase diagram data [61Poo, 71Lan, Elliott] failed to reproduce the phase diagram. First, the observed miscibility gap is nearly vertical, compared to calculations, which tended to place a critical point at temperatures slightly above the peritectic temperature. Second, calculations of the (βTi) transus do not show the strong negative curvature of the observed transus, and hence, calculations of the eutectoid temperature tend to lie as much as 50 °C above the observed eutectoid reaction. For temperatures above 500 °C, the optimizations lead to phase boundaries similar to most other Ti systems without compounds.

The approximation of zero solubility of Ti in (αPu), (βPu), and (γPu) is not adequate to reproduce the phase diagram in this system. The expectation implicit in this assumption is that, because the homogeneity ranges of (αTi) and (δPu) are restricted in the relevant temperatures ranges, the invariant temperatures would fall close to the pure metal transformation points. This expectation

Fig. 3---Detail of the (δ'Pu) Region

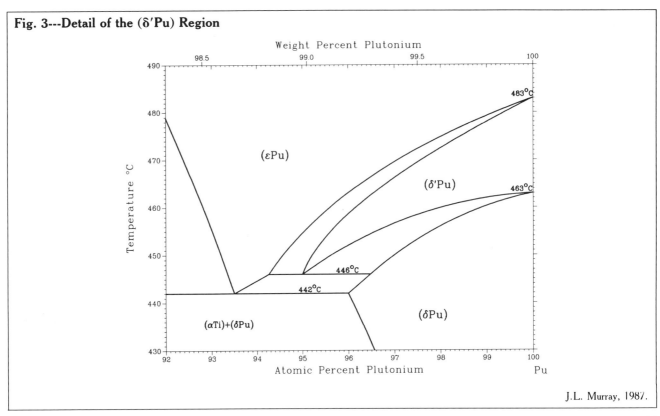

J.L. Murray, 1987.

Fig. 4---Approximate Thermodynamic Calculation of the Ti-Pu Phase Diagram

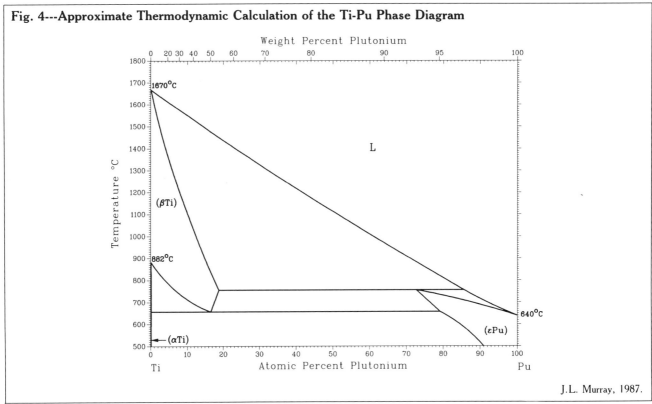

J.L. Murray, 1987.

was not realized, because of the very low heats of transformation of pure Pu.

A more detailed calculation of the Ti-Pu system is at present prohibited by the lack of thermochemical data and accurate data for most of the phase boundaries. However, the regular solution parameters obtained in the present rough calculations are consistent with the expected values for Ti-lanthanide and Ti-actinide systems.

Table 5 Gibbs Energies of the Ti-Pu System

Properties of the pure elements

$G^0(\text{Ti,L}) = 0$

$G^0(\text{Pu,L}) = 0$

$G^0(\text{Ti,bcc}) = -16\,234 + 8.368\,T$

$G^0(\text{Pu,bcc}) = -3\,347 + 3.6659\,T$

$G^0(\text{Ti,cph}) = -20\,585 + 12.134\,T$

$G^0(\text{Pu,cph}) = -4\,477 + 8.14\,T$

$G^0(\text{Ti,fcc}) = -17\,238 + 12.134\,T$

$G^0(\text{Pu,fcc}) = -4\,477 + 5.1944\,T$

$G^0(\text{Ti},\delta'\text{Pu}) = -12\,000 + 8.368\,T$

$G^0(\text{Pu},\delta') = -3\,933 + 4.4441\,T$

$G^0(\text{Pu},\gamma) = -4\,561 + 5.3373\,T$

$G^0(\text{Pu},\beta) = -6\,402 + 9.1568\,T$

$G^0(\text{Pu},\alpha) = -9\,247 + 16.3960\,T$

Excess Gibbs energies of the solution phases

$B(\text{L}) = 10\,158$

$B(\text{bcc}) = 17\,559$

$B(\text{cph}) = 31\,245$

$B(\text{fcc}) = 16\,677$

$B(\delta'\text{Pu}) = 13\,730$

Note: Values are given in J/mol and J/mol · K.

Cited References

57Wal: M.B. Waldron, "Phase Diagrams of Plutonium Alloys Studied at Harwell," *The Metal Plutonium*, A.S. Coffinberry and W.N. Miner, Ed., University of Chicago Press, Chicago, IL, 225-239 (1957). (Equi Diagram; Experimental)

***61Ell** R.O. Elliott and A.C. Larson, "δ′ Plutonium," *The Metal Plutonium*, A.S. Coffinberry and W.N. Miner, Ed., University of Chicago Press, Chicago, IL, 265-280 (1961).

***61Poo** D.M. Poole, M.G. Bale, P.G. Mardon, J.A.C. Marples, and J.L. Nichols, "Phase Diagrams of Some Plutonium Alloy Systems," in *Plutonium, 1960*, E. Grison W.B.H. Lord, and R.D. Fowler, Ed., Cleaver-Hume Press, Ltd., London, 267-280 (1961). (Equi Diagram; Experimental)

64Dav L.G.T. Davy and J.S. White, "The Kinetics of the β to α Transformation in Plutonium-Titanium," *J. Nucl. Mater.*, 12(2), 221-225 (1964). (Meta Phases; Experimental)

***65Kut** V.I. Kutaitsev, N.T. Chebotarev, I.G. Lebedev, M.A. Andrianov, V.N. Konve, and T.S. Menshikova, "21 Phase Diagrams of Plutonium with the Metals of Groups IIA, IVA, VIIIA, and IB," Plutonium 1965, Proc. Third Int. Conf. on Plutonium, London, 420-449 (1965). (Equi Diagram, Meta Phases; Experimental)

67Lip R.G. Liptai and R.J. Friddle, "The Effects of High Pressure on the Phase Equilibria of Some Plutonium Alloys," *J. Nucl. Mater.*, 21, 114-116 (1967). (Equi Diagram; Experimental)

70Lan A. Languille, C. Remy, and C. Calais, "Effect of the Allotropic β → α Transformation and of Oxygen on the Segregation of Plutonium in Dilute ZrPu and TiPu Alloys," *J. Nucl. Mater.*, 37, 139-152 (1970). (Meta Phases; Experimental)

***71Lan** A. Languille, "The Plutonium-Titanium System: Phase Diagram and Diffusion of Plutonium in Beta Titanium," *Mem. Sci. Rev. Met.*, 68(6), 435-441 (1971). (Equi Diagram; Experimental)

73Ell R.O. Elliott and A.M. Russell, "Retention of a Metastable Bcc ε-Pu Solid Solution: ε-Pu (Ti)," *J. Mater. Sci.*, 8, 1325-1330 (1973). (Meta Phases; Experimental)

The Rb-Ti (Rubidium-Titanium) System

85.4678   47.88

By C.W. Bale

Equilibrium Diagram

The assessed Rb-Ti phase diagram is shown in Fig. 1. As part of a project to determine the corrosive effects of Rb on commercial refractory and nonrefractory metals, [62You] measured the equilibrium concentrations of Mo–0.5 wt.% Ti alloys in liquid Rb at 760, 927, and 1093 °C. The results were scattered and are reproduced in Fig. 2. The maximum solubility was reported as approximately 0.0007 at.% Ti at 1093 °C, and the minimum solubility was approximately 0.0002 at.% Ti at 929 °C (not at 760 °C). [62You] also reported oxygen impurity levels of 18 to 64 wt.ppm oxygen at 760 °C, 18 to 65 wt.ppm oxygen at 927 °C, and 4 to 6 wt.ppm oxygen at 1093 °C. It is quite possible that the oxygen impurity enhanced the corrosion of Ti (at 760 °C, for example), because rubidium titanates should be extremely stable [82Koh].

The following expression was calculated by least-squares regression applied to the results of [62You]:

$$\log \text{at.\% Ti} = -3.467 + 16/T \quad \text{(Eq 1)}$$

where T is in K.

Equation 1 is represented by the solid line in Fig. 2. This expression is very approximate; the second coefficient in Eq 1 should have a negative value. If Fig. 2 approaches true equilibrium conditions, the actual solubility of pure Ti in Rb is much higher, because the results in Fig. 2 correspond to Ti in a dilute solid solution of Mo.

There are no other published data for this system. By analogy with the other alkali metal–Group IVA systems, it is probable that the system is almost completely immiscible in both the solid and liquid states. If the solubilities are very limited, it follows from thermodynamic considerations that the univariant temperatures in the phase diagram are virtually identical with the transition temperatures of the pure components.

From vapor-pressure calculations, at 2000 °C the composition of the gas phase at 1 atm is estimated to be approximately 0.04 at.% Ti, in equilibrium with almost pure liquid Ti.

The crystal structure data for the elements are summarized in Table 1.

Cited References

62You: P.F. Young and R.V. Arabian, "Determination of Temperature Coefficient of Solubility of Various Metals in Rubidium and the Corrosive Effects of Rubidium on Various Alloys at Temperatures from 1000 to 2000 °F," USAEC Contract AT (04-3)-368; AGN-8063 (1962). (Equi Diagram; Experimental)

82Koh: R. Kohli, "A Thermodynamic Assessment of the Behavior of Cesium and Rubidium in Reactor Fuel Elements," *Material Behavior and Physical Chemistry in Liquid Metal Systems*, Plenum Press, New York 345-350 (1982). (Equi Diagram; Theory)

Fig. 1---Assessed Rb-Ti Phase Diagram

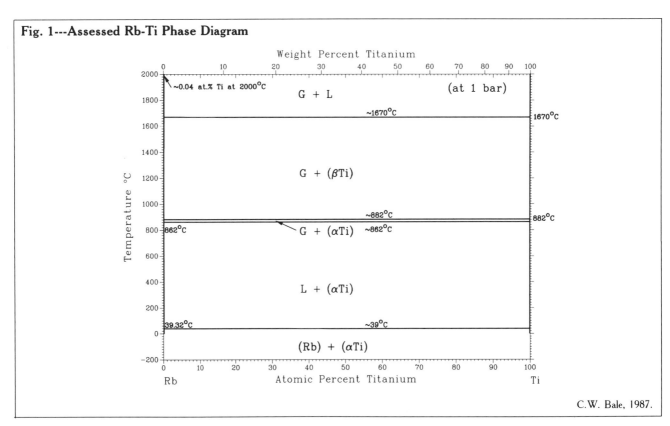

C.W. Bale, 1987.

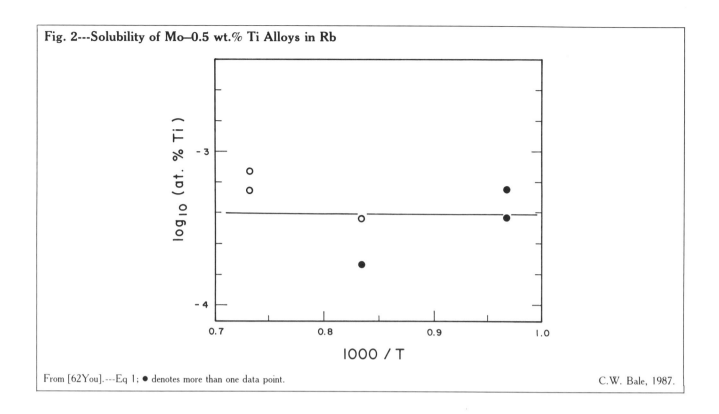

Fig. 2---Solubility of Mo–0.5 wt.% Ti Alloys in Rb

From [62You].---Eq 1; ● denotes more than one data point.   C.W. Bale, 1987.

Table 1 Crystal Structures and Lattice Parameters of the Rb-Ti System

Phase	Composition, at.% Ti	Pearson symbol	Space group	Struktur-bericht designation	Proto-type	Lattice parameters, nm a	c	Comment	Reference
(Rb)	0.0	cI2	Im3m	A2	W	0.5703	...	At RT	[King1]
(αTi)	100	hP2	P6$_3$/mmc	A3	Mg	0.29503	0.46836	At RT	[King2]
(βTi)	100	cI2	Im3m	A2	W	0.33065	...	Above 900 °C	[King2]

The RE-Ti (Rare Earth-Titanium) Systems

By J.L. Murray

Phase diagrams for the binary systems of Ti with La, Ce, Nd, Eu, Gd, Er, Sc, and Y can be found in this compilation. They are characterized by the absence of intermetallic compounds and quite limited mutual solubilities in the solid solutions, as well as miscibility gaps in the liquid solutions. Some of the liquid phase miscibility gaps are entirely metastable; others appear in the equilibrium diagrams. Several monotectoid reactions were predicted by a thermodynamic analysis of the limited available experimental data, but have not been reported in the experimental literature.

The diagrams of the binary systems of Ti with Pr, Pm, Sm, Tb, Dy, Ho, Tm, and Lu can thus be predicted in their qualitative features with some confidence. They contain no compounds and are characterized by limited solid solubilities and liquid phase miscibility gaps (either equilibrium or metastable). The prediction of liquid immisci-

bility in the equilibrium diagram and quantitative details of the diagrams would require sufficient thermodynamic information on several Ti-RE systems for systematic trends over the rare earth (RE) series to be deduced. However, thermodynamic information is not sufficiently accurate for such quantitative predictions to be made.

The reader is referred to [86Gsc] for data on properties of the pure rare earths, such as crystal structures, lattice parameters, and allotropic transformation temperatures.

Cited Reference

86Gsc: K.A. Gschneidner, Jr. and F.W. Calderwood, in *Handbook of the Physics and Chemistry of Rare Earths*, Vol. 8, K.A. Gschneidner, Jr. and L. Eyring, Ed., North-Holland Physics Publishing, Amsterdam (1986).

The Re-Ti (Rhenium-Titanium) System

186.207   47.88

By J.L. Murray

Equilibrium Diagram

The Ti-Re phase diagram has been studied only in one laboratory [58Sav, 59Sav, 69Sav] and only in limited composition and temperature ranges. In view of the incompleteness of the investigations and the difficulties intrinsic to working with this system, the phase diagram must still be considered uncertain. A tentative phase diagram is shown in Fig. 1, and its special points are listed in Table 1.

The equilibrium solid phases of the Ti-Re system are:

- The cph (αTi) and (Re) solid solutions. The cph structure is the equilibrium form of pure Re and pure Ti below 882 °C.
- The high-temperature bcc (βTi) solid solution. The maximum solubility of Re in (βTi) is approximately 50 at.%.
- The ordered compound Ti_5Re_{24} (composition 82.8 at.% Re) with a structure with 58 atoms per unit cell, isomorphous with αMn [55Trz]. The presence of Ti_5Re_{24} was confirmed by X-ray diffraction and hardness studies [58Sav]. There is no evidence for any significant homogeneity range.

Liquidus and Solidus.

The melting temperature of pure Re is 3186 °C [Melt]. [58Sav] determined melting temperatures by the "drop method;" the solidus temperature was judged to be that at which a hole in the sample completely filled with liquid. For Ti-rich compositions, the data, although scattered, define a (βTi) solidus between 1670 and about 2000 °C. For the high melting temperatures at Re-rich compositions, the drop method provided only a rough estimate of the solidus. The melting points determined by [58Sav] are shown in Fig. 1. The data were interpreted as evidence for a peritectic reaction L + Ti_5Re_{24} ⇄ (βTi) at 2025 °C and a peritectic reaction L + (Re) ⇄ Ti_5Re_{24} at 2750 °C. There appears to have been some microscopic evidence for the peritectic reactions. This interpretation, however, is inconsistent with the datum near 40 at.% Re; the phase diagram predicts that this alloy would melt at 2025 °C. An additional compound would have to intervene between (βTi) and Ti_5Re_{24} for that alloy to melt at such a high temperature. X-ray diffraction results do not suggest the existence of another compound; according to [57Phi], the

(βTi) solution is homogeneous at 1200 °C to about 50 at.% Re. Therefore, the peritectic reactions proposed by [58Sav, 59Sav] are tentatively accepted in the assessed diagram.

[59Sav] studied the (βTi)/(αTi) equilibria using metallographic, X-ray, and dilatometric methods; the experimental results are summarized in Fig. 1. [59Sav] also estimated the solubility of Re in (αTi) to be approximately 0.03 at.% at 750 °C. In a study of the decomposition of metastable (βTi), [61Age] verified that a 6 at.% Re alloy lies within the two-phase (βTi) + (αTi) region at 400 °C.

The solubility of Ti in (Re) was determined metallographically by [69Sav] at four temperatures, and the data are shown in Fig. 1.

Metastable Phases

Metastable phases are formed from (βTi) during quenching. In Ti-rich alloys, the cph phase (α'Ti) can form martensitically, and at slightly higher Re contents, an orthorhombic distortion of the cph structure (α''Ti) is formed. ω phase forms as an intermediate phase in the decomposition of (βTi) into (αTi) at low temperatures.

[58Bag] observed the distorted (α'''Ti) martensite in an as-quenched 2.8 at.% Re alloy. For all Ti-transition metal systems, the start temperature (M_s) of the martensitic bcc to cph transformation decreases with increasing transition metal content. M_s has not been determined experimentally as a function of composition for the Ti-Re system. [59Sav] found that the bcc phase could be retained metastably during quenching in alloys containing more than 3.6 at.% Re; thus, M_s reaches room temperature at approximately that composition.

The ω phase is formed either during quenching of alloys of sufficient Re content that (αTi) is not formed during quenching, or during aging of metastable bcc alloys at temperatures between approximately 300 and 450 °C. During aging at temperatures above approximately 450 °C, the equilibrium (αTi) + (βTi) usually forms directly. During aging, the composition of ω becomes more Ti-rich, approaching a "saturation composition," which coincides approximately with the composition at which ω is most easily produced in as-quenched alloys. [58Bag] observed ω + (βTi) in an as-quenched 4 at.% Re alloy. In a 5.3 at.% Re alloy, only ω phase was found. [61Age] produced ω + (βTi) in a 6 at.% Re alloy during aging at 400 °C.

Table 1 Special Points of the Assessed Ti-Re Phase Diagram

Reaction	Compositions of the respective phases, at.% Re			Temperature, °C	Reaction type
L + Ti_5Re_{24} ⇄ (βTi)	~ 42	82.8	~ 50	~ 2025	Peritectic
L + (Re) ⇄ Ti_5Re_{24}	~ 81	82.8	~ 97	~ 2750	Peritectic
L ⇄ (βTi)		0		1670	Melting point
(βTi) ⇄ (αTi)		0		882	Allotropic transformation
L ⇄ (Re)		100		3186	Melting point

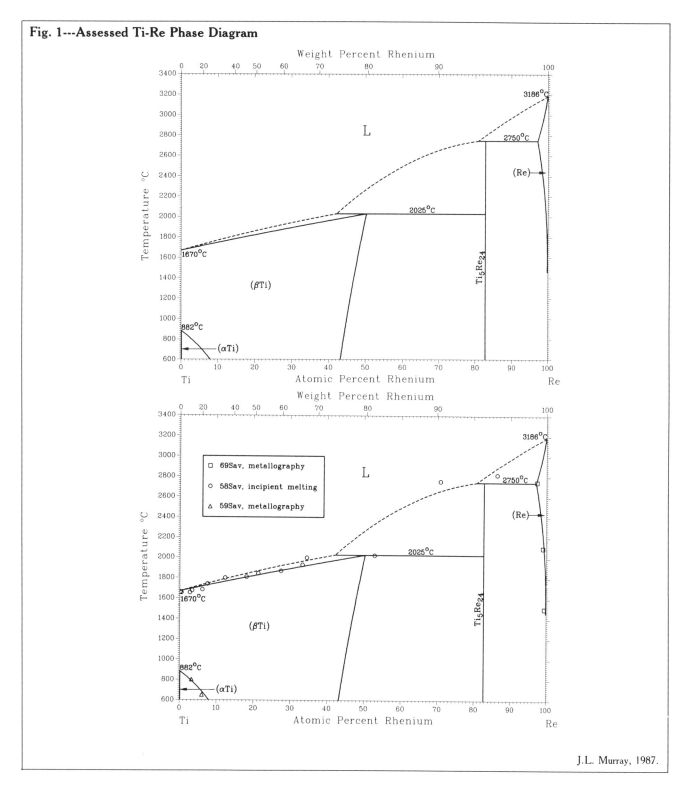

Fig. 1---Assessed Ti-Re Phase Diagram

J.L. Murray, 1987.

Crystal Structures and Lattice Parameters

The structures of the equilibrium and metastable phases of the Ti-Re system are summarized in Table 2. Lattice parameter data are given in Table 3.

Thermodynamics

Experimental determinations of high-temperature thermodynamic properties of the Ti-Re system have not yet been made. [70Kau] calculated the diagram in the regular solution approximation using theoretical estimates of excess Gibbs energies. The major discrepancy between calculation and experiment was that [70Kau] obtained a eutectic reaction L $\rightleftarrows$ Ti$_5$Re$_{24}$ + (Re) and congruent melting of Ti$_5$Re$_{24}$. The purpose of the present calculations is to test what properties of the Gibbs energies would be required to obtain the observed peritectic reaction L + (Re) $\rightleftarrows$ Ti$_5$Re$_{24}$.

261

Table 2 Crystal Structures of the Ti-Re System

Phase	Homogeneity range, at.% Re	Pearson symbol	Space group	Struktur- bericht designation	Proto- type	Reference
(αTi)	0 to 0.2	$hP2$	$P6_3/mmc$	$A3$	Mg	[Pearson2]
(βTi)	0 to 50	$cI2$	$Im3m$	$A2$	W	[Pearson2]
χ	82.8	$cI58$	$I43m$	$A12$	αMn	[55Trz]
(αRe) ...	97.5 to 100	$hP2$	$P6_3/mmc$	$A3$	Mg	[Pearson2]
α″	(a)	$oC4$	$Cmcm$	$A20$	αU	[58Bag]
ω	(a)	$hP3$	$P6/mmm$ or $P\bar{3}m1$	...	...	[58Bag]

(a) Metastable.

Table 3 Lattice Parameters of Ti-Re Phases

Phase	Reference	Composition, at.% Re	Lattice parameters, nm	
			a	c
(βTi)	[57Phi]	25	0.3199	...
		50	0.3104	...
		...	0.3104	...
(αRe) ...	[69Sav]	99.2	0.27555	0.4447
Ti$_5$Re$_{24}$ ..	[55Trz]	82.8	0.9609	...
	[61Buc]	83	0.9595	...
	[61Mat]	82.8	0.9587	...

Table 4 Ti-Re Thermodynamic Properties

Gibbs energies of the pure components [70Kau]

$G^0(Ti,L) = 0$
$G^0(Re,L) = 0$

$G^0(Ti,bcc) = -16\ 234 + 8.368\ T$
$G^0(Re,bcc) = -27\ 196 + 10.042\ T$

$G^0(Ti,cph) = -20\ 585 + 12.134\ T$
$G^0(Re,cph) = -28\ 870 + 8.346\ T$

$G(Ti_5Re_{24}) = -65\ 996 + 13.441\ T$

Excess Gibbs energies

$B(L) = -87\ 500$
$B(bcc) = -84\ 000$

Note: Values are given in J/mol and J/mol · K.

Because of the absence of thermochemical data and the sketchiness of the phase diagram data, optimization calculations can be done only in the simplest approximation. Ti and Re were assumed to be mutually insoluble in the cph form. Ti$_5$Re$_{24}$ was assumed to be strictly stoichiometric. The liquid and bcc Gibbs energies were modeled in the regular solution approximation:

$$G(i) = G^0(Ti,i)\ (1 - x) + G^0(Re,i)\ x + RT[x \ln x + (1 - x) \ln (1 - x)] + B(i)\ x\ (1 - x)$$

where i is the phase; x is the atomic fraction of Re; G^0 is the pure component Gibbs energies; and $B(i)$ is the regular solution interaction parameter. The pure element properties are taken from [70Kau].

Rough estimates of the regular solution bcc and liquid Gibbs energies can be made from the (βTi) transus and the (Re) liquidus, respectively. From the phase diagram, the parameters are estimated as: $B(bcc) \equiv -60$ to 100 kJ/mol and $B(L) \equiv >0$ kJ/mol.

On the other hand, from the linear behavior of the (βTi) liquidus, the excess Gibbs energies are expected to be nearly equal and to show little deviation from the regular solution approximation. There appears, therefore, to be some inconsistency in the reported phase diagram.

The parameter $B(bcc)$ was chosen as -84 kJ/mol, in conformity with theoretical estimates and the trend of Ti-transition metal systems. The Gibbs energies of the liquid and Ti$_5$Re$_{24}$ were optimized with respect to the phase diagram data. The calculated reactions are:

L ⇌ Ti$_5$Re$_{24}$ + (Re) at 2764 °C, $x(L) = 49$ at.% Re

L ⇌ Ti$_5$Re$_{24}$ + (βTi) at 1997 °C, $x(L) = 90$ at.% Re

L ⇌ Ti$_5$Re$_{24}$ at 2826 °C

Excess Gibbs energies (see Table 4) are close to those previously proposed by [70Kau]. No further calculation is presently justified in view of the scarcity of input data.

It is recommended for future experimental work that the peritectic reaction L + (Re) ⇌ Ti$_5$Re$_{24}$ be confirmed.

Cited References

55Trz: W. Trzebiatowski and J. Niemiec, "The Structure of Re$_{24}$Ti$_5$," *Roczn. Chem., 29,* 277-283 (1955) in Polish. (Crys Structure; Experimental)

57Phi: T.V. Philip and P.A. Beck, "CsCl-Type Ordered Structures in Binary Alloys of Transition Elements," *Trans. AIME, 209,* 1269-1271 (1957). (Crys Structure; Experimental)

58Bag: Yu.A. Bagariatskii, G.I. Nosova, and T.V. Tagunova, "Factors in the Formation of Metastable Phases in Titanium-Base Alloys," *Dokl. Akad. Nauk SSSR, 122,* 593-598 (1958) in Russian; TR: *Sov. Phys. Dokl., 3,* 1014-1018 (1958). (Meta Phases; Experimental)

58Sav: E.M. Savitskii and M.A. Tylkina, "Alloys of Rhenium with Refractory Metals (Mo, Ti, Zr, Ta, Ni, Co, Cr, W, Mn)," *Zh. Neorg. Chem., 3*(1), 815 (1958) in Russian; TR: *Russ. J. Inorg. Chem., 3*(1), 338-415 (1958). (Equi Diagram; Experimental)

59Sav: E.M. Savitskii, M.A. Tylkina, and Yu.A. Zotev, "Phase Diagram of the Rhenium-Titanium System," *Zh. Neorg. Chem., 4*(2), 702-703 (1959) in Russian; TR: *Russ. J. Inorg. Chem., 4*(3), 319-320 (1958). (Equi Diagram; Experimental)

61Age: N.V. Ageev, O.G. Karpinskii, and L.A. Petrova, "Mechanism of the Decomposition of the β-Solid Solution in Titanium-Rhenium Alloys," *Zh. Neorg. Chem., 6*(1), 251 (1961) in Russian; TR: *Russ. J. Inorg. Chem., 6*(1), 127-128 (1961). (Meta Phases; Experimental)

61Buc: E. Bucher, F. Heiniger, and J. Mueller, "Superconductivity and Paramagnetism of Complex Transition Metal Phases," *Hev. Phys. Acta, 34,* 843-858 (1961). (Crys Structure; Experimental)

61Mat: B.T. Matthias, V.B. Compton, and E. Corenzwit, "Some New Superconducting Compounds," *J. Phys. Chem. Solids, 191*(1-2), 130-133 (1961). (Crys Structure; Experimental)

69Sav: E.M. Savitskii, M.A. Tylkina, and O.Kh. Khamidov, "Investigation of the Solid Solubility of Transition Metals in Rhenium and Some Properties of Their Alloys," *Izv. Akad. Nauk SSSR, Met.,* (4), 200-208 (1969) in Russian; TR: *Russ. Metall.,* (4), 130-135 (1969). (Equi Diagram; Experimental)

70Kau: L. Kaufman and H. Bernstein, *Computer Calculation of Phase Diagrams,* Academic Press, New York (1970). (Thermo; Review)

The Rh-Ti (Rhodium-Titanium) System

102.0955   47.88

By J.L. Murray

Equilibrium Diagram

There are qualitative disagreements about the Ti-rich portion of the Ti-Rh phase diagram, and two alternate versions of the relevant phase equilibria are given in the assessed diagram (Fig. 1) and in the list of salient points given in Table 1. On the Ti-rich side, the solid curves are based on the work of [66Rau], and the dotted curves are based on the work of [72Ere] and [75Sht]. All investigators agreed on the existence of Ti_2Rh, $\beta TiRh$, and $TiRh_3$. The quantitative features about which there is fair agreement are the $(\alpha Ti)/(\beta Ti)$ boundaries, the homogeneity range of $TiRh_3$, part of the (Rh) solvus, and the congruent melting point of $\beta TiRh$. The most serious discrepancies concern the extent of the single-phase (βTi) field.

Fig. 1---Assessed Ti-Rh Phase Diagram

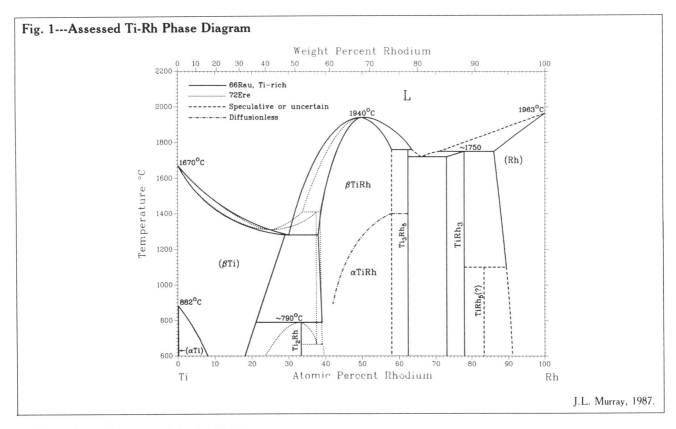

J.L. Murray, 1987.

Table 1 Special Points of the Ti-Rh System

Reaction	Compositions of the respective phases, at.% Rh			Temperature, °C	Reaction type
$(\beta Ti) + \beta TiRh \rightleftarrows Ti_2Rh$	21	39	33.3	790	Peritectoid
$L \rightleftarrows \beta TiRh$	...	50	...	1940	Congruent
$L + (Rh) \rightleftarrows TiRh_3$	71	78	86	1750	Peritectic
$L \rightleftarrows (\beta Ti)$		25		1310	Congruent [72Ere]
$L \rightleftarrows Ti_3Rh_5 + TiRh_3$	66	62.5	73	~ 1720	Eutectic (?)
$\beta TiRh + Ti_3Rh_5 \rightleftarrows \alpha TiRh$	~ 58	62.5	~ 58	~ 1400	Peritectoid
$L \rightleftarrows Ti_3Rh_5 + TiRh_3$	~ 66	~ 62.5	73	~ 1760	Eutectic
$TiRh_3 + (Rh) \rightleftarrows TiRh_5$	78	~ 89.5	83.3	~ 1100	Peritectic
$L \rightleftarrows (\beta Ti) + \beta TiRh$	30	29	38.0	1280	Eutectic [68Rau]
$(\beta Ti) \rightleftarrows Ti_2Rh + \beta TiRh$	37.5	33.3	39	~ 665	Eutectoid [72Ere]
$L + \beta TiRh \rightleftarrows (\beta Ti)$	30	40	39	~ 1400	Peritectic [72Ere]
$L \rightleftarrows (Rh)$		100		1963	Melting point
$L \rightleftarrows (\beta Ti)$		0		1670	Melting point
$(\beta Ti) \rightleftarrows (\alpha Ti)$		0		882	Allotropic transformation

263

The observed solid phases of the Ti-Rh system are:

- The cph (αTi), bcc (βTi), and fcc (Rh) solid solutions. The cph and bcc structures are the equilibrium forms of pure Ti below and above 882 °C, respectively. The maximum solubility of Rh in (αTi) is low and that in (βTi) has not yet been definitively determined. The maximum solubility of Ti in (Rh) is approximately 14 at.%.
- The stoichiometric compound Ti_2Rh, with the $MoSi_2$ structure, stable below about 790 °C.
- Two compounds with wide homogeneity ranges about TiRh stoichiometry. The high-temperature form, βTiRh, has the CsCl structure; it transforms on cooling to αTiRh, which has the tetragonal AuCu structure.
- The essentially stoichiometric compound Ti_3Rh_5, which has a structure that is isomorphous with Ge_3Rh_5.
- $TiRh_3$, which has the Cu_3Au ordered fcc structure. $TiRh_3$ has a homogeneity range of 73 to ~78 at.% Rh.
- A possible equilibrium compound with composition near 83 at.% Rh ($TiRh_5$) was identified by [66Rau].

A Ti_2Ni-type phase is often found in Ti-transition metal alloys. It is an equilibrium phase of several Ti systems (e.g., Ti-Fe), but it is usually a ternary phase stabilized by oxygen impurities. [60Nev] showed that this phase does not appear in the binary Ti-Rh system, but that it does appear in ternary Ti-Rh-O alloys.

Ti-Rich Liquidus and Solidus and Extent of the (βTi) Field.

The addition of Rh to (βTi) lowers its melting point. There are experimental discrepancies concerning the extent of the single-phase (βTi) field and whether (βTi) melts congruently or by a eutectic reaction.

[72Ere] claimed that the maximum solubility of Rh in (βTi) is at least 37.5 at.% Rh. They did not observe CsCl superlattice lines in X-ray diffraction patterns in alloys containing less than 37.5 at.% Rh; the X-ray results were supported by metallographic evidence that the same alloys were single phase after annealing at 1000 °C. [72Ere] found no well-defined two-phase (βTi) + TiRh region; they hypothesized a very narrow two-phase region between the observed (βTi) and βTiRh alloys. Based on solidus and metallographic data, [72Ere] concluded that the liquidus and solidus have a minimum at 1310 °C and about 25 at.% Rh.

[66Rau], however, presented metallographic evidence that a 30 at.% Rh alloy was two phase at 1270 °C; their as-cast microstructures clearly supported a eutectic reaction at about 30 at.% Rh. [66Rau] reported the compositions of (βTi) and βTiRh in equilibrium at 1100 °C as 26 and 37 at.% Rh, respectively, based on lattice parameter data.

The thermal analysis data of [66Rau] and [72Ere] were quite similar, as illustrated in Fig. 2. The eutectic isotherm of [66Rau] (1280 ± 20 °C) corresponds closely to the congruent melting point of [72Ere] (1310 ± 20 °C). Because both [72Ere] and [66Rau] presented positive evidence in favor of their versions of the diagram, the discrepancy cannot be resolved simply by reinterpretation of the data. Both versions are therefore shown in Fig. 1.

Ti_2Rh.

[64Sch1] first noted the presence of a $MoSi_2$ structure in Ti-Rh, and they designated the phase $Ti_{70}Rh_{30}$. (Note that the phase designated Ti_2Rh by [64Sch1] had the CsCl structure and was not the $MoSi_2$-type phase.) [72Ere]

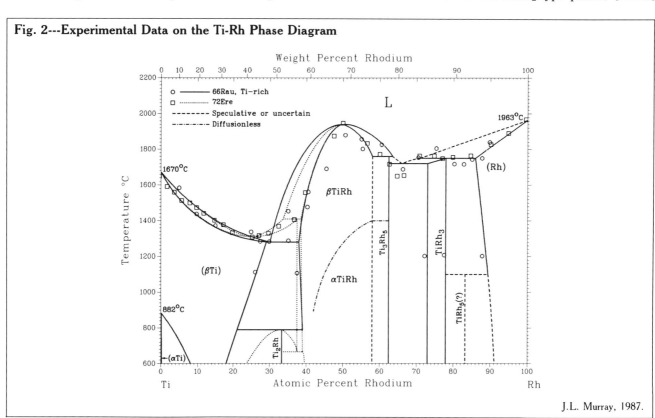

Fig. 2---Experimental Data on the Ti-Rh Phase Diagram

J.L. Murray, 1987.

found single-phase alloys only at the stoichiometric Ti₂Rh composition, and these metallographic results are preferred to the X-ray data [64Sch1].

Ti₂Rh is not stable above 790 °C [66Rau, 72Ere]. [66Rau] represented the reaction at 790 °C as peritectoid in type. [72Ere] represented it as a congruent solid-state transformation from (βTi). This follows from the disagreement between these authors concerning the extent of the single-phase (βTi) field. Neither [72Ere] nor [66Rau] presented any direct evidence for the type of the transformation.

(αTi)/(βTi) Boundaries.

The solubility of Rh in (αTi) is low (<0.07 at.%). [64Rau] originally reported the compositions of (αTi) and (βTi) in equilibrium at 700 °C as 2 to 3 at.% and 5 to 6 at.%, respectively, using X-ray fluorescence microprobe analysis of annealed specimens. [72Ere] also reported a single-phase (αTi) alloy at about 2.5 at.% Rh and 800 °C. [66Rau], however, revealed that the solubility of Rh in (αTi) is much lower, between 0.07 and 0.08 at.% Rh at 600 °C. [66Rau] attributed the discrepancy to the grain size of the two-phase alloys, which was too small for application of microprobe analysis. The lower solubility is also more consistent with the negligible effect of Rh on the lattice parameters of (αTi) [64Rau].

[72Ere] found the (βTi) transus to lie at approximately 6 to 7 at.% Rh at 600 °C, using standard metallographic techniques. [66Rau] (metallographic and X-ray studies) reported that the (βTi) transus lies between 8 and 12 at.% Rh at 700 °C. In Fig. 1, the (αTi)/(βTi) boundaries are drawn from the results of a thermodynamic calculation. The calculated composition of (βTi) in equilibrium with (αTi) at 600 °C is 8.0 at.% Rh, in satisfactory agreement with both experimentally determined values.

Composition Range 40 to 60 at.% Rh.

Three structures have been observed in alloys in the composition range 35 to 60 at.% Rh—the CsCl structure in alloys containing less than 45 at.% Rh [66Rau], the AuCu structure in alloys containing more than 45 at.% Rh [64Ram, 66Rau, 72Ere], and also at 55 at.% Rh, an orthorhombic distortion of CsCl with the NbRu-type structure [64Ram, 66Rau].

The NbRu-type phase is not included in the assessed diagram as an equilibrium phase. A complete structure determination for several well-equilibrated high-purity alloys near 55 at.% Rh is needed before any definitive conclusions can be drawn about the existence of a NbRu-type phase. In the Zr-Rh system, a compound designated Zr₃Rh₅ was also identified by the same investigators [66Rau] as having the NbRu structure. The structural and phase equilibrium study of [69Gie] showed that this structure does not occur in the binary Zr-Rh system at all. Finally, the NbRu structure has not been rigorously determined, so that identification of any structure as "NbRu-type" is ambiguous.

[66Rau] and [72Ere] agreed that the microstructures of alloys containing more than 45 at.% Rh showed signs of transformation during quenching. The assessed diagram is based on the interpretation given by [72Ere]; the high-temperature form of the equiatomic compound has the CsCl structure, and only at less than 45 at.% Rh can the CsCl structure be retained during quenching. [66Rau], however, believed that the CsCl structure belongs to a distinct phase near stoichiometry Ti₃Rh₂ and that the high-temperature βTiRh phase must have yet another structure. Still, it should be emphasized that the present

interpretation does not contradict metallographic and X-ray observations reported by [66Rau].

The interpretation by [72Ere] is preferred, because neither investigator observed (CsCl + $L1_0$) two-phase alloys and because the CsCl structure appears in all analogous Ti-based systems as the high-temperature equiatomic phase with a low-temperature modification (see Ti-Ir, Ti-Pt, and Ti-Pd). In several systems, the transformation of the CsCl structure was observed directly by high-temperature X-ray diffraction.

The two-phase field (αTiRh + βTiRh) cannot be drawn on the basis of the available information. The dot-dashed curve should be interpreted only as a representation of the observed diffusionless transformation temperatures and not as an equilibrium phase boundary. The uncertainty in its position is about ±50 °C, based on data of [66Rau] and [72Ere].

On the Rh-rich end of the single-phase αTiRh field, there are discrepancies between the two studies, because [72Ere] did not take the phase Ti₃Rh₅ into account. The work of [66Rau] is therefore used to place this boundary at 58 ± 2 at.% Rh between 1500 and 1760 °C.

Ti₃Rh₅.

The first studies of Ti₃Rh₅ led to some confusion about this region of the phase diagram; [64Ram] reported a phase Ti₃Rh₅ with a hexagonal structure similar to that of TiNi₃. [66Rau] also identified the structure as hexagonal, adding on the basis of thermal analysis and X-ray diffraction, that the compound probably has a very narrow homogeneity range and melts congruently near the congruent melting temperature of βTiRh. [66Ere] and [72Ere] reported that alloys of the Ti₃Rh₅ composition are single phase, but did not verify the existence of Ti₃Rh₅.

[69Gie] investigated the series of alloys—Ti₃Rh₅, Zr₃Rh₅, and Hf₃Rh₅, and made complete structural determinations of Ti₃Rh₅ and Hf₃Rh₅ by X-ray powder diffraction. Ti₃Rh₅ has a unique orthorhombic structure isomorphous with Ge₃Rh₅. The structure is related to, but not identical with, TiNi₃.

Ti₃Rh₅ is therefore included as a line compound in the assessed phase diagram. It is not known how this phase reacts with the liquid. [66Rau] hypothesized congruent melting at about 1900 °C and eutectic reactions with βTiRh and TiRh₃. If, however, there were a peritectic reaction βTiRh + L ⇌ Ti₃Rh₅, this would help to explain why the compound was overlooked in other investigations.

Rh-Rich Alloys.

[59Dwı] identified TiRh₃ as having the AuCu₃ structure. The existence of TiRh₃ has been verified by all subsequent investigations [64Ram, 66Ere, 66Rau, 72Ere].

Data on the homogeneity range of TiRh₃ are mutually consistent. [66Rau] gave 73 to 78 at.% Rh, determined from lattice parameters. [66Ere] and [72Ere] gave 73 to slightly less than 80 at.% Rh, based on metallographic and X-ray examination of annealed specimens. Both [66Rau] and [72Ere] reported little dependence of the homogeneity range on temperature. The homogeneity range is drawn as 73 to approximately 78 at.% Rh.

[66Rau] reported that TiRh₃ melts congruently at approximately 1760 °C and has a eutectic reaction with (Rh) at 1720 °C (±50 °C). [72Ere] claimed that it is formed by the peritectic reaction L + (Rh) ⇌ TiRh₃ at 1750 °C (±50 °C). An as-cast microstructure of an 81 at.% Rh alloy does not allow for a eutectic reaction [72Ere]. In the assessed diagram, the peritectic reaction is therefore shown. Both [66Rau] and [72Ere] reported that qualita-

Table 2 (Rh) Solvus

Reference	Composition, at.% Rh	Temperature, °C	Method/note
[66Rau] ...	84 ± 2	1720	Intersection of solidus with eutectic
	88	1200	Lattice parameters
[72Ere] ...	88 to 90	1200	Metallography, X-ray diffraction
	85 to 88	1700	
	86	1750	

tively the solidus is an approximately straight line between the melting point of Rh (1963 °C [Melt]) and the invariant reaction.

(Rh) Solvus and a Possible Compound TiRh₅.

Data on the (Rh) solvus are summarized in Table 2. In the temperature range 1700 to 1200 °C, agreement among various investigators is good. The maximum solubility of Ti in (Rh) is about 14 at.% Ti at 1750 °C.

[66Rau] identified an additional compound with a cubic structure in alloys containing 81 to 90 at.% Rh that had been annealed below 1100 °C. The existence of such a phase was not verified by [72Ere], and further attempts to identify its structure have not yet been made.

Similarly, the (Rh) solvus is uncertain below 1200 °C. [72Ere] found the solvus to lie at less than 88 at.% Rh at 600 °C, the only temperature below 1200 °C that they examined. The high solubility at 600 °C conflicts with the solvus data above 1200 °C; at 600 °C, it is expected that the observed solubility may be larger than the equilibrium solubility. Diffusion experiments indicated that the mutual solubilities of Ti and Rh are very low at 400 °C [75Deb].

Critical Experiments.

How wide is the (βTi) field? Because [66Rau] and [72Ere] presented contradictory experimental results for the same alloy, the most likely explanation is that one of them used higher purity alloys. Re-examination of as-cast and annealed microstructures in a 30 at.% Rh alloy, taking particular care to avoid contamination, should resolve the contradiction.

If the (βTi) field is wide, then a search for the two-phase [(βTi) + βTiRh] region is of theoretical interest. The use of annealed diffusion couples or temperature gradient specimens could resolve this question. It is possible that the (βTi)/βTiRh transition is higher than first

order and that [72Ere] did not find the two-phase region because it does not exist.

Examination of as-cast structures on either side of 62.5 at.% Rh would elucidate the reactions between Ti₃Rh₅ and the liquid.

Examination of as-cast structures near 80 at.% Rh would resolve the discrepancy between the proposed peritectic vs eutectic formation of (Rh). If the peritectic reaction type is verified, then a few careful solidus and liquidus measurements between 88 and 100 at.% Rh would be of interest.

TEM observations are needed to verify the extent of the single-phase (Rh) solid solution. The appearance of TiRh₅ suggests that apparently single-phase alloys may reveal more complex precipitation processes. A structure determination of TiRh₅ is needed.

Metastable Phases

In Ti-transition metal systems, metastable phases are formed from (βTi) during quenching; in Ti-rich alloys, the cph phase (α'Ti) can form martensitically, and at slightly higher transition metal content, the hexagonal ω phase forms as an intermediate phase in the decomposition of retained β into α at low temperatures. In the Ti-Rh system, the formation of supersaturated (αTi) has been verified, but not that of ω phase [66Rau].

Crystal Structures and Lattice Parameters

The structures of equilibrium and metastable phases of the Ti-Rh system are summarized in Table 3. Lattice parameter measurements are presented in Tables 4 and 5. For a discussion of the structures of the equiatomic phases, see Equilibrium Diagram.

Thermodynamics

The only experimental thermodynamic work on the Ti-Rh system is on formation of molecules in the gaseous state [74Coc]. Ti-Rh Gibbs energies were estimated in the regular solution approximation by [70Kau], based primarily on theoretical considerations. The present calculations add subregular contributions, based on the present assessment of the phase diagram. In the absence of either thermochemical data or an accurate phase diagram, both sets of Gibbs energies must be considered very tentative estimates. (See Fig. 3).

Table 3 Crystal Structures of the Ti-Rh System

Phase	Homogeneity range, at.% Rh	Pearson symbol	Space group	Strukturbericht designation	Prototype	Reference
(αTi)	0 to 0.075	$hP2$	$P6_3/mmc$	$A3$	Mg	[Pearson2]
(βTi)	0 to 29	$cI2$	$Im3m$	$A2$	W	[Pearson2]
Ti₂Rh	33.3	$tI6$	$I4/mmm$	$C11_b$	MoSi₂	[64Sch1, 64Sch2]
βTiRh	~ 38 to 58	$cP2$	$Pm3m$	$B2$	CsCl	[66Rau]
αTiRh	~ 38 to 58	$tP2$	$Pm3m$	$L1_0$	AuCuI	[64Ram]
Ti₃Rh₅	62.5	$oP16$	$Pbam$	...	Ge₃Rh₅	[69Gie]
TiRh₃	73 to 78	$cP4$	$P4/mmm$	$L1_2$	AuCu₃	[59Dwi]
TiRh₃	~ 83.8	(a)	...	...	...	[66Rau]
(Rh)	86 to 100	$cF4$	$Fm3m$	$A1$	Cu	[Pearson2]

(a) Unknown.

Fig. 3---Phase Diagram Calculated Using Thermodynamic Parameters Listed in Table 6

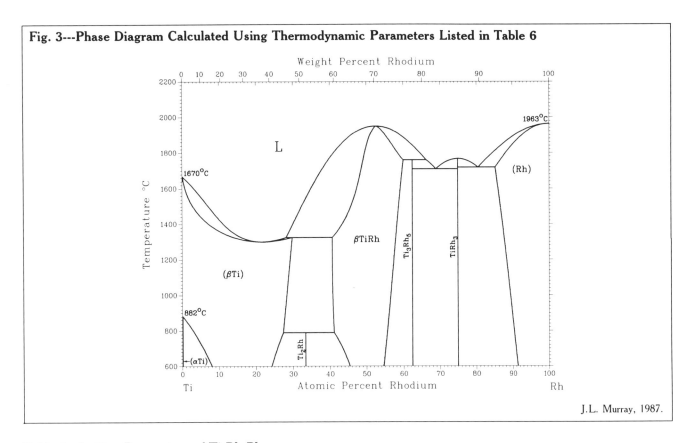

J.L. Murray, 1987.

Table 4 Lattice Parameters of Ti-Rh Phases

Phase	Reference	Composition, at.% Rh	Lattice parameters, (nm)	
			a	c
(βTi)	[62Buc]	5	0.3264	...
		10	0.3242	...
		12.5	0.3240	...
		15	0.3228	...
	[64Rau]	4	0.3245	...
		12	0.3223	...
	[66Rau]	12	0.3217	...
		15	0.3214	...
		20	0.3178	...
		25	0.3164	...
		27.5	0.3157	...
Ti$_2$Rh	[66Ere]	33.3	0.3078	0.9882
	[64Rau]	30	0.306	0.981
Ti$_3$Rh$_5$	[69Gie]	62.5	0.536	1.042
TiRh$_3$	[59Dwi]	75	0.3845	...
	[66Ere]	73 to 80	0.3823	...
	[66Rau]	75	0.3821	...
		77.5 to 80	0.3815	...
(Rh)	[66Rau]	95	0.3807	...

Thermodynamic Models.

The Gibbs energies of the solution phases are represented as:

$$G(i) = G^0(\text{Ti},i)\,(1 - x) + G^0(\text{Rh},i)\,x + RT\,[x \ln x + (1 - x) \ln (1 - x)] + x\,(1 - x)\,[B(i) + C(i)\,(1 - 2x)]$$

where i designates the phase; x is the atomic fraction of Rh; G^0 is the Gibbs energies of the pure components; and $B(i)$ and $C(i)$ are parameters in the polynomial expansion of the excess Gibbs energy. The properties of the pure components are taken from [70Kau].

βTiRh and αTiRh are for simplicity represented as a single Wagner-Schottky phase, TiRh, with Gibbs energy:

$$G(\text{TiRh}) = F(\text{TiRh}) + RT\,[x_{\text{Ti}} \ln x_{\text{Ti}} + x_{\text{Rh}} \ln x_{\text{Rh}} + v \ln v + s \ln s - (1 + v) \ln (1 + v)] + C(i)\,s + D(i)\,v$$

The Wagner-Schottky compound is conceptually resolved into Ti and Rh sublattices; v is the concentration of vacancies on the Ti sublattice; s is the concentration of substitutional Ti atoms on the Rh sublattice; x_{Ti} and x_{Rh} are the concentrations of Ti and Rh on their respective

Table 5 Lattice Parameters of Structures Near 50 at.% Rh

Reference	Composition, at.% Rh	Structure	Lattice parameters, nm			Note
			a	b	c	
[62Dwi]	50	(a)	0.2735	...	0.3679	Distorted CsCl
[66Ere]	50	(b)	0.296	0.286	0.341	$\alpha = 90°37'$
[64Ram, 64Sch1, 64Sch2]	35	CsCl	0.311	...	...	...
	50	CuAu	0.4173	...	0.354	
	55	NbRu	0.415	0.4111	0.340	
	57	CuAu	0.411			
[66Rau]	45	CuAu	0.427	...	0.333	...
	55	CuAu	0.3120	...	...	
	37.5	CsCl	0.3120	...	...	
	40	CsCl	0.3121	...	...	
	42.5	CsCl	0.3100	...	...	

(a) Tetragonal. (b) Monoclinic.

Table 6 Thermodynamic Properties of the Ti-Rh System

Properties of the pure components [70 Kau]

$G^0(\text{Ti,L}) = 0$
$G^0(\text{Ti,bcc}) = -16\ 234 + 8.368\ T$
$G^0(\text{Ti,cph}) = -20\ 585 + 12.134\ T$
$G^0(\text{Ti,fcc}) = -17\ 238 + 12.134\ T$

$G^0(\text{Rh,L}) = 0$
$G^0(\text{Rh,bcc}) = -11\ 841 + 12.761\ T$
$G^0(\text{Rh,cph}) = -18\ 117 + 8.996\ T$
$G^0(\text{Rh,fcc}) = -18\ 711 + 8.368\ T$

Excess Gibbs energies of the Ti-Rh solution phases

This evaluation	[70Kau] (calculated)
$B(\text{L}) = -161\ 000$ $C(\text{L}) = 0$	$B(\text{L}) = -150\ 783$
$B(\text{bcc}) = -180\ 000$ $C(\text{bcc}) = 40\ 000$	$B(\text{bcc}) = -148\ 038$
$B(\text{fcc}) = -131\ 000$ $C(\text{fcc}) = 40\ 000$	$B(\text{fcc}) = -146\ 658$
$B(\text{cph}) = -95\ 000$ $C(\text{cph}) = 40\ 000$	$B(\text{cph}) = -109\ 420$

Compound Gibbs energies

$G(\text{Ti}_2\text{Rh}) = -56\ 037 + 8.0\ T$
$G(\text{TiRh}) = -70\ 000 + 8.0\ T$
$G(\text{Ti}_3\text{Rh}_5) = -59\ 363 + 5.0\ T$
$G(\text{TiRh}_3) = -56\ 660 + 8.3\ T$

$C(\text{TiRh}) = 102\ 500$ $D(\text{TiRh}) = 47\ 000$

Note: Values are given in J/mol and J/mol · K.

sublattices; and x is the overall composition. All concentrations are referred to the total number of atoms. The concentrations of vacancies and substitutions are determined by minimizing the Wagner-Schottky Gibbs energy with respect to the variables s and v.

Ti_2Rh, Ti_3Rh_5, and TiRh_3 are represented as line compounds with compositions 33.333, 62.5, and 75 at.% Rh, respectively.

Thermodynamic Calculations.

For the purpose of the present calculations, it was assumed that (βTi) melts congruently, as reported by [72Ere]. Although [66Rau] reported a eutectic rather than a peritectic melting of (βTi), the (βTi) liquidus and solidus of [66Rau] extrapolate to a metastable congruent point and are not qualitatively very different from the [72Ere] results.

The regular solution Gibbs energies of [70Kau] were chosen as the starting point of the calculations, and the bcc subregular term was determined by the (βTi) liquidus and solidus. The subregular contributions to the three solid-solution phases were assumed to be equal. The Gibbs energy of TiRh_3 was determined by the (Rh) solvus and

the melting data. The Gibbs energy of TiRh was determined by the congruent melt and the extent of the single-phase region. The Gibbs energies of the compounds Ti_3Rh_5 and Ti_2Rh were determined by their three-phase equilibria (in the case of Ti_3Rh_5, by a rough guess at a peritectic temperature). None of the enthalpies or entropies thus determined is unreasonable, and the model appears to be a reasonable and self-consistent, if very approximate, representation of the thermodynamics.

In Table 6, the Gibbs energies used in the present calculation are compared to the regular solution Gibbs energies calculated by [70Kau].

Cited References

59Dwi: A.E. Dwight, "CsCl-Type Equiatomic Phases in Binary Alloys of Transition Elements," *Trans. AIME, 215,* 283-286 (1959). (Equi Diagram; Experimental)

60Nev: M.V. Nevitt, J.W. Downey, and R.A. Morris, "A Further Study of Ti$_2$Ni-Type Phases Containing Titanium, Zirconium or Hafnium," *Trans. AIME, 218,* 1019-1023 (1960). (Equi Diagram; Experimental)

62Buc: W. Buckel, G. Dummer, and W. Gey, "Superconductivity and Crystal Structure of Titanium-Rhodium Alloys," *Z. Ang.*

Phys., 14, 703-706 (1962) in German. (Crys Structure; Experimental)

62Dwi: A.E. Dwight and J.B. Darby, "Equiatomic Phases," U.S. Atomic Energy Comm. ANL-6677, 258-260 (1962). (Equi Diagram; Experimental)

64Ram: A. Raman and K. Schubert, "Structural Investigations on Some Alloy Systems Homologous and Quasi-Homologous to T4-T9," *Z. Metallkd., 55*(11), 704-710 (1964) in German. (Crys Structure; Experimental)

64Rau: Ch.J. Raub and G.W. Hull, "Superconductivity of Solid Solutions of Ti and Zr with Co, Rh, and Ir," *Phys. Rev., 133*(4A), A932-A934 (1964). (Crys Structure; Experimental)

64Sch1: K. Schubert, A. Raman, and W. Rossteutscher, "Structure Data for Intermetallic Phases (10)," *Naturwissenschaften, 51*(2), 506-507 (1964) in German. (Crys Structure; Experimental)

64Sch2: K. Schubert, H.G. Meissner, A. Raman, and W. Rossteutscher, "Structure Data for Intermetallic Phases (9)," *Naturwissenschaften, 51*(12), 287-288 (1964) in German. (Crys Structure; Experimental)

66Ere: V.N. Eremenko, T.D. Shtepa, and V.G. Sirotenko, "Intermediate Phases in Alloys of Titanium with Iridium, Rhodium, and Osmium," *Porosh. Metall., 42*(6), 68-72 (1966) in Russian; TR: *Sov. Powder Metall., 42*(6), 487-490 (1966). (Equi Diagram; Experimental)

66Rau: E. Raub and E. Roschel, "The Titanium-Rhodium Alloys," *Z. Metallkd., 57*, 546-551 (1966) in German. (Equi Diagram; Experimental)

69Gie: B.C. Giessen, R. Wang, and N.J. Grant, "New A_3B_5 Phases of the Titanium Group Metals with Rhodium," *Trans. AIME, 245*, 1207-1210 (1969). (Equi Diagram; Experimental)

70Kau: L. Kaufman and H. Bernstein, *Computer Calculations of Phase Diagrams,* Academic Press, New York (1970). (Thermo; Theory)

72Ere: V.N. Eremenko and R.D. Shtepa, "Phase Equilibria in the Binary Systems of Titanium with Ruthenium, Osmium, Rhodium, Iridium, and Palladium," *Colloq. Int. CNRS,* (205), 403-413 (1972) in German. (Equi Diagram; Experimental)

74Coc: D.L. Cocke and K.A. Gingerich, "Thermodynamic Investigation of the Gaseous Molecules TiRh, Rh_2, and Ti_2Rh by Mass Spectrometry," *J. Chem. Phys., 60*(5), 1958-1965 (1974). (Thermo; Theory)

75Deb: W.J. DeBonte, J.M. Poate, C.M. Melliar-Smith, and R.A. Levesque, "Thin-Film Interdiffusion. II. Ti-Rh, Ti-Pt, Ti-Rh-Au, and Ti-Au-Rh," *J. Appl. Phys., 46*(10), 4284-4290 (1975). (Equi Diagram; Experimental)

75Sht: T.D. Shtepa, "Interactions of Titanium with Platinum-Group Metals," *Fiz. Khim. Kondens. Sverkhtverd. Mater.,* 175-191 (1975) in Russian. (Equi Diagram; Experimental)

The Ru-Ti (Ruthenium-Titanium) System

101.07   47.88

By J.L. Murray

Equilibrium Diagram

The equilibrium solid phases of the Ti-Ru system are:

- The bcc (Ti) solid solution, based on the equilibrium solid phase of pure Ti above 882 °C. The maximum solubility of Ru in (Ti) is 25 at.% at the peritectic temperature of 1575 °C.
- The ordered bcc (CsCl) compound TiRu, which melts congruently at approximately 2130 ± 20 °C. Its homogeneity range is 45 to 53 at.% Ru.
- The cph solid solutions, (Ru) and (αTi). Pure Ti has the cph structure below 882 °C. The solubility of Ru in (αTi) is less than 0.1 at.%. The maximum solubility of Ti in (Ru) is about 14 at.% at 1825 °C.

The assessed phase diagram is shown in Fig. 1, and special points of the diagram are summarized in Table 1. The assessed diagram is drawn from the thermodynamic calculations, where they agree satisfactorily with the experimental data (see Thermodynamics).

Three determinations have been made of the Ti-Ru phase diagram [63Rau, 73Ere, 76Bor] (see also [75Sht]). Qualitatively, these investigations are in accord; quantitative discrepancies are probably caused by the difficulty in bringing alloys into equilibrium and the high temperatures involved, which cause severe oxygen contamination. The work of [73Ere] and [76Bor] are preferred, because iodide Ti was used for preparation of the alloys, which were arc- or levitation-melted.

Fig. 1---Assessed Ti-Ru Phase Diagram

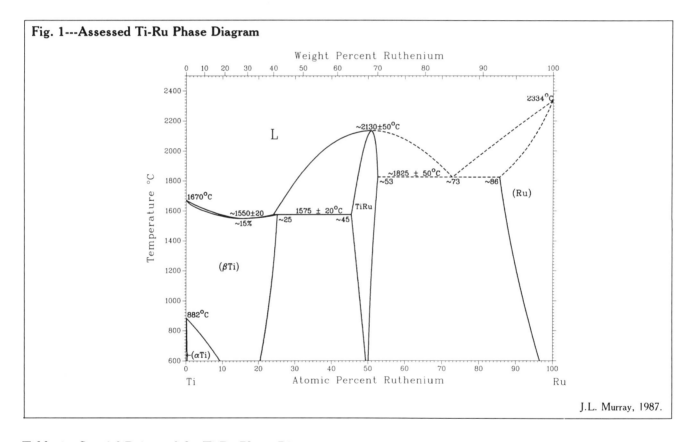

J.L. Murray, 1987.

Table 1 Special Points of the Ti-Ru Phase Diagram

Reaction	Compositions of the respective phases, at.% Ru			Temperature, °C	Reaction type
L ⇌ (βTi)		15		1550 ± 20	Congruent
L + TiRu ⇌ (βTi)	24	25	45	1575 ± 20	Peritectic
L ⇌ TiRu + (Ru)	53	73	86	1825 ± 50	Eutectic
L ⇌ TiRu		50		2130 ± 20	Congruent
L ⇌ (βTi)		0		1670	Melting point
(βTi) ⇌ (αTi)		0		882	Allotropic transformation
L ⇌ (Ru)		100		2334	Melting point

270

Ti-Rich Liquidus and Solidus.

The Ti-rich liquidus and solidus have a minimum at about 15 at.% Ru and 1550 ± 20 °C, and they terminate at the peritectic equilibrium with TiRu at 1575 ± 20 °C. The maximum solubility of Ru in (Ti) is 25 ± 2 at.% at the peritectic temperature. The assessed phase boundaries are based primarily on the melting point data of [73Ere] obtained using the Pirani method. According to the thermal analysis and microscopic work of [76Bor] and the optical pyrometric and microscopic work of [63Rau], the congruent point is about 50 °C lower than shown in Fig. 1. Experimental liquidus and solidus data are compared to the assessed diagram in Fig. 2. Ti-rich liquidus and solidus data are presented in Table 2.

The first version of the diagram [63Rau] showed a eutectic rather than a peritectic reaction at 1540 ± 10 °C. However, both [63Rau] and [73Ere] found that the liquidus and solidus become very flat at about 20 at.% Ru. [63Rau] apparently rejected the possibility of a congruent melt, because the microstructure of a 30 at.% Ru alloy annealed at 1540 °C appeared to be a coarsened eutectic. However, this microstructure can also be interpreted by the solid-state reaction implied by the assessed diagram. The as-cast microstructure published by [73Ere] strongly suggests the absence of a eutectic reaction. The version of [73Ere] is chosen, because it provides a self-consistent interpretation of all the experimental data; the minimum is clearly observed in the thermal analysis data, as well as the microstructures of both [63Rau] and [73Ere]. Therefore, the disagreement between [63Rau] and [73Ere] is considered to be satisfactorily resolved.

(αTi)/(βTi) and (βTi)/TiRu Boundaries.

[63Rau] placed an upper limit of 1 at.% on the solubility of Ru in (αTi). Experimental data on the extent of the single-phase (βTi) field are given in Table 3. All determinations of the (βTi)/(βTi) + TiRu boundary agree well. The solubility of Ru in (βTi) decreases with decreasing temperature, but in the temperature range where equilibrium can be achieved, no eutectoid reaction (βTi) ⇌ (αTi) + TiRu has been observed. The assessed phase diagram does not extend below 600 °C because of the difficulty in achieving equilibrium at lower temperature. [76Bor] examined alloys at 400 °C, for example, but these data are not included.

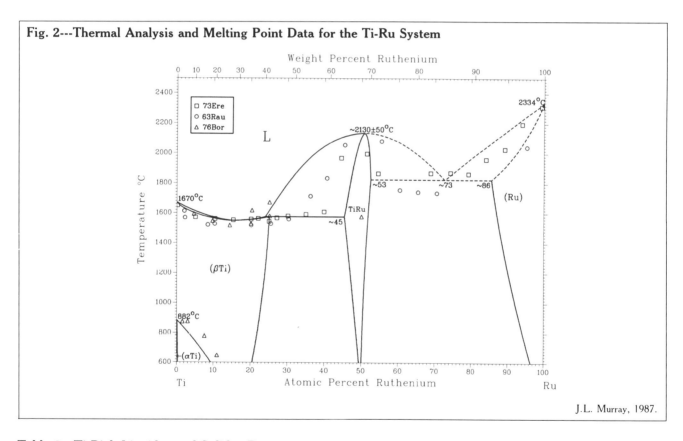

Fig. 2---Thermal Analysis and Melting Point Data for the Ti-Ru System

J.L. Murray, 1987.

Table 2   Ti-Rich Liquidus and Solidus Data

Reference	Reaction type	Composition, at.% Ru		Temperature, °C	Experimental method
		Liquidus	Solidus		
[63Rau]	Eutectic	27	24	1540 ± 10	Thermal analysis
		...	...	...	Metallography
[73Ere]	Congruent	15	...	1550	Thermal analysis
	Peritectic	24	25	1575	Metallography
[76Bor]	Congruent	8 to 11	...	< 1500	Thermal analysis
	Peritectic	23	...	1570	...

Table 3 Extent of the Single-Phase (βTi) Field

Reference	Composition, at.% Ru		Temperature, °C	Experimental method
	(βTi) transus	(βTi)/[TiRu + (βTi)]		
[63Rau]	10	20	600	Metallography,
	4 to 5	20	800	lattice parameters
	...	24 to 27	1540	
[73Ere]	...	21	1100	Metallography
	...	23	1400	
	...	27	1575	
[76Bor]	3.7	22 ± 2	800	Thermal analysis,
	6.3	23	600	metallography
		23	1570	

Table 4 Experimental Data on the Composition Range of TiRu

Reference	Composition, at.% Ru		Temperature, °C	Method/note
	Ti-rich	Ru-rich		
[63Rau]	50	...	2150(a)	Thermal analysis, liquidus
	45	52	1000	Lattice parameters
[73Ere]	50	...	2120(a)	Thermal analysis,
	< 45	51	1700	metallography
	> 45	...	1400	
[76Bor]	< 47	> 52	300 to 1100	X-ray, metallography

(a) Congruent melt.

In the (βTi)/(αTi) + (βTi) boundary, (βTi) transus, the scatter in the data is somewhat larger; the microscopic work of [63Rau] done with a greater number of alloys is weighted more heavily than that of [76Bor]. Thermal analysis measurements of [76Bor] are not used because they were carried out in air.

TiRu.

TiRu has the ordered CsCl structure [55Jor]. It melts congruently at 2130 ± 20 °C [73Ere, 2120 °C; 63Rau, ~2150 °C]. [63Rau] reported additional thermal analysis data for the liquidus, but because very large uncertainties (±50 °C) must be attached to these liquidus measurements, the assessed diagram is based only on the more accurate congruent melting point data, with the requirement of consistency with the (βTi) boundaries.

Experimental data on the homogeneity range of TiRu are given in Table 4. Based primarily on the lattice parameter data of [63Rau], the homogeneity range is 45 to 52 ± 1 at.% Ru.

Ru-Rich Liquidus, Solidus, and (Ru) Solvus.

The melting point of pure Ru is 2334 °C [Melt]. The liquidus and solidus terminate at a eutectic isotherm; the eutectic composition lies between 70 and 75 at.% Ru. The maximum solubility of Ti in (Ru) is 15 ± 1 at.%.

There is a rather large discrepancy in reported values of the eutectic temperature (1760 and 1855 °C, according to [63Rau] and [73Ere], respectively, see Table 5). The average of the two determinations (1810 ± 50 °C) was used as a preliminary estimate for the eutectic temperature. The calculated eutectic temperature is 1825 °C; considering the large uncertainty in this temperature, the calculated value agrees sufficiently well with the data that it was used to draw the assessed diagram (Fig. 1).

The (Ru) solvus was investigated only by [73Ere], using metallography of alloys heat treated at 1100, 1400, and 1700 °C in 5 at.% increments. The extrapolation of the solvus to 600 °C in Fig. 1 was made using the thermodynamic calculations described below.

Table 5 Experimental Data on the (Ru) Solvus

Reference	Composition, at.% Ru	Temperature, °C	Experimental method
[63Rau]	< 95	1000 to 1600	X-Ray
		1760(a)	Thermal analysis
[73Ere]	85	1855(a)	Thermal analysis,
	< 90	1400 to 1700	metallography
	> 90	1000	
[76Bor]	> 90	400 to 1000	Metallography, X-ray

(a) Eutectic temperature.

Metastable Phases

Martensites.

At low Ru content, (βTi) cannot be retained metastably during quenching, but transforms martensitically. At the lowest Ru content, the martensite, (α'Ti), has the cph structure. For Ru contents exceeding 2 at.%, an orthorhombic martensite (α″Ti) was observed [73Gus, 76Bor]. The structure of the orthorhombic martensite (α″Ti) is similar to that of αU and is a distortion of the cph structure. In alloys containing more than 2.4 at.% Ru, (βTi) does not transform to (α″Ti) or (α'Ti) during quenching.

ω Phase.

The ω phase appears as a transitory phase during the transformation of metastable (βTi) alloys to the equilibrium two-phase (αTi) + (βTi) condition. ω phase does not form if (βTi) transforms martensitically to supersaturated (αTi); if (βTi) alloys are aged at too high a temperature, the equilibrium cph phase precipitates directly and ω does not appear.

ω phase is of two types. Athermal (as-quenched) forms during rapid quenching of (βTi) phase alloys and has the same composition as the (βTi) matrix; aged (isothermal) ω forms during aging metastable (βTi) alloys at tempera-

Table 6 Crystal Structures of the Ti-Ru System

Phase	Homogeneity range, at.% Ru		Pearson symbol	Space group	Struktur-bericht designation	Prototype	Reference
(αTi)	0 to	>0.1	hP2	P6$_3$/mmc	A3	Mg	[Pearson2]
(βTi)	0 to	25	cI2	Im3m	A2	W	[Pearson2]
TiRu	45 to	53	cP2	Pm3m	B2	CsCl	[55Jor]
(Ru)	85 to	100	hP2	P6$_3$/mmc	A3	Mg	[Pearson2]
(α'Ti)	(a)		hP2	P6$_3$/mmc	A3	Mg	[73Gus, 76Bor]
(α''Ti)	(a)		oC4	Cmcm	A20	αU	[73Gus, 76Bor]
ω	(a)		hP3	P6/mmm	...	ωMnTi	[73Gus, 76Bor]

(a) Metastable.

Table 7 Ti-Ru Lattice Parameter Data

Phase	Reference	Composition, at.% Ru	Lattice parameters, nm		
			a	b	c
(α'Ti)	[73Gus]	1.19	0.2943	...	0.4691
(α''Ti)	[73Gus]	2.39	0.3015	0.4981	0.4666
(βTi)	[63Rau]	15	0.3194	...	...
		17.5	0.3196	...	...
		20	0.3186	...	...
	[74Gus](a)	3.8	0.3261	...	...
		5.5	0.3254	...	...
		6.4	0.3250	...	...
		7.2	0.3248	...	...
TiRu	[55Jor]	50	0.306	...	...
	[63Rau]	45	0.3076	...	...
		50	0.3067	...	...
(Ru)	[63Rau]	90	0.27135	...	0.4287
		95	0.27095	...	0.4308
	[Pearson2]	100	0.27058	...	...

(a) The original Russian article rather than the abridged translation must be consulted for information on Ti-Ru alloys.

tures usually between 200 and 450 °C, accompanied by composition changes. During aging, the composition becomes more Ti-rich, approaching a "saturation composition," which coincides approximately with the composition at which ω is most easily produced in as-quenched alloys.

Athermal ω was observed in Ti-Ru alloys containing between 3.8 and 10.5 at.% Ru [73Gus, 76Bor]; at higher Ru concentrations, only the bcc phase is found after quenching. Low-temperature aging experiments were not performed.

Crystal Structures and Lattice Parameters

The structures of the equilibrium and metastable phases of the Ti-Ru system are summarized in Table 6. Lattice parameter data are presented in Table 7.

Thermodynamics

Experimental thermodynamic data are not available for the Ti-Ru system. [70Kau] modeled the phase equilibria using the regular solution approximation. The present calculations were performed in order to model the congruent melting of (βTi), which cannot be done using the regular solution approximation. As estimated from other binary phase diagrams (e.g., Nb-Ru), the melting point of metastable bcc Ru is quite low compared to that of the equilibrium fcc Ru, between 1200 and 1300 °C. Therefore,

the bcc liquidus and solidus, having reached a minimum at about 15 at.% Ru, must, at yet higher Ru content, reach a maximum and then decrease to the metastable melting point. The regular solution model does not give rise to double extrema. The Gibbs energies used in the present calculation are compared in Table 8 to those calculated by [70Kau].

Based on the work of [70Kau], [79Lio] predicted the effect of radiation-induced disorder on the stability range of TiRu.

Thermodynamic Models.

The Gibbs energies of the solution phases are represented as:

$$G(i) = G^0(\text{Ti},i)(1-x) + G^0(\text{Ru},i)x + RT(x \ln x + (1-x)\ln(1-x) + B(i)x(1-x) + C(i)x(1-x)(1-2x)$$

where i designates the phase; x is the atomic fraction of Ru; G^0 is the Gibbs energies of the pure metals; and $B(i)$ and $C(i)$ are parameters in the polynomial expansion of the excess Gibbs energy. The pure metal Gibbs energies are from [70Kau].

TiRu was modeled with a Wagner-Schottky Gibbs energy:

$$G(\text{TiRu}) = G^0(\text{TiRu}) + RT[x_{\text{Ti}} \ln x_{\text{Ti}} + x_{\text{Ru}} \ln x_{\text{Ru}} + v \ln v + s \ln s - (1+v)\ln(1+v)] + C(\text{TiRu})s + D(\text{TiRu})v$$

Table 8 Thermodynamic Parameters of the Ti-Ru System

Properties of the pure elements

$G^0(\text{Ti,bcc}) = -16\,234 + 8.368\,T$
$G^0(\text{Ti,L}) = 0$
$G^0(\text{Ti,cph}) = -20\,585 + 12.134\,T$
$G^0(\text{Ru,bcc}) = -16\,652 + 11.715\,T$
$G^0(\text{Ru,L}) = 0$
$G^0(\text{Ru,cph}) = -21\,815 + 8.368\,T$

Excess Gibbs energy parameters (see text)

Present evaluation	Regular solution approximation [70Kau]
$B(\text{L}) = -47\,000$	$B(\text{L}) = -63\,137$
$C(\text{L}) = -15\,000$	
$B(\text{bcc}) = -60\,000$	$B(\text{bcc}) = -68\,580$
$C(\text{bcc}) = 6\,000$	
$B(\text{cph}) = -10\,000$	$B(\text{cph}) = -32\,807$

Gibbs energy of TiRu

$G^0(\text{TiRu}) = -39\,000 + 5.871\,T$

$C(\text{TiRu}) = 80\,000$
$D(\text{TiRu}) = 60\,000$

Note: Values are given in J/mol and J/mol · K.

The Wagner-Schottky compound is conceptually resolved into Ti and Ru sublattices; v is the concentration of vacancies on the Ti sublattice; s is the concentration of substitutional Ti atoms on the Ru sublattice; x_{Ti} and x_{Ru} are the concentrations of Ti and Ru on their respective sublattices; and x is the overall composition. All concentrations are referred to the total number of atoms. The concentrations of vacancies and substitutions are determined by minimizing the Wagner-Schottky Gibbs energy with respect to the variables s and v.

Present Thermodynamic Calculations.

The regular solution Gibbs energies [70Kau] were used as the starting point of the present calculations. Differences between the subregular contributions to the Gibbs energies were estimated by trial and error, based on the assessed (βTi) congruent melting point. Similarly, the Gibbs energy of TiRu was found by starting rough estimates of the enthalpy and entropy of fusion and by modifying Gibbs energy parameters by trial and error, based on the phase diagram.

The thermodynamic parameters listed in Table 8 imply a eutectic temperature of 1825 °C, a peritectic temperature of 1552 °C, a solubility of Ru in (βTi) of 20 at.% at 600 °C, and that the (Ru) solvus reaches 96 at.% at 600 °C. The calculated congruent melting point of (βTi) is 1538 °C and 17.5 at.% Ru. The calculated TiRu congruent melting point is 2130 °C, and the calculated homogeneity range is 45.6 to 52.6 at.% Ru. Thus, the calculated phase diagram agrees well with all the experimental data. For all phase boundaries except those that intersect the peritectic isotherm, the calculation is therefore used to draw the assessed diagram. The calculated and assessed diagrams are compared in Fig. 3. It must be noted, however, that because there is no thermodynamic data for this system the Gibbs energies presented here should be considered only as rough estimates.

Cited References

55Jor: C.B. Jordan, "Crystal Structure of ΓiRu and TiOs," *Trans. AIME, 203,* 832-833 (1955). (Crys Structure; Experimental)

63Rau: E. Raub and E. Roeschel, "Alloys of Ruthenium with Titanium and Zirconium," *Z. Metallkd., 54*(8), 455-462 (1963) in German. (Equi Diagram, Crys Structure; Experimental)

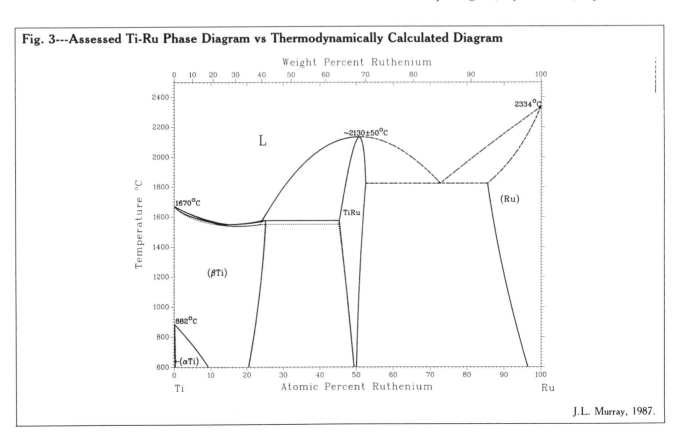

Fig. 3---Assessed Ti-Ru Phase Diagram vs Thermodynamically Calculated Diagram

J.L. Murray, 1987.

70Kau: L. Kaufman and H. Bernstein, *Computer Calculation of Phase Diagrams,* Academic Press, New York (1970). (Theory; Review)

73Ere: V.N. Eremenko, T.D. Shtepa, and V.G. Khoruzhaya, "The Ti-Ru Phase Diagram," *Izv. Akad. Nauk SSSR, Met.,* (2), 204-206 (1973) in Russian; TR: *Russ. Metall.,* (2), 155-156 (1973). (Equi Diagram; Experimental)

73Gus: L.N. Guseva, N.G. Boriskina, and L.K. Dolinskaya, "Metastable Phases in Titanium-Rich Ti-Ru Alloys," *Izv. Akad. Nauk SSSR, Met.,* (3), 215-217 (1973) in Russian; TR: *Russ. Metall.,* (3), 182-184 (1973). (Meta Phases, Crys Structure; Experimental)

74Gus: L.N. Guseva and L.K. Dolinskaya, "Metastable Phases in Titanium Alloys with Group VIII Elements Quenched from the

β-Region," *Izv. Akad. Nauk SSSR, Met.,* (6), 195-202 (1974) in Russian; TR: *Russ. Metall.,* (6), 155-159 (1974). (Meta Phases; Experimental)

75Sht: T.D. Shtepa, "Interactions of Titanium with Platinum-Group Metals," *Fiz. Khim. Knode4ns. Sverkhtverd. Mater.,* 175-191 (1975) in Russian. (Experimental)

76Bor: N.G. Boriskina and I.I. Kornilov, "The Ti-Ru Phase Diagram," *Izv. Akad. Nauk SSSR, Met.,* (2), 214-217 (1976) in Russian; TR: *Russ. Metall.,* (2), 162-165 (1976). (Equi Diagram; Experimental)

79Lio: K.-Y. Liou and P. Wilkes, "The Radiation Disorder Model of Phase Stability," *J. Nucl. Mater.,* 87, 317-330 (1979). (Meta Phases, Thermo; Theory)

The S-Ti (Sulfur-Titanium) System

32.06 47.88

By J.L. Murray

Equilibrium Diagram

The Ti-S system is very complex and requires further experimental study before a complete phase diagram can be constructed. The sulfides generally are prepared by direct reaction of the elements in sealed silica tubes at one or two elevated temperatures; quenched samples are examined by X-ray diffraction. Commonly used preparation techniques can introduce significant contamination; the predominant use of X-ray diffraction also creates an uncertainty as to whether apparently single-phase material is actually so. Systematic studies have not been performed on the temperature range of stability of phases in the TiS to TiS$_2$ region.

The assessed Ti-S phase diagram is shown in Fig. 1, and a partial listing of invariant reactions is given in Table 1. Most of the Ti-rich portion of the diagram is based on a single experimental study [67Ere]. The eutectic reaction L $\rightleftarrows$ (βTi) + Ti$_3$S, however, is supported by several independent studies. Similarly, the peritectic reaction L + TiS$_2$ $\rightleftarrows$ TiS$_3$ is well established. The remainder of the diagram, particularly the composition interval 45 to 70 at.% S, should be regarded as a map of observed structures as a function of composition. The equilibrium relationships among these phases are unknown. In the region labeled "polytypes," at least 13 distinct structures have been observed that differ only in the stacking sequences of close-packed S layers. In the region labeled "Ti$_{2.67}$S$_4$ and its superlattices," three related structures appear.

For the Ti-S system, crystal structure and phase equilibrium data are inseparable, and they will be treated together in this assessment. The structures believed to be equilibrium phases are given in Table 2. Lattice parameter data are listed in Table 3. Table 4 gives more detailed structural data for the phases in the TiS to TiS$_2$ region, including the polytype structures.

The phase nomenclature in the experimental literature is not uniform. In this assessment, the nomenclature

of [83Leg] is adopted. The stoichiometries are derived from the number of close-packed S layers in the structure and the occupancies of the Ti layers. Table 4 gives additional common nomenclatures as a guide to the experimental literature.

The assessment of thermodynamic data for the Ti-S system requires a treatment of the vapor phase of S and is thus beyond the scope of the present work.

Ti-Rich Phases (0 to 40 at.% S)

To date, [67Ere] has made the only detailed phase equilibrium study, using microscopy, thermal analysis, dilatometry, and X-ray diffraction. [67Ere] minimized contamination by preparing alloys by arc melting from iodide titanium and a master alloy. [53Gol], [57Ber], and [57Per] examined dilute alloys by microscopy and thermal analysis as part of a mechanical-properties study. Additional information on the Ti-rich phases was obtained from X-ray studies of alloys prepared by direct reaction of the elements in sealed silica tubes at elevated temperatures [37Bil, 54Hag, 57Hah, 58Bar, 59Hah]. In addition to the difficulties of determining the equilibrium diagram solely by X-ray methods, reaction of Ti with quartz tubes appears in almost every instance to have caused gross contamination of alloys and the formation of ternary phases. Some literature on titanium subsulfides identified in some steels has been omitted in the present evaluation, because these structures are stabilized by carbon or nitrogen (see [63Jel]).

The maximum solubility of S in (αTi) is about 0.02 at.% [53Gol, 57Ber, 67Ere], and the temperature of the (βTi) $\rightleftarrows$ (αTi) transition is not changed measurably by additions of S [57Ber, 67Ere]. Based on grain size studies, [57Ber] concluded that the solubility of S in (βTi) is somewhat less than in (αTi), and (Ti$_6$S + (βTi) $\rightleftarrows$ (αTi)) is therefore listed as a peritectoid-type reaction in Table 1. Based on X-ray diffraction studies, [57Hah] proposed a solubility of S in (αTi) of 23 at.%, but the diffraction data

Phase Diagrams of Binary Titanium Alloys

Fig. 1---Assessed Ti-S Phase Diagram

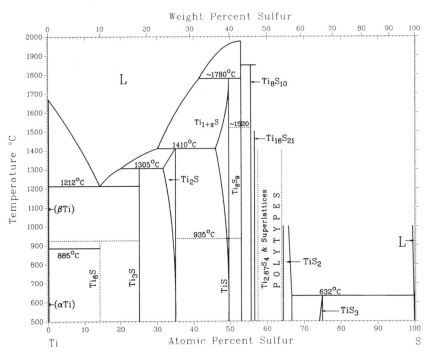

The phase diagram is extremely uncertain. Most solid lines have been used to enhance the clarity of the figure rather than to indicate the accuracy of the phase boundary. In the region between 58 and 64 at.% S, many structurally similar phases have been observed, and equilibrium relations are unknown. See text and tables for descriptions of these structures.

J.L. Murray, 1987.

Table 1 Special Points of the Assessed Ti-S Phase Diagram

Reaction	Compositions of the respective phases, at.% S			Temperature, °C	Reaction type
L ⇌ (βTi) + Ti₃S	14	0.01	25	1212	Eutectic
(βTi) + Ti₆S ⇌ (αTi)	0.01	~ 14	0.02	~ 885	Peritectoid
(βTi) + Ti₃S ⇌ Ti₆S	0.01	25	14	~ 925	Peritectoid
L + Ti₂S ⇌ Ti₃S	~ 20	~ 34	25	1305	Peritectic
L + Ti₁₊ₓS ⇌ Ti₂S	30	46	35	1410	Peritectic
Ti₁₊ₓS ⇌ TiS		49.7		935	Unknown
L + Ti₈S₉ ⇌ Ti₁₊ₓS	~ 41.5	~ 53	49.7	1780	Peritectic
L ⇌ Ti₈S₉		~ 53		~ 1975	Congruent
L + Ti₈S₉ ⇌ Ti₈S₁₀	~ 57	~ 53	~ 55.6	1850	Peritectic
L + TiS₂ ⇌ TiS₃	~ 100	~ 66.7	75	632	Peritectic
L ⇌ (βTi)	0			1670	Melting point
(βTi) ⇌ (αTi)	0			882	Allotropic transformation

can be reinterpreted as evidence for a hexagonal compound phase, Ti₆S [58Bar, 63Jel].

A eutectic reaction, L ⇌ (βTi) + Ti₃S, occurs at 14 at.% S and 1212 °C. The eutectic composition is taken from [67Ere] and [57Ber], and the temperature is taken from [67Ere] and [57Per]. Agreement among investigators is good (± 3 °C).

At more S-rich compositions (19 and 25 at.% S), [57Ber] found microscopic evidence for a peritectic reaction, and [67Ere] observed two peritectic reactions—at 20 at.% S and 1305 °C and at 30 at.% S and 1410 °C. Finally, thermal analysis effects in alloys containing up to 10 at.%

S were observed at 925 °C, but were not interpreted by [67Ere].

Using metallography and X-ray diffraction, [67Ere] found two intermetallic phases in the range 0 to 45 at.% S—tetragonal Ti₃S with a narrow homogeneity range and a complex low-symmetry phase Ti₂S with a range of about 30 to 35 at.% S at 1305 °C. [58Bar] reported an additional hexagonal phase of approximate composition Ti₆S, mentioned above as possibly explaining the very large range of (αTi) proposed by [57Hah]. According to [67Ere], who did not find Ti₆S, the discrepancy must be resolved by further work involving longer thermal treatments. Ti₆S is tentatively included in the present diagram, and the reaction at 925 °C [67Ere] is tentatively interpreted as the peritectoid decomposition of Ti$_6$S. The peritectic reactions at 1305 and 1410 °C are the melting points of Ti$_3$S and Ti$_2$S, respectively.

Monosulfides (40 to 53 at.% S)

In the composition range 40 to 53 at.% S, three structures have been established as equilibrium phases:

Table 2 Crystal Structures of the Ti-S System

Phase	Homogeneity range, at.% S	Pearson symbol	Space group	Strukturbericht designation	Prototype	Reference
(αTi)	0 to 0.02	hP2	P6$_3$/mmc	A3	Mg	[Pearson2]
(βTi)	0 to 0.01	cI2	Im3m	A2	W	[Pearson2]
Ti$_6$S	~ 14	(a)	...	...	...	[58Bar]
Ti$_3$S	25	t*24	...	...	...	[67Ere]
Ti$_2$S	31 to 35	(b)	...	...	...	[67Ere]
Ti$_{1+x}$S	46 to 49.7	hP2	P$\bar{6}$m2	...	...	[58Bar]
TiS	~ 49.7	hP4	P6$_3$/mmc	B8$_1$	NiAs	[67Ere, 58Bar]
Ti$_8$S$_9$	~ 52.6	hR18	R̄3m	...	...	[58Bar, 56Hah, 67Ere]
	53.1 to 54.6					[63Jac]
Ti$_8$S$_{10}$	~ 55.6	hP18	P6$_3$/mmc	...	...	[70Wie, 70Bec]
Ti$_{16}$S$_{21}$	~ 56.6	hR37.1	R̄3m	...	...	[70Wie]
Ti$_{2.67}$S$_4$	57.9 to 61.4	hP6.8	P6$_3$mc	...	...	[70Nor, 79Ono]
(4H)$_2$	59.8 to 60.3	mC40.14	Cc	...	...	[79Ono]
(4H)$_3$	...	mC59.8	Cc	...	...	[79Ban2, 79Ono]
Ti$_7$S$_{12}$	~ 62.8	hR19.1	R̄3m	...	...	[75Tro, 66Fli]
TiS$_2$	64.4 to 66.7	hP3	P$\bar{3}$m	C6	CdI$_2$	[75Tho, 59Jea]
TiS$_3$	~ 75	mP8	P2$_1$/m	...	...	[63Har, 75Fur]

(a) Hexagonal. (b) Unknown low symmetry.

Table 3 Ti-S Lattice Parameter Data

Phase	Lattice parameters, nm			Comment	Reference
	a	b	c		
Ti$_6$S	0.2967	...	1.450	...	[58Bar]
Ti$_3$S	0.9978	...	0.490	...	[67Ere]
Ti$_2$S	...	...	...	...	[67Ere]
Ti$_{1+x}$S	0.3287	...	0.6421	...	[58Bar]
	0.330	...	0.644	...	[54Hag]
	0.3272	...	0.6438	...	[67Ere]
	0.3299	...	0.6380	...	[58Bar]
Ti$_8$S$_9$	0.3425	...	2.6493	...	[58Bar]
	0.3147	...	2.645	...	[56Hah]
	0.3423	...	2.646	...	[67Ere]
	...	...	...	...	[63Jac]
Ti$_8$S$_{10}$	0.3429	...	2.893	...	[70Wie]
	0.3439	...	2.893	...	[70Bec]
Ti$_{16}$S$_{21}$	0.3441	...	6.048	...	[70Wie]
Ti$_{2.67}$S$_4$	0.342 to 0.3442	...	1.144 to 1.1431	...	[60Jea, 62Jea]
	0.34198	...	1.1444	...	[70Nor]
	0.3445	...	1.145	...	[57Wad]
	0.34385	...	1.14322	...	[79Ono]
	0.343	...	1.142	...	[58Mct]
	0.343	...	1.144	...	[56Hah]
(4H)$_2$	0.594395	1.02951	2.28583	...	[79Ono]
(4H)$_3$	1.03	0.592	3.49	...	[79Ban2]
	1.0238	0.59384	3.49245	...	[79Ono]
Ti$_7$S$_{12}$	0.3420	...	3.4326	...	[75Tro]
	0.3418	...	3.436	...	[66Fli]
TiS$_2$	0.34073	...	0.56953	...	[75Tho]
	0.339	...	0.570	...	[54Hag]
	0.34049	...	0.56912	...	[59Jea]
TiS$_3$	0.501	0.340	0.888	β = 97.74°	[58Mct]
	0.497	0.342	0.878	β = 97°10′	[58Jea, 62Jea]
	0.499	0.338	0.87784	β = 97.324°	[75Fur]
	0.499	0.338	1.76	β = 97.324°	[56Han]
	0.4973	0.3433	0.8714	β = 97.5°	[63Har]

Table 4 Equilibrium and Metastable Phases in the TiS to TiS$_2$ Region: Nomenclature and Descriptions

Phase	Ramsdell notation	Space group	Reference	Zhadanov symbol	Other nomenclatures
Composition wave structures					
Ti$_{16}$S$_{21}$ 21R		$R\bar{3}m$	[66Fli, 70Wie, 63Jea, 73Til]	$(2221)_3$	Ti$_3$S$_4$ [66Fli, 70Wie]
Ti$_8$S$_{10}$ 10H		$P6_3/mmc$	[66Fli, 70Wie, 70Bec]	$(221)_2$	Ti$_4$S$_5$ [66Fli, 70Wie, 63Jac]
Ti$_8$S$_9$ 9R		$R\bar{3}m$		$(21)_3$	Ti$_{(1-x)}$S [58Bar]
					HT-TiS [57Hah]
Stacking variant structures (polytypes)					
Ti$_{2.67}$S$_4$ 4H		$P6_3mc$	[70Nor, 62Jea, 57Wad, 79Ono]	(22)	Ti$_3$S$_4$ [56Hah]
Ti$_7$S$_{12}$ 12R		$R\bar{3}m$	[66Fli, 71Tro, 75Tro]	$(31)_3$	Ti$_{2+x}$S$_4$ [70Nor]
					Ti$_5$S$_8$ [66Fli, 73Til]
					Ti$_8$S$_{12}$ [71Tro]
	8H	$P\bar{3}m1$	[73Tro]	(3212)	...
	10H	$P\bar{3}m1$	[73Tro]	(321112)	...
	12H	$P6_3/mc$	[71Tro]	(312312)	...
	16H	...	[78Leg]	(32132221)	...
	18H	...	[78Leg]	(3212321112)	...
	24H	...	[71Tro]	...	...
	26H	$P\bar{3}m1$	[76Leg]	$(3212)_2(321112)$	...
	40H	...	[73Tro]	...	...
	120H	...	[77Leg]	...	...
	24R	$R3m$	[76Mor1, 76Mor2]	$(3221)_3$	...
	48R	$R3m$	[73Tro]	$(31231231)_3$	...
	696R	...	[73Tro]	...	...

- TiS with the hexagonal NiAs structure [49Ehr, 54Hag, 56Hah, 58Bar, 63Jac, 67Ere].
- Ti$_{1+x}$S with another hexagonal structure related to that of NiAs, believed to be a distinct phase with the WC structure [56Hah, 58Bar, 59Hah]. [63Jel] reported that crystallographic data are also consistent with a disordered NiAs structure. Conceptually, the disordered NiAs structures appear more likely, because Ti atoms occupy octahedral rather than tetrahedral sites in the structure. This is also the situation with other structures of this system and of other transition metal-S systems. [58Bar] designated various hexagonal structures TiS$_{1-x}$ I, II, and III and considered them distinct phases. [59Hah] designated the disordered NiAs-type structure Ti$_3$S$_2$.
- Ti$_8$S$_9$ has a rhombohedral crystal structure. Structurally, this phase should be classed among the series in the range TiS to TiS$_2$. However, it is discussed here because [56Hah] and [59Hah] incorrectly believed it to be the high-temperature form of TiS.

[59Abe] suggested that the composition range 40 to 57 at.% S spans a two-phase field, "Ti$_5$S$_4$" plus another phase with a wide homogeneity range. These results were based on measurements of the ratios of the H$_2$S/H$_2$ pressures when hydrogen gas was equilibrated with the sulfide. They are discounted, because the diffraction pattern of "Ti$_5$S$_4$" [59Abe] is different from any other reported in the literature [63Jel], and the S-rich phase with a wide homogeneity range clearly does not represent an equilibrium phase.

The claim [56Hah, 59Hah] that rhombohedral Ti$_8$S$_9$ is a high-temperature form of TiS was based on the observation of an irreversible transformation to Ti$_8$S$_9$ when TiS that had been prepared at 700 °C was heated above 1000 °C. [63Jac] demonstrated that the appearance of Ti$_8$S$_9$ during high-temperature heat treatment is due to shifts in overall composition by the loss of Ti in a reaction

with the silica tubes. [63Jac] showed that TiS is stable with respect to Ti$_8$S$_9$ to at least 1000 °C. Therefore, the range 33 to 43 at.% S attributed by [56Hah] and [59Hah] to Ti$_{1+x}$S should not be used. [63Jac] did not distinguish the two hexagonal NiAs-type structures. However, based on X-ray diffraction data, they assigned homogeneity ranges of 49.2 to 51.5 at.% S and 53.1 to 54.6 at.% S to Ti$_{1-x}$S and Ti$_8$S$_9$, respectively, at 1000 °C. [58Bar] reported that Ti$_8$S$_9$ has an approximate composition of 52.6 at.% S, based on structural, density, and X-ray fluorescence measurements.

The only other quantitative determination of the stability ranges was conducted by [67Ere]. They reported that the homogeneity range of hexagonal Ti$_{1-x}$S depends strongly on temperature, being very narrow at low temperatures and about 46 to 49.7 at.% S at 1400 °C. Ti$_8$S$_9$ has a composition of ~52 at.% S and undergoes a phase transition involving only a small enthalpy change at 1520 °C. Ti$_8$S$_9$ is the low-temperature form; the high-temperature form was not identified.

Melting points were determined by [65Fra] and [67Ere]. According to [65Fra], a sample of TiS melted at 1947 ± 20 °C. It is not known whether the composition was controlled sufficiently to distinguish the melting of TiS from that of Ti$_8$S$_9$. [67Ere] found that Ti$_{1+x}$S melts by the peritectic reaction L + Ti$_8$S$_9$ ⇌ Ti$_{1+x}$S at 1780 °C and that alloys containing 50 to 53 at.% S melted near ~2000 °C, roughly corresponding to the [65Fra] melting point of the monosulfide.

In the present assessment, the melting points of [67Ere] are used. The transformation at 935 °C is tentatively interpreted as the order-disorder transition $P6_3/mmc \rightarrow P6m$ in the hexagonal NiAs-type phase. The phase transition in Ti$_8$S$_9$ is indicated on the assessed phase diagram, but no structural interpretation can be offered at present. Because X-ray diffraction tends to overestimate the homogeneity ranges of the phases, they are shown as narrower than those given by [63Jac] and closer to the ranges determined microscopically by [67Ere]. Because the temperature dependence of the homogeneity range has not been established, Ti_8S_9 is shown as a line compound at 53 at.% S on the assessed diagram.

TiS to TiS₂ Region

TiS and TiS_2 have $B8_1$ (NiAs) and $C6$ (CdI_2) structures, respectively. Both structures contain close-packed layers of S atoms, with Ti atoms occupying octahedral sites between the close-packed layers. The two structures thus differ only in the occupations of metal atom sites. Nineteen distinct additional structures have been observed in the TiS and TiS_2 region (see Tables 2 through 4).

These structures were prepared at temperatures between 600 and 1000 °C. The available information on the composition ranges in which the phases were observed were derived from lattice parameter vs composition curves, the appearance of a second phase, crystal structure determinations, or vapor pressure measurements as a function of composition.

These structures are discussed in terms of the following structural classification:

- The occupations of the Ti layers can deviate from the simple alternation of full and partially full layers, and structures of this type can be described by concentration waves in the direction perpendicular to the close-packed plane. The structures of Ti_8S_9, Ti_8S_{10}, and $Ti_{16}S_{21}$ are described by concentration waves.
- Polytype structures are stacking variants of the S-layers, which can have very long periods (as many as 696 layers). The Ti layers alternate between full and partially full and are disordered within the plane. The possible space groups for polytype stacking variants are $P6_3mc$, $P3m1$, $P\bar{3}m1$, $R\bar{3}m$, and $R3m$. The most commonly observed polytype structures are $Ti_{2.67}S_4$ and Ti_7S_{12}.
- Ti atoms and vacancies can order within a partially occupied layer, and these structures will be referred to as the superlattice structures. These structures are based on $Ti_{2.67}S_4$ and are designated $(4H)_2$ and $(4H)_3$.

In this assessment, both the Ramsdell and the Zhadanov notations will be given for the various structures in Table 4. The Ramsdell notation gives the number of S layers and the symmetry class (hexagonal or rhombohedral), e.g., $12R$. The Zhadanov notation explicitly gives the stacking sequence of S layers. Cubic and hexagonal stackings are abbreviated as:

$$cch \equiv 3, \; ch \equiv 2, \; h \equiv 1$$

where h or c is assigned to a layer according to whether the two neighboring layers are the same or different from each other. For example, in the Zhadanov notation, the $12H$ structure ($BCABABCBACAC$) is (31231). In the following discussion, the number of layers refers to the S layers.

Structures Described by Concentration Waves

Ti₈S₉.

The nine-layer rhombohedral phase Ti_8S_9 was observed by [56Hah], [58Bar], and [67Ere]. Phase equilibrium data involving Ti_8S_9 have been discussed previously in connection with the monosulfides.

Ti₈S₁₀.

[34Pic] synthesized Ti_4S_5, and [63Jac] found a narrow single-phase region near this composition (55.6 at.% S),

Table 5 Experimental Data on the Stability Ranges of $Ti_{2.67}S_4$ and Related Phases

Temperature, °C	Homogeneity range, at.% S	Phase	Reference
800	57.6 to 59.2	$Ti_{2.67}S_4$	[63Ben]
	59.2 to 62.3	$(4H)_3$	
900	? to 49.6	$Ti_{2.67}S_4$ (SRO)	[82Sae2]
	59.7 to 60.6	$(4H)_2 + (4H)_3$	
	60.8 to 63.4	$Ti_{2.67}S_4$ (disordered)	
1000	57.9 to 59.8	$Ti_{2.67}S_4$	[63Ben]
	59.8 to 60.2	$(4H)_3$	
	60.2 to 61.4	$Ti_{2.67}S_4$	

but did not determine the structure. An unidentified low-symmetry phase was also reported by [58Mct] at about 54.5 at.% S. The hexagonal ten-layer structure is due to [66Fli], [70Bec], and [70Wie]. Using electron microscopy, [73Til] found only $Ti_{16}S_{21}$ in a sample ostensibly of Ti_8S_{10} stoichiometry.

Ti₁₆S₂₁.

$Ti_{16}S_{21}$ was reported to have a rhombohedral 21-layer structure by [66Fli] and [70Wie], who also established that this is the same structure previously indexed as monoclinic by [63Jea]. The homogeneity range was reported as 56.2 to 56.7 at.% S at 1000 °C [63Jac, 63Jea]; [73Til] found that the pure phase was almost perfectly ordered in a 55.6 at.% S sample, and [66Fli] and [70Wie] assigned the Ti_3S_4 stoichiometry, although crystals of that composition were not single phase. Based on the refined crystal structure, the composition 56.6 at.% S can be derived, which lies on the edge of the single-phase field [63Jac, 63Jea].

At Ti_3S_4 stoichiometry, [73Til] found two new phases with 11-layer and 15-layer structures that had not been reported previously.

Ti₂.₆₇S₄ and Superlattice Structures

Three predominant phases have been observed in the range 57.1 to 63.4 at.% S—$Ti_{2.67}S_4$ with the $(4H)$ structure and two $(4H)$-based superlattices, $(4H)_2$ and $(4H)_3$. The structure of the disordered phase $Ti_{2.67}S_4$ originally was assigned the symmetry $P6_3/mcc$ [56Hah] and later was correctly determined as $P6_3mc$ [57Wad, 58Mct, 70Nor]. In the long-range ordered superlattice structures, vacancies are ordered in the partially filled Ti planes, and stacking sequences $(4H)_2$ and $(4H)_3$ occur. Both $(4H)_2$ and $(4H)_3$ have the monoclinic symmetry Cc. The $(4H)_2$ structure was determined by [58Bar], [79Ban2], [79Ono], and [80Mor]; the $(4H)_3$ structure was determined by [79Ban1], [79Ono], and [81Tro]. In addition, [60Jea], [62Jea], and [63Ben] reported a superlattice involving only a tripling of the c axis. Based on re-examination of the powder patterns, [63Jel] identified this phase as $(4H)_3$.

Short-range ordered $Ti_{2.67}S_4$ also has been found [76Mor2, 78Mor, 82Sae2]. The ordering of Ti atoms and vacancies in the Ti plane can occur without the formation of 8- and 12-layer fully long-range ordered structures. [78Mor] postulated that the long-range order locks in at the stoichiometric composition $Ti_{2.67}S_4$.

$Ti_{2.67}S_4$ or one of its superlattices has been observed over the range 57.1 to 63.4 at.% S, but this approximate range does not define the single-phase field or the effect of temperature. Attempts at quantitative determination of the homogeneity ranges of $Ti_{2.67}S_4$, $(4H)_2$, and $(4H)_3$ were made by [62Jea], [63Ben], and [82Sae2] (see Table 5).

[82Sae2] noted that the $(4H)_2 + (4H)_3$ two-phase field was not distinguishable in their vapor pressure data, either because of the imprecision of the technique or because of transformations that occur during quenching. It appears that the superlattice structures are stable in a small range about stoichiometry (60 at.% S). However, at present, it is not possible to draw quantitative phase boundaries, but only to indicate the composition range in which these structures occur. Note that the polytype range and that of the superlattice structures overlap in Fig. 1.

Polytypes

Polytypes occur in the composition range 59 to 62 at.% S (see [83Leg] for a comprehensive review of polytype structures). Polytype structures are given in Table 4. The phases do not occur homogeneously, and it is a question of current interest as to whether polytypes should be considered distinct equilibrium phases. Thus, only the region in which the polytypes occur is indicated on the assessed phase diagram, and phase boundaries are not drawn. Some progress toward delineating possible equilibrium relations has been made by [77Leg], [78Leg], and [80Leg], who demonstrated that $2H$ and $4H$ single crystals underwent the transformations $2H \rightarrow 8H$ and $4H \rightarrow 16H$ after annealing at 820 °C.

The $12R$ polytype Ti_7S_{12} is frequently observed [73Til], and its structure was determined before the discovery of polytypism as a pervasive phenomenon in this system [66Fli]. [82Sae2] did not find Ti_7S_{12} in samples prepared by direct reaction of the elements, but only in samples prepared by a chemical transport reaction. [82Sae1] prepared single-phase Ti_7S_{12} at 62.3 at.% S. At other compositions between 60.1 and 63.7 at.% S, they found two-phase (Ti_7S_{12} + $Ti_{2.67}S_4$). Ti_7S_{12} appeared to undergo a phase transformation at about 100 °C, but the kinetics were too slow for any further elucidation.

[73Til] examined a 61.5 at.% S sample by TEM, finding Ti_7S_{12}, $Ti_{2.67}S_4$, and TiS_2. Coherent lamellae of some other stacking variants were found intergrown in Ti_7S_{12}, and it was noted that they could be Ti_7S_{12}, $Ti_{2.67}S_4$, or some new, unidentified phase.

The most frequently observed hexagonal polytypes are $8H$, $10H$, and $12H$.

Disulfide TiS_2

The existence of TiS_2 was recognized in early work [28Oft, 37Bil, 49Ehr]. TiS_2 has a homogeneity range that varies with temperature [59Jea, 62Jea, 75Tho, 77Mik], and there has been controversy concerning the extent of this range. At 800 and 1000 °C, [59Jea] and [62Jea] were unable to prepare TiS_2 containing more than 65.8 at.% S. On the Ti-rich side, they noted a discrepancy between the Ti-rich phase boundary determined from lattice parameters (63 at.% S) and that determined from the appearance of a second phase (64.4 at.% S). The latter value is in good agreement with the results of vapor pressure measurements [82Sae2] and is adopted in the assessed diagram.

[75Chi], [75Rie], and [75Tho] made further studies of structure and lattice parameters using alloys prepared at 600 °C and concluded that at 600 °C TiS_2 can indeed be prepared with the stoichiometric composition. Lattice parameters [75Tho] varied linearly with composition, except in the range 65.7 to 66.7 at.% S, where a plateau was attributed to the onset of a superlattice. Discrepancies in lattice parameters [70Tak, 75Tho] remain to be resolved and interpreted.

[77Mik] measured vapor pressures as a function of composition and temperature to project the phase diagram

for the condensed phases. In agreement with [75Tho], TiS_2 was reported as being single phase at 632 °C. The homogeneity range decreases with increasing temperature, qualitatively in agreement with [59Jea], [63Ben], and [75Tho]. [77Mik] attributed the deviations from stoichiometry at 800 to 1000 °C to the high S pressure (>20 bar at 800 °C) and noted that the pressure decreases rapidly with excess Ti content. The phase diagram reported by [77Mik] includes a two-phase $Ti_{1+x}S_2$ + TiS_6 region, but no $TiS_2 \rightleftarrows Ti_{1+x}S_2$ + TiS_3 reaction. It is not known how the two-phase field was derived from the P-T-x data; the assessed diagram shows a single-phase TiS_2 field.

Titanium Trisulfide

Numerous investigators have verified the existence of TiS_3 as an equilibrium phase with a restricted homogeneity range [37Bil, 56Hah, 58Jea, 58Mct, 63Har, 75Fur, 77Mik]. One contradictory report [54Sch] can be explained; the reaction temperature of 750 °C is above that of the peritectic reaction L + $TiS_2 \rightleftarrows TiS_3$, and TiS_3 is not a stable phase at any pressure at that temperature.

The stability range of TiS_3 was examined by [58Jea], [62Jea], and [77Mik]. [77Mik] measured the total pressure of the S vapor in equilibrium with the condensed phases over the range 400 to 800 °C, 10^3 to 10^6 Pa, and 60 to 75 at.% S. From the P-T-x equilibrium data, the phase diagram for the condensed phases was projected. The peritectic reaction L + $TiS_2 \rightleftarrows TiS_3$ occurs at 632 °C [77Mik], in good agreement with measurements of the temperature above which TiS_2 rather than TiS_3 is formed by direct reaction (625 ± 5 °C [62Jea]).

For compositions between 67 and 75 at.% S, [63Jea] measured the temperature at which precipitation of TiS_3 occurred during cooling from 700 °C and the temperature at which precipitation of TiS_3 disappeared during heating. This temperature should lie within the two-phase field TiS_2 + TiS_3, but does not necessarily establish the homogeneity range of either phase. From vapor pressure measurements, [77Mik] established three data points on the Ti-rich side of the homogeneity range. Based on these data, the Ti limit was determined to be 75 at.% S at 632 °C and 74.2 at.% S at 500 °C.

Cited References

28Oft: I. Oftedal, "X-Ray Investigation of SnS_2, TiS_2, $TiSe_2$, $TiTe_2$" *Z. Phys. Chem., 134,* 301-310 (1928) in German. (Equi Diagram; Experimental)

34Pic: M. Picon, "On the Sulfides of Titanium" *Bull. Soc. Chim.,1,* 919-926 (1934) in French. (Equi Diagram; Experimental)

37Bil: W. Biltz and P. Ehrlich, "The Sulfides of Titanium," *Z. Anorg. Chem., 234*(2), 97-116 (1937) in German. (Equi Diagram; Experimental)

49Ehr: P. Ehrlich, "X-Ray Investigation of Titanium Sulfides and Vanadium Monotelluride" *Z. Anorg. Chem., 260,* 13 (1949) in German. (Equi Diagram; Experimental)

53Gol: R.M. Goldhoff, H.L. Shaw, C.M. Craighead, and R.I. Jaffee, "The Influence of Insoluble Phases on the Machinability of Titanium," *Trans. ASM, 45,* 941-971 (1953). (Equi Diagram; Experimental)

54Hag: G. Hagg and N. Schonberg, "X-Ray Studies of Sulfides of Titanium, Zirconium, Niobium, and Tantalum" *Arkiv. Kemi, 7*(40), 371-380 (1954). (Equi Diagram, Crys Structure; Experimental)

54Sch: N. Schonberg, "The Tungsten Carbide and Nickel Arsenide Structures" *Acta Metall.,2,* 427-432 (1954). (Equi Diagram, Crys Structure; Experimental)

56Hah: H. Hahn and B. Harder, "On the Crystal Structure of Titanium Sulfides" *Z. Anorg. Chem., 288,* 241-256 (1956) in German. (Equi Diagram, Crys Structure; Experimental)

57Ber: L.W. Berger, D.N. Williams, and R.I. Jaffee, "The Effect of Sulphur on the Properties of Titanium and Titanium Alloys" *Trans. ASM, 49,* 300-314 (1957). (Equi Diagram; Experimental)

57Hah: H. Hahn and P. Ness, "On the Question of the Existence of Subchalcogenides of Titanium," *Naturwissenschaften,44,* 581 (1957) in German. (Equi Diagram; Experimental)

57Per: R.A. Perkins, "Discussion" of [57Ber],*Trans. ASM, 49,* 312-314 (1957). (Equi Diagram; Experimental)

57Wad: A.D. Wadsley, "Partial Order in the Non-Stoichiometric Phase $Ti_{2+x}S_4$ (0.2 < x < 1)," *Acta Crystallogr.,* 10, 715-716 (1957). (Crys Structure; Experimental)

58Bar: S.F. Bartram, "The Crystallography of Some Titanium Sulfides," thesis, Rutgers University, 136 p (1958);*Dissertation Abstr., 19,* 1216 (1958). (Equi Diagram, Crys Structure; Experimental)

58Jea: Y. Jeannin and J. Benard, "Structure and Stability of Titanium Trisulfides," *Compt. Rend., 246,* 614-617 (1958) in French. (Equi Diagram, Crys Structure; Experimental)

58Mct: F.K. McTaggart and A.D. Wadsley, "The Sulphides, Selenides, and Tellurides of Titanium, Zirconium, Hafnium, and Thorium," *Aust. J. Chem., 11,* 445-457 (1958). (Crys Structure; Experimental)

59Abe: R.P. Abendroth and A.W. Schlechten, "A Thermodynamic Study of the Titanium-Sulfur System in the Region $TiS_{1.93}$ to $TiS_{0.80}$, *Trans. AIME, 215,* 145-151 (1959). (Equi Diagram; Experimental)

59Hah: H. Hahn and P. Ness, "The Subchalcogenides of Titanium," *Z. Anorg. Chem., 302,* 17-36 (1959). (Equi Diagram; Experimental)

59Jea: Y. Jeannin and J. Benard, "Non-Stoichiometric Phase TiS_2: Stability Range and Defects," *Compt. Rend., 248,* 2875-2877 (1959) in French. (Experimental)

60Jea: Y. Jeannin, "Non-Stoichiometric Phase Ti_2S_3: Extent, Defects, and Ti_8S_{12} Superstructure," *Compt. Rend., 251,* 246-248 (1960) in French. (Equi Diagram, Crys Structure; Experimental)

*62Jea: Y. Jeannin, "Contribution to the Crystallochemistry of the Titanium-Sulfur System," *Ann. Chim., 7,* 57-83 (1962) in French. (Equi Diagram; Experimental)

63Ben: J. Benard and Y. Jeannin, "Investigations of Nonstoichiometric Sulfides. I. Titanium Sulfides, TiS_2 and Ti_2S_3," *Advan. Chem. Ser. 39,* Nonstoichiometric Compounds, American Chemical Society, Washington, DC, 191-203 (1963). (Equi Diagram, Crys Structure; Experimental)

63Har: H. Haraldsen, A. Kjekshus, E. Rost, and A. Steffensen, "On the Properties of TiS_3, ZrS_3, and HfS_3," *Acta Chem. Scand.,* 7, 1283-1292 (1963). (Equi Diagram, Crys Structure; Experimental)

63Jac: Y. Jacquin and Y. Jeannin, "The Titanium-Sulfur System near the Composition TiS," *Compt. Rend., 256,* 5362-5365 (1963) in French. (Equi Diagram, Crys Structure; Experimental)

63Jea: Y. Jeannin, "Existence of an Ordered Vacancy Superstructure Ti_3S_4," *Compt. Rend., 256,* 3111-3113 (1963) in French. (Equi Diagram, Crys Structure; Experimental)

63Jel: F. Jellinek, "Sulfides of the Transition Metals of Groups IV, V, and VI," *Arkiv. Kemi, 20*(26), 447-480 (1963). (Equi Diagram, Crys Structure; Review)

65Fra: H.F. Franzen and P.W. Gilles, "High-Temperature Vaporization and Thermodynamic Properties of Titanium Monosulfide," *J. Chem. Phys., 42*(3), 1033-1038 (1965). (Equi Diagram; Experimental)

66Fli: E. Flink, G.A. Wiegers, and F. Jellinek, "The System Titanium-Sulfur. I. The Structure of Ti_3S_4 and Ti_4S_5," *Rec. Trav. Chim., 85,* 869-872 (1966). (Equi Diagram, Crys Structure; Experimental)

*67Ere: V.N. Eremenko and V.E. Lismovnichii, "Structures and Properties of Titanium-Sulfur Alloys," *Khal'kogenidy (Kiev),* 69-78 (1967) in Russian. (Equi Diagram; Experimental)

70Bec: O. Beckmann, H. Boller, and H. Nowotny, "The Crystal Structures of Ta_2S_2C and Ti_4S_5 ($Ti_{0.81}S$)," *Monatsh. Chem., 101,* 945-955 (1970). (Crys Structure; Experimental)

70Nor: L.J. Norrby and H.F. Franzen, "Refinement of the Crystal Structure of Nonstoichiometric $Ti_{2+x}S_4$," *J. Solid State Chem., 2,* 36-41 (1970). (Crys Structure; Experimental)

70Tak: S. Takeuchi and H. Katsuta, "Characteristics of Nonstoichiometry and Lattice Defects of the TiS_2 Phase," *J. Jpn. Inst. Met., 34,* 758-763 (1970) in Japanese. (Crys Structure; Experimental)

70Wie: G.A. Wiegers and F. Jellinek, "The System Titanium-Sulfur. II. The Structure of Ti_3S_4 and Ti_4S_5," *J. Solid State Chem., 1,* 519-525 (1970). (Equi Diagram, Crys Structure; Experimental)

71Tro: E. Tronc and M. Huber, "Polytypism in the Ti-S System: 12H Polytype $TiS_{1.6}$," *C.R. Acad. Sci., Paris, 272,* 1018-1021 (1971) in French. (Crys Structure; Experimental)

73Til: R.J.D. Tilley, "An Electron Microscope Study of Some Titanium Sulphides," *J. Solid State Chem., 7,* 213-221 (1973). (Experimental)

73Tro: E. Tronc and M. Huber, "Polytypism in the Titanium-Sulfur System," *J. Phys. Chem. Solids, 34,* 2045-2058 (1973). (Equi Diagram, Crys Structure; Experimental)

75Chi: R.R. Chianelli, J.C. Scanlon, and A.H. Thompson, "Structure Refinement of Stoichiometric TiS_2," *Mater. Res. Bull., 10,* 1379-1382 (1975). (Equi Diagram, Crys Structure; Experimental)

75Fur: S. Furuseth, L. Brattas, and A. Kjekshus, "On the Crystal Structures of TiS_3, ZrS_3, $ZrSe_3$, HfS_3, and $HfSe_3$," *Acta Chem. Scand. A, 29,* 623-631 (1975). (Crys Structure; Experimental)

75Rie: C. Riekel and R. Schollhorn, "Structure Refinement of Nonstoichiometric TiS_2," *Mater. Res. Bull., 10,* 629-634 (1975). (Crys Structure; Experimental)

75Tho: A.H. Thompson, F.R. Gamble, and C.R. Symon, "The Verification of the Existence of TiS_2," *Mater. Res. Bull., 10,* 915-920 (1975). (Equi Diagram, Crys Structure; Experimental)

75Tro: P.E. Tronc, R. Moret, J.J. Legendre, and M. Huber, "Refined Structure of the 12 R Polytype Ti_8S_{12}," *Acta Crystallogr., B31,* 2800-2804 (1975). (Crys Structure; Experimental)

76Leg: J.L. Legendre and M. Huber, "A Method of Structural Resolution for Polytypes. Application to the Titanium-Sulfur Polytype 26H," *Acta Crystallogr., B32,* 3209-3213 (1976) in French. (Crys Structure; Experimental)

76Mor1: R. Moret and M. Huber, "Structure of a New Sulfur-Titanium Polytype: 24R," *Acta Crystallogr., B32,* 1302-1303 (1976) in French. (Crys Structure; Experimental)

76Mor2: R. Moret, M. Huber, and R. Comes, "Diffuse Scattering and Titanium Short-Range Order in $Ti_{1-x}S_2$," *Phys. Status Solidi (a), 38,* 695-700 (1976). (Crys Structure; Experimental)

77Leg: J.J. Legendre and M. Huber, "Formation of Titanium Sulphide Hexagonal Polytypes," *Acta Crystallogr., A33,* 971-975 (1977). (Crys Structure; Experimental)

*77Mik: J.C. Mikkelsen, "P-T-X Phase Diagram for Ti-S from 60 to 75 Atomic Percent Sulfur," *Nuovo Cimento, 38*(2), 378-386 (1977). (Equi Diagram; Experimental)

78Leg: J.J. Legendre and M. Huber, "Two New Polytype Structures in the Titanium-Sulfur System: 16H and 18H," *Acta Crystallogr., A34,* 982-986 (1978) in French. (Crys Structure; Experimental)

78Mor: R. Moret, E. Tronc, M. Huber, and R. Comes, "Two-Dimensional Model for Titanium Ordering in the $Ti_{1+x}S_2$ Polytypes," *Philos. Mag. B, 38*(2), 105-119 (1978). (Crys Structure; Experimental)

79Ban1: Y. Bando, M. Saeki, M. Onoda, I. Kawada, and M. Nakahira, "(4H)3-6C-Type Superstructure of $TiS_{1.51}$ as Revealed by High-Resolution Electron Microscopy," *Acta Crystallogr., B35,* 2522-2525 (1979). (Crys Structure; Experimental)

79Ban2: Y. Bando, M. Saeki, Y. Sekikawa, Y. Matsui, S. Horiuchi, and M. Nakahira, "(4H)2-4C Type Superstructure of $TiS_{1.4}$ as Determined by High-Resolution Electron Microscopy," *Acta Crystallogr., A35,* 564-569 (1979). (Crys Structure; Experimental)

79Ono: M. Onoda, M. Saeki, and I. Kawada, "Superstructure of Ti₂S₃(4H)," *Z. Anorg. Chem.*, 457, 62-74 (1979). (Equi Diagram, Crys Structure; Experimental)

80Leg: J.J. Legendre and M. Huber, "Generation of Polytypic Structures by Thermally Induced Expansion of Stacking Faults in a Titanium Sulphide Single Crystal," *Acta Crystallogr.*, 1031-1032 (1980). (Equi Diagram, Crys Structure; Experimental)

80Mor: R. Moret and E. Tronc, "Comments on (4H)2-4 C Type Superstructure of $TiS_{1.46}$ as Determined by High-Resolution Electron Microscopy," *Acta Crystallogr., B36*, 2854-2855 (1980). (Crys Structure; Experimental)

81Tro: E. Tronc and R. Moret, "Structural Aspects of the Metal-Metal Interactions in the $Ti_{1+x}S_2$ Materials," *J.*

Solid State Chem., 36, 97-106 (1981). (Crys Structure; Experimental)

82Sae1: M. Saeki and M. Onoda, "The Preparation of Ti_5S_8," *Bull. Chem. Soc. Jpn.*, 55, 113-116 (1982). (Equi Diagram, Crys Structure; Experimental)

*82Sae2: M. Saeki and M. Onoda, "The Phase Relation of the Titanium-Sulfur System," *Bull. Chem. Soc. Jpn.*, 55, 3144-3146 (1982). (Equi Diagram; Experimental)

83Leg: J.J. Legendre, R. Moret, E. Tronc, and M. Huber, "$Ti_{1+x}S_2$ Polytypes," in *Crystal Growth and Characterization of Polytype Structures*, Vol. 7, P. Krishna, Ed., Pergamon Press, New York, 309 (1983). (Equi Diagram, Crys Structure; Review)

The Sb-Ti (Antimony-Titanium) System

121.75   47.88

By J.L. Murray

Equilibrium Diagram

Structural data on the Ti-Sb system are given in Table 1. Based on sparse microstructural data, [Hansen] composed a schematic phase diagram showing equilibrium compounds Ti_4Sb, $TiSb$, and $TiSb_2$. The diagram was intended to convey that:

- TiSb and Ti_4Sb melt congruently and, hence, there occur at least two eutectic reactions on the Ti-side of TiSb [51Now2].
- The solubility of Ti in (Sb) is sufficiently low that the lattice parameter of Sb is not affected [51Now2].
- $TiSb_2$ forms peritectically from the melt. Specifically, the reaction L + TiSb ⇌ $TiSb_2$ occurs at 1010 °C, and the liquidus temperature is 1030 °C at

33.3 at.% Sb [60Dud, thermal analysis]. The addition of 1.3 at.% Ti raises the liquidus temperature from 630.755 °C (the melting point of pure Sb) to 680 °C [51Now2].

Although none of these points has been contradicted, four additional equilibrium phases have been identified since the [Hansen] compilation. This means that a more complicated reaction scheme is needed to describe the equilibria for Ti-rich alloys. In the composition range Ti_6Sb_5 to $Ti_{1.5}Sb$, alloys were found to be easily decomposed, either because of the presence of a high-temperature phase, or because of a phase stabilized by oxygen [58Aue]. At least one reported structure (Ti_3SbX in Table 1) is clearly an impurity-stabilized phase; it is irreversibly formed when alloys are melted in alumina crucibles [62Kje].

Table 1   Crystal Structures of the Equilibrium Phases of the Ti-Sb System

Phase	Composition, at.% Sb	Pearson symbol	Space group	Struktur-bericht designation	Proto-type	Lattice parameters, nm			Reference
						a	*b*	*c*	
(βTi)	0 to ?	cI2	Im3m	A2	W	...	...	...	(a)
(αTi)	0 to ?	hP2	P6₃/mmc	A3	Mg	...	...	...	(a)
Ti₄Sb	20.1 to ?	hP8	P6₃/mmc	D0₁₉	Ni₃Sn	0.5958	...	0.4808	[51Now1]
Ti₃Sb	25	cP8	Pm3n	A15	Cr₃Si	0.52186	...	...	[62Kje]
Ti₃SbX(b) ..	25	tI32	I4/mcm	D8ₘ	SbTi₃	1.0465	...	0.52639	[62Kje]
Ti₂.₅Sb	28.6	tP12	P4₂/mmc	...	Ti₂Bi	0.4008	...	1.4529	[58Aue]
Ti₅Sb₃	37.5	oP32	Pnma	...	...	1.0221	0.83548	0.71818	[77Ste]
						1.0172	0.8348	0.7135	[62Kje]
Ti₆Sb₅(c) ..	45.5	oP*	Pbam or Pba2	...	...	1.455	1.634	0.531	[62kje]
TiSb	50	hP4	P6₃/mmc	B8₁	AsNi	0.4070	...	0.6306	[51Now2]
						0.4115	...	0.6264	[62Kje]
TiSb₂	66.2 to 67.1	tI12	I4/mcm	C16	Al₂Cu	0.665315	...	0.580927	[75Don]
						0.6654	...	0.4806	[72Hav]
						0.6666	...	0.5817	[51Now2]
(Sb)	100	hR2	R3m	A7	As	0.43007	...	1.1222	[Pearson2]

(a) See text under "Ti (Titanium)." **(b)** An impurity-stabilized phase. **(c)** Designated $Ti_{1.2}Sb$ by [62Kje].

Fig. 1---Schematic Ti-Sb Phase Diagram

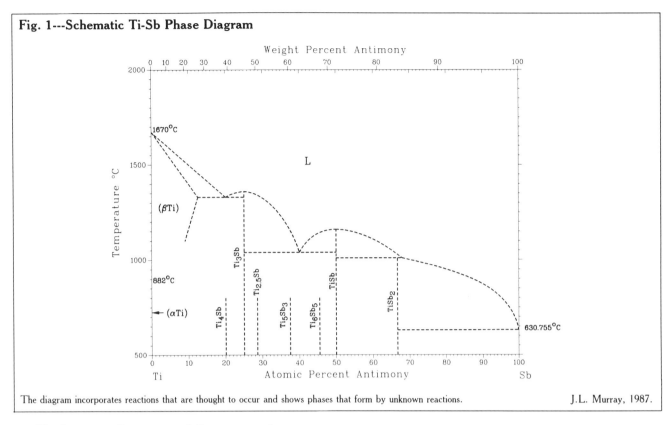

The diagram incorporates reactions that are thought to occur and shows phases that form by unknown reactions.

J.L. Murray, 1987.

The homogeneity ranges of the compounds are reported as narrow [51Now2]. [75Don] reported the homogeneity range of $TiSb_2$ as 66.2 to 67.1 at.% Sb, based on microscopic, X-ray, and density measurements. In summary, the Ti-Sb phase diagram can be drawn only as a schematic diagram, as shown in Fig. 1, and is in need of considerable further study.

Thermodynamics

Compounds are reported to be very stable [51Now2]. The heat of formation of TiSb from αTi and solid Sb was found to be -140.6 kJ/mol [59Sch], from the difference between heats of combustion between the compound and the unalloyed components. [78Alg] measured partial excess Gibbs energies of Ti for liquid alloys in the range 90 to 100 at.% Sb at 1130 K. In the limit $x_{Sb} \to 1$, the data can be described by an excess Gibbs energy function:

$$G^{ex}(L) = -74\ 760\ x(1 - x)\ \text{J/mol}$$

where x is the atomic fraction of Sb. The data are too scattered and the composition range too limited to permit a more detailed description. Phase equilibrium calculations are not feasible for this system because of insufficient experimental data.

Cited References

51Now1: H. Nowotny, R. Funk, and J. Pesl, "Crystal Chemistry Investigation of the Systems Mn-As, V-Sb, Ti-Sb," *Monatsh. Chem.*, *82*, 513-525 (1951) in German. (Crys Structure; Experimental)

51Now2: H. Nowotny and J. Pesl, "Investigation of the Titanium-Antimony System," *Monatsh. Chem.*, *82* 336-343 (1951) in German. (Crys Structure; Experimental)

58Aue: H. Auer-Welsbach, H. Nowotny, and A. Kohl, "Friction-Pyrophoric Titanium Alloys; Ti_2B, A New Structure Type," *Monatsh. Chem.*, *89*, 154-159 (1958) in German. (Equi Diagram; Experimental)

59Shc: S.A. Shchukarev, M.P. Morozova, and Li Miao-hsiu, "Enthalpy of Formation of Compounds of Titanium with Elements of the Main Subgroup of Group V," *Zh. Obshch. Khim.*, *29*, 2465 (1959) in Russian; TR: *J. Gen. Chem. USSR*, *29* 2427-2429 (1959). (Thermo; Experimental)

60Dud: L.D. Dudkin and V.I.Vaidanich, "The Nature of the Electrical Conductivity of Certain Compounds of Transition Metals with $CuAl_2$-Type Lattices," *Fiz. Tverd. Tela*, *2*(29), 404-405 (1960) in Russian; TR: *Soviet Phys.-Solid State*, *2* 377-378 (1960). (Equi Diagram; Experimental)

62Kje: A. Kjekshus, F. Gronvold, and J. Thorbjornsen, "On the Phase Relationships in the Titanium-Antimony System. The Crystal Structures of Ti_3Sb," *Acta Chem. Scand.*, *16*, 1493-1510 (1962). (Crys Structure; Experimental)

72Hav: E.E. Havinga, H. Damsma, and P. Hokkeling, "Compounds and Pseudo-Binary Alloys with the $CuAl_2$(C16)-Type Structure," *J. Less-Common Met.*, *27* 169-186 (1972). (Crys Structure; Experimental)

75Don: J.D. Donaldson, A. Kjekshus, D.G. Nicholson, and F. Rakke, "Properties of $TiSb_2$ and VSb_2," *J. Less-Common Met.*, *41* 255-263 (1975). (Crys Structure; Experimental)

77Ste: J. Steinmetz, B. Malaman, and B. Roques, "Structural Investigation of Antimonides V_3Sb_2 and Ti_5Sb_3," *C.R. Acad. Sci. Paris*, *284*, 499-502 (1977) in French. (Crys Structure; Experimental)

78Alg: M.M. Alger, "The Thermodynamics of Highly Solvated Liquid Metal Solutions," *Diss. Abstr. Int.*, *42*(11), 251 p (1982). (Thermo; Experimental)

The Sc-Ti (Scandium-Titanium) System

44.9559   47.88

By J.L. Murray

Equilibrium Diagram

The Ti-Sc phase diagram was studied by [61Sav] and [62Bea], and the present assessment (Fig. 1) is based on [62Bea]. The [61Sav] study was done with Sc of 96% purity. Thermal analysis and X-ray diffraction were used, with microscopic work limited to Ti-rich alloys. [61Sav] proposed that both β and α forms are immiscible up to a eutectic reaction at 1440 °C. They placed the αSc/βSc transition at 1450 °C, compared to the assessed value of 1337 °C. It is, therefore, certain that the diagram is not correct on the Sc-rich side. [62Bea], however, obtained

sufficient 99.8 wt%. Sc to perform a comprehensive metallographic study, as well as thermal analysis and melting point determinations by optical pyrometry. Ti and Sc undergo allotropic transformations between a low-temperature α cph form and a high-temperature β bcc form, Ti at 882 °C and Sc at 1337 °C, respectively. The continuous range of bcc β solutions in Ti-Sc alloys is the evidence by which it is known that pure βSc has the bcc structure. The terminal solutions are the only equilibrium solid phases of the system.

The important features of the phase diagram are summarized in Table 1, and crystal structure and lattice

Table 1 Special Points of the Assessed Ti-Sc Phase Diagram

Reaction	Compositions of the respective phases, at.% Sc			Temperature, °C	Reaction type
(βTi) ⇌ (αTi) + (αSc)	~ 7	~ 7	43	875 ± 8	Eutectoid
(βSc) ± (βTi) + (αSc)	53	~ 25	87.5	1050 ± 10	Monotectoid
β ⇌ (βTi) + (βSc)		~ 37		< 1100	Critical
L ⇌ (βTi, βSc)		~ 50		1300	Congruent
L ⇌ (βTi)		0		1670	Melting point
L ⇌ (βSc)		100		1541	Melting point
(βTi) ⇌ (αTi)		0		882	Allotropic transformation
(βSc) ⇌ (αSc)		100		1337	Allotropic transformation

Fig. 1---Assessed Ti-Sc Phase Diagram

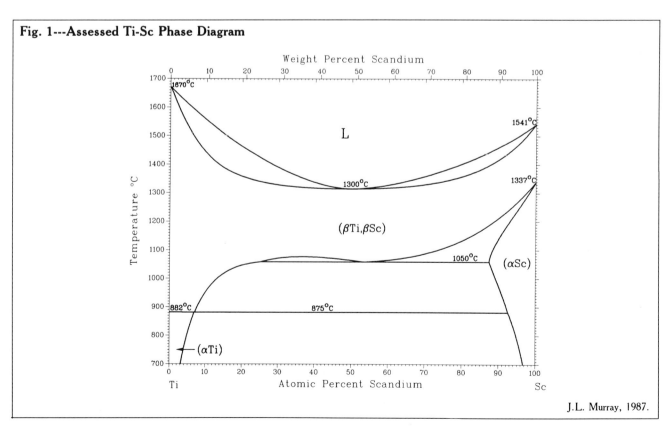

J.L. Murray, 1987.

284

Fig. 2---Assessed Phase Diagram vs Experimental Data

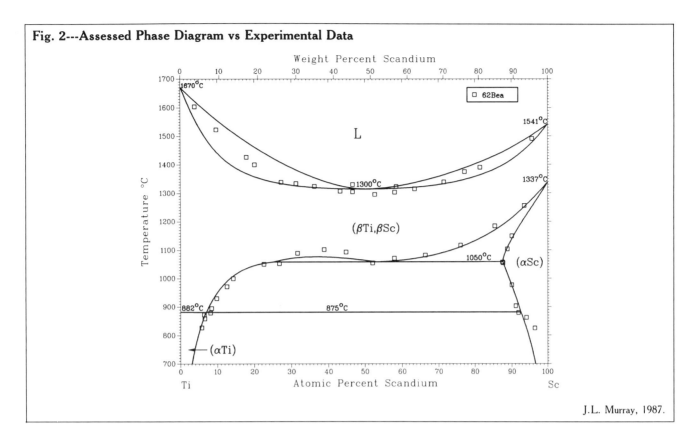

J.L. Murray, 1987.

Table 2 Crystal Structures of the Ti-Sc System

Phase	Composition range, at.% Sc	Pearson symbol	Space group	Struktur- bericht designation	Proto- type	Lattice parameters (a), nm	
						a	c
(αTi)	0 to 7	hP2	P6₃/mmc	A3	Mg	0.29512	0.46826(b)
(βTi,βSc)	0 to 100	cI2	Im3m	A2	W	0.3311(b)	...
(αSc)	87.5 to 100	hP2	P6₃/mmc	A3	Mg	0.33088	0.52680(c)

(a) Lattice parameters are for the pure elements. (b) See [Ti], βTi at 900 °C. (c) From [86Gsc].

parameter information is summarized in Table 2. Alloying lowers the melting temperature of (βTi) and (βSc) to a congruent melting point at about 1300 °C and 50 at.%. A monotectoid reaction (βTi,βSc) $\rightleftarrows$ (βTi) + (αSc) occurs at 1050 ± 10 °C. The critical point of the miscibility gap lies below 1100 °C. (βTi) disappears in the eutectoid reaction (βTi) $\rightleftarrows$ (αSc) + (αTi) at 875 °C. The maximum solubilities of Ti in (αSc) and Sc in (αTi) are about 12.5 at.% Ti and 7 at.% Sc, respectively.

The monotectoid and eutectoid temperatures and (βTi,βSc)/(αSc) boundaries were determined by thermal analysis on heating and cooling. For the three-phase reactions, the average of heating and cooling arrests were taken, and the uncertainties indicate the difference between the arrests on heating and cooling. For the solid-state reactions, the arrest on heating is probably closer to the equilibrium reactions, giving 1060 and 883 °C as possibly more accurate temperatures. Note that the uncertainty in the lower reaction temperature spans the difference between a eutectoid and peritectoid reaction type. The accuracy of the congruent melting point is about ±5 to 10 °C; the accuracy for alloys with a melting range is probably less. Based on the composition increments of

the alloys examined, the accuracy of the solvus is about ±1 at.%. The compositions of the phases in equilibrium at the monotectoid temperature are less well known (± 2 to 3 at.%), because they are based on extrapolations to the monotectoid temperature of curves with low slopes.

The phase diagram (Fig. 1) is the result of thermodynamic calculations, using Gibbs energies obtained by least-squares optimization with respect to the experimental data of [62Bea]. The calculated phase diagram is compared to the experimental data in Fig. 2. It lies within the experimental uncertainties, except for a small discrepancy in the (αSc) solvus.

Metastable Equilibria

The solubility of Ti in (αSc) can be extended to 41 at.% by rapid solidification [76Wan]. The extended solubility arises from solute trapping during the L → β → α transformation; the β → α transformation is martensitic. As an aid in the interpretation of these results, Fig. 3 includes, in addition to the calculated equilibrium diagram, the T_0 curve (i.e., the curve of equal Gibbs energy

Phase Diagrams of Binary Titanium Alloys

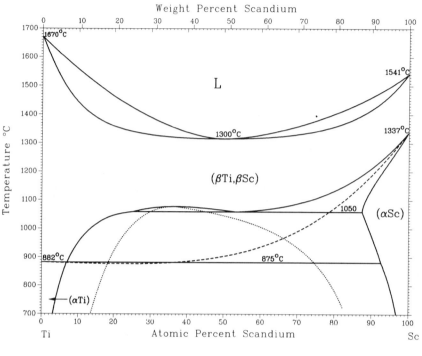

Fig. 3---Spinodal Curve of the (βTi,βSc) Phase (dotted) and The T_0 Curve for the bcc and cph Phases (dashed)

J.L. Murray, 1987.

for the cph and bcc phases) and the spinodal of the bcc miscibility gap.

Thermodynamics

Experimental thermochemical data are not available for the Ti-Sc system. However, approximate Gibbs energy functions can be constructed, because the two pure metals are isomorphous and because there is equilibrium data on the miscibility gaps in both solid phases.

Gibbs energies of the solution phases are represented as:

$$G(i) = (1 - x)\, G^0(\text{Ti},i) + x\, G^0(\text{Sc},i) + RT\, [x \ln x + (1 - x) \ln (1 - \text{Tulx})]\, x(1 - x)\, [B(i) + C(i)\, (1 - 2x) + D(i)\, (6x^2 - 6x + 1)]$$

where i designates the phase; x is the atomic fraction of Sc; $G^0(i)$ is the Gibbs energies of the pure metals; and $Bl(i)$, $C(i)$, and $D(i)$ are parameters in the polynominal expansion of the excess Gibbs energy. The Gibbs energies of the pure components are based on data in [Hultgren, E].

The input to the optimization calculations were the data shown in Fig. 2 and the three-phase equilibria. Optimizations were performed with and without excess entropies and with and without extension of the polynomials to include D parameters for the solid phases. The Gibbs energies listed in Table 3 represent the optimum number of parameters for fitting the data. Agreement between the calculated (αSc) solvus and the input data was not improved by including excess entropy terms.

Table 3 Gibbs Energies of the Ti-Sc System

Gibbs energies of the pure components

$G^0(\text{Ti,L})$	$=$	0
$G^0(\text{Sc,L})$	$=$	0
$G^0(\text{Ti,bcc})$	$=$	$-16\,234 + 8.368\,T$
$G^0(\text{Sc,bcc})$	$=$	$-14\,100 + 7.782\,T$
$G^0(\text{Ti,cph})$	$=$	$-20\,585 + 12.134\,T$
$G^0(\text{Sc,cph})$	$=$	$-18\,135 + 10.292\,T$

Interaction parameters for the solution phases

$B(\text{L}) = 15\,918$	$B(\text{bcc}) = 26\,431$	$B(\text{cph}) = 28\,396$
$C(\text{L}) = -1\,352$	$C(\text{bcc}) = 1\,011$	$C(\text{cph}) = 289$
	$D(\text{bcc}) = 2\,260$	$D(\text{cph}) = 2\,154$

Note: Values are given in J/mol and J/mol·K.

Cited References

61Sav: E.M. Savitskii and G.S. Burkhanov, "Equilibrium Diagram of the Scandium-Titanium System," *Zh. Neorg. Khim.*, 6(5), 1253-1255 (1961) in Russian; TR: *Russ. J. Inorg. Chem.*, 6(5), 642-643 (1961). (Equi Diagram; Experimental)

62Bea: B.J. Beaudry and A.H. Daane, "Sc-Ti System and the Allotropy of Sc," *Trans. AIME*, 224, 770-775 (1962). (Equi Diagram; Experimental)

76Wan: R. Wang, "Solubility and Stability and Liquid-Quenched Metastable Hcp-Solid Solutions," *Mater. Sci. Eng.*, 23(2-3), 135-140 (1976). (Meta Phases; Experimental)

86Gsc: K.A. Gschneidner, Jr. and F.W. Calderwood, in *Handbook of the Physics and Chemistry of Rare Earths*, Vol. 8, K.A. Gschneidner, Jr. and L. Eyring, Ed., North-Holland Physics Publishing, Amsterdam (1986). (Crys Structure; Review)

The Se-Ti (Selenium-Titanium) System

78.96   47.88

By J.L. Murray

Introduction

It is not presently possible to construct a phase diagram for the Ti-Se system. Table 1 summarizes crystal structure and lattice parameter data on the established phases, and Table 2 summarizes crystal structure data, some of which were recently shown to require reinterpretation.

Several phases with closely related crystal structures occur in the composition range TiSe to TiSe$_2$. TiSe$_2$ has the C6 (CdI$_2$) structure, and near-equiatomic alloys have structures closely related to that of hexagonal NiAs. Both the NiAs and the CdI$_2$ structures contain close-packed layers of Se, with the Ti atoms occupying the octahedral sites between the close-packed layers; the two structures differ only in the occupations of the octahedral sites. A continuous transition was originally thought to occur between the two structures [49Ehr], but in fact, several

distinct intermediate phases—Ti$_3$Se$_4$, Ti$_5$Se$_8$, and Ti$_8$Se$_9$—have been identified in this region. The reader is referred to the Ti-S system, which exhibits similar complex phase equilibria between TiS and TiS$_2$ and which has been more intensely studied. It is probable that further study will reveal additional phases in the Ti-Se system.

Ti-Rich Compositions

Titanium undergoes a transition from the high-temperature bcc (βTi) form to the low-temperature cph (αTi) form at 882 °C. According to [53Gol], the solubility of Se in (αTi) at 790 °C is about 0.15 at.%. At this composition, the first insoluble phase was observed metallographically in arc-melted dilute alloys of Mg-reduced (low-purity) Ti. The maximum solubility value should be taken as a rough estimate.

Table 1   Crystal Structures and Lattice Parameters of the Ti-Se System

Phase	Homogeneity range, at.% Se			Pearson symbol	Space group	Strukturbericht designation	Prototype	Lattice parameters, nm			Comment	Reference
								a	b	c		
(αTi)	0	to	0.15	hP2	P6$_3$/mmc	A3	Mg	...	...	...	...	[Pearson2]
(βTi)	0	to	?	cI2	Im3m	A2	W	...	...	...	...	[Pearson2]
Ti$_{1+x}$Se	48.7			oP8	Pcmn	...	...	0.6222	0.3494	0.6462	...	[61Gro]
TiSe(a)	...			hP4	P6$_3$/mmc	B8$_1$	NiAs	...	...	...	...	...
Ti$_8$Se$_9$	~ 52.9			hP12	R$\bar{3}$m	...	...	1.0684	...	0.6248	...	[72Bru]
Ti$_3$Se$_4$	55.0 to 57.6			mC14	C2/m	...	Cr$_3$Se$_4$	0.640	0.357	1.204	β = 90°42'	[62Che]
								0.635	0.357	1.201	β = 90°30'	[62Ber1]
Ti$_5$Se$_8$	60.9 to 62.0			(b)		...	...	0.619	1.194	0.3608	...	[63Ber]
αTiSe$_2$	?	to	66.7	hP24	P$\bar{3}$1c	...	...	...	...	...	...	[77Hol]
βTiSe$_2$	?	to	66.7	hP3	P$\bar{3}$m1	C6	CdI$_2$	0.35401	...	0.60083	...	[76Rie]
								0.3535	...	0.6004	...	[58Mct]
								0.350	...	0.6007	...	[28Oft]
								0.3548	...	0.5998	...	[49Fur]
TiSe$_3$	~ 75			(c)	...	...	...	...	...	...	...	[75Fur]

(a) A dubious structure identification, see Table 2 and text. (b) Monoclinic. (c) Unknown.

Table 2   Reported Structure Data on Controversial Hexagonal Phases

Phase	Composition range, at.% Se	Pearson symbol	Space group	Lattice parameters, nm		Reference
				a	b	
"Simple" NiAs-type structures						
Ti$_{1-x}$Se	51.2	hP4	P6$_3$/mmc	0.3571	0.6301	[61Gro]
Ti$_3$Se$_2$	33 to 40	...	...	0.3574	0.6322	[59Hah]
Ti$_2$Se$_3$	60	...	...	0.3595	0.5994	[58Mct]
TiSe	50	...	...	0.3566	0.6232	[49Ehr]
(2a,2c) NiAs superlattices						
TiSe$_{1.2}$	54.6 to 56.5	hP16	(a)	0.357 (×2?)	0.6175 (×2?)	[61Gro]
	50	...	...	0.715	1.200	[58Mct]
	50	...	...	0.7149	1.199	[59Hah]
(4a,4c) NiAs superlattices						
βTiSe	...	...	(b)	0.720	2.741	[59Hah]

(a) Hexagonal. (b) Trigonal.

Phase Diagrams of Binary Titanium Alloys

Alloys containing less than 50 at.% Se are extremely difficult to equilibrate and are highly reactive [59Hah, 58Mct, 61Gro]. [49Ehr] suggested that there may be a subselenide, Ti_2Se. [59Hah] attributed the composition range 33 to 40 at.% Se to a phase denoted Ti_3Se_2 with the NiAs structure. [61Gro] found that a 37.5 at.% Se sample was two-phase and that many X-ray diffraction lines were not from metallic Ti. [61Gro], therefore, postulated the existence of a subselenide of unknown composition and structure.

[61Gro] reported that the 37.5 at.% Se sample was partly melted after heating to 1550 °C. This is the only reported Ti-Se melting point.

Ti-Se samples generally are prepared either by direct reaction of the components at elevated temperature or, for studies of higher Se-content compounds, by vapor transport. Reaction with Si and oxygen causes significant loss of Ti, as well as the probable formation of ternary phases. The exclusive use of X-ray diffraction to determine phase equilibria can be misleading in this system (see, Ti-S). Therefore, it is currently not known whether a subselenide exists.

Intermetallic Compounds (45 to 65 at.% Se)

As mentioned above, the structures occurring in the range near 50 at.% are closely related to the hexagonal NiAs structure. Table 2 summarizes data on phases reported to have simple NiAs structures or superlattices based on the NiAs structure. There are discrepancies among the compositions attributed to these phases as well as the dimensions of the unit subcells, although there is some agreement about the superlattice structures.

[49Ehr] considered the range 50 to 66.7 at.% Se to be a single-phase field with a continuous transition between the NiAs and CdI_2 structures. Because the structures have different space groups, this is impossible on the basis of symmetry considerations alone. [61Gro] recognized symmetry changes from hexagonal to monoclinic to hexagonal with increasing Se content, but did not distinguish separate phases. Similarly, [62Ber1] found a discontinuity in the slope of the lattice parameters, but did not attribute it to the formation of a distinct phase.

Additional phases—Ti_8Se_9, Ti_3Se_4, and Ti_5Se_8—were discovered by [62Che], [62Ber1], and [72Bru]. Alloys were chemically analyzed and homogeneity ranges were determined by the lattice parametric method and by the appearance of a second crystal structure. [72Bru] was able to reinterpret previous results on the proposed hexagonal NiAs-type structures [61Gro, 58Mct, 59Hah]. [72Bru] found that the 50 at.% alloy contained a two-phase mixture of $Ti_1 + xSe$ and Ti_8Se_9 and that the diffraction patterns previously attributed to the superlattice $(2a, 2c)$ should be reinterpreted as belonging to a two-phase mixture of Ti_8Se_9 and Ti_3Se_4.

Diselenide and Triselenide

The CdI_2 structure of the diselenide is well established [28Oft, 49Ehr, 62Ber2, 58Mct, 76Rie]. Recent work has also uncovered, at low temperatures, a higher than first-order phase transition to a superlattice structure with $a = 2a_0$, and $c = 2c_0$ [76Woo, 77Hol]. The high- and low-temperature forms are designated $\beta TiSe_2$ and $\alpha TiSe_2$, respectively.

The $\beta TiSe_2$ structure was refined by [76Rie], for a stoichiometric sample. Based on lattice parameter measurements, [62Ber2] reported the homogeneity range of $\beta TiSe_2$ as 58.7 to 66.2 at.% Se at 1000 °C. The Ti-rich side of this homogeneity range, however, does not take into account the Ti_5Se_8 phase.

At stoichiometry, the higher order $\beta TiSe_2 \rightleftarrows \alpha TiSe_2$ transition occurs at 202 K [80Vat]. [76Woo] observed the transition at 150 ± 5 K, but noted that excess Ti decreases the transition temperature.

The triselenide $TiSe_3$ was prepared by [75Fur], but the crystals were unsuitable for structural determination.

Cited References

28Oft: I. Oftedal, "X-Ray Investigation of SnS_2, TiS_2, $TiSe_2$, $TiTe_2$," *Z. Phys. Chem., 134*, 301-310 (1928) in German. (Equi Diagram, Crys Structure; Experimental)

49Ehr: P. Ehrlich, "X-Ray Investigation of Titanium Sulfides and Vanadium Monotelluride," *Z. Anorg. Chem., 260*, 13 (1949) in German. (Equi Diagram, Crys Structure; Experimental)

53Gol: R.M. Goldhoff, H.L. Shaw, C.M. Craighead, and R.I. Jaffee, "The Influence of Insoluble Phases on the Machinability of Titanium," *Trans. ASM, 45*, 941-971 (1953). (Equi Diagram, Crys Structure; Experimental)

57Sch: A. Schneidner and K.H. Imhagen, "The Temperature Dependent Lattice Parameters of the NiAs Phase," *Naturwissenschaften, 44*, 324 (1957) in German. (Crys Structure; Experimental)

58Mct: F.K. McTaggart and A.D. Wadsley, "The Sulphides, Selenides, and Tellurides of Titanium, Zirconium, Hafnium, and Thorium," *Aust. J. Chem., 11*, 445-457 (1958). (Equi Diagram, Crys Structure; Experimental)

59Hah: H. Hahn and P. Ness, "The Subchalcogenides of Titanium," *Z. Anorg. Chem., 302*, 17-36 (1959) in German. (Equi Diagram, Crys Structure; Experimental)

61Gro: F. Gronvold and F.J. Langmyhr, "X-Ray Study of Titanium Selenides," *Acta Chem. Scand., 15*(10), 1949-1962 (1961). (Equi Diagram, Crys Structure; Experimental)

62Ber1: P. Bernusset and Y. Jeannin, "The Non-Stoichiometric Phase Ti_3Se_4: Preparation, Homogeneity Range, Lattice Parameters," *Compt. Rend., 255*, 2973-2974 (1962) in French. (Equi Diagram, Crys Structure; Experimental)

62Ber2: P. Bernusset and Y. Jeannin, "Non-Stoichiometric Phase $TiSe_2$: Homogeneity Range, Variation of Lattice Parameter," *Compt. Rend., 255*, 934-936 (1962) in French. (Equi Diagram, Crys Structure; Experimental)

62Che: M. Chevreton and F. Bertaut, "Titanium and Vanadium Selenides and Titanium Telluride," *Compt. Rend., 255*, 1275-1277 (1962) in French. (Equi Diagram, Crys Structure; Experimental)

63Ber: P. Bernusset, "On the Existence of a Non-Stoichiometric Phase Ti_5Se_8," *Compt. Rend., 257*, 2840-2842 (1963) in French. (Equi Diagram, Crys Structure; Experimental)

72Bru: S. Brunie and M. Chevreton, "Structural Study of the Titanium Selenide Ti_8Se_9. Epitaxy Between Ti_8Se_9 and Ti_3Se_4," *C.R. Acad. Sci., 274*, 278-281 (1972) in French. (Equi Diagram, Crys Structure; Experimental)

75Fur: S. Furuseth, L. Brattas, and A. Kjekshus, "On the Crystal Structures of TiS_3, ZrS_3, $ZrSe_3$, $ZrTe_3$, HfS_3, and $HfSe_3$," *Acta Chem. Scand., A29*, 623-631 (1975). (Equi Diagram, Crys Structure; Experimental)

76Rie: C. Riekel, "Structure Refinement of $TiSe_2$ by Neutron Diffraction," *J. Solid State Chem., 17*, 389-392 (1976). (Equi Diagram, Crys Structure; Experimental)

76Woo: K.C. Woo, F.C. Brown, W.L. McMillan, R.J. Miller, M.J. Schaffman, and M.P. Sears, "Superlattice Formation in Titanium Diselenide," *Phys. Rev. B, 14*(8), 3242-3247 (1976). (Equi Diagram, Crys Structure; Experimental)

77Hol: J.A. Holy, K.C. Woo, M.V. Klein, and F.C. Brown, "Raman and Infrared Studies of Superlattice Formation in $TiSe_2$," *Phys. Rev. B, 16*(8), 3628-3637 (1977). (Equi Diagram, Crys Structure; Experimental)

80Vat: H.P. Vaterlaus, S. Ansermet, M. Py, and F. Levy, "Electron Diffraction and Infrared Study of the Semimetallic Layered Compound $Ti_{1-x}Se_2$," *Solid State Comm., 35*, 925-929 (1980). (Equi Diagram, Crys Structure; Experimental)

The Si-Ti (Silicon-Titanium) System

28.0855   47.88

By J.L. Murray

Equilibrium Diagram

The equilibrium solid phases of the Ti-Si system are:

- The cph (αTi), bcc (βTi), and diamond cubic (Si) solid solutions. (Si) dissolves essentially no Ti [52Han, 70Sve].
- Stoichiometric Ti_3Si, with the Ti_3P structure. [65Ros] first observed Ti_3Si, and [70Sve] verified its existence and determined its phase boundaries.
- Ti_5Si_3, with a homogeneity range of about 4 at.% about stoichiometry [52Han, 59Age, 70Sve]. Ti_5Si_3 was originally considered the most Ti-rich intermetallic phase [Hansen].
- Ti_5Si_4, a tetragonal phase near 45 at.% Si. It was first observed by [62Sch] and [66Bru], who tentatively assigned it a $D8_8$-type structure and the stoichiometry Ti_6Si_5. [70Sve] also observed a tetragonal phase to which they attributed stoichiometry Ti_5Si_4, based on the structural isomorphism with Zr_5Si_4. Based on phase equilibrium work [70Sve], Ti_6Si_5 and Ti_5Si_4 are tentatively identified as a single-phase Ti_5Si_4 in the present assessment.
- Stoichiometric TiSi, formed from the melt by a peritectic reaction. TiSi is extremely brittle, and single crystals have been prepared only by directional solidification [59Age].
- Essentially stoichiometric $TiSi_2$, the most Si-rich intermetallic phase. The existence of a thermal effect in $TiSi_2$ at about 1200 °C [53Gla] suggests the possibility of an allotropic transformation in $TiSi_2$.

Comprehensive studies of the Ti-Si diagram were made by [52Han] and [70Sve]. Both used thermal analysis, metallography, and X-ray diffraction; [52Han] in addition performed incipient melting studies. Additional work limited to Ti-rich alloys was done by [54Sut1], [54Sut2], [62Pyl], [63Luz], and [78Pli]. [53Gla] measured the melting point of $TiSi_2$.

The assessed diagram is shown in Fig.1, summarized in Table 1, and compared with experimental data in Fig. 2. Experimental data on the invariant reactions and the (αTi)/(βTi) equilibria are given in Tables 2 and 3. There is good general agreement among all studies ($\pm$ 15 °C) on the positions of the eutectic reactions, the congruent melting points, and the (αTi)/(βTi) boundaries. Agreement on these features is sufficiently good that the assessed values are averages of the various experimental determinations.

Concerning phase equilibria involving Ti_3Si and Ti_5Si_4 [70Sve], there are marked disagreements; [52Han] did not show either of these compounds. The two diagrams of [52Han] and [70Sve] predict different phases appearing

---

Fig. 1---Assessed Ti-Si Phase Diagram

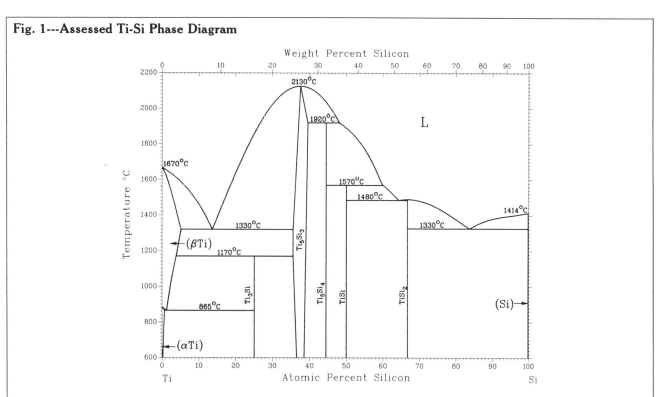

Except for the homogeneity range of Ti_5Si_3, the diagram was drawn from the present calculations. Temperatures of invariant reactions may differ slightly from the tabulated values, in all cases within the experimental uncertainty.

J.L. Murray, 1987.

Phase Diagrams of Binary Titanium Alloys

Table 1 Special Points of the Assessed Ti-Si Phase Diagram

Reaction	Compositions of the respective phases, at.% Si			Temperature, °C	Reaction type
(βTi) ⇌ (αTi) + Ti₃Si	1.1	0.5	25	865	Eutectoid
(βTi) + Ti₅Si₃ ⇌ Ti₃Si	3.5	37.5	25	1170	Peritectoid
L ⇌ (βTi) + Ti₅Si₃	13.5	4.7	37.5	1330	Eutectic
L ⇌ Ti₅Si₃		37.5		2125	Congruent
L + Ti₅Si₄ ⇌ TiSi	60	44.4	50	1570	Peritectic
L + Ti₅Si₃ ⇌ Ti₅Si₄	48.1	37.5	44.4	1920	Peritectic
L ⇌ TiSi + TiSi₂	64.2	50	66.7	1480	Eutectic
L ⇌ TiSi₂		66.7		1500	Congruent
L ⇌ TiSi₂ + (Si)	83.8	66.7	100	1330	Eutectic
L ⇌ (βTi)		0		1670	Melting point
(βTi) ⇌ (αTi)		0		882	Allotropic transformation
L ⇌ (Si)		100		1414	Melting point

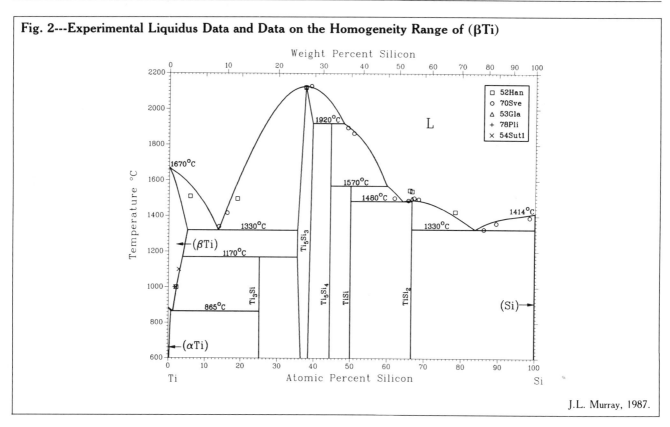

Fig. 2---Experimental Liquidus Data and Data on the Homogeneity Range of (βTi)

J.L. Murray, 1987.

in as-cast and heat treated alloys over a wide temperature and composition range.

The assessed diagram is based on [70Sve]. This judgment is based on: (1) the probable purity of the alloys, (2) the fact that the phases of composition 25 and 45 at.% Si were also observed by [65Ros] and [62Sch], respectively, and (3) the present thermodynamic analysis of the system.

Metastable Phases

A (βTi) → (αTi) massive transformation was shown to exist for alloys containing up to 1.1 at.% Si [77Pli, 78Pli]. From continuous cooling experiments, the start temperature, M_s, for the martensitic transformation was also found to be: 808 and 814 °C for 0.68 and 1.1 at.% Si alloys, respectively.

[78Pol] prepared an amorphous 20 at.% Si alloy by rapid solidification. The crystallization temperature was reported as 594 °C. Rapidly solidified alloys containing 15 and 20 at.% Si were crystalline.

Crystal Structures and Lattice Parameters

Crystal structure data are summarized in Table 4, and lattice parameters of the compounds are listed in Table 5. There are discrepancies in the reported structures of Ti₅Si₄, TiSi, and TiSi₂. [60Aro] assessed crystal structure data for TiSi and TiSi₂. He found the data to be too meager for any definite conclusions about the TiSi structure, or to be able to determine whether there are two modifications of this phase. The C49 structure reported for TiSi₂ [56Cot] belongs to a ternary Ti-Si-Al phase rather than to the binary compound [61Bru, 64Sch].

Table 2 Experimental Data on Invariant Equilibria Involving the Liquid

Phase	Composition, at.% Si			Temperature, °C	Reference
L ⇌ (βTi) + Ti$_5$Si$_3$	13.7	5	~ 36	1340 ± 10	[70Sve]
	13.7	~ 5	~ 37	1330 ± 5	[52Han]
	...	5.3	...	1320 ± 6	[54Sut1, 54Sut2]
L ⇌ TiSi$_2$ + (Si)	85.8	66.7	100	1330 ± 5	[52Han]
	86	66.7	100	1330 ± 10	[70Sve]
L ⇌ Ti$_5$Si$_3$	...	37.5	...	2120 ± 15	[52Han]
	...	37.5	...	2130 ± 20	[70Sve]
L ⇌ TiSi + TiSi$_2$	64.1	50	66.7	1470 ± 10	[70Sve]
	64.2	50	66.7	1490 ± 5	[52Han]
L ⇌ TiSi$_2$	...	67.7	...	1540	[52Han]
	...	66.7	...	1500	[70Sve, 53Gla]
L + Ti$_5$Si$_3$ ⇌ TiSi	56.3	37.5	50	1760 ± 15	[70Sve]
L + Ti$_5$Si$_4$ ⇌ TiSi	60	44.4	50	1570 ± 10	[70Sve]
L + Ti$_5$Si$_3$ ⇌ Ti$_5$Si$_4$	48.1	37.5	44.4	1920 ± 20	[70Sve]

Table 3 Experimental Data on the (αTi)/(βTi) Equilibria

Reference	Temperature, °C	Composition, at.% Si	
		(αTi)	(βTi)
Eutectoid reaction			
[54Sut1, 54Sut2]	855 ± 7	~ 0.8	1.1
[52Han]	860 ± 10	0.5 to 0.9	1.5
[70Sve]	860 ± 5	0.65	1.1
[78Pli]	872 ± 5	< 0.65	1.3
Additional (αTi) solvus data			
[54Sut1, 54Sut2]	650	< 0.4	...
[63Luz]	600 to 700	0.5 to 0.8	...
	840	0.8 to 1.2	...
[62Pyl]	850	0.7	...
	800	0.6	...
	600	0.5	...

Table 4 Crystal Structures of the Equilibrium Phases of the Ti-Si System [Pearson2]

Phase	Composition, at.% Si	Pearson symbol	Space group	Struktur- bericht designation	Prototype
(αTi)	0 to 0.5	hP2	P6$_3$/mmc	A3	Mg
(βTi)	0 to 3.5	cI2	Im3m	A2	W
Ti$_3$Si	25	tP32	P4$_2$/n	...	Ti$_3$P
Ti$_5$Si$_3$	35.5 to 39.5	hP16	P6$_3$/mcm	D8$_8$-type	Mn$_5$Si$_3$
Ti$_5$Si$_4$	44.4	tP36	P4$_1$2$_1$2	...	Si$_4$Zr$_5$
Ti$_6$Si$_5$	45.5	(a)	...	...	...
TiSi	50	oP8	Pmm2	...	TiSi
	50	oP8	Pnma	B27	BFe
TiSi$_2$	66.7	oF24	Fddd	C54	Si$_2$Ti
(Si)	100	cF8	Fd3m	A4	C(diamond)

(a) Tetragonal, related to σ (D8$_8$).

Thermodynamics

Heats of formation of the liquid were measured by [81Esi] in the composition range 60 to 100 at.% Si at 2000 K. Experimental data are compared with the present calculations in Fig. 3.

The phase diagram was calculated by [73Kau] in the range 0 to 70 at.% Si and by [79Kau] over the whole range. In both calculations, the input experimental data were heats of formation of the compounds and the diagram proposed by [52Han]. The calculation by [79Kau] differs from the input data on the Si-rich side.

The present calculations were undertaken in order to include the liquid phase heat of mixing data and to examine the discrepancy between [52Han] and [70Sve].

The Gibbs energies of the solution phases are represented as:

Table 5 Ti-Si Lattice Parameter Data

Phase	Composition, at.% Si	Lattice parameters, nm			Reference
		a	*b*	*c*	
(αTi)	0 to 0.5	...	...	...	[Pearson2]
(βTi)	0 to 3.5	...	...	...	[Pearson2]
Ti₃Si	25	1.039	...	0.517	[64Sch]
		1.0206	...	0.5069	[70Sve]
Ti₅Si₃	35.5 to 39.5	0.7465	...	0.5162	[50Pie]
		0.7429	..	0.51392	[59Swa]
		0.7448	...	0.5114	[59Now]
		0.7431	...	0.5135	[70Sve]
		0.7449 to	...	0.5149 to	[66Bra]
		0.7454		0.5144	
Ti₅Si₄	44.4	0.7133	...	1.2977	[70Sve]
Ti₆Si₅	45.5	0.7133	...	1.2977	[62Sch]
TiSi	50	0.3618	0.4973	0.6492	[57Age, 59Age]
		0.6544	0.3638	0.4997	[61Bru]
		0.6537	0.3638	0.5013	[61Bru]
TiSi₂	66.7	0.8253	0.4783	0.8540	[39Lav, 53Now]
		0.8279	0.4819	0.8568	[64Duf]
		0.8235	0.4775	0.8549	[66Bru]
		0.826719	0.480005	0.8550511	[77Jei]
(Si)	100	0.54307	...	...	[Pearson2]

Fig. 3---Heats of Mixing of the Liquid Phase vs the Present Calculations

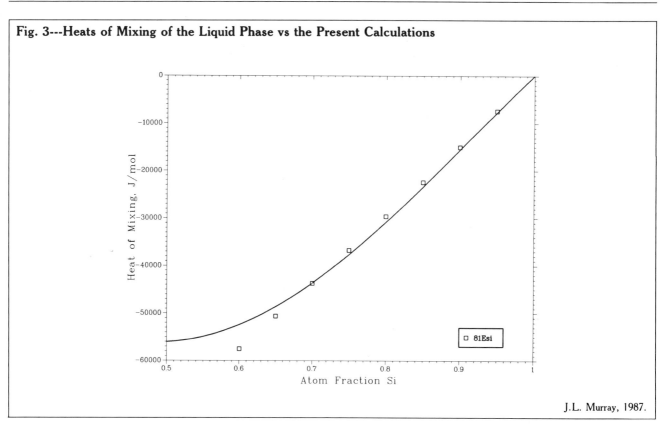

J.L. Murray, 1987.

$$G(i) = G^0(\text{Ti},i)\,(1 - x) + x\,G^0(\text{Si},i) + RT\,[x \ln x + (1 - x)\ln(1 - x)] + x(1 - x)\,[B_j(i)\,P_j\,(1 - 2x)]$$

where i designates the phase; x is the atomic fraction of Si; G^0 is the Gibbs energies of the pure metals, and P_j is the jth Legendre polynomial in the expansion of the excess Gibbs energy, with $B_j(i)$ as the coefficient. The intermetallic phases Ti₃Si, TiSi₂, Ti₅Si₃, and Ti₅Si₄ were assumed to be line compounds, and zero solubility range was also attributed to solid (Si). Pure element Gibbs energies are from [79Kau].

Input data for the optimizations were heats of mixing [81Esi], heats of formation [Hultgren, B], and the invariant reactions. In preliminary calculations, invariant points according to [52Han] were used. Although the heats of mixing and invariant reactions could be reproduced, the calculated (βTi) liquidus and solidus disagreed with the data. This can be understood as due to having more thermochemical data for Si-rich compositions. None of the attempts to weight the data differently succeeded in improving the fit to the phase diagram, and it was tentatively concluded that using

Table 6 Thermodynamic Properties of the Ti-Si System

Properties of the pure elements

$G^0(\text{Ti,L})$ = 0
$G^0(\text{Si,L})$ = 0

$G^0(\text{Ti,bcc})$ = $-16\,234 + 8.368\,T$
$G^0(\text{Si,bcc})$ = $-6\,276 + 10.46\,T$

$G^0(\text{Ti,cph})$ = $-20\,585 + 12.134\,T$
$G^0(\text{Si,cph})$ = $-418 + 12.13\,T$

$G^0(\text{Si})$ = $-50\,626 + 30.0\,T$

Excess Gibbs energies of the solution phases

$B(\text{L},0)$ = $-204\,457 - 5.2354\,T$
$B(\text{L},1)$ = $-17\,313 - 5.1910\,T$
$B(\text{L},1)$ = $39\,223$

$B(\text{bcc},0)$ = $-158\,920 - 20.9796\,T$
$B(\text{cph},0)$ = $-178\,060 - 3.8666\,T$

Gibbs energies of the compounds

$G(\text{Ti}_3\text{Si})$ = $-73\,861 + 12.95\,T$
$G(\text{Ti}_5\text{Si}_3)$ = $-96\,609 + 11.285\,T$
$G(\text{Ti}_5\text{Si}_4)$ = $-100\,431 + 13.08\,T$
$G(\text{TiSi})$ = $-120\,731 + 26.3785\,T$
$G(\text{TiSi}_2)$ = $-102\,245 + 25.2465\,T$

Note: Values are given in J/mol and J/mol · K.

the [52Han] diagram introduced inconsistencies among the input data.

With regard to thermochemical properties, the [70Sve] diagram differs greatly from that of [52Han]. According to [70Sve], the Ti_5Si_3 liquidus is metastable below the peritectic reaction at 1920 °C, and the liquid is considerably more stable with respect to Ti_5Si_3. Calculations suggest that this difference was the source of discrepancies in the preliminary calculations. The diagram of [70Sve] is easily reproduced within the experimental uncertainty. The results of the calculation are given in Table 5 and were used to draw the assessed diagram (Fig. 1).

Cited References

39Lav: F. Laves and H.J. Wallbaum, "Crystal Structures of Ni_3Ti and Si_2Ti," *Z. Kristallogr., 101*, 78-93 (1939) in German. (Crys Structure; Experimental)

51Pie: P. Pietrokowsky and P. Duwez, "Crystal Structure of Ti_5Si_3, Ti_5Ge_3, Ti_5Sn_3," *Trans. AIME, 191*, 772-773 (1951). (Crys Structure; Experimental)

*52Han: M. Hansen, H.D. Kessler, and D.J. McPherson, "The Titanium-Silicon System," *Trans. ASM, 44*, 518-538 (1952). (Crys Structure; Experimental)

53Gla: F.W. Glaser and D. Moskowitz, "Electrical Measurements at High Temperatures as an Efficient Tool for Thermal Analysis," *Powder Met. Bull., 6*(6), 178-185 (1953). (Equi Diagram; Experimental)

53Now: H. Nowotny, H. Schroth, R. Kieffer, and F. Benesovsky, "Constitution of Several Silicide Systems with Transition Metals," *Monatsh. Chem., 84*, 579-584 (1953) in German. (Crys Structure; Experimental)

54Sut1: D.A. Sutcliffe, "Titanium-Silicon Alloys," *Met. Treat., 21*, 191-197 (1954). (Equi Diagram; Experimental)

54Sut2: D.A. Sutcliffe, "Titanium-Silicon Alloys," *Rev. Met., 51*, 524-536 (1954) in German. (Equi Diagram; Experimental)

56Cot: P.G. Cotter, J.A. Kohn, and R.A. Potter, "Physical and X-Ray Study of the Disilicides of Titanium, Zirconium, and Hafnium," *J. Am. Ceram. Soc., 39*, 11-12 (1956). (Crys Structure; Experimental)

57Age: N. Ageev and V. Samsonov, "X-Ray Determinations of the Crystal Structures of TiSi and TiGe," *Dokl. Akad. Nauk SSSR, 112*(5), 853-855 (1959) in Russian. (Crys Structure; Experimental)

59Age: N.V. Ageev and V.P. Samsonov, "An X-Ray Study of the Crystal Structures of Titanium Silicides and Germanides," *Zh. Neorg. Khim., 4*(7), 1590-1595 (1959) in Russian; TR: *Russ. J. Inorg. Chem., 4*(7), 716-719 (1959). (Crys Structure; Experimental)

59Now: H. Nowotny, H. Auer-Welsbach, J. Bruss and A. Kohl, "An Investigation of the Mn_5Si_3 Structure ($D8_8$-type)," *Monatsh. Chem., 90*, 15-23 (1959) in German. (Crys Structure; Experimental)

59Swa: H.E. Swanson, N.T. Gilfrich, M.I. Cook, R. Stinchfield, and P.C. Parks, "Titanium Silicide, Ti_5Si_3 (hexagonal)," *Nat. Bur. Stand. (U.S.) Cir., 539* VIII, 66-65 (1959). (Crys Structure; Experimental)

60Aro: B. Aronsson, "Borides and Silicides of the Transition Metals," *Arkiv. Kemi., 16*(36) 379-423 (1960). (Crys Structure; Review)

61Bru: C. Brukl, H. Nowotny, O. Schob, and F. Benesovsky, "The Crystal Structures of TiSi, Ti(Al,Si)_2 and Mo(Al,Si)_2," *Monatsh. Chem., 92*, 781-788 (1961) in German. (Crys Structure; Experimental)

62Pyl: E.N. Pylaeva and M.A. Volkova, "Solubility of Silicon in α-Titanium," *Titan Splav. Akad. Nauk SSSR Inst. Met., 7*, 74-77 (1962) in Russian. (Equi Diagram; Experimental)

62Sch: O. Schob, H. Nowotny, and F. Benesovsky, "The Ternary (Titanium, Zirconium, Hafnium) – Aluminum Silicon," *Planseeber Pulvernet, 10*, 65-71 (1962) in German. (Crys Structure; Experimental)

63Luz: L.P. Luzhnikov, V.M. Novikova, and A.P. Mareev, "Solubility of β-Stabilizers in α-Titanium," *Metalloved. Term. Obrab. Met., (2)*, 13-16 (1963) in Russian; TR: *Met. Sci. Heat Treat., (2)*, 78-81 (1963). (Equi Diagram; Experimental)

64Duf: W.J. Duffin, E. Parthe, and J.T. Norton, "The Structure of (Ti,Re) Si_2," *Acta Crystallogr., 17*, 450-451 (1964). (Crys Structure; Experimental)

64Sch: K. Schubert, A. Raman, and W. Rossteutscher, "Structure Data on Metallic Phases," *Naturwissenschaften, 51* 506-507 (1964) in German. (Crys Structure; Experimental)

65Ros: W. Rossteutscher and K. Schubert, "Structure Determination for Several T4...5-B4...5 Systems," *Z. Metallkd., 56*(11), 813-822 (1965) in German. (Equi Diagram; Experimental)

66Bru: C.E. Brukl, U.S. Air Force Tech. Doc. Rep. AFML-TR-65-2, Part II, Vol VII (May 1966). (Equi Diagram, Crys Structure)

*70Sve: V.N. Svechnikov, Yu.A. Kocherzhisky, L.M. Yupko, O.G. Kulik, and E.A. Shishkin, "Phase Diagram of the Titanium-Silicon System," *Dokl. Akad Nauk SSSR, 193*(2), 393-396 (1970) in Russian. (Equi Diagram; Experimental)

73Kau: L. Kaufman and H. Nesor, "Phase Stability and Equilibria as Affected by the Physical Properties and Electronic Structure of Titanium Alloys," *Sci. Technol. Appl. Titanium, Proc. Int. Conf., R.I. Jaffee, Ed., Vol. 2*, 773-800 (1973). (Thermo; Theory)

77Jei: W. Jeitschko, "Refinement of the Crystal Structure of TiSi_2 and Some Comments on Bonding in TiSi_2 and Related Compounds," *Acta Crystallogr. B, 33*, 2347-2348 (1977). (Meta Phases; Experimental)

77Pli: M.R. Plichta, J.C. Williams, and H.I Aaronson, "On the Existence of the β to α-m Transformation in the Alloy Systems Ti-Ag, Ti-Au, Ti-Si," *Metall. Trans. A, 8*, 1885-1892 (1977). (Meta Phases; Experimental)

78Pli: M.R. Plichta, H.I. Aaronson, and J.H. Perepezko, "The Thermodynamics and Kinetics of the β to α-m Transformation in Three Ti-X Systems," *Acta Metall., 26*, 1293-1305 (1978). (Equi Diagram; Experimental)

78Pol: D.E. Polk, A. Calka, and B.C. Giessen, "The Preparation and Thermal and Mechanical Properties of New Titanium Rich Metallic Glasses," *Acta Metall., 26*, 1097-1103 (1978). (Meta Phases; Experimental)

79Kau: L. Kaufman, "Coupled Phase Diagrams and Thermochemical Data for Transition Metal Binary Systems–VI," *Calphad, 3*(1), 45-76 (1979). (Thermo; Theory)

81Esi: Yu.O. Esin, M.G. Valishev, A.F. Ermankov, P.V. Gel'd,

and M.S. Petrushevskiy, "Enthalpy of Formation of Liquid Binary Alloys of Vanadium and Titanium with Silicon," *Izv. Akad. Nauk SSSR, Met.*,(2), 95-96 (1981) in Russian; TR: *Russ. Metall.*, (2), 71-72 (1981). (Thermo; Theory)

The Sn-Ti (Tin-Titanium) System

118.69   47.88

By J.L. Murray

Equilibrium Diagram

The equilibrium solid phases of the Ti-Sn system are: (1) the low-temperature cph (αTi) and high-temperature bcc (βTi) solid solutions; (2) the ordered hexagonal Ti_3Sn phase, whose homogeneity range has been the subject of dispute; (3) ordered hexagonal phases Ti_2Sn and Ti_5Sn_3, of which the phase equilibria involving these compounds have received only cursory study; (4) high- and low-temperature forms of the most Sn-rich intermetallic phases---βTi_6Sn_5 and αTi_6Sn_5---for which the nomenclature used here conforms to the convention that the low-temperature form is designated α, and reverses the previously used nomenclature [64Vuc, Pearson2]; and (5) (βSn), which dissolves essentially no Ti.

The assessed diagram, Fig. 1, has been constructed from the thermodynamic calculations presented below, except for the homogeneity ranges of Ti_3Sn and Ti_2Sn, and the $\beta Ti_6Sn_5 \rightarrow \alpha Ti_6Sn_5$ transition. In the calculations, these phases were modeled as line compounds. This assessment tentatively resolves the controversy concerning the peritectoid reaction (βTi) + $Ti_3Sn \rightleftharpoons$ (αTi) in favor of its existence.

Ti-Rich Liquidus and Intermetallic Compounds.

The assessed three-phase equilibria and congruent melting points are given in Table 1 and experimental data are shown in Fig. 2. For reactions in the range 25 to 50 at.% Sn, the only experimental data are those of [57Pie]. [57Pie] made incipient melting studies using optical

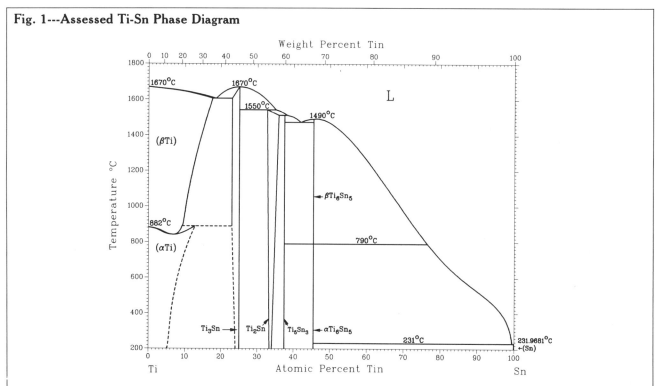

Fig. 1---Assessed Ti-Sn Phase Diagram

The figure is based on thermodynamic calculations. Drawn invariant temperatures may differ slightly (within experimental uncertainty) with the temperatures listed in Table 1.

J.L. Murray, 1987.

Table 1 Special Points of the Assessed Ti-Sn Phase Diagram

Reaction	Composition of the respective phases, at.% Sn			Temperature, °C	Reaction type
L ⇌ (βTi) + Ti₃Sn	18.5	17.5	~ 25	1605	Eutectic
L ⇌ Ti₃Sn		25		1670	Congruent
(βTi) + Ti₃Sn ⇌ (αTi)	~ 9.5	23	~ 12.5	890	Peritectoid
(βTi) ⇌ (αTi)		6.7		842	Congruent
L + Ti₃Sn ⇌ Ti₂Sn	35.5	25	33.3	1550	Peritectic
L + Ti₂SN ⇌ Ti₅Sn₃	38	33.3	37.5	~ 1510	Peritectic
L ⇌ Ti₅Sn₃ + βTi₆Sn₅............	42	37.5	45.5	~ 1475	Eutectic
L ⇌ Ti₆Sn₅		45.5		~ 1490	Congruent
βTi₆Sn₅ ⇌ αTi₆Sn₅		45.5		790	Unknown
L ⇌ αTi₆Sn₅ + (Sn)	~ 99.5	45.5	~ 99.98	231	Eutectic
L ⇌ (βTi)		0		1670	Melting point
(βTi) ⇌ (αTi)		0		882	Allotropic transformation
L ⇌ (Sn).....................		100		231.9681	Melting point

$L \rightleftarrows (\beta Ti) + Ti_3Sn$
$L \rightleftarrows Ti_3Sn$
$(\beta Ti) + Ti_3Sn \rightleftarrows (\alpha Ti)$
$(\beta Ti) \rightleftarrows (\alpha Ti)$
$L + Ti_3Sn \rightleftarrows Ti_2Sn$
$L + Ti_2SN \rightleftarrows Ti_5Sn_3$
$L \rightleftarrows Ti_5Sn_3 + \beta Ti_6Sn_5$
$L \rightleftarrows Ti_6Sn_5$
$\beta Ti_6Sn_5 \rightleftarrows \alpha Ti_6Sn_5$
$L \rightleftarrows \alpha Ti_6Sn_5 + (Sn)$

Fig. 2---Experimental Liquidus Data vs the Assessed Diagram

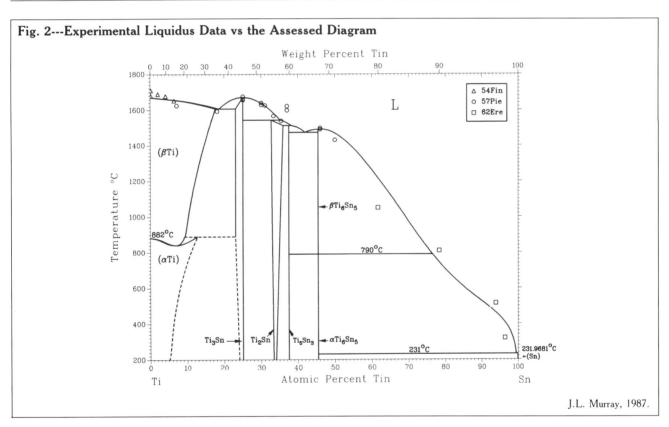

J.L. Murray, 1987.

pyrometry, with estimated uncertainties in the temperature measurement of 10 to 20 °C. Thus, the congruent melting temperatures of Ti₃Sn and βTi₆Sn₅ were determined as 1663 ± 20 °C and 1494 ± 20 °C, respectively. Similarly, the eutectic reaction L ⇌ (βTi) + Ti₃Sn was located at 1607 ± 20 °C; [55Mcq] found 1590 °C from metallographic studies, in good agreement with [57Pie]. [57Pie] located the peritectic temperature (Ti₃Sn + L ⇌ Ti₂Sn) as 1552 °C based on metallographic work. The remaining two reactions, Ti₂Sn + L ⇌ Ti₅Sn₃ and L ⇌ Ti₅Sn₃ + βTi₆Sn₅, were investigated by metallographic examination of as-cast and slowly cooled specimens. The temperatures of these reactions are thus only rough estimates. From the same metallographic study it was de-

duced that Ti₂Sn has a homogeneity range of 32.7 to 35.9 at.% Sn at high temperatures, but only a narrow range at low temperatures. [57Pie] attempted to synthesize single-phase Ti₅Sn₃ but was unsuccessful.

Sn-Rich Alloys.

[57Pie] found by X-ray diffraction that alloys containing more than 45 at.% Sn were mixtures of βTi₆Sn₅ and (βSn). The lattice parameters of (βSn) and melting temperatures of Sn-rich alloys did not differ detectably from those of pure βSn.

[62Ere] determined the Sn-rich liquidus and the types of the invariant reactions at 790 °C and 231 °C. Differential thermal analysis (DTA) data for the liquidus are

Fig. 3---Detail of the Ti-Rich Solid Phase Equilibria

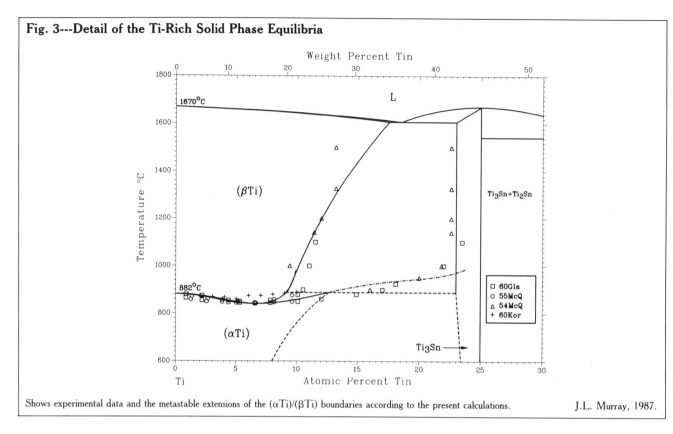

Shows experimental data and the metastable extensions of the (αTi)/(βTi) boundaries according to the present calculations.

J.L. Murray, 1987.

Table 2 Experimental Data on the (αTi)/(βTi)/Ti₃Sn Equilibria

Reference	(βTi) composition, at.% Sn	Reaction temperature, °C	Reaction type	Experimental method
[54Fin]	8.4	940	Peritectoid	Metallography
[52Wor]	8	885	Peritectoid	Metallography
[57Pie]	9.5	864	Eutectoid	Metallography
[60Kor]	10	890	Peritectoid	Thermal analysis
[55Mcq]	(αTi) Ti₃Sn ⇌ continuously		...	Metallography
[60Gla]	(αTi) Ti₃Sn ⇌ continuously		...	Metallography, resistivity

was determined that the melting point of Sn is lowered by 1 to 2 °C with addition of Ti.

[62Ere] tentatively identified an exothermic reaction at 790 °C with a polymorphic transformation in Ti₆Sn₅. That such a transformation exists was independently discovered in the X-ray diffraction study of [64Vuc]. The assignment of the hexagonal structure to the high-temperature form and the orthorhombic structure to the low-temperature form is based first on the lower symmetry of the hexagonal form, and second on the predominance of βTi₆Sn₅ and the observation of αTi₆Sn₅ only in relatively Sn-rich alloys [61Sam, 57Pie, 64Vuc].

Equilibria of (αTi), (βTi), and Ti₃Sn.

(βTi) transforms congruently to (αTi) at 6.7 at.% Sn and 842 °C [52Wor, 54Mcq, 55Mcq, 60Gla, 60Kor]. Neither [54Fin] nor [57Pie] showed the congruent transformation, but both examined too coarse a temperature interval to have been able to distinguish the minimum. Experimental data are compared to the assessed boundaries in Fig. 3. Resistivity measurements

of [60Kor] made on heating disagreed quantitatively with the others, being consistent with a congruent point at 860 °C. The scatter in this heating data suggests 860 °C is an upper bound on the congruent point. Therefore, the data of [60Gla], [54Mcq], and [52Wor], which are in close agreement, were used in this assessment, though obtained by several techniques: resistivity (cooling), hydrogen pressure, and metallography, respectively.

The maximum solubility of Sn in (αTi) and hence the existence of the peritectoid reaction (βTi) + Ti₃Sn ⇌ (αTi) have also been the subject of controversy. Reported reactions are summarized in Table 2. [55Mcq] and [60Gla] believed the single-phase (αTi) solution to be considerably more extensive than was reported elsewhere [52Wor, 54Fin, 57Pie, 60Kor], attributing reports of two-phase microstructures to a failure to achieve equilibrium. [55Mcq] used extremely long annealing times, but did not work the alloys to hasten equilibration. Similarly, [60Gla] used levitation melting to make unusually homogeneous alloys. Neither [55Mcq] nor [60Gla] found evidence for a two-phase (αTi) + Ti₃Sn field. [55Mcq] speculated that the

Table 3   Crystal Structures and Lattice Parameters of the Equilibrium Phases of the Ti-Sn System

Phase	Equilibrium composition range, at.% Sn			Pearson symbol	Space group	Struktur-bericht designation	Prototype	Lattice parameters, nm			Reference
								a	b	c	
(βTi)	0	to	>7.5	$cI2$	$Im3m$	$A2$	W	...	...	...	[Pearson2]
(αTi)	0	to	17	$hP2$	$P6_3/mmc$	$A3$	Mg	...	...	...	[Pearson2]
Ti$_3$Sn	23	to	25	$hP8$	$P6_3/mmc$	$D0_{19}$	Ni$_3$Sn	0.5916	...	0.4764	[52Pie]
Ti$_2$Sn	32.7	to	35.9	$hP6$	$P6_3/mmc$	$B8_2$	InNi$_2$	0.4653	...	0.570	[57Pie]
Ti$_5$Sn$_3$	37.5			$hP16$	$P6_3/mcm$	$D8_8$-type	Mn$_5$Si$_3$	0.8049	...	0.5454	[59Now, 51Pie]
βTi$_6$Sn$_5$ ...	45.5			$hP22$	$P6_3/mmc$	...	αTi$_6$Sn$_5$	0.922	...	0.569	[64Vuc]
					$P\bar{3}1c$	...	...	0.924$\bar{8}$	...	0.589	[63Sch]
αTi$_6$Sn$_5$ ...	45.5			$oI44$	$Immm$	...	Nb$_6$Sn$_5$	1.693	0.9144	0.5735	[64Vuc]
(Sn)	99.98	to	100	$tI4$	$I4_1/amd$	$A5$	βSn	0.58315	...	0.31814	[Pearson2]

ordering was continuous, and both [55Mcq] and [60Gla] showed the (αTi) and Ti$_3$Sn phase fields joining continuously.

It can be shown on the basis of symmetry alone, however, that the cph → D0$_{19}$ ordering cannot occur by a higher-order transition along a line of critical points. We must therefore conclude that equilibrium was actually more closely approached by [52Wor], [54Fin], [57Pie] and [60Kor]. It is proposed that under the conditions studied by [55Mcq] and [60Gla] incoherent precipitation of Ti$_3$Sn in (αTi) was suppressed, and that they measured the metastable extension of the (αTi)/(βTi) boundary. We also suggest that TEM could be used to shed more light on the source of these discrepancies.

Reported invariant temperatures range between 864 °C [57Pie] and 940 °C [54Fin]. The value 890 °C obtained from thermal analysis and metallographic studies of [60Kor] is preferred.

Table 4   Lattice Parameters of (αTi) [60Cou]

Composition, at.% Sn	Lattice parameters, nm	
	a	c
6	0.2955	0.4731
8	0.2957	0.4735
10	0.2957	0.4744
12	0.2957	0.4754

[52Wor], [55Mcq], [60Gla], and [60Kor] agreed in placing the homogeneity range of Ti$_3$Sn at about 23 to 25 at.% Sn. [52Wor], [57Pie], and [60Kor] found that the (αTi)/((αTi) + Ti$_3$Sn) boundary is essentially vertical between the peritectoid reaction and 700 °C. To construct the assessed diagram, thermodynamic calculations were used to extrapolate this boundary to lower temperatures.

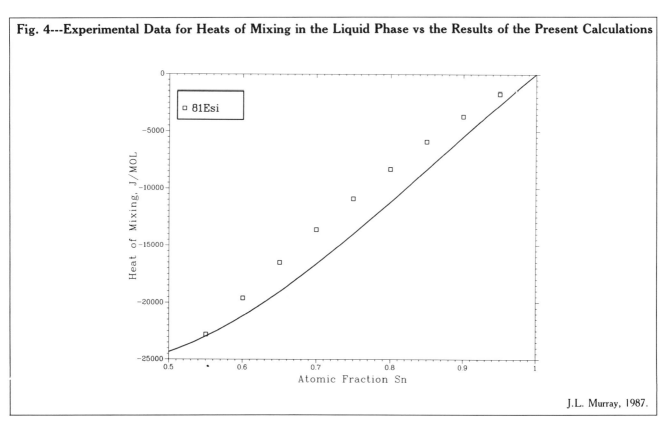

Fig. 4---Experimental Data for Heats of Mixing in the Liquid Phase vs the Results of the Present Calculations

□ 81Esi

J.L. Murray, 1987.

Table 5 Experimental Data on Ti Partial Excess Gibbs Energy in the Liquid

Reference	Temperature, °C	Composition, at.% Sn	Partial Gibbs energy, J/mol
[78Alg]	850	84.9	− 38 849
		88.3	− 43 265
		88.5	− 36 753
		91.1	− 37 743
		97.5	− 35 857
		98.2	− 37 743
		98.6	− 34 656
		98.9	− 33 262

Metastable Phases

The start temperature of the martensitic (βTi) → (αTi) transformation (M_s) was measured by [60Sat]. In the range 2 to 9 at.% Sn, M_s is 740 to 760 °C, apparently independent of composition.

Crystal Structures and Lattice Parameters

Crystal structures and lattice parameters for the equilibrium phases are summarized in Table 3. Two symmetries have been reported for βTi$_6$Sn$_5$; $P6_3/mmc$ [64Vuc] is preferred.

Lattice parameters of hexagonal (αTi) are given separately in Table 4. (αTi) lattice parameters are from [60Cou], who showed that hydrogen contamination causes a decrease, particularly in the c parameters.

Thermodynamics

[55Kub] attempted to measure heats of formation of the intermetallic compounds, but very little alloying occurred at 1000 °C and no heat effect was observed.

[81Esi] measured enthalpies of mixing in the liquid state at 2000 K, over the composition range 50 to 100 at.% Sn. Extrapolation of the data to pure Ti would predict a very large negative heat of mixing near pure Ti, and it is therefore believed that the data for Sn-rich compositions are more accurate. Data are compared to our calculation in Fig. 4.

[78Alg] and [80And] measured Ti activities in liquid Sn, [80And] using nitrogen-nitride phase equilibria and [78Alg] using an emf cell technique. The results of [80And] are consistent with a limiting value of $RT(\ln \gamma_{Ti})$ of −29 900 J/mol. Data of [78Alg] for $RT(\ln \gamma_{Ti})$ are given in Table 5.

[73Geg] measured Ti and Sn activities in bcc (βTi) by the Knudsen cell technique: they reported a temperature-independent excess Gibbs energy of −92 500 ± 3800 $x_{Ti}x_{Sn}$ J/mol.

Calculation of the Phase Diagram

[73Kau] calculated the Ti-rich side of the phase diagram using the regular solution approximation for all the solution phases. Because of intrinsic limitations of the regular solution model, the congruent (βTi) $\rightleftarrows$ (αTi) transition was not reproduced. The availability of thermodynamic data for the liquid now makes it possible to use more detailed models to reproduce the details of the phase diagram.

In the present calculation, the Gibbs energies of the solution phases are represented as:

Table 6 Ti-Sn Thermodynamic Properties

Gibbs energies of the pure elements

G^0(Ti,L) = 0
G^0(Sn,L) = 0

G^0(Ti,bcc) = − 16 234 + 8.368 T
G^0(Sn,bcc) =   6.276 T

G^0(Ti,cph) = − 20 585 + 12.134 T
G^0(Sn,cph =   − 418 + 6.694 T

G^0(Sn)   = − 7 031 + 13.913 T

Excess Gibbs energies of the solution phases

B(L,0) = − 97 252
B(L,1) = − 45 236

B(bcc,0) = − 109 000
B(bcc,1) = − 45 000

B(cph,0) = − 154 500
B(cph,1) = − 47 420
B(cph,2) = 114 767
B(cph,3) = − 67 397

Gibbs energies of the compounds

G(Ti$_3$Sn) = − 51 000 + 10.0 T
G(Ti$_2$Sn) = − 45 046 + 5.7369 T
G(Ti$_5$Sn$_3$) = − 38 808 + 1.9804 T
G(Ti$_6$Sn$_5$) = − 43 109 + 4.4646 T

Note; Values are given in J/mol and J/mol·K.

$$G(i) = (1 - x)\, G^0(\text{Ti},i) + x\, G^0(\text{Sn},i) + RT[x \ln x + (1 - x) \ln (1 - x)] + x(1 - x)\, [B(i,j) + P_j(1 - 2x)]$$

where i designates the phase; x, is the atomic fraction of Sn; and $G^0(i)$ is the Gibbs energies of the pure metals. $P_j f1$ is the jth Legendre polynomial used to expand the excess Gibbs energy, with $B(i,j)$ as coefficient. The intermetallic phases---Ti$_3$Sn, Ti$_2$Sn, Ti$_5$Sn$_3$, and Ti$_6$Sn$_5$—were modeled as line compounds, and zero solubility range was also attributed to (βSn).

Input data for optimization calculations were: heat of mixing data for the liquid [81Esi]; the (αTi)/(βTi) phase boundaries [60Gla]; the Sn-rich liquidus [62Ere]; and the three-phase equilibria and congruent melting points [57Pie]. As discussed above, we interpreted the results of [60Gla] and [55Mcq] as pertaining to the metastable extensions of the (αTi)/(βTi) boundaries.

Preliminary calculations were done in the subregular solution approximation. Based on the good agreement between the limiting values of excess partial Gibbs energies and heats of mixing for the Sn-rich liquid, temperature-independent excess functions were used for the solution phases. In this approximation, some conflict was found between the liquidus data and partial excess Gibbs energies. Thermodynamic functions (Table 6) were chosen that reproduced the liquidus.

In addition to requiring a congruent (αTi)/(βTi) transformation, we required that the metastable extensions of the (αTi)/(βTi) boundaries become nearly vertical as the composition Ti$_3$Sn is approached. We also hypothesized that the broadening of the two-phase (αTi)/(βTi) field beyond the congruent point reflects a complicated composition dependence of the cph Gibbs energy, which reflects the effect of local ordering.

This hypothesis appears to be justified by the calculations. In the subregular solution model, both the composition of the critical point and the width of the two-phase region agreed only qualitatively with experiment.

Additional terms in the composition expansion were then determined by the metastable $(\alpha Ti)/[(\alpha Ti)+(\beta Ti)]$ boundary. We found that not only could the shape of the metastable boundary be reproduced, but also that the position of the congruent point was shifted to precise agreement with the assessed value. We concluded that the interpretation of the phase equilibria described in [55Mcq] and [60Gla] as metastable equilibria leads to a self-consistent model for the thermodynamics of this system.

Cited References

51Pie: P. Pietrokowsky and P. Duwez, "Crystal Structure of Ti_5Si_3, Ti_5Ge_3, and Ti_5Sn_3," *Trans. AIME, 191*, 772-773 (1951). (Crys Structure; Experimental)

52Pie: P. Pietrokowsky, "Crystal Structure of Ti_3Sn," *Trans. AIME, 194*, 211-212 (1952). (Crys Structure; Experimental)

*52Wor: H.W. Worner, "The Structure of Titanium-Tin Alloys in the Range 0-25 at.% Tin," *J. Inst. Met., 81*, 521-529 (1952-1953). (Equi Diagram; Experimental)

*54Fin: W.L. Finlay, R.I. Jaffee, R.W. Parcel, and R.C. Durstein, "Tin Increases Strength of Ti-Al Alloys Without Loss in Fabricability," *J. Met., 6*, 25-29 (1954). (Equi Diagram; Experimental)

*54Mcq: A.D. McQuillan, "A Study of the Behaviour of Titanium-Rich Alloys in the Titanium-Tin and Titanium-Aluminum Systems," *J. Inst. Met., 83*, 181-184 (1954-1955). (Equi Diagram; Experimental)

55Kub: O. Kubaschewski and W.A. Dench, "The Heats of Formation in the Systems Titanium-Aluminum and Titanium-Iron," *Acta Metall., 3*, 339-346 (1955). (Thermo; Experimental)

*55Mcq: M.K. McQuillan, "The Constitution of the Titanium-Tin System in the Region 0-25 at.% Tin," *J. Inst. Met., 84*, 307-312 (1955-1956). (Equi Diagram; Experimental)

*57Pie: P. Pietrokowsky and E.P. Frink, "A Constitution Diagram for the Alloy System Titanium-Tin," *Trans. ASM, 49*, 339-358 (1957). (Equi Diagram; Experimental)

59Now: H. Nowotny, B. Auer-Welsbach, J. Bruss, and A. Kohl, "Contribution on the Mn_5Si_3 ($D8_8$-Type) Structure," *Monatsh Chem., 90*, 15-23 (1959) in German. (Crys Structure; Experimental)

60Cou: A. Coucoulas and H. Margolin, "The Influence of Hydrogen on the Lattice Parameters of Ti-Sn Alloys," *Trans. AIME, 218*, 958-959 (1960). (Crys Structure; Experimental)

*60Gla: V.V. Glazova and N.N. Kurnakov, "Equilibrium in the Titanium-Tin System," *Dokl. Akad. Nauk SSSR, 134*, 1087-1090 (1960) in Russian; TR: *Proc. Acad. Sci. USSR, Chem. Sect., 134*, 1129-1133 (1961). (Equi Diagram; Experimental)

*60Kor: I.I. Kornilov and T.T. Nartova, "Phase Diagram of the Titanium-Tin System," *Zh. Neorg. Khim., 5*, 622-629 (1960) in Russian; TR: *Russ. J. Inorg. Chem., 5*(3), 300-303 (1960). (Equi Diagram; Experimental)

60Sat: T. Sato, S. Hukai, and Y. Huang, "The M_s Points of Binary Titanium Alloys," *J. Aust. Inst. Met., 5*(2), 149-153 (1960). (Meta Phases; Experimental)

61Sam: V.P. Samsonov, "Production of Single Crystals of the Compound Ti_6Sn_5 and Its Crystal Structure," *Tr. Gos. Nauchn.-Issled. Proek. Inst. Obrab. Tsvetn. Metall.*, (20), 136-142 (1961) in Russian. (Equi Diagram; Experimental)

*62Ere: V.N. Eremenko and T.Ya. Velikanova, "The Tin-Titanium System in the Tin-Rich Region," *Zh. Neorg. Khim., 7*, 1750-1752 (1962) in Russian; TR: *Russ. J. Inorg. Chem., 7*(7), 902-904 (1962). (Equi Diagram; Experimental)

63Sch: K. Schubert, K. Frank, R. Gohle, A. Maldonado, H.G. Meissner, A. Raman, and W. Rossteutscher, "Structure Data on Metallic Phases," *Naturwissenschaften, 50*, 41 (1963) in German. (Crys Structure; Experimental)

*64Vuc: J.H. Vucht, H.A. Brunning, H.C. Donkersloot, and A.H. Mesquita, "The System Vanadium-Gallium," *Phillips Res. Rep., 19*, 407-421 (1964). (Crys Structure; Experimental)

73Geg: H.L. Gegel and M. Hoch, "Thermodynamics of Alpha-Stabilized Ti-X-Y Systems," Sci. Technol. Appl. Titanium, Proc. Intl. Conf., Vol. 2, R.I. Jaffee, Ed., 923-933 (1973). (Thermo; Experimental)

73Kau: L. Kaufman and H. Nesor, "Phase Stability and Equilibria as Affected by the Physical Properties and Electronic Structure of Titanium Alloys," Sci. Technol. Appl. Titanium, Proc. Intl. Conf., Vol. 2, R.I. Jaffee, Ed., 773-800 (1973). (Thermo; Theory)

78Alg: M.M. Alger, "The Thermodynamics of Highly Solvated Liquid Metal Solutions," thesis, Mass. Inst. Tech. (1978); *Diss. Abstr. Int., 42*(11), 251 (1982). (Thermo; Experimental)

80And: R.N. Anderson and G.S. Selvaduray, "The Thermodynamics of Liquid Titanium Alloys," *Titanium '80*, Sci. and Technol., Vol. 4, Kyoto, Japan, 3009-3018 (1980). (Thermo; Experimental)

*81Esi: Yu.O. Esin, M.G. Valishev, A.F. Ermakov, P.V. Gel'd, and M.S. Petrushevskii, "The Enthalpies of Formation of Liquid Cobalt-Titanium and Tin-Titanium Alloys," *Zh. Fiz. Khim., 55*, 747-748 (1981) in Russian; TR: *Russ. J. Phys. Chem., 55*(3), 417-418 (1981). (Thermo; Theory)

The Sr-Ti (Strontium-Titanium) System

87.62   47.88

By J.L. Murray

Table 1   Special Points of the Assessed Ti-Sr Phase Diagram

Reaction	Compositions of the respective phases, at.% Sr			Temperature, °C	Reaction type
(βTi) + (LSr) ⇌ (αTi)	(0.0059)	99.967	(0.019)	~ 882	Peritectic
(LSr) ⇌ (αTi) + (βSr)	99.983	(<0.019)	~ 100	~ 769	Eutectic
(βTi) ⇌ (αTi)		0		882	Allotropic transformation
L ⇌ (βSr)		100		769	Melting point

Table 2   Crystal Structures of the Ti-Sr System [Pearson2]

Phase	Homogeneity range, at.% Sr	Pearson symbol	Space group	Struktur-bericht designation	Prototype
(βTi)	0 to ~ 0.0059	cI2	Im3m	A2	W
(αTi)	0 to 0.019	hP2	P6₃/mmc	A3	Mg
(βSr) .... ~ 100		cI2	Im3m	A2	W

Fig. 1---Assessed Ti-Sr Phase Diagram

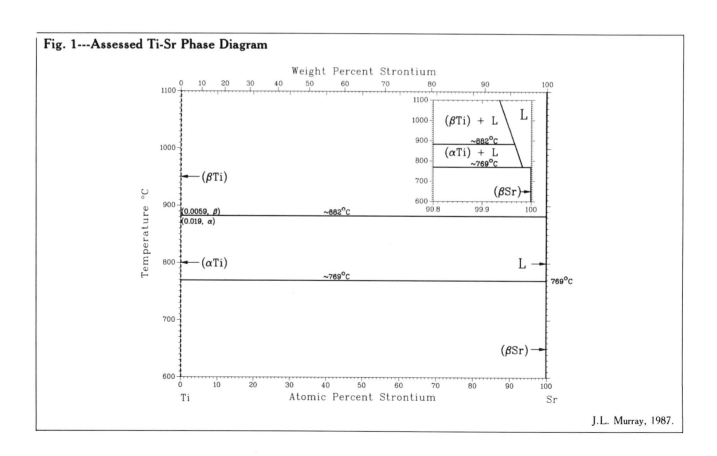

J.L. Murray, 1987.

300

[78Ali] measured solubilities of Ti in liquid Sr at several temperatures and found them to be very low. From the limited liquid solubility, it is deduced that mutual solid solubilities are extremely low and that no intermetallic compounds occur in the Ti-Sr system.

[78Ali] held Ti isothermally in liquid Sr for 1.5 to 3 h under pressure in helium. The solubility of Ti in liquid Sr was determined by chemical analysis of several samples of the liquid layer. Microstructures were also reported to have been examined. Alloys were prepared from "degassed Ti used as a getter, and ... strontium of analytical purity."

[78Ali] reported solubilities of Ti in liquid Sr and Ba at 800, 900, and 1000 °C in the form of equations and figures. However, values of the two do not correspond. Assuming the figures to be correctly labeled, the present evaluator concluded that the equations for Sr and Ba had been reversed, and that, in the equation, temperature was given in °C and composition x in atomic fraction of Sr. Then the equation for the solubility is:

$$\log x = -2.0265 - 1279/T$$

Using 7431 J/mol for the heat of fusion of Sr [84Cha] and the van't Hoff relation, it is estimated that the temperature of the eutectic reaction $L \rightleftarrows (\alpha Ti) + (\beta Sr)$ lies about 0.1 °C below the melting point of Sr, 769 °C [Melt].

Based on the experimental liquid solubility data, [78Ali] estimated solubilities of Sr in (αTi) and (βTi) at 882 °C as 0.019 and 0.0059 at.% Sr, respectively, using a method that was not described. These values are shown in the assessed phase diagram, (Fig. 1), but it is not recommended that much reliance be placed on solid solubility values obtained by an unknown method from unconfirmed liquid solubility data. The experimental data are shown in Fig. 2.

Special points of the phase diagram are listed in Table 1, and crystal structure data for the pure components are listed in Table 2.

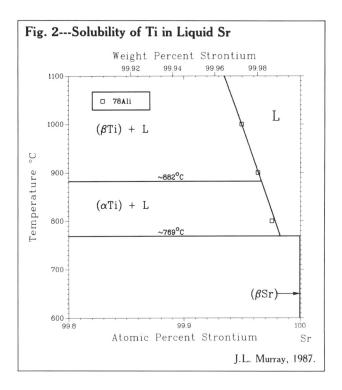

Fig. 2---Solubility of Ti in Liquid Sr

J.L. Murray, 1987.

Cited References

78Ali: F.N. Alidzhanov, A.V. Vakhobov, and T.D. Dushanbe, "Ba-Ti and Sr-Ti Phase Diagrams," *Izv. Akad. Nauk SSSR, Met.*,(2) 223-224 (1978); TR: *Russ. Metall.*, (2), 177-178 (1978). (Equi Diagram; Experimental)

84Cha: M.W. Chase, "Heats of Transformation of the Elements," *Bull. Alloy Phase Diagrams*, 4(1), 123-124 (1984). (Thermo; Compilation)

The Ta-Ti (Tantalum-Titanium) System

180.9479   47.88

By J.L. Murray

Equilibrium Diagram

The Ti-Ta phase diagram (Fig. 1 and Table 1) is of the simple isomorphous type, but quantitative data are lacking on the liquidus. Data on the solid phase boundaries are both mutually contradictory and inconsistent with thermodynamic properties of pure Ti. Thermodynamic calculations therefore played a major role in the construction of the assessed diagram. The justification for preferring the calculated diagram and the method of constructing it will be discussed in detail in the following sections.

The equilibrium solid phases of the Ti-Ta system are:

- The bcc (βTi,Ta) solid solution, in which there is complete mutual solubility at temperatures above the (βTi) ⇌ (αTi) transition of pure Ti (882 °C).
- The low-temperature cph (αTi) solid solution, which exhibits a retrograde solubility. The sugges-

tion [68Kal] that an equilibrium compound Ti₃Ta occurs was based on anomalies in physical properties and has not been verified by any microstructural or crystal structure study.

Liquidus and Solidus.

In Fig. 2, melting point data [52Sum, 53May, 65Bru, 67Bud] are compared with the assessed solidus and thermodynamically predicted liquidus. [53Sum] measured the temperature of the solid-liquid interface and suggested that these data represented points about midway between the solidus and liquidus. On the Ti-rich side, the data of [53May] were obtained by optical observation of melting; other data were obtained from microscopic examination of quenched alloys. [65Bru], [67Bud], and [69Rud] obtained solidus data by optical methods.

Although all investigators agreed that the melting points vary approximately linearly with composition, dis-

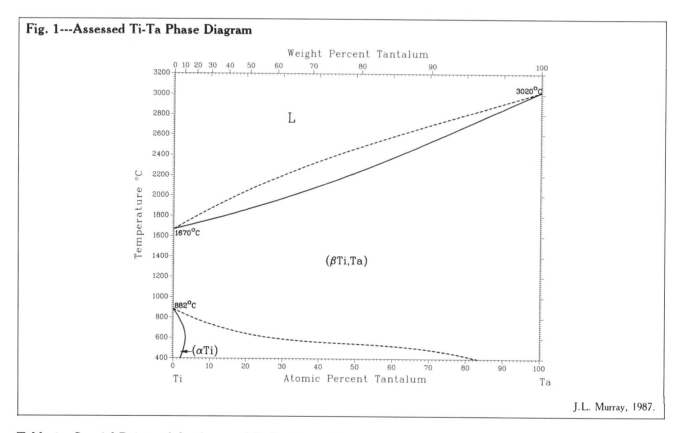

Fig. 1---Assessed Ti-Ta Phase Diagram

J.L. Murray, 1987.

Table 1   Special Points of the Assessed Ti-Ta Phase Diagram

Reaction	Composition, at.% Ta	Temperature, °C	Reaction type
L ⇌ (βTi)	0	1670	Melting point
(βTi) ⇌ (αTi)	0	882	Allotropic transformation
L ⇌ (Ta)	100	3020	Melting point
(βTi,Ta) ⇌ (βTi) + (Ta)	~ 50	~ 450	Critical point (metastable)

302

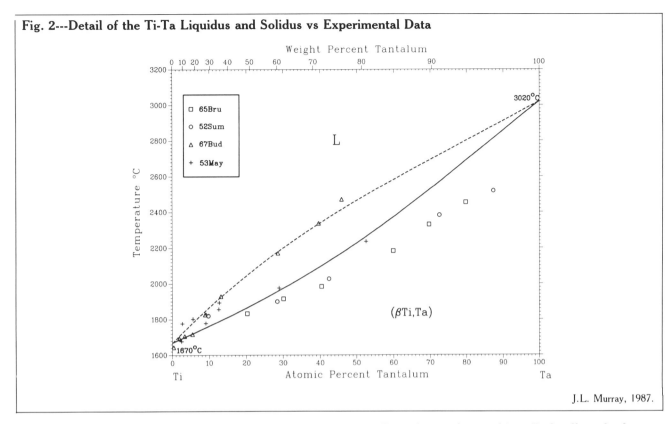

Fig. 2---Detail of the Ti-Ta Liquidus and Solidus vs Experimental Data

J.L. Murray, 1987.

crepancies in the melting temperatures reach nearly 500 °C for Ta contents exceeding 40 at.%. Alloy contamination at high temperature is thus a predominating effect. Because of the large discrepancies, the assessed solidus is based on the pure metal melting points and the approximate linearity of the melting point curve. The linearity implies that excess Gibbs energies of the liquid and bcc phases are approximately equal. The width of the two-phase field is determined primarily by the ideal entropy contribution to the Gibbs energies at these high temperatures, and it is influenced only slightly by the magnitudes of the excess Gibbs energies. Thus, with respect to the solidus, the predicted liquidus can be considered reliable to within about ±2 at.% at a given temperature.

(αTi) Solvus and (βTi) Transus.

The (αTi) solvus [52Sum, 53May, 67Bud] and the (βTi) transus [53May, 67Bud, 69Nik] were examined by metallographic and X-ray techniques, supplemented by resistivity [53May] and physical and mechanical-property [67Bud] measurements. Experimental data are shown in Fig. 3. There is good agreement that the maximum solubility of Ta in (αTi) is about 3 ± 0.2 at.% at 600 °C. A disagreeing report that the solubility of Ta in (αTi) is at least 5 at.% [65Rau] is attributed to failure to reach equilibrium.

The [67Bud] data points for the (βTi) transus indicate the most Ta-rich two-phase alloy found at each given temperature. Note the discrepancy between the resistivity (lowest temperature datum) and the metallographic data of [53May]. There are not only discrepancies concerning the position of the transus, but also discrepancies in the curvature of the boundary.

There are two expected sources of experimental error for the (βTi) transus—(1) contamination of the alloys by interstitial impurities and (2) precipitation of (αTi) during

insufficiently rapid quenching. Both effects lead to an erroneous high-temperature (βTi) transus. The latter effect has been shown to be important in both the Ti-V and Ti-Nb systems, which are chemically similar to Ti-Ta. The absence of Ti-Ta compounds strongly suggests that excess Gibbs energies are positive and that the (βTi) transus lies above a metastable (βTi) miscibility gap. Although the experimental evidence is not conclusive, the appearance of two bcc phases in tempered (βTi,Ta) tends also to support the existence of a metastable miscibility gap with approximate tie line compositions of 20 and 70 at.% Ta at 400 °C [72Byw2]. The closeness of the lattice parameters of bcc Ti and Ta and the symmetrical miscibility gap tie line estimates imply that the bcc excess Gibbs energy is not strongly asymmetrical. Gibbs energy functions were constructed that reproduce the (αTi) solvus and the approximate metastable bcc miscibility gap, and the calculated boundaries are shown in Fig. 1 and 3.

Comparison with chemically related system indicates that the (βTi) transus is geometrically of the type calculated in the present assessment. In particular, near the Ti-rich side, the two-phase field can be extrapolated from the van't Hoff value in the dilute limit. It is estimated that the temperature of the (βTi) transus is uncertain by not more than about ±25 °C at 10 at.% Ta; the uncertainty on the Ta-rich side of the diagram is unknown, but considerably greater. Thus, the boundary is drawn with a dashed curve; further experimental phase diagram and/or thermochemical data are needed.

Metastable Phases

Metastable phases are formed from (βTi) during quenching. In Ti-rich alloys, the cph phase (α'Ti) can form martensitically; at slightly higher Ta contents, a distorted cph phase (α″Ti) is formed, and at sufficiently high Ta content, (βTi) can be retained. ω phase forms as an

Fig. 3---Detail of the Ti-Ta (βTi)/(αTi) Boundaries vs Experimental Data

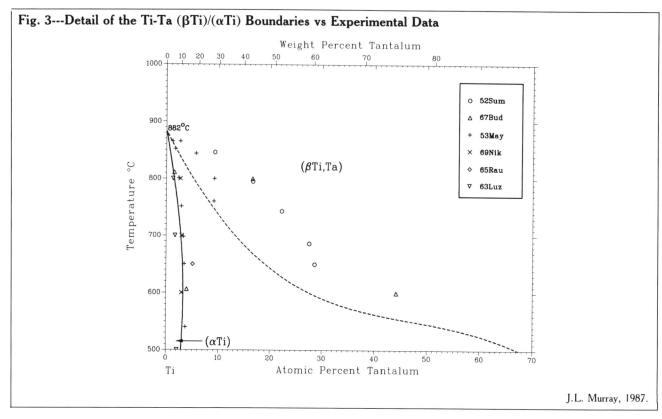

J.L. Murray, 1987.

intermediate phase in the decomposition (βTi) → (αTi) at low temperatures.

Metastable bcc Miscibility Gap.

As mentioned above, there is some electron diffraction evidence for the formation of two bcc phases in a 21 at.% Ta alloy tempered at 400 °C [72Byw2]. Based on lattice parameters, the composition of the two phases were estimated as 20 and 70 at.% Ta. It is also possible, however, that inhomogeneous precipitation was responsible for the appearance of two bcc phases.

Martensites.

The martensite transformation of (βTi) was reported to be partly suppressed in alloys containing more than 9 at.% Ta [52Sum] or 15 at.% Ta [53May] and fully suppressed in alloys containing more than 14 at.% Ta [53Sum], 15 at.% Ta [58Bag], or 21 at.% Ta [53May]. [72Byw1] showed that prior heat treatment can influence the product structures and also that samples which showed no optical evidence of martensite may nevertheless be fully transformed.

In alloys containing up to 7 at.% Ta, the martensite has the cph structure. In alloys containing more than 7 at.% Ta, the martensite has an orthorhombic structure, which may be viewed as distorted cph [63Bor, 66Bro, 71Byw1].

The start temperature of the martensite transformation, M_s, was measured by [53Duw] for alloys containing up to 5 at.% Ta. M_s varies approximately linearly with composition and reaches 750 °C at 5 at.% Ta.

ω Phase.

In Ti-Ta alloys, ω phase is not found in quenched specimens of any composition, but only forms during tempering of the bcc phase near 400 °C [58Bag, 72Byw2].

ω particles are coherent with the bcc matrix and ellipsoidal in shape, indicative of low misfit between the ω phase and the bcc matrix [72Byw2, 73Ika]. A 23 at.% Ta alloy forms ω phase when tempered at 400 °C, but forms equilibrium (αTi) + (βTi) directly when tempered at 600 °C [72Byw2]. [72Byw2] noted that a metastable congruent (βTi) → ω transformation at −14 at.% Ta and −600 °C with a wide two-phase (ω + (βTi)) field would be consistent with their observations.

Crystal Structure and Lattice Parameters

Equilibrium and metastable crystal structures of Ti-Ta alloys are summarized in Table 2. ω phase may also have a trigonal symmetry $P\bar{3}m1$, rather than the hexagonal $P6/mmm$ symmetry listed. Detailed structure determinations have not been performed for ω in the Ti-Ta system. In Fig. 4 and 5, respectively, room-temperature lattice parameters of (βTi,Ta) [53Sum] and of the (α'Ti) and (α"Ti) martensites [72Byw1] are shown.

Thermodynamics

No experimental thermodynamic data for Ti-Ta alloys are available. Figure 1 is the result of a regular solution calculation. The Gibbs energies of the equilibrium phases are represented as:

$$G(i) = G^0(\text{Ti},i)(1 - x) + G^0(\text{Ta},i)x + RT[x \ln x + (1 - x)\ln(1 - x)] + B(i)x(1 - x)$$

where i designates the phase; x is the atomic fraction of Ta; G^0 is the Gibbs energies of the pure components; and $B(i)$ is the coefficient of the excess Gibbs energy in the regular solution approximation. The lattice stability parameters G^0 are from [70Kau], but are slightly modified to reproduce the melting point of pure Ta.

Table 2 Crystal Structures of the Equilibrium Phases of the Ti-Ta System

Phase	Homogeneity range, at.% Ta	Pearson symbol	Space group	Struktur-bericht designation	Prototype	Reference
(αTi)	0 to 3.6	hP2	P6₃/mmc	A3	Mg	[Pearson2]
(βTi,Ta)	0 to 100	cI2	Im3m	A2	W	[Pearson2]
(α'Ti)	(a)	hP2	P6₃/mmc	A3	Mg	[72Byw1]
(α"Ti)	(a)	oC4	Cmcm	A20	αU	[72Byw1]
ω	(a)	hP3	P6/mmm	...	ωMnTi	[Pearson2]
			or P3̄m1	...	ωCrTi	[Pearson2]

(a) Metastable.

$P6_3/mmc$, $Im3m$, $P6_3/mmc$, $Cmcm$, $P6/mmm$, $P\bar{3}m1$

Fig. 4---Room-Temperature Lattice Parameters of the bcc Phase

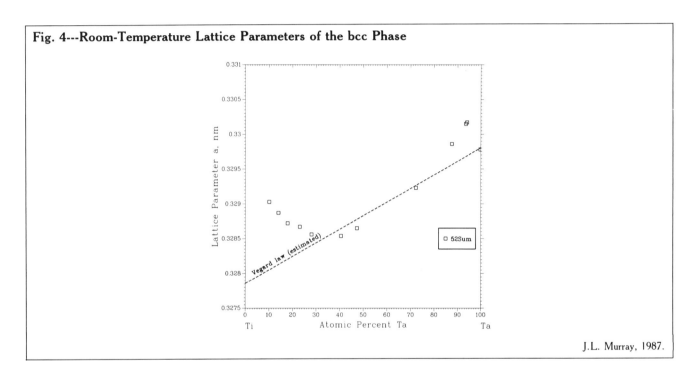

J.L. Murray, 1987.

Fig. 5---Room-Temperature Lattice Parameters of the Martensites, (α'Ti), and (α"Ti)

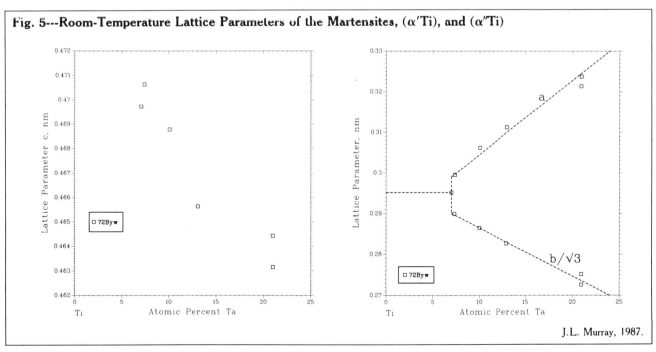

J.L. Murray, 1987.

Table 3 Thermodynamic Properties of the Ti-Ta System

Properties of the pure elements

G^0(Ti,L) = 0
G^0(Ta,L) = 0

G^0(Ti,bcc) = $-16\,234 + 8.368\,T$
G^0(Ta,bcc) = $-27\,363 + 8.309\,T$
G^0(Ti,cph) = $-20\,585 + 12.134\,T$
G^0(Ta,cph) = $-21\,087 + 11.715\,T$

Ti-Ta excess Gibbs energies

B(L)   = 12 417
B(bcc) = 12 000
B(cph) = 12 500

Note: Values are given in J/mol and J/mol · K.

In the absence of thermochemical data and accurate phase diagram data, least-squares optimization techniques are not suitable for treating this system, and trial and error was used. The Gibbs energy of the bcc phase is based on an estimate of the "submerged" metastable miscibility gap. From the available phase diagram data, the critical point must lie below 500 °C. Based on the Gibbs energy of the bcc phase, those of the liquid and cph phases are determined from the solidus and (αTi) solvus, respectively. Thermodynamic properties of Ti-Ta as determined in the present assessment are summarized in Table 3.

Cited References

*52Sum: D. Summers-Smith, "The Constitution of Tantalum-Titanium Alloys," *J. Inst. Met.*, *81*, 73-76 (1952-53). (Equi Diagram; Experimental)

53Duw: P. Duwez, "The Martensite Transformation Temperature in Titanium Binary Alloys," *Trans. ASM*, *45*, 934-940 (1952). (Meta Phases; Experimental)

*53May: D.J. Maykuth, H.R. Ogden, and R.I. Jaffee, "Titanium-Tungsten and Titanium-Tantalum Systems," *Trans. AIME*, *197*, 231-237 (1953). (Equi Diagram; Experimental)

58Bag: Yu.A Bagariatskii, G.I. Nosova, and T.V. Tagunova, "Factors in the Formation of Metastable Phases in Titanium-Base Alloys," *Dokl. Akad. Nauk SSSR*, *122*, 593-596 (1958) in Russian; TR: *Sov. Phys. Dokl.*, *3*, 1014-1018 (1958). (Meta Phases; Experimental)

63Bor: B.A. Borok, E.K. Novikova, L.S. Golubeva, R.P. Shchego-leva, and N.A. Ruch'eva, "Dilatometric Investigation of Binary Alloys of Titanium," *Metall. Term. Obrab. Met.*, (2), 32-36 (1963) in Russian; TR: *Met. Sci. Heat Treat.*, (2), 94-98 (1963). (Meta Phases; Experimental)

63Luz: L.P. Luzhnikov, V.M. Novikova, and A.P. Mareev, "Solubility of β Stabilizers in α-Titanium," *Metalloved. Term. Obrab. Met.*, (2), 13-16 (1963) in Russian; TR: *Met. Sci. Heat Treat.*, (2), 78-81 (1963). (Equi Diagram; Experimental)

*65Bru: C.E. Brukl *et al.*, unpublished work, cited in [69Rud] (1965). (Equi Diagram; Experimental)

65Rau: C.J. Raub and U. Zwicker, "Superconductivity of α-Titanium Solid Solutions with Vanadium, Niobium, and Tantalum," *Phys. Rev.*, *137A*, A142-A143 (1965). (Equi Diagram; Experimental)

66Bro: A.R.G. Brown and K.S. Jepson, "Physical Metallurgy and Mechanical Properties of Ti-Nb Alloys," *Mem. Sci. Rev. Metall.*, *63*(6), 575-584 (1966) in French. (Meta Phases; Experimental)

*67Bud: P.B. Budberg and K.K. Shakova, "Phase Diagram of the Titanium-Tantalum System," *Izv. Akad. Nauk SSSR Neorg. Mat.*, *3*(4), 656-660 (1967) in Russian; TR: *Russ. J. Inorg. Mater.*, *3*(4), 577-580 (1967). (Equi Diagram; Experimental)

68Kal: G.P. Kalinin and O.P. Eljutin, "The Anomalies of the Properites of Titanium Alloys with β Stabilizing Elements," *Metalloved. Term. Obrab. Met.*, (4), 52-54 (1968) in Russian; TR: *Met. Sci. Heat Treat.*, (4), 301-302 (1968). (Equi Diagram; Experimental)

*69Nik: P.N. Nikitin and V.S. Mikheyev, "Solubility of Tantalum in α-Titanium," *Fiz. Met. Metalloved.*, *28*(6), 1127-1129 (1969) in Russian; TR: *Phys. Met. Metallogr.*, *28*(6), 190-192 (1969). (Equi Diagram; Experimental)

69Rud: E. Rudy, "Compilation of Phase Diagram Data," USAF Tech. Rept. AFML-TR-65-2, Part V (1969). (Equi Diagram; Experimental)

70Kau: L. Kaufman and H. Bernstein, *Computer Calculation of Phase Diagrams*, Academic Press, New York (1970). (Thermo; Theory)

72Byw1: K.A. Bywater and J.W. Christian, "Martensitic Transformations in Titanium-Tantalum Alloys," *Philos. Mag.*, *25*, 1249-1273 (1972). (Meta Phases; Experimental)

73Byw2: K.A. Bywater and J.W. Christian, "Precipitation Reactions in Titanium-Tantalum Alloys," *Philos. Mag.*, *25*, 1275-1289 (1972). (Meta Phases; Experimental)

73Ika: H. Ikawa, S. Shin, M. Miyagi, and M. Morikawa, "Some Fundamental Studies on the Phase Transformation from β to α Phase in Titanium Alloys," Sci. Tech. Appl. Titanium, Proc. Int. Conf., R.I. Jaffee, Ed., 1545 (1973). (Meta Phases; Experimental)

The Tc-Ti (Technetium-Titanium) System

(98) 47.88

By J.L. Murray

Equilibrium Diagram

For ten Ti-Tc alloy compositions, [76Koc] examined the phases present in as-cast samples annealed at 700, 1000, or 1500 °C, and [62Dar] examined the phases present in several alloys near the compositions of the two compounds of the system. A systematic study of the details of the phase relationships of the Ti-Tc system has not yet been made. Based on the work of [76Koc] and [62Dar], the equilibrium solid phases of the Ti-Tc system are:

- The bcc (β) solid solution, based on the equilibrium solid phase of pure Ti above 882 °C.
- The cph (αTi) and (Tc) solid solutions. The cph phase is the equilibrium solid of pure Tc and the equilibrium phase of pure Ti below 882 °C.
- The CsCl-type ordered bcc compound TiTc. TiTc probably has a wide homogeneity range; it appears in single-phase alloys containing 67 at.% Tc.

- An ordered χ phase with narrow homogeneity range about 85 at.% Tc. χ has the αMn structure with 58 atoms per unit cell.

Although the equilibrium phases of the Ti-Tc system have been identified experimentally, the phase boundaries have not. The Ti-Tc system is similar to the Ti-Re system in that it forms the αMn compound, but different in that Re alloys generally do not tend to form CsCl compounds. The CsCl compound is an equilibrium phase of the Ti-Fe, Ti-Ru, Ti-Os, and Ti-Co systems. In alloy chemistry, Tc resembles Ru in forming αMn phases only with Cr-group elements, whereas Mn and Re form αMn phases with elements of both the V and Cr groups.

The assessed diagram (Fig. 1) and Table 1 summarize the experimental data on the system. A rough schematic sketch of possible phase relations is included, based on analogy with chemically related systems. The sketch is primarily intended as a guide.

Table 1 Experimental Data for the Ti-Tc Phase Diagram

Reference	Composition, at.% Tc	Temperature, °C	Structure	Method/note
[62Dar]	87.5, 85.7	As-cast	χ	X-ray
[76Koc]	97	As-cast	(αTc)	X-ray, optical metallography
	95, 87.5	1500	(αTc) + χ	
	85	1500	χ	
	75, 67, 50	As-cast		
	67	1000	TiTc	
	50	700	TiTc	
	50, 25, 15	As-cast	β	

Table 2 Crystal Structures of the Ti-Tc System [Pearson2]

Phase	Homogeneity range, at.% Tc	Pearson symbol	Space group	Strukturbericht designation	Prototype
(αTi)	0 to ?	hP2	P6₃/mmc	A3	Mg
(βTi)	0 to ~ 50	cI2	Im3m	A2	W
TiTc	~ 50	cP2	Pm3m	B2	CsCl
χ	~ 85	cI58	I4̄m	A12	αMn
(Tc)	~ 96 to 100	hP2	P6₃/mmc	A3	Mg

Table 3 Lattice Parameters of Ti-Tc Alloys

Phase	Reference	Composition, at.% Tc	Lattice parameter, nm	Condition
β	[76Koc]	50	0.3098	As-cast
		50	0.3091	Annealed (700 °C)
		25	0.3181	As-cast
		15	0.3221	As-cast
TiTc	[76Koc]	50	0.3083	Annealed (1000 °C)
		50	0.3091	Annealed (700 °C)
	[62Dar]	50	0.3110 ± 0.0005	Annealed (700 °C)
χ	[76Koc]	85	0.9512	Annealed (1500 °C)
	[62Dar]	87.5	0.9579 ± 0.0001	As-cast

Fig. 1---Experimental Data on the Ti-Tc System

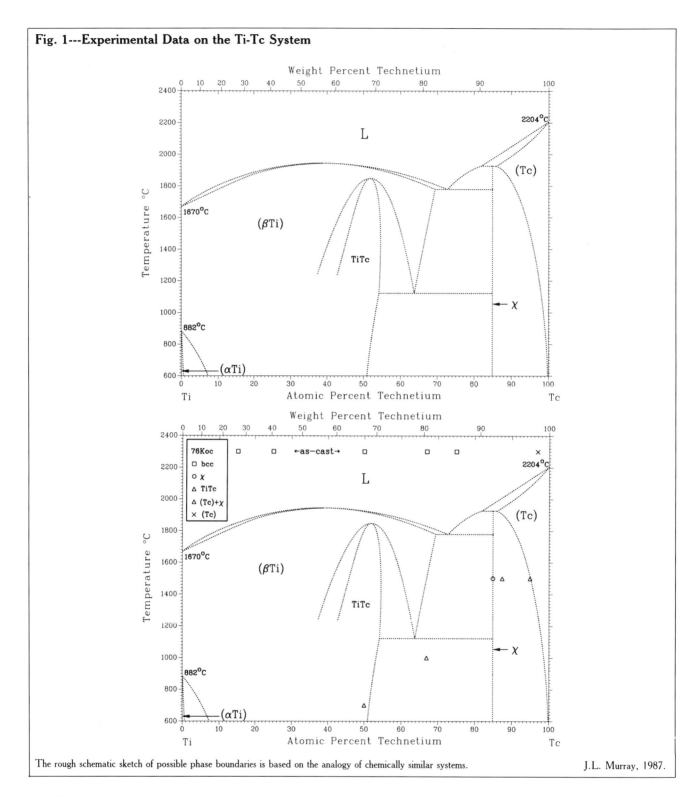

The rough schematic sketch of possible phase boundaries is based on the analogy of chemically similar systems.

J.L. Murray, 1987.

Liquidus and Solidus.

The melting point of pure Tc is 2204 °C [Melt] and that of Ti is 1670 °C. Related systems that may be a guide to the possible behavior of Ti-Tc are Ti-Ru, Ti-Os, and especially Ti-Re. All these metals have the cph structure and higher melting points than Ti. In these systems, the liquidus and solidus on the Ti-rich side are nearly flat, and the two-phase region is narrow. Addition of Re in-

creased the melting point of Ti; addition of Ru and Os lower it to a shallow minimum.

(αTi)/(βTi) Equilibria.

Neither the (βTi) transus nor the (αTi) solvus has been studied experimentally in Ti-Tc alloys. Addition of any of the transition metals to Ti stabilizes (βTi) with respect to (αTi). Based on comparison of the solubilities of Fe, Ru, and Re in (αTi) (0.05, 0.05, and 0.1 respectively), the maximum solubility of Tc in (αTi) can be estimated to be probably less than 0.5 at.% Tc.

TiTc.

Samples containing 50 and 67 at.% Tc exhibit the disordered bcc structure in the as-cast form and exhibit the CsCl ordered structure after annealing at 700 °C [76Koc]. This suggests that TiTc undergoes an ordering transformation in the solid state. The CsCl compounds in related Ti systems, e.g., Ti-Ru or Ti-Os, melt congruently at about 2150 °C.

χ and the (Tc) Solvus.

[76Koc] observed one single-phase χ alloy at 85 at.% Tc. χ may have some homogeneity range which includes 85 at.% Tc. It is not known whether χ forms by the peritectic reaction L + (Tc) $\rightleftarrows$ χ or by the eutectic reaction L $\rightleftarrows$ χ + (Tc).

[76Koc] found that an as-cast 97 at.% Tc sample had a single-phase (Tc) structure, but that 95 at.% Tc samples were two-phase (Tc) + χ, both in the as-cast form and after annealing at 1500 °C.

Crystal Structures and Lattice Parameters

The structures of the equilibrium and metastable phases of the Ti-Tc system are summarized in Table 2, and lattice parameter data are given in Table 3.

Cited References

62Dar: J.B. Darby, D.J. Lam, L.J. Norton, and J.W. Downey, "Intermediate Phases in Binary Systems of Technetium-99 with Several Transition Elements," *J. Less-Common Met., 4* 558-563 (1962). (Equi Diagram; Experimental)

76Koc: C.C. Koch, "Superconductivity in the Technetium-Titanium Alloy System," *J. Less-Common Met., 44,* 177-181 (1976). (Crys Structure, Equi Diagram; Experimental)

The Te-Ti (Tellurium-Titanium) System

127.60   47.88

By J.L. Murray

Introduction

The Ti-Te system bears marked similarities to the Ti-Se and Ti-S systems, but has been studied less thoroughly. Most of the experimental work was done by X-ray diffraction; there are many discrepancies concerning the number of equilibrium phases, their structures, and their homogeneity ranges. Comparison with the Ti-Se and Ti-S systems indicates that there are experimental difficulties in determining both the compositions of the phases and their structures. The most recent work on the Ti-Te system is a vapor pressure study [66Suz], which demonstrates that this system is more complex than X-ray diffraction studies had previously indicated. However, [66Suz] was not able to determine the structures of the new phases, or to establish a correspondence with previously determined structures for most phases. Therefore, no phase diagram has been constructed for this assessment.

Ti-Based Alloys.

[53Gol] reported that the solubility of Te in Mg-reduced (low-purity) cph (αTi) annealed for 10 min at 790 °C is 0.18 at.%. This value should be considered a rough estimate.

X-Ray Diffraction Studies of Intermetallic Phases.

Between 50 and 66.7 at.% Te, the Ti-Te system was originally considered to have a single phase with a wide homogeneity range and a continuous transition between the related NiAs and CdI_2 structures [49Ehr]. [62Raa] also considered this composition range to span a single-phase field $Ti_{2-x}Te_2$, but reported a monoclinic rather than a hexagonal structure between 55.4 and 59.2 at.% Te. The monoclinic structure was verified by [62Che] and assigned the Ti_3Te_4 stoichiometry. Thus, as in the Ti-S and Ti-Se systems, it is established that the range 50 to 66.7 at.% Te contains several distinct phases.

[57Hah] and [58Mct] attributed a superlattice structure similar to the NiAs structure to TiTe. For Ti-Se alloys, it has been shown that the hexagonal NiAs superlattice structure should be reinterpreted as two phases, but corresponding work has not been performed for Ti-Te alloys.

It has been established that the Ti-Te system includes at least one subtelluride with a tetragonal crystal structure. The phases Ti_2Te [59Hah], Ti_3Te_2 [57Hah], and Ti_5Te_4 [61Gro, 62Raa] have been reported. According to [61Gro], comparison of the powder photograph data indicates that the phases previously identified as Ti_2Te and Ti_3Te_2 were identical with Ti_5Te_4. Alloys containing less than 50 at.% Te are extremely reactive and difficult to equilibrate [58Mct]. Discrepancies in observed stoichiometries of subtellurides probably can be attributed to loss of Ti by reaction with SiO_2.

In Table 1, crystal structures are listed essentially as reported in the original literature, except where several reported phases clearly have the same crystal structure. Thus, the monoclinic and hexagonal $Ti_{2-x}Te_2$ phases reported by [62Raa] are listed as Ti_3Te_4 and $TiTe_2$, respectively.

Vaporization Study.

The most recent work on the Ti-Te system is a vaporization study by the Knudsen effusion technique [66Suz]. Phase boundary data were derived from breaks in the mass versus time data from the samples in the Knudsen cell. The following phases were found: (αTi), (βTi), Ti_2Te, Ti_5Te_4, TiTe, Ti_2Te_3, Ti_4Te_7, $Ti_{10}Te_{19}$, and $TiTe_2$. It was proposed that the addition of Te stabilizes (αTi) with respect to (βTi) and that the solubility of Te in (αTi) extends to about 25 at.%. Based on metallographic data [53Gol], the solubility of Te in (αTi) is quite low; consequently, the phase identified by [66Suz] as (αTi) may

Phase Diagrams of Binary Titanium Alloys

Table 1 Crystal Structures and Lattice Parameters of the Ti-Te System

Phase	Homogeneity range(a), at.% Te			Pearson symbol	Space group	Strukturbericht designation	Proto-type	Lattice parameters, nm			Comment	Reference
								a	b	c		
(αTi)	0	to	~0.18	$hP2$	$P6_3/mmc$	$A3$	Mg	...	...	...	...	[Pearson2]
(βTi)	0	to	?	$cI2$	$Im3m$	$A2$	W	...	...	...	...	[Pearson2]
Ti_2Te(b) ...		...		(c)	...	...	...	...	...	...	...	[59Hah]
Ti_3Te_2(b) ..		...		(c)	...	...	...	1.434	...	0.3580	...	[57Hah]
Ti_5Te_4		44.4		$tI18$	$I4/m$	...	Te_4Ti_5	1.0164	...	0.37720	...	[61Gro]
								1.0164	...	0.37720	...	[62Raa, 66Suz]
TiTe	40	to	?	$hP16$	$\sim P6_3/mmc$(?)	$\sim B8_1$	~NiAs	...	0.3842	0.6402	...	[49Ehr]
								0.772	...	1.265	...	[58Mct]
								0.7704	...	1.2626	...	[57Hah]
Ti_3Te_4	55.0 to		57.6	$mC14$	$C2/m$	...	Cr_3Se_4	0.682	0.385	1.266	$\beta = 90°28'$	[62Che]
	55.4 to		59.2					0.6954 to 0.6840	0.3836 to 0.3850	1.2716 to 1.2658	$\beta = 90.63°$ to 90.42°	[62Raa]
$TiTe_2$ .....~	60	to	66.7	$hP3$	$P\bar{3}m1$	$C6$	CdI_2	0.3782	...	0.654	...	[28Oft]
								0.3764	...	0.6526	...	[49Ehr]
								0.376	...	0.648	...	[58Mct]
								0.3755	...	0.646	...	[66Suz]
								0.3884 to 0.3766	...	0.6348 to 0.6491	...	[62Raa]

(a) The homogeneity ranges are very uncertain. (b) According to [61Gro], this phase is Ti_5Te_4. (c) Tetragonal.

Table 2 Ti-Te Vaporization Data [66Suz]

Phase	Homogeneity range, at.% Te		Temperature, °C	Enthalpy of formation(a) $(\Delta_f H)$, kJ/mol	Entropy of formation(a) $(\Delta_f S)$, kJ/mol · K
(βTi)	0	to 7	1250	...	...
(αTi)	14	to 25	1250	...	...
Ti_2Te	32	to 35	1250	$- 79.5 \pm 10$	4.6 ± 10
Ti_5Te_4	39.8	to 45.1	1250	$- 113.0 \pm 17$	1.7 ± 17
TiTe	46.8	to 57.6	750 to 1250	$- 125.5 \pm 21$	$- 9.7 \pm 21$
Ti_2Te_3	59.0	to 60.9	750	$- 147.3 \pm 33$	$- 36.4 \pm 33$
Ti_4Te_7	61.8	to 64.3	450 to 750	$- 148.5 \pm 38$	$- 44.8 \pm 38$
$Ti_{10}Te_{19}$	~ 65.5		450	$- 149.4 \pm 42$	$- 43.1 \pm 42$
$TiTe_2$	~ 66.7		450	$- 150.6 \pm 42$	$- 44.4 \pm 41$

(a) In the reaction $Ti(s) + rTe(s) \rightarrow TiTe_r(s)$.

therefore be an additional subtelluride. Ti_4Te_7 and $Ti_{10}Te_{19}$ are new phases identified by [66Suz].

[66Suz] verified the structures of Ti_5Te_4 and $TiTe_2$ by X-ray diffraction. For other samples, the diffraction patterns were very complex, and it was not possible to verify the vacuum balance results by structure analysis.

[66Suz] calculated vapor pressures assuming that, for 0 to 50 at.% Te, Te was the only species present in the vapor, and that, for 50 to 100 at.% Te, Te_2 was the only species present. The standard heat and entropy of formation data listed in Table 2 refer to the reaction:

$$Ti(s) + rTe(s) \rightarrow TiTe_r(s)$$

Cited References

28Oft: I. Oftedal, "X-Ray Investigation of SnS_2, TiS_2, $TiSe_2$, $TiTe_2$," *Z. Phys. Chem., 134*, 301-310 (1928) in German. (Equi Diagram, Crys Structure; Experimental)

49Ehr: P. Ehrlich, "X-Ray Investigation of Titanium Sulfides and Vanadium Monotelluride," *Z. Anorg. Chem., 260*, 13 (1949) in German. (Equi Diagram, Crys Structure; Experimental)

53Gol: R.M. Goldhoff, H.L. Shaw, C.M. Craighead, and R.I. Jaffee, "The Influence of Insoluble Phases on the Machinability of Titanium," *Trans. ASM, 45*, 941-971 (1953). (Equi Diagram, Crys Structure; Experimental)

57Hah: H. Hahn and P. Ness, "On the Question of the Existence of Subchalcogenides of Titanium," *Naturwissenschaften, 44*, 581 (1957) in German. (Equi Diagram, Crys Structure; Experimental)

58Mct: F.K. McTaggart and A.D. Wadsley, "The Sulphides, Selenides, and Tellurides of Titanium, Zirconium, Hafnium, and Thorium," *Aust. J. Chem., 11*, 445-457 (1958). (Equi Diagram, Crys Structure; Experimental)

59Hah: H. Hahn and P. Ness, "The Subchalcogenides of Titanium," *Z. Anorg. Chem., 302*, 17-36 (1959) in German. (Equi Diagram, Crys Structure; Experimental)

61Gro: F. Gronvold, A. Kjekshus, and F. Raaum, "The Crystal Structure of Ti_5Te_4," *Acta Crystallogr., 14*, 930-934 (1961). (Equi Diagram, Crys Structure; Experimental)

62Che: M. Chevreton and F. Bertaut, "Titanium and Vanadium Selenides and Titanium Telluride," *Compt. Rend., 255*, 1275-1277 (1962) in French. (Equi Diagram, Crys Structure; Experimental)

62Raa: F. Raaum, F. Gronvold, A. Kjekshus, and H. Haraldsen, "The Titanium-Tellurium System," *Z. Anorg. Chem., 317*, 91-104 (1962) in German. (Equi Diagram, Crys Structure; Experimental)

*66Suz: A. Suzuki and P.G. Wahlbeck, "Vaporization Study of the Titanium-Tellurium System," *J. Phys. Chem., 70*, 1914-1923 (1966). (Equi Diagram, Thermo; Experimental)

The Th-Ti (Thorium-Titanium) System

232.0381   47.88

By J.L. Murray

The assessed Ti-Th phase diagram is shown in Fig. 1, and special points of the diagram are summarized in Table 1. Crystal structure data are summarized in Table 2. Ti-Th alloys solidify by a eutectic reaction with a rod eutectic morphology; the Ti-Th phase diagram is applied in the production of aligned Nb-Ti superconducting filaments [80Ped].

The equilibrium solid phases of the Ti-Th system are the solutions based on the structures of the pure components:

- The low-temperature cph (αTi) solution
- The low-temperature fcc (αTh) solution

- The high-temperature bcc solutions, (βTi) and (βTh)

The constitution of the Ti-Th system was examined by [56Car] and [80Ped]. [56Car] surveyed the entire system using alloys prepared from ~99.95 wt.% Th and 99.5 wt.% Ti. Experimental data are shown in Fig. 2. X-ray and metallographic examination showed that there are no intermetallic compounds, and no detectable mutual solubilities in the solid state. The eutectic temperature was measured by optical observation of incipient melting.

[56Car] metallographically determined the eutectic composition as 59.9 ± 1.2 at.% Th. By chemical analysis

Table 1 Special Points of the Assessed Ti-Th Phase Diagram

Reaction	Compositions of the respective phases, at.% Th			Temperature, °C	Reaction type
L $\rightleftarrows$ (βTi) + (αTh)	57.5	~ 0	~ 100	1190	Eutectic
L + (βTh) $\rightleftarrows$ (αTh)	~ 78	~ 100	~ 100	~ 1363	Unknown
(βTi) + (αTh) $\rightleftarrows$ (αTi)	~ 0	~ 100	~ 0	~ 882	Unknown
L $\rightleftarrows$ (βTi)	0			1670	Melting point
(βTi) $\rightleftarrows$ (αTi)	0			882	Allotropic transformation
L $\rightleftarrows$ (βTh)	100			1755	Melting point
(βTh) $\rightleftarrows$ (αTh)	100			1360	Allotropic transformation

Fig. 1---Assessed Ti-Th Phase Diagram

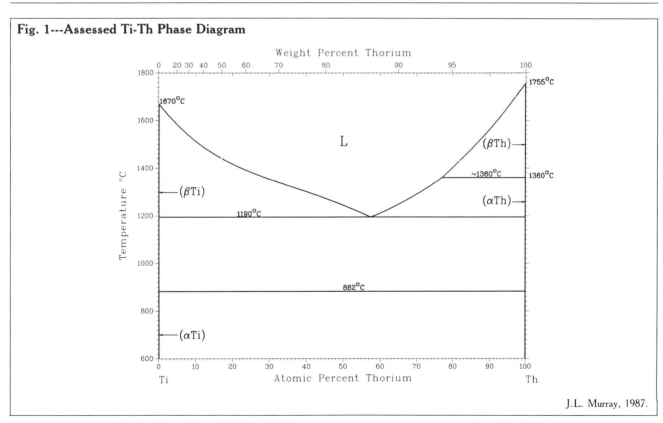

J.L. Murray, 1987.

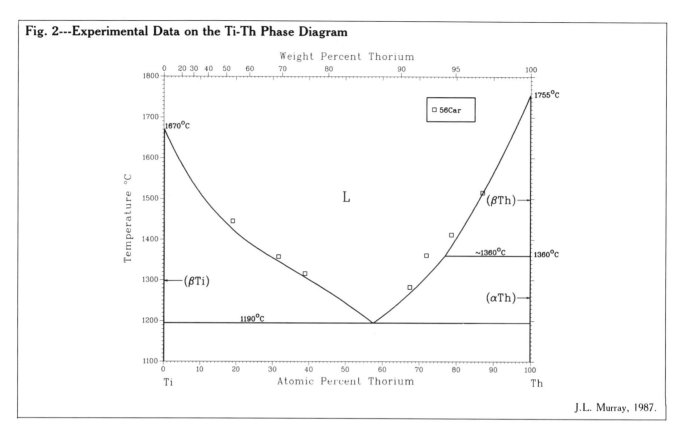

Fig. 2---Experimental Data on the Ti-Th Phase Diagram

J.L. Murray, 1987.

Table 2 Crystal Structures of the Ti-Th System

Phase	Pearson symbol	Space group	Struktur-bericht designation	Proto-type	Lattice parameters(a), nm		Comment
					a	c	
(αTi)	hP2	P6₃/mmc	A3	Mg	0.29512	0.46826	
(βTi)	cI2	Im3m	A2	W	0.33065	...	At 900 °C
(βTh)	cI2	Im3m	A2	W	0.411	...	At 1450 °C
(αTh)	cF4	Fm3m	A1	Cu	0.50845	...	

(a) Lattice parameters are for the pure elements. Ti parameters are from [Ti]; Th parameters are from [81Gsc].

Table 3 Gibbs Energies of the Ti-Th System

Properties of the pure elements

$G^0(\text{Ti,L})$ = 0
$G^0(\text{Th,L})$ = 0
$G^0(\text{Ti,bcc})$ = $-16\,234 + 8.368\,T$
$G^0(\text{Th,bcc})$ = $-16\,121 + 7.937\,T$
$G^0(\text{Ti,cph})$ = $-20\,585 + 12.134\,T$
$G^0(\text{Th,fcc})$ = $-18\,857 + 9.610\,T$

Excess Gibbs energy of the liquid

$B(\text{L})$ = 16 798

$C(\text{L})$ = − 4 853

Note: Values are given in J/mol and J/mol · K.

of a eutectic alloy produced by directional solidification, [80Ped] found 57.5 ± 0.8 at.% Th, using Ti of 99.94 wt.% purity. The assessed value is 57.5 at.% Th.

[56Car] determined liquidus points by holding eutectic alloys in Ti crucibles at temperatures in the (liquid + solid) phase field and chemically analyzing the composition of the portion of the alloy that melted. This technique avoids contamination of alloys due to interaction with a crucible.

The assessed diagram was drawn from a thermodynamic calculation of the system. The solid phases were modeled as mutually immiscible, with Gibbs energies determined from heats and temperatures of transformation [Hultgren,E]. The excess Gibbs energy of the liquid was represented as:

$$G^{ex}(\text{L}) = x(1 - x)\,[B(\text{L}) + C(\text{L})\,(1 - 2x)]$$

where x is the atomic fraction of Th, and $B(\text{L})$ and $C(\text{L})$ are temperature-independent coefficients describing the excess Gibbs energy.

The excess Gibbs energy of the liquid was determined by least-squares optimizations of the parameters, using liquidus data and the eutectic point as input. The resulting Gibbs energy is given in Table 3.

The liquidus data [56Car] are consistent with the thermodynamic properties of the elements, and a reasonable model for the excess Gibbs energy of the liquid can be constructed based on these data. This indicates that the liquidus may be accurate to within ± 1 at.%. The mutual solid solubilities, however, should be re-examined using high-purity Ti before it is assumed that Ti and Th have undetectably low mutual solid solubilities.

Cited References

*56Car: O.N. Carlson, J.M. Dickinson, H.E. Lunt, and H.A. Wilhelm, "Thorium-Columbium and Thorium-Titanium Alloy Systems," *Trans. AIME, 206,* 132-136 (1956). (Equi Diagram; Experimental)

*80Ped: T.E. Pedersen, M. Noack, and J.D. Verhoeven, "Eutectic Alloys of Thorium-Niobium and Thorium-Titanium," *J. Mater. Sci., 15,* 2115-2117 (1980). (Equi Diagram; Experimental)

81Gsc: K.A. Gschneidner, Jr. and F.W. Calderwood, "Critical Evaluation of Binary Rare Earth Phase Diagrams," IS-RIC-PR-1, Rare-Earth Information Center, Iowa State University, Ames, IA (1981). (Crys Structure; Review)

The Ti-U (Titanium-Uranium) System

47.88   238.0289

By J.L. Murray

Equilibrium Diagram

The Ti-U equilibrium diagram was investigated by [51Sey], [53Buz], [54Udy], [55Kna], and [61Add]. The diagram proposed by [53Buz] differs from other proposed diagrams in the number of equilibrium phases and invariant reactions. The disagreements can be attributed plausibly in part to the contamination of the alloys used by [53Buz] during heat treatments and in part to misinterpretation of microstructures that arise from nonequilibrium transformations during quenching. Among [54Udy], [55Kna], and [61Add], there are some quantitative but no qualitative disagreements. The most pronounced discrepancies involve the homogeneity ranges of the phases and the position of the eutectoid point on the Ti-rich side of the diagram.

The assessed Ti-U phase diagram is shown in Fig. 1, and special points of the diagram are summarized in Table 1. The assessed diagram was obtained by optimizing Gibbs energy functions with respect to select experimental phase diagram data.

Fig. 1---Assessed Ti-U Phase Diagram

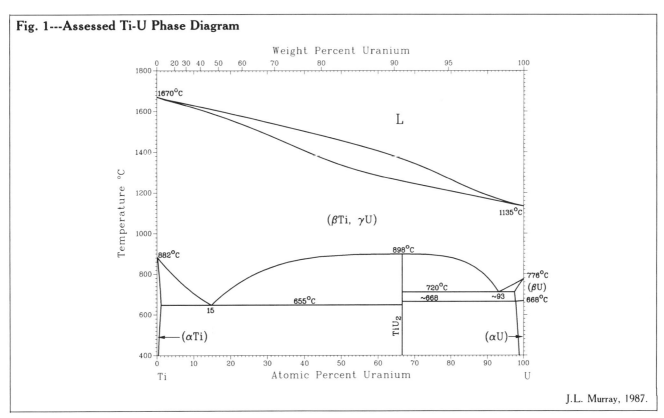

J.L. Murray, 1987.

Fig. 2---Experimental Liquidus and Solidus Data vs the Assessed Diagram

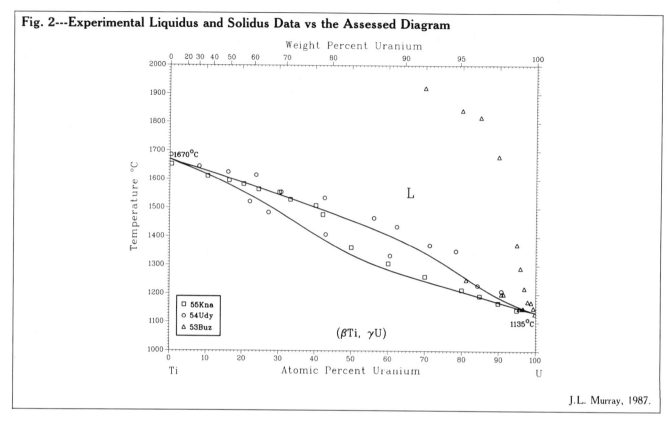

J.L. Murray, 1987.

Table 1   Special Points of the Assessed Ti-U Phase Diagram

Reaction	Compositions of the respective phases, at.% U			Temperature, °C	Reaction type
$(\beta Ti) \rightleftarrows (\alpha Ti) + TiU_2$	15	~ 1	66.7	655 ± 5	Eutectoid
$(\gamma U) \rightleftarrows (\beta U) + TiU_2$	~ 93	~ 98	66.7	720 ± 5	Eutectoid
$(\beta U) \rightleftarrows (\alpha U) + TiU_2$	~ 98	~ 98	66.7	~ 667	Eutectoid
$(\beta Ti, \gamma U) \rightleftarrows TiU_2$	66.7			898 ± 5	Congruent
$L \rightleftarrows (\beta Ti)$	0			1670	Melting point
$(\beta Ti) \rightleftarrows (\alpha Ti)$	0			882	Allotropic transformation
$L \rightleftarrows (\gamma U)$	100			1135	Melting point
$(\gamma U) \rightleftarrows (\beta U)$	100			776	Allotropic transformation
$(\beta U) \rightleftarrows (\alpha U)$	100			668	Allotropic transformation

The equilibrium solid phases of the system are:

- The bcc solid solution (βTi, γU), with complete mutual solubility above 898 °C.
- The low temperature cph (αTi) solid solution.
- The low temperature (αU) solid solution, whose structure is closely related to that of (αTi).
- (βU) with the A_b structure, related to that of σ phases.
- The compound TiU_2 with an ordered hexagonal structure closely related to that of ω phase. TiU_2 has a narrow homogeneity range.

Liquidus and Solidus.

Experimental data [53Buz, 54Udy, 55Kna] on the liquidus and solidus are compared with the assessed diagram in Fig. 2. Liquidus points are derived from thermal analysis [53Buz] or visual observation of melting [54Udy, 55Kna]. Solidus points are derived from metallographic examination of heat treated and quenched specimens. The very high liquidus temperatures [53Buz] are due to contamination of alloys by beryllia crucibles, exacerbated by holding alloys in the liquid phase field for several hours. [54Udy] noted similar high melting points in their preliminary work ([51Sey]).

The liquidus and solidus data of [55Kna] were used for the optimization of Gibbs energies. [54Udy] is in substantial agreement with [55Kna].

Solid Phase Equilibria.

Experimental data on solid phase equilibria [51Sey, 53Buz, 54Udy, 55Kna, 58Mur, 61Add] are shown in Fig. 3. [61Add] used microprobe analysis of equilibrated diffusion couples. Other work was done by metallography and

Fig. 3---Experimental Data Pertaining to the Solid Phase Equilibria vs the Assessed Diagram

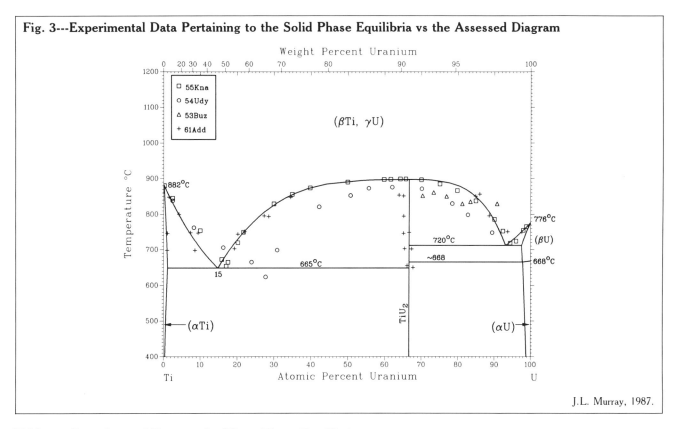

J.L. Murray, 1987.

Table 2   Experimental Data on the Three-Phase Equilibria

Reaction	Compositions, at.% U			Temperature, °C	Reference
$(\beta Ti) \rightleftarrows (\alpha Ti) + TiU_2$	15	0.7 to 0.8	66.7	655	[55Kna]
	28	...	60	620	[54Udy]
	15	~ 1	65	650	[61Add]
$(\gamma U) \rightleftarrows (\beta U) + TiU_2$	~ 96	~ 98.5	66.7	723	[55Kna]
	~ 95	~ 97	...	718	[53Buz]
	...	~ 97	~70	~ 650	[51Sey, 54Udy]
	...	- 97.5	...	...	[58Mur]
$(\beta U) \rightleftarrows (\alpha U) + TiU_2(a)$	> 99	...	66.7	~ 667	[55Kna]
	~ 99	~ 96	...	667	[53Buz]
	~ 99	~ 96	70	667	[51Sey, 54Udy]
	~ 97.5	99 to 99.55	...	...	[58Mur]

(a) According to [53Buz] and [54Udy], this is a peritectoid reaction and pure βU → αU at about 650 °C. Hence, according to these authors, the solubility of Ti is greater in (αU) than in (βU) at 667 °C.

X-ray diffraction, supplemented by dilatometry [55Kna]. The studies of [53Buz] and [58Mur] were restricted to U-rich alloys, and [58Mur] studied primarily mechanical properties.

The assessed diagram is based on [55Kna] and [61Add]. The major discrepancies involve: the number of equilibrium intermetallic phases [53Buz], the homogeneity range of TiU₂ [51Sey, 54Udy], the composition and temperature of the eutectoid points [54Udy, 55Kna, 61Add], and the solubility of Ti in (αU), (βU), and (γU) [54Udy, 55Kna, 58Mur].

[53Buz] claimed that an additional phase of distinct but unknown structure appeared in equilibrium with (γU) in the composition range 82 to 90 at.% U. A banded microstructure was interpreted as evidence for the two-

phase equilibria. The postulated composition range of the two-phase field corresponds to the range in which a metastable phase of monoclinic symmetry appears during quenching in single-phase (γU) alloys. It is concluded that the error of [53Buz] was one of interpretation and that γU and βTi are completely miscible at sufficiently high temperature. The congruent transformation $(\beta Ti, \gamma U) \rightleftarrows TiU_2$ occurs at 898 ± 5 °C [54Udy, 55Kna, 61Add].

[54Udy] showed a 10 at.% homogeneity range for TiU₂. [55Kna] found neither metallographic evidence for any measurable homogeneity range nor any significant variation of the TiU₂ lattice parameters as a function of composition. [60Add] and [61Add] verified that TiU₂ has a narrow range. [70Tom] found that (βTi,γU) decomposes during quenching to form a coherent hexagonal phase of

315

Fig. 4a---Lattice Parameters of Supersaturated (βTi,γU)

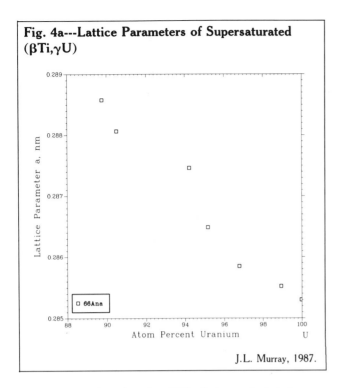

J.L. Murray, 1987.

Fig. 4b---Lattice Parameters of Supersaturated (βTi,γU)

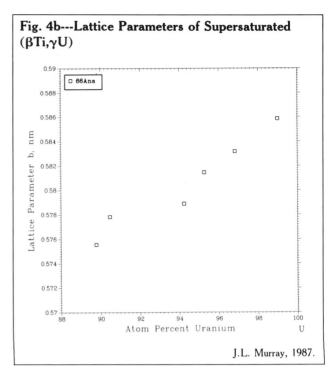

J.L. Murray, 1987.

structure closely related to TiU_2. It is concluded that an apparently wide TiU_2 field may be a property of the metastable rather than the stable equilibrium diagram.

The solubility of U in (αTi) was measured by microprobe analysis of annealed diffusion couples [61Add] and by autoradiography [55Kna]. Results were 0.9 and 0.7 to 0.8 at.% U, respectively, and the calculations presented below give 1.0 at.% U in good agreement.

There is considerable scatter in data on the (βTi)/[(βTi) + (αTi)] boundary. The eutectoid reaction is sluggish, and based on the low value of the reported eutectoid temperature and the relatively few alloys examined in this region, the data of [54Udy] can be rejected in favor of [55Kna] and [61Add]. The remaining discrepancy between [55Kna] and [61Add] is tentatively resolved by thermodynamic calculation (see Fig. 3).

On the U-rich side, sources of experimental error are (1) loss of Ti through combination with interstitial impurities, (2) the influence of contamination on the transformation points of pure U, and (3) the formation of several metastable phases from the solid solution during quenching.

The assessed invariant temperatures are 720 ± 3 °C [53Buz, 54Udy, 55Kna] and approximately 667 °C; the assessed solid solubilities conform with the lower estimates (see Tables 1 and 2).

Metastable Phases

Metastable phases are formed from (βTi,γU) during quenching. On the Ti-rich side, alloys containing up to 15 at.% U transform martensitically to the low-temperature cph form [55Kna], and it is likely that coherent or "diffuse" TiU_2 forms at compositions where (βTi) can be retained during quenching (15 to 30 at.% U) [55Kna].

Particular attention has been given to phases formed metastably from (γU) and (βU). [56Har] reported that, with pure enough alloys and high enough cooling rates, alloys containing more than 7 at.% Ti can be retained in

Fig. 4c---Lattice Parameters of Supersaturated (βTi,γU)

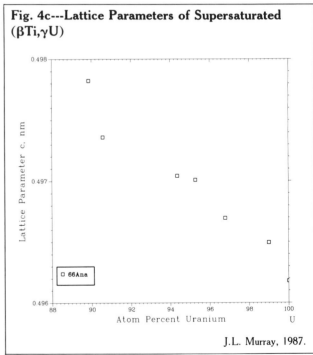

J.L. Murray, 1987.

the bcc structure. It is more usually reported that 90 to 100 at.% U alloys quenched from the bcc field transform martensitically to the orthorhombic αU structure [55Kna, 66Ana].

Orthorhombic martensites are designated α', α_a', or α_b', depending on their microstructures [66Ana]. In the range 2 to 5.5 at.% Ti, an acicular microstructure (α_a') appears; in the range 5.5 to 10.5 at.% Ti, a banded microstructure appears, which was discussed with reference to the equilibrium diagram [53Buz, 54Udy, 55Kna]. By examining dilute ternary Ti-U-Mo alloys, [66Ana] were able to deduce that α' and α_a' are the result of the

Table 3 Crystal Structures and Lattice Parameters of the Ti-U System

Phase	Homogeneity range, at.% U	Pearson symbol	Space group	Struktur-bericht designation	Proto-type	Lattice parameters, nm			Comment	Reference
						a	b	c		
(αTi)	0 to ~ 1	$hP2$	$P6_3/mmc$	$A3$	Mg	0.29511	...	0.46843	...	[Pearson2]
(βTi,γU) ...	0 to 100	$cI2$	$Im3m$	$A2$	W	0.33065	...	...	Ti at 900 °C	[Pearson2]
						0.3524	...	...	U at 805 °C	[Pearson2]
α_b''(a)	~ 11	(b)	...	...	...	0.2886	0.5733	0.4970	β = 90.68°	[66Ana]
TiU$_2$	66.7	$hP3$	$P6/mmm$	$C32$	AlB$_2$	0.4828	...	0.2847	...	[54Kna]
						0.4817	...	0.2844	...	[54Udy]
						0.483	...	0.2847	...	[70Tom]
						0.4834	...	0.2847	...	[71Bar]
(βU)	~ 98 to 100	$tP30$	$P4_2nm$	A_b	βU	1.0759	...	0.5656	At 720 °C	[81Gsc]
(αU)	~ 98 to 100	$oC4$	$Cmcm$	$A20$	αU	0.29537	0.58695	0.49548	At 25 °C	[81Gsc]

(a) Metastable. (b) Monoclinic.

transformation (βU) → (αU), but that the α_b' martensite is formed directly from the bcc phase. At 11 at.% Ti, the martensite becomes distorted to a monoclinic structure designated α_b'', and in the range 11.4 to 17 at.% Ti, α_b'' coexists with γ_0, a tetragonal phase. γ_0 can be obtained in single-phase alloys containing more than 17 at.% Ti [66Ana]. [68Bas] investigated the approach to equilibrium during the annealing of metastable orthorhombic phases. [81Nor] reported continuous cooling curves for the martensitic transformation and investigated the effect of ternary additions.

In quenched bcc alloys of composition near TiU$_2$ stoichiometry, equilibrium TiU$_2$ forms coherent precipitates in the bcc matrix [70Tom]. Coherent TiU$_2$ is also referred to as "diffuse TiU$_2$" [55Kna]. Fully ordered TiU$_2$ can be transformed at 100 °C to the metastable disordered bcc structure by neutron irradiation [71Bar].

Crystal Structures and Lattice Parameters

Lattice parameters of metastable supersaturated (βTi,γU) and (αU) solution phases are shown in Fig. 4 and 5. Crystal structure data for equilibrium and metastable phases and additional lattice parameter data are given in Table 3. Structures of pure U are from [81Gsc].

Thermodynamics

Measurements have not yet been made of thermodynamic properties of Ti-U alloys; however, certain qualitative features of the phase diagram allow a tentative thermodynamic analysis. The complete miscibility of Ti and U in (βTi,γU), the single compound of structure similar to ω, and the narrow melting range all indicate that the system is nearly ideal. That the excess Gibbs energies are positive is suggested first by the analogy between this system and Ti-Cr, but more importantly by the shape of the bcc/(bcc + TiU$_2$) phase boundaries near the congruent point. These phase boundaries are very flat, and the flatness implies very small curvature in the Gibbs energy function for (βTi,γU). That is, the congruent point of the equilibrium diagram must lie close to the critical point of a metastable (βTi,γU) miscibility gap. The constraint that a miscibility gap cannot appear in the equilibrium diagram, however, places an upper limit on the positive excess Gibbs energy of (βTi,γU). Thus, from the qualitative features of the phase diagram, it is estimated that a regular solution approximation of the excess Gibbs energy of (βTi,γU) is: $(18\,000 \pm 2000)\,x(1 - x)$ J/mol,

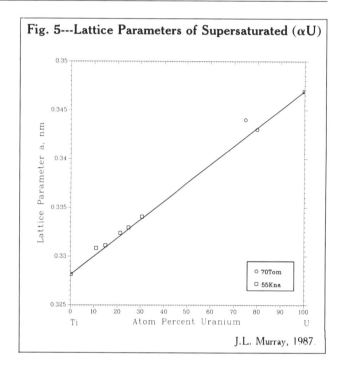

Fig. 5---Lattice Parameters of Supersaturated (αU)

J.L. Murray, 1987

where x is the atomic fraction of U. Gibbs energies of the other phases are determined relative to that of (βTi,γU). Excess entropies cannot be determined, but in a nearly ideal system, these terms are expected to be relatively unimportant.

In summary, because of the accidental near-coincidence of a critical point and a congruent point in this system, approximate thermodynamic functions can be obtained from the phase diagram.

Calculation of the Phase Diagram.

Gibbs energies of the solution phases are represented as:

$$G(i) = G^0(\text{Ti},i)\,(1 - x) + x\,G^0(\text{U},i) + RT\,[x \ln x + (1 - x) \ln (1 - x)] + x(1 - x)\,[B(i) + C(i)\,(1 - 2x)]$$

where i designates the phase; x is the atomic fraction of U; G^0 is the Gibbs energies of the pure metals; and $B(i)$ and $C(i)$ are coefficients in an expansion of the excess Gibbs energy. TiU$_2$ is represented as a stoichiometric line compound. Thermodynamic functions are listed in Table 4. Where possible, pure element properties are based on data

Table 4 Gibbs Energies of the Ti-U System

Properties of the pure elements and compound

$G^0(\text{Ti,L})$ = 0
$G^0(\text{U,L})$ = 0
$G^0(\text{Ti,bcc})$ = $-16\,234 + 8.368\,T$
$G^0(\text{U,bcc})$ = $-8\,519 + 6.0545\,T$
$G^0(\text{Ti,cph}(\alpha))$ = $-20\,585 + 12.134\,T$
$G^0(\text{U,}\alpha)$ = $-16\,103 + 13.594\,T$
$G^0(\text{Ti,}\beta\text{U})$ = $-21\,000 + 15.0\,T$
$G^0(\text{U,}\beta\text{U})$ = $-13\,311 + 10.627\,T$
$G(\text{TiU}_2)$ = $-12\,313 + 6.162\,T$

Interaction parameters of the solution phases

$B(\text{L})$ =	17 000	$B(\text{bcc})$ =	18 078
$C(\text{L})$ =	$-$ 1 458	$C(\text{bcc})$ =	$-$ 2 425
$B(\alpha)$ =	31 216	$B(\beta\text{U})$ =	25 400
$C(\alpha)$ =	3 257		

Note: Values are given in J/mol and J/mol · K.

in [Hultgren,E]; the Gibbs energies of Ti in the βU and αU structures and of cph U had to be roughly estimated. To estimate the lattice stabilities, the structurally similar (αU) and (αTi) were modeled as a single phase, designated α in Table 4. The lattice stability of Ti in the βU structure was roughly estimated and has no important effect on the diagram, as long as large extrapolations are not made of the (βU) phase boundaries.

Optimizations of Gibbs energies were performed with the regular solution term of either the liquid or the bcc phase held fixed. Phase diagram data of [55Kna] were used as input to the optimizations. The solubility of Ti in (βU) was required to be less than 2.5 at.%, but invariant temperatures were weighted more heavily than solubility data. The assessed diagram (Fig. 1, 2, and 3) is taken from the present calculations. The calculation has allowed provisional resolution of the discrepancies among [54Udy], [55Kna], and [61Add] on the Ti-side of the diagram.

Cited References

51Sey: A.W. Seybolt, D.E. White, F.W. Boulger, and M.C. Udy, *Reactor Sci. Technol., 1,* 118-119 (1951). (Equi Diagram; Experimental)

53Buz: R.W. Buzzard, R.B. Liss, and D.P. Fickle, "Titanium-Uranium System in the Region 0 to 30 at.% of Titanium," *J. Res. Nat. Bur. Stand., 50*(4), 209-214 (1953). (Equi Diagram; Experimental)

54Kna: A.G. Knapton, "The Crystal Structure of TiU₂," *Acta Crystallogr., 7,* 457-458 (1954). (Crys Structure; Experimental)

***54Udy:** M.C. Udy and F.W. Boulger, "Uranium-Titanium Alloy System," *Trans. AIME, 200,* 207-210 (1954). (Equi Diagram; Experimental)

***55Kna:** A.G. Knapton, "The System Uranium-Titanium," *J. Inst. Met., 83,* 497-504 (1954-55). (Equi Diagram; Experimental)

56Har: A.G. Harding, "The Constitution of Titanium Alloys," *J. Inst. Met., 84,* 532-535 (1956). (Meta Phases; Experimental)

58Mur: D.J. Murphy, "Some Properties of Uranium-Low Titanium Alloys," *Trans. ASM, 50,* 884-904 (1958). (Equi Diagram; Experimental)

60Add: Y. Adda and J. Philibert, "Diffusion in the Uranium-Titanium System," *Acta Metall., 8,* 700-710 (1960) in German. (Equi Diagram; Experimental)

61Add: Y. Adda, M. Beyeler, A. Kirianenko, and F. Maurice, "Determination of Equilibrium Diagrams by Diffusion in the Solid State," *Mem. Sci. Rev. Met., 58*(9), 716-724 (1961) in German. (Equi Diagram; Experimental)

66Ana: M. Anagnostidis, R. Baschwitz, and M. Colombie, "Metastable Phases in Uranium-Titanium Alloys," *Mem. Sci. Rev. Met., 63*(2), 163-168 (1966) in German. (Meta Phases; Experimental)

68Bas: R. Baschwitz, M. Colombie, and M. Foure, "Study of Annealing of Metastable Orthorhombic Phases in Uranium-Titanium Alloys Containing 4.8 and 9.5 at.% Ti," *J. Nucl. Mater., 28,* 246-256 (1968). (Meta Phases; Experimental)

70Tom: R.D. Tomlinson, J.M. Silcock, and J. Burke, "The Isothermal Decomposition of γ-Phase Uranium-Titanium Alloys," *J. Inst. Met., 98,* 154-160 (1970). (Meta Phases; Experimental)

71Bar: Z. Baran, J. Bazin, J. Bloch, and A. Accary, "The Effect of Neutron Irradiation on the Compound U₂Ti," *J. Nucl. Mater., 38,* 67-76 (1971) in German. (Meta Phases; Experimental)

81Gsc: K.A. Gschneidner, Jr. and F.W. Calderwood, "Critical Evaluation of Binary Rare Earth Phase Diagrams," IS-RIC-PR-1, Rare Earth Information Center, Iowa State University, Ames, IA (1981). (Crys Structure; Review)

81Nor: W.G. Northcutt, "Quenching Rate Effects on Microstructure and Phase Transformations of Uranium and Phase Transformations of Uranium-Titanium-X Ternary Alloys," Metallurgical Technology of Uranium and Uranium Alloys, Vol. 1, Physical Metallurgy of Uranium and Uranium Alloys, Gatlinburg, TN, May 26-28 (1981)

The Ti-V (Titanium-Vanadium) System

47.88   50.9415

By J.L. Murray

Equilibrium Diagram

The equilibrium solid phases of the Ti-V system are:
(1) cph (αTi), in which the maximum solubility of V is
about 3 at.%; and (2) bcc (βTi,V) with a complete range of
solid solubility. The (βTi,V) phase is referred to as β; the
designations (βTi) and (V) are used to distinguish the
phase separated alloys.

Important points of the assessed diagram (Fig. 1) are
summarized in Table 1. A miscibility gap in the β phase
gives rise to a monotectoid reaction (βTi) $\rightleftarrows$ (αTi) + (V),
based on the recent work of [80Nak]. This contradicts
previous assessments [Hansen, Elliott, Shunk, 81Mur], in
which the β miscibility gap was shown as entirely metastable and the β/[β+(αTi)] boundary as a single curve.

Similar conflicts occur with regard to the Ti-Mo
system. The present evaluator reassessed the Ti-Mo and
Ti-V systems together, and based on confirmation of the
miscibility gap in Ti-Mo, with reassessment of the original
data, the miscibility gaps in both systems have been
accepted in the present evaluation.

Gibbs energies were optimized with respect to selected experimental data, and Fig. 1 is the result of the
thermodynamic calculations discussed below. The calculations agree with select data within the experimental
uncertainty.

Liquidus/Solidus.

The liquidus has not been determined experimentally. [52Ade] and [69Rud] determined the solidus by

Table 1   Special Points of the Assessed Ti-V System

Reaction	Compositions of the respective phases, at.% V			Temperature, °C	Reaction type
(βTi) $\rightleftarrows$ (V) + (αTi)	18	~ 80	2.7	675	Monotectoid
(βTi,V) $\rightleftarrows$ (βTi) + (V)		~ 50		850	Critical point
L $\rightleftarrows$ (βTi,V)		32		1605	Congruent
L $\rightleftarrows$ (βTi)		0		1670	Melting point
L $\rightleftarrows$ (V)		100		1910	Melting point
(βTi) $\rightleftarrows$ (αTi)		0		882	Allotropic transformation

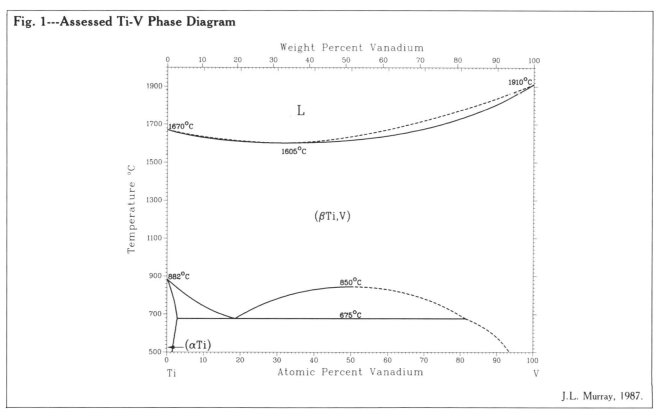

Fig. 1---Assessed Ti-V Phase Diagram

J.L. Murray, 1987.

319

Table 2   Experimental Solidus Data

Reference	Composition, at.% V	Temperature, °C
[52Ade]	0	1720
	28.7(a)	1620
	50	1610
	75	1690
[69Rud]	0	1668
	20	1630
	35(a)	1604
	40	1607
	55	1634
	70	1671
	85	1783

(a) Congruent point.

incipient melting techniques. The results of [69Rud] are preferred, because pure alloys were used. The liquidus and solidus have a congruent minimum at approximately 32 at.% V and 1605 ± 5 °C. Experimental data are summarized in Table 2 and compared to the assessed diagram in Fig. 2.

Solubility of V in (αTi).

Experimental studies of the solubility of V in (αTi) are summarized in Table 3 and data are plotted in Fig. 3. [51Mcq], [63Luz], and [73Shu] found a maximum solubility of about 1 at.% V, whereas other investigators found larger solubilities of 2 to 4 at.% V. [52Ade] compared the solubility of V in iodide Ti with that in less pure ("Process A") Ti; the solubility is decreased by the presence of impurities. Observed solubilities are higher when alloys have been severely cold worked before annealing [77Mol]. The assessed diagram is based on [52Ade] and [77Mol].

The calculated solubility at the monotectoid temperature is 2.7 at.% V.

β Phase and Monotectoid Reaction.

The assessed (βTi,V) miscibility gap and β/(α + β) boundary are based on [61Erm] and [80Nak] for Ti-rich alloys and on [81Nak] for alloys containing more than 20 at.% V. Experimental data are shown in Table 4 and plotted in Fig. 3.

For alloys containing less than 20 at.% V, the observed temperature of the β/(α + β) boundary is raised by oxygen contamination or insufficiently rapid quenching after annealing [61Erm]; therefore, the data of [61Erm] and [80Nak] are preferred. The β/(α + β) boundary determined by hydrogen pressure measurements [51Mcq] is lower than that determined by conventional methods for unknown reasons.

Evidence for the monotectoid reaction comes from resistivity measurements on heating and cooling, supported by X-ray diffraction of both quenched and annealed specimens [80Nak]. The monotectoid temperature is 675 °C, based on the heating curves. The miscibility gap was determined from the cooling data, obtained in the composition range 10 to 50 at.% V.

[80Nak] contradicts previous work [74Mol, 75Mar]. Conflicting observations may be due to the use of longer anneals at higher temperatures by [80Nak]. The conflict cannot be resolved by thermodynamic analysis. Based on thermodynamic analyses of the Ti-rich part of the phase diagram, [61Kri] predicted an equilibrium miscibility gap, but [61Har] predicted only a metastable miscibility gap. The present calculations show that the experimental data in the range 0 to 18 at.% V can be fitted within the uncertainty either by an equilibrium or by a metastable miscibility gap.

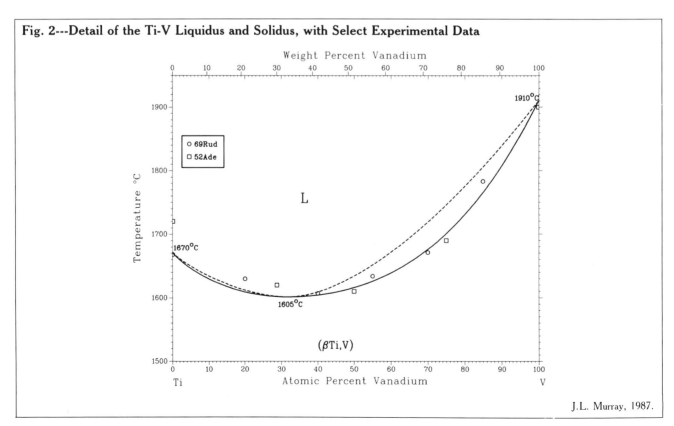

Fig. 2---Detail of the Ti-V Liquidus and Solidus, with Select Experimental Data

J.L. Murray, 1987.

Fig. 3---Detail of the (αTi)/(βTi,V) Boundaries, with Select Experimental Data

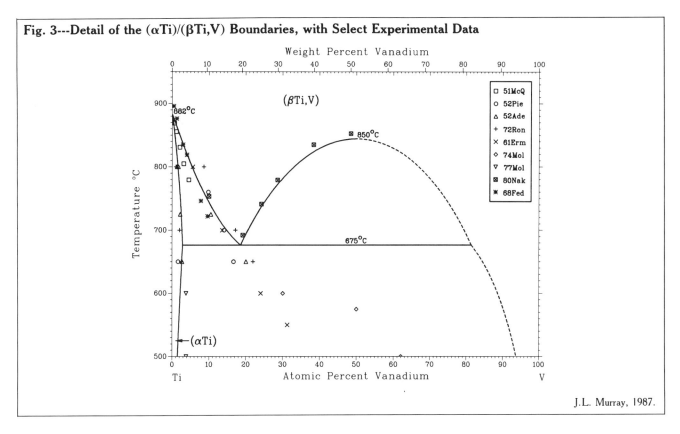

J.L. Murray, 1987.

Table 3 Solubility of V in (αTi)

Reference	Composition, at.% V	Temperature, °C	Experimental technique/comment
[51Mcq]	0.5 to 1	700	Metallography
[52Pie]	~ 1.5	650	Metallography
[52Ade]	1.65	800	Metallography
	2.6	750	(cold worked alloys)
	3.0	700	
[63Luz]	0.5 to 1.2	500, 600, 750	Short anneals
[65Rau]	< 2.5	650	Unpublished research
[66Gla]	2.8 to 3.3	650 to 700	...
[70Ron]	~ 1	800	Microprobe analysis
	~ 2	700	of annealed diffusion couples
	2 to 3	650	
[73Shu]	< 0.85	650 to 700	Paramagnetic susceptibility, modulus of elasticity
[77Mol]	3.7	600	Metallography, X-ray, hardness
	3.7	500	(cold worked alloys)
	2.5	400	

Metastable Phases

Metastable phases are formed from the β phase. In Ti-rich alloys, the hexagonal phase can form martensitically during quenching (α'), or a distorted hexagonal phase (α″) can form similarly in more V-rich alloys. At sufficiently high V content, the β phase can be retained metastably. The ω phase also appears as an intermediate phase in the decomposition of β into the equilibrium (α + β) assemblage. Stress-induced transitions in this system have been examined by several workers [70Kou, 72Oka, 73Wil, 75Kua, 79Oka, 83Men].

Spinodal Decomposition of Metastable (βTi,V).

Spinodal decomposition of unstable β alloys occurs within the coherent spinodal, which is depressed from the chemical spinodal by elastic energy contributions to the Gibbs energy. According to the present rough estimates, the elastic energy is approximately equal in magnitude to the chemical excess Gibbs energy, and spinodal decomposition should be observable only at low temperatures.

Several investigators reported the formation of two β phases after heating the supercooled β solid solution to 120 to 280 °C [61Har, 61Kri, 70Fed2, 72Fed, 74Mol].

Table 4 Experimental Data on the (βTi)/[(βTi) + (αTi)] Boundary

Reference	Composition, at.% V	Temperature, °C	Experimental technique
[51Mcq]	0.5	870	Hydrogen pressure
	1.0	855	
	2.0	830	
	3.0	805	
	4.4	780	
[52Ade]	1.6	800	Metallography
	10.4	725	
	20.0	650	
[52Pie]	16.6	650	Metallography, X-ray,
	14.0	700	thermal analysis
	9.7	760	
[61Erm]	5.5	800	Metallography (rapid quench)
	13.5	700	
	24.0	600	
	31.2	550	
[70Ron, 72Ron]	8.5	800	Microprobe analysis of annealed
	17.1	700	diffusion couples
	21.9	650	
[74Mol]	62	500	Metallography, X-ray, microhardness
	50	575	electrical resistivity
	30	600	
[80Nak]	9.9	753	Resistivity (cooling)

Using dilatometry and X-ray techniques, [61Har] observed the formation of two bcc phases in a 19 at.% V sample aged at 120 to 200 °C; V contents were estimated as 10.4 and 21 at.% V, respectively.

[68Hic] showed that the transformation seen by [61Har] could be attributed to inhomogeneous precipitation during the early stages of ω formation. [68Bla] found broadening, but not splitting, of the bcc phase diffraction lines after aging at 200 °C and reached no conclusion about whether there are two β phases. [70Fed2] found no splitting of the β X-ray diffraction lines after heating samples to 200 °C, but found a splitting after heating to 280 °C.

Martensite Formation.

Depending on the V content, a cph (α') or an orthorhombic (α'') structure may form martensitically during quenching from the β phase. The α' martensite forms in the Ti-rich alloys. The martensite becomes distorted to the α'' structure at 9 at.% V according to [58Bag] and [63Bor], but at 5 at.% V according to [75Age], [78Gus], and [82Gus].

For compositions greater than about 10 at.% V, β or (β + ω) begin to be retained following the quench [58Bag, 71Mcc, 75Col, 80Lei]; and above 14 to 15 at.% V [58Bag, 61Hul, 78Gus], the martensitic α' and α'' phases disappear. The martensite transformation start temperatures, M_s, were measured by [53Duw], [60Sat], [70Hua], and [75Col]. The studies using the techniques of thermal analysis [60Sat, 53Duw] and magnetic susceptibility [75Col] are essentially in agreement. Decomposition of martensites during heating has also been studied [68Fed1, 68Fed2, 70Fed1, 70Fed2, 71Kon, 73Fed, 74Lya]. [68Fed], [72Fed], and [73Fed] reported start and end temperatures A_s and A_f, for the reverse martensitic transformation. Data on M_s, A_s and A_f are shown in Fig. 4.

Metastable ω Phase.

[58Sil] identified the ω structure as one of hexagonal symmetry $P6/mmm$. In the Ti-V system, the ω phase can become distorted to a trigonal structure of symmetry $P\bar{3}m1$ [58Bag, 58Sil, 72Sas]. The ω phase forms in the metastably retained β phase either during quenching, or after aging at temperatures up to about 500 °C [55Bro, 68Hic]. ω particles are coherent with the matrix for all particle sizes observed [68Bla].

As-quenched (or athermal) ω is found coexisting with β in alloys containing more than about 11 to 14 at.% V [68Hic, 71Mcc, 58Bag]. With increasing V content, the ω diffraction lines broaden and disappear; the transition from small ω particles to short-range order is not a clear one. The maximum V content for formation of as-quenched ω was reported as 15 at.% V [58Bag], about 25 at.% V [80Lei], and about 20 at.% V [60Age, 75Age]. According to [68Hic], athermal ω is not found in 19 and 25 at.% V alloys, but according to [71Mcc], ω is found in alloys containing up to about 50 at.% V.

Start temperatures for the athermal β → ω transformation were measured by [55Bro, 72Ika, 73Pat]. Transition temperatures lie between room temperature and 400 °C. As a function of composition they are scattered and depend on cooling rate. [73Pat] showed that oxygen contamination can depress the start temperature for ω formation below room temperature; this is a possible reason for the discrepancy in the reported composition range of the ω phase.

During low-temperature aging, the composition of ω changes until it reaches a metastable equilibrium value. This composition is 13.5 to 15 at.% [68Hic] and independent of aging temperature. At 400 °C, (αTi) begins to precipitate from β + ω after approximately 20 h of aging. At temperatures about 500 to 550 °C, (αTi) forms directly from retained β without the intermediate precipitation of ω phase.

ω is a high-pressure equilibrium phase of pure Ti [Ti]. The combined effect of pressure and alloying on ω formation was studied by [80Voh, 81Lei, 81Min, 81Voh]. [81Min] used pressures up to 25 GPa, [81Lei] to 9.2 GPa, [81Voh] to 8 GPa. The pressure at which (αTi) transforms

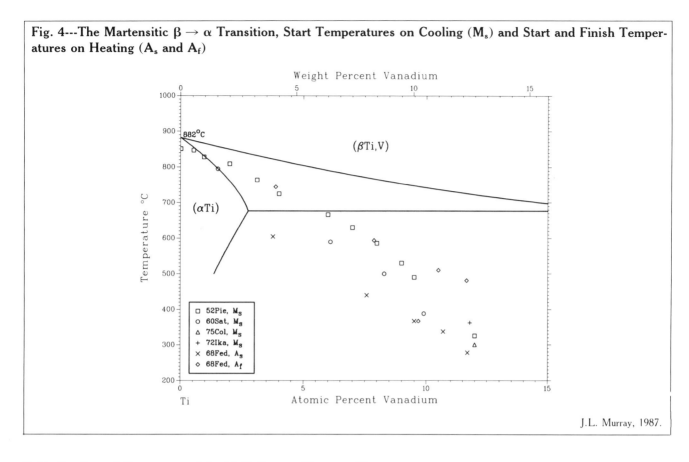

Fig. 4---The Martensitic β → α Transition, Start Temperatures on Cooling (Mₛ) and Start and Finish Temperatures on Heating (Aₛ and A_f)

J.L. Murray, 1987.

Table 5   Crystal Structures of the Ti-V System [Pearson2]

Phase	Composition range, at.% V	Pearson symbol	Space group	Struktur-bericht designation	Prototype
(αTi)	0 to ~ 3	hP2	P6₃/mmc	A3	Mg
α'(a)	0 to 5	hP2	P6₃/mmc	A3	Mg
(βTi,V)	0 to 100	cI2	Im3m	A2	W
α"(a)	5 to 15	oC4	Cmcm	A20	αU
ω(a)	11 to ~ 50	hP3	P6/mmm or P3̄m1	...	ωCrTi

(a) Metastable.

to ω decreases with increasing V content [81Min, 81Voh]. The transformation proceeds via an intermediate β phase [80Voh]. Application of pressure also induces the β → ω transition, with a two-phase (ω + β) region extending from about 18 to 30 at.% V [80Lei, 81Min]. For V content greater than 30 at.%, ω phase was not observed at any pressure. [81Voh] suggested that the equilibrium pressure/composition diagram (at an unspecified temperature) is of the peritectoid type, with ω the high-pressure phase of both constituents. However, a eutectoid type of reaction is equally suggested by the experiments, and an approximately vertical (ω + β) phase field is more consistent with the failure to observe a β → ω transition in pure V.

To draw a P-x diagram, one must take into account coherency effects and the effect of temperature (i.e., the β miscibility gap) on the phase equilibria. Also, the uncertainties in the pure Ti pressure-temperature diagram must be considered. Thus, the effect of pressure on the

Ti-V diagram cannot be quantitatively described at the present time.

Crystal Structures and Lattice Parameters

Crystal structure data are summarized in Table 5. Lattice parameters of the cph and bcc phases are plotted in Fig. 5 and 6, respectively. The break in the bcc lattice parameters at about 10 at.% V represents the composition at which the β phase cannot be fully retained during quenching. Extrapolation of the β lattice parameter to pure Ti provides an estimate of the lattice parameter of hypothetical room temperature bcc Ti. Lattice parameters of the cph phase are for the martensite phase. For the α" martensite in a 19 at.% V alloy, [83Men] gives lattice parameters of $a = 0.2589$ nm, $b = 0.4200$ nm, and $c = 0.3838$ nm. Lattice parameters of the ω phase as a function of overall alloy composition are given in Table 6. The data of [68Hic] are for aged ω; the data of [80Lei] are for ω phase formed during high-pressure soaking.

323

Fig. 5---Lattice Parameters of the cph Solid Solution

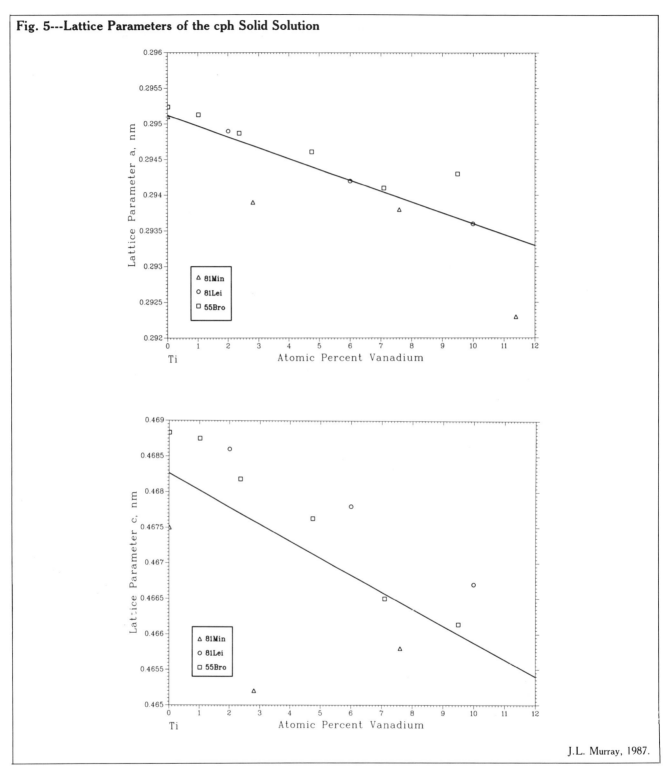

J.L. Murray, 1987.

[80Lei] analyzed lattice parameter data to produce the following expressions for the molar volumes of the various phases (x refers to composition in atomic fraction V):

$$V_{\omega}(x) = 10.47\,(1 - 0.141\,x)\ \text{cm}^3/\text{mol}\quad V_{\beta}(x) =$$
$$10.60\,(1 - 0.157\,x)\ \text{cm}^3/\text{mol}\quad V_{\alpha}(x) =$$
$$10.66\,(1 - 0.214\,x)\ \text{cm}^3/\text{mol}$$

Thermodynamics

Experimental Work.

[71Rol] and [72Rol] measured activities of V and Ti at high temperatures in the β phase by the Knudsen cell technique; they found the β excess Gibb energy to be temperature independent and to follow an $x(1 - x)$ composition dependence. The value of the regular solution interaction parameter based on the experimental activity

324

Table 6 Lattice Parameters of the ω Phase

Reference	Composition, at.% V	Lattice parameters, nm a	c
[68Hic]	14.3	0.4598	0.2818
	16.0	0.4597	0.2817
	17.2	0.4595	0.2815
	19.6	0.4590	0.2814
	22.8	0.4586	0.2811
[80Lei]	6	0.4604	0.2819
	10	0.4594	0.2813
	14	0.4586	0.2810
	20	0.4573	0.2804

data is 7615 ± 500 J/mol. This value implies that the critical temperature is 185 °C and disagrees with phase diagram studies. [70Rub] also reported measurements of V activities in the β phase, but a reference state was not defined, and the authors considered the results to be approximate only.

[68Kal] and [75Mar] found anomalies in the physical properties at 25 at.% V and on this basis, [68Kal] claimed that there is short range order at 25 at.% V. This interpretation is also inconsistent with the phase diagram.

Equilibrium Diagram Calculations.

Several calculations of regular solution parameters of the α and β phases have been made from tie line data from the Ti-rich part of the phase diagram [61Kri, 70Kau, 75Che1,75Che2, 80Lei]. Such calculations do not lead to definitive results, because these phase boundaries are strongly dependent only on the difference between the regular solution parameters and are relatively insensitive to the critical temperature chosen for the bcc miscibility gap.

Table 7 Thermodynamic Parameters

Lattice stability parameters of Ti and V [70Kau]

$G^0(\text{Ti,L}) = 0$
$G^0(\text{V,L}) = 0$

$G^0(\text{Ti,bcc}) = -16\,234 + 8.368\,T$
$G^0(\text{V,bcc}) = -18\,242 + 8.368\,T$

$G^0(\text{Ti,cph}) = -20\,585 + 12.134\,T$
$G^0(\text{V,cph}) = -11\,966 + 11.715\,T$

Excess Gibbs energy parameters

Reference	B(cph)	B(bcc)	B(L)
[61Kri]	29 288	18 828	...
[70Kau]	11 125	11 125	3 644
[75Che1,75Che2]	11 506	11 297	...
[80Lei]	(a)	...	...
This evaluation	18 997	18 581	13 177

Note: Values are given in J/mol; T in K.
(a) B(cph) − B(bcc) = 21 673 − 23 T.

The present calculations (Fig. 1) therefore rely on the miscibility gap data [81Nak]. The free energies of the phases are represented as regular solutions:

$$G(i) = (1 - x)\,G^0(\text{Ti}, i) + G^0(\text{V},i)x + RT[x \ln x + (1 - x)\ln(1 - x)] + B(i)x(1 - x)$$

where i designates the phase; x is the atomic fraction of V; G^0 is the Gibbs energies of the pure components; and $B(i)$ is the regular solution interaction parameter. The lattice stability parameters were taken from [70Kau] and were not varied in the optimizations.

Fig. 6---Lattice Parameters of the bcc Solid Solution

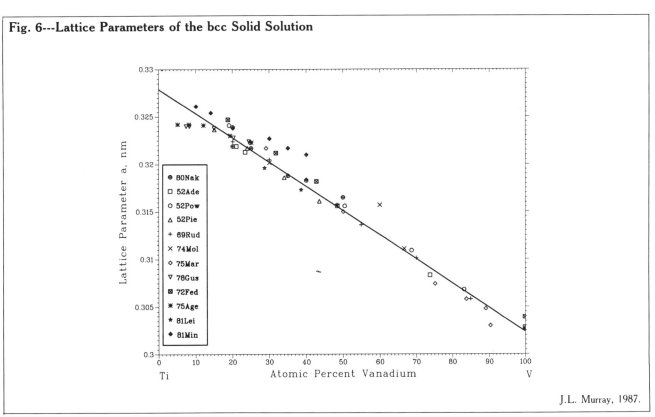

J.L. Murray, 1987.

Phase Diagrams of Binary Titanium Alloys

The results of the present calculations are summarized in Table 7. The regular solution model was found adequate to represent the thermodynamics of this system. The shape of the β miscibility gap and the position of the melting minimum depend on deviations from the regular solution, and they are reproduced within the experimental uncertainty by the present calculations.

Coherent Equilibria.

The elastic energy associated with coherency stresses is required for two aspects of phase equilibria in this system: (1) to predict the composition and temperature range in which spinodal decomposition occurs in the β phase, and (2) to model the β to ω phase equilibria. The contribution of the elastic terms to the Gibbs energy, in the regular solution model, is

$$-x(1-x)\,2e^2E/(1-v)$$

where E is the Young's modulus; v, is the Poisson ratio; and e is $d \ln(a)/dx$, where a is the lattice parameter. This leads to an excess Gibbs energy of about $-19\,000\,x(1-x)$ J/mol. Comparing this with the excess Gibbs energy derived from the incoherent equilibrium diagram, it can be seen that this rough calculation predicts that spinodal decomposition will not occur in this system.

[80Lei] estimated lattice stability parameters and some interaction energies for the metastable ω phase. However, elastic energies must also be included in a treatment of the separation of coherent ω phase particles. The Gibbs energies derived from the incoherent diagram lead to the impossibility of calculating phase equilibria with the unstable β phase inside the spinodal, where β alloys are in fact observed in coherent metastable equilibrium with ω. This problem was not recognized by [80Lei], because they calculated only T_0 curves of equal Gibbs energies of the β and ω phases and not tielines.

Cited References

51Mcq: A.D. McQuillan, "The Effect of the Elements of the First Long Period on the α-β Transformation in Ti," *J. Inst. Met., 80*, 363-368 (1951). (Equi Diagram; Experimental)

52Ade: H.K. Adenstedt, J.R. Pequignot, and J.M. Raymer, "The Titanium-Vanadium System," *Trans. AIME, 44*, 990-1003 (1952). (Equi Diagram; Crys Structure; Experimental)

52Pie: P. Pietrokowsky and P. Duwez, "Partial Titanium-Vanadium Phase Diagram," *Trans. AIME, 194*, 627-630 (1952). (Equi Diagram, Meta Phases; Experimental)

52Pow: R.M. Powers and H.A. Wilhelm, WADC Technical Rept. 54-502, 42 (1952). (Equi Diagram; Experimental)

53Duw: P. Duwez, "The Martensite Transformation Temperature in Titanium Binary Alloys," *Trans. AIME, 45*, 935-940 (1953). (Meta Phases; Experimental)

55Bro: F.R. Brotzen, E.L. Harmon, and A.R. Troiano, "Decomposition of β Titanium," *Trans. AIME, 203*, 413-419 (1955). (Meta Phases; Experimental)

58Bag: Yu.A. Bagariatskii, G.I. Nosova, and T.V. Tagunova, "Factors in the Formation of Metastable Phases in Titanium-Base Alloys," *Dokl. Akad. Nauk SSSR, 122*, 593-596 (1958) in Russian; TR: *Sov. Phys. Dokl., 3*, 1014-1018 (1958). (Meta Phases; Experimental)

58Sil: J.M. Silcock, "An X-Ray Examination of the ω Phase in TiV, TiMo and TiCr Alloys," *Acta Metall., 6*, 481-493 (1958). (Meta Phases; Crys Structure; Experimental)

60Age: N.V. Ageev and L.A. Petrova, "Stability of the β Phase in Titanium-Vanadium Alloys," *Zh. Neorg. Khim., 5*, 615 (1960) in Russian; TR: *Russ. J. Inorg. Chem., 5*, 295-298 (1960). (Meta Phases, Crys Structure; Experimental)

60Sat: T. Sato, S. Hukai, and Y.C. Huang, "The M_s Points of Binary Titanium Alloys," *J. Aust. Inst. Met., 5(2)*, 149-153 (1960). (Meta Phases; Experimental)

61Erm: F. Ermanis, P.A. Farrar, and H. Margolin, "A Reinvestigation of the Systems Ti-Cr and Ti-V," *Trans. AIME, 221*, 904-908 (1961). (Equi Diagram; Experimental)

61Har: E.L. Harmon and A.R. Troiano, "β Transformation Characteristics of Titanium Alloyed with Vanadium and Aluminum," *Trans. AIME, 53*, 43-53 (1961). (Equi Diagram, Meta Phases; Experimental)

61Hul: J.K. Hulm and R.D. Blaugher, "Superconducting Solid Solution Alloys of Transition Elements," *Phys. Rev., 123*, 1569 (1961). (Equi Diagram; Experimental)

61Kri: O. Krisement, "Precipitation in β Ti-V Alloys," *Z. Metallkd., 52*, 695-704 (1961) in German. (Thermo; Experimental)

63Bor: B.A. Borok, E.K. Novikova, L.S. Golubeva, R. Shchegoleva, and N.A. Ruch'eva, "Dilatometric Investigation of Binary Alloys of Titanium," *Metalloved. Term. Obrab. Met.*, (2), 32-36 (1963) in Russian; TR: *Met. Sci. Heat Treat.*, (2), 94 (1963). (Equi Diagram; Experimental)

63Luz: L.P. Luzhnikov, V.M. Novikova, and A.P. Mareev, "Solubility of β-Stabilizers in α-Titanium," *Metalloved. Term. Obrab. Met.*, (2), 13 (1963) in Russian; TR: *Met. Sci. Heat Treat.*, (2), 78 (1963). (Equi Diagram; Experimental)

65Rau: Ch.J. Raub and U. Zwicker, "Superconductivity of α-Titanium Solid Solutions with Vanadium, Niobium, and Tantalum," *Phys. Rev. A, 137(1)*, A142-A143 (1965). (Equi Diagram; Experimental)

66Gla: V.V. Glazova, "Titanium Alloys," *Metallurgiya* (1966). (Equi Diagram; Experimental)

68Bla: M.J. Blackburn and J.C. Williams, "Phase Transformations in Ti-Mo and Ti-V Alloys," *Trans. AIME, 242*, 2461-2468 (1968). (Meta Phases; Experimental)

68Fed1: S.G. Fedotov, K.M. Konstantinov, and Ye.P. Sinodova, "Disintegration of Titanium-Vanadium Martensite on Continuous Heating," *Fiz. Met. Metalloved., 25*, 860-866 (1968) in Russian; TR: *Phys. Met. Metallogr.*, 104-110 (1968). (Meta Phases; Experimental)

68Fed2: S.G. Fedotov, K.M. Konstantinov, and Ye.P. Sinodova, "Decomposition of Titanium-Vanadium Martensite During Heating," *Titanium Alloys for Modern Technology*, N.P. Sazhin, Ed., NASA TT F-596, 187-193 (1968). (Meta Phases; Experimental)

68Hic: B.S. Hickman, "Precipitation of the ω Phase in Titanium-Vanadium Alloys," *J. Inst. Met., 96*, 330-337 (1968). (Meta Phases; Experimental)

68Kal: G.P. Kalinin and O.P. Elyutin, "The Anomalies of the Properties of Titanium Alloys with β Stabilizing Elements," *Metalloved. Term. Obrab. Met.*, (4), 52-54 (1968) in Russian; TR: *Met. Sci. Heat Treat.*, (4), 301-302 (1968). (Equi Diagram; Experimental)

69Rud: E. Rudy, "Compilation of Phase Diagram Data," USAF Tech. Rept. AFML-TR-65-2, Part V (1969). (Equi Diagram; Experimental)

70Fed1: S.G. Fedotov and K.M. Konstantinov, "Effect of the Heating Rate on the Decomposition of Titanium-Vanadium Martensite," *Dokl. Akad. Nauk SSSR, 191(6)*, 1270-1273 (1970) in Russian; TR: *Sov. Phys. Dokl., 15*, (4), 410-413 (1970). (Meta Phases; Experimental)

70Fed2: S.G. Fedotov and K.M. Konstantinov, "Decomposition of Unstable β Solid Solution of Titanium with 18 wt.% Vanadium," *Dokl. Akad. Nauk SSSR, 192(3)*, 555-558 (1970) in Russian; TR: *Sov. Phys. Dokl., 15(4)*, 516-519 (1970). (Meta Phases; Experimental)

70Hua: Y.C. Huang, S. Suzuki, H. Kaneko, and T. Sato, "Continuous Cooling Transformation of β-Phase in Binary Titanium Alloys," Sci. Technol. Appl. Titanium, Proc. Int. Conf., R.I. Jaffee, Ed., 695 (1970). (Equi Diagram; Experimental)

70Kau: L. Kaufman and H. Bernstein, *Computer Calculation of Phase Diagrams*, Academic Press, New York, (1970). (Meta Phases; Theory)

70Kou: M.K. Koul and J.F. Breedis, "Phase Transformation in β Isomorphous Titanium Alloys," *Acta Metall., 18*, 579-588 (1970). (Meta Phases; Experimental)

70Ron: G.N. Ronami, S.M. Kuznetsova, S.G. Fedotov, and K.M. Konstantinov, "Determination of the Phase Boundaries in Ti Systems with V, Nb, and Mo by the Diffusion-Layer Method,"

Vest. Mosk. Univ. Fiz., 25(2), 186-189 (1970) in Russian; TR: *J. Moscow Univ. Phys., 25*(2), 55-57 (1970). (Equi Diagram; Experimental)

70Rub: A.N. Rubtsov, Yu.G. Olesov, V.I. Cherkashin, and A.B. Suchkov, "Activity of Al, V and Cr in Binary Alloys with Titanium," *Izv. Akad. Nauk SSSR, Met.,* (6), 84-87 (1970) in Russian; TR: *Russ. Metall.,* (6), 56-58 (1970). (Thermo; Experimental)

71Kon: K.M. Konstantinov, S.G. Fedotov, and G.D. Shnyrev, "Phase Transformations During Rapid Heating of Titanium-Vanadium Martensite," *Izv. Akad. Nauk SSSR, Met.,* (1), 172-175 (1971) in Russian; TR: *Russ. Metall.,* (1) 114-117 (1971). (Meta Phases; Experimental)

71Mcc: K.K. McCabe and S.L. Sass, "The Initial Stages of the ω Phase Transformation in Ti-V Alloys," *Philos. Mag., 23,* 957-970 (1971). (Meta Phases; Experimental)

71Rol: E.J. Rolinski, M. Hoch, and C.J. Oblinger, "Determination of Thermodynamic Interaction Parameters in Solid V-Ti Alloys Using the Mass Spectrometer," *Metall. Trans., 2,* 2613-2618 (1971). (Thermo; Experimental)

72Fed: S.G. Fedotov and K.M. Konstantinov, "Structural Features and Instability of β Solid Solutions in the System Ti-V," *New Constr. Mater. Ti, Akad Nauk SSSR,* 37-41 (1972) in Russian. (Crys Structure; Experimental)

72Ika: H. Ikawa, S. Shin, and M. Morikawa, "The Studies Continuous Cooling Transformation in Binary α + β and Metastable β Titanium Alloys," *J. Jpn. Inst. Met., 41*(4) 394-402 (1972) in Japanese. (Meta Phases; Experimental)

72Oka: M. Oka, C.S. Lee, and K. Shimizu, "Transmission Electron Microscopy Study of Face-Centered Orthorhombic Martensite in Ti–12.6% V Alloy," *Metall. Trans., 3,* 37-45 (1972). (Meta Phases; Experimental)

72Rol: E.J. Rolinski, M. Hoch, and C.J. Oblinger, "Determination of Thermodynamic Interaction Parameters in Solid V-Ti-Cr Alloys Using the Mass Spectrometer," *Metall. Trans., 3* 1413-1418 (1972). (Thermo; Experimental)

72Ron: G.N. Ronami, S.M. Kuznetsova, S.G. Fedotov, and K.M. Konstantinov, "Determination of Phase Equilibria of Superconducting Alloys Using Electron Beam Microanalysis," *Kristallogr. Technik., 7,* 615 (1972). (Equi Diagram; Experimental)

72Sas: S.L. Sass and B. Borie, "The Symmetry of the Structure of the ω Phase in Zr and Ti Alloys," *J. Appl. Crystallogr., 5,* 236-238 (1972). (Crys Structure; Experimental)

73Fed: S.G. Fedotov "Peculiarities of Changes in Elastic Properties of Titanium Martensite," *Sci., Technol. Appl. Titanium, Proc. Int. Conf.,* R.I. Jaffee, Ed., 871-881 (1973). (Meta Phases; Experimental)

73Pat: N.E. Paton and J.C. Williams, "The Influence of Oxygen Content on the Athermal β-ω Transformation," *Scr. Metall., 7,* 647-650 (1973). (Meta Phases; Experimental)

73Shu: V.M. Shushkanov, L.S. Moroz, V.V. Obukhovskiy, N.P. Kapitonova, and N.V. Ivanova, "Solubility of Vanadium in α-Titanium," *Izv. Akad. Nauk SSSR, Met.,* (4) 221-224 (1973) in Russian; TR: *Russ. Metall.,* (4), 156-158 (1973). (Equi Diagram; Experimental)

73Wil: J.C. Williams, "Kinetics and Phase Transformations," *Sci. Technol. Appl. Titanium, Proc. Int. Conf.,* R.I. Jaffee, Ed., 1433 (1973). (Meta Phases; Experimental)

74Lya: V.S. Lyasotskaya, B.A. Kolachev, S.G. Fedotov, and K.M. Konstantinov, "Transformations in Quenched Ti-V Alloys During Heating," *Stali Splavy Tsvet Met., Kuibyshev,* 136-143 (1974) in Russian. (Meta Phases; Experimental)

74Mol: V.V. Molokanov and P.B. Budberg, "Phase Structure of Ti-V Alloys," *Dokl. Akad. Nauk SSSR, 215* (5), 1125-1127 (1974) in Russian; TR: *Sov. Phys. Dokl., 49*(3), 237-239 (1974). (Equi Diagram; Experimental)

75Age: N.V. Ageev, L.N. Guseva, and L.K. Dolinskaya, "Metastable Phases in Quenched Ti-Mo and Ti-V Alloys and the Influence of Small Quantities of Oxygen on Them," *Izv. Akad. Nauk SSSR, Met.,* (4), 151 (1975) in Russian; TR: *Russ. Metall.,* (4), 113-115 (1975). (Meta Phases; Experimental)

75Che1: D.B. Chernov and A.Ya. Shinyaev, "Use of a Regular Solution Model for the Analysis of Phase Equilibria in the Titanium-Vanadium System," *Zh. Fiz. Khim., 48,* 762-763 (1975) in Russian; TR: *Russ. J. Phys. Chem., 49*(3), 445-445 (1975). (Thermo; Theory)

75Che2: D.B. Chernov and A.Y. Shinyayev, "Calculation of the Interaction Parameters of Titanium with Vanadium, Niobium and Molybdenum," *Izv. Akad. Nauk SSSR, Met.,* (5), 212-219 (1975) in Russian; TR: *Russ. Metall.,* (5), 167-172 (1975). (Thermo; Theory)

75Col: E.W. Collings, "Magnetic Studies of ω Phase Precipitation and Aging in Titanium-Vanadium Alloys," *J. Less-Common Met., 39,* 63-90 (1975). (Meta Phases; Experimental)

75Kua: T.S. Kuan, R.R. Ahrens, and S.L. Sass, "The Stress-Induced ω Phase Transformation in Ti-V Alloys," *Metall. Trans. A, 6,* 1767-1774 (1975). (Meta Phases; Experimental)

75Mar: L.V. Marchukova, N.M. Matveeva, and I.I. Kornilov, "Phase Equilibria and Certain Properties of V-Ti Alloys," *Izv. Tsvetn. Metall.,* (3), 131-134 (1975) in Russian. (Equi Diagram; Experimental)

77Mol: V.V. Molokanov, D.B. Chernov, and P.B. Budberg, "Solubility of Vanadium in α-Ti," *Metalloved. Term. Obrab. Met.,* (8), 60-61 (1977) in Russian; TR: Met. Sci. Heat Treat., (8), 704-705 (1977). (Equi Diagram; Experimental)

78Gus: L.N. Guseva and L.K. Dolinskaya, "Formation Conditions of Athermal ω Phase in Alloys of Ti with Transition Elements," *Krist. Strukt. Svoistva Met. Splavov,* 59-63 (1978) in Russian. (Meta Phases; Experimental)

79Oka: M. Oka and Y. Taniguchi, "332 Deformation Twins in Ti–15.5% V Alloy," *Metall. Trans. A, 10,* 651-653 (1979). (Meta Phases; Experimental)

80Lei: Ch. Leibovitch and A. Rabinkin, "Metastable Diffusionless Equilibria in Ti-Mo and Ti-V Systems under High Pressure," *Calphad, 4*(1), 13-26 (1980). (Meta Phases; Theory)

80Nak: O. Nakano, H. Sasano, T. Suzuki and H. Kimura, "Phase Separation in Ti-V Alloys," Titanium '80 Proc. 4th Int. Conf. on Titanium, Kyoto, Japan, Kimura, H. and Izumi, O. eds., Vol. 2, 2889-2895 (1980). (Equi Diagram; Experimental)

80Voh: Y.K. Vohra, S.K. Sikka, E.S.K. Menon and R. Krishnan, "Direct Evidence of Intermediate State During Alpha-Omega Transformation in Ti-V Alloy," *Acta Metall., 28,* 683-685 (1980). (Meta Phases; Experimental)

81Lei: Ch. Leibovitch, A. Rabinkin, and M. Talianker, "Phase Transformations in Metastable Ti-V Alloys Induced by High Pressure Treatment," *Metall. Trans. A, 12,* 1513-1519 (1981). (Meta Phases; Experimental)

81Min: L.-C. Ming, M.H. Manghanani, and K.W. Katahana, "Phase Tranformations in the Ti-V System under High Pressure up to 25 GPa," *Acta Metall., 29,* 479-485 (1981). (Meta Phases; Experimental)

81Mur: J.L. Murray, "The Ti-V(Titanium-Vanadium) System," *Bull. Alloy Phase Diagrams, 2*(1), 48-55 (1981). (Meta Phases; Experimental)

81Voh: Y.K. Vohra, E.S.K. Menon, S.K. Sikka, and R. Krishnan, "High Pressure Studies on a Prototype ω Forming Alloy System," *Acta Metall., 29,* 457-470 (1981). (Meta Phases; Experimental)

82Gus: L.N. Guseva and L.K. Dolinskaya, "Metastable Phases in Quenched Titanium Alloys with Transition Elements" *Titanium and Ti Alloys, Scientific and Technological Aspects, 2,* J.C. Williams and A.F. Delov, Ed., 1559-1565 (1982). (Meta Phases; Experimental)

83Men: E. Sarath Kumar Menon and R. Krishnan, "Phase Transformations in Ti-V Alloys," *J. Mater. Sci., 18,* 365-374 (1983). (Meta Phases; Experimental)

The (Ti-W) Titanium-Tungsten System

47.88   183.85

By J.L. Murray

Equilibrium Diagram

The equilibrium solid phases of the Ti-W system are: (1) the low-temperature cph (αTi) solid solution; and (2) the bcc (βTi,W) solid solution. Ti and W exhibit complete mutual solid solubility in (βTi,W) between a critical point at 1250 °C and the solidus. The assessed Ti-W diagram is shown in Fig. 1, and the special points are summarized in Table 1.

(βTi,W) Miscibility Gap.

This assessment differs significantly from [Hansen], [Elliott], and [Shunk]. The evaluation by [Hansen] relied on the diagram proposed by [53May], in which a bcc miscibility gap is shown persisting up to a peritectic reaction, L + (W) ⇄ (βTi), at 1880 °C. However, by

metallographic and X-ray studies, [68Rud] showed that a complete range of (βTi,W) solutions exists above a critical point at about 1250 °C and that there is no peritectic reaction.

Previous results were attributed to insufficient homogenization of arc-melted alloys. [68Rud] estimated compositions of coexisting bcc phases as 20 and 50 at.% W at 1100 °C based on lattice parameter data. A thermodynamic calculation of the phase diagram is used in this assessment to extrapolate the miscibility gap to the W side of the system.

Liquidus and Solidus.

Experimental studies of the liquidus and solidus were made by [53May], [54Now], [68Rud], and [71Ole]. Because the proposed peritectic reaction was shown to be incorrect,

Table 1   Special Points of the Ti-W System

Reaction	Compositions of the respective phases, at.% W			Temperature, °C	Reaction type
(βTi) ⇄ (W) + (αTi)	9	∼ 70	0.2	740 ± 20	Monotectoid
(βTi,W) ⇄ (βTi) + (W)		33		1250	Critical point
L ⇄ (βTi)		0		1670	Melting point
(βTi) ⇄ (αTi)		0		882	Allotropic transformation
L ⇄ (W)		100		3422	Melting point

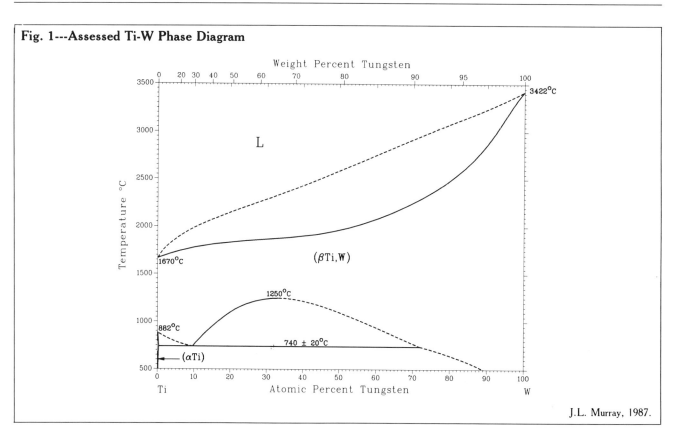

Fig. 1---Assessed Ti-W Phase Diagram

J.L. Murray, 1987.

Fig. 2---Experimental Data on the Liquidus and Solidus

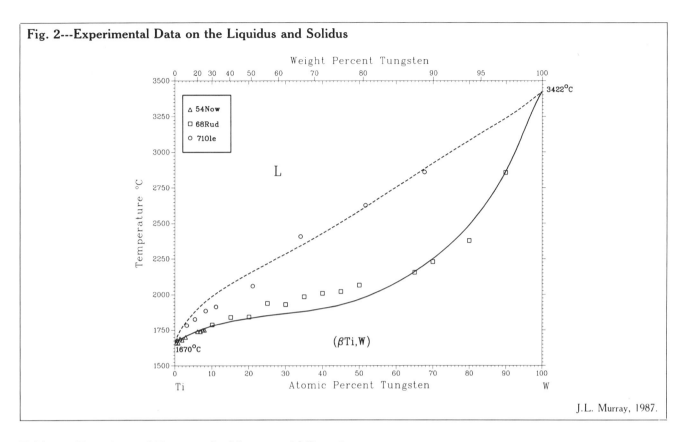

J.L. Murray, 1987.

Table 2   Experimental Data on the Monotectoid Reaction

| | Monotectoid point | | |
Reference	Composition, at.% W	Temperature, °C	Experimental technique
[53May]	9	715	Metallography
[58Bag1]	9	725	Metallography, X-ray
[62Img]	8.7	735	Metallography
[71Ole]	9	760	DTA (heating)

high-temperature data of [53May] are not used in the present assessment. In the range 0 to 10 at.% W, however, the work of [53May] is consistent with the other determinations.

Data of [54Now], [68Rud], and [71Ole] are plotted in Fig. 2. [68Rud] and [71Ole] used the incipient melting technique and high-temperature differential thermal analysis, respectively, to determine the solidus. [54Now] gave no details on the experimental methods used to determine the liquidus. The temperatures of sample collapse [68Rud] lie 50 to 100 °C below the liquidus data of [54Now]. [68Rud] estimated that the equilibrium liquidus may lie as much as 200 °C above the temperature of sample collapse.

The assessed phase boundaries represent the results of optimizations of Gibbs energies with respect to the data of [54Now], [68Rud], and [71Ole]. Although the solid-state equilibria could be reproduced using simple temperature-independent subregular solution models for the excess quantities, the liquid/solid equilibria could not. Large excess entropy terms had to be included in order to reproduce the liquidus and solidus within the experimental uncertainty, as shown in Fig. 1 and 2. At this time, it is not certain whether these excess entropy terms are

realistic, or whether some real inconsistency between the liquidus and the solidus data is indicated.

Monotectoid and (αTi)/(βTi) Equilibria.

Reported values of the monotectoid composition and temperature are given in Table 2. The reaction (βTi) ⇄ (αTi) + (W) is sluggish, and the discrepancy between thermal analysis arrests on heating [71Ole] and metallographic studies [53May] is 45 °C. Moreover, data on the (βTi) transus do not appear to be consistent with the miscibility gap and the monotectoid point. These data are shown in Fig. 3. From [53May], only data from iodide Ti-based alloys are shown. The work of [58Bag1] is not shown, because they showed the bcc → cph transition of pure Ti above 950 °C, indicating that alloys were contaminated. [61Img, see Elliott] found the (βTi) transus to be nearly a straight line between the monotectoid point and the transformation in pure Ti, in agreement with the monotectoid reaction and the present thermodynamic calculations.

The assessed monotectoid reaction has benn placed at 740 ± 20 °C, somewhat below the heating arrests [71Ole], but above the values based on microscopic studies. The assessed (βTi) transus is based on [61Img].

Fig. 3---Experimental Data on the Equilibria of the Solid Phases

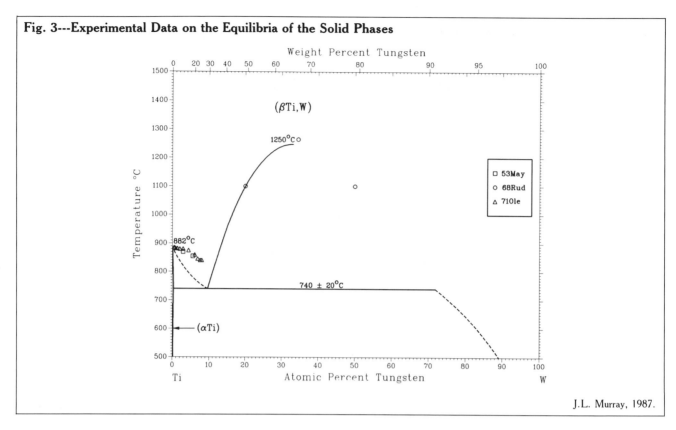

J.L. Murray, 1987.

Table 3   Crystal Structures of the Ti-W System [Pearson2]

Phase	Composition range, at.% W	Pearson symbol	Space group	Struktur-bericht designation	Prototype
(αTi)	0 to 0.2	$hP2$	$P6_3/mmc$	$A3$	Mg
α′	0 to 2	$hP2$	$P6_3/mmc$	$A3$	Mg
(βTi,W)	0 to 100	$cI2$	$Im3m$	$A2$	W
α″	2 to 5.5	$oC4$	$Cmcm$	$A20$	αU
ω	6 to 10	$hP3$	$P6/mmm$	...	ωMnTi

Using iodide Ti-based alloys, [53May] determined the maximum solubility of W in (αTi) as 0.2 at.% at 715 °C. [71Ole] verified that the solubility must be less than about 0.3 at.%.

Metastable Phases

Martensites.

Bcc (βTi,W) cannot be retained during quenching in alloys containing less than about 6 at.% W [53Duw, 53May, 71Ole]. For alloys containing 0 to 2 at.% W, the cph structure (α′) is formed martensitically; for alloys containing 2 to 6 at.% W, the martensite has an orthorhombic structure that is a distortion of the cph lattice [58Bag2, 58Bag3, 70Wil, 71Mor].

[53Duw] measured the start temperature, M_s, of the martensite transformation as:

Composition, at.% W	M_s,°C
4.0	600
6.2	400

[70Wil] compared the tempering behavior of the two martensites. The decomposition process proceeds as:

$$\alpha'' \rightarrow (\alpha Ti) + \alpha'(\text{W-enriched}) \rightarrow (\alpha Ti) + (\beta Ti) \text{ (equilibrium)}$$

ω Phase.

The ω phase is ubiquitous in Ti-transition metal alloys in the range where the bcc phase can be retained during quenching. The reader is referred to reviews of Ti-Cr or Ti-V, for example, for detailed studies of the structure and properties of ω phase. ω formation in the Ti-W system was reported by [58Bag2, 58Bag3] and [73Ika]. ω forms either during quenching (athermally) or during annealing in the approximate range 30 to 450 °C. ω precipitates are coherent with the bcc matrix. In the Ti-W system, the precipitates have the ellipsoidal morphology associated with low misfit of the two structures [73Ika].

Crystal Structures and Lattice Parameters

The structures of the equilibrium and metastable phases of the Ti-W system are summarized in Table 3. Lattice parameter measurements for (βTi,W) alloys are presented in Table 4, and lattice parameters for the

Table 4 (βTi,W) Lattice Parameters

Reference	Composition, at.% W	Lattice parameter, nm
[68Rud]	15	0.324
	25	0.3222
	35	0.3208
	40	0.3200
	45	0.3193
	50	0.3184
	55	0.3181
	65	0.3170
	70	0.3167
	80	0.3164
	90	0.3164
[58Bag3]	7	0.3264

metastable hexagonal and orthorhombic phases are presented in Table 5.

Thermodynamics

Excess thermodynamic quantities have not yet been determined experimentally for the Ti-W system. Previous thermodynamic analyses of the system [75Che, 75Kau, 79Les], were based on phase diagram data. The thermodynamic parameters used in the present calculation are compared in Table 6 to the parameters used by [75Kau] and [75Che].

Calculation of the Phase Diagram.

The Gibbs energies of solution phases are represented as:

$$G(i) = G^0(\text{Ti},i)(1-x) + x\,G^0(\text{W},i) + RT\,[x \ln x + (1-x)\ln(1-x)] + x(1-x)\,[B(i) + C(i)(1-2x)]$$

where i designates the phase; x is the atomic fraction of W; G^0 is the Gibbs energies of the pure metals; and $B(i)$ and $C(i)$ are coefficients in an expansion of the excess Gibbs energies. The lattice stability parameters are from [70Kau] and have been modified only to reproduce the melting point of W.

The presence of the miscibility gap in the bcc phase provides a means of estimating its excess Gibbs energy. Four points on the miscibility gap were chosen from the data of [68Rud], and an asymmetric temperature-dependent excess Gibbs energy reproduced these points.

The Gibbs energies of the liquid and of (αTi) were then determined with respect to that of the bcc solution. Constraining the maximum solubility of W in (αTi) to be 0.2 at.% leads to a calculated monotectoid point at 9.5 at.% and 740 °C, in good agreement with the experimental data.

Calculation of the liquidus and solidus led to some difficulties in this system. If it is assumed that the excess entropy of the liquid is small in magnitude and a symmetric function of composition, then the calculation can reproduce the observed liquidus or the observed solidus, but not the very wide two-phase region. The Gibbs energy

Table 5 Lattice Parameters of the Martensites

Phase	Reference	Composition, at.% W	Lattice parameters, nm		
			a	b	c
α′	[58Bag3]	1.3	0.2883	...	0.4716
		2.0	0.2909	...	0.4693
α″	[70Wil]	4	0.3000	0.4995	0.4655
	[58Bag3]	2.9	0.2906	0.4916	0.4669
		3.4	0.2771	0.4800	0.4673
		4.3	0.2993	0.4964	0.4662

Table 6 Thermodynamic Parameters

Properties of the pure elements

$G^0(\text{Ti,L}) = 0$
$G^0(\text{W,L}) = 0$

$G^0(\text{Ti,bcc}) = -16\,234 + 8.368\,T$
$G^0(\text{W,bcc}) = -30\,543 + 8.266\,T$

$G^0(\text{Ti,cph}) = -20\,585 + 12.134\,T$
$G^0(\text{W,cph}) = -22\,175 + 8.266\,T$

Properties of the solution phases

Parameter	This evaluation	[75Kau]	[75Che]
B(bcc)	14 653 + 4.32 T	3900 + 3.98 T	15 690
C(bcc)	− 947 + 5.37 T	3.98 T	7 322
B(L)	124 120 − 41.84 T	9623 + 3.98 T	...
C(L)	41 236 − 19.88 T	3347 + 3.98 T	...
B(cph)	41 531	6300 + 3.98 T	...
C(cph)	...	3.98 T	...

Note: Values are given in J/mol and J/mol · K.

function used to generate the assessed diagram is given in Table 6. The present evaluation does not believe that it realistically represents the partitioning of Gibbs energy into enthalpy and entropy contributions. The Gibbs energies of [75Kau], for example, are thermodynamically more plausible, but they do not reproduce the observed liquidus and solidus. The lack of experimental thermodynamic data, the uncertainty of the phase diagram, and the very large temperature range over which Gibbs energies are approximated by simple models preclude further progress at this time.

Cited References

53Duw: P. Duwez, "The Martensite Transformation Temperature in Titanium Binary Alloys," *Trans. ASM, 45,* 934-940 (1953). (Meta Phases; Experimental)

53May: D.J. Maykuth, H.R. Ogden, and R.I. Jaffee, "Titanium-Tungsten and Titanium-Tantalum Systems," *Trans. AIME, 197,* 231-237 (1953). (Equi Diagram; Experimental)

54Now: H. Nowotny, E. Parthe, R. Kieffer, and F. Benesovsky, "The Ternary Titanium-Tungsten-Carbon System," *Z. Metallkd., 45,*(3), 97-99 (1954) in German. (Equi Diagram; Experimental)

58Bag1: Yu.A. Bagariatskii, G.I. Nosova, and T.V. Tagunova, "Study of the Phase Diagrams of the Alloys Titanium- Chromium, Titanium-Tungsten, and Titanium-Chromium-Tungsten, Prepared by the Method of Powder Metallurgy," *Zh. Neorg. Khim., 3*(3), 777-785 (1958) in Russian; TR: *Russ. J. Inorg. Chem., USSR, 3*(3) 330-341 (1958). (Equi Diagram; Experimental)

58Bag2: Yu.A. Bagariatskii, G.I. Nosova, and T.V. Tagunova, "Factors in the Formation of Metastable Phases in Titanium-Base Alloys," *Dokl. Akad. Nauk SSSR, 3,* 1014 (1958) in Russian; TR: *Sov. Phys. Dokl., 3,* 1014-1018 (1958). (Metal Phases; Experimental)

58Bag3: Yu.A. Bagariatskii, T.V. Tagunova, and G.I. Nosova, "Metastable Phases in Titanium-Base Alloys," *Coll. Prob. Metallogr. Phys. Met., 5,* 210-234 (1958) in Russian. (Meta Phases; Experimental)

*61Img: A.G. Imgram, D.N. Williams, R.A. Wood, H.R. Ogden, and R.I. Jaffee, "Metallurgical and Mechanical Characteristics of High-Purity Titanium-Base Alloys," WADC Tech. Rept. 59-595, Part II (1961). (Equi Diagram; Experimental)

*68Rud: E. Rudy and St. Windisch, "Revision of the Titanium-Tungsten System," *Trans. AIME, 242,* 953-954 (1968). (Equi Diagram; Experimental)

70Kau: L. Kaufman and H. Bernstein, *Computer Calculation of Phase Diagrams,* Academic Press, New York (1970). (Thermo; Theory)

70Wil: J.C. Williams and B.S. Hickman, "Tempering Behavior of Orthorhombic Martensite in Titanium Alloys," *Metall. Trans., 1,* 2648-2650 (1971). (Metal Phases; Experimental)

71Mor: H.A. Moreen, "Discussion of Tempering Behavior of Orthorhombic Martensite in Titanium Alloys," *Metall. Trans., 2,* 2953 (1971). (Meta Phases; Experimental)

*71Ole: S.V. Oleynikova, T.T. Nartova, and I.I. Kornilov, "Examination of the Structure and Properties of Ti-rich Ti-W Alloys," *Izv. Akad. Nauk SSSR, Met.,* (3), 192-196 (1971) in Russian; TR: *Russ. Metall.,* (3), 130-133 (1971). (Equi Diagram; Experimental)

73Ika: H. Ikawa, S. Shin, M. Miyagi, and M. Morikawa, "Some Fundamental Studies on the Phase Transformation from β Phase to α Phase in Titanium Alloys," Sci. Technol. Appl. Titanium, Proc. Int. Conf., R.I. Jaffee, Ed., 1545-1556 (1973). (Experimental)

75Che: D.B. Chernov, A.Ya. Shinyaev, and E.M. Savitskii, "Calculation of Phase Diagrams of Titanium with β-Isomorphos Elements," *Dokl. Akad. Nauk SSSR, 225*(2) 390-392 (1975) in Russian. (Thermo; Theory)

75Kau: L. Kaufman and H. Nesor, "Calculation of Superalloy Phase Diagrams Part IV," *Metall. Trans. A, 6,* 2123-2131 (1975). (Thermo; Theory)

79Les: A.G. Lesnik, V.V. Nemonshkalenko, and A.A. Ovcharenko, "Computer Calculation of Phase Diagrams of Some Binary Alloys in the Subregular Solution Approximation," *Akad. Nauk Ukr. SSR, Metallofiz., 75,* 20-31 (1979) in Russian. (Thermo; Theory)

The Ti-Y (Titanium-Yttrium) System

47.88   88.9059

By J.L. Murray

Equilibrium Diagram

Ti and Y both undergo allotropic transformation from a low-temperature cph form to a high-temperature bcc form, Ti at 882 °C and Y at 1478 °C. The system has no intermetallic compounds. The solid phases have only small homogeneity ranges, and experimental studies [61Bar, 62Lun, 62Sav, 68Bea] were limited to Ti- and Y-rich alloys. There is agreement among investigators about the topology of the diagram, but there are discrep-

ancies exceeding the experimental precision in the quantitative details, particularly in the eutectic temperature.

The assessed invariant temperatures and other special points of the phase diagram are summarized in Table1. The assessed Ti-Y phase diagram (Fig. 1) is the result of the thermodynamic calculations described below and reproduces the input data within the experimental uncertainty. The liquid miscibility gap and the monotectic reaction are predictions of the thermodynamic analysis

Table 1 Special Points of the Assessed Ti-Y Phase Diagram

Reaction	Compositions of the respective phases, at.% Y			Temperature, °C	Reaction type
$(\beta Y) \rightleftarrows L + (\alpha Y)$	~ 99	~ 94	99	1440	Catatectic(a)
$L \rightleftarrows (\beta Ti) + (\alpha Y)$	80	~ 1	98	1355	Eutectic
$L_1 \rightleftarrows L_2 + (\beta Ti)$	~ 30	~ 80	~ 2	1370(b)	Monotectic
$(\beta Ti) \rightleftarrows (\alpha Ti) + (\alpha Y)$	~ 0.1	~ 0.1	~ 99.8	~ 870	Eutectoid
$L \rightleftarrows L_1 + L_2$		~ 56		~ 1545(b)	Critical point
$L \rightleftarrows (\beta Ti)$		0		1670	Melting point
$L \rightleftarrows (\beta Y)$		100		1522	Melting point
$(\beta Ti) \rightleftarrows (\alpha Ti)$		0		882	Allotropic transformation
$(\beta Y) \rightleftarrows (\alpha Y)$		100		1478	Allotropic transformation

(a) "Inverted peritectic." (b) Predicted by the present thermodynamic calculations.

Fig. 1---Assessed Ti-Y Phase Diagram

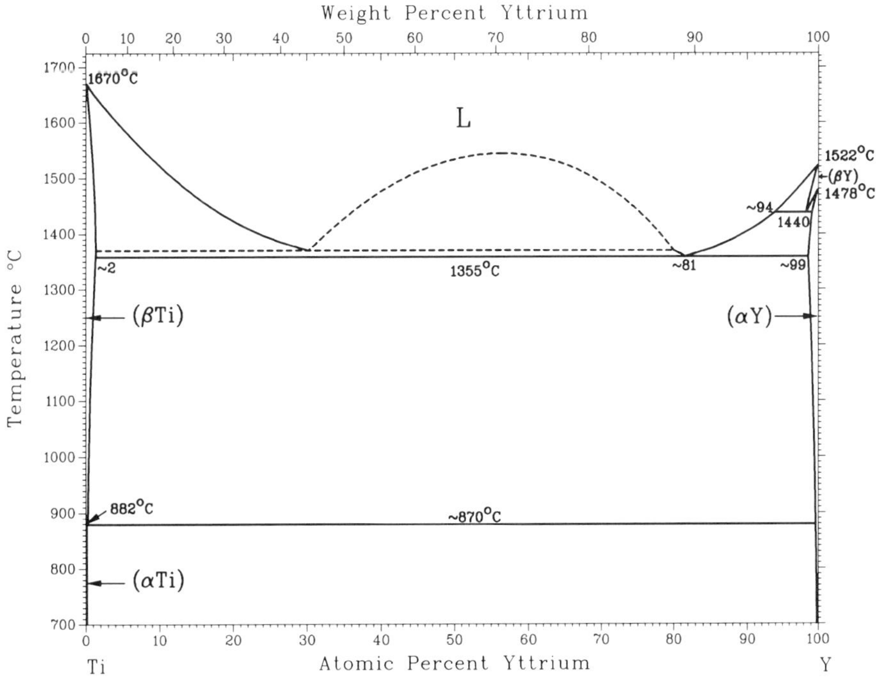

The diagram, which has been calculated, may differ slightly, but within the experimental uncertainty, from values listed in Table 1.   J.L. Murray, 1987.

333

and appear in a composition range for which data have not been reported in the literature. The detailed shape of the miscibility gap is not proposed as a quantitative description of the system, but it is thought that because the calculations provide sufficient evidence for some unmixing in the liquid state, the miscibility gap should be shown on the assessed diagram.

Table 2   Experimental Data on the Ti-Y Diagram

Reference	Composition, at.% Y			Temperature, °C
	(βTi)	L	(αY)	
L ⇄ (βTi) + (αY)				
[60Lov]	...	84.5	...	1420
[61Bar]	~ 0.5	78.3	98.2	1330
[62Lun]	0.5	79.8	98.7	1385
[62Sav]	~ 1.1	...	...	1400
[68Bea]	...	81.3	> 98	1355

Reference	Temperature, °C
(βY) ⇄ L + (αY)	
[61Bar]	1480
[62Lun]	1490
[68Bea]	1440

Reference	Composition, at.% Y	Temperature, °C
	(βTi)	
(βTi) ⇄ (αTi) + (αY)		
[60Lov]	...	800
[61Bar]	...	880
[62Lun]	0.2	870
[62Sav]	1.1	890

[60Lov] determined invariant temperatures by differential thermal analysis and estimated liquidus temperatures by optical pyrometry. The results are quite scattered. No estimates of the limited solid solubilities could be made based on X-ray diffraction and metallographic work because all alloys examined were two phase.

[61Bar] determined invariant melting temperatures by the incipient melting technique, liquidus compositions by chemical analysis of the equilibrated liquid, solid solubilities by microscopy, and the effect of Y on the (βTi) → (αTi) transition by thermal analysis. Both the Ti and the Y used to make alloys were rich in interstitial impurities, particularly oxygen.

[62Lun] produced microscopic evidence that the (βY) → (αY) reaction with the liquid is of the catatectic ("inverted peritectic") type. The solubilities of Y in (αTi) and (βTi) were found to be about 0.1 and 0.2 at.% at the eutectoid temperature (870 °C) by metallography. The Y used by [62Lun] contained 1750 ppm (by wt.?) of O_2 and 1000 ppm (wt.?) Zr, and the melting temperature of Y was reported as 1552 °C, compared to the accepted value of 1522 °C [86Gsc].

[68Bea] verified the catatectic reaction and determined the Y-rich part of the diagram by thermal analysis. [68Bea] used distilled Y that was very pure with respect to nonmetallic impurities. [62Sav] investigated the Ti-rich part of the diagram, but did not describe experimental details.

Numerical data on the invariant temperatures and solubilities are compared in Table 2 and liquidus data [61Bar, 68Bea] are compared in Fig. 2. As input data for thermodynamic optimizations, the three-phase equilibria, liquidus and solidus data from [68Bea], liquidus data from [61Bar], and low-temperature solubility data from [62Lun] were used.

Fig. 2---Y-Rich Region of the Ti-Y Phase Diagram, with Select Experimental Data

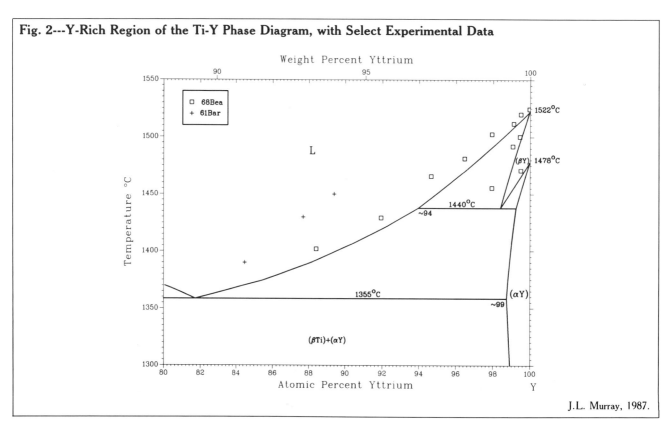

J.L. Murray, 1987.

Table 3 Crystal Structures of the Ti-Y System

Phase	Homogeneity range, at.% Y		Pearson symbol	Space group	Struktur- bericht designation	Proto- type	Lattice parameters(a), nm		Reference
							a	c	
(αTi)	0 to	~ 0.01	hP2	$P6_3/mmc$	A3	Mg	0.29511	0.46843	[Pearson2]
(βTi)	0 to	~ 1	cI2	$Im3m$	A2	W	0.33065	...	[Pearson2]
(βY) .... ~ 99 to		100	cI2	$Im3m$	A2	W	0.407	...	[86Gsc]
(αY) .... ~ 99 to		100	hP2	$P6_3/mmc$	A3	Mg	0.36482	0.57318	[86Gsc]

(a) Lattice parameters are for the pure elements.

Metastable Phases

[76Wan] used TEM to examine splat-cooled Ti-Y alloys for evidence of metastable extension of the solid solubility of Ti in Y. The observed maximum solubility was 4 to 5 at.% Ti, not significantly larger than the present estimate of the equilibrium solubility of Ti in (βY).

Crystal Structures

Crystal structure data for Ti-Y are listed in Table 3. Data on the lattice parameters as a function of composition are not available.

Thermodynamics

Although experimental thermochemical data are not available for the Ti-Y system, approximate Gibbs energies can be constructed from the phase diagram data. The accuracy of Gibbs energies, however, is limited by the accuracy of that input data. In the present calculation, a number of polynomial expansions of the Gibbs energies and a number of optimization strategies were examined in order to separate artifacts of the fitting procedure from those features of the system which are model independent. In the present calculations, Gibbs energies of the solution phases are represented as:

$$G(i) = (1 - x) G^0(Ti,i) + x G^0(Y,i) + RT(x \ln x + (1 - x) \ln (1 - x)] + x(1 - x) [B(i) + C(i) (1 - 2x)]$$

where i designates the phase; x is the atomic fraction of Y; G^0 is the Gibbs energies of the pure metals; and $B(i)$ and $C(i)$ are the parameters describing the excess Gibbs energies. The pure component properties are based on data in [Hultgren,E].

Preliminary optimizations were based on the regular solution model and the interaction parameters of all three phases were simultaneously varied. An equilibrium miscibility gap appeared in the liquid phase. The invariant temperatures were reproduced, but the solubility of Ti in (αY) and (βY) was too low and all the phase boundaries of the Y side of the diagram were displaced to the Y side of the data. A subregular solution model is therefore required. The resulting Gibbs energies with which the assessed diagram was calculated are given in Table 4. Excess entropies cannot be determined from the available data, but do not appear to be necessary to model the phase equilibria.

Table 4 Estimated Gibbs Energies of the Ti-Y System

Gibbs energies of the pure components

$G^0(Ti,L)$ = 0
$G^0(Y,L)$ = 0

$G^0(Ti,bcc)$ = $- 16\,234 + 8.368\ T$
$G^0(Y,bcc)$ = $- 11\,397 + 6.349\ T$

$G^0(Ti,cph)$ = $- 20\,585 + 12.134\ T$
$G^0(Y,cph)$ = $- 16\,388 + 9.200\ T$

Excess Gibbs energies

$B(L)$ = 29 688	$B(bcc)$ = 55 666	$B(cph)$ = 73 302
$C(L)$ = $- 2\,708$	$C(bcc)$ = 5 370	$C(cph)$ = 14 931

Note: Values are given in J/mol and J/mol · K.

A further series of calculations was performed assuming various values of the miscibility gap critical temperature, to examine whether the monotectic reaction is an artifact of the model used. With decreasing excess terms, the miscibility gap sinks below the equilibrium liquidus, and the eutectic composition rapidly approaches Ti-rich concentrations. It is concluded that at the eutectic temperature there is substantial unmixing in the metastable liquid. For this reason, the calculated monotectic reaction is shown in Fig 1.

Cited References

*61Bar: D.W. Bare and O.N. Carlson, "Phase Equilibria and Properties of Yttrium-Titanium Alloys," *Trans. ASM, 53*, 1-11 (1961). (Equi Diagram; Experimental)

60Lov: B. Love, "The Metallurgy of Yttrium and the Rare- Earth Metals, Part I, Phase Relationships," WADD Tech. Report 60-74, Part I (1960).

*62Lun: C.E. Lundin and D.T. Klodt, "Phase Equilibria of the Group IVA Metals with Yttrium," *Trans. AIME, 224*, 367-372 (1962). (Equi Diagram; Experimental)

*62Sav: E.M. Savitskii and G.S. Burkhanov, "Phase Diagrams in Titanium with Rare Earth Metals," *J. Less-Common Met., 4*, 301-314 (1962). (Equi Diagram; Experimental)

*68Bea: B.J. Beaudry, "The Effect of Titanium on the Melting Point and Transition Temperature of Yttrium," *J. Less-Common Met., 14*, 370-372 (1968). (Equi Diagram; Experimental)

76Wan: R. Wang, "Solubility and Stability of Liquid- Quenched Metastable H.C.P. Solid Solutions," *Mater. Sci. Engr., 7*(3), 135-140 (1976). (Meta Phases; Experimental)

86Gsc: K.A. Gschneidner, Jr. and F.W. Calderwood, in *Handbook of the Physics and Chemistry of Rare Earths*, Vol. 8, K.A. Gschneidner, Jr. and L. Eyring, Ed., North-Holland Physics Publishing, Amsterdam (1986). (Crys Structure; Review)

The Ti-Zn (Titanium-Zinc) System
47.88   65.38

By J.L. Murray

Equilibrium Diagram

The difference between the melting point of Ti (1670 °C [Melt]) and the boiling point of Zn (907 °C [Hultgren,E]) prevents determination of a standard metallurgical phase diagram, except for Zn-rich compositions. The extreme Zn-rich part of the diagram (the eutectic reaction) has been studied extensively, because additions of Ti to Zn act as a grain refiner [75Leo, 76Leo] and promote creep resistance in rolled alloys. Beyond 0.7 at.% Ti, the liquidus has been examined between 418 and 1000 °C, and a series of peritectic reactions occur. In addition, several intermetallic compounds have been found whose phase relations are unknown. The effect of Zn on the (βTi) to (αTi) transformation is entirely unknown. The existence of several parallel crystal structures suggests that analogy with the better known systems Fe-Zn and Co-Zn may be a useful guide to the alloy chemistry of Ti-Zn.

The assessed diagram (Fig. 1) is based on thermal analysis data and the observed compound phases; the phase relations are a possible construction based on incomplete information. Data on the special points of the diagram are summarized in Table 1. Figures 2 and 3 show details of the eutectic region and liquidus, respectively.

Eutectic Reaction.

The solubility of Ti in (Zn) is very low. [44And] estimated it to be 0.01 to 0.02 at.% Ti at 300 °C by

metallographic examination. [62Hei] estimated it to be less than 0.0006 at.% Ti by the same technique. The lower boundary is used in Fig. 1.

The eutectic reaction L $\rightleftarrows$ (Zn) + $Zn_{15}Ti$ occurs at 418.5 ± 0.1 °C. Data on the L/(Zn) phase boundaries are summarized in Table 2. Reported eutectic compositions are between 0.16 and 0.63 at.% Ti [41Geb, 44And, 62Hei, 61Pel, 66Ren, 72Spi]. The reason for discrepancies in reported eutectic compositions is that this eutectic is of the faceted-nonfaceted type [76Leo]. The coupled eutectic zone—the composition and temperature range in which eutectic microstructures can appear—extends to the Ti-rich side of the equilibrium eutectic composition [72Spi, 73Got, 75Leo, 76Leo].

A value 0.21 ± 0.01 at.% Ti for the eutectic composition is preferred, based on the following considerations:

- Van't Hoff's law gives 0.18 at.% Ti as an estimate for the eutectic composition, if zero solubility of Ti in (Zn) is assumed. Because the solubility is actually finite, the Ti content of the eutectic point may be greater than, but is probably not less than, the value 0.18 at.% Ti. This suggests that the lowest value 0.16 at.% Ti [44And] probably lies to the Zn-side of the eutectic.

The eutectic composition proposed by [62Hei], [62Pel], and [66Ren] were based on observations of eutectic microstructures, or on extrapolation of the liquidus to

Fig. 1---Provisionally Assessed Zn-Ti Phase Diagram

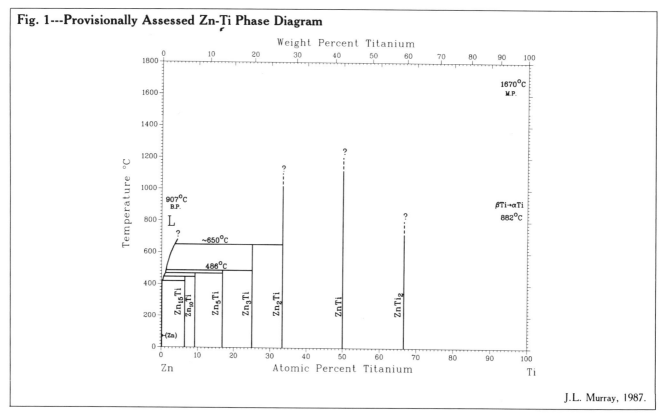

J.L. Murray, 1987.

Table 1 Special Points of the Assessed Zn-Ti System

Reaction	Compositions of the respective phases, at.% Ti			Temperature, °C	Reaction type
L $\rightleftarrows$ (Zn) + Zn$_{15}$Ti	0.21	~ 0.0006	6.3 ± 0.4	418.6 ± 0.1	Eutectic
L + Zn$_{10}$Ti(?) $\rightleftarrows$ Zn$_{15}$Ti	~ 0.45	9.1	6.3	445	Peritectic
L + Zn$_5$Ti(?) $\rightleftarrows$ Zn$_{10}$Ti(?)	~ 1	16.7	9.1	468	Peritectic
L + Zn$_3$Ti(?) $\rightleftarrows$ Zn$_5$Ti(?)	~ 1	25	16.7	486	Peritectic
L + Zn$_2$Ti(?) $\rightleftarrows$ Zn$_3$Ti(?)	~ 3.5	33.3	25	650	Peritectic
L $\rightleftarrows$ (Zn)		0		419.58	Melting point
G $\rightleftarrows$ L		0		907	Boiling point
L $\rightleftarrows$ (βTi)		100		1670	Melting point
(βTi) $\rightleftarrows$ (αTi)		100		882	Allotropic transformation

Fig. 2---Detail of the Eutectic Reaction

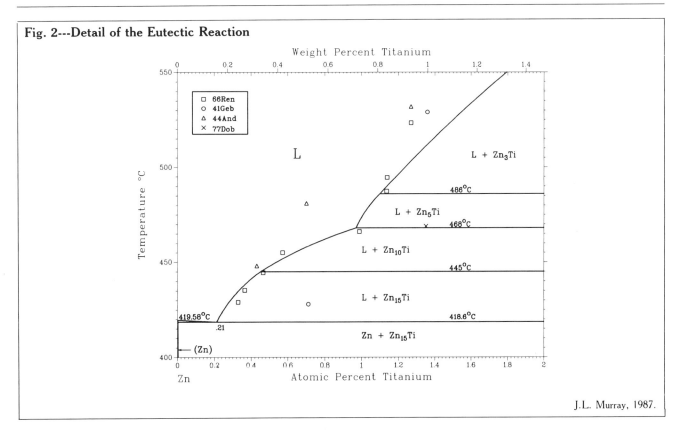

J.L. Murray, 1987.

418.5 °C. Both methods tend to overestimate the eutectic composition, the former because the coupled eutectic zone is skewed to the Ti-rich side and the latter because of undercooling. [72Spi] observed primary Zn$_{15}$Ti in a slowly cooled 0.22 at.% Ti sample, and the eutectic composition must therefore be less than or equal to 0.22 at.% Ti.

Intermetallic Compounds and Their Structures.

The compounds that have been proposed as equilibrium phases of the Zn-Ti system are Zn$_{15}$Ti, Zn$_{10}$Ti, Zn$_5$Ti, Zn$_3$Ti, Zn$_2$Ti, ZnTi, and ZnTi$_2$. All investigators have agreed that Zn$_{15}$Ti is the most Zn-rich compound. Its composition was determined by [44And], who also found that it decomposes by a peritectic reaction. [63Pio] compared the structure to that of FeZn$_{13}$ and CoZn$_{13}$. [81Sai] found, by a single-crystal X-ray study, that the structure of Zn$_{15}$Ti is a variant of the FeZn$_{13}$ structure. The details of the structure are best described by the stoichiometry Zn$_{16}$Ti.

[54Pie] prepared Zn$_2$Ti and Zn$_3$Ti and determined their structures as $L1_2$ and (possibly) $C14$, respectively. The existence of Zn$_2$Ti and Zn$_3$Ti and the structure of Zn$_3$Ti were verified by [62Hei] and [63Pio]. [62Hei] did not verify that Zn$_2$Ti has the $C14$ structure, but reported that it has a hexagonal symmetry.

Between the Zn$_{15}$Ti and Zn$_3$Ti, at least one additional phase occurs; it has been identified as Zn$_{10}$Ti [44And, 63Pio], but also as Zn$_5$Ti [62Hei]. [62Hei] proposed the stoichiometry Zn$_5$Ti, based on the observation of two phases—Zn$_{15}$Ti and Zn$_5$Ti—in an 11.9 at.% Ti alloy. [63Pio] proposed the stoichiometry Zn$_{10}$Ti, based on chemical analysis of phases present in diffusion couples and X-ray analysis. [63Pio] related the structure of Zn$_{10}$Ti to that of corresponding δ_1 phases in Fe-Zn and Co-Zn. δ_1 had a distorted γ-brass structure and was accompanied by another phase with a γ-brass-like structure. By analogy with the Fe-Zn and Co-Zn systems, it is probable that more than one structural variant of γ-brass appears in

337

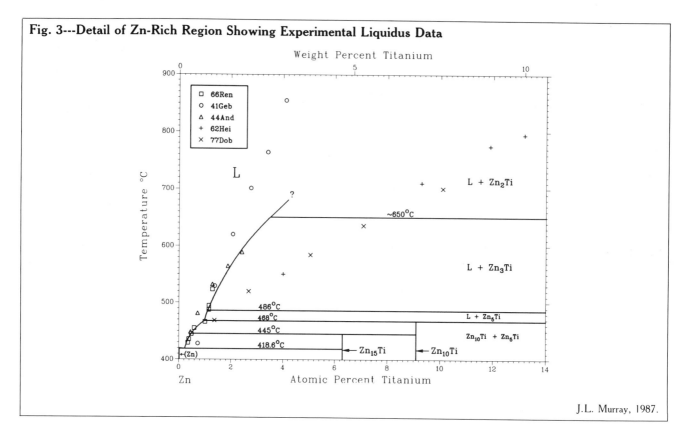

Fig. 3---Detail of Zn-Rich Region Showing Experimental Liquidus Data

J.L. Murray, 1987.

Table 2 Experimental Data on the Eutectic Reaction

Reference	Solubility of Ti in (Zn), at.% Ti	Eutectic Composition, at.% Ti	Eutectic Temperature, °C	Experimental technique
[41Geb]	< 0.027	0.61	418	Thermal analysis, metallography
[44And]	0.015	0.16	418.5	...
[62Hei]	< 0.0006	0.31	...	Chemical analysis of eutectic in zone-melted material
[62Pel]	...	0.25	...	Eutectic microstructure
[66Ren]	...	0.26	...	Thermal analysis, metallography
[72Spi]	...	0.22	...	Appearance of primary $Zn_{15}Ti$

Table 3 Experimental Data on the Peritectic Reactions

Reference	Reaction temperature, °C	Reaction attributed to arrest
[62Hei]	445	$L + Zn_5Ti \rightarrow Zn_{15}Ti$
[66Ren]	446	$L + ZnTi_? \rightarrow Zn_{15}Ti$
[41Geb]	460	$L + ZnTi_? \rightarrow Zn_{15}Ti$
[66Ren]	468	...
[41Geb]	490	...
[44And]	485	$L + Zn_{10}Ti \rightarrow Zn_{15}Ti$
[66Ren]	486	

Zn-Ti and that discrepancies in stoichiometry will be resolved by determination of the composition ranges of the phases. Both $Zn_{10}Ti$ and Zn_5Ti are tentatively included in Fig. 1.

[62Hei] verified the CsCl structure of ZnTi, originally found by [39Lav]. [63Sch] found a compound $ZnTi_2$ with the $MoSi_2$ structure. Structure and lattice parameter data are listed in Table 3.

Peritectic Reactions and the Liquidus.

Above the eutectic temperature, at least three peritectic reactions occur in the composition range 0 to 20 at.% Ti. Thermal analysis data are summarized in Table 4, as are the reactions to which each author attributed the peritectic arrests. It can be seen that each of the three reactions observed by [66Ren] at 446, 468, and 486 °C has also been observed by an independent investigation, although the interpretations differed greatly. The thermal analysis work of [66Ren] is therefore used to construct the peritectic reactions.

Liquidus data (primarily from thermal analysis experiments) fall into two groups: data of [41Geb], [44And], and [66Ren] rise steeply, reaching 600 °C at about 2.7 at.% Ti. Data of [62Hei] and [77Dob] lie considerably lower in temperature (see Fig. 3). The data of [62Hei] and [77Dob] are uncertain by at least ±0.5 at.%, having been digitized from published figures, but it is clear that they are not consistent with the first set. The present evaluator has

Table 4 Crystal Structures and Lattice Parameters of the Zn-Ti System

Phase	Homogeneity range, at.% Ti	Pearson symbol	Space group	Struktur-bericht designation	Prototype	Lattice parameters, nm			Reference
						a	*b*	*c*	
(Zn)	0 to 0.0006	hP2	$P6_3/mmc$	A3	Mg	0.26649	...	0.49468	[Pearson2]
$Zn_{15}Ti$	6.3	oC68	Cmcm	...	$Zn_{15}Ti$	0.77207	1.14497	1.17559	[81Sai]
$Zn_{10}Ti$	9.1	(a)	...	...	...	...	...	...	[63Pio]
Zn_5Ti	16.7	(b)	...	...	...	...	...	...	...
Zn_3Ti	25	cP4	Pm3m	$L1_2$	$AuCu_3$	0.39322	...	...	[54Pie]
						0.3023(c)	...	...	[39Lav]
						0.39324	...	...	[63Pio]
						0.3958	...	...	[62Hei]
Zn_2Ti(d)	33.3	hP12	$P6_3/mmc$	C14	$MgZn_2$	0.5064	...	0.8210	[54Pie]
ZnTi	50	cP2	Pm3m	B2	CsCl	0.3146	...	...	[62Hei]
$ZnTi_2$	66.7	tI6	I4/mmm	$C11_b$	$MoSi_2$	0.3036	...	1.06776	[63Sch]
(αTi)	? to 100	hP2	$P6_3/mmc$	A3	Mg	0.29512	...	0.46826	[Ti]
(βTi)	? to 100	cI2	Im3m	A2	W	0.33065(e)	...	...	[Pearson2]

(a) Distorted γ-brass. (b) Unknown. (c) Impure starting materials were used. (d) C14 was suggested as a possible structure by [54Pie], but rejected by [62Hei]. (e) At 900 °C.

provisionally chosen the liquidus data of [44And] and [66Ren] for this assessment, based on reinterpretation of a two-phase microstructure described in [62Hei]. The discrepancy, however, is not satisfactorily explained.

[62Hei] located the peritectic decomposition of Zn_5Ti between 600 and 700 °C by annealing and quenching experiments. Similarly, [44And] estimated a temperature slightly above 600 °C as the peritectic decomposition temperature of $Zn_{10}Ti$. [62Hei] proposed that Zn_3Ti melts congruently at about 900 °C based on the occurrence of eutectic in as-cast microstructures.

The provisional diagram was constructed from these rather disparate observations, based on the following considerations. The highest peritectic temperature was placed at 650 °C, and the liquidus was extrapolated through the data of [44And] to that temperature. Reactions were then assigned to all the observed peritectic arrests, using Zn_5Ti and $Zn_{10}Ti$ as distinct phases. The first compound that could melt congruently is, thus, Zn_2Ti. The compound phases were drawn as line compounds; it is possible that Zn_5Ti, $Zn_{10}Ti$, and ZnTi will be found to have substantial homogeneity ranges.

Cited References

39Lav: F. Laves and H.J. Wallbaum, "Crystal Chemistry of Titanium Alloys," *Naturwissenschaften, 27,* 674-675 (1939) in German. (Crys Structure; Experimental)

41Geb: E. Gebhardt, "Alloy Systems of Zinc with Titanium and Zirconium," *Z. Metallkd., 33,* 355-357 (1941) in German. (Equi Diagram; Experimental)

*44And: E.A. Anderson, E.J. Boyle, and P.W. Ramsey, "Rolled Zinc-Titanium Alloys," *Trans. AIME, 156,* 278-286 (1944). (Equi Diagram; Experimental)

54Pie: P. Pietrokowsky, "A Cursory Investigation of Intermediate Phases in the Systems Ti-Zn, Ti-Hg, Zr-Zn, Zr-Cd, and Zr-Hg by X-Ray Powder Diffraction Methods," *Trans. AIME, 200,* 219-226 (1954). (Equi Diagram; Experimental)

61Pel: M.E. Pelzel, "Structure of Zinc-Copper-Titanium Alloys," *Metall., 15,* 881-883 (1961) in German. (Equi Diagram; Experimental)

62Hei: W. Heine and U. Zwicker, "Alloy System Zinc-Titanium," *Z. Metallkd., 53,* 380-385 (1962) in German. (Equi Diagram; Experimental)

63Pio: W. Piotrowski, "Phases of the Zinc-Titanium System," *Z. Nauk. Politech. Lodz. Mech.,* (10), 33-41 (1963). (Equi Diagram; Experimental)

63Sch: K. Schubert, K. Frank, R. Gohle, A. Maldonado, H.G. Meissner, A. Raman, and W. Rossteutscher, "Structure Data on Metallic Phases (8)," *Naturwissenschaften, 50,* 41 (1963) in German. (Crys Structure; Experimental)

64Wen: Z. Wendorff and W. Piotrowsk, "AuCu₃-Type Phases in Zn Systems," *Hutnik, 31,* 246-249 (1964) in Polish. (Crys Structure; Experimental)

*66Ren: E.H. Rennhack, "Redetermined Zinc-Rich Portion of the Zn-Ti System," *Trans. AIME, 236,* 941-942 (1966). (Equi Diagram; Experimental)

72Spi: J.A. Spittle, "The Effects of Composition and Cooling Rate on the As-Cast Microstructures of Zn-Ti Alloys," *Metallography, 5,* 423-447 (1972).

73Got: S. Goto, K. Esashi, S. Koda, and S. Morozumi, "Structure-Controlling of Zn-Ti Hyper-Eutectic Alloys by Unidirectional Solidification," *J. Jpn. Inst. Met., 37*(4), 466-473 (1973) in Japanese. (Equi Diagram; Experimental)

75Leo: G.L. Leone, P. Niessen, and H.W. Kerr, "The Mechanism of Grain Refinement During Solidification of Zn-Ti Base Alloys," *Metall. Trans. B, 6,* 503-511 (1975). (Equi Diagram; Experimental)

76Leo: G.L. Leone and H.W. Kerr, "Grain Structures and Coupled Growth in Zn-Ti Alloys," *J. Crystal Growth, 32,* 111-116 (1976). (Equi Diagram; Experimental)

77Dob: R. Dobrev, V. Dimova, and I. Georgiev, "Study of Liquid Zinc Alloys with Some Refractory Elements," *Materialozn. Teknol., 5,* 40-44 (1977) in Russian. (Crys Structure; Experimental)

81Sai: M. Saillard, G. Develey, and C. Becle, "The Structure of $TiZn_{16}$," *Acta Crystallogr. B, 37,* 224-226 (1981). (Crys Structure; Experimental)

The Ti-Zr (Titanium-Zirconium) System

47.88   91.22

By J.L. Murray

Equilibrium Diagram

In both Ti and Zr, there is an allotropic transformation from the high-temperature bcc β form to the low-temperature cph α form. The solid solution phases (βTi,βZr) and (αTi,αZr) are referred to as β and α, respectively, in the present assessement. The phase diagram is characterized by two congruent transformations, L ⇄ β and β ⇄ α. The assessed Ti-Zr diagram is shown in Fig. 1, and its special points are summarized in Table 1. The assessed diagram is the result of thermodynamic calculations that reproduce the experimental data within the estimated uncertainty.

Recent work [82Auf] has uncovered several inconsistencies in the details of the phase boundaries. The congruent β ⇄ α transition was previously placed at 545 °C [66Far, 71Cha]. However, preliminary dilatometric studies [77Etc] indicated that for a 50 at.% alloy the β ⇄ α transition occurred near 600 °C. Thermodynamic data tend to support the diagram of [82Auf]. In this assessment, the positions of the α/β phase boundaries are consistent with the observations of [77Etc] and [82Auf].

Solidus and Liquidus.

Incipient melting determinations were made by [39Fas], [51Hay], and [69Rud]. The experimental data are compared to the assessed diagram in Fig. 2. The melting points reported by [39Fas] and [51Hay] are high compared to those found by [69Rud], and their value for the melting point of Ti (1720 °C) disagrees with the accepted value

Table 1   Special Points of the Assessed Ti-Zr Phase Diagram

Reaction	Composition, at.% Zr	Temperature, °C	Reaction type
L ⇄ (βTi,βZr)	38 ± 2	1540 ± 15	Congruent
(βTi,βZr) ⇄ (αTi,αZr)	52 ± 2	605 ± 10	Congruent
L ⇄ (βTi)	0	1670	Melting point
(βTi) ⇄ (αTi)	0	882	Allotropic transformation
L ⇄ (βZr)	100	1855	Melting point
(βZr) ⇄ (αZr)	100	863	Allotropic transformation

Fig. 1---Assessed Ti-Zr Phase Diagram

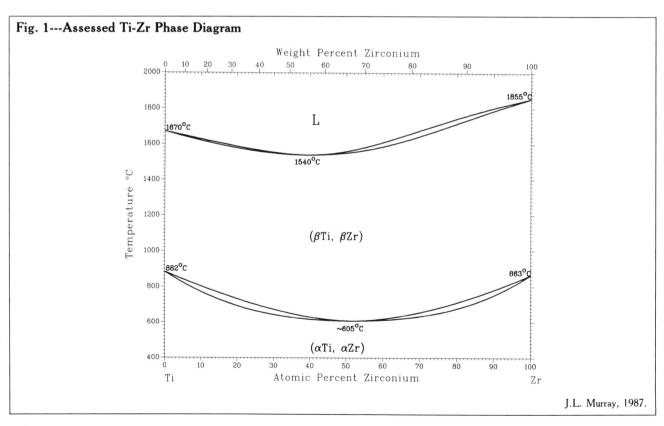

J.L. Murray, 1987.

Ti-Zr System

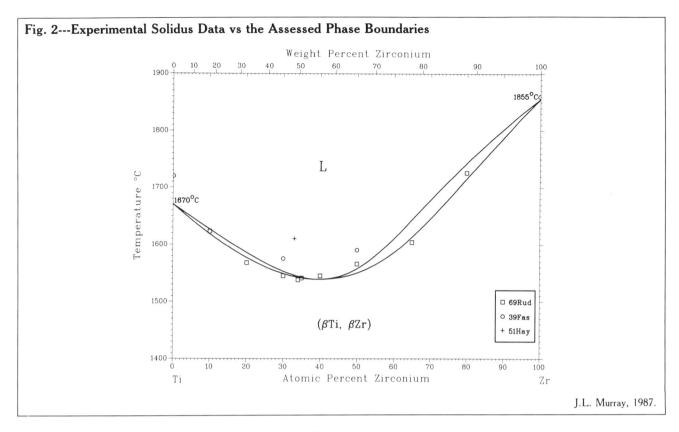

Fig. 2---Experimental Solidus Data vs the Assessed Phase Boundaries

J.L. Murray, 1987.

Table 2 Experimental Determinations of the β/α Phase Boundaries

Reference	Congruent temperature, °C	Experimental method
[82Auf]	608 ± 10	Microcalorimetry (β → α, 599 °C)
		(α → β, 616 °C)
[77Etc]	600 ± 15	Dilatometry (β → α, 586 °C)
		(α → β, 621 °C)
[71Cha]	~ 530	Hydrogen pressure method
		(minimum by interpolation)
[66Far]	~ 535	Metallography (minimum by interpolation)
[66Kup]	520	Thermal analysis, metallography
[61Enc]	None	Metallography, X-ray diffraction
[62Img]	...	Metallography, Ti-rich alloys only
[51Hay]	~ 535	Resistivity (cooling), ~ 50 °C hysteresis
[39Fas]	~ 545	Resistivity (cooling)

(1670 °C [Melt]); this suggests oxygen contamination of the alloys. The solidus data of [69Rud] were used as input for thermodynamic calculations, and his value for the congruent melting temperature was also accepted. Note that the melting point of pure Zr as determined by [69Rud] was 1876 °C, compared to the presently accepted value of 1855 °C [Melt].

α/β Boundaries.

Experimental determinations of the α/β phase boundaries are summarized in Table 2, and the data are compared with the assessed diagram in Fig. 3. The results fall into three categories:

- [61Enc] reported that (βTi) is strongly stabilized by addition of Zr, and that no congruent point is observable above 500 °C. [66Far], however, argued plausibly that the observations of [61Enc] were the effect of failure to remove the original transforma-

tion microstructure during heat treatment in the α field.

- [39Fas], [51Hay], [66Far], [66Kup], and [71Cha] reported that the congruent β ⇌ α transition occurs at about 545 °C. The transition temperature was determined by a variety of experimental techniques, all of which posed experimental difficulties in the range 40 to 70 at.% Zr.
- [77Etc] and [82Auf] reported that the congruent β ⇌ α transition occurs at about 605 ± 15 °C. The measurements pertain to the martensitic transformation at cooling and heating rates of 300 °C/h, with hysteresis of about 35 °C.

There are experimental difficulties in measuring the β/α boundaries in the range 40 to 70 at.% Zr by classical techniques. Because of the retention of the original transformation microstructure in alloys heat treated in the α

341

Fig. 3---Experimental Data on the α/β Boundaries vs the Assessed Phase Boundaries

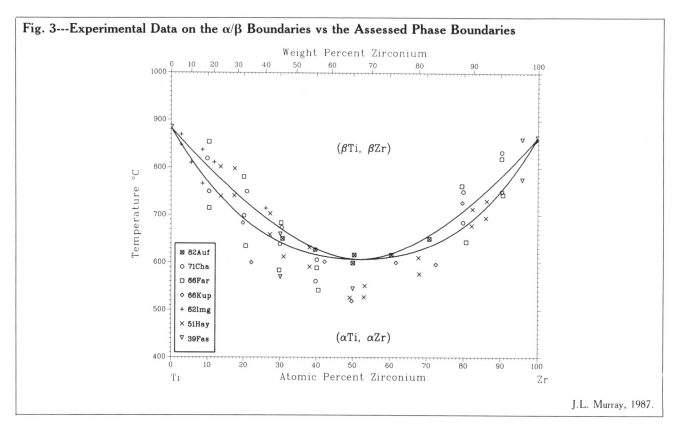

J.L. Murray, 1987.

field, metallographic results can be misleading, except in the composition range where two-phase structures can be obtained. Resistivity measurements on cooling and heating showed large (50 °C) hysteresis, and only cooling curves were used for the reported phase boundaries [39Fas, 51Hay], thus the equilibrium transition temperature was probably underestimated.

For the congruent transformation temperature, the dilatometric and microcalorimetric work of [77Etc] and [82Auf] is used for three reasons. First, three alloys of different purities were used (the highest purity alloys were made from Ti and Zr of less than 100 ppm oxygen), and precautions were taken to avoid contamination during the long experiments. Second, transition points above 600 °C have been tentatively verified in this laboratory by DSC [84Shu]. Third, the microcalorimetric work yielded heats of transformation as well as transition temperatures. The measured heat of transformation is somewhat more consistent with the higher congruent temperature.

The discrepancy between [77Etc] and [82Auf] and earlier work occurs only in the range 35 to 65 at.% Zr. Outside this range, however, there are some discrepancies concerning the width of the two-phase field [66Far, 66Kup, 71Cha]. The effect of oxygen and nitrogen contamination is to increase the width of the observed two-phase field [82Auf] and to raise the transition temperatures of the pure metals. The width of the two-phase field in the binary system is therefore less than that generally observed. In the present thermodynamic calculations, the phase boundaries were required to lie inside the experimentally observed two-phase field.

Ordered Low-Temperature Phase.

[66Kup] found a hardness minimum in the range 30 to 65 at.%, which they interpreted as evidence for an ordered phase in this composition range. They supported

this interpretation with observations of superlattice lines in X-ray diffraction patterns. There are no data on either the structure of an ordered phase, or on its equilibrium boundaries. There would probably be great experimental difficulties in demonstrating the existence of an equilibrium ordered phase in this system, because of the difficulty of attaining equilibrium at the appropriate temperatures.

Metastable Phases

Martensitic Transformation.

β can transform martensitically to the cph phase (α') across the entire composition range [70Hua]. In other Ti systems (e.g., Ti-Ta and Ti-V), an orthorhombic distortion of the hexagonal phase can also be formed martensitically; this form has not been observed in the Ti-Zr system [66Bro].

Martensite transformation temperatures were determined by [52Duw], [60Gri], [70Hau], [77Etc], and [82Auf]. The work of [77Etc] and [82Auf] at a cooling rate of 300 °C/h were discussed above in connection with the equilibrium diagram. The work of [52Duw], [60Gri], and [70Hau] was done at higher cooling rates and gave lower transformation temperatures with a larger difference between the transformations on heating and cooling.

[70Hua] measured the start temperatures of the β → α transformation on cooling and on heating (M_s and β_s, respectively) over the entire composition range. At the congruent composition, M_s and β_s were found to be 475 and 645 °C, respectively. Both [60Gri] and [52Duw] found the minimum M_s temperature to be about 485 °C.

The internal structure of martensite plates in 9 to 32.3 at.% Zr alloys was studied by [73Ban]. No retained β was found in any of the alloys studied. The 9 at.% Zr alloy exhibited a lath martensite structure with few twins; the other alloys formed plates and were profusely twinned.

ω Phase.

The metastable ω phase has been reported to occur in Ti-Zr alloys as an intermediate phase in the decomposition of β → α [57Hat, 60Gri]. It appears, however, that the precondition for ω phase formation (the retention of metastable β during quenching) is not fulfilled in sufficiently pure specimens [59Hat, 69Sas, 70Hua].

Retention of some β phase to room temperature has been reported in the following composition ranges:

Reference	Composition, at.% Zr	Comment
[51Hay]	18.4 to 44.1	Quenched from >800 °C
[52Duw]	20 to 80	...
[60Gri]	40 to 80	Found with ω

Based on thermal analysis, [60Gri] reported that the temperature of the ω → β transformation is about 450 °C. [69Sas] made a dark field electron microscopy study of ω phase in a 75 at.% Zr sample that had been deliberately contaminated with oxygen. The ω phase formed during quenching from the β region. In as-quenched alloys, ω appeared in the form of thin rods, and rods that were composed of equiaxed particles aligned in the $<111>_\beta$ direction.

Crystal Structures and Lattice Parameters

Data on the crystal structures of equilibrium and metastable phases of this system are summarized in Table 3. Room-temperature lattice parameters for (αTi,αZr) are shown as a function of composition in Fig. 4. Because the bcc phase can only be retained to room temperature in alloys contaminated by gaseous impurities and because martensitic α′ or ω phase may coexist with retained β, bcc lattice parameters [51Hay, 52Duw] are not given.

The ideal ω structure is hexagonal with the symmetry P_6/mmm. In some alloys, the ω phase becomes distorted to a trigonal structure of symmetry $P\bar{3}m1$. Lattice parameters of the ω phase in a 50 at.% are $a = 0.3051$ nm and $c = 0.4859$ nm [57Hat].

Thermodynamics

Experimental Data.

[71Pey] measured the activity of Ti in two Zr-rich liquid alloys. The results are consistent with an ideal solution, but the experimental uncertainty does not per-

mit derivation, of either the sign or the magnitude, of the deviation from ideality.

[82Auf] measured the enthalpy of the β → α transformation in a 50 at.% alloy, finding −2520 J/mol on cooling and 2460 J/mol on heating.

Calculations of the Diagram.

Previous calculations of the diagram were made by [59Kau, 70Kau]. [59Kau] considered only the bcc/cph equilibria. [70Kau] calculated the complete diagram in the regular solution approximation. The present calculation of the diagram includes more recent data and addresses the experimental discrepancies. The present results are very similar to those of [59Kau] or [70Kau]; small changes in the Gibbs energies were made in order to reproduce the diagram within the experimental uncertainty. The results of the present and previous studies are compared in Table 4.

The Gibbs energies of the bcc, cph and liquid phases are represented as:

$$G(i) = (1 - x) G^0(Ti,i) + G^0(Zr,i) x + RT[x \ln x + (1 - x) \ln (1 - x)] + B(i) x(1 - x)$$

where i designates the phase, x is the atomic fraction of Zr, G^0 is the Gibbs energies of the pure components, and $B(i)$ are the temperature-dependent parameters describing the excess Gibbs energies. The approximation will be termed "regular solution" only when a temperature-independent $B(i)$ is used. The Gibbs energies of the pure components are derived from the heats of transformation and transformation temperatures [Hultgren,E].

Present Calculations.

Input data to the calculations are the solidus data [69Rud], cph/bcc phase boundary data in the vicinity of the congruent point [82Auf], and cph/bcc phase boundary data away from the congruent point [66Far, 71Cha]. For comparison, two calculations were performed. The first assumed that the congruent bcc ⇄ cph transition occurs at 545 °C; the second assumed that the transition occurs at 605 °C. In both calculations, the cph/bcc phase boundaries were assumed to be straddled by the experimental data (Fig. 3), except in the vicinity of the congruent point.

In the first calculation, it was found that excess entropy contributions were required to simultaneously satisfy the two requirements. In the second calculation, a regular solution approximation could be used to reproduce the diagram. According to the two calculations, the heats of the bcc ⇄ cph transformation are −1360 and −3197 J/mol, respectively. The somewhat better agreement of the latter value with the measured heat of transformation suggests that the second calculation more nearly approximates the thermodynamics of the system.

Table 3 Crystal Structures of the Ti-Zr System [Pearson2]

Phase	Homogeneity range, at.% Zr	Pearson symbol	Space group	Strukturbericht designation	Prototype
(βTi,αZr)	0 to 100	cI2	Im3m	A2	W
(αTi,αZr)	0 to 100	hP2	P6₃/mmc	A3	Mg
α′	(a)	hP2	P6₃/mmc	A3	Mg
ω	(a)	hP3	P6/mmm or P3̄m1	...	ωCrTi

(a) Metastable.

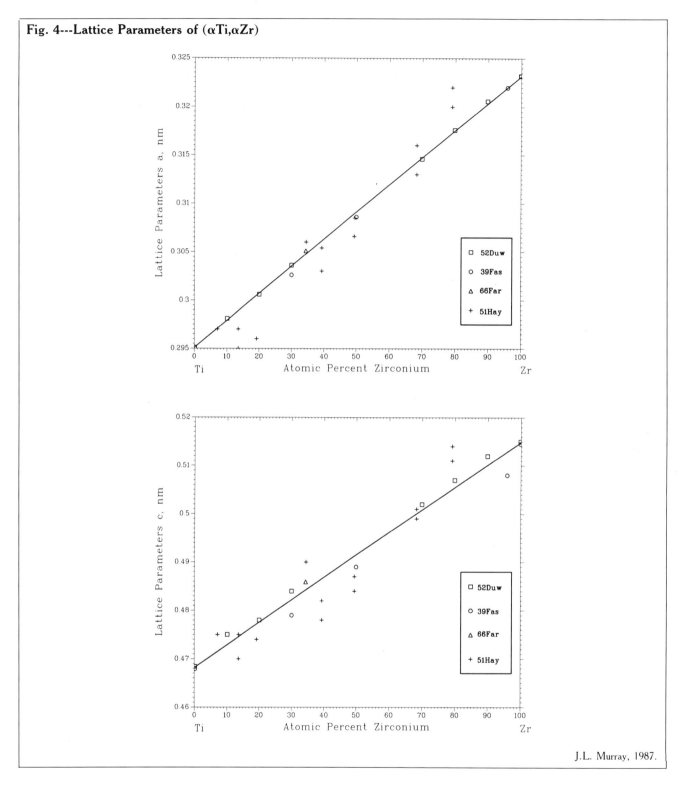

Fig. 4---Lattice Parameters of (αTi,αZr)

J.L. Murray, 1987.

Similar observations can be made regarding the liquidus and solidus boundaries; specifically, an excess entropy is required to match the incipient melting data. However, assuming a regular solution model, the calculated phase diagram differs very little from experiment, about 40 °C at the most Zr-rich compositions. To produce the assessed liquidus and solidus, the excess entropy term was included. For the purpose of extrapolation to ternary systems, the regular solution approximation may be more realistic, and the parameters of the regular solution Gibbs energy are therefore also tabulated in Table 4.

The calculated compositions of the L → bcc and bcc → cph congruent points are 39.5 and 51.9 at.% Zr, respectively. The latter differs slightly from the value usually assumed (50 at.%) but because the transformations experimentally must occur almost isothermally over a wide (10 at.%) range, there is no conflict between the calculations and the observations.

Table 4 Thermodynamic Properties of the Ti-Zr System

Properties of the pure components

$G^0(Ti,L) = 0$
$G^0(Zr,L) = 0$
$G^0(Ti,bcc) = -16\,234 + 8.368\,T$
$G^0(Zr,bcc) = -16\,895 + 7.939\,T$
$G^0(Ti,cph) = -20\,585 + 12.134\,T$
$G^0(Zr,cph) = -20\,868 + 11.438\,T$

Reference	B(cph)	B(bcc)	B(L)
Excess Gibbs energies			
[70Kau]	8749	3740	$-3\,138$
This evaluation (regular solution) ..	7600	3740	$-3\,100$
This evaluation (Fig. 1)	7600	3740	$-27\,620 - 13.4\,T$

Note: Values are given in J/mol; T in K.

Finally, it should be noted that, strictly speaking, only excess Gibbs energy differences are determined by the present calculations. The regular solution parameters of the phases may differ by several kJ/mol, as long as the differences are fixed.

Cited References

39Fas: J.D. Fast, "The Transition Point Diagram of the Zirconium-Titanium System," *Rec. Trav. Chim., 58*, 973-983 (1939). (Equi Diagram; Experimental)

51Hay: E.T. Hayes, A.H. Roberson, and O.G. Paasche, "Zirconium-Titanium System: Constitution Diagram and Properties," U.S. Bur. Mines Rept. Invest. No. 4826 (1951). (Equi Diagram; Experimental)

52Duw: P. Duwez, "Allotropic Transformation in Titanium-Zirconium Alloys," *J. Inst. Met., 80*, 525-527 (1952). (Meta Phases; Experimental)

57Hat: B.A. Hatt, J.A. Roberts, and G.I. Williams, "Occurrence of the Metastable ω Phase in Zirconium Alloys," *Nature, 180* (1957). (Meta Phases; Experimental)

59Hat: B.A. Hatt and G.I. Williams, "Some Observations on the Retention of the β Phase in Quenched Zr–50 at.% Ti Alloys," *Acta Metall., 7*, 682-706 (1959). (Meta Phases; Experimental)

59Kau: L. Kaufman, "The Lattice Stability of Metals---I. Titanium and Zirconium," *Acta Metall., 7*, 575-587 (1959). (Thermo; Theory)

60Gri: V.N. Gridnev, V.I. Trefilov, and V.N. Minakov, "Martensitic Transitions in the System Titanium-Zirconium," *Dokl. Akad. Nauk SSSR, 134*(6), 1334-1336 (1960) in Russian; TR:

Sov. Phys. Dokl., 5(6), 1094-1096 (1960). (Meta Phases; Experimental)

61Enc: E. Ence and H. Margolin, "Observations on the Ti-Zr System," *Trans. AIME, 221*, 205-206 (1961). (Equi Diagram; Experimental)

62Img: A.G. Imgram, D.N. Williams, and H.R. Ogden, "Tensile Properties of Binary Titanium-Zirconium and Titanium-Hafnium Alloys," *J. Less-Common Met., 4*, 217-225 (1962). (Equi Diagram; Experimental)

66Bro: A.R.G. Brown and K.S. Jepson, "Physical Metallurgy and Mechanical Properties of Titanium-Niobium Alloys," *Mem. Sci. Rev. Metall., 63*(6), 575-584 (1966). (Meta Phases; Experimental)

*66Far: P.A. Farrar and S. Adler, "On the System Titanium-Zirconium," *Trans. AIME, 236*, 1061-1064 (1966). (Equi Diagram; Experimental)

66Kup: V.V. Kuprina, V.B. Bernard, A.T. Grigor'ev, and E.M. Sokolovskaya, "Investigation of Solid State Transformations in Alloys of the Titanium-Zirconium," *Vest. Mosk. Univ. Khim., 21*(5), 69-73 (1966) in Russian; TR: *J. Moscow Univ. Chem., 21*(5), 396-400 (1966). (Equi Diagram; Experimental)

*69Rud: E. Rudy, "Compendium of Phase Diagram Data," Tech. Rep. AFML-TR-65-2, Part V, Wright Patterson Air Force Base (1969). (Equi Diagram; Experimental)

69Sas: S.L. Sass, "The ω Phase in a Zr–25 at.% Ti Alloy," *Acta Metall., 17*, 813-820 (1969). (Meta Phases; Experimental)

70Hua: Y.C. Huang, S. Suzuki, H. Kaneko, and T. Sato, "Thermodynamics of the M_s Points in Titanium Alloys," Sci. Technol. Appl. Titanium, Proc. Int. Conf., R.I. Jaffee, Ed., 691-693 (1970). (Meta Phases; Experimental)

70Kau: L. Kaufman and H. Bernstein, *Computer Calculations of Phase Diagrams*, Academic Press, New York (1970). (Thermo; Theory)

*71Cha: D. Chatterji, M.T. Hepworth, and S.J. Hruska, "On the System Ti-Zr," *Metall. Trans., 2*, 1271-1272 (1971). (Equi Diagram; Experimental)

71Pey: Sh.I. Peyzulayev, V.V. Sumin, V.N. Bykov, and L.K. Popova, "Activities of Titanium and Iron in Binary Alloys with Zirconium," *Izv. Akad. SSSR Met.*, (4), 144-148 (1971) in Russian; TR: *Russ. Metall.*, (4), 98-102 (1971). (Thermo; Experimental)

73Ban: S. Banerjee and R. Krishnan, "Martensitic Transformation in Zr-Ti Alloys," *Metall. Trans., 4*, 1811-1819 (1973). (Meta Phases; Experimental)

*77Etc: J. Etchessahar and J. Debuigne, "Study of the Allotropic Transformation in Equiatomic Titanium-Zirconium Alloys; Influence of Purity of the Materials and Nitrogen on the Phase Transition," *Mem. Sci. Rev. Metall., 74*(3), 195-205 (1977). (Equi Diagram, Meta Phases; Experimental)

*82Auf: J.P. Auffredic, E. Etchessahar, and J. Debuigne, "The Ti-Zr Phase Diagram: Microcalorimetric Study of the α to β Transition," *J. Less-Common Met., 84*, 49-64 (1982). (Equi Diagram, Meta Phases, Thermo; Experimental)

84Shu: R.D. Shull and A.J. McAlister, private communication (1984).